D00022776

ASH DEPOSITS AND CORROSION DUE TO IMPURITIES IN COMBUSTION GASES

EDITOR

Richard W. Bryers

Foster Wheeler Development Corporation
Livingston, New Jersey

HEMISPHERE PUBLISHING CORPORATION

Washington London

In Association With

McGRAW-HILL INTERNATIONAL BOOK COMPANY

New York St. Louis San Francisco Auckland Bogotá
Düsseldorf Johannesburg London Madrid Mexico
Montreal New Delhi Panama Paris São Paulo
Singapore Sydney Tokyo Toronto

"These proceedings are based upon the International Conference on Ash Deposits and Corrosion from Impurities in Combustion Gases, which was sponsored by the Engineering Foundation at the New England College, Henniker, New Hampshire on June 26–July 1, 1977. The views presented here are not necessarily those of the Engineering Foundation, 345 East 47th Street, New York, New York 10017."

ASH DEPOSITS AND CORROSION DUE TO IMPURITIES IN COMBUSTION GASES

1 2 3 4 5 6 7 8 9 0 D O D O 7 8 3 2 1 0 9 8

Library of Congress Cataloging in Publication Data

International Conference on Ash Deposits and
Corrosion from Impurities in Combustion Gases,
New England College, 1977.
Ash deposits and corrosion due to impurities
in combustion gases.

"Sponsored by the Engineering Foundation."
Includes index.
1. Steam-boilers—Corrosion—Congresses.
2. Gas-turbines—Corrosion—Congresses. 3. Combustion gases—Congresses. 4. Fly ash—Congresses.
I. Bryers, Richard W. II. Engineering Foundation, New York. III. Title.
TJ390.I57 1977 621.4'023 78-7001
ISBN 0-89116-074-4

ENGINEERING FOUNDATION CONFERENCE

PROCEEDINGS

OF THE

INTERNATIONAL CONFERENCE ON ASH DEPOSITS AND CORROSION FROM IMPURITIES IN COMBUSTION GASES

Held at

New England College
Henniker, New Hampshire

June 26-July 1, 1977

Co-Sponsored by

The American Society of Mechanical Engineers (ASME)
Research Committee on Corrosion and Deposits from Combustion Gases

Electric Power Research Institute (EPRI)
United States Department of Energy (DOE)

CONTENTS

PREFACE

External corrosion of boiler tubes and gas-turbine blades caused by the inorganic matter in fuels has been a problem for many years. The first indication of fireside deposit problems dates back to the formation of soot on heat-receiving surfaces. Improvements in combustion as a result of increasing residence time, raising temperature levels, and providing sufficient turbulence for a good mixing eventually eliminated the soot deposit problem. As steaming capacity and temperature increased, the flue gas temperatures and gas velocities rose. Ash particles were torn loose from the beds and deposited on tube surface immersed in the flue gases, forming deposit problems. Soot blowers gradually came into being as a means of removing deposits while on line. Pulverized coal was introduced during the mid-1920s creating an even greater need for soot blowers.

The first literature on corrosion from combustion gases appeared about 1942 when wall tubes began failing at an alarming rate in boiler furnaces. Accounts of internal corrosion and deposits continued through the 1950s. The Boiler Availability Committee was formed in 1942 to correlate the work of many independent investigators and groups, such as the British Coal Utilisation Research Association (BCURA), the Fuel Research Station of the Department of Scientific and Industrial Research, and the Central Electricity Generating Board (CEGB). During its 21 years the Boiler Availability Committee acted as a center for exchanging data on experimental results for manufacturers, operators, and researchers. Mayer has listed the bulletins and technical papers of the Boiler Availability Committee that are available for distribution.[1] In 1963 the CEGB sponsored a conference at their Marchwood Research Laboratories as a forum for the worldwide experts in this field to review their achievements. The conference resulted in publication of *The Mechanism of Corrosion by Fuel Impurities* containing 44 papers of related subject matter, which remains an excellent reference source.[2]

In 1951 the American Society of Mechanical Engineers, recognizing the need for research on fireside problems, formed the Research Committee on Corrosion and Deposits from Combustion Gases. The committee, consisting of 21 members and representing research laboratories, fuel suppliers, and manufacturers and operators of steam-generating and gas turbine equipment, was charged with the responsibility of: (1) improving and increasing the knowledge of the technology used in combatting corrosion and deposits from combustion gases, (2) surveying the literature and publications, and critically analyzing such materials to increase and coordinate the teachings for best use in the field of corrosion and deposits by combustion gases, (3)

[1] J. Mayer, *J. Inst. Fuel*, 35:86, 1965.

[2] *Proceedings of the Marchwood Conference, The Mechanism of Corrosion by Fuel Impurities*, eds. M. R. Johnson and D. J. Littler, Butterworths, London, 1963.

collecting, organizing, and publishing in usable form the best technical data available in the sphere of the committee's activity, and (4) encouraging, organizing, supervising, and conducting research within the "scope" of the committee's activity and cooperating with the ASME and other societies both domestic and foreign in conducting research on basic engineering problems in the field of corrosion and deposits by combustion gases.

In 1958 the committee sponsored an extensive review of the status of fireside problems through the Battelle Memorial Institute, which resulted in the publication of a valuable reference on fireside deposits and corrosion.[3] In 1962 the committee initiated a research program at Battelle that continued until 1968 on the basic problems related to corrosion and deposits. The literature files from which the above report was prepared were maintained and expanded at Battelle during the 6-year study of the fundamentals of corrosion that was conducted for the committee—and since completion of the study—to a total of more than 1500 technically evaluated references. These references were abstracted and indexed by the committee in 1970.[4]

Out of the early work of the Boiler Availability Committee, the ASME Research Committee on Ash Deposits and Corrosion from Impurities of Combustion Gases, and independent investigations of such researchers as W. T. Reid, R. Corey, H. Crossley, and many others too numerous to mention came the identification of the alkali iron trisulfates and a series of chemical reactions that led to the formation of a liquid film at metal temperatures. Later, the same compounds were found on superheater tubes where metal loss occurred at a much greater rate than could be explained on the basis of gas-phase oxidation.

All liquid attack thus far observed in coal-fired steam generators has been associated with the presence of compounds of alkali and sulfur. Temperature ranges in which accelerated corrosion takes place can be related to the melting temperatures of those various compounds. In the temperature range from about 550 to 900°F, in the presence of a reducing atmosphere, corrosion might be attributed to the pyrosulfates $Na_2S_2O_7$ and $K_2S_2O_7$. The alkali-iron trisulfates $Na_3Fe(SO_4)_3$ and $K_3Fe(SO_4)_3$ have been mainly responsible for the severe loss of metal in superheaters and reheaters where the metal temperature is in the range of 1100 to 1300°F. The following sequence of events is generally accepted as representing the actions occurring during the corrosion process[5,6]:

- Sodium and potassium are volatilized in part from the mineral matter in the high-temperature flame forming Na_2O and K_2O.

[3] M. W. Nelson et al., *A Review of Available Information on Corrosion and Deposits in Coal- and Oil-Fired Boilers and Gas Turbines*, ASME, New York, 1959.

[4] M. W. Nelson et al., *Corrosion and Deposits from Combustion Gases—Abstracts and Index*, prepared by Battelle Memorial Institute, ASME, New York, 1970.

[5] W. T. Reid, *External Corrosion and Deposits—Boilers and Gas Turbines*, Elsevier, New York, 1971.

[6] R. W. Borio and R. P. Hensel, Coal-Ash Composition as Related to High Temperature Fireside Corrosion and Sulfur Oxide Emission Control, ASME Paper No. 71-WA/CD-4, Winter Annual Meeting, Washington, D.C., 1971.

- During combustion, pyrites are dissociated thermally and, with the organic sulfur in the coal, react with oxygen, forming mostly SO_2 and some SO_3.
- The Na_2O and K_2O then react with SO_3 either in the gas stream or on the tube to form sodium or potassium sulfate, a low-melting material that deposits on the tube. This low-melting material attracts other ash particles, eventually building up a moderately thick deposit on the tube.
- Sulfur dioxide present in the flue gas is catalytically oxidized to sulfur trioxide on iron oxide surfaces present as tube scale. Sulfur trioxide reaches near equilibrium concentrations in localized areas near the tube ($\simeq$ 1000 ppm).

A difference of opinion centers around the source of Fe_2O_3 in the alkali-iron trisulfate and therefore the exact mechanism of tube wastage. Some investigators have indicated the source of Fe_2O_3 to be the coal ash and have shown that migration of the complex sulfate does occur, resulting in tube wastage according to the following reaction:

$$10Fe + 2Na_3Fe(SO_4)_3 \rightleftharpoons 3Fe_3O_4 + FeS + 3Na_2SO_4$$

Others indicate that tube scale is a more likely form of Fe_2O_3 resulting in a reaction where the complex sulfate is a product of the wastage reaction as follows:

$$3Na_2SO_4 + Fe_2O_3 + 3SO_3 \rightleftharpoons 2Na_3Fe(SO_4)_3$$

The third and highest temperature range is associated with the alkali sulfates Na_2SO_4 and K_2SO_4, which are highly corrosive when molten. The temperature range extends from 1530°F, which is the minimum melting point from the system $Na_2SO_4 \cdot K_2SO_4$, to about 1950°F, the melting point for pure K_2SO_4. Alkali sulfate corrosion has not been a problem in steam generators because the temperatures at which it would occur are not reached by metal parts in steam generators.[5,6]

Somewhat different conditions occur in gas turbines. The trisulfates cannot exist in combustion atmospheres at temperatures above 1250°F, but the alkali sulfates Na_2SO_4 and K_2SO_4 are molten at the working temperatures of gas turbine blades and can lead to metal loss. Vanadium compounds likewise cause trouble. Sodium and vanadium form compounds that melt between 950 and 1600°F.

Vanadium salts catastrophically attack the metal surface by dissolving the normally protective oxide layer and assisting in the transport of oxygen to the pure metal surface. With gas turbines, corrosion is minimized by controlling the impurities in the products of combustion through careful selection of fuel, by water washing or centrifuging some types of contaminated fuels, or by using additives. Deposits formed with some fuels and with most additives limit the length of the run of a gas turbine before washing is required.[6]

The role of chlorine in the corrosive attack is still uncertain. It has been indicated that chlorides will not condense as an independent phase in the deposit and the HC1 will have little effect on any metal surface it contacts. Furthermore, the absorption of gaseous chloride by the deposits will not enhance their long-term corrosiveness. It is recognized, however, that chlorides may deposit on tube surfaces due to short residence time or poor combustion.

The presence of chlorides as an independent phase on a surface by whatever mechanism it is introduced, may be taken as an indication of enhanced rates of attack on the metal. If the general atmosphere permeating the deposit is oxidizing, then the chloride can play an active role in the corrosive process. If it is reducing, the atmosphere may cause enhanced rates of corrosion, via destruction of any protective oxide layer, sulfidation of the metal, etc. The chloride itself may play little or no direct part.[7]

Fourteen years have elapsed since the last major international conference on fireside deposits and corrosion due to impurities in combustion gases. During that time an energy crisis has developed. A greater emphasis has been placed on the burning of coal and fuels of poor quality with increasing amounts of troublesome impurities. The greater efforts to control environmental emissions have stimulated the use of such previously untried fuels as lignites and subbituminous coals with new and unfamiliar impurities and refuse. In addition to new fuels, new energy conversion processes are being investigated with their attendant fireside problems.

Included among the new processes are fluidized bed combustion, gasification, combustion of refuse, and magnetohydrodynamics (MHD). The ASME Research Committee on Ash Deposit and Corrosion from Impurities in Combustion Gases felt the time was right to sponsor a second major international conference on fireside problems.

Assistance was solicited from the Engineering Foundation Conferences because the format they normally follow provides an opportunity for free and informal exchange of information, opinions, and ideas. Financial assistance was provided by the co-sponsors of the conference—the Electrical Power Research Institute (EPRI) and the Department of Energy (DOE). The conference consisted of ten sessions spanning $4\frac{1}{2}$ days. The sessions were scheduled in the morning and evening leaving the afternoons available for meetings, group discussions, and recreation. Henniker, New Hampshire was selected as the conference site so that participants were removed from the press of daily business into congenial surroundings where informal discussions could take place between those actively engaged in fireside investigations. The east coast location was selected as the most convenient place for delegates to meet from such distant countries as Australia, Germany, and Israel.

Each session was devoted to a specific topic. The topics were assigned in sequence such that the fireside problems could be examined from their origin in the mineral matter found in coal and oil through the process of combustion on to postcombustion, high-temperature reactions. A session was devoted to remedial efforts for conventional energy conversion technology. The conference was concluded with sessions dealing with advanced energy conversion processes, including fluidized-bed combustion, the combustion of refuse, and MHD. The traditional format of individual technical presentations, followed by open discussion, was used during each sesseion to exchange information. This publication is the accumulation of the 44 individual presentations made at the conference.

[7] A. J. B. Cutler, W. D. Halstead, J. W. Laxton, and C. G. Stevens, The Role of Chloride in the Corrosion Caused by Flue Gases and Their Deposits, ASME Paper No. 70-WA/CD-1, Winter Annual Meeting, New York, 1970.

The proceedings represent the beginning of a new era for the technology of fireside deposits and corrosion. The differences in mineral composition between lignite and subbituminous coals and the more familiar bituminous coals are responsible for different fireside problems despite similarities in elemental content. Although there are many similarities in the deposit formation and corrosion while firing coal and refuse, there are significant differences, which lead researchers to believe that fireside problems are due to chlorides and heavy metals such as lead and zinc, as well as the alkalis and sulfur. Unusual corrosion patterns encountered in fluidized-bed combustion make it impossible to interpret new problem areas completely in terms of past experience in conventional equipment.

This volume covers a very broad matrix of fireside technology, presenting an in-depth coverage of specific problem areas. Many problems remain unresolved. Perhaps these papers will provide the stimulus to fill the void in knowledge and serve as a reference in executing future research.

As chairman of the conference, I would like to thank the authors for their efforts in preparing the formal presentations appearing in this text. I would also like to thank the 98 delegates from 8 nations fro setting aside their regular business chores long enough to attend the conference and for making meaningful contributions to the discussions. On behalf of the ASME Research Committee on Ash Deposits and Corrosion from Impurities in Combustion Gases, I express our appreciation to the Engineering Foundation for making it possible to hold this conference, and especially to Dr. S. S. Cole and Harold Comerer of the Foundation for their efforts in implementing the administrative details. Financial support from the cosponsoring agencies EPRI and DOE permitted us to extend the coverage of technical matters discussed at the conference and made possible this publication. Finally, special thanks are directed to each of the session chairmen for organizing the technical sessions.

Richard W. Bryers
Conference Chairman

CONFERENCE ORGANIZING COMMITTEE

Chairman: Richard W. Bryers
Foster Wheeler Development Corporation
John Blizard Research Center
12 Peach Tree Hill Road
Livingston, New Jersey 07039

Session Chairmen

Dr. H. J. Gluskoter	Illinois State Geological Survey, Natural Resources Building, Urbana, Illinois 61801
Dr. G. K. Lee	Department of Energy, Mines & Resources, Fuels Research Center, 555 Booth Street, Ottawa, Ontario, Canada K1A OG1
R. E. Barrett	Battelle Columbus Laboratories, 505 King Avenue, Columbus, Ohio 43201
Dr. D. L. Keairns	Westinghouse Electric Corporation, R&D Center, 1310 Beulah Road, Pittsburgh, Pennsylvania 15235
Dr. J. S. Wilson	ERDA—Morgantown Energy Research Center, P.O. Box 880, Morgantown, West Virginia 26505
H. vonE. Doering	General Electric Company, 1 River Road #53-316, Schenectady, New York 12345
J. E. Radway	Basic Chemicals, 2532 St. Clair Avenue, Cleveland, Ohio 44114
C. O. Velzy	Charles R. Velcy Associates, Inc., 355 Main Street, Armonk, New York 10504
Dr. H. H. Krause	Battelle Columbus Laboratories, 505 King Avenue, Columbus, Ohio 43201
Dr. W. E. Young	Westinghouse Electric Company, R&D Center, 1310 Beulah, Road, Pittsburgh, Pennsylvania 15235
R. C. Corey	U.S. ERDA, FE/CCU/Syn Fuels/4th Floor/Rm 4103, 20 Massachusetts Avenue, N.W., Washington, D.C. 20545
W. T. Reid	Consultant, Energy Conversion, 2470 Dorset Road, Columbus, Ohio 43221

THE OCCURRENCE OF MINERAL MATTER IN COAL AND OIL

Session Chairman: **H. J. Gluskoter**
Illinois State Geological Survey

AN INTRODUCTION TO THE OCCURRENCE OF MINERAL MATTER IN COAL

HAROLD J. GLUSKOTER

Illinois State Geological Survey
Urbana, Illinois, USA 61801

ABSTRACT

Interest in trace elements and mineral matter in coals, especially the potentially detrimental effects of these materials on equipment and on the environment during coal utilization and their impact on availability of energy, has recently increased.

The phrase *mineral matter in coal* usually refers to mineral phases or species in coal and to all chemical elements in coal that are generally considered inorganic, even if they are present in organic combination. The principal minerals in coals are aluminosilicates (kaolinite, illite, and mixed-layer clay minerals); sulfides (primarily pyrite and marcasite); carbonates (calcite, siderite, and ankerite or ferroan dolomite); and quartz.

INTRODUCTION

The phrase *mineral matter in coal* is usually meant to include all of the inorganic noncoal material found in coal as mineral phases and the elements in coal that are considered inorganic. The definition thus includes all elements in coal except carbon, hydrogen, oxygen, nitrogen, and sulfur. Four of these five "organic" elements are also found in coals in inorganic combination as part of the mineral matter. Carbon is present in carbonates {$Ca(Fe,Mg)CO_3$}; hydrogen in free water and water of hydration; oxygen in water, oxides, carbonates, sulfates, and silicates; and sulfur in sulfides (primarily FeS_2) and sulfates.

As methods of utilizing coal have become more sophisticated and the amounts of coal being used at single locations have increased, the detrimental effects of mineral matter in coal during coal utilization have assumed greater importance. The amount of mineral matter in coals varies, but is usually significantly large with respect to any method of coal utilization. In a study of 65 Illinois coals, Rao and Gluskoter (1) found that the mineral matter content ranged from 9.4 to 22.3 percent (corresponding to a range of ash content from 7.3 percent to 15.8 percent). A recent study of 114 whole coal samples from the Illinois Basin reported a mean value of 15 percent for the mineral-matter content of the coals (2). O'Gorman and Walker (3) found an even larger range (9.05 percent to 32.26 percent) in 16 "whole coal" samples from a wide distribution of locations in North America. The mineral matter in the coal is mined along with the coal and is introduced with the coal into each utilization process unless it is partially removed by cleaning. In 1976 the total coal production in the United States was estimated to be 665×10^6 tons. Assuming that 15 percent is a reasonable estimate of the average value of mineral-matter content in North American coals, approximately 100×10^6 tons of this normally unwanted foreign material is included in the total coal produced.

Interest in mineral-matter content of coal has intensified since electric power plants began to expand and to increase the temperatures at which boilers operate. Fireside boiler-tube fouling and corrosion, which become increasingly severe with a rise in temperature, are related to the sulfur, chlorine, alkali, and ash content of the coals (4). The recent widespread public interest in air pollution and water pollution has made both the coal consumer and the producer aware of the need to know more about the constitution of mineral matter in coal and of the products and by-products of the mineral matter produced during combustion. Much of this interest has been directed to the forms of sulfur in coal and coal refuse, to the sulfur oxides formed during combustion of the coal, and to the sulfates that form when coal oxidizes. There has been a concomitant demand for data relating to the origin, distribution, and reactions of sulfur in coal and for data on the trace elements in coals, particularly their concentrations and distribution, volatility, and potential effects on the environment.

The mineral matter poses several problems for the conversion of coal during gasification, liquefaction, and the production of clean solid fuels. Not only the removal and disposal of the mineral matter, but also the possible chemical effects such as catalyst poisoning, which may occur during methanation of gas from coal, are important considerations.

Not all of the interest in mineral matter in coals is stimulated by detrimental effects during coal utilization; coal might also be a source of beneficial elements and materials. Uranium has been produced from lignite; germanium and sulfur could be produced from coal; and coal ash has been used for construction materials such as brick, lightweight aggregate, and paving material for roads.

MINERALS IN COAL

Many distinct mineral phases in coals have been reported; however, most of these occur only sporadically or in trace amounts. The overwhelming majority of the minerals in coal are in one of four groups: aluminosilicates, sulfides, carbonates, and silica (quartz) (table 1).

Aluminosilicates—Clay Minerals

Clay minerals are the most common inorganic constituents of coals and of the strata associated with the coals. In fact, much of the work on clays is concerned with the clay minerals in these strata rather than with the coals. The clay minerals most commonly reported in coals are illite $\{(OH)_4K_2(Si_6\ Al_2)Al_4O_{20}\}$, kaolinite $\{(OH)_8Si_4Al_4O_{10}\}$, and mixed-layer illite-montmorillonite. In their investigation of 65 coals from the Illinois Basin, Rao and Gluskoter determined the mean value of clay content in the mineral matter to be 52 percent (1). O'Gorman and Walker (3) and Ward (5) also found that clay minerals constituted the greater part of the mineral matter in most of the coals they studied.

Sulfides and Sulfates

Pyrite is the dominant sulfide mineral in coal. Marcasite has also been reported in many different coals. Pyrite and marcasite are dimorphs, minerals that are identical in chemical composition (FeS_2), but different in crystalline form; pyrite is cubic and marcasite is orthorhombic. Other sulfide minerals sometimes found in significantly large amounts in coals are sphalerite (ZnS) and galena (PbS).

TABLE 1. MINERALS IDENTIFIED IN COALS

Mineral	Chemical formula
Clay minerals	
Montmorillonite	$Al_2Si_4O_{10}(OH)_2 \cdot x\ H_2O$
Illite-sericite	$KAl_2(AlSi_3O_{10})\ (OH)_2$
Kaolinite	$Al_4Si_4O_{10}(OH)_8$
Halloysite	$Al_4Si_4O_{10}(OH)_8$
Chlorite (Prochlorite, Penninite)	$Mg_5Al(AlSi_3O_{10})\ (OH)_8$
Mixed-layer clay minerals	
Sulfide minerals	
Pyrite	FeS_2
Marcasite	FeS_2
Sphalerite	ZnS
Galena	PbS
Chalcopyrite	$CuFeS_2$
Pyrrhotite	$Fe_{1-x}S$
Arsenopyrite	FeAsS
Millerite	NiS
Carbonate minerals	
Calcite	$CaCO_3$
Dolomite	$(Ca,Mg)CO_3$
Siderite	$FeCO_3$
Ankerite (Ferroan dolomite)	$(Ca,Fe,Mg)CO_3$
Witherite	$BaCO_3$
Sulfate minerals	
Barite	$BaSO_4$
Gypsum	$CaSO_4 \cdot 2H_2O$

Anhydrite	$CaSO_4$
Bassanite	$CaSO_4 \cdot \frac{1}{2}H_2O$
Jarosite	$(Na,K)Fe_3(SO_4)_2(OH)_6$
Szomolnokite	$FeSO_4 \cdot H_2O$
Rozenite	$FeSO_4 \cdot 4H_2O$
Melanterite	$FeSO_4 \cdot 7H_2O$
Coquimbite	$Fe_2(SO)_3 \cdot 9H_2O$
Roemerite	$FeSO_4 \cdot Fe_2(SO_4)_3 \cdot 12H_2O$
Mirabilite	$Na_2SO_4 \cdot 10H_2O$
Kieserite	$MgSO_4 \cdot H_2O$
Sideronatrite	$2Na_2O \cdot Fe_2O_3 \cdot 4SO_3 \cdot 7H_2O$
Chloride minerals	
Halite	$NaCl$
Sylvite	KCl
Bischofite	$MgCl_2 \cdot 6H_2O$
Silicate minerals	
Quartz	SiO_2
Biotite	$K(Mg,Fe)_3(AlSi_3O_{10})(OH)_2$
Zircon	$ZrSiO_4$
Tourmaline	$Na(Mg,Fe)_3Al_6(BO_3)_3(Si_6O_{18})(OH)_4$
Garnet	$(Fe,Ca,Mg)_3(Al,Fe)_2(SiO_4)_3$
Kyanite	Al_2SiO_5
Staurolite	$Al_4FeSi_2O_{10}(OH)_2$
Epidote	$Ca_2(Al,Fe)_3Si_3O_{12}(OH)$
Albite	$NaAlSi_3O_8$
Sanidine	$KAlSi_3O_8$
Orthoclase	$KAlSi_3O_8$
Augite	$Ca(Mg,Fe,Al)(Al,Si)_2O_6$
Hornblende	$NaCa_2(Mg,Fe,Al)_5(SiAl)_8O_{22}(OH)_2$
Topaz	$Al_2SiO_4(OH,F)_2$
Oxide and Hydroxide minerals	
Hematite	Fe_2O_3
Magnetite	Fe_3O_4
Rutile	TiO_2
Limonite	$FeO \cdot OH \cdot nH_2O$
Goethite	$FeO \cdot OH$
Lepidocrocite	$FeO \cdot OH$
Diaspore	$AlO \cdot OH$
Phosphate minerals	
Apatite (Fluorapatite)	$Ca_5(PO_4)_3(F,Cl,OH)$

Sulfates are uncommon in fresh and unweathered coals. Pyrite, however, is susceptible to oxidation and at room temperature decomposes to several phases of iron sulfate minerals. Iron sulfate mineral phases found to be associated with Illinois coals under oxidizing conditions are szomolnokite, rozenite, melanterite, coquimbite, roemerite, and jarosite (6). Two separate studies of coals from the Illinois Basin reported that sulfides constituted 25 percent (1) and 24 percent (5) of the mineral matter.

Carbonates

Carbonate minerals generally exhibit a wide range of composition because of the extensive solid solution of Ca, Mg, Fe, Mn, and other elements that is possible within them. This wide range of mineral compositions also applies to the carbonate minerals in coals. The relatively pure end members, calcite ($CaCO_3$) and siderite ($FeCO_3$), have frequently been reported; however, the most common carbonate minerals in the majority of coals in the world are dolomite ($CaCo_3 \cdot MgCO_3$) and ankerite ($2CaCO_3 \cdot MgCO_3 \cdot FeCO_3$).

Coals from different parts of the world have different carbonate mineralogies. Calcite is nearly the only carbonate mineral observed in coals of the Illinois Basin (1, 5), whereas ankerite is the dominant carbonate mineral in coals from Britain (7, 8); siderite, ankerite, and calcite are common in Australian coals (9).

Silica (Quartz)

Quartz is ubiquitous in coals. Rao and Gluskoter reported that, on the average, quartz made up 15 percent of the mineral matter in coals they studied from the Illinois Basin (1), and Ward reported a range of 8 to 28 percent (5). O'Gorman and Walker found that quartz ranged from 1 to 20 percent in 16 whole coal samples from various parts of the United States (3).

MINERAL MATTER IN COAL AND IN HIGH-TEMPERATURE COAL ASH

The mineral matter of coal can be determined neither qualitatively nor quantitatively from the ash that is formed when the coal is oxidized. Normal high-temperature ashing of coal at 750°C, the temperature designated by ASTM standards (10), results in a series of reactions in all the mineral groups contained in the coal except quartz.

All clay minerals in coal contain water bound within their lattices. Kaolinite contains 13.96 percent, illite 4.5 percent, and montmorillonite 5 percent of bound water. In addition, the montmorillonite in the mixed-layer clays also contains interlayer or adsorbed water. All of the water is lost during the high-temperature ashing.

During high-temperature ashing, the iron sulfide minerals are oxidized to Fe_2O_3 and SO_2. Some of the SO_2 may remain in the ash in combination with calcium, but much is lost. If all of the SO_2 were emitted during ashing, 33 percent of the weight of pyrite or of marcasite in the original sample would be lost.

The $CaCO_3$ is calcined to lime (CaO) during high-temperature ashing, with a loss of CO_2 resulting in a 44 percent reduction in the weight of the original calcite.

The changes in the mineral-matter content in coal during ashing have long been recognized. A number of workers have suggested schemes for calculating the true amount of mineral matter from the determinations made during the chemical analyses of coal. One of the first schemes, and still the most widely used, was developed by Parr, who considered only the total sulfur and ash contents in developing the conversion formulas (11). A more sophisticated method was suggested by King, Maries, and Crossley (12); the "KMC" method takes into account both the loss of CO_2 for carbonates and the decomposition of the chlorides in addition to the variables considered by the Parr formula. Further modifications of these techniques have been suggested by Brown, Caldwell, and Fereday (13); Pringle and Bradburn (7); and Millot (14). The problem of converting raw data from chemical analyses to a pure coal basis by subtracting the calculated mineral matter has recently been investigated by Given et al. (15, 16).

METHODS FOR ANALYZING MINERALS IN COAL

Separation of Minerals from Coal

For many studies, workers have analyzed minerals (i.e., naturally occurring inorganic substances with definite chemical composition and ordered atomic arrangement, not chemical elements) that were picked by hand from coal seams, from coal partings, or cleats or that were separated from the coal by methods based on differences in specific gravity between the coal and the minerals in the coal. As a first step, these studies were important contributions; however, it soon became apparent that the large amount of inseparable mineral matter in coals resulted in incomplete analytical data. The next stage in this research was to ash coal at 300°C to 500°C, temperatures below that of normal combustion, or at room temperatures in an oxygen stream. Although these investigations provided additional information, they were inadequate in that the oxidation of many of the minerals in coals accompanied the oxidation of the organic fraction.

Within the past decade the technique of electronic (radio-frequency) low-temperature ashing has been applied to investigations of mineral matter in coal. In a low-temperature asher, oxygen is passed through a radio-frequency field and a discharge takes place. Activated oxygen formed in this way passes over the coal sample, and the organic matter oxidizes at relatively low temperatures-usually less than 150°C (17). The effects of low-temperature ashing and of the oxidizing gas stream upon minerals in coal are minimal. No oxidation of mineral phases present has been reported, and the only phase changes observed were those to be expected at a temperature of 150°C and a pressure of 1 torr. Therefore, most of the major mineral constituents of coals, including pyrite, kaolinite, illite, quartz, and calcite, are unaffected by the radio-frequency ashing. The results of studies of mineral matter in coal by this method can be found in Gluskoter (18); Estep, Kovach, and Karr (19); Wolfe (20); O'Gorman and Walker (3); Rao and Gluskoter (1); and Ward (5).

Identification of Minerals in Coal

After the low-temperature mineral-matter residue has been obtained by radio-frequency ashing, a variety of instrumental techniques, which are reviewed below, can be applied to identify the minerals and to determine their concentrations.

1. The best-developed, most inclusive, and probably most reliable method for distinguishing minerals in coal is x-ray diffraction analysis, which has been used extensively by Gluskoter (17), Wolfe (20), O'Gorman and Walker (3), Rao and Gluskoter (1), Ward (5), and Paulson et al. (21). Greater precision of the technique would increase its value; nevertheless, it has been used in quantitative mineral analyses with some success.
2. Estep, Kovach, and Karr (19) used infrared absorption bands in the region 650 to 200 cm^{-1} to analyze minerals in low-temperature ash quantitatively as well as qualitatively. O'Gorman and Walker (3) have also applied this technique.
3. Differential thermal analyses (DTA) of minerals in a high-temperature coal ash have been reported by Warne (22, 23, 24). Thermal methods have also been applied to the mineral-matter fraction of American coals by O'Gorman and Walker (3, 25).
4. The popularity of electron microscopy as a mineralogical research tool has increased rapidly, but as yet has only limited application to the identification of minerals in coals. Dutcher, White, and Spackman (26) described a limited investigation that demonstrated the use of the electron probe in analyzing mineral matter in coal. Scanning electron microscopy with an energy-dispersive x-ray system accessory has been used to a limited extent to study minerals obtained from the low-temperature ashing of coal by Gluskoter and Ruch (27); Gluskoter and Lindahl (28); Ruch, Gluskoter, and Shimp (29); and Russell (30).
5. A study of Fe^{57} Mössbauer spectra in coals published by Lefelhocz, Friedel, and Kohman (31) demonstrated the validity of applying Mössbauer spectroscopy to coal and suggested the presence of high-spin iron (II) in sixfold

coordination in several samples.

6. Mass spectrometric investigations of isotopes in coal and coal minerals have also been very limited. Rafter (32) published sulfur isotope data on 27 New Zealand coal samples but drew no conclusions. Smith and Batts (33) determined the isotopic composition of sulfur in a number of Australian coals and concluded that the origin of the organically combined sulfur, the depth of penetration of sea water into underlying coal measures, and factors controlling reduction of sulfates to sulfides by biogenic residues could be deduced from isotopic data on sulfur in coal.

Investigation of Minerals in Coal in Situ

The coal petrographer uses the optical microscope, usually illuminating the specimen with reflected light, to characterize the organic fraction (macerals) in coals. Coal petrography has reached a high degree of precision, particularly during the past quarter of a century; however, coal petrographic techniques have not been nearly so successful when used to study mineral matter in coal. Because of its high reflectance and its abundance in coals, pyrite (FeS_2) is the most suitable for study with a microscope. The use of petrographic techniques for determining the minerals in coal is examined in the sections of *Coal Petrology* authored by Mackowsky (34).

OBSERVATIONS ON GEOCHEMISTRY OF MINERAL MATTER IN COAL

The investigation and interpretation of the "mineral matter in coal" is complicated by the variety of physical and chemical conditions that existed in the environment in which the coal-forming materials were deposited and in which the coal was formed. The system with which we are concerned, the formation of mineral matter in coal, was at relatively low temperature and low pressure with a large number of component phases. The system is an open one, and many of the components are mobile. A further complication is that the system has been active and may have been in the process of altering at any time since its genesis (approximately 300 million years for coals deposited during the Pennsylvanian Period). Interpretations of the system are possible in spite of these complications, for the minerals associated with coal (and with all other sedimentary rocks) are the results not of random deposition, but of a definite set of biological, chemcial, and physical conditions that combined to provide an environment in which the minerals could be deposited or in which they could form.

The mineral matter and the ash in coal have often been informally classified as *inherent* (stemming from the plant material in the coal swamp) or as *adventitious* (added after the deposition of the plant material in the swamp). This classification is at times misleading and difficult to apply, especially to minerals that are contemporaneous with the peat swamp but not incorporated by the plants.

Some standard terms applied to sediments and sedimentary rocks are applicable to coal minerals. Minerals that were transported by water or wind and deposited in the coal swamp are called *allogenic* or *detrital*. All minerals that formed within the coal swamp, in the peat, or in the coal are called *authigenic*. The term *syngenetic* is used for the minerals that were formed at the same time as the coal, and the term *epigenetic* refers to those formed later (cleat fillings, for example).

CHEMCIAL ANALYSES OF MINERAL MATTER AND TRACE ELEMENTS IN COAL

Inasmuch as mineral matter has been broadly defined to include all "inorganic" elements in coals, the chemical characterization of mineral matter involves the determination of a large number of elements. In general, chemical analyses of geological materials have progressed from the "wet chemical" methods to sophisticated instrumental methods. The major elements in the mineral constituents of coal (Si, Al, Ti, Ca, Mg, Fe, P, S, Na, K) are the same as those in silicate rocks and are often determined by x-ray fluorescence spectroscopy and flame photometry.

The most popular techniques used to determine minor and trace elements in coals are optical-emission and atomic-absorption spectroscopy. Neutron activation analysis is an excellent technique for determining many elements; however, its use is limited by the need for a neutron source (usually an atomic reactor). X-ray fluorescence spectroscopy, ESCA (electron spectroscopy for chemical analyses), and spark source mass spectroscopy have also been sucessfully applied to the analyses of some minor and trace elements in coal.

Until recently, chemical analysis of ash produced from coal at relatively high temperatures was the standard approach and was the method used in an article on cadmium in coal that was published 125 years ago (35). Using a high-temperature ash sample has the limitation that volatile elements may not be detected because they are lost during combustion. Another limitation which applies especially to analyses for trace and minor elements is that no coal standards have been available until very recently.

Recent comprehensive investigations involving a large number of coal samples and the determination of many elements, including trace elements, have been undertaken by the U.S. Geological Survey (36, 37), the U.S. Bureau of Mines (38), the Illinois State Geological Survey (2, 29), and the Pennsylvania State University (3). These studies, which produced large amounts of data, have been published so recently that they are not cited in any but the most recent review articles or bibliographies on trace elements in coals.

ACKNOWLEDGMENTS

This article has borrowed heavily from an earlier one published as chapter 1, "Trace Elements in Fuel", Advances in Chemistry Series 141, edited by S. P. Babu, American Chemical Society, p. 1-22. It appears here as an updated revision.

Research into mineral matter in coal at the Illinois State Geological Survey is currently receiving partial support through Grant R804403 from the Fuel Process Branch, Industrial Environmental Research Laboratory, Environmental Protection Agency, Research Triangle Park, North Carolina. This support is gratefully acknowledged.

ANNOTATED BIBLIOGRAPHY

1) Rao, C. Prasada, and Harold J. Gluskoter, 1973, Occurrence and distribution of minerals in Illinois coals: Illinois State Geological Survey Circular 476, 56 p.

2) Gluskoter, H. J., R. R. Ruch, W. G. Miller, R. A. Cahill, G. B. Dreher, and J. K. Kuhn, 1977, Trace elements in coal: Occurrence and distribution: Illinois State Geological Survey Circular 499, 154 p.

3) O'Gorman, J. V., and P. L. Walker, Jr., 1972, Mineral matter and trace elements in U.S. coals: Office of Coal Research, U.S. Department of the Interior, Research and Development Report No. 61, Interim Report No. 2, 184 p.

4) Crossley, H. E., 1963, The Melchett Lecture for 1962: A contribution to the development of power stations: Journal of the Institute of Fuel, v. 36, no. 269, p. 228-239.

5) Ward, Colin R., 1977, Mineral matter in the Springfield-Harrisburg (No. 5) Coal Member in the Illinois Basin: Illinois State Geological Survey Circular 498, 35 p.

6) Gluskoter, Harold J., and Jack A. Simon, 1968, Sulfur in Illinois coals: Illinois State Geological Survey Circular 432, 28 p.

7) Pringle, W. J. S., and E. Bradburn, 1958, The mineral matter in coal II—The composition of the carbonate minerals: Fuel, v. 37, no. 2, p. 166-180.

8) Dixon, K., E. Skipsey, and J. T. Watts, 1970, The distribution and composition of inorganic matter in British coals. Part 3: The composition of carbonate minerals in the coal seams of the East Midlands coal fields: Journal of the Institute of Fuel, v. 43, no. 354, p. 229-233.

9) Kemezys, Michell, and G. H. Taylor, 1964, Occurrence and distribution of minerals in some Australian coals: Journal of the Institute of Fuel, v. 37, no. 284, p. 389-397.

10) American Society for Testing and Materials, 1973, ASTM standard D1374-73, in 1973 annual book of ASTM standards, part 19, Philadelphia, p. 438-439.

11) Parr, Samuel W., 1928, The classification of coal: University of Illinois Engineering Experiment Station Bulletin No. 180, 62 p.

12) King, J. G., M. B. Maries, and H. E. Crossley, 1936, Formulae for the calculation of coal analyses to a basis of coal substance free of mineral matter: Journal of the Society of Chemical Industry, v. 57, p. 277-281.

13) Brown, R. L., R. L. Caldwell, and F. Fereday, 1952, Mineral constituents of coal: Fuel, v. 31, no. 3, p. 261-273.

14) Millot, J. O., 1958, The mineral matter in coal I—The water of constitution of the silicate constituents: Fuel, v. 37, no. 1, p. 71-85.

15) Given, P. H., 1969, Problems of coal analysis: The Pennsylvanian State University Report SROCR-9, submitted to the U.S. Office of Coal Research under Contract No. 14-01-0001-390, 40 p.

16) Given, P. H., D. C. Gronauer, William Spackman, H. L. Lovell, Alan Davis, and Bimal Biswas, 1975, Dependence of coal liquefaction behaviour on coal characteristics: 2. Role of petrographic samples: Fuel, v. 54, no. 1, p. 40-49.

17) Gluskoter, H. J., 1965, Electronic low-temperature ashing of bituminous coal: Fuel, v. 44, no. 4, p. 285-291.

18) Gluskoter, H. J., 1967, Clay minerals in Illinois coals: Journal of Sedimentary Petrology, v. 37, no. 1, p. 205-214.

19) Estep, Patricia A., John J. Kovach, and Clarence Karr, Jr., 1968, Quantitative infrared multicomponent determination of minerals occurring in coal: Analytical Chemistry, v. 40, no. 2, p. 358-363.

20) Wolfe, D. F., 1969, Noncombustible mineral matter in the Pawnee coal bed, Powder River County, Montana: Unpublished M.S. thesis, Montana College of Mineral Science and Technology, Butte, Montana, 64 p.

21) Paulson, L. E., W. Beckering, and W. W. Fowkes, 1972, Separation and identification of minerals from Northern Great Plains Province lignite: Fuel, v. 51, no. 3, p. 224-227.

22) Warne, S. St. J., 1965, Identification and evaluation of minerals in coal by differential thermal analysis: Journal of the Institute of Fuel, v. 38, no. 292, p. 207-217.

23) Warne, S. St., J., 1970, The detection and identification of the silica minerals quartz, chalcedony, agate, and opal, by differential thermal analysis: Journal of the Institute of Fuel, v. 43, no. 354, p. 240-242.

24) Warne, S. St., J., 1975, An improved differential thermal analysis method for the identification and evaluation of calcite, dolomite, and ankerite in coal: Journal of the Institute of Fuel, v. 48, p. 142-145.

25) O'Gorman, James V., and Phillip L. Walker, Jr., Thermal behaviour of mineral fractions separated from selected American coals: Fuel, v. 52, no. 1, p. 71-79.

26) Dutcher, R. R., E. W. White, and William Spackman, 1964, Elemental ash distribution in coal components—Use of the electron probe: Proceedings of the 22nd Ironmaking Conference, Iron and Steel Division, Metallurgical Society, American Institute of Mining Engineers, New York, p. 463-483.

27) Gluskoter, Harold J., and R. R. Ruch, 1971, Iron sulfide minerals in Illinois coals: Geological Society of America, Abstracts with Programs, v. 3, no. 7, p. 582.

28) Gluskoter, Harold J., and Peter C. Lindahl 1973, Cadmium: Mode of occurrence in Illinois coals: Science, v. 181, no. 4096, p. 264-266.

29) Ruch, R. R., H. J. Gluskoter, and N. F. Shimp, 1973, Occurrence and distribution of potentially volatile trace elements in coal: An interim report: Illinois State Geological Survey Environmental Geology Note 61, 43 p.

30) Russell, Suzanne J., 1977, Characterization by scanning electron microscopy of mineral matter in residues of coal liquefaction: Illinois Institute of Technology, Scanning Electron Microscopy, v. 1, p. 95-100.

31) Lefelhocz, J. F., R. A. Friedel, and T. P. Kohman, 1867, Mössbauer spectroscopy of iron in coal: Geochimica et Cosmochimica Acta, v. 31, no. 12, p. 2261-2273.

32) Rafter, T. A., 1962, Sulfur isotope measurements on New Zealand, Australian, and Pacific Islands sediments, in M. L. Jensen {ed.}, Biogeochemistry of sulfur isotopes, Proceedings of the National Science Foundation Symposium at Yale University, April 1962, p. 42-60.

33) Smith, J. W., and B. D. Batts, 1974, The distribution and isotopic composition of sulfur in coal: Geochimica et Cosmochimica Acta, v. 38, no. 1, p. 121-133.

34) Mackowsky, M.-Th., 1975, Minerals and trace elements occurring in coals: in E. Stach et al., Coal Petrology: Gebrüder Borntraeger, Berlin, p. 121-132. This review is concerned with the petrographic characteristics of the mineral phases in coal, which are classified genetically and divided into groups by time of genesis.

35) Liebig, J., and H. Kopp, 1849, Jahresbericht über die Fortschritte der reinen, pharmaceutischen, und technischen Chemie, Physic, Mineralogie, und Geologie für 1847-1848, Giessen, p. 1120.

36) Swanson, V. E., 1972, Composition of coal, southwestern United States: U.S. Geological Survey, Southwest Energy Study, Coal Resources Work Group, Part II, 61 p.

37) Swanson, V. E., J. H. Medlin, J. R. Hatch, S. L. Coleman, G. H. Wood, S. D. Woodruff, R. T. Hildebrand, 1976, Collection, chemical analysis, and evaluation of coal samples in 1975: U.S. Geological Survey Open File Report 76-468, 503 p.

38) Kessler, T., A. G. Sharkey, Jr., and R. A. Friedel, 1973, Analysis of trace elements in coal by spark-source mass spectrometry: U.S. Bureau of Mines Report of Investigations No. 7714, 8 p.

39) Watt, J. D., 1968, The physical and chemical behaviour of the mineral matter in coal under the conditions met in combustion plant, Part I. The occurrence, origin, identity, distribution and estimation of the mineral species in British coals: Leatherhead, Surrey, England, British Coal Utilization Research Association, Literature Survey, 121 p. This excellent reveiw of minerals in coal and of the chemical composition of coals does not, as the title may suggest, limit itself to British coals; a great deal of attention is given to the rest of Europe and to North America. A large section is devoted to the methods of determining the amount of mineral matter in coal; another section is concerned with the methods of identifying mineral species in coals.

40) Nicholls, G. D., 1968, The geochemistry of coal-bearing strata, in Murchison, Duncan G., and T. Stanley Westoll, eds., Coal and coal-bearing strata: Edinburgh and London: Oliver and Boyd, p. 267-307. Many data on trace elements in coal are included. The emphasis is on the geochemistry of the trace elements; other topics are the levels of concentrations of trace elements in coals, the organic (or inorganic) affinities of the trace elements, and the geochemical controls of associations of elements.

41) Mackowsky, M.-Th., 1968, Mineral matter in coal, in Murchison, Duncan, G. and T. Stanley Westoll, eds., Coal and coal-bearing strata: Oliver and Boyd, Edinburgh and London, p. 309-321. Mackowsky discusses the mineral phases found in coals and differentiates those phases with geologically different genetic histories. A number of the mineral phases are shown in a series of photomicrographs of polished sections of coals.

42) Williams, F. A., and C. M. Cawley, 1963, Impurities in coal and petroleum, in Johnson, H. R., and D. J. Littler, eds., The mechanism of corrosion by fuel impurities: Butterworths London, p. 24-67. Williams and Cawley describe the major minerals and the elemental composition of coals, methods of removing impurities from coal, and the effects

ANALYSIS OF MINERAL MATTER IN COALS OF THE RUHR UNDER GASIFICATION CONDITIONS

H. PLOGMANN and M.-TH. MACKOWSKY

Bergbau-Forschung GmbH
Frillendorfer Strasse 351
4300 Essen 13
Federal Republic of Germany

ABSTRACT

The mineral matter in coal from the Ruhr region of Germany consists primarily of a mixture of clay minerals, carbonates, iron sulfide, and quartz. The melting behavior of the mineral matter in combustion processes, tested by conventional means, defines softening temperatures and sintering temperatures that are important indicators of the temperature range at which agglomeration or caking of ash may cause the formation of deposits.

Gasification experiments were performed with an apparatus that allowed the simulation of allothermal fluid bed gasification conditions. Although the experiments were carried out at bed temperatures at which the highest temperatures of the heater could not reach the ash sintering temperatures, deposits were frequently formed at the heater. Subsequent experiments under gasification conditions with steam and pressure were found to increase melt formation and lower viscosity.

INTRODUCTION

The growing awareness that fossil fuel reserves are not inexhaustible has led to the search for more efficient utilization of fossil fuels, including those of poor quality. Efforts are being made throughout the world to improve existing processes for coal utilization and to develop new methods to overcome corrosion and environmental damage caused by impurities in the fuels. The application of fluidized bed techniques with dry ash removal plays an important part in technologies developed for combustion as well as for gasification; however, fluidized bed techniques may be applied only to an upper temperature limit that must be determined by examining the melting behavior and corrosiveness of mineral matter. Much work on the behavior of mineral matter in combustion processes has been done, but less is known of the special influences involved in gasification. We report here the results of a study made in conjunction with gasification experiments of the effects of steam and pressure on the mineral matter in coals of the Ruhr.

Composition

The mineral matter in coals from the Ruhr district is composed essentially of a mixture of clay minerals, carbonates, iron sulfide, and quartz. In order to show their contributions to ash analysis, the most common minerals and their formulas have been compiled in table 1.

The clay minerals, almost completely represented by illite and kaolinite and generally forming about 60 to 80 percent of the mineral matter, are the main contributors of Al and Si to elemental ash analysis, in which the Si content may be augmented by a small amount of quartz, generally not exceeding 10 percent by weight of the mineral matter. Other aluminosilicates such as chlorite and feldspar, which are often present but only in small amounts, do not affect ash analysis significantly. Apart from kaolinite, which exhibits a relatively constant composition, and quartz, the silicates are known to be of variable composition, as shown by their structural formulae in table 1. When common X-ray powder analysis is used, illite cannot be distinguished from sericite (or muscovite), to which it is connected by compositional transitions and which may be considered an end member of a composition range. The sign $\underline{M}$ in the formula stands essentially for Fe and Mg. Thus, illite-sericite is the most important alkali-bearing constituent and may introduce about 7 percent potassium oxide and up to 8 percent of its weight in iron oxide into ash analysis. Most of the iron content results from the carbonate siderite or the iron sulfide pyrite, whereas the marcasite modification of FeS_2 is rarely observed. The carbonates calcite and dolomite are also the main source of the alkaline-earth contents; a small amount of MgO may also be contributed by the above-mentioned complex silicates. Soluble salts, such as NaCl, do not play an important role in German coals.

Table 1: Mineral matter in coals of the Ruhr

Mineral	Formula
Illite	$K_{X+Y}(Al_{4-Y}M^{2+}_{Y})\ (Si_{8-X}Al_{X})O_{20}(OH)_4$
Sericite	$K_2\ \ Al_4\ \ (Si_6\ Al_2)O_{20}(OH)_4$
Kaolinite	$Al_2O_3 \cdot 2SiO_2 \cdot 2H_2O$
Chlorite	$(Mg_{6-X-Y}Fe^{2+}_{Y}Al_X)\ (Si_{4-X}Al_X)O_{10}(OH)_8$
Feldspar	$(K,Na)_XCa_{1-X}Al_{2-X}Si_{2+X}O_8$
Quartz	SiO_2
Calcite	$CaO \cdot CO_2$
Siderite	$FeO \cdot CO_2$
Dolomite	$CaO \cdot MgO \cdot 2CO_2$
Ankerite	$CaO \cdot (Mg,Fe,Mn)O \cdot 2CO_2$
Pyrite	FeS_2
Marcasite	FeS_2
Anhydrite	$CaO \cdot SO_3$
Apatite	$9CaO \cdot 3P_2O_5 \cdot Ca(OH,F,Cl)_2$

Table 2 presents the common range of composition of ashes from upgraded coals of the Ruhr and some ash analyses from the coal materials used in the experiments reported here. Some data from these analyses fall outside of the common range of composition because only certain grain-size fractions were used for the fluidized bed experiments.

In comparison to the original mineral matter, ash analyses of this type lack structural bonded water, carbon dioxide (from carbonates), and a variable amount of sulfur, all of which are lost during ashing and which would have to be determined in subsequent examinations; however, a calculation of the mineral matter, even from a completed ash analysis, is not possible because of the number and chemical variability of the mineral species. On the other hand, a quantitative X-ray phase analysis suffers from ill-defined crystallinity of clay minerals. A combination of X-ray and chemical analyses allows a semiquantitative estimation of the mineral constituents.

Table 2: Common range of coal ash composition and ash analyses of coal and chars used in experiments

Mineral	Common range (%)	Percentage by weight of coal G	Percentage by weight of char H	Percentage by weight of char L	Percentage by weight of char L 172
SiO_2	32.0-55.0	47.3	36.7	41.4	34.6
Al_2O_3	24.0-37.0	26.4	24.2	25.9	22.5
TiO_2	0.5-1.5	1.1	1.0	0.9	0.9
Fe_2O_3	9.0-27.0	16.7	27.9	16.9	30.1
CaO	0.0-9.0	1.7	3.5	3.7	3.3
MgO	1.0-4.0	n.d.	0.9	1.2	n.d.
K_2O	0.5-5.5	4.0	2.3	3.2	2.5
Na_2O	0.5-1.5	n.d.	2.0	1.1	n.d.
SO_3	0.5-9.0	1.5	2.4	3.6	2.5

The mineral matter of coal G was estimated to contain about 50 percent illite, 25 percent kaolinite, and 5 percent quartz; the remainder consisted of almost equal amounts of dolomite, siderite, and pyrite. Because the chars were prepared from bituminous coals at 800°C, their original minerals, apart from quartz, decomposed. Of the clay minerals, some relic X-ray-reflections of illite could still be detected. Only magnetite and pyrrhotite (FeS), which result from the decomposition of siderite and pyrite, occurred as new crystalline products.

Melting Behavior

To evaluate the melting behavior of mineral matter in combustion processes, a conventional, standardized method for testing ash fusibility, DIN 51730, comparable to an ASTM method, is generally applied. The softening temperature defined by this test, accepted as the point of initiation of melting, as well as the sintering temperature, which is not standarized by DIN but which is considered to be the temperature at which shrinkage of the ash specimen first occurs, are of special interest because they indicate the temperature range at which agglomeration or caking of ash may cause the formation of deposits. These temperatures for 60 ashes are shown in figure 1. At process temperatures, which reach about 950° to 1000°C depending on the process, only some coal ashes would be considered problematical because of their melting behavior. In addition, the temperatures reflect the familiar lowering of temperatures in a reducing atmosphere caused by the iron content's acting as a flux in the lower valance state.

We found that mineral matter characterized by a high content of illite as the dominant clay mineral and relatively high contents of iron-bearing minerals, especially pyrite and calcite, yielded ashes with low melting temperatures. It is known and can be substantiated with the aid of phase diagrams that a specific ratio of components is necessary to reach a low melting point. This was confirmed by earlier trials to lower the melting temperatures of an ash by stepwise-increasing additions of lime or iron oxide, with the result that an excess of additives had the reverse effect. The influence of illite on melting behavior results from its alkali content. At about 850°C the lattice of illite is destroyed, and, as proposed by Grim and Bradley (1), alkalies react with the adjacent SiO_4-tetrahedron layer of the structure to form a glass at about 950°C. The effects of pyrite and calcite on low melting may be clarified by a recent publication of Koch et al. (2), who observed the formation of a melt even at 900°C during the reaction of CaO and FeS in the presence of metallic iron. The first stages of melting of mineral matter under strong reducing conditions appear to substantiate this observation; however, simple 2- or 3-phase systems may explain only the initial stages of local melting, which rapidly interact with other surrounding components.

The temperatures yielded by the ash fusibility test may serve only as rough support in evaluating melting behavior under gasification conditions, which differ strongly from conditions of gasification with steam under pressure.

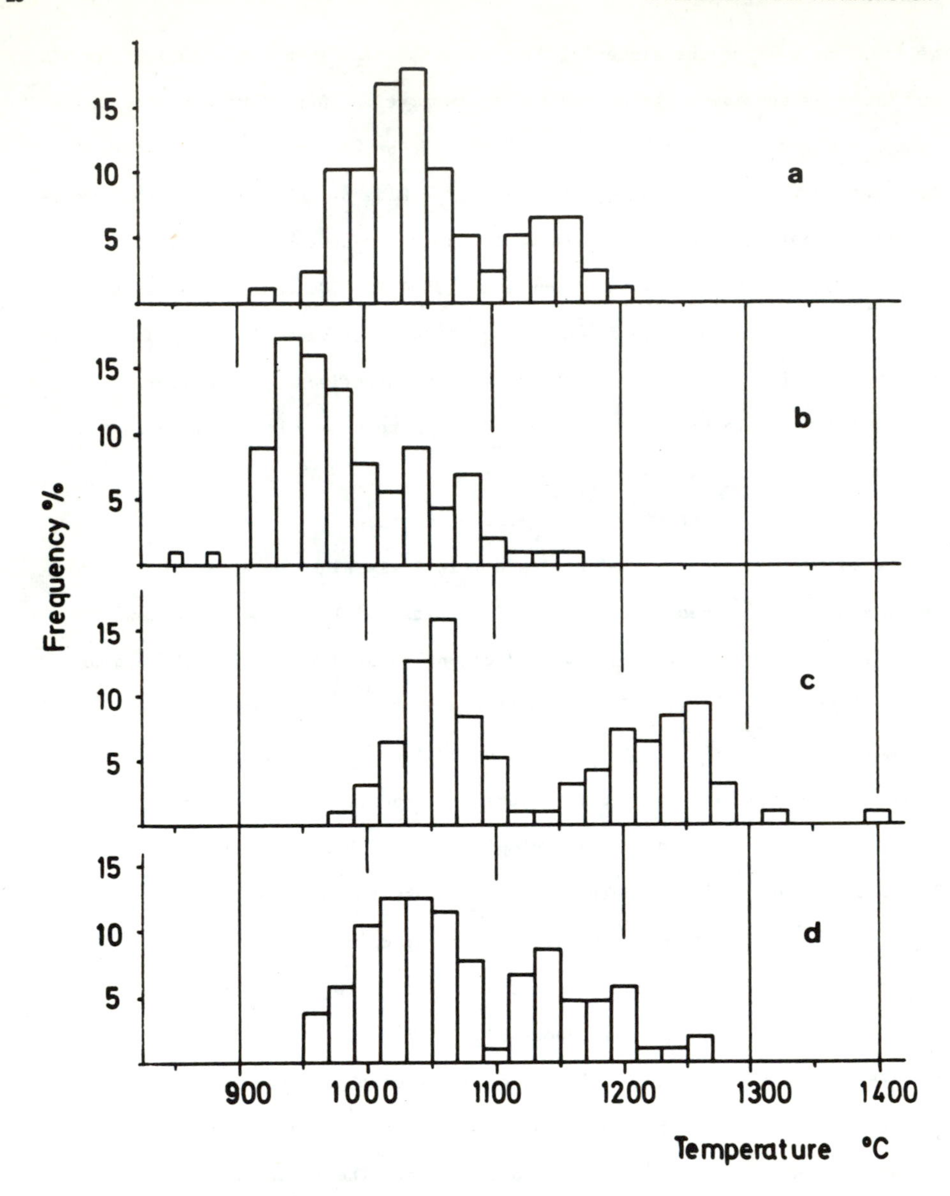

Figure 1: Ash fusibility of 60 samples: (a) sintering temperatures, oxidizing atmosphere (air); (b) sintering temperatures, reducing atmosphere $(CO:C)_2$=2:1); (c) softening temperatures, oxidizing atmosphere (air); (d) softening temperatures, reducing atmosphere ($CO:CO_2$=2.1)

A systematic study of these influences on melting behavior of coal ash is not known to us, but from experimental petrology it is clear that silicate melts under high water-vapor pressures may absorb water, which causes a lowering of melting temperature, a change of equilibrium ratios of the components in the melt, and a lowering of viscosity of the melt. The lowering of melting temperature in certain silicate systems reaches a rate of $1^{o}K$ per 0.1 MN/m^2 (1 bar) water vapor pressure, which seems severe enough to discern in gasification experiments. (3, 4)

Deposit Formation in Gasification Experiments

The gasification experiments were performed with an apparatus that allowed the simulation of allothermal fluidized bed gasification conditions; an electrically heated spiral of incology, which served as a heat exchanger, was immersed into the fluid bed. These experiments offered the opportunity to study the behavior of mineral matter during gasification at a pressure of 4 MN/m^2 (40 bars) and temperatures in the range of 800° to $900^{o}C$. At stationary conditions, the temperature difference between the surface of the heater and the center of the cylindrical fluid bed, which measured 10 cm in diameter, was 40° to $70^{o}K$, depending on mass transfer and entering temperature of steam; but because the heater also served as temperature control of the fluidized bed, short-term overheating of the heater could occur, and increase the temperature difference up to about $120^{o}K$.

Though the gasification experiments were carried out at bed temperatures at which the highest temperatures of the heater could not reach the ash sintering temperatures determined by the standard test at reducing atmosphere, deposits were frequently formed at the heater. A microscopic study of these deposits exhibited the same principally heterogeneous structure in each (fig. 2). That the deposits consisted of a melt filled with unmolten ash particles showed clearly the effect of a temperature gradient because the inclusions became increasingly molten towards the heater's surface. Even the melt generally was not homogenous, as indicated by different, locally concentrated crystallites in the glassy groundmass. New phases formed in the deposits are listed in table 3.

Table 3: Crystalline phases of heater deposits

Mineral	Formula
Magnetite	$FeO \cdot Fe_2O_3$
Hercynite	$FeO \cdot Al_2O_3$
Fayalite	$2FeO \cdot SiO_2$
Anorthite	$CaO \cdot Al_2O_3 \cdot 2SiO_2$
Gehlenite	$2CaO \cdot Al_2O_3 \cdot SiO_2$
Clinopyroxene	$CaO \cdot (Mg,Fe)O \cdot 2SiO_2$
Mullite	$3Al_2O_3 \cdot 2SiO_2$
Spurrite	$4CaO \cdot 2SiO_2 \cdot CaCO_3$
Nepheline	$Na_2O \cdot Al_2O_3 \cdot 2SiO_2$

Apart from spurrite, all phases crystallized from the melt, but occurred only in the following associations, in order of frequency:

1. magnetite + hercynite
2. magnetite + hercynite + feldspar
3. magnetite + hercynite + feldspar + fayalite
4. magnetite + hercynite + mullite
5. magnetite + hercynite + fayalite + gehlenite
6. magnetite + hercynite + clinopyroxene + gehlenite
7. magnetite + hercynite + nepheline

Additional phases in small amounts that could not be identified sometimes occurred.

The amount of Ca-bearing phases was sometimes so large in comparison to the amount in ash analysis that it suggested Ca-enrichment of melt. This could be proved qualitatively by analyzing melted crusts, which were chosen so as to be fairly free of inclusions. In addition, a slight elevation of iron content was found when the iron content of the ash was relatively low. Moreover, the described phase associations suggested a strong over-heating above fluid bed temperature. This was evident by directly comparing ash particles of the residue from the fluid bed with those of the deposits. These particles were sintered except for a small part, which seemed to have been softened slightly because the particles exhibited rounded pores. Apart from the spinels, magnetite, and hercynite, ash particles had developed only fayalite as a new phase, formed by reaction in the solid state.

Whether the observed overheating is cause or consequence of deposit formation cannot be determined because the heater serves also as temperature control. Thus, once a deposit is initiated, the restricted heat transfer would cause this overheating automatically.

Melting Experiments at Gasification Conditions

As for the initiation of deposit formation, extreme overheating of the heater could not be excluded without any doubt; whether gasification conditions caused these deposits by lowering the melting temperature of ashes could not be determined. Therefore, melting experiments with mineral matter, isolated from chars by radio-frequency low-temperature ashing, were carried out at gasification conditions at $4MN/m^2$ (40 bars) pressure and a simulated gasification atmosphere consisting of a product gas mixture of 60 percent H_2, CO and CO_2 at 17.5 percent each, and 5 percent by volume CH_4, which was passed through a saturizer to acquire about 50 percent steam. Parts of a mineral sample were tempered for one hour at several temperatures. The specimens were then compared microscopically with those tempered at the same temperatures, but at normal pressure and reducing atmosphere corresponding to the DIN test.

The ash of char L 172 is this standard test exhibited initial shrinkage in the range of 960° to 1010^oC. Softening occurred slowly, over a wide temperature range, beginning at about 1040^oC until at 1100^oC a distinct deformation was noticeable, thus demonstrating the accuracy that may be expected of values yielded by this test, which allows a deviation up to 30^oK for softening temperature and repetition by the same observer and up to 50^oK with different equipment and observers. After several repetitions of the test, first shrinkage was fixed at 970^oC and softening at 1060^oC. Samples were tempered for one hour at 990°, 1040°, and 1100^oC. After this procedure, the 1040^oC specimen showed only shrinkage, but no deformation, while the 1100^oC specimen was markedly rounded. Under the microscope (with reflected light), the 990^oC specimen appeared only slightly sintered, whereas the 1040°C specimen, shown in figure 3, exhibited a slight plasticity, noticeable by the beginning of rounding of irregularly shaped pores. In the 1100°C specimen (figure 4), rounding of pores is nearly complete and homogenization has proceeded markedly. In contrast, figure 5 shows a 1040°C specimen that was tempered for the same time under the gasification conditions described above. The viscosity revealed here corresponds closely to that of the 1100°C specimen of the standard test. In contrast to the standard test, in which formed specimens are used, a loose package of powders was preferred in the experiments under gasification conditions in order to offer a good contact with the gas atmosphere; this accounts for the formation of large pores shown in figure 6. Figure 7 shows that under these conditions, even at 990°C when the standard test specimen exhibited only slight sintering, considerable amounts of partial melts were formed. Even at 940°C (figure 8), some particles revealed a slight softening or viscous flow. Phase analyses of both series did not differ in species formed, but differed in their ratios. With increasing temperature, spinel formation increased, whereas formation of fayalite decreased, feldspar and mullite being unchanged. Under gasification conditions, however, quartz, except for scarce relics, had almost already disappeared at 990°C, whereas it decreased in the standard test only slightly even at 1040°C. Pyrrhotite (FeS) had almost disappeared at 940°C under gasification atmosphere, but considerable relics were found at 1040°C. This can be observed in figure 6, in which pyrrhotite appeared as an inclusion within the melt. It can be concluded that inclusion by rapid melting protected pyrrhotite against further reaction with the atmosphere. During the heating of mineral matter under the reducing conditions of the standard test, pyrrhotite did not decompose even at 1040°C.

The same experiments were performed also for the mineral matter of coal G and led to the same result: In comparison with the melting behavior revealed by the standard test, gasification conditions at corresponding temperatures caused an increase of melt formation and a general lowering of viscosity. It is difficult, however, to establish in this way a reliable point of lowering of the beginning of fusion, because under microscopic examination a formerly liquid spot is hard to distinguish from an amorphous solid. But with regard to the formation of agglomerates or deposits, the general lowering of viscosity seems more efficient and this lowering of corresponding stages of viscosity may be estimated by the experiments to be on the order of $50^{o}K$. This lowering seems insufficient to explain the formation of deposits at the heater of the gasification test apparatus, so overheating of the heater must also have taken place. Evidence in some cases showed that overheating was not the only cause, however.

Corrosion Phenomena by Gas Transfer

It is known that the solubility of solids in steam under pressure at a given temperature may reach the multiple of that corresponding to the vapor pressure of the solids, for which SiO_2 is an especially good example (5). Deposits temporarily formed in the steam feed pipe and at the sieve plate at the bottom of the fluid bed made it evident that this transport of inorganic components by steam may also play an important role in causing corrosion in gasification processes. Analyses of these deposits revealed that they consisted partly of condensates and partly of particles introduced mechanically by the gas stream. The latter could be derived from materials of the steam superheater vessel and are of little interest here. The condensates consisted predominantly of crystalline SiO_2 phases, quartz, and cristobalite, which were derived from a ceramic insolation material used in the steam superheater. But of special interest was the occurrence of Ca and alkalies, which were found to result from temporarily bad feed-water quality. After this was detected, the occasional deposits at the heater, characterized by nepheline when rich in sodium and others enriched by Ca, could be related to feed-water quality. In the case of alkalies it is clear they may cause deposit formation at the heater; whether this is so in the case of Ca may only be supposed. Evidence for the Ca-transport by the vapor phase was also found when deposits of a brown coal char rich in Ca were studied under the microscope.

Figure 9 shows one of the quartz grains of the ash that had developed reaction rims of spurrite, a Ca-bearing phase (see table 3), thus demonstrating an initial stage of quartz transormation by Ca-attack of the atmosphere.

Conclusion

The examples outlined demonstrate the specific influences that gasification conditions, expecially pressure and steam, have on the behavior of mineral matter in coal. These influences have proved to be important, and, for better evaluation, their study should be continued. The formation of deposits at the heater shows the complex interlocking of different influences of a technical system that must be considered in evaluating the alteration of mineral matter by such a system; phases may result not only from the composition of the system, but also from influences inherent in the system.

BIBLIOGRAPHY

1) Grim, R. E., Bradley, W. F.: "Dehydration and rehydration of illite." Am. Min. Vol. 33 (1948) 50.

2) Koch, K., Trömel, G., Laspeyres, J.,: "Entschwefelung von Eisenschmelzen über die Schlackenphase unter oxidierenden Bedingungen." Arch. Eisenhüttenwes. 48 (1977) Nr. 3 pp. 133-138.

3) Schröcke, H.,: "Grundlagen der magmatogenen Lagerstättenbildung." Ferdinand Enke Verlag, Stuttgart 1973. p. 58 ff.

4) Khitarov, N. J., Kadik, A. A.: "Water and Carbon Dioxide in Magmatic Melts and Peculiarities of the Melting Process." Contr. Mineral. and Petrol. 41 (1973) 3, 205-215.

5) Morey, G. W.: "The Solubility of Solids in Gases." Economic Geology, Vol. 52, (1957) pp. 225-251.

6) Mackowsky, M.-Th.: "Mineral Matter in Coal." Coal and Coal-Bearing Strata, ed. Murchison and Westoll, Oliver & Boyd, Edinburgh, (1968) pp. 309-321.

7) Gumz, W., Kirsch, H., Mackowsky, M.-Th.: "Schlackenkunde." Springer Verlag, Berlin-Göttingen-Heidelberg (1958).

8) Mitchell, R. S., Gluskoter, H. J.: "Mineralogy of ash of some American coals: Variations with temperature and source." Fuel, Vol. 55 (April 1976) pp. 90-96.

9) Natesan, K.: "Corrosion-Erosion Behaviour of Materials in a Coal-Gasification Environment." Corrosion-NACE, Vol. 32, H. 9 (1976) pp. 364-370.

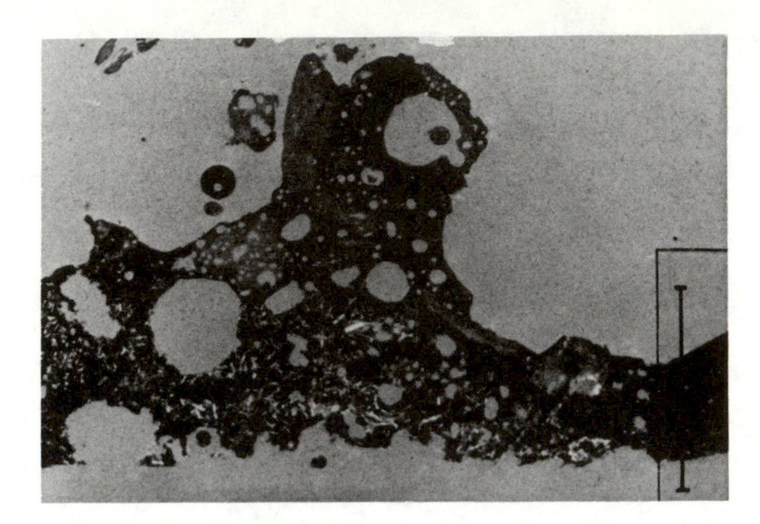

Figure 2: Deposit from heater, loosened from base (straight side), exhibiting heterogeneous structure; melt including slightly altered mineral particles. Scale: 1mm; transmitted light.

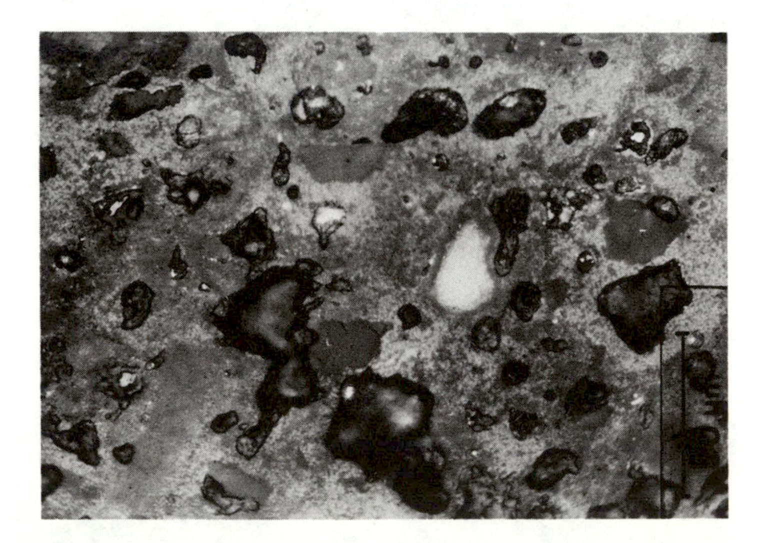

Figure 3: Ash of char L 172; tempered 1 h; 1040°C; reducing atmosphere; rounding of irregularly shaped pores is beginning (reflected light).

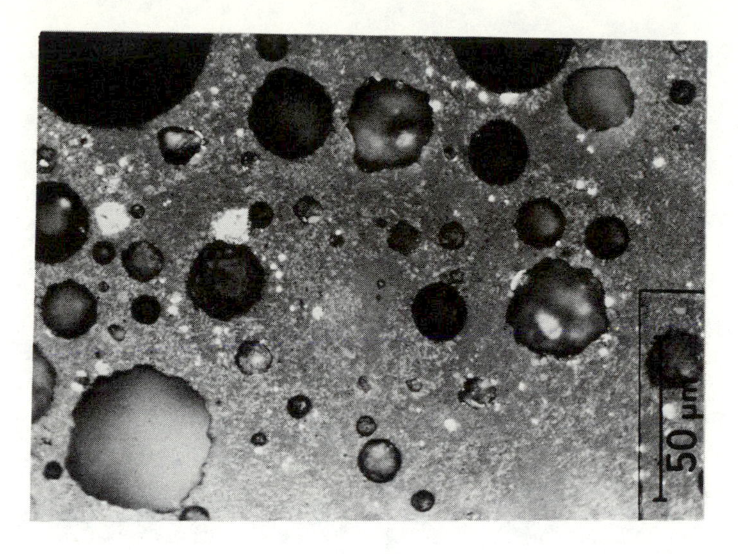

Figure 4: The same ash as figure 2; tempered 1 h; 1100°C; reducing atmosphere; roundness of pores nearly complete. Good homogenization of groundmass. Bright spots are FeS. Fine grey spots are crystalline phases, predominantly spinels.

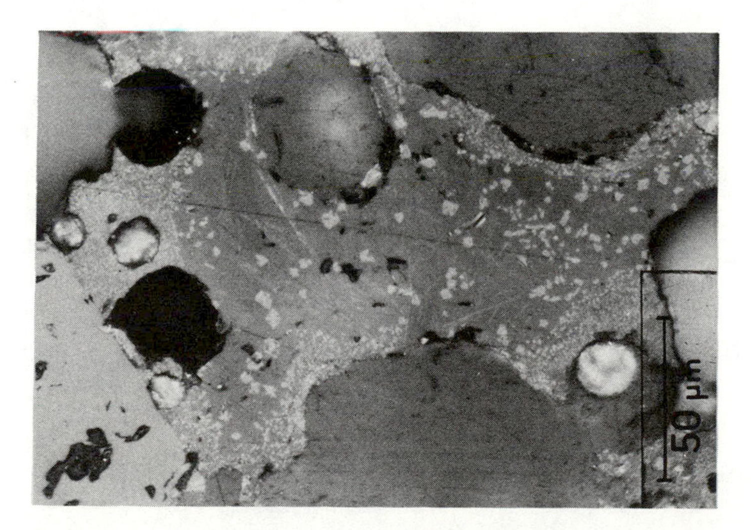

Figure 5: Mineral matter of char L 172 (isolated by RF-LTA): tempered 1 h; 1040°C; gasification atmosphere (H_2O-H_2-CO_2-CO-CH_4 = 50-30-8.7-8.7-2.6 percent by volume); almost completely molten part; at the corner: large quartz relic; spinels and mullite (laths) in the melt.

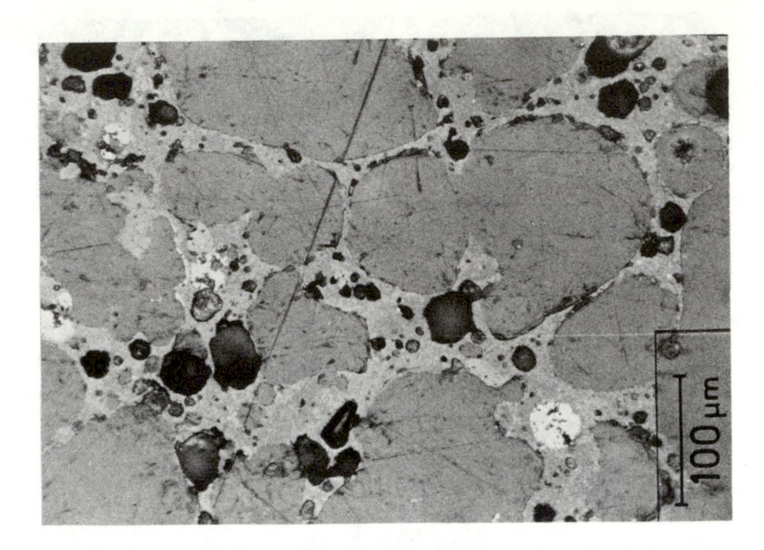

Figure 6: The same specimen as figure 5, but half the magnification; large pores (embedding medium) by loose powder usage; bright spots are FeS.

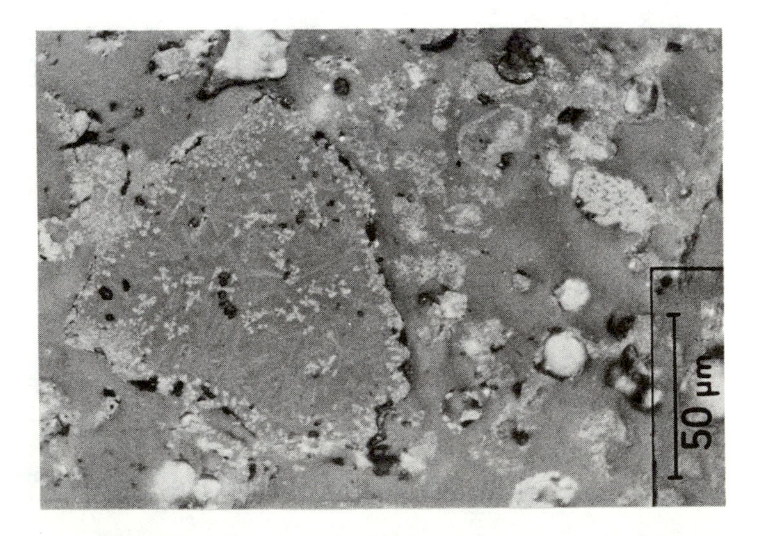

Figure 7: The same mineral matter as figure 5; tempered 1 h; 990°C; gasification atmosphere; partial melt with crystals of spinels. (Lath-shaped crystals are skeletohedral spinels.)

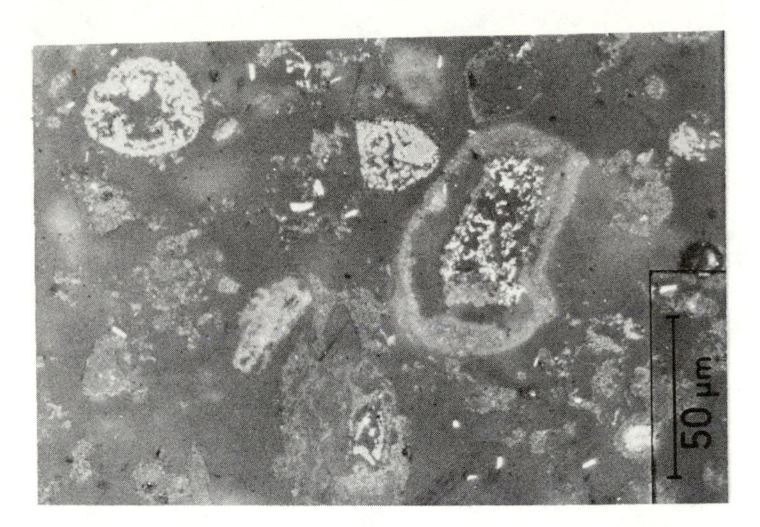

Figure 8: The same mineral matter as figure 5; tempered 1h; 940°C; gasification atmosphere; initial stages of softening of clay minerals.

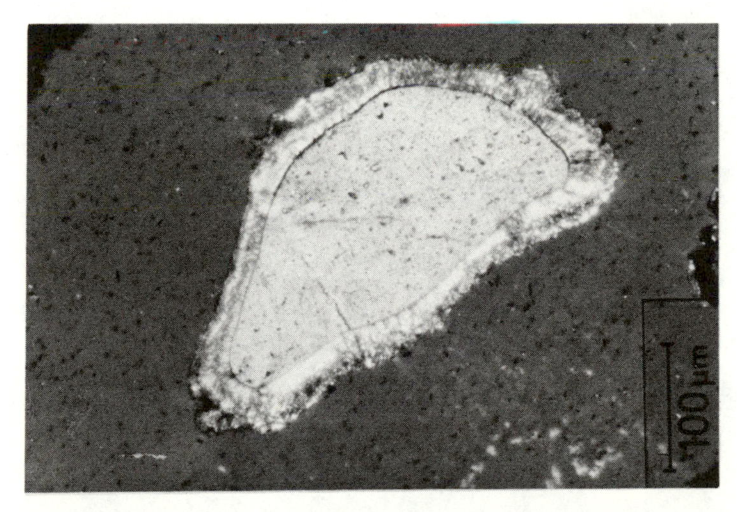

Figure 9: Quartz crystal with a reaction rim of spurrite (5 CaO · 2 SiO_2 · CO_2); formed in a brown coal char ash deposit at about 800-850°C during gasification.

VARIATIONS IN ORGANIC CONSTITUENTS OF SOME LOW RANK GOALS

ROBERT N. MILLER

Department of Geosciences, Geochemistry, and Mineralogy Section
The Pennsylvania State University
University Park, Pennsylvania, USA 16802

PETER H. GIVEN

Department of Material Sciences, Fuel Science Section
The Pennsylvania State University
University Park, Pennsylvania, USA 16802

ABSTRACT

Twenty to thirty elements have been determined at many levels in 4 drill cores and in various specific gravity fractions of a fifth sample. The data have been analyzed statistically. A number of elements (e.g. Ca, Na, K, Sr, Mg, Mn) are present in ion exchangeable form attached to organic acid groups. Other elements (e.g. Al, Ti, Be, V) appear to be present partly as chelated organometallic complexes decomposable by dilute mineral acid. The major part of the Mo in the samples is associated with the mineral pyrite. Another group of elements (e.g. Si, Al, K, Mg, Zr) are evidently associated with detrital minerals. Variations of elements within seams are discussed. Some practical implications of these findings are pointed out.

INTRODUCTION

It has become increasingly evident that a knowledge of the nature and distribution of inorganic matter in coals is of vital importance when considering any coal for practical applications. It is certainly well known that in combustion processes using coal as an energy source, the presence of the mineral matter causes severe problems of corrosion and fouling or the emission of pollutants. Coal conversion technology is likewise plagued with problems owing to the presence of the mineral matter, for example, the poisoning of catalysts or the formation of solid deposits in reactors.

With regard to the inorganic species they contain, lignites and subbituminous coals present a special case. They are known to have a large amount of reactive oxygen-containing functional groups that participate in ion exchange and chelation reactions and can thereby hold inorganic species. This mode of occurrence can largely determine the amount and kind of inorganic residues found in utilization processes, which is not true of bituminous coals or coals of higher rank. Inasmuch as there has been a renewed interest in the potential of coal as a major contributor to this nation's energy requirements, utilization processes are likely to become dependent on the vast reserves of low rank coals in the Western United States.

Geochemical studies of coal offer potentialities for interpreting what inorganic species are found in coal in such a way that predictions concerning the usefulness of particular coals can be made. Since coal is primarily a carbonaceous sediment, derived mainly from the organs and tissues of the higher plants, the mineral matter generally represents a minor fraction; nevertheless it is chemically significant. Suites of inorganic elements in coals are related to the specific geological circumstances that prevailed at particular sites; that is, to the nature of the rocks undergoing erosion in the area. Therefore, in order to realize the potentialities referred to above, extensive geological and geochemical studies of coal basins are needed, to reach an understanding of how and why particular suites of inorganic materials are present in coals. A generalization that holds for one basin is unlikely to hold for others.

Coals originate in peat swamps or marshes which are waterlogged systems where ground waters transport mineral particles and soluble ions. Peats, and low rank coals formed from them by metamorphism, contain reactive organic functional groups of various kinds (e.g. carboxyl and hydroxyl). Consequently, the inorganic constituents of low rank coals can be of several types:

1. As discrete mineral phases originating as detrital grains or as authigenic species formed *in situ*;

2. As elements present as ion exchangeable cations on carboxyl groups or clays, emplaced by entrapment from circulating ground waters;

3. As elements chelated in stable organometallic complexes.

Extensive studies of inorganics in bituminous coals have been made in several areas[1,2,3] but less complete accounts are available for low rank coals. Most detailed accounts of the inorganic constituents in coals treat only the minerals that occur in coals[4,5,6] but some studies are available which have concerned themselves with ion exchange[7,8] and with the association of elements with the organic matter[9,10] in low rank coals. Extensive studies of certain elements in American lignite ashes have also been published[11], but little information on elements as they exist naturally in low rank coals is available. Little attention has been given either to the geochemical relationships of mineral matter in low rank coals or to vertical variations in seam profiles. Therefore, this study is concerned with the mineral matter geochemistry of some low rank coals of the Northern Great Plains and the Gulf Coast Provinces. Vertical variations of inorganic elements in seams have been studied, and the analytical scheme was planned so as to obtain a better idea of the organic-inorganic associations in coals.

METHODS OF ANALYSIS

Four drill cores, including roof and floor, from Alabama (2), Texas (1), and Wyoming (1) were subjected to detailed inorganic analyses. A channel sample of a North Dakota lignite was also included in the study. All drill cores were subdivided into a number of small sections (80-150 mm), the size of each depending on the thickness of the seams. Total ash, sulfatic and pyritic sulfur were determined on each section by standard ASTM procedures. Analyses were made for 30 elements on each section in the total ash, in certain extracts, and in the insoluble residues. The instrumental techniques used were conventional emission spectroscopy, atomic absorption, and argon plasma arc emission spectroscopy. In order to get some understanding of how elements might be bound in coals, the samples were put through a simple fractionation scheme to obtain materials as follows:

1) Soluble in 1.0N ammonium acetate (pH=7, 25°C), which contain ion-exchangeable cations and soluble salts[12];

2) Soluble in 1.0N hydrochloric acid (25°C), which contain the cations of carbonates and some organometallic elements;

3) Insoluble in ammonium acetate and hydrochloric acid, which are primarily the silicate minerals and pyrite.

The extracts were analyzed in solution. The residue was ashed, fused with lithium metaborate, dissolved in dilute nitric acid and then analyzed in solution according to published procedures[13]. The channel sample was crushed and a size fraction of 80x200 mesh (Tyler) was obtained and subjected to a series of specific gravity separations using centrifugation in order to obtain fractions of varying mineral matter content. Chemical fractions of each gravity fraction were obtained as above. Major mineral phases present in the samples were determined qualitatively by X-ray diffraction on an ash produced by low temperature plasma oxidation[14].

RESULTS AND DISCUSSION

Specific gravity fractionation has proved to be a valuable means of differentiating organophilic elements from mineral-forming ones[15], especially when coupled with the chemical fractionation scheme. An example is shown in figure 1 which includes the data for Na and K. Most of the sodium in the coal is ion exchangeable, and little is found in any form in the heavy fraction. On the other hand, although most of the K in the lighter fractions is ion exchangeable, increasing proportions in the heavier fractions are found in the acid insoluble residue. Clearly, some K is associated with carboxyl groups in the organic matter, but a similar amount is tightly bound with a mineral phase, in this case the clay mineral illite. Since K is tightly held in illites, this comparison suggests one reason why fouling in the combustion of lignites should correlate well with total Na content, but not with total K content.

The distributions of the alkaline earth metals (Sr, Ca, and Mg) are similar to that of Na, being almost entirely present in ion exchangeable form. A number of other elements (e.g. Be, V, Y, Yb, Sc, Cr; also, acid soluble Ti, Cu and Zn) are concentrated in the fractions of lower specific gravity. It is not surprising that many of these elements should be associated with organic matter, presumably in chelated form[9,15,16]. But it is somewhat surprising that part of the aluminum behaves similarly. Figure 2 shows that the amount of Al in the acid soluble material decreases as the specific gravity of the fraction increases. The rest of the aluminum is in the acid insoluble form in the higher specific gravity fractions. The inference is that some of the Al is chelated to organic matter.

The extraction studies of the drill cores indicate that a considerable portion (50-80%) of the inorganic matter of these low rank coals is either ion exchangeable or soluble in dilute acid under mild conditions and concentrations of such elements vary little through the seam. Any major variations in the total mineral content are presumably a result of the variable input of detrital material (clays and sand) during deposition. These points are illustrated by a number of examples.

Figures 3 and 4 show rather flat profiles for ion exchangeable calcium within the coal seams. Minor variations of ion exchangeable Ca in figure 3 correlate with variations in the petrographic composition of the organic matter; the various macerals presumably contain varying amounts of acidic functional groups capable of ion exchange. Calcium is, by far, the predominant ion exchangeable cation, representing 1-2.5% of dry coal at all levels in the samples studied. 75-95% of the total calcium was removed by ammonium acetate. Profiles similar to those of calcium were generally found for the ammonium acetate soluble fractions of Mg, Sr, Ba, Mn, Na, and K. It is thought that circulating ground waters are the determining factors setting the levels of concentration in ion exchangeable form. Since calcium is generally the major cation found in most ground waters and because its ion exchange properties favor the replacement of monovalent cations, it is not surprising that this element is the predominant ion exchange species. There is little doubt that exchangeable Ca is responsible for the build-up of calcite in coal liquefaction reactors when low rank coals are used.

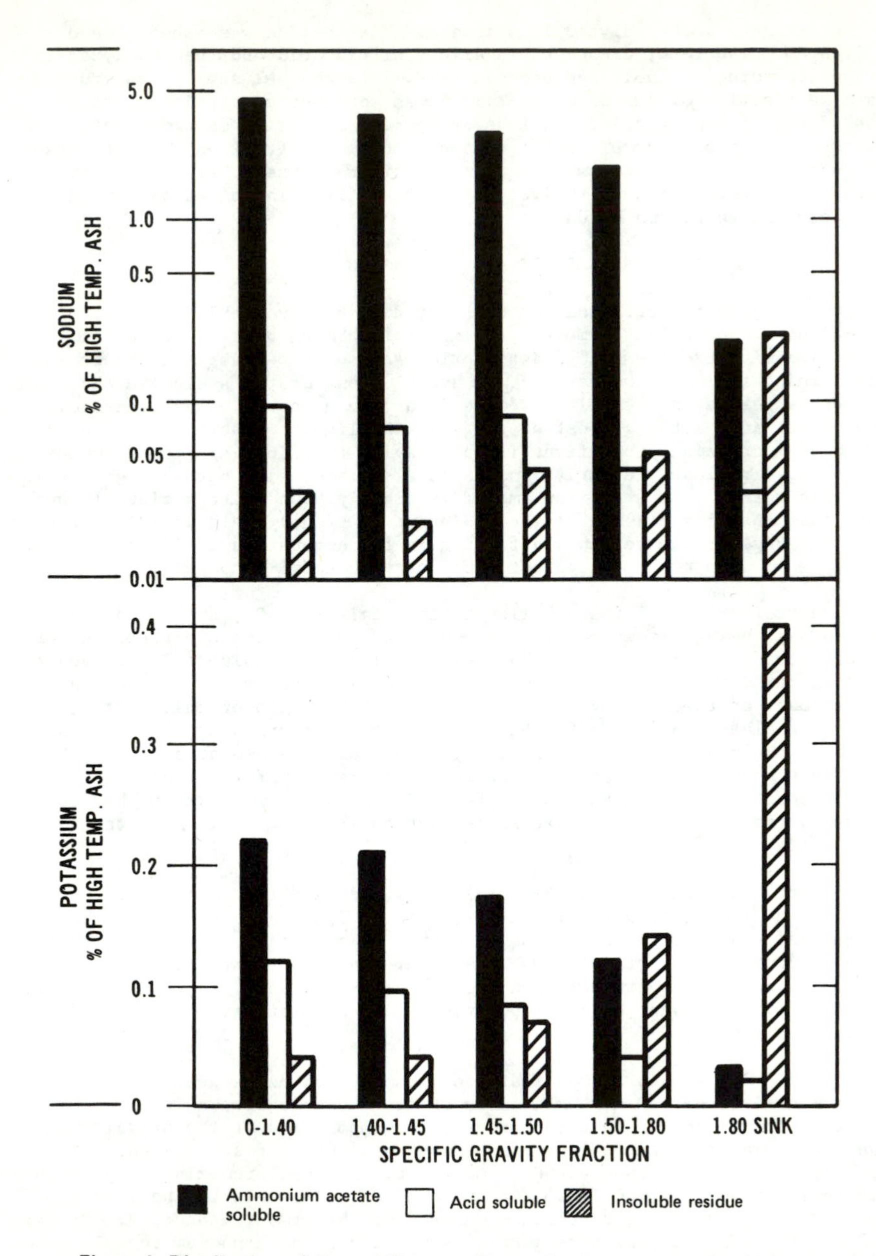

Figure 1. Distribution of Na and K in specific gravity fractions of a North Dakota lignite.

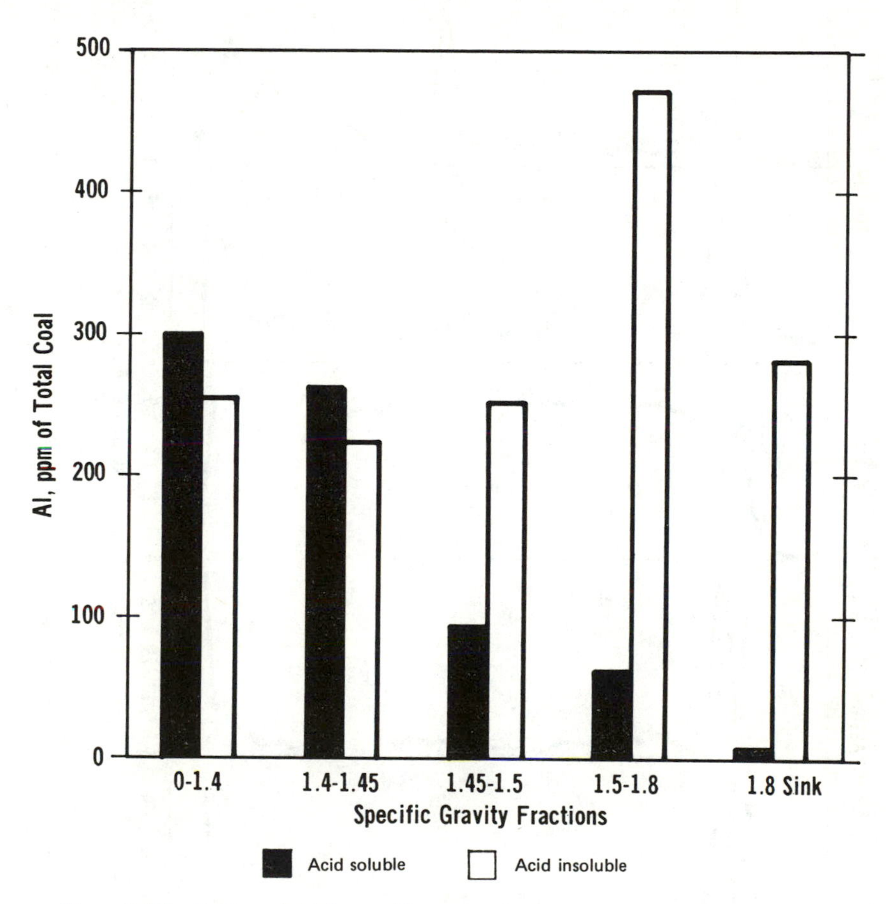

Figure 2. Distribution of A1 in specific gravity fractions of a North Dakota lignite.

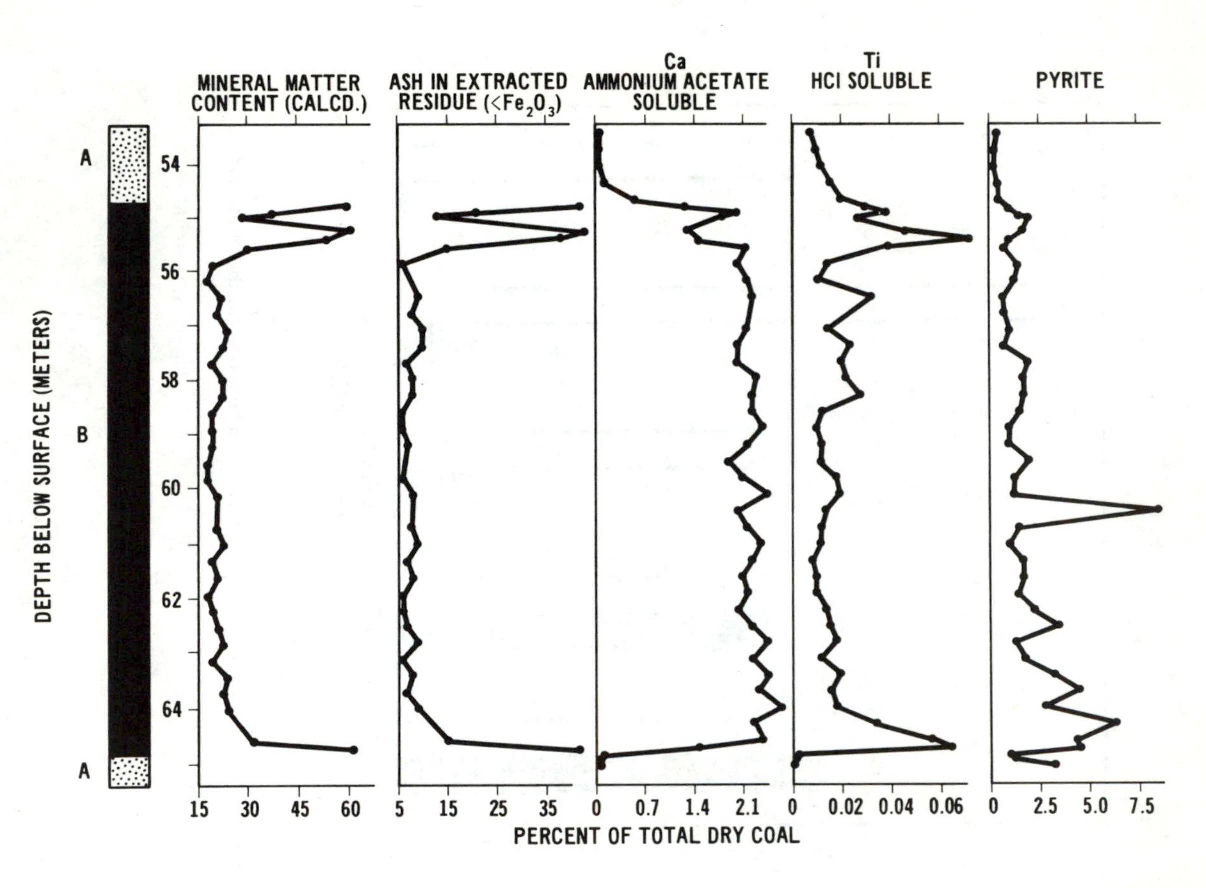

Figure 3. Profiles of some inorganic components in an eastern Alabama lignite. (A) Sandstone roof and floor; (B) Coal.

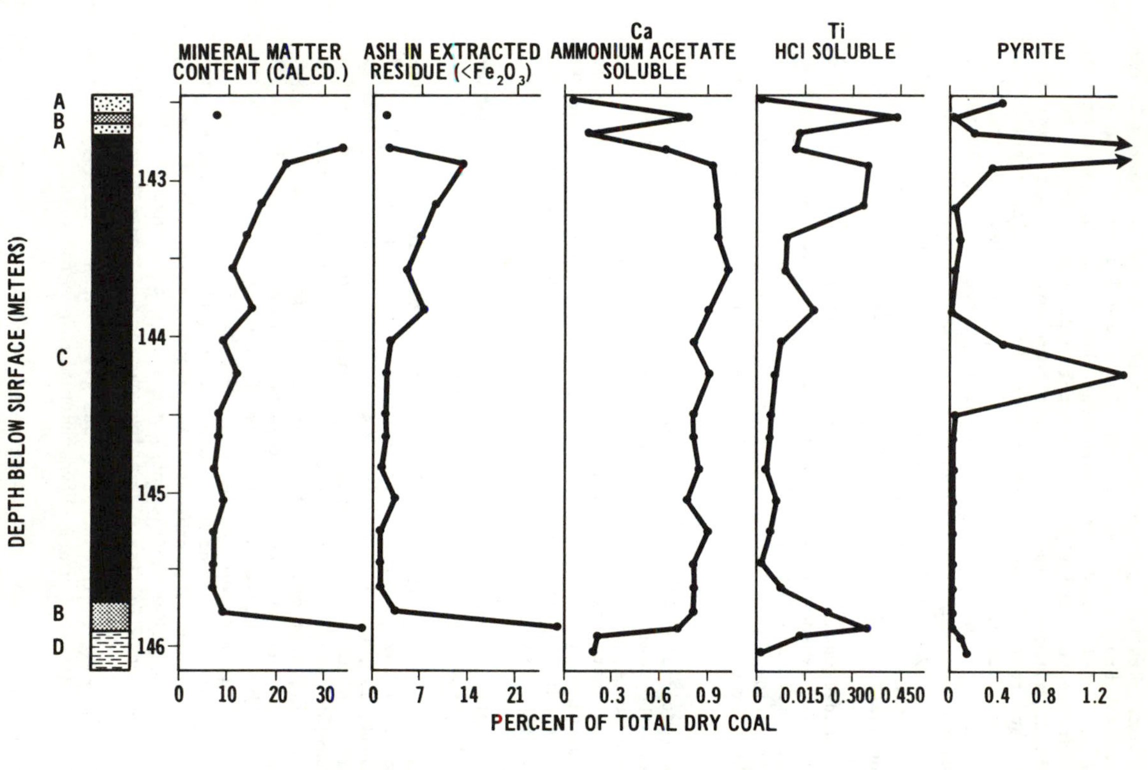

Figure 4. Profiles of some inorganic components in a Wyoming subbituminous coal. (A) Sandstone roof; (B) Bright coal; (C) Dull coal; (D) Mudstone floor.

The picture with Ti is interesting. The solubility of major proportions of titanium in acid leads one to believe that this portion is combined with the organic matter rather than in mineral phases, since the common Ti minerals (rutile, anatase, brookite, ilmenite, sphene) are all insoluble in acid under mild conditions. Acid soluble Ti peaks near the margins of the seams, but is low or zero in the roof and floors adjacent to the seams. On the other hand, acid insoluble Ti, although also peaking near seam margins, is relatively high in the floor and roof. These differences provide further indications that organic matter is involved in trapping Ti in acid soluble form. The presence of organically held Ti is interesting because it is presumably only in this form that it exercises its well-known poisoning effect on coal liquefaction catalysts. Also, Ti molecularly dispersed in organic complexes will yield very finely divided oxide on combustion of the coal, so that it will be in a state of high catalytic activity for the oxidation of SO_2 to SO_3.

The inorganic species in the acid-insoluble residues will consist mainly of detrital minerals and the authigenic mineral, pyrite. There is no reason why the input of detritus should correlate with pyrite formation, and therefore in seeking correlations between the ash yielded by detrital minerals and other parameters, the ferric oxide yielded by pyrite was subtracted from the total ash yield.

The corrected ash yields from the insoluble residues showed major variations through the seams, especially near margins and major partings. X-ray diffraction studies of the low temperature ash from these high mineral matter zones indicated that they were characterized by the presence of large amounts of clay (kaolinite and illite mainly) and quartz. Evidently there were transitional phases at the start and termination of the accumulation of peat, when the input of inorganic sediment was relatively large.

Pyrite contents, as well, show considerable variability through the seams and when they are exceptionally high, the mineral is present in nodular form.

With data available for up to 30 elements at many levels in each core, interpretation of the data was a formidable problem. Therefore, "principal component analysis", using a computer, was applied to the data. This is a statistical technique, and is a specially valuable tool when one has a large number of variables and when it can be assumed that the variables are not all truly independent; that is, that there are underlying processes or phenomena that make some of the variables interrelated. Figure 5 shows the results of principal component analysis used to classify the elements in the chemical fractions from one coal into various factors, each of which indicate parallel variations in the concentration of various elements with depth. Each factor is independent of the other. The statistical loading factors (which are more or less equivalent to correlation coefficients) are all in the range 0.5-0.96. The several trace elements in figure 5, which are shown in parentheses, were not included in the principal component analysis, but an independent correlation matrix indicated that their concentrations gave statistically significant correlations with the other elements shown under the same factor.

Obviously, when elements are classified in this way, one must attempt to produce a geochemical reason why the groups are as they are; that is, one must try to identify the underlying processes associated with each of the factors. Factors 1 and 6 explain the major variations in the total mineral matter content of the coal, appearing to be related to the input of detritus during deposition. Factor 2 is obviously associated with the trapping of cations by carboxyl groups. There is evidence to show that pyrite in this coal has become oxidized by weathering, and Factor 3 appears to be associated with this phenomenon. On the other hand, Factor 5 is associated with the initial formation of sulfides. Pyrite is, of course, the most abundant of these found in coals and its formation is widely recognized to be a result of the bacterial reduction of sulfate ion to H_2S. It is known that in the presence of H_2S, molybdate ion can be converted

PRINCIPAL COMPONENT ANALYSIS of an EASTERN ALABAMA LIGNITE

FACTORS					
1	2	3	4	5	6
Res. K	AmOAc sol. Sr	HCl sol. Fe	HCl sol. Ca	Res. Mn	Ash
Res. Ti	AmOAc sol. Mg	Sulfate	HCl sol. Sr	Pyrite	Res. Si
HCl sol. Ti	AmOAc sol. Mn	Res. Fe	HCl sol. Mg	Res. Fe	Res. Al
Res. Mg	AmOAc sol. Ca	Pyrite	HCl sol. Mn	(Mo)	
–AmOAc Ba	HCl sol. Mn			(Yb)	
Res. Ca	AmOAc sol. K				
(Zr)	(Be)				
(Sc)					

Figure 5. Principal component analysis of an eastern Alabama lignite.

to thiomolybdate, which in time can decompose to MoS_3 or MoS_2. The pyrite-molybdenum association was found in all samples studied. Moreover the presence of Mo was almost entirely restricted to the heaviest specific gravity fraction of the North Dakota channel sample, as was the pyrite. The acid soluble cations in Factor 4 are the most common cations of carbonate minerals, and this group may be associated with the formation of these minerals; however, the absolute amounts of these minerals must be low (e.g. acid soluble Ca represents less than 10% of the total calcium).

The data for trace elements in the seams indicated that many of these (e.g. Be, Ge, Yb, Y, Sc) tend to be concentrated at seam margins, while others (e.g. Cu, V) appeared to have fairly uniform distributions throughout the seams (but in one core V peaked near the margins). Concentration levels of many elements differ widely from one seam to another, presumably owing to differences in the availability of the elements in the source rocks in proximity to the coal seams either during or after deposition of the coals. Entrapment by the organic matter from circulating ground waters appears to be responsible for the relatively high marginal concentrations of many elements.

Little clay but unusually high concentrations of many trace elements were found in the 8 cm section of coal (the "rider") which was parted by sand from the main seam shown in figure 4. On the other hand, most trace elements were either low or undetectable in the roof and parting. These factors strongly suggest that the organic matter was involved in some manner in the trapping of many trace elements. Figure 6 shows data for a 2.10-sink specific gravity fraction of the rider, and the results of the chemical fractionation procedure applied to the fraction that floated at 2.10 S.G. Pb and Sn were the only elements that were significantly concentrated in the heavy fraction. Averaging the percentages shown in figure 6 for the acid soluble portions of the 12 elements, we obtain a mean content of 46%. It should be noted that the mean contains a value of zero for Sn and values approaching 100% for V and Be. Be is known to form stable complexes with organic carboxyl groups[17] which are soluble in dilute acid. Although a large part of the vanadium in petroleums is present as vanadyl porphyrins, these substances are most probably not present in coals. It must be assumed, then, that vanadium is being held in another form, not yet identified but probably organic. It should be recalled that V_2O_5 is an excellent catalyst for the oxidation of SO_2, so that the presence of V in coals in this form or a form convertible to V_2O_5, would be of practical significance.

SUMMARY AND CONCLUSIONS

A detailed study of a large number of elements has been made for several low rank coal seams. For many elements, considerable variations with depth in the seams have been found, and this is likely to be so with other coals. It has been shown that parallel variations of different suites of elements are related to independent geochemical processes that were active at different stages of peat accumulation or after burial. It is evident, particularly with lignites, that elements may be held in a number of different forms.

These observations may have a number of practical implications. For example, if it is desired to know the mean concentration of any element in a coal seam, it is obviously vital that a sample representative of the whole seam be submitted for analysis. Inasmuch as particularly high concentrations of some elements and high total mineral matter contents were found at seam margins, it might be desirable, by selective mining, to reject the margins from the feedstock to a conversion process or any other process in which these elements are deleterious in one way or another to the process. Also, it must be recognized that where an element occurs in different forms, it may be desirable to know in which form it actually is present in any coal, because this may determine whether or not its presence is deleterious. The point already made with regard to titanium, that if present as an organic complex it will be released on combustion as very finely dispersed oxide, applies, of course, to any element similarly held.

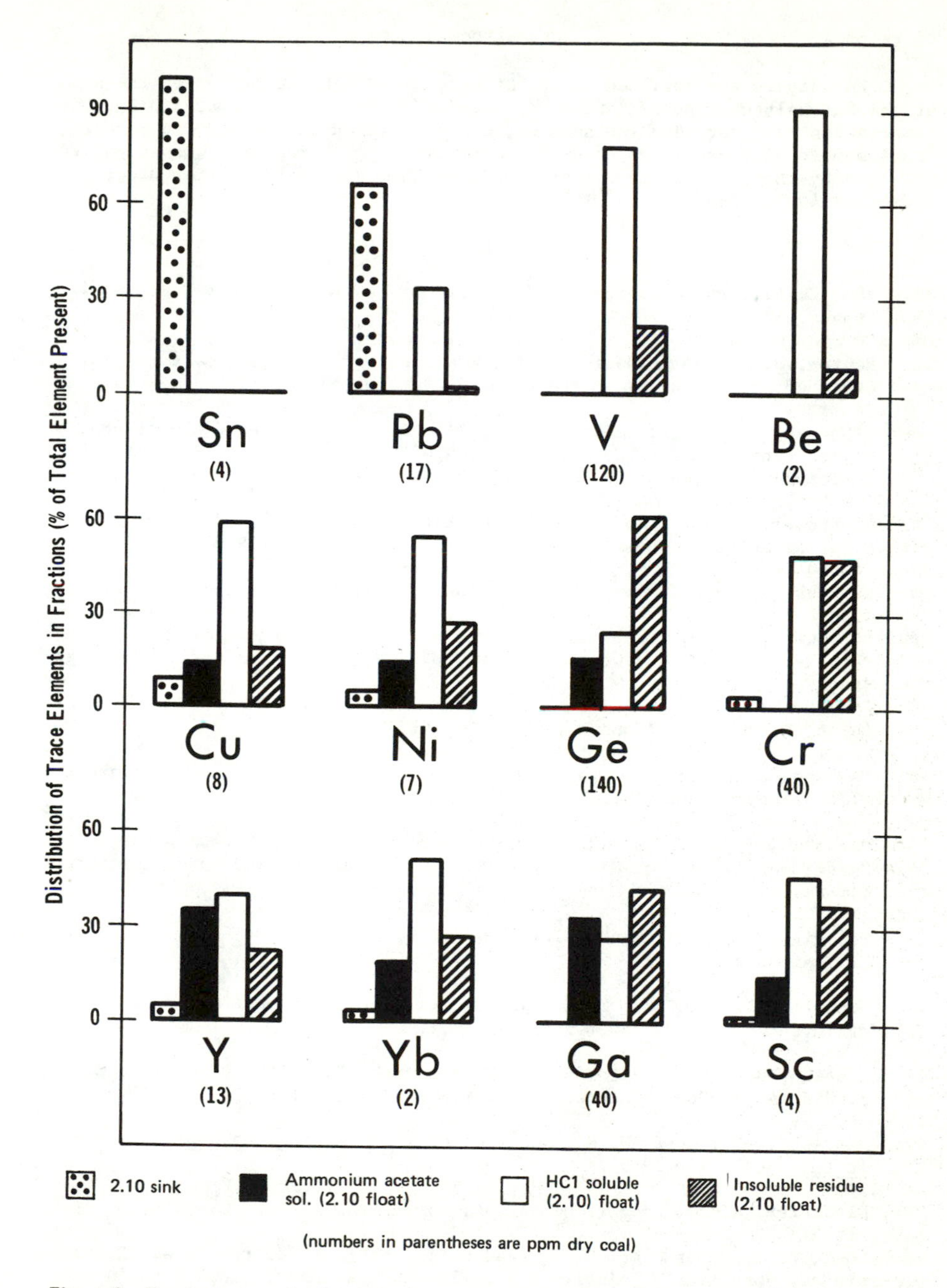

Figure 6. Trace element distribution of an upper section of a Wyoming subbituminous coal seam.

ACKNOWLEDGMENTS

The authors are indebted to Mr. Richard F. Yarzab who performed the statistical analyses. Appreciation is extended to Dr. William Spackman and his associates for providing the samples, three of which were taken from the PSU-ERDA sample base assembled under his direction. This research was supported in part by the U.S. Energy Research and Development Administration under contract FE(49-18)-2494.

REFERENCES

1. Rao, C. P., and Gluskoter, H. J., Illinois St. Geol. Survey Circular 476, Urbana, Ill. (1973).

2. Renton, J. J., and Hidalgo, R. V., Coal Geology Bulletin No. 4, W. Va. Geol. and Econ. Survey, Morgantown, W. Va. (1975).

3. O'Gorman, J. V., and Walker, P. L., Jr., Research and Development Rept. No. 61, Interim Rept. No. 2, to Office of Coal Research, U.S. Dept. of Interior, under Contract No. 14-01-0001-390 (1971).

4. Gluskoter, H. J., Chap. 1, in "Trace Elements in Fuel", ed. Babu, S. I., p. 1, Advances in Chem. Series 141, Amer. Chem. Soc. (1975).

5. Kemezys, M., and Taylor, G. N., J. Inst. Fuel, 37, 389 (1964).

6. Mackowsky, M.-Th., Chap. 13, in "Coal and Coal-Bearing Strata", eds. Murchison, D. G., and Westoll, T. S., p. 309, Oliver and Boyd, Edinburgh (1968).

7. Durie, R. A., Fuel (London), 40, 407 (1961).

8. Paulson, L. E., and Fowkes, W. W., U.S. Bureau of Mines Rept. of Invest. 7176, 18 pp. (1968).

9. Nicholls, G. D., Chap. 12, in "Coals and Coal-Bearing Strata", eds. Murchison, D. G., and Westoll, T. S., p. 269, Oliver and Boyd, Edinburgh (1968).

10. Zubovic, P., Stadnichenko, T., and Sheffey, N. B., U.S. Geol. Survey Bull. 1117-A, 58 pp. (1961).

11. Sondreal, E. A., Kube, W. R., and Elder, J. L., U.S. Bureau of Mines Rept. of Invest. 7158, 94 pp. (1968).

12. Chapman, H. D., in "Methods of Soil Analyses, Part II", ed. Black, C. A., p. 891, Amer. Soc. of Agron. Inc., Madison, Wisc. (1965).

13. Medlin, J. H., Suhr, N. H., and Bodkin, J. B., Atomic Absorption Newsletter, 8(2), 25 (1969).

14. Gluskoter, H. J., Fuel (London), 44, 285 (1965).

15. Ruch, R. R., Gluskoter, H. J., and Shimp, N. F., "Occurrence and Distribution of Potentially Volatile Trace Elements in Coal", Final Rept., Environmental Geol. Notes, (72), 96 pp., Ill. St. Geol. Survey (1974).

16. Zubovic, P., Chap. 13, in "Coal Science", Advances in Chem. Series 55, Am. Chem. Soc., p. 221 (1966).

17. Sidgwick, N. V., "The Chemical Elements and their Compounds", Vol. 1, p. 213, Oxford Univ. Press, Oxford (1952).

THE NATURE OF METALS IN PETROLEUM FUELS AND COAL-DERIVED SYNFUELS

R. H. FILBY, K. R. SHAH, and F. YAGHMAIE

Nuclear Radiation Center
Washington State University
Pullman, Washington, USA 99164

ABSTRACT

Petroleum is a complex mixture consisting mainly of aliphatic, alicyclic and aromatic hydrocarbons but also containing variable amounts of metal containing compounds including metalloporphyrins. Nickel and vanadium are normally the most abundant metals in petroleum with Ni ranging from 0.001 to 500 μg/g and V ranging from 0.01 to 1500 μg/g in crude oils. Both Ni and V are distributed in porphyrin (Ni and V) and non-porphyrin forms. The nature of the non-porphyrin Ni and V is poorly defined but is associated with the high molecular weight asphaltene sheet structure. Trace elements, other than Ni and V, are present in oils at much lower concentrations and appear to be associated only with the high asphaltene/resin fraction of the oil.

The trace element contents of shale oil and coal-derived Solvent Refined Coal are lower than the parent oil shale and coal but (except for Ni and V) are generally higher than the trace element contents of residual fuel oils used for power generation.

INTRODUCTION

Crude oils from which petroleum products are made vary greatly in composition depending on age, geological location, and history. Hydrocarbons, alkanes, cycloalkanes, aromatics and mixed hydrocarbons are the principal constituents of crude oils, normally exceeding 75% by weight of the total crude oil. Small but significant amounts of non-hydrocarbon constituents, including sulfur, oxygen, nitrogen and metal compounds, are normally present but vary widely depending on the type of petroleum. In general, the younger asphaltic oils contain larger amounts of non-hydrocarbons than do light highly paraffinic oils. The hydrocarbons found in a typical light and a typical heavy asphaltic crude oil (1) are shown in Table I. The fuel components obtained by refining of a medium crude oil are shown in Table II; however, it should be noted that increased yields of certain components (e.g. gasoline) can be obtained by changing refining conditions.

The small amounts of metal containing compounds are important for several reasons. Metallic elements such as Ni, V, As, Cu, Fe, and alkali metals are highly detrimental to cracking and reforming catalysts, either changing the selectivity of the catalyst or poisoning it. Also, V and alkali metals present in residual fuel oil are extremely corrosive to refractories used in furnaces. Some metallic species in petroleum, e.g. vanadium porphyrins, are volatile and may appear as undesirable components of distillate fractions or may result in toxic emissions. In recent years the emphasis on pollution control and control of emissions from power plants has focused attention on the Hg, Se, As, Ni, V and S contents of residual fuel oils. Although the metal containing species of petroleum are important for the above reasons, of equal importance is the geochemical information they provide concerning the origin and subsequent geological transformations of petroleum.

TABLE I

Composition of Heavy and Light Crude Oils (1)

	Percent of Total Crude	
	Light Oil	Heavy Oil
Normal paraffins (alkanes)	23.3	0.95
Branched chain alkanes	12.8	3.20
Cycloalkanes or naphthenes	41.0	19.2
Aromatics (mono- and polycyclic)	6.4	9.5
Naphtheno-aromatics or mixed	8.1	27.9
Hydrocarbons (include S compounds)		
Resins]	8.4	23.1
Asphaltenes]		16.5
TOTAL	100	100

TABLE II

Components of a Medium Crude Oil (1)

	Carbon Numbers and Type		B.P.°C	Percent of Crude
Gasoline	C_5-C_{10}	a, ia	180	25
Kerosene	C_{11}-C_{13}	a, ia, ca, ar	180-250	10
Gas Oil	C_{14}-C_{15}	a, ia, ca, ar	250-300	15
Light Oil Distillate	C_{18}-C_{25}	a, ia, ca, ar	300-400	20
Lube Oil Distillate	C_{26}-C_{35}	a, ia, ca, ar	400-500	10
Residue (fuel oil, etc.)	C_{36}-C_{60}	a, ia, ca, ar, r, as	500	20

a = alkanes, straight chain
ia = isoalkane, branched chain
ca = cycloalkanes
ar = aromatics
r = resins
as = asphaltenes

TRACE ELEMENTS IN PETROLEUM

Abundances of Trace Elements in Crude Oils

Several early studies reported the presence of trace elements in the ashes of crude oils (2,3) but the first quantitative data were presented by Shirey (4) who determined 13 metallic elements in the ashes of crude oils. Most of the work reported after 1931 concerned the two most abundant metals in crude oils, Ni and V, although several studies reported qualitative or semiquantitative data for other elements (5-7). With the exception of Ni and V, however, few accurate, precise, and representative abundance data for trace elements in petroleum are available. Several authors have reviewed the trace element data on crude oils in order to determine their environmental significance and to correlate trace element concentrations with geological or geochemical factors (5,8,9). The available data indicate that the metal contents of crude oils vary greatly. The values for Ni and V range from 0.001 to 500 μg/g for Ni and from 0.01 to 1500 μg/g for V (9,10). Magee et al. (11) have pointed out the need for better analyses of crude oils, particularly those oils that comprise the bulk of U.S. production and imports.

The main reason for the lack of accurate data on trace elements in crude oils is that sensitive and precise analytical methods for the determination of extremely low metal concentrations have only recently been developed. The conventional analytical methods used in the inorganic analysis of petroleum have been reviewed recently by Milner (12) and McCoy (13) and the most frequently used techniques have been emission spectrography, X-ray fluorescence spectrography, and colorimetry. Because of the low concentrations of most trace elements in oils (generally less than 10 μg/g) the analytical techniques require preconcentration of the trace elements before analysis and this is often accomplished by ashing large volumes of the crude oil. The ashing procedure is subject to loss of volatile metal-containing compounds, e.g. Ni and V porphyrins, and several other elements may form volatile compounds during combustion. Examples are As, Sb, and Hg halides, and SeO_2. The ashing procedure may also lead to erroneous results by introducing contaminant elements from reagents or equipment used in the chemical methods. Recently atomic absorption, plasma arc spectroscopy, and polarography have been applied to selected elements in petroleum but general methods of trace element analysis have not been developed. Neutron activation analysis has been used extensively for multielement analysis of petroleum because the high sensitivity of the technique allows direct measurement on the untreated oil. Several studies have been reported recently from this laboratory (14-18), by Al-Shahristani and Al-Atiya (19), and by Cahill (20).

One problem faced in attempting to determine the trace element composition of crude oils is that there has been no representative sampling of oils and that many reported values in the literature are from unusual or atypical oils. Hence the abundance data should be interpreted carefully.

Table III shows the concentrations of S, Ni, V and Na in 6 crude oils. Four of these are from producing wells, one is a sample of Athabasca Tar Sand Crude obtained by solvent extraction and one is a very atypical oil, probably from the Cymric field in California. This California oil is included because it illustrates "abnormal" trace element abundances and because it was used in studies of the distribution of trace elements among petroleum components. Table IV shows values of 15 other trace elements in the same set of oils. It is apparent that wide variations in trace element contents are found. The very high value for V in Boscan is unusual, although most Venezuelan crudes are high in V. The high Hg value for the California oil comes from an area subject to Hg mineralization (Cymric field). It should also be pointed out that most metal abundances are less than 1 ppm (1 μg/g) and are much lower than those values found in coal (see later). The large variations in trace elements shown by the crude oils in Tables III and IV are not unusual, and are largely a function of oil type, e.g. heavy young asphaltic oils are high in trace metals whereas old or light paraffinic oils are generally very low in trace metals. The same variations may be found within an oil producing area as shown in Table V which shows average values for some trace elements in oils from the W. Canada

TABLE III

Nickel, Vanadium, Sodium, and Sulfur in Crudes

Element	Crude Oil 1	2	3	4	5	6
S (%)	0.44	3.9	<2	1.28	0.35	0.98
Ni	<5	72	117	12	5.9	94
V	0.02	178	1120	53	9.5	7.5
Na	0.04	21	25	0.6	1.1	11

KEY:
1. Alberta, Canada (Devonian) (18)
2. Athabasca Tar Sand crude - Canada (18)
3. Boscan, Venezuela (17)
4. Iranian light crude (21)
5. Nigerian crude (21)
6. California (Tertiary) heavy crude (17)

TABLE IV

Trace Elements in Some Crude Oils (ng/g)

Element	Crude Oil 1	2	3	4	5	6
Fe (μg/g)	<0.1	254	0.33	1.4	37	73
Co	6.8	2000	11	300	198	13000
Zn	-	-	41	324	2600	9300
Cu	-	-	17	32	210	930
Cr	<5	1700	19	17	380	634
Se	<3	520	39	58	370	360
As	1.5	320	4.8	95	1200	660
Sb	<0.01	28	<0.1	<0.1	273	52
Cs	0.9	69	-	-	-	-
Rb	<10	720	-	-	-	-
Sc	0.1	190	-	0.07	4.4	8.8
Eu	0.2	23	0.3	0.21	-	-
Br	8.0	104	91	16	-	-
Cl	3800	8000	8700	2270	-	-
Hg	<20	<20	5.6	<20	139	21000

Basin. These oils range from Devonian to Upper Cretaceous in age of producing formation and in general the younger heavy asphaltic oils are those high in trace elements.

Other fuels derived from petroleum, e.g. gasoline, kerosene, diesel oil and residual fuel oil (RFO) have trace element compositions that reflect the composition of the crude oil from which they were derived and the distillate or residual nature of the product. As discussed later, during refining most of the trace elements in crude oil are found in the RFO, except for volatile porphyrins which may appear in distillate cuts. Hence the trace element concentrations of distillate oils tend to be very low compared to RFO and the RFO values tend to vary as the crude oils. Table VI shows some values for four RFO's. Table VII shows similar data for a furnace oil, diesel oil, and RFO from a typical refinery crude oil. These data show clearly that the concentrations in the RFO are larger than in the crude indicating that most trace elements concentrate in the RFO. The concentrations of such elements as Ni, V, Fe, Cu, Cr, etc., are very low in the distillate products.

Forms of Trace Elements in Petroleum

To explain the behavior of trace elements during refining and for the geochemical interpretation of trace element data in crude oils it is necessary to know in what forms the trace elements occur. Except for Ni and V such knowledge is largely lacking. The metals may be present in oils as inorganic particulate matter (discrete mineral grains, adsorbed on clay minerals, etc.), in emulsified formation waters, introduced as drilling fluids or corrosion inhibitors, or present as organometallic complexes. Nickel and vanadium in crude oils have been extensively studied because of the presence of Ni and V porphyrins which are thought to have been derived from chlorophyll (or hemoglobin), thus indicating a biogenic origin for petroleum. It is now recognized (22,23) that significant amounts of Ni and V are present in crude oils which cannot be accounted for by metalloporphyrins. The non-porphyrin Ni and V is present in the asphaltic component and either occupies sites bounded by heteroatoms (N, S, O) or is present in metalloporphyrins strongly associated by π-π bonding to the asphaltene aromatic sheets (24). The origin of this non-porphyrin Ni and V is not known. It may have been incorporated in the asphaltene structure from the original organic source material, replaced other metal cations in the asphaltenes, or it may have been incorporated by complexation from aqueous or solid phases during the migration or maturation of petroleum. Sugihara et al. (23) have postulated that the non-porphyrin V in asphaltenes was the source of V introduced into the porphyrin structure during the conversion of chlorophyll in the source material to the metalloporphyrins found in oil. The fact that V and Ni may occur in porphyrins and non-porphyrins forms may explain some of the conflicting data concerning the correlations of V/Ni ratios with age of reservoir rock.

Very little is known of the nature of metals other than Ni and V in crude oils. No well documented organometallic compounds of metals other than Ni and V have been found. In several studies it was demonstrated that the metals occurred in oil soluble form and some authors have noted the association of metals with the asphaltic fraction of petroleum (25,26).

Filby (17) studied the trace element distribution in a California Tertiary oil. This oil is very high in trace metals and is not typical, but does allow separated components of the oil to be analyzed. In this study the crude oil was washed and filtered to remove suspended mineral matter and emulsified water and the oil was then separated into three components as shown in Figure 1. Each component was analyzed by neutron activation analysis for 9 trace elements. The results are shown in Table VIII. The data show clearly that the resins and asphaltenes contain the major fraction for each trace element, as has been shown for Ni and V in other studies. To further understand how these elements are distributed within each fraction, gel permeation chromatography (GPC) was used to separate molecular weight fractions of each component. The distribution of arbitrary molecular weight fractions, 300-1000, 1000-4000, and 4000-8000, among each petroleum component is shown in Table IX. As expected, the relative amounts of the four molecular weight fractions are different for the three

TABLE V

General Statistics for 22 Elements Determined in 88 Crude Oils (Total Oil Basis), Alberta, Canada (18)

Element (conc.)	Detection Limit	Number of Samples Above Detection Limits	Average (conc.)	Maximum (conc.)
S (%)	0.05	88	0.83	3.88
V (ppm)	0.1	84	13.6	177
Cl (ppm)	-	87	39.3	1010
Na (ppm)	-	85	3.62	64.7
Fe (ppm)	0.1	41	10.8	254
Ni (ppm)	0.1	69	9.38	74.1
Zn (ppb)	25	79	459	5920
Co (ppb)	0.2	84	53.7	2000
I (ppb)	10	51	719	9000
Mn (ppb)	3	78	100	3850
Se (ppb)	3	62	51.7	517
Hg (ppb)	2	39	50.9	399
Cs (ppb)	0.5	43	4.27	68.5
Br (ppb)	2	82	491	12500
As (ppb)	2	71	111	1990
Au (ppb)	-	4	0.438	1.32
Sb (ppb)	0.01	24	6.22	34.8
Cr (ppb)	5	42	93.3	1680
Rb (ppb)	10	9	148	720
Sc (ppb)	0.05	51	7.76	199
Eu (ppb)	0.05	50	0.935	23.2
Ga (ppb)	-	-	-	445

TABLE VI

Trace Elements in Some Residual Fuel Oils

Element (ng/g)	Residual Fuel Oil 1	2	3	4
Fe (μg/g)	4.6 ± 4.5	-	3.3	21.0
Co	320 ± 170	-	400	59.2
Zn	1000 ± 840	1070	244	2324
Cu	-	-	-	330
Cr	70 ± 50	78.1	164	147
Se	90 ± 32	42.4	130	100
As	55 ± 40	55.7	105	160
Sb	4 ± 3	-	8.5	-
Cs	-	-	-	4.7
Rb	-	-	-	-
Sc	2 ± 7	0.23	0.6	0.35
Eu	2 ± 3	1.22	-	1.12
Br	220 ± 100	497	<100	232
Hg	-	4.0	-	3.3
Cl (μg/g)	40 ± 38	63	25.3	26.5
Ni (μg/g)	-	12.5	-	21.7
V (μg/g)	87 ± 58	11.7	49.5	29.2
Na (μg/g)	33 ± 23	59.1	19.3	29.1
K (μg/g)	-	3.6	-	1.9

[1] Average of 13 RFO's reported by Cahill (20)
[2] Canadian crude oil feedstock (21)
[3] Heavy Venezuelan (20)
[4] Canadian crude oil feedstock (21)

TABLE VII
Trace Element Concentrations in Products Derived from a Crude Oil (21)

Element (μg/g)	Crude	Bunker C	Furnace Oil	Diesel
Fe (μg/g)	0.64	21.0	<0.1	<0.1
Co	5.4	59.2	51	63
Zn	76	2324	17	<70
Cu	43.4	330	<5	7.1
Cr	-	147	<5	<5
Se	22.5	100	11.8	5.8
As	34.7	160	9.5	4.3
Sb	<1.0	<1.0	<1.0	<1.0
Cs	-	4.7	0.7	0.9
Rb				
Sc	0.03	0.35	0.03	0.04
Eu	1.03	1.12	0.13	0.13
Br	218	232	5.6	5.6
Hg	<4	3.3	<4	<4
Cl (μg/g)	16.2	26.5	3.8	1.25
Ni (μg/g)	2.1	21.7	0.08	0.11
V (μg/g)	2.8	29.2	0.40	0.25
Na (μg/g)	9.1	29.1	0.002	0.002
S (%)	0.32	2.85	0.27	0.12

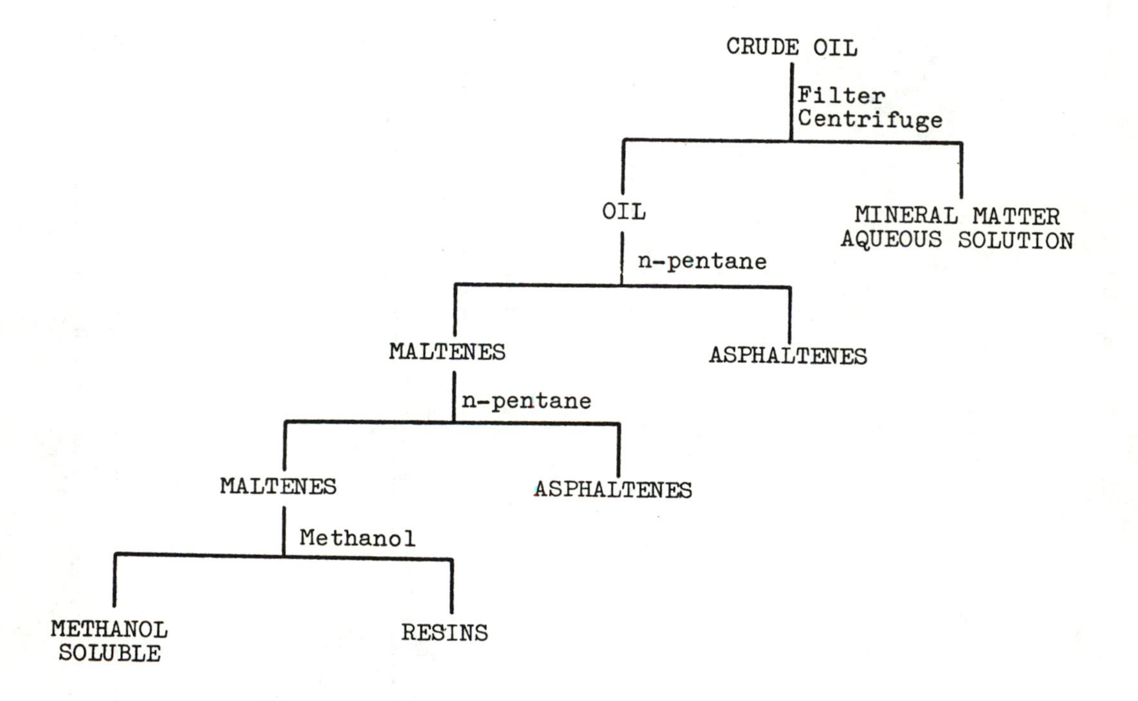

Fig. 1.--Schematic Diagram of Separation Scheme for Oil Components.

TABLE VIII

Distribution of Trace Elements in Components of California Crude Oil (17)

Concentration (g/g)	Crude Oil	Methanol Soluble	Resins	Asphaltenes
% Crude Oil	100	57.5	37.5	4.99
Ni	93.5	7.21	147	852
Co	12.7	0.8	10.7	122
Fe	73.1	1.95	66.4	895
Hg	21.2	0.686	29.6	140
Cr	0.634	0.300	0.894	7.54
Zn	9.32	0.74	8.86	109
Cu	0.897	0.500	0.32	3.02
Sb	0.0517	0.0033	0.0130	1.22
As	0.656	0.546	0.290	2.25

TABLE IX

Distribution of MW Fractions of Oil Components (17)

Molecular Weight Fraction	Percentage of Component		
	MeOH Soluble	Resins	Asphaltenes
300 - 1000	93.8	29.4	11.0
1000 - 4000	---*	21.2	23.0
4000 - 8000	---	49.4	50.6
8000 - 22000	---	0	15.2
TOTAL	100	100	100

*Note: In MeOH soluble fraction, 6.2% of fraction has MW > 1000 but was not separated.

components. The methanol-soluble fraction contains only 6.2% high molecular-weight material whereas the resins and asphaltenes contain 71% and 89% high molecular-weight material respectively. For both resins and asphaltenes the 4000-8000 molecular-weight fraction is the most abundant and the heaviest molecular-weight fraction (8000-22000) was found only in the asphaltenes (14.2%). These findings are consistent with the nature of the asphaltenes and resins as described by Yen (24) and others (27). The asphaltenes contain 11% of low molecular-weight material which is consistent with the idea that small polar molecules are associated with the asphaltene micelles during precipitation but which are separated in the GPC procedure. The fact that the resins and the asphaltenes contain material of the same molecular weight range indicates the similar nature of these materials and the difference in precipitation behavior may be attributed to the higher polarity of the asphaltenes as suggested by Witherspoon and Winneford (32)

The concentrations of 9 trace elements in each of the molecular-weight fractions of the resins and the asphaltenes are shown in Tables X and XI. Vanadium was not determined in the separated molecular-weight fractions because of the small amount of material available for analysis. For the asphaltenes, the highest trace element concentrations, with the exception of Ni and Sb, are found in the highest molecular-weight fraction. The highest Ni was found in the 300-1000 fraction but this is due entirely to Ni porphyrins which are associated with the asphaltenes (17). Antimony exhibits a behavior contrary to that of all the other elements studied. The highest Sb content is found in the lowest molecular-weight fraction (11.0 μg/g) and the Sb concentration decreases with increasing molecular weight of the asphaltene fraction. This suggests that Sb is contained in small molecules of high polarity that are insoluble in n-pentane and which precipitate with the asphaltenes. The nature of the Sb compounds is unclear and further work is necessary. Possible compounds are the alkyl and aryl stibines R_xSbH_{3-x}. Of the other elements in the asphaltene fractions, Cr, Cu and Co show increasing concentrations with increasing molecular weight of the fraction and this indicates that these elements are associated with the large asphaltene molecules. The highest concentrations of Fe, Hg, Zn, and As are found in the highest molecular-weight fraction of the asphaltenes but these elements show higher concentrations in the 300-1000 molecular-weight fraction than in the 1000-4000 fraction. For Fe, Hg, and Zn this may indicate the presence of porphyrin complexes or other organometallic compounds. Iron porphyrins have been identified in shale oil (28) and Zn coproporphyrin occurs in animal feces (29) but neither of these metalloporphyrins has been found in crude oil, however. Mercury forms porphyrin complexes with mesoporphyrin (30) but the complexes are decomposed by water. The extraction of the crude oil sample with distilled water removed insignificant amounts of Hg thus indicating that Hg is not present in the oil as a porphyrin complex. Arsenic also shows a higher concentration in the low molecular-weight fraction than in the 1000-4000 fraction. This is consistent with the conclusion that part of the As in crude oil is present in low molecular-weight compounds such as alkyl or aryl arsines R_xAsH_{3-x} (17).

TABLE X

Distribution of Trace Elements in Asphaltenes (17)

	Fraction 1 300-1000	Fraction 2 1000-4000	Fraction 3 4000-8000	Fraction 4 8000-22000
% of Crude	11.0	23.0	50.6	15.2
Element		Concentration (μg/g)		
Ni	1327.0	189.0	984.0	1060.0
Co	2.67	0.30	167.0	176.0
Fe	480.0	368.0	867.0	1934.0
Hg	72.0	20.9	90.0	350.0
Cr	0.77	4.80	9.12	19.6
Zn	112.0	52.0	103.0	225.0
Cu	0.34	1.50	4.0	7.20
Sb	11.0	0.91	0.35	0.10
As	0.850	0.620	1.90	6.60

TABLE XI

Distribution of Trace Elements in Resins (17)

	Fraction 1 300-1000	Fraction 2 1000-4000	Fraction 3 4000-8000
% of Crude	29.4	21.2	49.4
Element	Concentration of Element (μg/g)		
Ni	206	110	80.2
Co	4.37	10.0	24.9
Fe	30.1	24.0	236
Hg	22.0	44.0	72.0
Zn	3.31	11.0	27.0
Cr	0.310	0.800	2.96
Sb	0.043	0.0026	0.0054
As	0.407	0.200	0.200
Cu	<0.20	<0.20	1.30

The resin fractions show trace element patterns similar to those observed for the asphaltenes, although the elemental concentrations in the resin fractions are lower than in the corresponding asphaltene fractions.

The conclusions reached in this study are that trace elements may occur in the organic matrix of petroleum in several ways - as discrete organometallic species (e.g. Ni and V porphyrins), bound to the high molecular-weight asphaltene sheet structures, or in small molecules associated with the high molecular-weight sheets. Obviously much more work is needed to determine the nature of metals in petroleum and the oil studied by Filby (17) represents only one atypical petroleum.

Trace Elements in Synfuels

A number of liquid products derived from coal and oil shale are being proposed as substitutes for petroleum fuels or feedstocks. Solvent Refined Coal (SRC), Synthoil, H-Coal and Shale Oil are low sulfur, low ash fuels which may possibly be used and which may replace residual fuel oil as a boiler fuel. The coal products are obtained by dissolving coal, hydrogenating it under elevated temperature and pressure and either distilling off the product (H-Coal) or filtering off the mineral residue (Synthoil and SRC). Shale oil is obtained by retorting oil shale, either in-situ or in retorts. The trace element contents of these new materials is of interest for the same reasons advanced for petroleum. In the case of the coal products, the starting material has a much higher trace element content than most crude or residual fuel oils. Also, there is the possibility that toxic organometallic compounds may be formed under the highly reactive conditions of coal hydrogenation.

Table XII shows trace element data for shale oil (21), Solvent Refined Coal (31), parent coal (31) and the "average" RFO reported by Cahill (20). As can be seen from Table XII, residual fuel oil is generally lower in trace metals than SRC, except for Ni, V and Na. Shale oil is generally lower in trace elements for most elements than residual fuel oil, except for the important element As. Shale oil has a very high As concentration and may involve organoarsenicals. The forms of trace elements in SRC have not been determined and the high concentration of Ti is of particular interest. Work is currently underway in this laboratory to determine the chemical form of Ti and other elements in SRC as obviously these are important for environmental and industrial considerations. What can be seen by examining Table XII is that the reduction in trace element content from coal to SRC is not the same for all elements and the volatile elements Cl, As, Se, Hg, Cr and Sb appear to be depleted to a lesser extent than do typically refractory elements such as Fe, Sc, Eu. Whether this difference is due to reactions of volatile metal species with the reactive coal matrix undergoing hydrogenation or whether the difference is due to different forms in the original coal is not clear at this time.

TABLE XII

Trace Elements in Shale Oil and Solvent Refined Coal Compared to Coal and Residual Fuel Oil (ng/g)

Element	Shale Oil (21)	Coal (31)	SRC (31)	Residual Fuel Oil (20)
Fe (μg/g)	5.35	17300	68	4.6
Co	96.3	3700	310	320
Zn	73.0	-	-	1000
Cu	-	-	-	-
Cr	21.9	14000	2680	70
Se	252	1530	148	90
As	15600	13600	1390	55
Sb	4.38	500	74	4
Cs		890	23	-
Rb		22400	570	-
Sc		2800	130	2
Eu		292	13	2
Br	9.3	3510	3950	220
Hg	75.0	56	25	4.0*
Cl (μg/g)	-	260	160	40
Ni (μg/g)	0.88	20.0	2.7	12.5*
V (μg/g)	-	30.1	4.63	87
Na (μg/g)	1.0	148	9.55	33
Ti (μg/g)	-	530	465	-
S (%)	-	4.34	0.74	0.32*

*Values for Canadian Residual Fuel Oil

LITERATURE CITED

1. Bestougeff, M. A., in "Fundamental Aspects of Petroleum Geochemistry," B. Nagy and U. Colombo, editors, Elsevier, Amsterdam, 1967, Chapter 3.
2. Hackford, J. E., J. Inst. Petrol. Tech. 8, 193 (1922).
3. Ramsay, A., Ibid. 10, 87 (1924).
4. Shirey, W. B., Ind. Eng. Chem. 23, 1151 (1931).
5. Hodgson, G. W., Bull. Amer. Assoc. Petrol. Geol. 38, 2537 (1954).
6. Erickson, R. L., Myers, A. T., and Horr, C. A., Ibid. 2200 (1954).
7. Bonham, L. C., Ibid. 40, 897 (1956).
8. Hyden, H. J., U.S. Geol. Survey Bull. 1100 1 (1961).
9. Sugihara, J. M., Amer. Petrol. Inst. Proc., Div. Sci. Tech. Sec. 8 42 30 (1962).
10. Nellensteyn, F. J., J. Inst. Petrol. Tech. 10, 311 (1924).
11. Magee, E. M., Hall, H. J., and Varga, G. M., "Potential Pollutants in Fossil Fuels," EPA Report EPA-R2-73-249 (1973).
12. Milner, O. I., "Analysis of Petroleum for Trace Elements," Pergamon Press, Oxford, 1963.
13. McCoy, J. W., "The Inorganic Analysis of Petroleum," Chemical Publishing Co., Inc., N. Y., 1962.
14. Shah, K. R., Filby, R. H., and Haller, W. A., J. Radioanal. Chem. 6, 185 (1970).
15. Shah, K. R., Filby, R. H., and Haller, W. A., J. Radioanal. Chem. 6, 413 (1970).
16. Filby, R. H. and Shah, K. R., in "Role of Trace Metals in Petroleum," T. F. Yen, editor, Chapter 5, Ann Arbor Science Publishers, Ann Arbor (1975).
17. Filby, R. H., in "Role of Trace Metals in Petroleum," T. F. Yen, editor, Chapter 2, Ann Arbor Science Publishers, Ann Arbor (1975).
18. Hitchon, B., Filby, R. H., and Shah, K. R., in "Role of Trace Metals in Petroleum," Chapter 6, Ann Arbor Science Publishers, Ann Arbor (1975).
19. Al Shahristani, H., and Al-Atiya, M. J., Geochim. Cosmochim. Acta 36, 929 (1972).
20. Cahill, R. A., A Study of Trace Element Distribution in Petroleum, M.S. Thesis, University of Maryland, 1974.
21. Filby, R. H., unpublished data.
22. Yen, T. F., Erdman, T. G., and Saraceno, A. J., PREPRINTS, Div. Petrol. Chem. ACS 6 (3) B53 (1961).
23. Sugihara, J. M., Branthaver, J. F., Wu, G. Y., and Weatherbee, C., PREPRINTS, Div. Petrol. Chem. ACS 15 (2) C5 (1970).
24. Yen, T. F., PREPRINTS, Div. Petrol. Chem. ACS 17, (4) F120 (1972).
25. Kotova, A. V., Tokareva, L. N., and Berkovskii, V. G., Tr. Inst. Khim. Prir. Solei. Akad. Nauk. Kuz., SSR No. 1, 83 (1970).
26. Katchenkov, S. M. and Flegentova, E. I., Vestsi. Akad. Navuk. Belarus. SSR. Ser. Khim. Navuk. 95 (1970).
27. Witherspoon, P. B. and Winneford, R. S., in "Fundamental Aspects of Petroleum Geochemistry," B. Nagy and U. Colombo, editors, Elsevier, Amsterdam, 1967, Chapter 6.
28. Moore, J. W. and Dunning, H. N., Ind. Eng. Chem. 47, 1440 (1955).
29. Falk, J. E., "Porphyrins and Metalloporphyrins," Elsevier, Amsterdam, 1964.
30. Hodgson, G. W., Flores, J., and Baker, B. L., Geochim. Cosmochim. Acta 33, 532 (1969).
31. Filby, R. H., Shah, K. R., and Sautter, C. A., J. Radioanal. Chem. 37, 693 (1977).

THE BEHAVIOR OF IMPURITIES DURING COMBUSTION

Session Chairman: **G. K. Lee**
Canada Centre for Mineral Energy Technology
Co-Chairman: **T. D. Brown**
Canada Centre for Mineral Energy Technology

MICROSCOPIC INVESTIGATION OF THE BEHAVIOUR OF INORGANIC MATERIAL IN COAL DURING COMBUSTION

M. SHIBAOKA and A. R. RAMSDEN

CSIRO Minerals Research Labs.
Division of Mineralogy
North Ryde, N.S.W., 2113, Australia

ABSTRACT

The behaviour of mineral matter during the combustion of coal is discussed on the basis of (a) electron microscopic observations of samples taken from a pilot-scale pulverized-fuel-fired furnace and (b) controlled combustion experiments on individual coal particles heated under a hot-stage light microscope and recorded by high-speed cine photography.

The process of ash formation is influenced mainly by the chemical composition and thermal properties of the mineral matter and its mode of occurrence in the coal. Other factors involved include interaction between the inorganic and organic constituents of the burning coal particles and the physical and chemical environments through which these particles pass within the furnace.

INTRODUCTION

Study of the behaviour of inorganic material during the combustion of coal is essential to an understanding of the processes related to formation of fly-ash and boiler deposits and their associated electrostatic precipitation and corrosion problems. Direct observation of these processes is difficult but we have been attempting to provide a qualitative picture of coal combustion based on electron microscopic studies of samples taken at various stages during the burning of pulverized coal in a pilot-scale furnace and also on high-speed cine photography of individual coal particles burning in a hot-stage light microscope.

Knowledge of the average composition of a coal ash is not itself sufficient to predict, for example, the efficiency of electrostatic precipitation of the fly-ash. This is because the inorganic material in coal is not uniformly distributed either within the seam or even within an individual lump. Consequently, its behaviour in individual particles of pulverized coal is complex, depending upon the mode of occurrence (whether present in finely disseminated form or as relatively large inclusions), and also on the thermal behaviour of the coal. Moreover, the surface composition of fly ash will almost certainly be completely different from its bulk composition (1).

Another important factor to which relatively little attention has so far been given concerns the influence of the organic material itself on ash formation during combustion. The process of coal combustion is not as simple as that of pure carbon. Furthermore the physical and chemical environments, such as temperature and availability of oxygen, experienced by the particles vary greatly within a furnace. These factors, together with the wide range of melting points and/or decomposition temperatures likely to be exhibited by a heterogeneous assemblage of mineral matter, further complicate the behaviour of inorganic material during combustion.

These problems will be discussed at the Symposium with the aid of various photomicrographs taken in the electron microscope and a 16 mm movie film taken under the light microscope.

PREVIOUS WORK

The combustion behaviour of coal particles, particularly differences in the behaviour of the various coal microlithotypes, has been discussed by many authors, such as Newall and Sinatt (2,3), Alpern *et al.* (4), Littlejohn (5), Street *et al.* (6), Field (7) and Szpindler (8,9,10). Ivanova and Babii (11,12) and Shibaoka (13,14) have observed the combustion of coal particles directly under the microscope using high speed cine photography.

Typical fly-ash particles from British bituminous coals have been described microscopically by Hamilton and Jarvis (15) and classified by Watt and Thorne (16). Raask (17) investigated the occurrence of ash cenospheres in fly-ashes but, until the development of a technique for direct sampling from a pulverized-coal-fired surface by Ramsden (18,19), little work had been done on the processes of fly-ash formation itself. Subsequently, similar techniques have been used by Bonafede and Drew (20), Szpindler (8,9,10) and Pietzner and Schiffers (21). The significance of the microscopic features of fly-ash in relation to electrostatic precipitation have been discussed by Paulson and Ramsden (22).

SAMPLING AND MICROSCOPIC EXAMINATION

Two methods of investigation have been used in the present work. In one, fly-ash and partly burnt coal particles were collected from the gas stream of the combustion chamber and flue of CSIRO's technical-scale pulverized-coal-fired rig by means of a water-cooled probe (Fig. 1) (18). A bituminous coal from the Great Northern Seam, N.S.W., Australia, with a typical composition of moisture 2.9%, ash 13.2% and volatile matter 31.6% (air dried), was used. The particles collected were examined by transmission and scanning electron microscopy and by light microscopy. Some data obtained in a similar manner by Szpindler (8,9,10) during the C15 trials (1969) of the International Flame Foundation are also used in this paper.

In the other method, single coal particles (0.4-0.7 mm and 0.05-0.1 mm in diameter) and also some thin-sections of coal (0.015-0.02 mm) were heated rapidly in a hot-stage microscope (13,14) and the behaviour of the various microscopic constituents (macerals) in the coal were recorded during the combustion process using a high-speed cine camera. In some experiments the effect of combustion on the physical structure of the macerals and minerals was also studied by terminating combustion before completion and examining the residue under an electron microscope.

INORGANIC MATTER IN COAL

Mackowsky (23) has summarized the occurrence of mineral matter in coals in general, and Kemezys and Taylor (24) have reported on the occurrence and distribution of mineral matter in some Australian coals.

Fig.1 Water-cooled probe to collect particles from the gas stream of the p.f. furnace

(a) Collecting grid of stainless steel gauze
Diameter of disc = 3 mm
(b) Water-cooled probe with collecting grid in position
(c) Detail of the collecting grid in position close to a
Pt/Pt - 13% Rh thermocouple

From the viewpoint of combustion, the mineral matter in coal can be classified into two categories : (1) finely disseminated, generally submicroscopic, particles, and (2) relatively coarse mineral inclusions. The first category includes the inorganic constituents of the original coal-forming plants as well as fine detrital particles deposited during sedimentation. The second category comprises detrital grains of quartz, clay and various heavy minerals washed in with the coal-forming material and also authigenic constituents such as carbonates, pyrite and marcasite formed during diagenesis.

X-ray diffraction charts of some Australian coals show prominent peaks due to kaolinite. However, this clay mineral is almost always accompanied by other inorganic constituents, including other clay minerals of low crystallinity, and also iron and calcium carbonates that have a fluxing effect during combustion. Consequently, the thermal behaviour of the actual inorganic material in the coals may differ significantly from that of kaolinite alone.

The behaviour of relatively coarse mineral inclusions such as quartz and carbonates of authigenic origin is more or less independent of the presence of other inorganic matter because they occur as isolated grains having little opportunity for interation.

In practice, therefore, the inorganic materials within individual coal particles constitute independent reaction systems within the combustion furnace. However some limited reaction between ash particles and atmospheric gases may also take place, and coalescence of molten ash released from several different particles may also generally be expected to occur. After solidification of fly-ash particles away from the combustion zone, some constituents of the flue gas are adsorbed and absorbed by their glassy surface.

FINELY DISSEMINATED MINERAL MATTER

When particles of vitrinite and exinite in bituminous and sub-bituminous coals are heated rapidly, they release volatile matter and quickly swell to form char cenospheres. Vitrinite from semi-anthracite and the huminite of brown coal may also swell but less significantly.

The walls of the char cenospheres are very thin and consequently perforate and burn out very rapidly (Fig. 2). If the local temperature is relatively low, thin films of fine ash are formed along the retreating edge of the combustion fronts (Fig. 3). These films are often lined by partly molten ash grains. Such films, however, are not commonly observed because, as mentioned later, the cenospheres tend to burn off in the high temperature zone of the furnace. If, as is commonly the case, the temperature is high enough to fuse the ash, then submicroscopic ash droplets are formed just behind the combustion fronts (Fig. 4). These coalesce into larger and larger droplets as the combustion fronts advance and merge (Fig. 5), and, provided the temperature within the burning cenosphere remains high enough, any such droplets released locally inside the burning particle will either coalesce further with other droplets or, if no longer molten, will tend to stick to other droplets to form aggregates.

In contrast to the behaviour of vitrinite and exinite reported above, particles of inertinite do not fuse and swell during rapid heating. Instead, combustion takes place only at the outer surface of the particle and the advance of the combustion fronts is essentially planar. This results in a relatively slow burn-off of inertinite particles with ash droplets formed mainly on the external surfaces of the particles (Fig. 6). The lateral movement of such ash droplets along the combustion fronts may be less free than in the case of ash formed within cenospheres and consequently there is less opportunity for coalescence and growth. Instead the droplets tend to be dispersed into the gas stream to form the finest fractions of the fly-ash.

Generally, the local temperatures at the microscopic combustion fronts of burning coal particles may be expected to be higher than those in the surrounding gas stream due to the heat generated at the fronts. Locally, this could be high enough to fuse disseminated mineral matter and produce spherical ash particles even where the gas temperature is below the melting points of the mineral matter. However, as mentioned later, in some parts of the furnace the temperatures may not be high enough to fuse the ash.

MINERAL INCLUSIONS

Much of the mineral matter present as discrete inclusions can be freed from coal during pulverization; some, however, always remains. Because these inclusions are not intimately mixed with the carbonaceous matter, they will not in general be directly affected by the burning coal (in contrast to the marked effects on disseminated mineral matter) and the temperatures reached by individual inclusions may be lower than their melting points or dissociation temperatures, with the consequent production of angular and irregularly-shaped fly-ash particles.

Secondary minerals such as quartz and carbonates often fill the pore spaces in fibrous fusinite. The melting point of these minerals is high, so that while they will be externally heated by combustion of the fusinite itself, they may yet not fuse. There may be some fluxing effects on these inclusions from the disseminated ash in the fusinite but such effects are likely to be limited due to the relatively small areas of contact between the two materials.

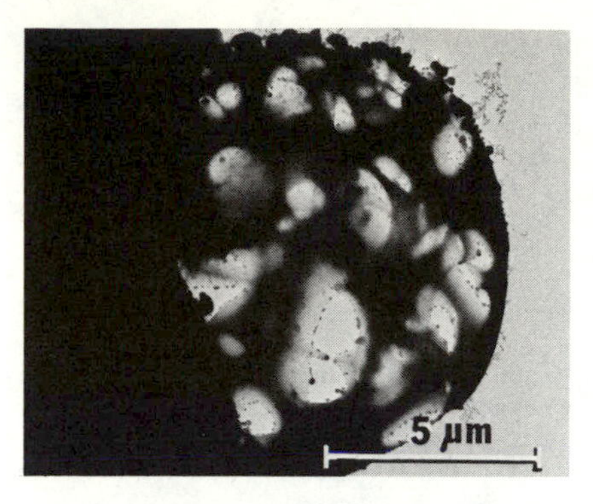

Fig.2 Partly reacted, devolatilized cenosphere from pulverized bituminous coal

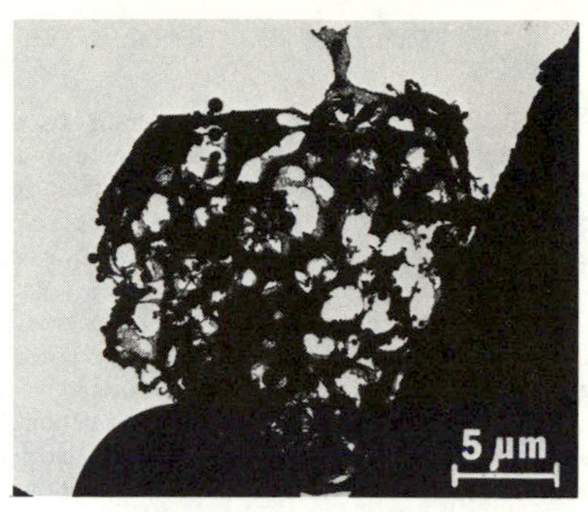

Fig.3 Cenosphere showing very thin films of fine ash formed along the retreating edge of the combustion front at a relatively low temperature

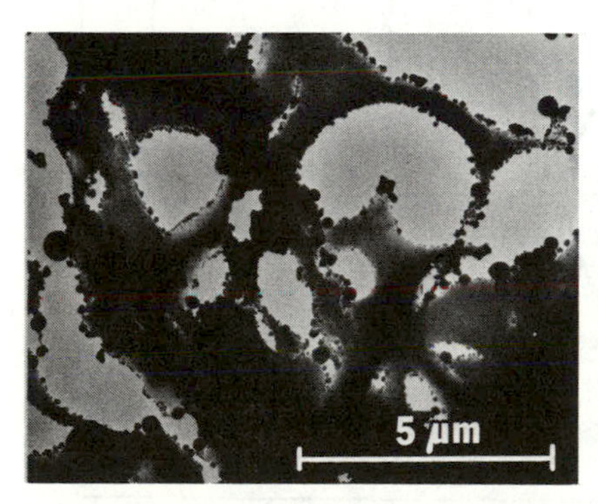

Fig.4 Fragment of a cenosphere showing fused ash droplets formed just behind the combustion front at a high temperature

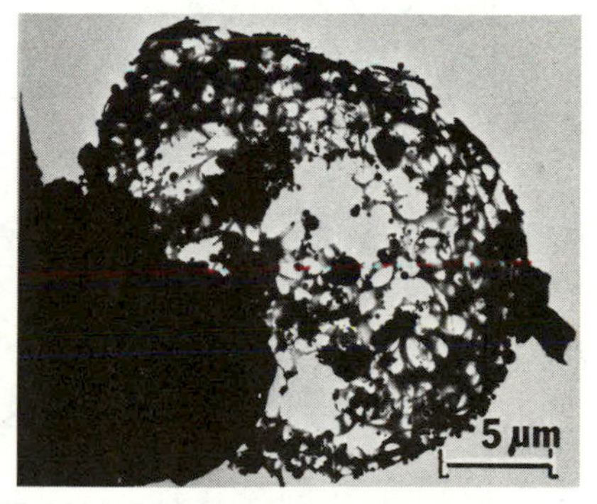

Fig.5 Partly reacted cenosphere with large ash droplets formed by coalescence of finer ash droplets

Fig.6 Fine submicron ash droplets typically formed from inertinite particles

BEHAVIOUR OF MINERAL MATTER IN TWO TEMPERATURE ZONES

When pulverized coal is burned in a furnace, two combustion zones (which may overlap to a greater or lesser extent) are formed in the flame. The first zone arises through combustion of the volatile matter given off immediately after the coal particles are injected into the furnace. The second zone is formed by combustion of the residual chars.

It is worth noting that only a small portion of the furnace reaches a really high temperature, 1600°C for instance. Generally, if a coal of low calorific value (e.g. high ash coal) or a brown coal is burned, the temperatures shown in Fig. 7 would be lower. Although the actual distribution of the two combustion zones would be different for a semianthracite fuel, the basic pattern should nevertheless be essentially the same.

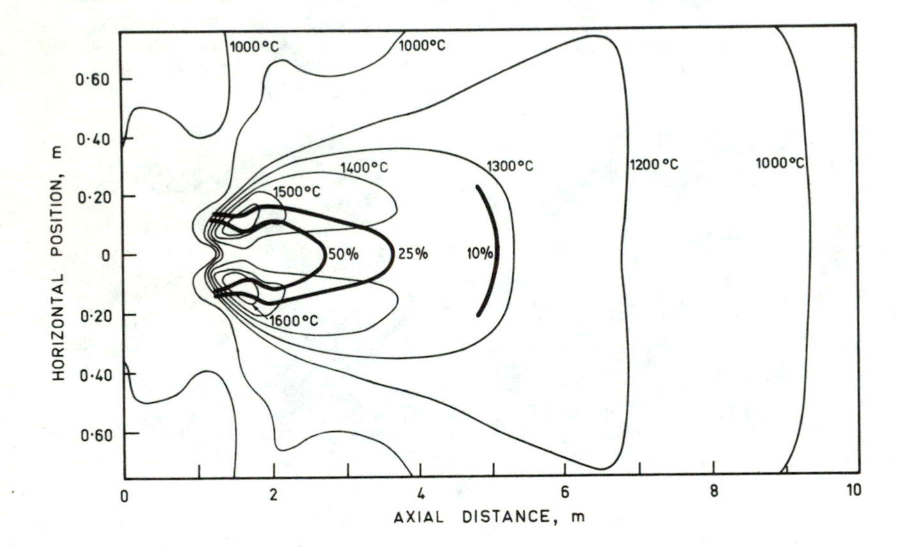

Fig.7 Variations in temperature (——) and combustible material content (——) in a p.f. furnace (8,9,10)

In the case of char cenospheres formed from vitrinite particles, burning takes place both at internal as well as external surfaces (25) and they tend to burn off quickly within the relatively high temperature zone of a furnace. Because of the rapid combustion, the temperature of the cenospheres will rise quickly and, as mentioned before, may be sufficiently high at the combustion fronts to form spherical ash droplets which tend to grow through coalescence with others. These droplets solidify, or at least tend to become less fluid, soon after the char is consumed by combustion and they then move into the cooler part of the furnace.

The combustion of inertinite particles is slow compared with that of vitrinite and thus may continue from the high temperature zone into the low temperature zone. If the inertinite particle burns off entirely within the high temperature zone, then the finely disseminated mineral matter within it may be fused. However, where the particle continues to burn in the low temperature zone, not all the inorganic material may be fused or, if fused, may not be sufficiently fluid to allow for coalescence of the fine ash droplets. In this case, very fine fly-ash particles are produced. A similar result may also be expected to occur for any coal particle that is exceptionally rich in mineral matter because of slow rate of combustion.

In the example given in Fig. 7, about 10% of the char is burned in a region where the temperature is less than 1300°C. Most mineral matter, with melting points above this, will not fuse in this zone (unless subjected to locally higher temperatures within cenospheres) but will produce very fine ash particles. If the furnace temperature is low due to, for example, high moisture or mineral matter content of the coal, then the region in which low temperature fine ash is formed will be extended, leading to production of larger numbers of very fine fly-ash fractions. High concentrations of such very fine particles (less than 5 μm) in fly-ash decreases the efficiency of collection in electrostatic precipitation (22).

Isolated mineral inclusions entering the furnace as liberated particles will be exposed directly to a high temperature and may be fused or at least softened as they pass through the high temperature zone. Some of the larger spherical or rounded ash particles that can often be seen embedded in char particles sampled from the volatile combustion zone may have such an origin. However, most of the large mineral particles probably undergo only partial softening in this high temperature zone due to their large thermal capacity and the relatively short residence time.

To further complicate the picture, some mineral particles may actually be ejected from a swelling vitrinite particle and subsequently fuse in the gas stream of the high temperature zone. In other cases, external mineral particles may be encapsulated in a rapidly swelling vitrinite particle or become partly coated by molten coaly matter and thereby be protected from the early high temperatures. In consequence, they may undergo only partial fusion in the high temperature zone even though their melting points are relatively low.

As each individual coal particle behaves as a separate reaction system during combustion, each fly-ash particle will be different in its detailed chemical composition. Some fly-ash particles also will have a heterogeneous composition either because of incomplete homogenization during fusion or because they have formed through coalescence of ash droplets of widely different compositions at a relatively low temperature.

In our experience with Australian coals, the following factors are likely to be the main cause for the formation of very fine fly ash particles :

High ash fusion point
High mineral matter in the coal
High inertinite content in the coal.

Under any of these conditions relatively more mineral matter is converted to ash at temperatures lower than or close to its melting point. If two or more of these conditions occur together, then the concentration of very fine fly-ash particles in the gas stream may become significantly high and may cause serious precipitation problems.

CONCLUSIONS

It is concluded that the process of ash formation during combustion of pulverized coal is complex and dependent upon a variety of factors, of which the most important are the chemical composition and thermal properties of the mineral matter and its mode of occurrence in the coal. In detail, the inorganic constituents in each coal particle behave as essentially independent reaction systems; however, certain generalizations can be made:

1. Vitrinite particles form cenospheres which burn off rapidly in the high temperature zone of the furnace. Combustion takes place at both internal and external surfaces in such particles and disseminated mineral matter in them tends to fuse just behind the combustion fronts into submicron ash droplets which then increase in size by progressive coalescence as the particle burns out. In this way, well rounded fly-ash particles ranging from submicron to

several hundred microns in size are released into the furnace.

2. Inertinite particles do not form cenospheres and burn off relatively slowly in both the low temperature as well as the high temperature zones of the furnace. This results in the release of a large proportion of very fine fly-ash particles into the gas stream. Where combustion continues into the low temperature zone of the furnace such very fine ash particles will be less well fused.

3. The behaviour of discrete mineral inclusions during combustion can vary considerably depending upon their size, thermal capacities and fusion points and also at what stage and for how long they are exposed to high temperatures. In general, however, such inclusions produce the coarser, and usually only partly fused, fractions of fly-ash irrespective of whether they originate in vitrinite or intertinite particles.

4. The proportion of very fine ash produced from disseminated mineral matter during combustion increases if the fusion point of the mineral matter is high and/or the coal particle is rich in mineral matter and intertinite, because relatively more mineral matter is converted to ash at temperatures below its fusion point.

5. Further complications in the production of fly-ash can be envisaged where fused or partly fused ash becomes coated by molten organic matter and thereby temporarily protected from the high temperatures and where aggregation of ash particles occurs in the gas stream, both circumstances leading to the production of irregular ash particles of highly variable composition.

ACKNOWLEDGEMENTS

The authors are indebted to Mr G. à Donau Szpindler for information obtained during the C15 trials in 1969 of the International Flame Foundation and wish to thank Mr I.W. Smith and C.A.J. Paulson for valuable suggestions during preparation of the paper.

BIBLIOGRAPHY

1. Collin, P.J., Symposium on Changing Technology of Electrostatic Precipitation, Institute of Fuel, Adelaide (1974).

2. Newall, H.E., and F.S. Sinnatt, Fuel, 3, 424 (1924).

3. Newall, H.E., and F.S. Sinnatt, Fuel, 5, 335 (1926).

4. Alpern, B., P. Courbon, J. Plateau, and G. Tissandier, J. Inst. Fuel, 33, 399 (1960).

5. Littlejohn, R.F., J. Inst. Fuel, 40, 128 (1967).

6. Street, P.J., R.P. Weight, and P. Lightman, Fuel, 48, 343 (1969).

7. Field, M.A., Combust. Flame, 14, 237 (1970).

8. Szpindler, G. à Donau, CSIRO Division of Mineral Chemistry, Investigation Report 87 (1971).

9. Szpindler, G. à Donau, CSIRO Division of Mineral Chemistry, Investigation Report 89 (1972).

10. Szpindler, G. a Donau, CSIRO Division of Mineral Chemistry, Investigation Report 94 (1972).

11. Ivanova, I.P., and V.L. Babii, Thermal Engineering (Teploenergetika) 13, No 4, 70 (1966).

12. Ivanova, I.P., and V.L. Babii, Thermal Engineering (Teploenergetika), 13, No 5, 100 (1966).

13. Shibaoka, M., J. Inst. Fuel, 42, 59 (1969).

14. Shibaoka, M., Fuel, 48, 285 (1969).

15. Hamilton, E.M., and H.D. Jarvis, "The Identification of Atmospheric Dust by Use of the Microscope", Central Electricity Generating Board, London (1963).

16. Watt, J.D., and D.J. Thorne, J. Appl. Chem., 15, 585 (1965).

17. Raask, E., J. Inst. Fuel, 41, 339 (1968).

18. Ramsden, A.R., J. Inst. Fuel, 41, 451 (1968).

19. Ramsden, A.R., Fuel, 48, 121 (1969).

20. Bonafede, G., and W.M. Drew, State Electricity Commission of Victoria, Planning and Investigation Department, Scientific Division, Report MR-181 (1969).

21. Pietzner, H., and A. Schiffers, VGB-Speisewassertagung, 30 (1972).

22. Paulson, C.A.J., and A.R. Ramsden, Atmos. Environ., 4, 178 (1970).

23. Macowsky, M.-Th., In: Murchison, D.G., and T.S. Westoll (eds.): "Coal and Coal-bearing Strata", p. 309, Oliver and Boyd, Edinburgh (1968).

24. Kemezys, M., and G.H. Taylor, J. Inst. Fuel, 37, 389 (1964).

25. Ramsden, A.R., and I.W. Smith, Fuel, 47, 253 (1968).

RESEARCH INTO THE COMBUSTION AND FOULING BEHAVIOUR OF BROWN COALS

KLAUS R. G. HEIN

Rheinisch-Westfalisches
Electrizitatswerk AG RWE
Krupstrasse-43 Essen 1
West Germany

Introduction

In the Federal Republic of Germany energy production to a large extent is based on brown coals, hence fuels with a low calorific value. The majority of the 120 million tons raw brown coals mined anually is used for electrical energy generation in thermal power stations with individual boiler capacities up to 19oo Mg/h steam. The moisture content of the fuel varies from 57 to 6o %, the non-combustible content from 2 to 2o %. Furthermore, the composition of this non-combustible matter showes extreme variations.

On a dry coal basis the concentrations of the main components (expressed as p.p.m.) Na, K, Ca, Mg, Fe, Si, Al, S can vary in vertical direction in every seam by one to two powers of ten. This ash is partly bound organically within the carbon matrix - up to 2.5 % of the total ash content - whereas higher percentages occur as sandy and clay-type mineral impurities in the seams.

Due to the characteristics of the fuel particularly the composition

---

[1] This research work was financially supported by the Government of Northrhine-Westphalia.

of the non-combustible matter, severe difficulties arise for the operation of the power station. In particular material erosion in the fuel preparation and grinding systems, fouling of the heat transfer surfaces in the combustion chamber, and variation of the particulate emission values has been experienced. Whilst the first problem is mainly a question of materials, and the latter can be dealt with by proper percipitator construction, fouling remains still a largely unknown process. Up to now the only practical method to deal with the difficulties due to fouling is the use of blowers. Water and/or steam blowers are common in modern boilers.

In order to find other means to minimize fouling and slagging in brown coal fired boilers the physical, chemical and mineralogical processes involved have to be known. Although many attempts have been made throughout the world a clear picture has not yet been reached.

Therefore, a several years' research project has been started recently in the Federal Republic of Germany by the Rheinisch-Westfälisches Elektrizitätswerk AG jointly with the Rheinische Braunkohlenwerke AG. The project ist entirely devoted to the three interconnected problems of ignition, combustion and subsequent fouling of low calorie fuels, in particular brown coal. For this purpose a research rig (Fig. 1) with a throughput of 5oo kg/h raw brown coal has been errected.

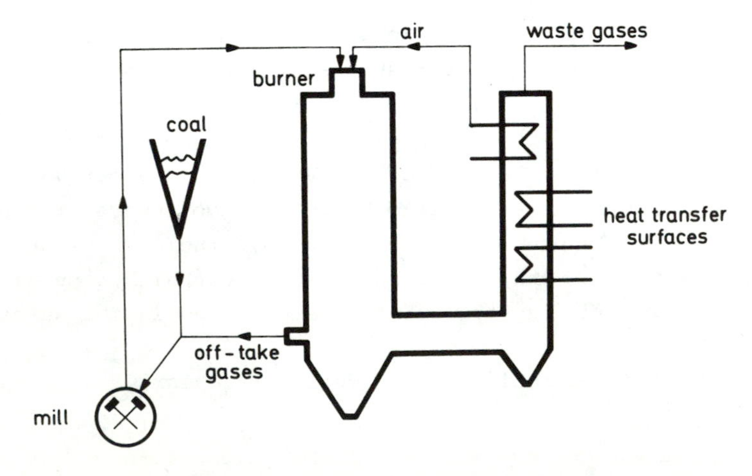

Fig. 1 Research rig Niederaußem

The rig allows the main industrial parameters such as particle size distribution, excess air level, velocities at the burner exit, and combustion air temperature to be varied. By probing the flame and taking samples of coal, fly ash, and deposits it has been found possible to correlate fuel characteristics and furnace conditions with the observed fouling behaviour.

Results

Although the research project is still in progress several important results have been reached already.

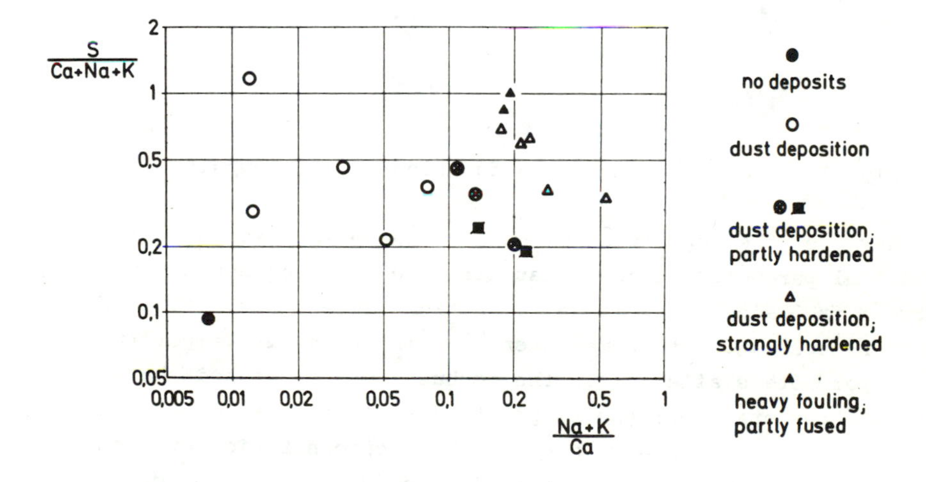

Fig. 2 Observed fouling tendencies (Molar ratios of raw coal)

When burning different brown coals from different cuts and seams a change in the fouling tendencies can be observed (Fig. 2): Increasing sulfur- and alkali-content and decreasing calcium concentration in the non-combustible matter of the fuel leads to enhanced fouling tendencies. Measuring the deposited mass per unit time and surface at the inner walls of the combustion chamber (Fig. 3) the same tendencies are found. In addition silica also seems to worsen the fouling problem.

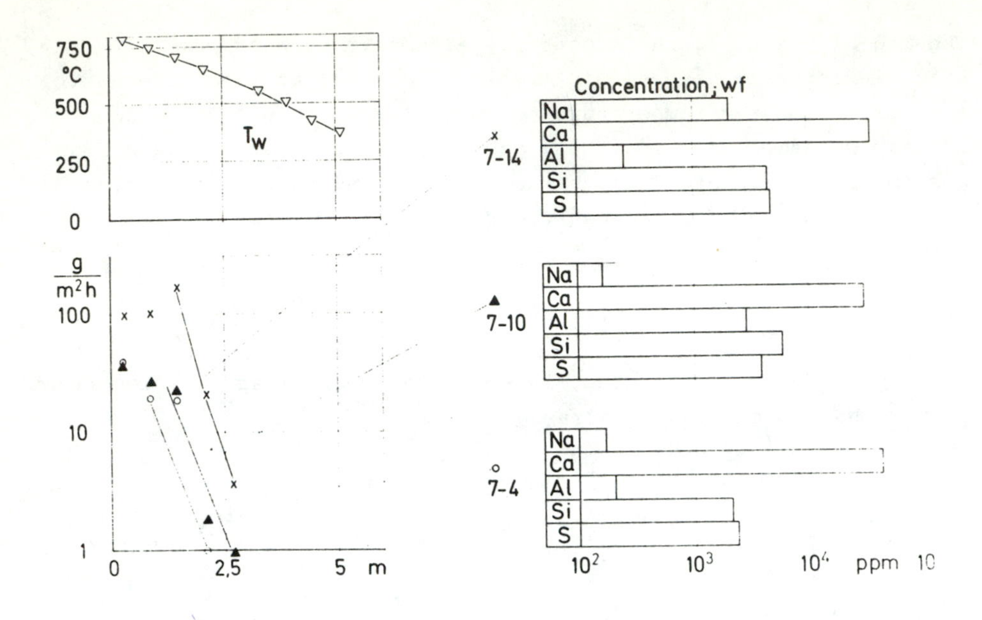

Fig. 3 Deposit formation as function of ash compsition

Furthermore, the deposition mechanism can be influenced by the technical parameters like excess air level, burner dependent mixing between fuel and combustion air, particle size distribution of the incoming coal and - to some extent - combustion air temperature. These parameters affect (via the combustion process and the subsequent heat transfer) the temperature field in the combustion chamber, hence the local temperature and concentration gradients of gas and solid which are key influences on the chemical reactions in the flame and the deposits.

Secondly, the excess air level and the mixing between coal and air were found to be the most important operational parameters. They determine the local oxygen partial pressure, the second key influence on the chemical reactions involved. The results in figures 4 and 5 show not only the expected shift in the flame temperatures (Tg), but indicate also a strong reduction of the deposits per unit time and area at the combustion chamber walls with increased excess air levels. A possible explaination is offered by a detailed investigation into the sulfur bound in the fly ash

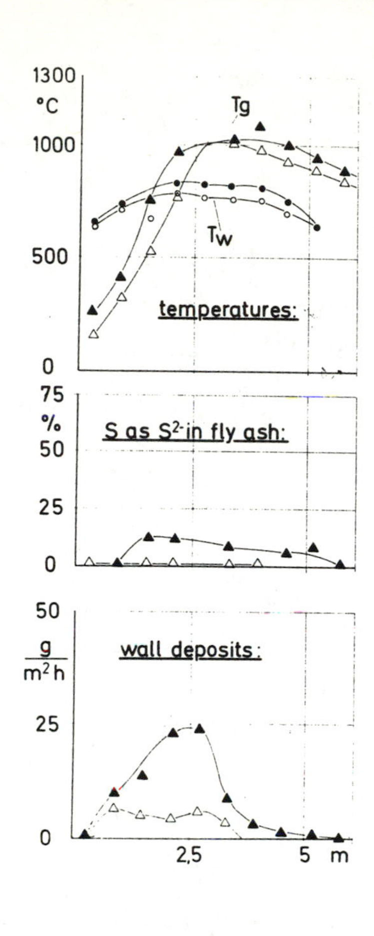

	△ ○	▲ •
flame	9/9	9/10
excess air	50%	25%
swirl	-	
air outlet	cyl.	

Fig. 4
Influence of
excess air level

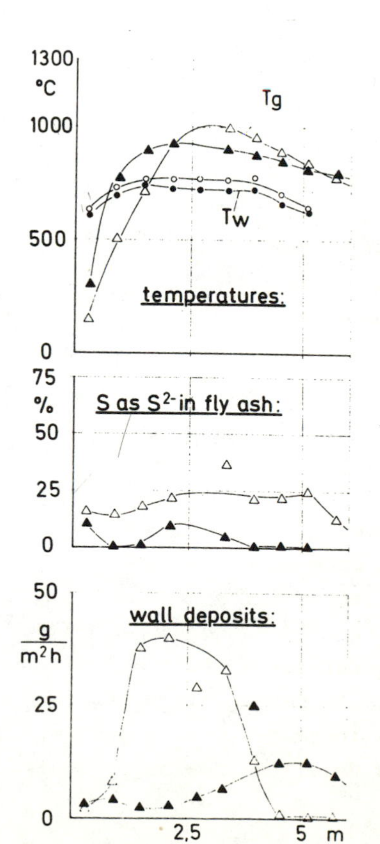

Fig. 5
Influence of
Excess air level

	△ ○	▲ •
flame	9/7	9/8
excess air	0%	65%
swirl	-	
air outlet	cyl.	

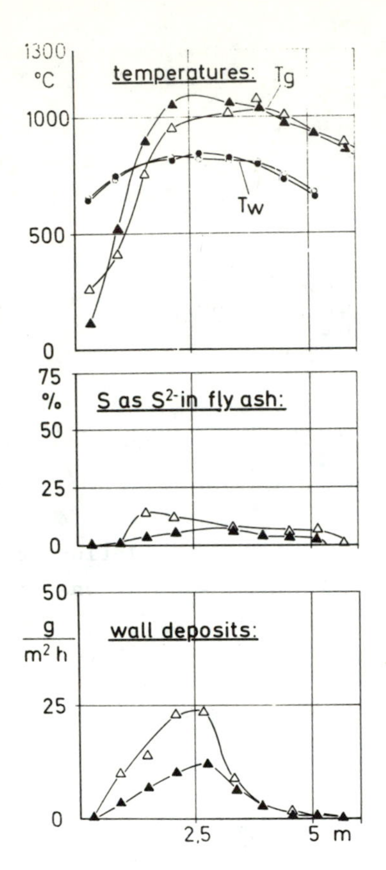

	△ ○	▲ ●
flame	9/10	9/11
excess air	25%	
swirl	-	+
air outlet	cyl.	

Fig. 6 Influence of Swirl

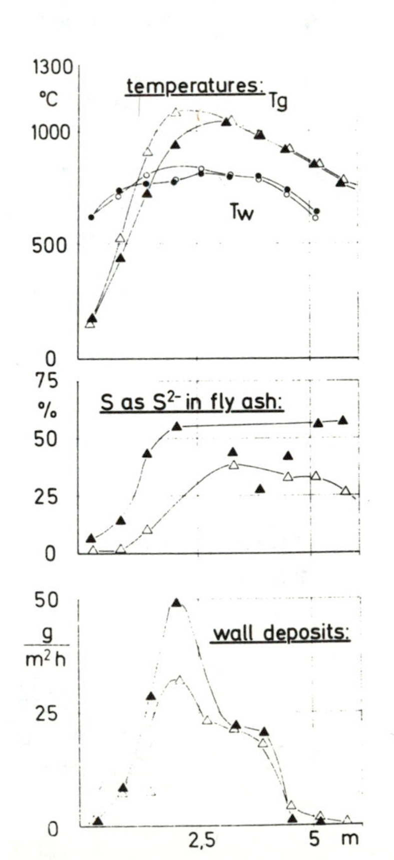

Fig. 7 Influence of burner exit geometry

	△ ○	▲ ●
flame	9/5	9/6
excess air	0%	
swirl	-	
air outlet	div.	cyl.

samples taken locally from the flame. With higher excess - air levels and an increased partial pressure of oxygen in the flame the percentage of sulfur fixed in the fly ash as sulfides (S^{-2}) is reduced. The sulfides in an oxide mixture are known to have generally lower melting points than the comparable sulfates; they are intermediate products which neither exist in the original raw coal nor can be detected in the deposits. Even the analysis of the fly ash samples has to be done as soon as possible after flame probing since these sulfides are not stable under ambient atmospheric conditions.

Improved mixing between fuel and combustion air also leads to elevated oxygen partial pressures early in the flame. This effect can be created either by using swirl in the combustion air (Fig.6) or by choosing a diverging air nozzle instead of the usual cylindrical air exit (Fig. 7). In these cases the explanation above also applies. The obvious reduction of the sulfide percentage in the fly ash samples leads to a subsequent reduction of the deposits at the combustion chamber walls.

Conclusions

Fouling of the heat transfer surfaces of boilers in which fuels of low calorific value are burnt causes severe difficulties for power station operation. The deposit formation is strongly dependent on the composition of the fuel ash. The phenomenon of fouling can be influenced by operational parameters like excess air level and mixing between fuel and combustion air.

Further research is necessary to quantify the discussed tendencies in order to optimize boiler operation and ultimately improve power station availability.

ASH FOULING STUDIES OF LOW-RANK WESTERN U.S. COALS

E. A. SONDREAL, G. H. GRONHOVD, P. H. TUFTE, and W. BECKERING

Energy Research and Development Administration
Grand Forks Energy Research Center
Grand Forks, North Dakota, USA 58201

ABSTRACT

The mineral content in lignite and subbituminous coals in the Western U.S. is distinctly different from that found in the bituminous coals of the East and Midwest. Most of the mineral content in lignite consists of alkali cations bound to the organic structure of the coal and of finely dispersed silica and aluminum silicates. Much of the low-rank Western U.S. coal contains an appreciable sodium content which tends to flux the inherent mineral matter and cause serious ash fouling problems in power boilers. The ERDA Grand Forks Energy Research Center (GFERC) has studied ash fouling both in full-scale boiler tests and in a 34 kg/hr (75-lb/hr) test furnace. The most consistent results have been that the degree of fouling is a direct function of the sodium and ash contents of these coals. Silica in the ash tends to increase fouling and calcium is beneficial. A fouling mechanism for North Dakota lignite has been proposed.

INTRODUCTION

Ash fouling of heat transfer surfaces is the most serious operating problem of boilers fired on Western U.S. lignite and subbituminous coal. Problem coals cause rapid ash deposition which can force repeated unscheduled shutdowns. The formation of bonded deposits in the convection section of the boiler is the principal problem in burning low-rank Western coals. Wall slag in the furnace can normally be removed by soot blowing, including the use of water blowers in recent years. However, in burning high-sodium coals, rapid formation of hard bonded deposits on convection surfaces is very difficult to control by on-line cleaning.

In commercial practice, fouling has been reduced by constructing larger boilers having a low heat release per unit volume and by very extensive soot blowing capabilities. Evaluation of fouling potentials in the 34 kg/hr (75-lb/hr) pc-fired laboratory furnace at GFERC has assisted in establishing boiler design specifications for new coals. Selective mining and blending have also been used to alleviate fouling by burning low sodium fuel during the periods of highest electrical load, during winter months.

Development of furnace designs for burning highly fouling lignites has been an evolutionary process, which is still continuing. Changes in design for the purpose of reducing NO_x emissions may be detrimental to fouling, which will pose a new challenge to boiler manufacturers and utilities.

Research at GFERC has included field testing in operating boilers, operation of a laboratory furnace, and studies on the mechanism of the fouling process. Remedial methods studied have involved either the removal of sodium from the coal by ion exchange or the use of additives to reduce the action of the sodium during combustion. Basic supporting studies on ash fouling at GFERC include: 1) continuing characterization of the mineral matter in Western coals, 2) examination of the microscopic structure and analysis of boiler deposits, 3) determination of the microscopic melting temperatures of fly ash particles and deposit fragments, and 4) studies on volatility of mineral constituents from coal and ash. Specialized analytical tools available to this work include X-ray fluorescence, X-ray diffraction, an electron microprobe, a scanning and analyzing electron microscope, a heated stage microscope, and a graphite furnace attached to an atomic absorption unit.

FIELD TESTS ON ASH FOULING IN OPERATING BOILERS

Starting in 1965, GFERC has performed tests on ash fouling in operating boilers burning North Dakota lignites for the purpose of relating the rate of fouling to the ash characteristics and powerplant operating conditions. The principal method of testing has involved the use of air cooled probes to obtain direct measurements of rates of ash deposition. In addition, statistical methods have been used to correlate observations on the severity of fouling with coincidental changes in coal or ash analysis or operating conditions.

A series of probe tests was conducted on a 53 MW pc-fired boiler in 1965 and 1966 on four selected lots of lignite containing nominally 2, 4, 6, and 8 pct Na_2O as percent of ash, and 6 to 7 pct ash in the as-received coal (1,2). Rates of fouling at three different probe locations are shown in Figure 1 as a function of percent Na_2O in ash. These results demonstrate the paramount effect of sodium content and the importance of temperature.

Recent probe tests (1976) were conducted to determine relative fouling rates in cyclone firing versus pulverized coal firing when burning the same high fouling North Dakota lignite (3). Tests were conducted on identical lignite burned in both a 216-MW pulverized coal-fired boiler and a 440-MW cyclone-fired boiler. Averaged results for 15 tests on the pc-fired boiler and 8 tests on the cyclone-fired unit indicated that the fouling rate for cyclone firing was only about half that for pc firing. This finding agreed with the past opinion of lignite-burning utilities, which held that cyclone firing should significantly reduce fouling because of the smaller quantity of fly ash passing up through the furnace. Operation of the first cyclone-fired boiler (212 MW) using lignite had been virtually free of fouling; however, this unit was operated on low sodium fuel.

In probe tests, a slightly greater sodium enrichment was observed in the deposits formed in the cyclone-fired unit as compared to the pc-fired unit. At different levels of temperature or sodium content from those experienced in these tests, this sodium enrichment could possibly cause aggravated fouling in a cyclone-fired boiler. Two new 440-MW cyclone-fired units burning North Dakota lignite have experienced instances of severe fouling involving extremely hard deposits which are difficult to remove by soot blowing. Thus, while the rate of deposition as measured by short-term probe testing indicated an important advantage for cyclone firing, the overall problem may be too complex to be adequately assessed by this method. An important limitation of this method of study is that deposit aging, hardness, and effectiveness of soot blowing are not evaluated.

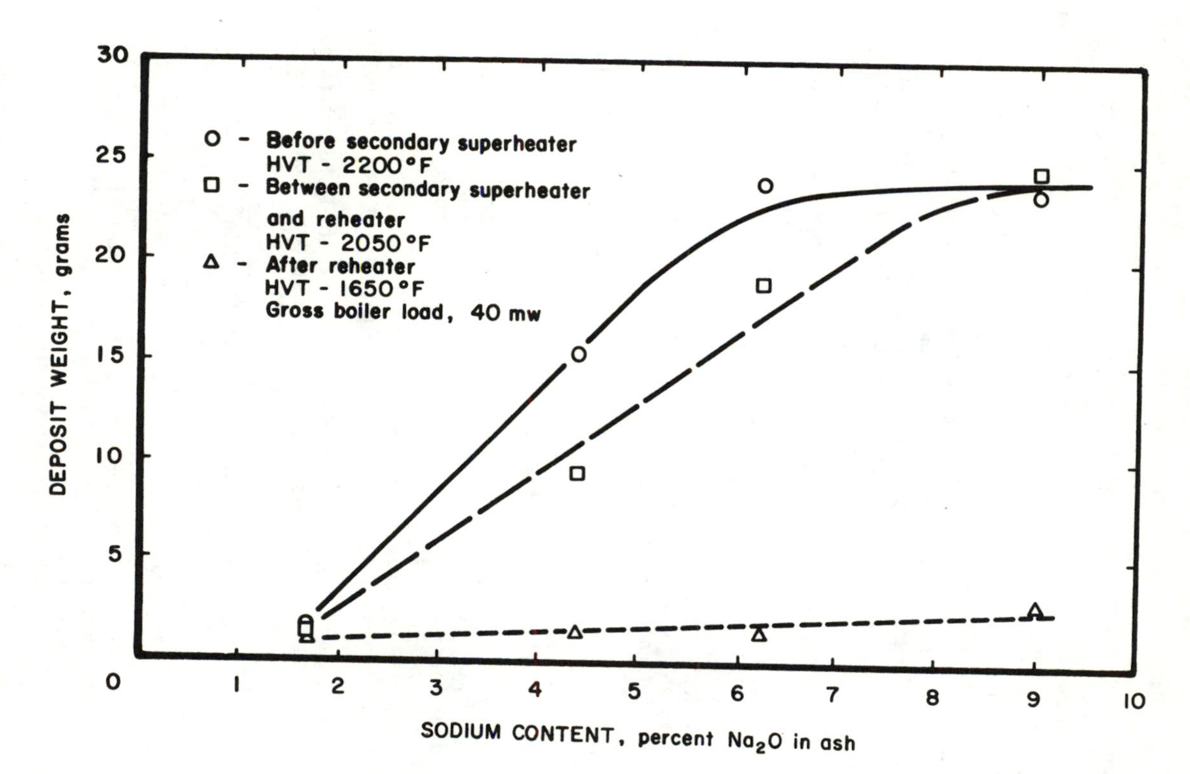

Figure 1. Ash deposition rates as a function of sodium content at three locations in a 53-MW boiler.

As an alternative method of study on cyclone firing (1976), statistical methods were used to relate observations on boiler fouling conditions in a 440-MW boiler to coal ash content, the ash analysis, and operating parameters, principally boiler load (3). Observations on boiler fouling condition were made daily through 32 viewing ports on the boiler, assigning a numerical score from 1 to 4 according to the extent that deposits were formed and not removed by soot blowing. Operating data logged each day included boiler load, air heater inlet temperature, flue gas oxygen, and ID fan suction pressure. Daily representative coal samples were analyzed for percent ash and elemental ash composition. The effects of sodium and ash, as the principal coal related properties, are shown in Figure 2 for intermediate boiler loads in the range of 360 to 385 MW. The predominant effect of sodium on boiler fouling is again well illustrated, and higher ash contents are shown to correspond to higher fouling ratings. Of numerous multiple regressions performed on these data, the one best representing the effects of the major coal related properties was as follows:

$$\text{Fouling rating} = 0.38\ Na_2O + 0.006\ SiO_2 - 0.008\ CaO + 0.062\ \text{Ash} + 0.0037,$$

where oxide constituents are expressed as percent of ash and "Ash" is on a dry coal basis. This equation, which was generated on 19 data sets collected over a period of two months operation, had a regression coefficient (R) of 0.986. The effects represented by the terms in the equation are all consistent with findings obtained in field tests or in the laboratory fouling furnace at GFERC. However, the statistical approach did not identify any previously unrecognized effect. The principal limitation of this method of study is that the observed fouling at any one time is not only a function of the operating conditions at that time, but also of all previous fouling since the last boiler cleaning.

MODELING ASH FOULING IN THE GFERC FURNACE

Much of the research on ash fouling at GFERC has been performed using a pulverized-coal fired, laboratory furnace. Layout and an overall view of the furnace and panel board are shown in Figures 3 and 4. The bottom-fired combustion chamber is 76 cm (30 in.) in diameter by 2.4 meters (8 ft) high and is refractory lined. The volumetric heat release rate is about 120,100 kg-cal/hr/m^2 (13,500 Btu/hr/cu ft), which is similar to that in a commercial boiler.

Test coal is prepulverized to a fineness of approximately 80 pct through 200 mesh in a hammermill and charged to a volumetric feeder. Coal is fed at a nominal rate of 34 kg/hr (75 lb/hr). Combustion air is preheated by flue gas in a tubular heat exchanger and then brought to final temperature in an electric air heater. When burning pulverized lignite containing 30-pct moisture, air is preheated to from 370 to 480° C (700 to 900° F). The pulverized coal is fed into the throat of a venturi in the primary air line to the burner. Heated secondary air is introduced into the burner, and heated tertiary air is added through two tangential ports located in the furnace wall about 30 cm (1 ft) above the burner cone. The percentages of the total air used as primary, secondary, and tertiary are about 10, 30, and 60, respectively.

Flue gas at 1093° C (2000° F) passes out of the furnace into a 25 cm (10-in.) square duct which is also refractory lined. Three probe banks located in the furnace exit duct are designed to simulate superheater or reheater surfaces in a commercial boiler. The first probe bank is vertical and the second and third are horizontal. The probes, three at each bank, are 4.2 cm (1.66 in.) OD type 304 stainless steel pipes and are cooled by compressed air. Normally the metal temperatures for probe banks 1, 2, and 3 are controlled at 538, 538, and 427° C (1000, 1000, and 800° F), respectively. The gas velocity between the tubes at the first probe bank is about 7.6 m/sec (25 fps). Normal gas temperatures entering the first, second and third banks are 1093, 982, and 871° C (2000, 1800 and 1600° F).

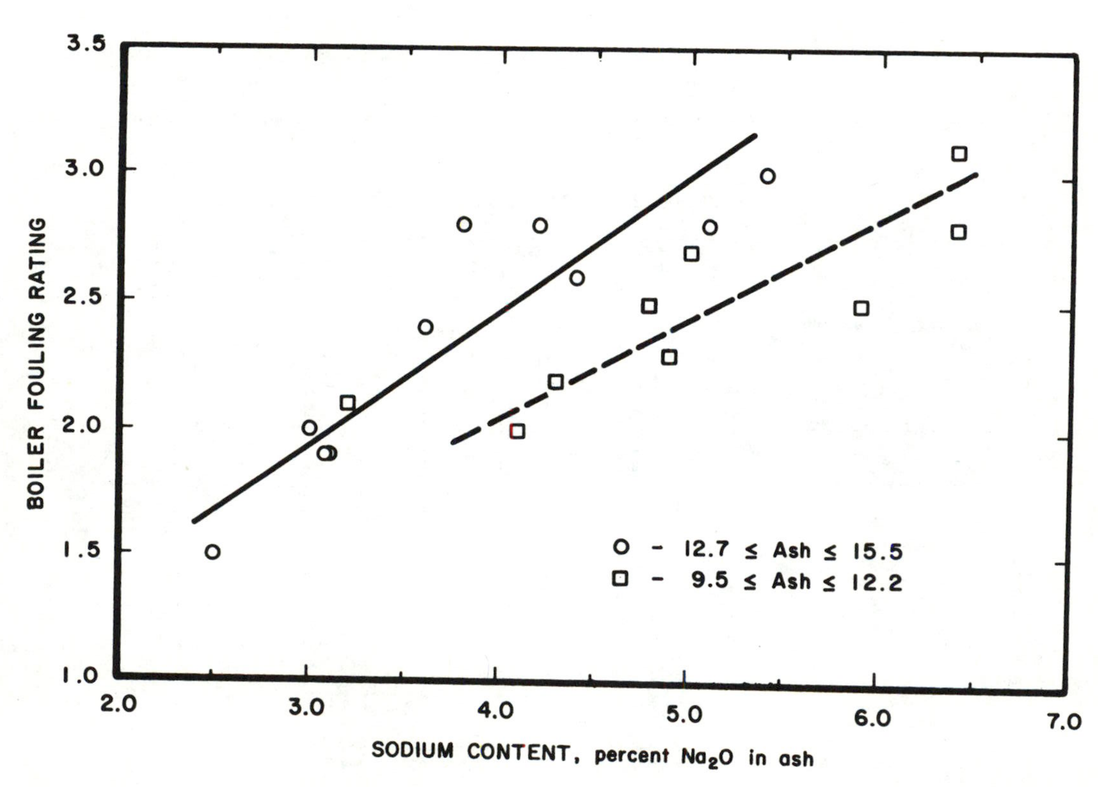

Figure 2. Boiler fouling rating versus sodium content in a 440-MW boiler. Each rating represents 32 averaged observations on a scale of 1, for a clean boiler, to 4 for deposits which could not be removed by soot blowing or hand lancing.

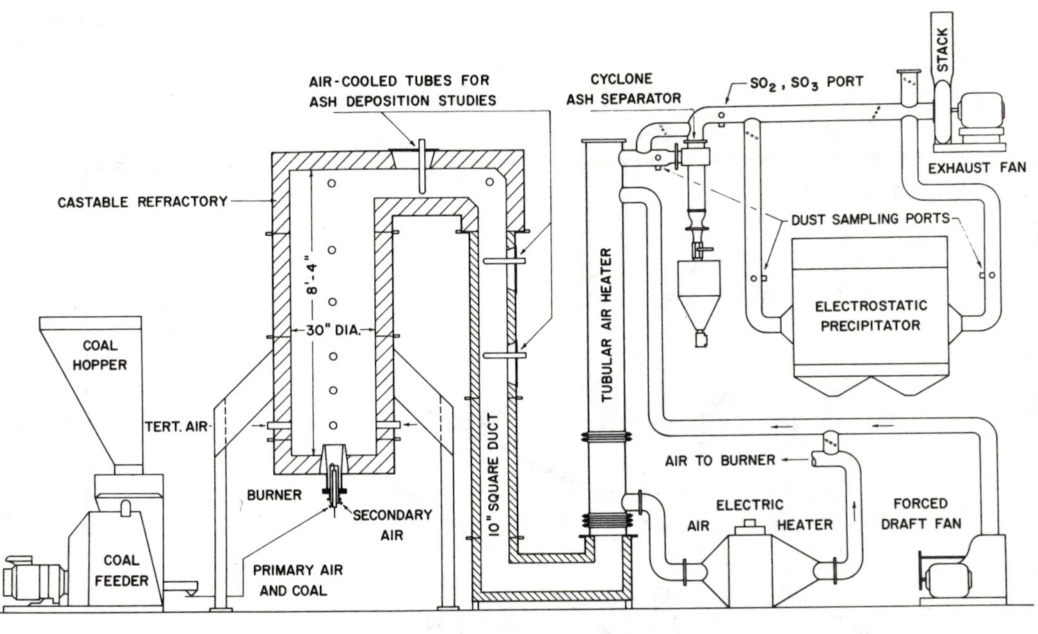

Figure 3. Schematic of the 34 kg/hr (75-lb/hr) test furnace at the Grand Forks Energy Research Center.

Figure 4. Overall view of the test furnace at the Grand Forks Energy Research Center.

After leaving the probe duct, the flue gases pass through the tubular air heater and either a cyclone ash separator or a plate-type electrostatic precipitator before being discharged to the stack.

The standard test procedure used to evaluate the relative fouling rates of test coals includes a 1-1/2 hour preheat on natural gas, a 1-hour transition period burning both gas and coal, and a 5-1/4-hour test period burning only coal. This schedule permits completion of a test in one 8-hour shift. Steady state conditions are maintained for the last three hours of the test.

After cooling, ash deposits are removed from the test probes, weighed and analyzed. Figure 5 shows the deposits formed on the upstream side of the tubes when burning high fouling and low fouling lignites. The following classification of fouling potential was adopted after much testing and experience:

Deposit weight on first probe bank	Relative fouling potential
Below 150 grams	low
150 - 300 grams	medium
Above 300 grams	high

The test furnace has to date been used to test over 100 coals in 400 tests, with greatest emphasis on North Dakota lignites, followed by Montana and Wyoming subbituminous coals and Texas lignites.

Correlation of Fouling Rates (4,5)

The many indices proposed to judge the quality of coal with respect to fouling and slagging have been recently summarized by Winegartner (6). Most have been developed for a particular coal supply and a specific type of firing. Their reliability is questionable outside the range of conditions for which they were developed.

The parameters which have predictive value for fouling by Western U.S. subbituminous coal and lignite are principally the ash content of the coal, the sodium content in the ash, and to a lesser extent the silica and calcium contents in the ash. Other indices may be illustrative of trends or principles, but they do not have predictive value for Western U.S. coals.

The effect of sodium in tests run on a North Dakota lignite (7) in the 75-lb/hr test furnace is shown in Figure 6. In these tests, the sodium level of a single coal was varied by ion exchange or addition of sodium acetate. As in field tests, fouling first increased sharply as sodium content was increased and then plateaued at higher levels. An increase in the flue gas temperature at the probe also markedly increased fouling (4,5).

The percent of ash in the coal is also a major factor in fouling. Figure 7 shows the exponential form of the increase in deposition occurring with an increase in ash content for subbituminous coals; in these tests sodium content in the ash was controlled at three levels by ion exchange or addition of sodium acetate. In practical terms, the fouling rate remained nominally low regardless of ash content when sodium content was low (1 pct of ash), but fouling increased rapidly to a very high level as ash increased at intermediate (5 pct) to high (10 pct) sodium levels. The same trends were observed in statistical studies conducted in the field (3).

The ash constituents besides sodium having important effects on ash fouling are calcium and silica. The effect of a difference in calcium over a range of sodium from 1 to 14 pct of ash may be seen in Figure 8, where the upper curve represents a low calcium lignite and the lower curve a high calcium lignite. Higher calcium acts to decrease fouling.

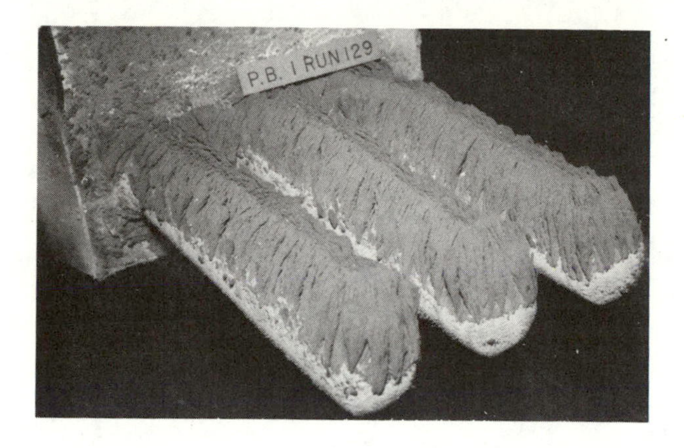

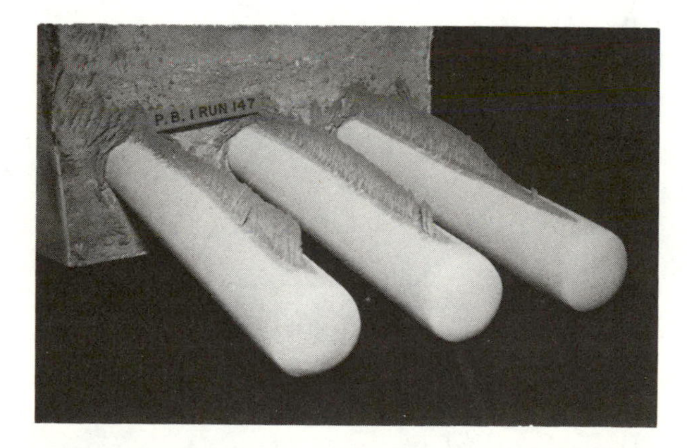

Figure 5. Comparison of furnace probe deposits from (top photograph) a high fouling lignite (6.5 pct Na_2O in ash), deposit weight 315 grams; and deposits from (bottom photograph) a low fouling lignite (0.5 pct Na_2O in ash), deposit weight 35 grams. Ash percentages: High fouling—10.28 pct dry basis. Low fouling—11.76 pct dry basis.

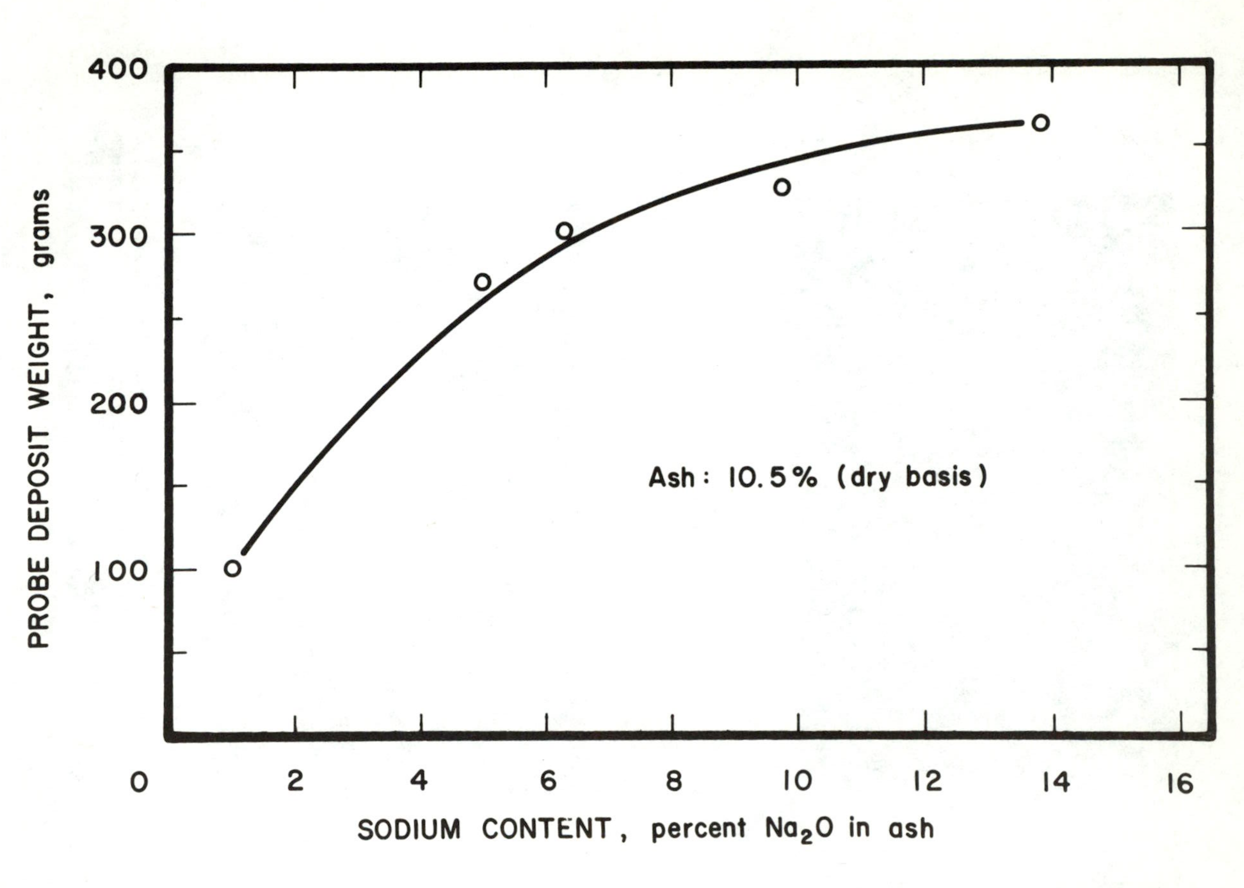

Figure 6. Probe deposit weight as a function of sodium content in ash, North Dakota lignite, using the GFERC test furnace.

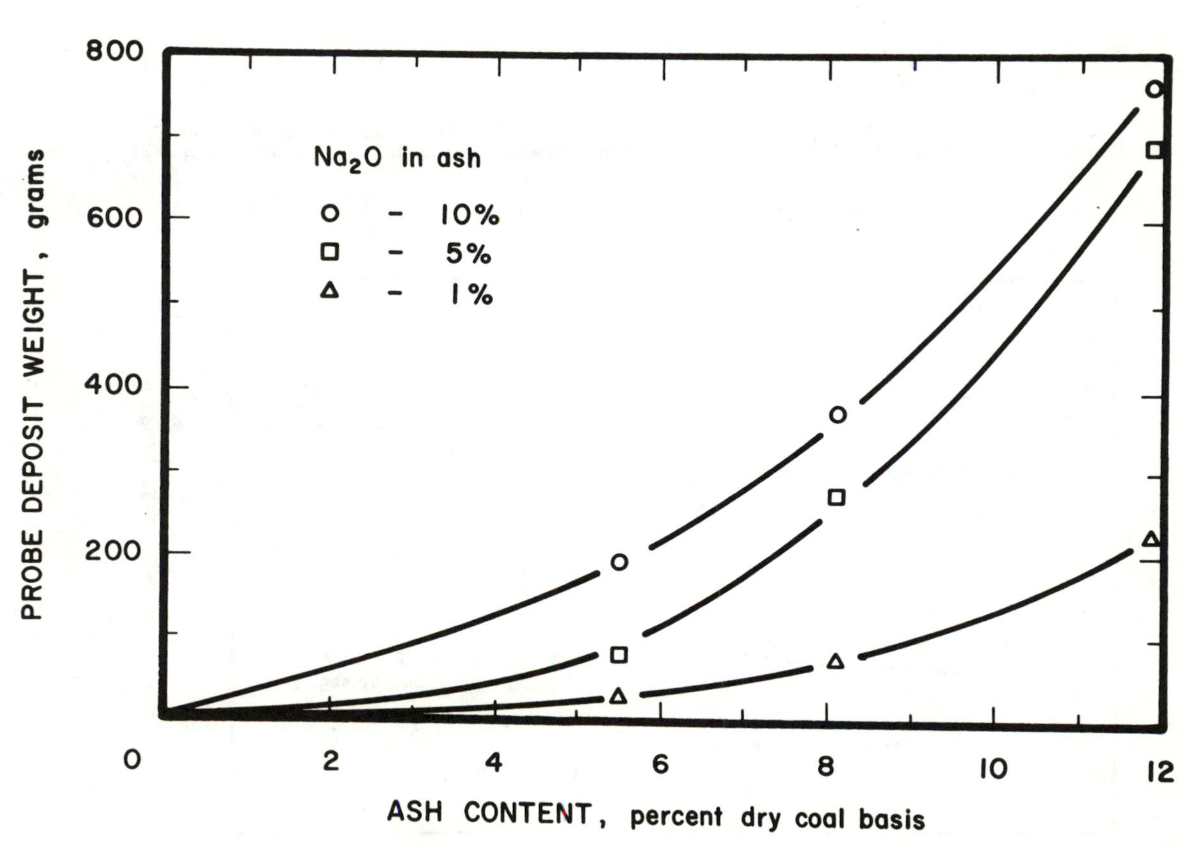

Figure 7. Probe deposit weight as a function of ash content for three subbituminous coals at three sodium levels, using the GFERC test furnace.

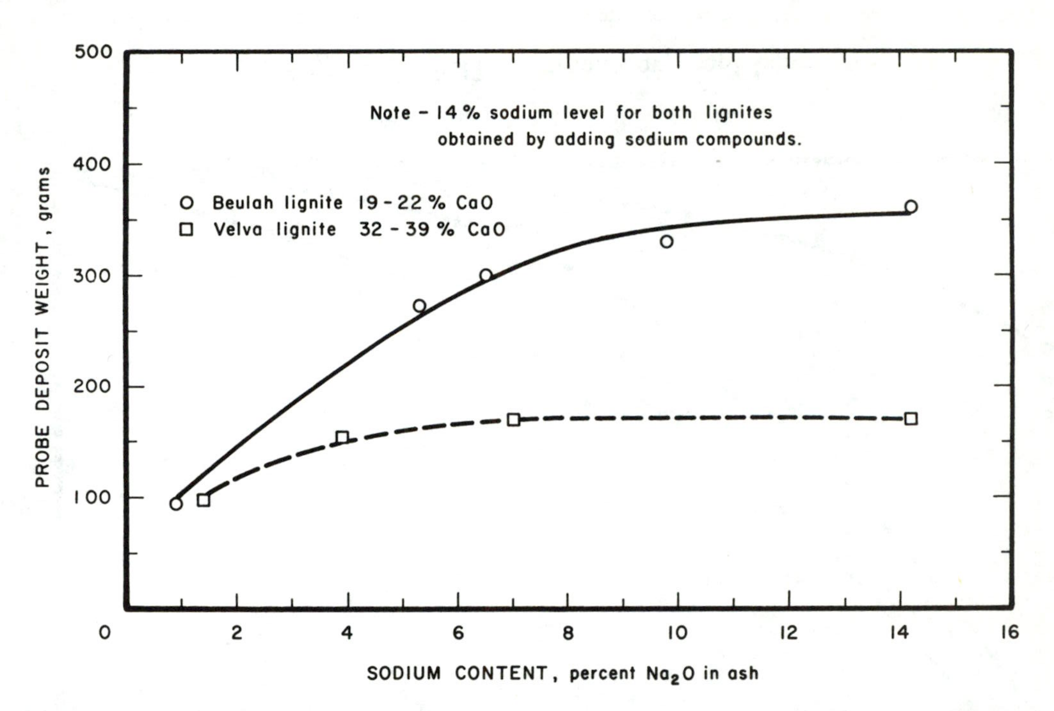

Figure 8. Comparison of fouling rates between high and normal calcium lignites as a function of sodium content, using the GFERC test furnace.

The addition of calcium compounds in tests on lignite has not produced the beneficial effect predicted by Figure 8. In order to explain this, it is necessary to consider the form of the calcium and its dispersion in the coal. It is hypothesized that for the calcium to be effective it must be inherent in the coal, that is, be bound organically to the coal molecule as inherent ash. In this form it is effective in raising the fusion temperature of the inherent ash which reacts with sodium to form the partial melt phase that causes high fouling. This will be discussed in greater detail in the section dealing with the mechanism of fouling.

Silica has been found statistically to contribute to high fouling. High silica in Western coals is associated with high ash, and the separate effects of silica and total ash have not been delineated. It has been observed, however, that high silica samples result in higher rates of deposition in the GFERC test furnace than would be predicted based on sodium content alone. This is reflected in the regression equation which best describes the GFERC test furnace data for North Dakota lignite:

$$\text{Probe wt in grams} = 47\ SiO_2 + 364\ Na_2O - 87$$

where the oxide constituents are expressed as pct of dry coal.

ASH COMPOSITION AND FUSION CHARACTERISTICS

Ranges of ash analyses for North Dakota lignite, Montana subbituminous coal, and Eastern U.S. bituminous coal are shown in Table 1 (8,9,10). A decrease in coal rank is usually associated with increased concentrations of CaO, MgO, Na_2O, and SO_3, with reduced concentrations of SiO_2 and Al_2O_3.

TABLE 1. Typical limits of ash composition of United States bituminous coals and lignites[a/]

	Eastern bituminous coal, pct of oxide (8)	Montana subbituminous coal, pct of oxide (9)	North Dakota lignite, pct of oxide (10)
Silica, SiO_2	20 to 60	19 to 52	11 to 28
Aluminum oxide, Al_2O_3	10 to 35	10 to 27	8 to 14
Ferric oxide, Fe_2O_3	5 to 35	2 to 18	2 to 16
Calcium oxide, CaO	1 to 20	3 to 28	18 to 31
Magnesium oxide, MgO	0.3 to 4	1 to 8	2 to 9
Sodium oxide, Na_2O	0.2 to 3	0.1 to 7	1.4 to 7
Potassium oxide, K_2O	0.2 to 4	0.0 to 1.4	0.2 to 0.6
Titanium oxide, TiO_2	0.5 to 2.5	0.1 to 1.3	0.2 to 0.6
Phosphorous pentoxide, P_2O_5	0.0 to 3	0.0 to 0.9	0.0 to 0.6
Sulfur trioxide, SO_3	0.1 to 12	5 to 21	12 to 27
Number of samples	260	125	212

a/ Ranges shown represent approximately 95 pct of the total variation in the data considered.

Despite differences in composition, the ranges of fusibilities for the different coal ranks do not differ significantly, as shown in Table 2 (9,11). However, the effect of changes in ash analysis on fusibility is different in the various coals. Lignite ash is high in the major basic constituents calcium, magnesium, and sodium and relatively low in the acidic constituents silica and alumina. The fusion temperature of lignite ash is lowered by increasing silica content and is raised by increasing calcium and magnesium contents. Bituminous coal ash, by contrast, is high in the acid constituents, and the effect of changes due to variations in major individual constitutents tends to be reversed (11). ASTM fusion temperatures do not correlate with fouling tendency for Western U.S. coals, but their trends are useful in interpreting the probable melting behavior of matrix materials in bonded deposits.

TABLE 2. - ASTM fusion temperatures

Ash Type	Softening Temperatures °C (°F)		
	Low	Average	High
North Dakota lignite ashes (11)	1,077 (1,970)	1,272 (2,322)	1,499 (2,730)
Montana subbituminous ashes (9)	1,060 (1,940)	1,204 (2,200)	1,588 (2,890)
Eastern U.S. bituminous ashes (11)	1,060 (1,940)	1,290 (2,355)	1,610 (2,930)

Few published data are available on viscosity of lignite-type ash. Duzy (12) has correlated the temperature of 250 poise viscosity with the percentage of basic constituents and the dolomite ratio and has shown that the trends are the same as for the ASTM Softening Temperature. Similarly, for bituminous coal ash the temperature of 25 N s/m^2 (250 poise) viscosity follows the same trend as the ASTM Softening Temperature with respect to correlation with silica ratio (11). Therefore, in the absence of viscosity data, general trends in viscosity can be inferred from ASTM fusion data. However, the correspondence is not precise and it cannot be used reliably to characterize individual samples of coal ash.

MINERAL FORMS IN WESTERN U.S. COALS

Three distinctive modes of mineral occurrence have been observed in the North Dakota lignites and Montana and Wyoming subbituminous coals studied at GFERC (13,14). About 35 to 70 pct occurs in an organically bound form. An additional 15 to 30 pct is present as mineral matter separable in a CCl_4 float-sink separation. And, approximately 20 to 40 pct is present as finely divided nonseparable clay or SiO_2 particles.

Ion Exchangeable Mineral Content

Elements that are bound to the organic matter in lignite included Na, K, Ca, Mg, Al, and Fe (14). These cations exist in part as metallic salts of the carboxylic acid groups and hydroxyl groups present in humic acids. Being salts of weak acids, they can be readily exchanged with the hydrogen ion to form the parent acid. Consequently, washing lignite with a dilute acid solution (0.1N HCl) results in substitution of hydrogen for the mineral cations. It is also possible to replace the monovalent cations Na^+ and K^+ by the divalent ions Ca^{++} and Mg^{++}, which in turn can be displaced by the trivalent cations Fe^{+3} and Al^{+3}.

Microprobe Scans

At GFERC samples of polished coal have been scanned using an AMR/3 electron microprobe at 15 kV to determine the distribution of elemental composition. Typical microprobe scans, for lignites from mines located at Beulah, North Dakota, and Savage, Montana, are shown in Figure 9.

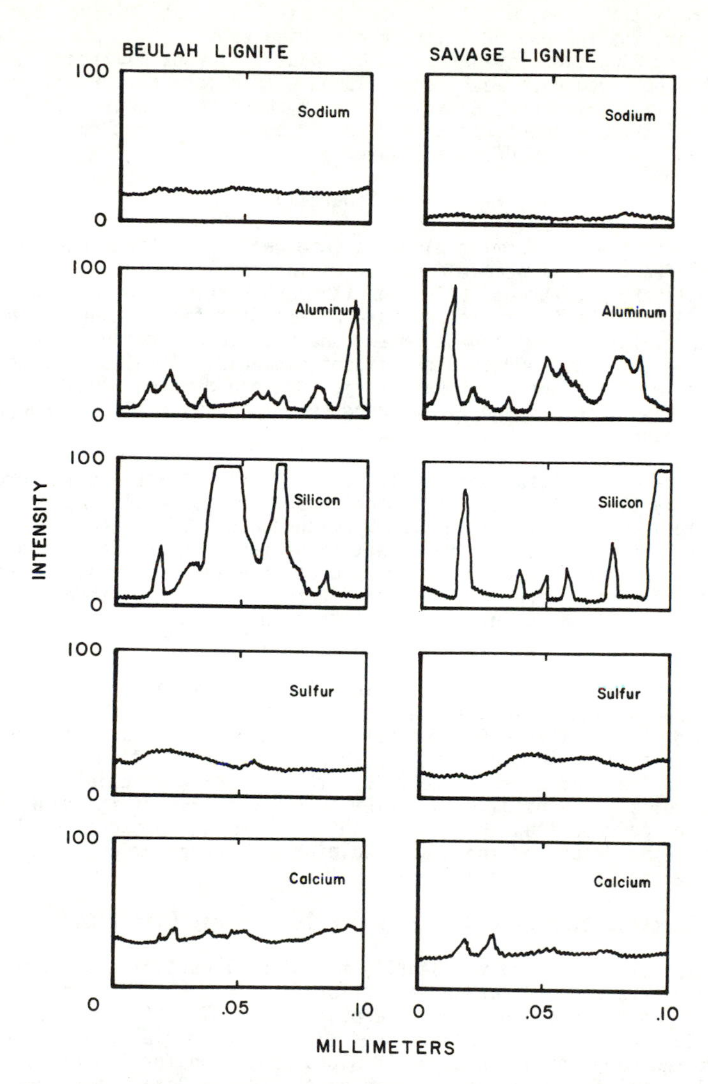

Figure 9. Distribution of mineral elements in several North Dakota lignites from electron microprobe scanning.

The scans for sodium, calcium, magnesium, and sulfur indicate an essentially uniform distribution, with few sharp peaks to indicate mineral particles. However, the traces for silica and alumina exhibit many sharp peaks, indicating SiO_2 and clay particles of about 3 microns diameter. All of the elements observed here would be included within most pulverized coal particles, which typically have a median size of about 50 microns. The juxtaposition of these elements is considered to be an important factor in the mechanism of fouling, as discussed later.

Float-Sink Separation

Separation of the ash-forming elements into separable (sink) and unseparable (float) fractions is accomplished by mixing coal that has been pulverized to nominally 80 pct through 200-mesh with carbon tetrachloride (specific gravity of 1.6) and allowing it to separate by standing for 16 hours. The float and sink fractions are then ashed and analyzed to determine the distribution of the total ash and the constituent elements. Results of such separations performed on one North Dakota lignite, one Montana lignite, two Montana subbituminous coals, and two Wyoming subbituminous coals are shown in histograms in Figure 10.

The portions of the total ash appearing in the sink fractions (Figure 10) are approximately 10 pct for the two Montana subbituminous coals, 20 pct for the two lignites, and 30 pct for the two Wyoming coals. Of interest to ash fouling is the fact that only a minor fraction of the mineral content in these coals was separated from the coal particles upon grinding to the size used in pulverized coal firing; therefore, the major fraction remained "inherent" in the coal particles and could intimately interact in the combustion of the coal particle.

The separations of the constituent elements in ash into float and sink fractions are shown in the first seven sets of histograms in Figure 10. The major portion of the sink fraction consists of oxides of silicon, aluminum and iron. These three oxides account for 70 pct of sink fraction in the two North Dakota lignites and approximately 90 pct of the sink material in the subbituminous coals. Essentially no sodium is found in any of the sink fractions, which corroborates the finding that sodium is ion exchangeable from the organic structure of the coal. Calcium and magnesium are also greatly enriched in the float fraction.

Identification of Separable Minerals by X-Ray Diffraction

X-ray diffraction is used to identify the mineral particles occurring in a sink fraction separated in carbon tetrachloride from coal sized to 50 by 100 mesh. X-ray diffraction patterns are obtained by placing the particle on the tip of a 0.3 mm glass capillary mounted in a 114.6 mm dia Debye-Sherrer powder camera and exposing it to copper K alpha radiation for 1.5 hr at 40 kW and 15 mA. A nickel filter is used to remove the copper Beta radiation.

Greater than 90 pct of the separable minerals from the previous six coal samples from North Dakota, Montana, and Wyoming were found to be aluminum silicates and quartz. Isolated particles of pyrite, calcite, dolomite, and hematite were also observed. Occurrences of minerals in individual samples are given in Table 3. Only one sample, a lignite, was found to have a high content of pyrite in the sink fraction. However, the sink fractions contained the major portion of the iron for all samples (Fig. 10). Iron particles present in the sink fraction were very small and therefore difficult to isolate and identify.

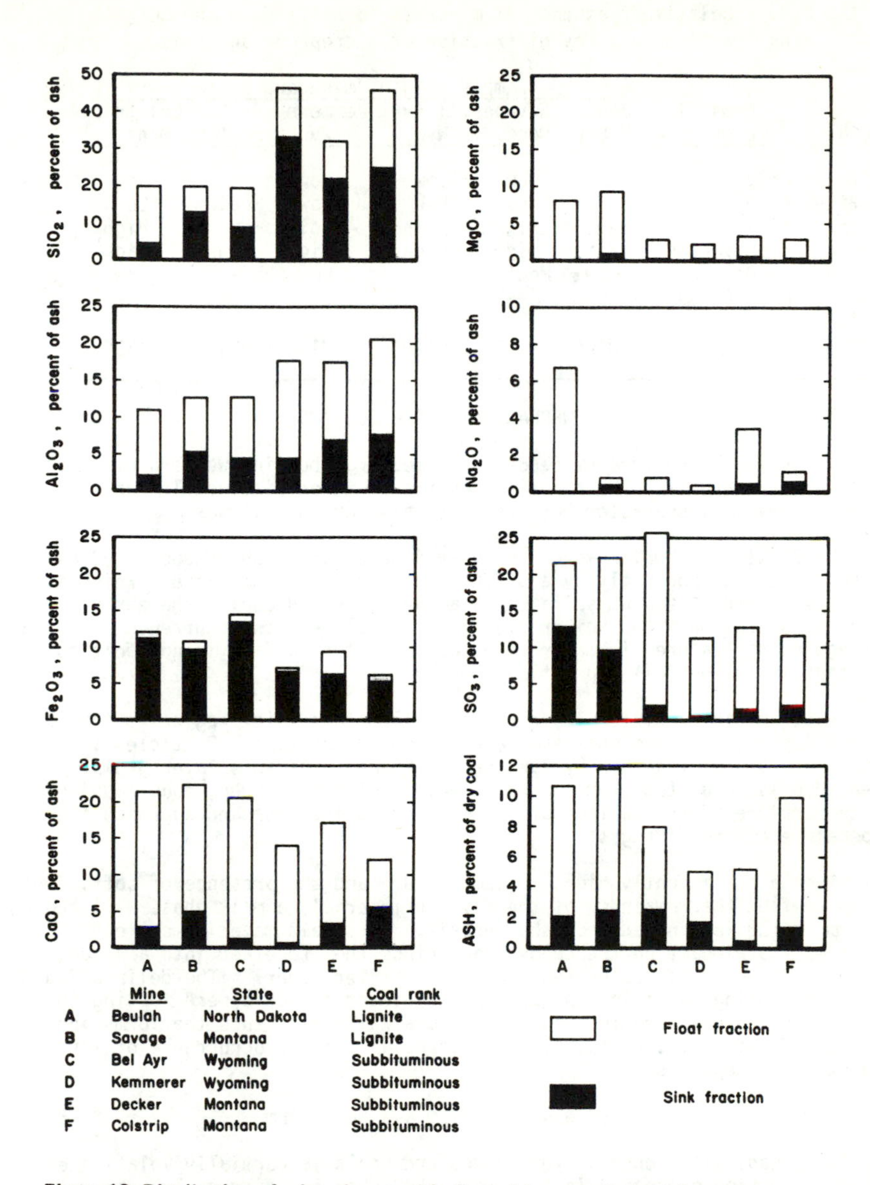

Figure 10. Distribution of mineral content in float-sink separations performed on some western U.S. coals.

TABLE 3. - Relative frequency of minerals identified in the CCl_4 sink fraction by X-ray diffraction or microprobe analysis

Mineral	Chemical formula	Sample origin, mine and state: Beulah N.Dak.	Savage Mont.	Bel Ayr Wyo.	Kemmerer Wyo.	Decker Mont.	Colstrip Mont.
Pyrite	FeS_2	High	Low	Low	Low	Low	Low
Hematite	Fe_2O_3	High	Low	Low	Low	Medium	Low
Quartz	SiO_2	Medium	High	High	Medium	High	High
Nacrite	$Al_2Si_2O_5(OH)$	Medium	High	High	High	High	High
Calcite	$CaCO_3$	Low	Medium	--	Low	--	Low
Dolomite	$CaMg(CO_3)_2$	Low	--	--	--	Low	--
Aluminum Silicates		High	High	High	High	High	High

THE MECHANISM OF ASH FOULING

The typical structure of deposits formed from burning Western U.S. coals is shown in Figure 11. The deposit consists of three distinct layers which differ in physical character but are quite similar in analyses.

The first thin "white layer" of very fine powdery ash is deposited all around the tube, apparently by a diffusional process. This layer, which is usually enriched in sodium sulfate, is always observed during the early period of operation after boiler cleanup. Therefore, its occurrence is not a distinguishing feature of low or high fouling rates, and it is not felt to be important in the overall deposition process.

Next, an "inner sinter layer" a few millimeters thick begins to form by initial impaction on the upstream face of the boiler tube. Particles in this deposit are bonded together by surface stickiness. As this layer grows, its outer edge is insulated from the relatively cool boiler tube, thus causing the temperature of the surface of the deposit to increase and approach the temperature of the flue gas.

Given a sufficiently high gas temperature and the presence of sufficient sodium to flux the remainder of the fly ash material, a melt phase will begin to form at the leading edge of the deposit. This melt material collects particles that impact on the deposit and binds them together into a strong bulk deposit which is designated the "outer sinter layer." The delineation of an ash fouling mechanism to explain the occurrence of severe fouling in burning Western coals centers on the factors which influence the formation of this melted matrix material, which is essential for the occurrence of large, high-strength deposits.

Reactions of Sodium and Potassium

The organically bound sodium in Western coals is partially volatilized in the combustion process. In tests at GFERC, heating either fly ash or sodium sulfate above 1316° C (2400° F) has resulted in substantial loss of sodium (Figure 12). Volatilization from the organic structure of the coal would be expected to occur more easily than revolatilization from glassy fly ash particles. Sodium that is not volatilized in the flame is retained in the complex ash particle produced from non-volatilized inherent ash. The portion of sodium retained and not volatilized is not known, but it is believed to be significant. Below 982° C (1800° F), the volatilized sodium is condensed into the entrained fly ash in a manner which causes the finer fraction of the fly ash to be substantially enriched in sodium (2). Sodium enrichment also occurs in boiler deposits which are formed at lower temperatures

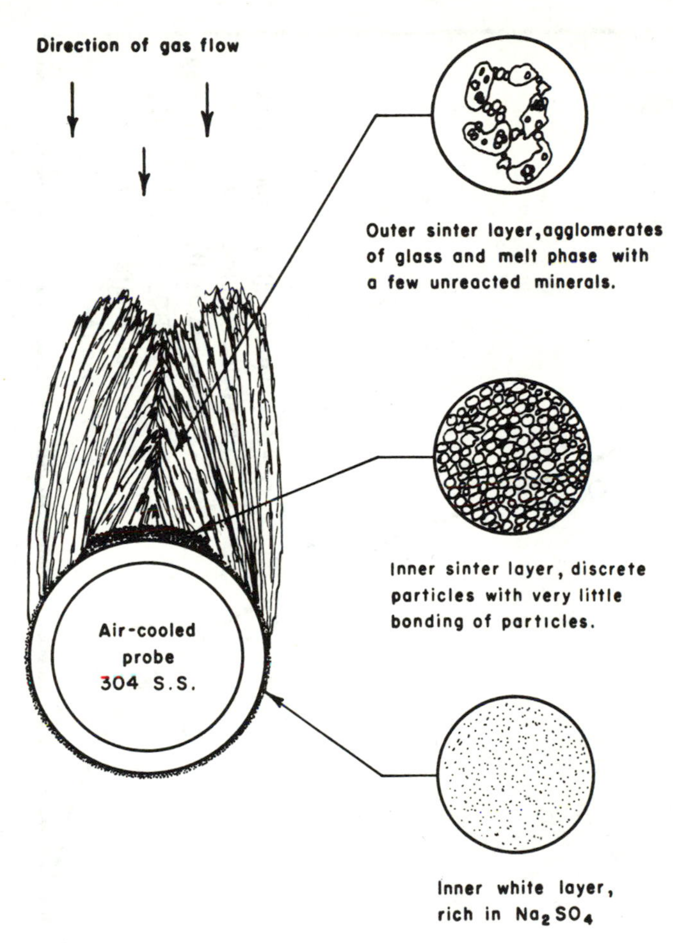

Figure 11. Physical structure of a typical deposit from the test furnace at the Grand Forks Energy Research Center when burning a highly fouling North Dakota lignite.

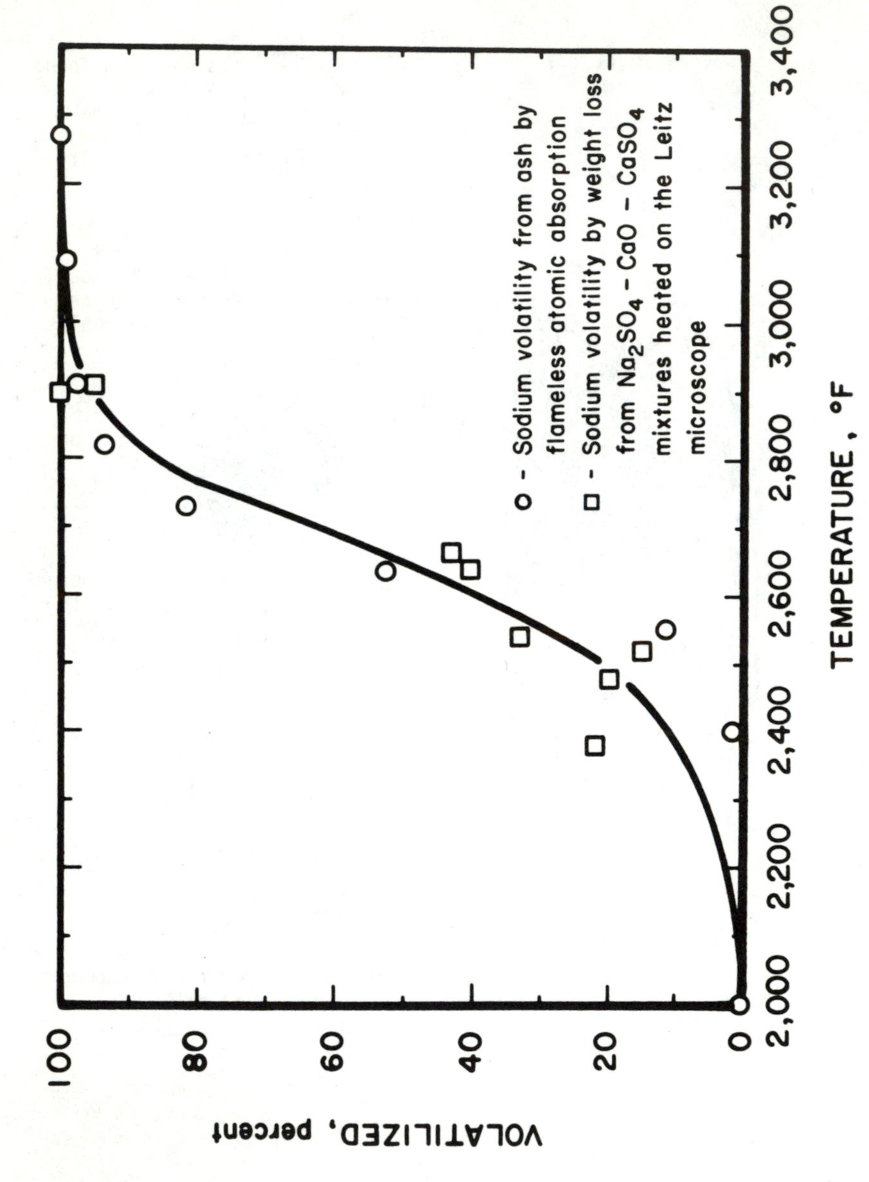

Figure 12. Weight loss for sodium in fly ash or Na_2SO_4 as a function of temperature.

rather than higher temperatures. In the end, the roles of the volatilized sodium and the retained sodium are the same, since both react to flux the ash and worsen fouling.

The form of the sodium entering the combustion process has not been found to significantly alter the severity of fouling in burning Western coals. Changes in level of sodium brought about by ion exchange (which alters the amount of Na held on the coal structure) and by addition of $NaC_2H_3O_2$, NaOH, NaCl, Na_2CO_3, or Na_2SO_4 have had essentially the same effect on the severity of fouling. Water soluble sodium in lignite does not correlate with fouling.

It has been shown at GFERC that addition of potassium as an organo-metallic compound causes fouling in the same manner as sodium. Organically bound potassium in coal would be expected to react similarly. However, most of the potassium in Western coals is bound to the clay, and in this form and in the small amounts present it does not influence fouling to an appreciable extent.

Reactions of Sulfur

It has been suggested by some investigators (15,16,17) that SO_3 is important in the process of forming bonded fouling deposits. However, at GFERC the severity of fouling has not been correlated with coal sulfur content; and fly ash has been shown to be fluxed by sodium in the form of the hydroxide, chloride, or carbonate.

Most of the sulfur in the coal is released as SO_2 in the flame, whether the sulfur occurs in the coal in organic or pyritic form. However, some of the organic sulfur, which is intimately dispersed with alkaline cations in a Western coal, may be retained directly in the fly ash particles derived from the inherent ash. The small amount of sulfur originally present as sulfate may also be either retained or volatilized. A portion of the volatized sulfur is reacted back with the alkaline constituents in the fly ash at lower temperatures. Tests on powerplants burning Western coals have indicated that from 10 to 40 pct of the total sulfur ends up in the fly ash, depending largely on the amount of sodium in the ash. In fluidized-bed combustion, where the average temperature is much lower, 816° C (1500° F), a large fraction of the sulfur may be retained on the ash from some highly alkaline coals (18). High retentions are also observed in laboratory ashing at nominally 800° C (1472° F) (11).

It has been observed that ash deposits harden as they age in the furnace. This has been attributed to "sulfating" of the deposits, which is believed to give rise to dense, hard deposits over a long period of time. This phenomena has not been studied in any sigificant detail for Western U.S. coals.

Mixing in Flames

It has already been pointed out that most of the mineral matter in Western coals is retained as inherent ash within individual pulverized coal particles. There is nevertheless ample opportunity for variations in analysis between different fly ash particles and their agglomerates due to diversity in the original distribution of mineral matter and to the volatilization and condensation processes. These variations will be more or less important depending on the amount of mixing of molten ash occurring in the flame.

Preliminary experiments on mixing of fly during combustion in the GFERC test furnace have involved the addition of a tracer, a pulverized copper mineral (cuprite), to the primary air stream along with the pulverized coal. Cuprite has a melting temperature, 1233° C (2255° F), that is similar to the fusion temperature of the ash, and it was ground and screened to the same size consist as that of the fly ash. Measurements performed on the fly ash produced indicated that some copper was contained in most of the fly ash particles, at least on the surface, where the electron microprobe could be used for detection. This was interpreted as evidence that material was transferred from particle to particle by coalescence and redispersal in the high-temperature region of the flame. However, since the experiment did not provide quantitative data on the distribution of copper and some slight mixing could have occurred by adherence of the finer cuprite to coal particles in the primary air stream, the findings are not conclusive. Further experiments are required to establish the importance of mixing.

Formation of Matrix Bonding Material

The matrix which bonds strong deposits together is believed to be formed by the fluxing action of sodium on the inherent ash (19). Not all inherent ash, however, is fluxable owing to diversity in elemental composition. The "matrix parent" is only that portion having an elemental composition in a range that is fluxable. The empirically determined effects of silica and calcium are believed to reflect differences in proneness to fluxing.

In the literature, matrix material has been variously suggested to be Na_2SO_4, $CaSO_4$, or a eutectic of sodium, calcium, and sulfate. In studies on Western U.S. coals at GFERC (19), the matrix has been identified to be a complex crystalline mineral specie, Melilite, having as its important constituents sodium, calcium, and silica. Identification is based on X-ray diffraction patterns and elemental analysis. Other melt species may also occur, but the Melilite is believed to have particular significance, at least for the deposits formed by Western coals.

The formation of Melilite is not perfectly understood. It has been shown in the studies at GFERC that sodium must be present to produce this matrix material found in high fouling deposits, and that calcium and silica are the other principal constituents. There is not sufficient Ca and Si in the extraneous mineral matter to account for the large amount of bonding Melilite that can be formed. However, since Ca and Si do occur in quantity and in juxtaposition in the inherent mineral content, it is believed that it is primarily the direct fluxing of the inherent ash by sodium which produces the Melilite. An alternative hypothesis offered by Hein (20) is that the Melilite is formed when particles having a high SiO_2 content are dissolved in a Na-Ca-SO_4 eutectic. Both reactions may occur, and the effect would be the same in either case.

The deposit building process then involves the continuous formation of a Melilite melt phase to which unfluxed extraneous or inherent ash is added as inert aggregate. The size and strength of the deposit depend on both the amount of bonding material and the amount of aggregate. A high-sodium, low-ash North Dakota lignite would form a Melilite-rich deposit that is moderately large and very hard; continuing boiler operation would build up very troublesome deposits in this situation. A moderate-sodium, high-ash Texas lignite would form a very large deposit of moderate strength; continuing operation might be better or worse than for the North Dakota lignite depending on the effectiveness of on-line cleaning. A low-sodium, high-ash Montana subbituminous coal would tend to form deposits having too little strength to permit extensive growth, and boiler operation should be relatively good. These generalizations do not take into account differences in furnace design, which would be made to compensate for ash behavior, or long-term reactions which could affect the hardness of deposits. Good judgment in design and operation of boilers for fouling coals will remain an important part of the solution to the problem.

The direct solutions to the fouling problem suggested by the mechanism are either the removal of sodium from the coal or the introduction of an additive which interrupts the fouling mechanism. To interrupt the mechanism, an additive must either combine with the sodium to produce a non-fluxing product or combine with the matrix parent to render it non-fluxable. Additives being tested at GFERC include materials in both categories.

REMEDIAL MEASURES

Over the past years considerable work has been done at GFERC to seek remedies for fouling by Western U.S. coals. The two principal methods studied have been coal washing to remove sodium and additives to reduce the effect of the sodium during combustion.

Removal of Sodium by Ion Exchange

Sodium can be removed from lignite by washing with dilute acid or solutions containing suitable concentrations of divalent ions (Ca^{++} or Mg^{++}) or trivalent ions (Fe^{+3} or Al^{+3}). The ion exchange reactions have been studied at GFERC in bench scale and beaker tests (14,21), and in a pilot plant apparatus (22). The variables investigated include different reagents, reagent sodium stoichiometry, reagent concentration, coal particle size and moisture content, and agitation. Settling properties and removal of moisture by centrifugation have also been studied.

Tests conducted to date have shown that a major fraction of the sodium in lignite can be removed by ion exchange treatment with wash water containing a few hundred ppm Ca^{++}. It has been further shown that lignite crushed to a commercially feasible size of 1.3 cm (1/2 in.) by zero can be effectively treated using a 2-hour residence time in a continuous process. Loss of coal fines is acceptably low; recovery of fines and dewatering of treated coal appear to be technically feasible.

Process development (22) for removal of sodium from lignite has been pursued in the pilot plant apparatus shown in Figure 13. In this device, lignite was conveyed through an inclined trough with a drag type conveyor, and wash solution was circulated in the countercurrent direction. The capacity of the apparatus was nominally 27 kg coal/hr (60 lbs/hr). Coal residence time ranged from 1 to 4 hours, and liquid residence time from 6 to 30 minutes. Coal tested was sized by crushing and screening to minus 1.3 cm (1/2 in.) by zero. Chemical reagents tested included lime, cement, calcium chloride, calcium sulfate, and sulfuric acid; the calcium chloride and sulfuric acid were shown to be most effective. Sodium removal ranged from a trace to 77 pct in treating lignites originally containing approximately 8 pct ash and 7 pct Na_2O in the ash. A typical condition for high removal was treatment with an excess of calcium chloride solution containing 570 ppm Ca^{++} for 2 hours or longer. Removal was shown to depend on calcium concentration in the range of 0 to 1900 ppm Ca^{++}, but this was partially a stoichiometric limitation based on the quantity of wash solution used. The ratio of sodium ions removed to calcium ions absorbed ranged from 1.3 to 2.7. Sodium removal after 4 hours treatment in calcium chloride solution containing 214 ppm Ca^{++} increased from 58 pct to 77 pct as the stoichiometric ratio of calcium to sodium was changed from 0.32 to 1.28 (based on an equivalence of 1 calcium ion to 2 sodium ions). Maximum removal was approached in 2 hours. Approximately 3 pct of the coal solids were lost in the treating solution; however, these solids settled rapidly at the ionic strength of the wash solution, and coal solids could be recovered in a suitably designed settling tank. Surface water adhering to the treated coal was effectively removed by centrifuging for 5 to 15 minutes.

Future work to develop a commercial process for the removal of sodium should involve 1) additional bench-scale tests to establish more complete data on kinetics and equilibria, 2) conceptual design and evaluation of process economics, 3) construction of a process development unit to treat nominally 20 tons/day, which would permit a one week combustion test to be run on a 50 megawatt boiler using approximately one month's production from the wash plant, and 4) investigation of problems relating to disposal of sodium enriched water from a wash plant.

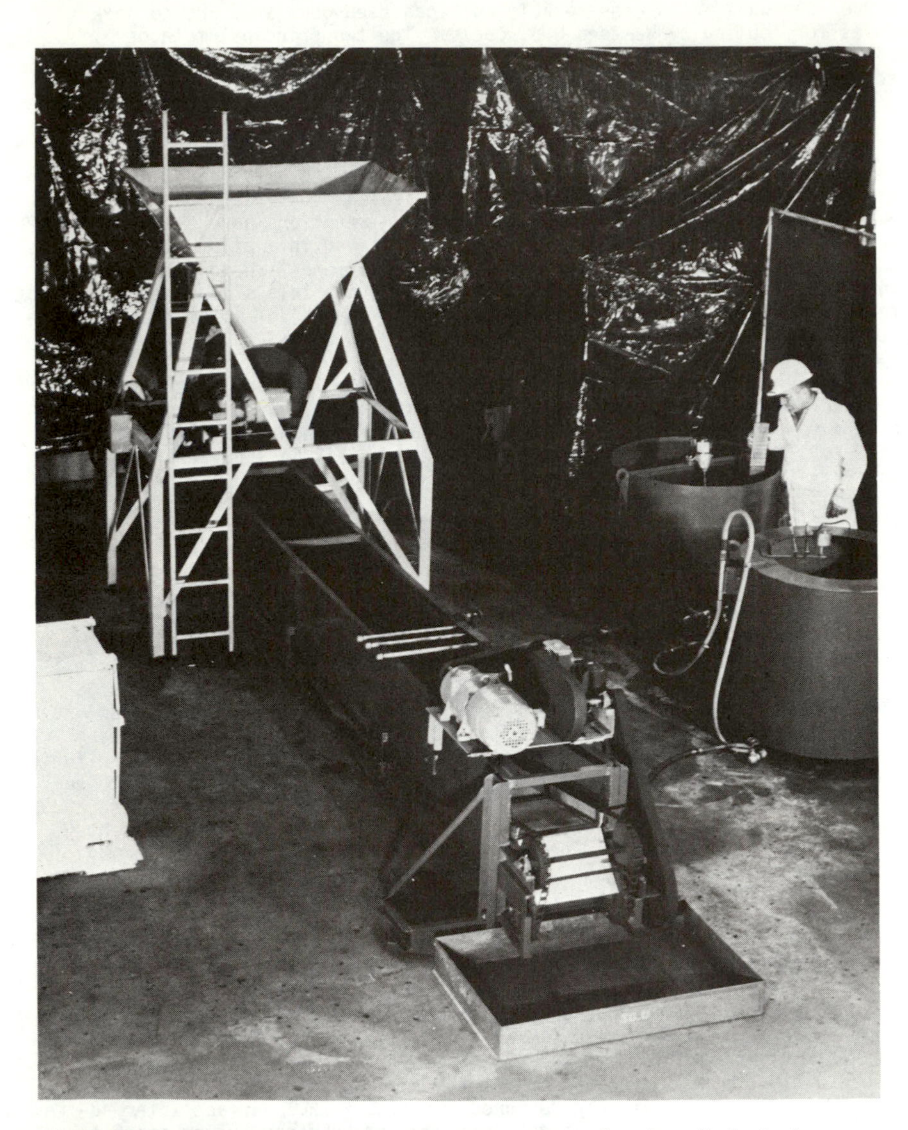

Figure 13. Pilot plant process equipment used to remove sodium from lignite by ion exchange, capacity 27 kg coal/hr (60 lb coal/hr).

Additives

Additives are being tested in the GFERC laboratory furnace, looking both at the additive and the extent of dispersion. Additives containing calcium and magnesium have had little effect on the rate of deposition, but deposit strength has been reduced; these elements act by reducing the fluxing tendency of the matrix parent. Aluminum compounds have been tested to determine if sodium and aluminum would react to form high-melting products that would tie up sodium in a harmless form; these additives have not reduced either the size or the strength of deposits. Further tests with tracers will be conducted to establish more definitely the extent of mixing between molten ash and additive particles in the turbulent flame.

FUTURE WORK

A total solution to the problem of burning highly fouling coals in conventional boilers will require the development of either a practical process for removing sodium or an effective additive. Process development for sodium removal should be expedited to learn if a breakthrough is possible. Short of a breakthrough, some improvement will be obtained through continuing refinement of boiler design, with the added requirement that changes in combustion conditions for reducing NO_x emission not be allowed to aggravate fouling. Planning and selection of boilers for new low-grade coals will continue to require prior measurement of fouling tendency, which can be provided by a test furnace such as the 34 kg/hr (75-lb/hr) GFERC furnace when full-scale tests cannot be arranged.

Future research on Western U.S. coals will shift towards measurements on deposit strength and adhesion, rather than deposition rate only. This is particularly important in work on additives, where additives have little effect on the rate of deposition but do affect strength. Long-term tests will be required to study strength of deposits.

Field studies on effect of excess air and fuel/air distribution are urgently needed to guide changes in boiler design pertaining to NO_x control for burning highly fouling coals.

Basic research is needful: 1) to better understand the mechanism of fouling and the significance of different mineral forms, 2) to learn more about the reactions of mineral constituents during combustion, and 3) to obtain data on the thermal conductivity and emissivity of differing deposits.

In the future, an alternative solution to the ash fouling problem will be offered by the development of fluidized-bed combustion (FBC) for Western U.S. coals. At temperatures below 982° C (1800° F), which are typical of FBC, fouling would be expected to be greatly reduced and perhaps eliminated.

ACKNOWLEDGEMENTS

The field tests reported were conducted with the cooperation and assistance of several utilities and mining companies, whose contribution is gratefully acknowledged. Also, the authors wish to acknowledge the assistance of Mr. Arthur L. Severson and Mr. Francis J. Schanilec for their work on a mineral identification and furnace operation at the Grand Forks Energy Research Center.

REFERENCES

1. Gronhovd, G.H., R.J. Wagner, and A.J. Wittmaier, "A Study of the Ash Fouling Tendencies of a North Dakota Lignite as Related to Its Sodium Content," Transaction of the Society of Mining Engineers, September 1967, pp. 313-322.

2. Gronhovd, G.H., A.E. Harak, and L.E. Paulson, "Ash Fouling Studies on North Dakota Lignite," Bureau of Mines IC 8376, May 1968, pp. 76-94.

3. Sondreal, E.A., S.J. Selle, P.H. Tufte, V.H. Menze, and V.R. Laning, "Correlation of Fireside Boiler Fouling with North Dakota Lignite Ash Characteristics and Powerplant Operating Conditions," presented at the American Power Conference, Chicago, IL, April 1977.

4. Gronhovd, G.H., A.E. Harak, and P.H. Tufte, "Ash Fouling and Air Pollution Studies Using a Pilot Plant Test Furnace," Bureau of Mines IC 8471, 1970, pp. 69-88.

5. Gronhovd, G.H., W. Beckering, and P.H. Tufte, "Study of Factors Affecting Ash Deposition from Lignite and Other Coals," presented at the Annual Meeting of the American Society of Mechanical Engineers, July 1969, Los Angeles, CA, Paper No. 69-WA/CD-1, 9 pp.

6. Winegartner, E.C., "Coal Fouling and Slagging Parameters," American Society of Mechanical Engineers, No. H00086, New York, NY, 1974, 34 pp.

7. Tufte, P.H., G.H. Gronhovd, E.A. Sondreal, and S.J. Selle, "Ash Fouling Potentials of Western Subbituminous Coal as Determined in a Pilot Plant Test Furnace," presented at the American Power Conference, April 1976, Chicago, IL 15 pp.

8. Selvig, W.A. and F.W. Gibson, "Analyses of Ash from United States Coals," Bureau of Mines Bulletin 567, 1956, 33 pp.

9. Cooley, S.A. and R.C. Ellman, "Analyses of Coal and Ash from Lignites and Subbituminous Coals of Eastern Montana," Energy Resources of Montana, 22nd Annual Publication of the Montana Geological Society, June 1975, pp. 143-158.

10. Sondreal, E.A., W.R. Kube, and J.L. Elder, "Analysis of the Northern Great Plains Province Lignites and Their Ash: A Study of Variability," Bureau of Mines RI 7158, 1968, 94 pp.

11. Sondreal, E.A. and R.C. Ellman, "Fusibility of Ash from Lignite and Its Correlation with Ash Composition," ERDA Report of Investigation GFERC/RI-75/1, 1975, 121 pp.

12. Duzy, A.F., "Fusibility-Viscosity of Lignite Type Ash," presented at the Annual Meeting of the American Society of Mechanical Engineers, Chicago, IL, November 1965, ASME Paper No. 65-WA/FU-7, 8 pp.

13. Paulson, L.E., W. Beckering, and W.W. Fowkes, "Separation and Identification of Minerals from Northern Great Plains Province Lignite," Fuel, Vol. 51, July 1972, p. 224.

14. Paulson, L.E. and W.W. Fowkes, "Changes in Ash Composition of North Dakota Lignite Treated by Ion Exchange," Bureau of Mines RI 7176, 1968, 18 pp.

15. Bishop, R.J., "The Formation of Alkali-Rich Deposits by a High-Chlorine Coal," Journal of the Institute of Fuel, Vol. 41, February 1968, pp. 51-65.

16. Coats, A.W., D.J.A. Dear, and D. Penfold, "Phase Studies on the Systems Na_2SO_4-SO_3, K_2SO_4-SO_3, and Na_2SO_4-K_2SO_4-SO_3, "Journal of the Institute of Fuel, Vol. 41, March 1968.

17. Grant, K. and J.H. Weymouth, "The Nature of Inorganic Deposits Formed During the Use of Victorian Brown Coal in Large Industrial Boilers," Journal of the Institute of Fuel, Vol. 35, No. 255, April 1962, pp. 154-160.

18. Goblirsch, G.M. and E.A. Sondreal, "Fluidized Combustion of North Dakota Lignite," presented at the ERDA-University of North Dakota Lignite Symposium, May 1977.

19. Tufte, P.H. and W. Beckering, "A Proposed Mechanism for Ash Fouling Burning Northern Great Plains Lignite," Journal of Engineering for Power, July 1975, pp. 407-412.

20. Hein, Klaus. Unpublished data from studies at Rheinisch Westfolisches Elektrizetatswerk AG in Essen, West Germany.

21. Crystal, J.T., "The Removal of Sodium from Lignite by Ion Exchange," Masters Thesis, University of North Dakota, Grand Forks, ND, August 1970.

22. Paulson, L.E. and R.C. Ellman, "Reduction of Sodium in Lignite by Ion Exchange in a Pilot Plant Study," ERDA RI in preparation at the Grand Forks Energy Research Center.

AN INVESTIGATION OF ACID SMUT FORMATION IN AN INDUSTRIAL BOILER PLANT

A. B. HEDLEY and P. PAPAVERGOS

Department of Chemical Engineering and Fuel Technology
University of Sheffield
Sheffield S1 3JD, England

1. ABSTRACT

A boiler plant over forty years old consisting of five water tube boilers originally coal fired but converted to residual fuel oil firing in 1972 subsequently started emitting acidic carbonaceous smuts from the chimney stack. This problem was investigated experimentally by flue gas sampling for sulphur oxides and particulate solids at different points in the system and relating the results to operating conditions. It was concluded that acid smut particles could be formed by interaction of sulphur oxides and smoke particles within the flue gases at temperatures higher than the sulphuric acid dewpoint. The distribution of smut fall out in the neighbourhood of the chimney stack was also reported.

2. INTRODUCTION

It is known that all plants burning residual fuel oil tend to emit finely divided solid carbonaceous particles which fall to the ground in the neighbourhood of the chimney. If burners are correctly designed and operated, the amount of material in the stack gases should not exceed 0.20% by weight of the oil burnt (1), but when the burners are dirty or badly adjusted this figure can easily rise to 0.50% or higher. Sulphur oxides originating from the sulphur in the oil can combine with these carbonaceous solids somewhere in the system resulting in the emission of acidic smuts from the chimney.

This paper is concerned with an investigation of the extent of acid smut formation and emission from an industrial power plant, originally constructed as a chaingrate coal fired installation over 40 years ago and converted to residual fuel oil firing in 1972. Soon after the commissioning of the oil firing arrangements, complaints from various individuals and adjacent works were received claiming damage to

(i) clothing especially synthetic fibres such as nylon;
(ii) cellulose finish on parked automobiles within the region;
(iii) roofs of surrounding buildings.

Tests were carried out at the plant in an attempt to determine the extent of the fall out problem as well as the origin and cause of smut formation in the plant.

3. POWER PLANT DESCRIPTION

The power plant consisted of five oil fired boilers, of which four (Boilers 1 to 4) were identical, each capable of producing a maximum of about 40,000 Lbs/hr (18.15 Tonnes/hr) steam (see FIG.1) and one (Boiler 5) of larger steam load capacity of about 60,000 Lbs/hr (27.21 Tonnes/hr) at 380 p.s.i.g. (25.85 atm.) and 600-650° F (316°-342° C). All oil fired boilers at the power plant were converted coal fired chaingrate stoker boilers, which during conversion had the water tubes at the rear walls of the boilers removed and substituted with refractory, and two steam assisted 'Y' jet atomizing burners fitted (Boiler 5 was equipped with three such burners, but during this investigation was out of order due to maintenance).

The residual fuel oil was preheated to a temperature of 220°-238° F (104°-114° C) prior to entering the burner gun, where atomisation at the burner tip was by means of superheated steam at 110 p.s.i.g. (7.48 atm.) and at about 419° F (215° C) measured before entry to the burner gun. The rate of steam consumption was about 1.50% of the fuel oil burnt by weight. The fuel oil pressure was variable (hence the flow rate) according to the steam load fluctuations of the boiler as determined by the control equipment. The fuel oil pressure fluctuation range was 100-140 p.s.i.g. (6.80-9.52 atm.), and the respective fuel oil rate range was 1212 — 2535 lbs/hr. (0.55-1.15 Tonnes/hr).

All boilers were equipped with steam heated air preheaters located between the forced draught fans and the Ljungstrom regenerative air-heaters; raising the ambient air temperature to an average temperature of 221° F (105° C) before entering the Ljungstrom heaters. The flue gases of boilers 1 to 4, after being discharged from each individual boiler's I.D. fan, entered the main flue duct prior to the 165 ft. (50.3 m) brick chimney, as shown in FIG.2. Boiler No. 5 had its own steel chimney separated from the main flue duct by a closed damper. During the investigation, one of the boilers (No. 4) was separated from the master control, equipped with its own new control system and was operated as a constant load experimental boiler.

The preliminary measures which had been adopted by the company in an attempt to control the acid smut emissions were as follows:

- (i) Improvement on the poor ignition conditions were made, by using less excess combustion air during light up; together with modifications of the flame failure protection system in order to reduce the number of boiler trips out.
- (ii) More frequent cleaning of the atomizing burner tips.
- (iii) All grit arrestors left from the coal-fired operation had been removed, in order to minimize possible tramp air inleakage and soot build-up.
- (iv) Oxygen analysers were installed at the economiser flue gas inlets, upon which the combustion control was based.
- (v) The burner throat diameter was reduced by 2 inches, in an attempt to improve mixing in the flame gases especially at low loads, by increasing the combustion air velocity and hence the swirler turbulence.

Finally (vi) Ammonia gas was injected into the flue gases at the discharge duct of the I.D. fan of each boiler in an attempt to neutralize the SO_3 vapour as well as the condensed sulphuric acid on the soot particles.

Even after the above modifications, the periodical acid smut emissions were present and various complaints were still received on many occasions.

4. MEASUREMENT OF ACID SMUT FALL OUT FROM CHIMNEY

Shallow wooden trays approximately 25 x 40 cm (10" x 15") were placed 60 cm above ground level at convenient locations in the works at distances of up to 350 metres from the base of the brick chimney. Moist congo red indicator paper was used to line the bottom of each tray. The falling acid smuts left a blue mark on this paper according to their size, even if they were blown away from the surface in strong wind conditions. To keep the indicator papers moist over a 24 hour period they were soaked in a 50% glycerine solution. Every 24 hours the papers were collected and a smut count taken,to establish the rate and range of fall out with respect to wind and weather conditions.

The results were expressed as "smuts/m^2/day" and being in three size ranges namely large; 0.60-2.50 cm mean diameter, medium; 0.30-0.60 cm mean diameter and small; 0.10-0.30 cm mean diameter. The peak fall out concentration varied with regard to wind velocity and distance from the chimney base, and FIGS. 3, 4 show typical results. These results are averaged from many tests taken continuously over several months duration. It was concluded that the incidence of smut emission could be classified into five categories:

25 - 150 smuts/m^2/day	-	very light emission
150 - 300 smuts/m^2/day	-	light emission
300 - 600 smuts/m^2/day	-	medium emission
600 - 1000 smuts/m^2/day	-	heavy emission
> 1000 smuts/m^2/day	-	very heavy emission.

Complaints could be expected from outside bodies and adjacent works even during "light emission" conditions if the acid content of the smuts proved to be high (i.e. $\sim$ 10% free SO_3 by weight).

5. SAMPLING OF THE FLUE GASES

The boiler plant was always in continuous operation 24 hrs/day throughout the year. The load fluctuations under normal conditions could vary between 100,000 lbs/hr (45.36 tonnes/hr) and 60,000 lbs/hr (27.21 tonnes/hr). In order to have conditions as near constant as possible No. 4 boiler was used as a test boiler and operated at different but known constant load conditions, with plant variations in steam demand being met by the other boilers. Gas and solid samples were taken at the inlet to the economiser, the air heater outlet, and the main flue duct just prior to chimney base. Samples from the latter were therefore subject to total load fluctuations and variable conditions prevailing in the other boilers.

5.1 Oxygen concentration measurements

Oxygen concentrations at the inlet to the economiser were continuously measured by means of a Westinghouse oxygen probe. A portable analyser was used to measure oxygen concentrations in the flue gas at the inlet and outlet to the Ljungstrom air heater,whilst simultaneously measuring static pressure differences across the air heater. The results indicated that air inleakage across the air heater averaged the equivalent of approximately 50% excess air indicating that in some instances the oxygen concentration in the flue gas stream rose from 1% at the inlet to 9% at the outlet. Tests on the other boilers showed the same severe problem due to bad seals. The air heater seals of Boiler No. 3 were replaced with new ones and set as suggested by the manufacturers and the leakage was reduced from over 60% excess air to 18% excess air. After the first shut down of the boiler the leakage was found to be about 23%. Experience over several years at this plant proved that all air heaters of this type had suffered from much higher leakages than the 10% predicted by the suppliers regardless of the efforts involved to control it. The flue gas temperature on leaving the air heater therefore could often be as low as 160° C. (320° F).

5.2 Oxides of Sulphur measurement

A direct manual method used was (2) (3) employing a sintered glass filter at 90° C to collect condensed SO_3 and mist and then absorbing the remaining SO_2 in hydrogen peroxide solution.

After both gases had been sampled at the economiser flue gas inlet section, the average conversion of SO_2 to SO_3 was found to be below 4%, under normal boiler operating conditions. Under poor combustion conditions (i.e. excess combustion air operation and/or malfunctioning of the boiler control system), this percentage of oxidation was found to be as high as 6% at the same sampling position. When the boiler was carefully controlled, by means of a new control panel,the percentage of oxidation was reduced considerably to as low as 0.05%, and the corresponding SO_3 concentration was found to be about 1 p.p.m.

The percent oxidation of SO_2 to SO_3 increased either when the steam load increased, or when the excess combustion air increased, or both. The extent of oxidation was found to increase considerably as the steam load rose to values higher than 30,000 Lbs/hr (13.61 Tonnes/hr) for various percent O_2 levels measured at the economiser inlet section, as shown in FIG.5. The percent oxidation of SO_2 to SO_3 as a function of the oxygen concentration as measured at the sampling position of the economiser inlet section, was generally found to bear a straight line relationship (as shown in FIG.6); on occasions under high load conditions this straight line relationship was disturbed by the sudden production of excessive quantities of SO_3, resulting in high oxidation values of over 6%. It was not immediately clear why this should have occurred. However an analysis of the operational data indicated that there were occasions when combustion conditions would suddenly deteriorate due to momentary increases in the oil flowrate thus necessitating an increased air supply by the control system to prevent smoke formation. At high loads this increased air supply would have reduced combustion chamber residence times even further with the possibility of quenching SO_3 levels at a higher than usual value (4). It was observed that these conditions produced the highest gas temperatures at the economiser inlet sampling point of 393° C with corresponding SO_3 levels of approximately 120 p.p.m.

When oxides of sulphur were sampled at the Ljungstrom air heater outlet the percentage oxidation of SO_2 to SO_3 was also found to be below 4% on average, i.e. of the same magnitude as those values found from the measurements at the economiser inlet section, when the boiler was operating under normal steam load conditions. The air-heater leakage dilutes the flue gases consequently the concentration of each flue gas constituent falls. Since the percentage oxidation of the SO_2 to SO_3 is a ratio of concentrations, this remains unaffected by the dilution effects of the inleaking air, as shown in FIG.7, which is similar to FIG.5). The amount of the air-heater leakage (about 50% excess air), did not affect the amount of oxidation of SO_2 to SO_3, and since no further oxidation of SO_2 to SO_3 took place after the economiser inlet, catalytic oxidation could not have been taking place to increase the gas SO_3 content.

Sampling at the main flue just prior to the chimney stack entry for sulphur oxides with and without the ammonia injection system in operation gave results typical of that shown in table 1.

Table 1.

Run	Specifications	SO_3 p.p.m.	SO_2 p.p.m.
1.	Ammonia being added	0.6	700
2.	" " "	0.5	873
3.	" " "	0.6	848
4.	No addition of ammonia	15.7	899
5.	" " " "	19.6	812
6.	" " " "	25.3	740

It can be seen from the low SO_2 values that severe dilution due to air inleakage to the flues was occurring, nevertheless the ammonia was quite effective in reducing the flue gas SO_3 contents. Chemical analysis of a deposit formed around the ammonia injection point of Boiler 4 proved to consist mainly of ammonium bisulphate (80% water soluble with a free acidity of 14.4 SO_3% w/w).

5.3 Solids sampling

All solids sampling was carried out under isokinetic conditions according to the British Standards method (5). The test results are summarised in table 2. The trend of increasing unburnt carbon carryover with decreasing combustion air is shown in Fig.8. It can be seen that under normal operating conditions Boiler 4 is emitting particulate matter at a rate of about 0.16% by weight of the oil burnt measured at the economiser. By means of a sticky slide technique and using a microscope, the particles in the gas stream at this point were observed to be very fine and less than 20 µm diameter. The solid samples taken from the sampling probe were similarly very fine in all the test runs, showing no signs of agglomeration or bonding between particles. An analysis of the free acidity of these samples from the economiser inlet showed that this varied from 0.98 to 2.00 (expressed as SO_3 % w/w) where the oxygen concentration and gas temperature were 0.67% and 2.50% and 334° C and 269° C respectively.

The solids samples taken from the flue gas outlet of the Ljungstrom air heater could be divided into two categories, namely agglomerated and non agglomerated. Although the mass rate of flow of the agglomerated particles was similar to the results of samples from the economiser inlet i.e. ∿ 0.16% by weight of the oil burnt, the particles were all within the size range 1.5-3.5 mm. The temperatures and oxygen concentrations at the sampling point being ∿ 175° C and 9.0% respectively with the corresponding oxygen concentration at the economiser inlet being ∿ 0.9%. The free acidity of these agglomerated particles (% SO_3 w/w) varying between 1.85 and 2.39 showing little change from the acidity of the finer particles taken from the economiser inlet. This implies that under these particular operating conditions, even though considerable air dilution was taking place across the air heater, the only significant changes that were occurring was particle agglomeration.

Table 2

Particulate Emission Results from Solid Sampling

Test No.	Flue Gas Temp. at Samp. Pt. °C	Flue Gas Velocity (m/s)	FLUE GAS FLOW (m^3/s) Thro' Duct	At S.T.P.	RATE OF EMISSION (kg/hr)	% dust to fuel oil
...............ECONOMISER FLUE GAS INLET OF BOILER NO. 4..................						
1	337	3.1	8.29	3.71	1.47	0.15
2	334	3.2	8.55	3.85	1.80	0.19
3	338	2.8	7.48	3.34	1.40	0.16
4	338	2.6	6.95	3.11	0.82	0.10
5	269	3.6	9.62	4.85	0.15	0.21
..............AIR-HEATER FLUE GAS OUTLET OF BOILER NO. 4.................						
1	175	8.2	6.76	4.12	1.67	0.17
2	172	9.0	7.42	4.55	0.92	0.08
3	177	8.4	6.92	4.20	1.14	0.12
4	175	8.8	7.25	4.42	1.10	0.11
5	174	6.6	5.44	3.32	0.97	0.11
6	163	8.3	6.84	4.28	0.43	0.06
7	160	8.2	6.76	4.26	0.38	0.07
............MAIN FLUE GAS DUCT PRIOR TO BRICK STACK ENTRY................						
1	165	3.2	14.86	9.26	6.56	0.25
2	168	4.2	19.51	12.08	5.98	0.21
3	168	6.5	30.19	18.69	5.71	0.22
4	170	6.2	28.80	17.75	6.86	0.26
5	166	5.0	23.22	14.44	4.89	0.19
6	167	5.6	26.01	16.14	36.90	1.40
7	168	5.2	24.15	14.95	5.14	0.19
8	167	5.7	26.48	16.43	4.88	0.18

When non agglomerated particles were sampled at the same air heater outlet i.e. $< 20\ \mu m$ diameter it was found that their acid content was always very high (over 6% SO_3 w/w) the gas temperature and oxygen concentration at the sampling point being of the order of 160° C and 15.0% respectively. However it was felt more important to note that this occurred under conditions of very low boiler load (∿ 13,000 lbs steam/hr) where only one burner gun was operating during the tests and the control system was unable to maintain combustion with low excess oxygen in the combustion chamber, consequently oxygen levels measured at the inlet to the economiser were between 8 - 10% O_2. This obviously favoured more SO_3 formation and accounted for the fact that the rate of solids emission was more than halved to 0.07% by weight of the oil burnt. Under these conditions either the fine particles were adsorbing SO_3 or H_2SO_4 vapour above the dewpoint, or oxidation of SO_2 to SO_3 was taking place on the surface of the particles. Solid samples taken in the main flue are influenced by the operating conditions of the other boilers. Table 2 indicates that the rate of emission generally is slightly worse than that coming from No. 4 Boiler alone. During run 6 however deposit detachment from the walls of the duct was obviously taking place since the rate of emission increased by an order of magnitude to 36.9 kg/hr (1.4% by weight of the oil burnt). The size of the particles sampled under these conditions was up to 2.5 cm in diameter with a free acidity of over 6% SO_3 w/w. These were obviously conditions when severe acid smut fall out from the chimney was taking place.

The majority of samples at this point however indicated dust loadings of the order of 5 kg/hr being approximately 0.2% by weight of the oil burnt. The sampled particles were found to be mixtures of very fine particles similar to those collected at the economiser inlet i.e. ∼ 20 µm diameter and particles within a size range of approximately 2.5 - 6.0 mm diameter. The acidity of these mixtures being variable between 0.86% and 4.8% SO_3 w/w. The oxygen content at the sampling point varied between 11% and 17%, indicating that further air inleakage was taking place through the walls of the brick flue due to the chimney draft.

5.4 Gas temperature measurements

In addition to gas temperature measurements being taken at each point of sampling, continuous recording of gas temperature at the base of the chimney stack allowed comparison of these recorded values with periods of severe acid smut fall out as assessed by the collection trays. These gas temperatures fluctuated considerably and this was attributed to variations in the quantity of the inleaking air mixing with the hot flue gases. Similar measurements were taken of the axial gas temperature in the top half of the chimney stack by lowering a thermocouple down the chimney on the end of a 10 lb weight. Again considerable fluctuations were observed as shown in FIG.9. No explanation could be offered for this except either variations in the quantity of inleaking air and or the displacement of the thermocouple and weight from a true axial path.

FIG.10 shows a typical gas temperature variation at the base of the stack during a period when heavy acid smut fall out occurred that day between 6 a.m. and 9 a.m. It can be seen that this followed several hours of elevated flue gas temperature implying that this was the initiating step resulting in deposit detachment and subsequent acid smut emission.

6. CONCLUSIONS

Poor atomisation and combustion,conditions can result in increased unburnt carbon particles being carried out of the combustion chamber. At high boiler load this problem was accentuated. Attempts to overcome this by using more combustion air only marginally reduced the carbon carry over by increased particle burn away,but greatly increased the SO_3 content of the combustion gases and the acidity of the carbon particles emitted.

The acidity of the carbon particles could increase as they were transported by the combustion gases through the system. This implied that sulphuric acid vapour was being adsorbed on the particles and/or SO_2 was being oxidised to SO_3 on the particles at temperatures above the acid dewpoint of the gases. Severe air inleakages taking place in the air heater and following brick flues, did not increase the percentage oxidation of SO_2 to SO_3 in the flue gases however. Ammonia injection after the induced draft fan of each boiler was effective in almost eliminating the SO_3 content of the flue gases but did not effectively reduce the free acidity of the soot particles. From the economiser onwards there was evidence of particle agglomeration. If these particles were carried out of the chimney stack they would constitute an acid smut fall out problem. If they were deposited in the flues or stack they could become detached at a later date as much larger flakes. The subsequent size of such flakes or particles constituting an acid smut fall out problem, would depend on factors causing detachment in the flues and aerodynamic factors causing flake break up in the flue and chimney as well as in the surrounding air outside. In any installation where there is no evidence of sulphuric acid condensation on carbon coated flue surfaces at temperatures below the dewpoint, being responsible for deposit detachment and subsequent emission of acid smuts from the chimney, it must be concluded that such smuts originate from the combustion chamber. Here, as a result of poor burner design and/or operation, unburnt carbon particles could be produced under excess oxygen conditions which, in the presence of sulphur oxides, result in particle agglomeration and increased acidity, as they are carried through the flue system. Deposition and subsequent intermittent detachment due to load and temperature fluctuations affecting adhesion, may exacerbate the eventual emission and fall out problems.

7. ACKNOWLEDGEMENTS

The authors wish to thank Imperial Chemical Industries Ltd. for their generous provision of facilities and financial support of this work.

8. REFERENCES

1. Clean Air Regulation 1971 (Emission of Grit & Dust from Furnaces) Statutory Instrument 1971, No. 162 H.M.S.O.
2. British Standards Institution Doc.No. 71/42839.
3. Goksøyr, H. & Ross K. Shell Research Ltd. TRC paper 1962 "SO_3 measurement in Combustion Gases".
4. Hedley, A.B. "Factors affecting the formation of SO_3 in Combustion Gases", J.Inst.Fuel, April, 1967, pp.142-151.
5. British Standard BS 3405 1971, Appendix D.

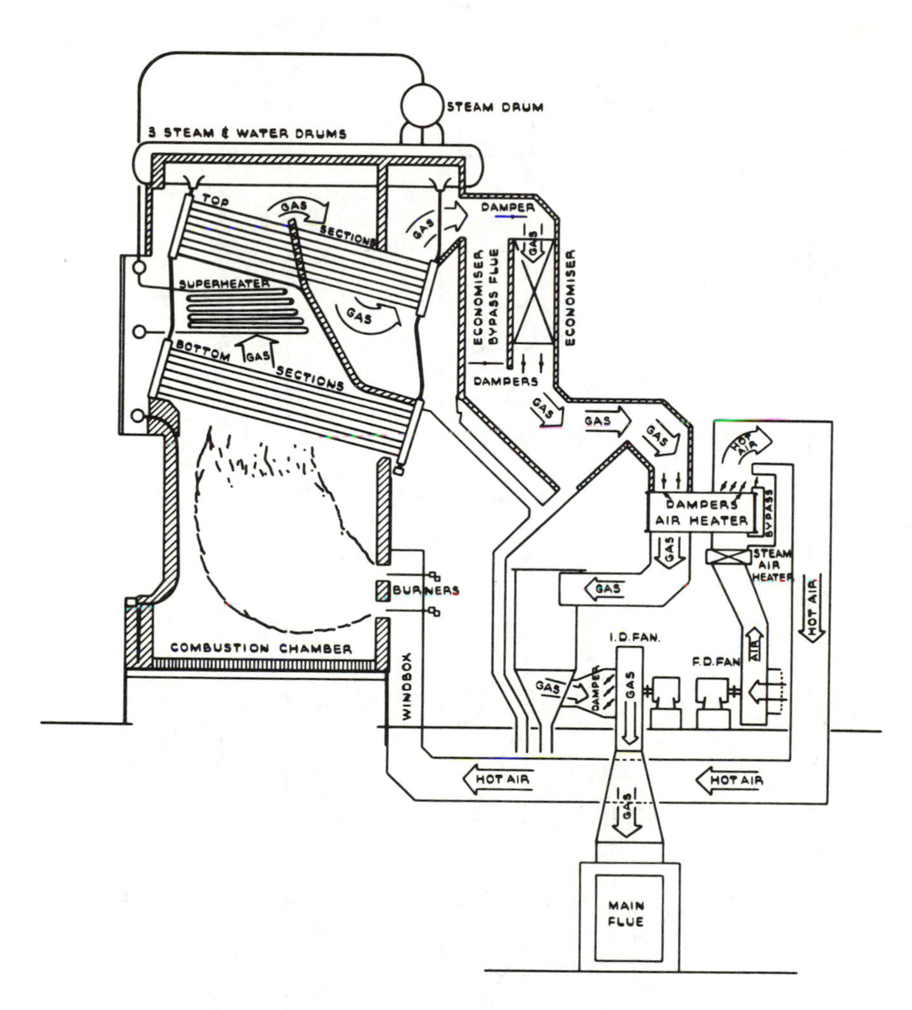

FIG. 1
BOILERS 1 TO 4 – OIL FIRING
DIAGRAM OF GAS & AIR CIRCUITS.

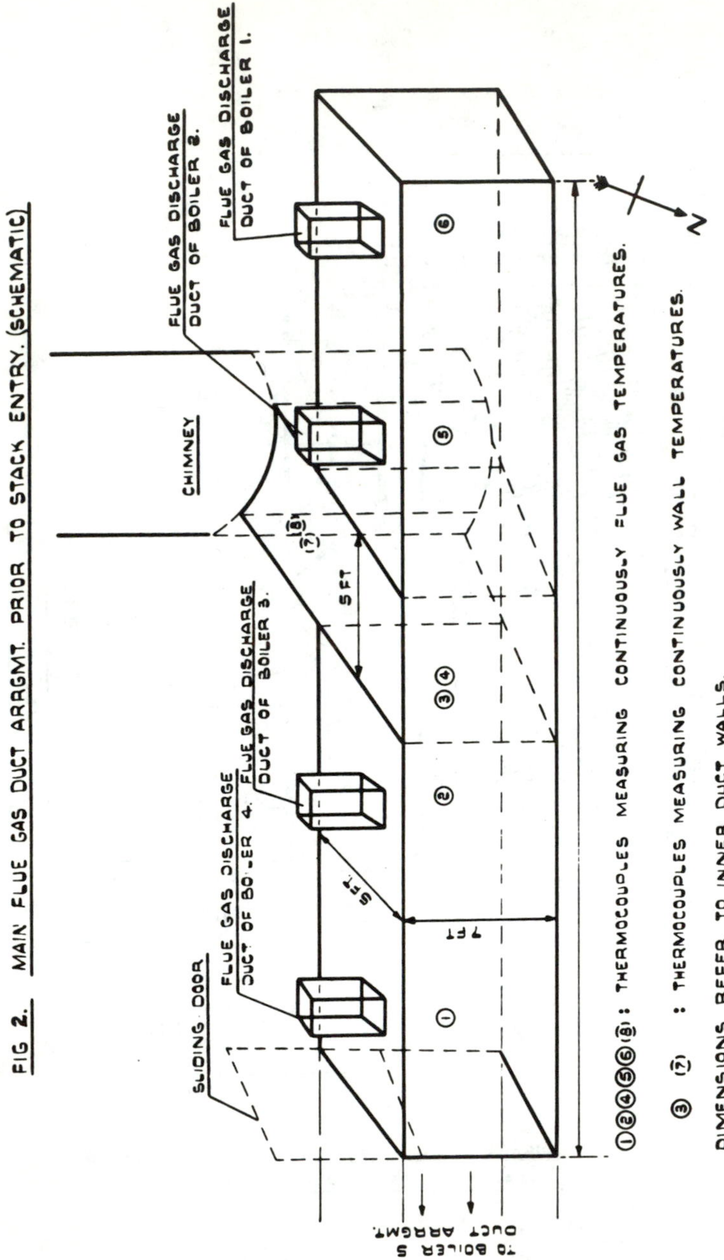

FIG 2. MAIN FLUE GAS DUCT ARRGMT. PRIOR TO STACK ENTRY. (SCHEMATIC)

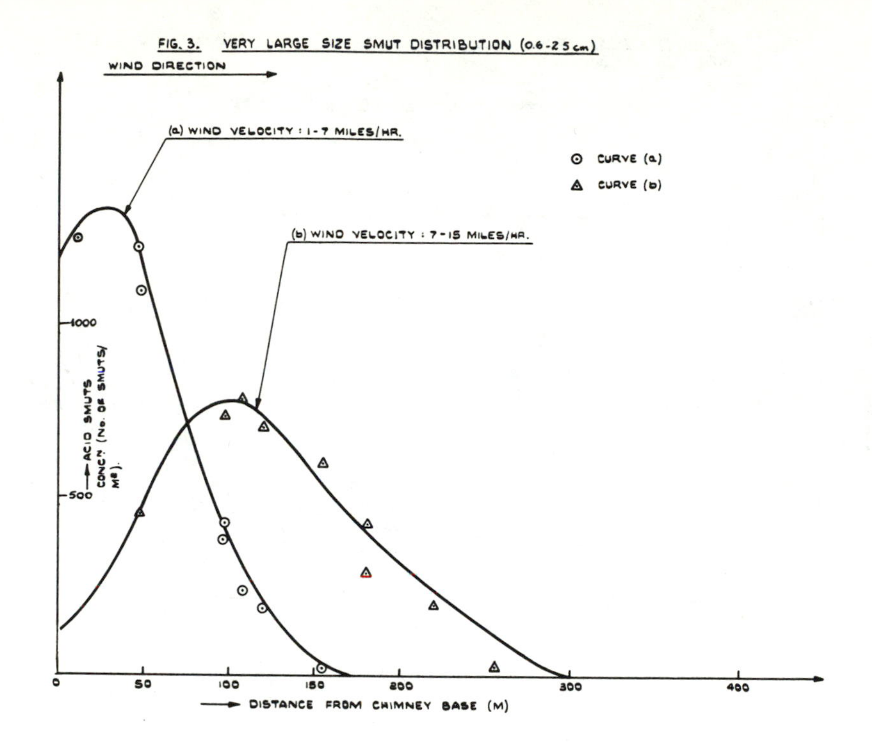
FIG. 3. VERY LARGE SIZE SMUT DISTRIBUTION (0.6-2.5 cm)
WIND DIRECTION
(a) WIND VELOCITY : 1-7 MILES/HR.
(b) WIND VELOCITY : 7-15 MILES/HR.
CURVE (a)
CURVE (b)
ACID SMUTS CONCN (No. OF SMUTS/M²).
1000
500
0
50
100
150
200
300
400
DISTANCE FROM CHIMNEY BASE (M)

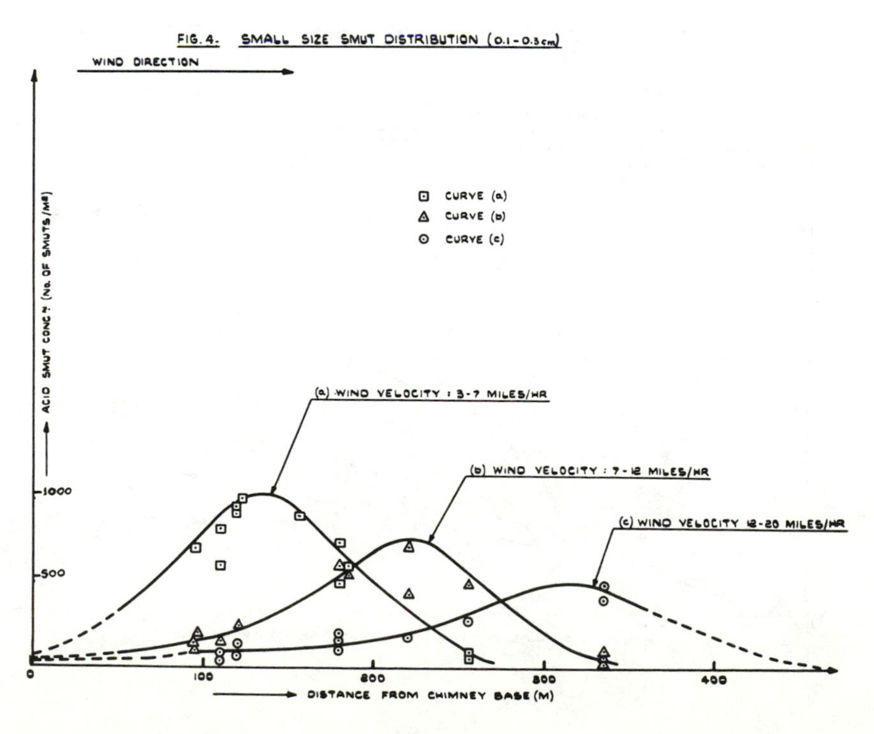
FIG. 4. SMALL SIZE SMUT DISTRIBUTION (0.1-0.3 cm)
WIND DIRECTION
CURVE (a)
CURVE (b)
CURVE (c)
ACID SMUT CONCN (No. OF SMUTS/M²)
(a) WIND VELOCITY : 3-7 MILES/HR
(b) WIND VELOCITY : 7-12 MILES/HR
(c) WIND VELOCITY 12-20 MILES/HR
1000
500
0
100
200
300
400
DISTANCE FROM CHIMNEY BASE (M)

FIG. 5. % OXIDATION OF SO_2 TO SO_3 VS. STEAMLOAD (Lbs/hr.) – AT ECONOMISER FLUE GAS INLET OF BOILER No. 4.

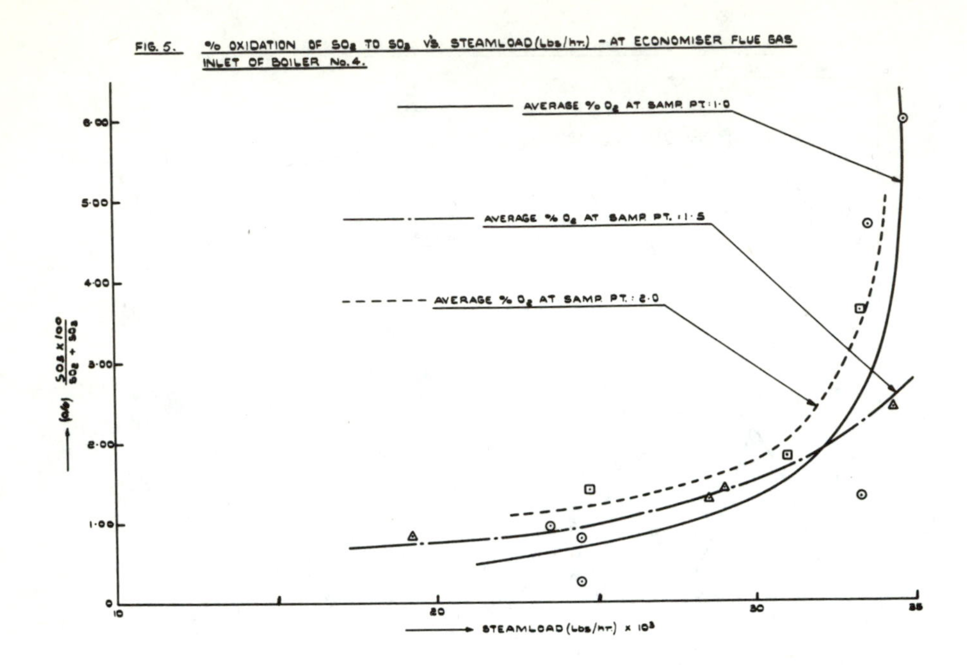

FIG. 6. % OXIDATION OF SO_2 TO SO_3 VS. % O_2 CONTENT IN THE FLUES – AT ECONOMISER FLUE GAS INLET OF BOILER No. 4.

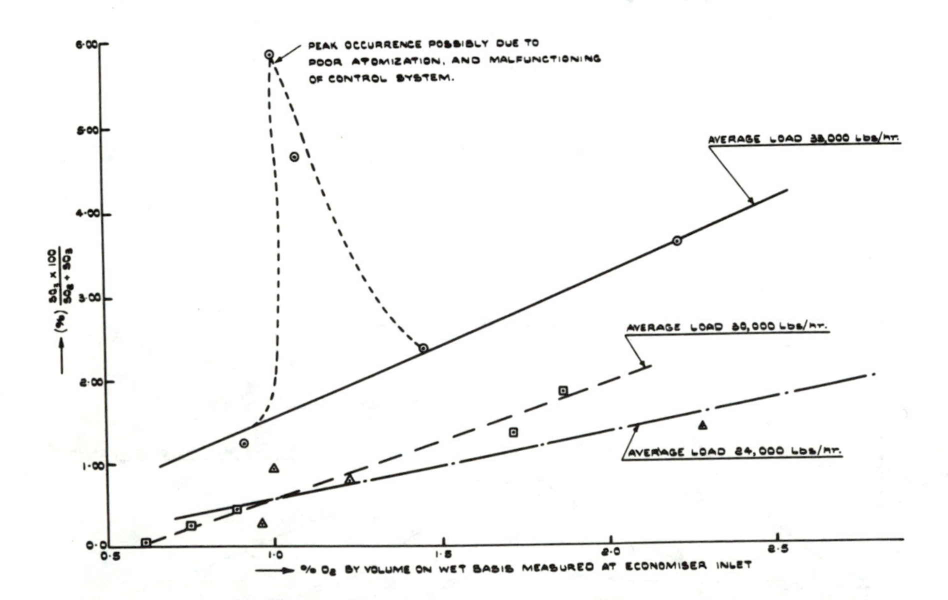

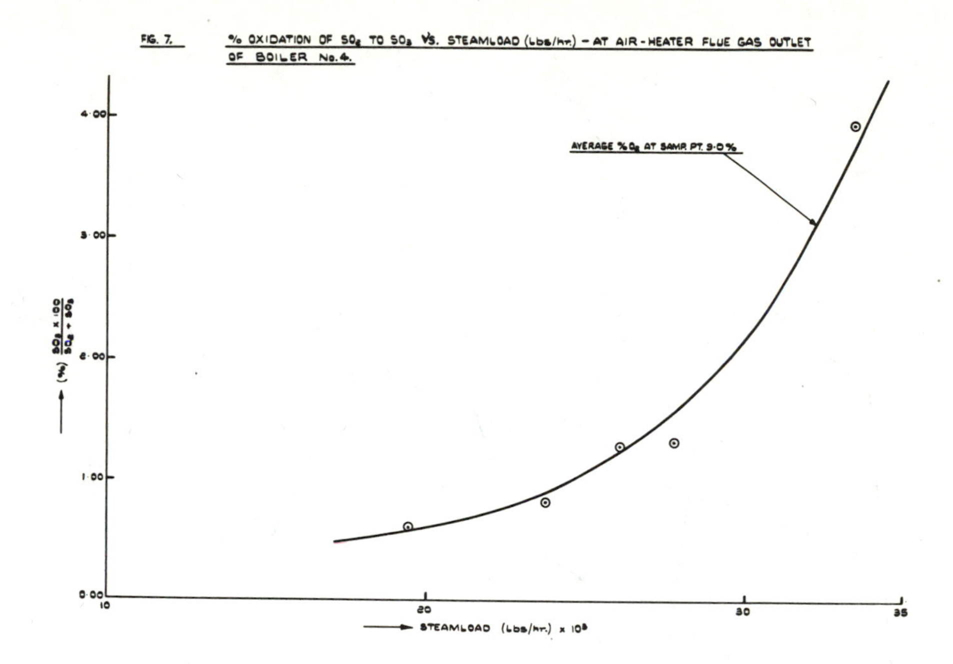
FIG. 7.
% OXIDATION OF SO2 TO SO3 VS. STEAMLOAD (Lbs/hr.) – AT AIR-HEATER FLUE GAS OUTLET OF BOILER No. 4.
AVERAGE % O2 AT SAMP. PT. 3·0 %
(%) SO3 x 100 / SO2 + SO3
4·00
3·00
2·00
1·00
0·00
10
20
30
35
STEAMLOAD (Lbs/hr.) x 10³

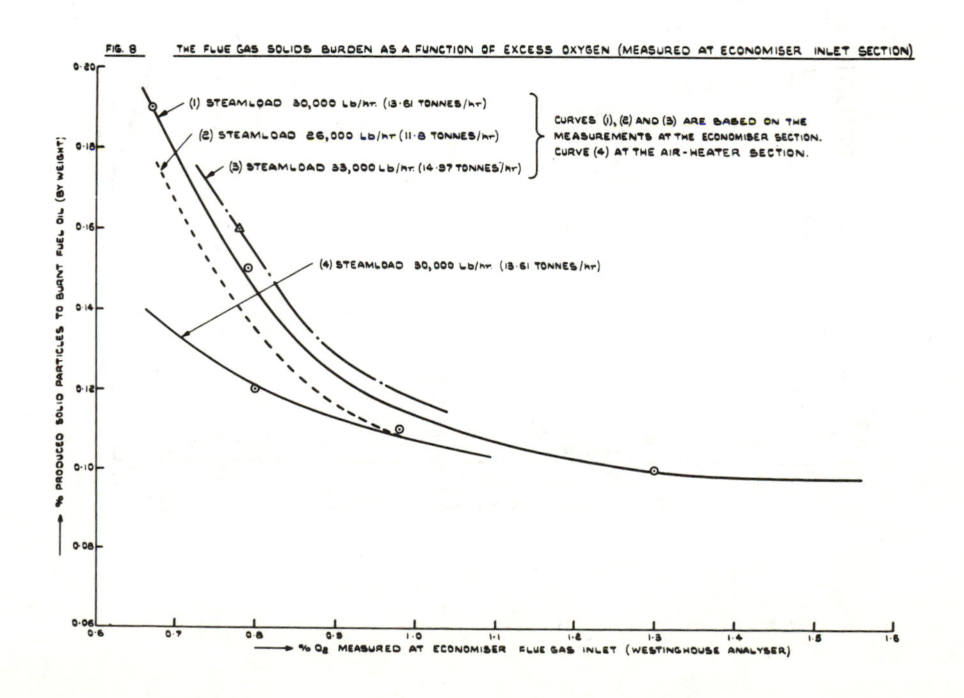
FIG. 8
THE FLUE GAS SOLIDS BURDEN AS A FUNCTION OF EXCESS OXYGEN (MEASURED AT ECONOMISER INLET SECTION)
(1) STEAMLOAD 30,000 Lb/hr. (13·61 TONNES/hr)
(2) STEAMLOAD 26,000 Lb/hr (11·8 TONNES/hr)
(3) STEAMLOAD 33,000 Lb/hr. (14·97 TONNES/hr)
CURVES (1), (2) AND (3) ARE BASED ON THE MEASUREMENTS AT THE ECONOMISER SECTION.
CURVE (4) AT THE AIR-HEATER SECTION.
(4) STEAMLOAD 30,000 Lb/hr (13·61 TONNES/hr)
% PRODUCED SOLID PARTICLES TO BURNT FUEL OIL (BY WEIGHT)
0·20
0·18
0·16
0·14
0·12
0·10
0·08
0·06
0·6
0·7
0·8
0·9
1·0
1·1
1·2
1·3
1·4
1·5
1·6
% O2 MEASURED AT ECONOMISER FLUE GAS INLET (WESTINGHOUSE ANALYSER)

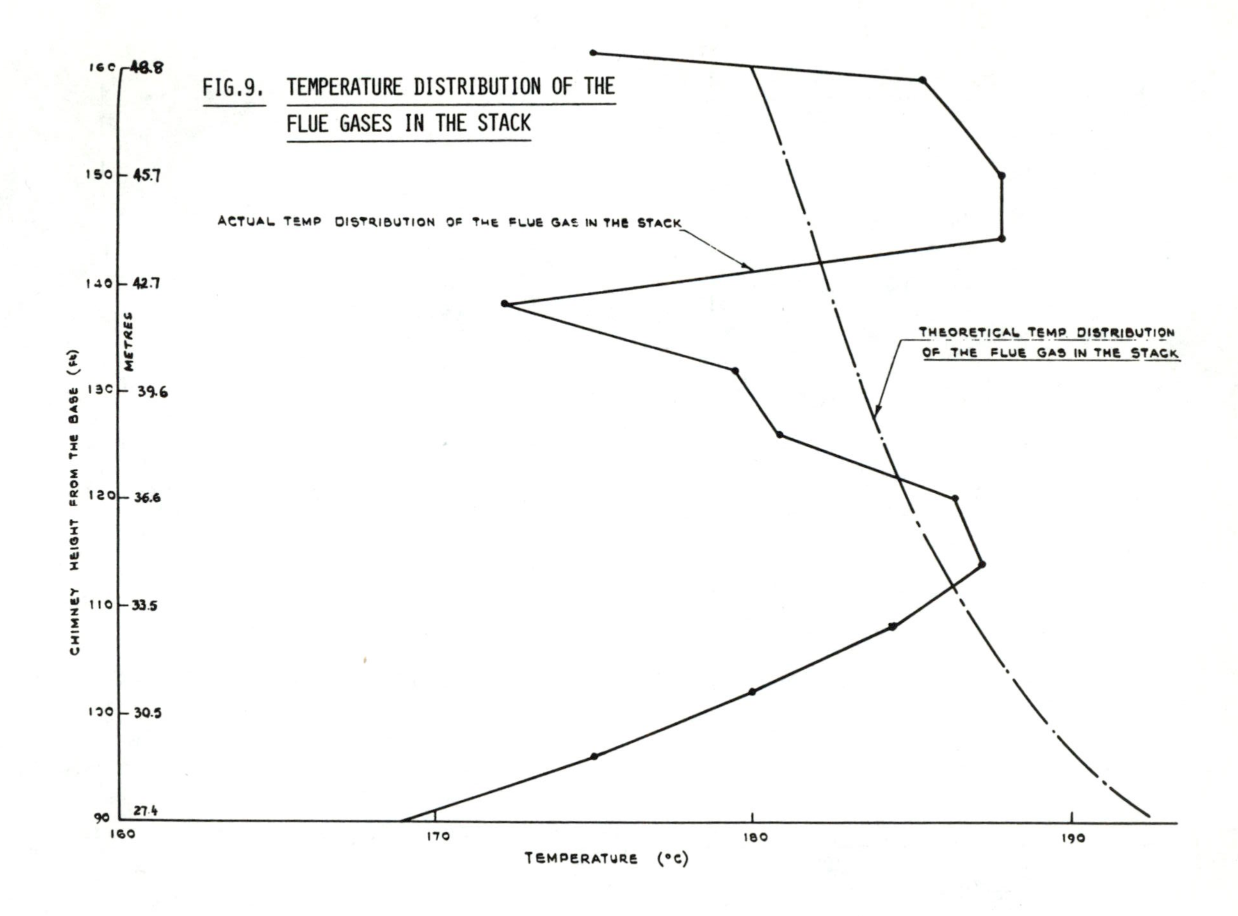

FIG.9. TEMPERATURE DISTRIBUTION OF THE FLUE GASES IN THE STACK

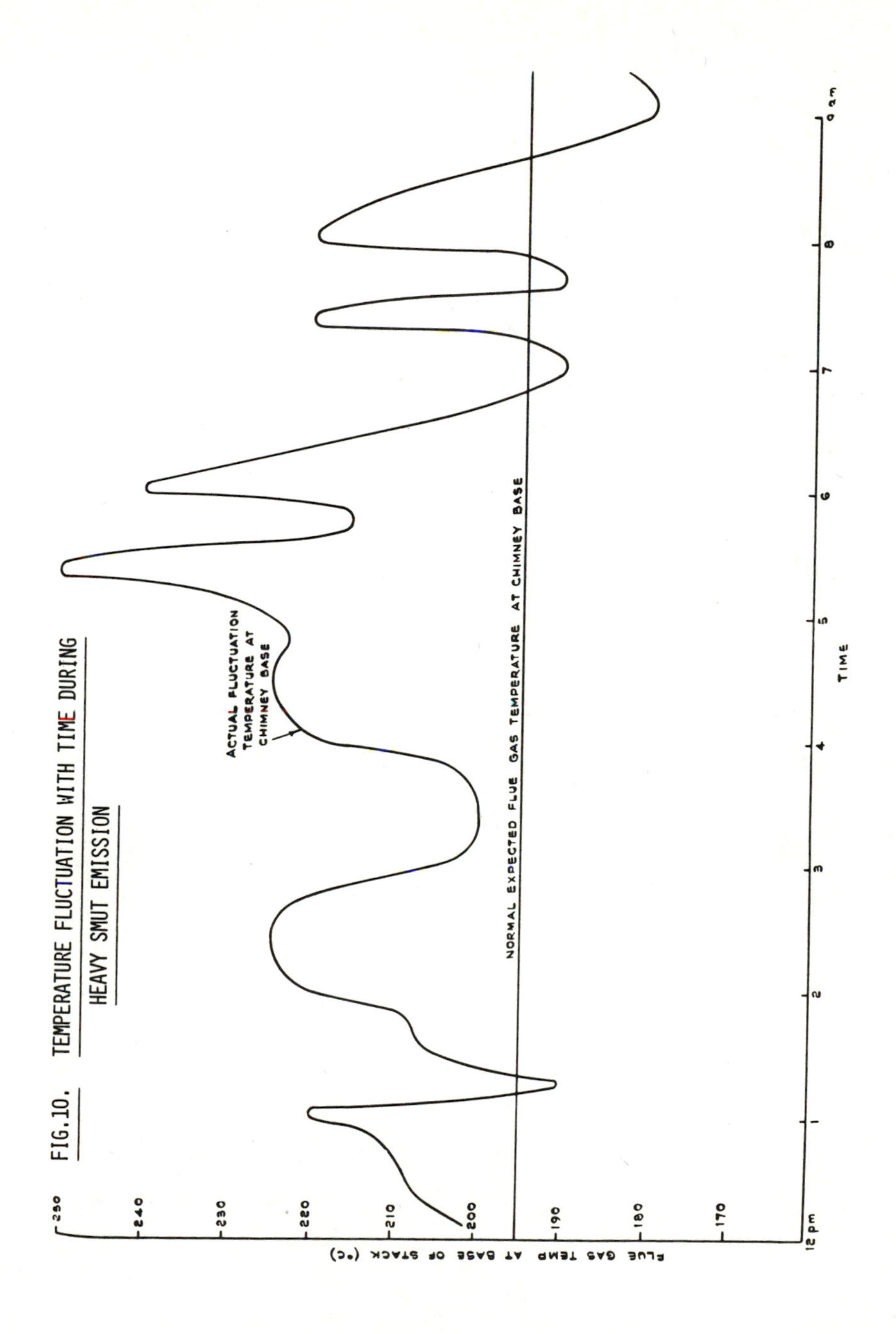

FIG.10. TEMPERATURE FLUCTUATION WITH TIME DURING HEAVY SMUT EMISSION

THE FATE OF FUEL NITROGEN—IMPLICATIONS FOR COMBUSTOR DESIGN AND OPERATION

M. P. HEAP, T. L. CORLEY, C. J. KAU, and T. J. TYSON

Energy and Environmental Research Corporation
8001 Irvine Boulevard
Santa Ana, California, USA 92705

ABSTRACT

The processes occurring during the combustion of fuels containing trace quantities of bound nitrogen are reviewed. Pyrolysis, heterogeneous and homogeneous reactions are considered. Experimental information on the interaction of fuel sulfur and fuel nitrogen compounds is presented. Limit case studies which consider chemistry to be the controlling phenomenon in fuel NO formation are presented for advanced power cycles burning coal-derived fuels.

INTRODUCTION

National security requires that the United States reduces its dependency upon imported crude oil, thus necessitating the need for energy conservation and the increased utilization of coal as a raw fuel. Economic and environmental considerations are key factors, both in conversion of conventional plants to coal-firing and in the production and utilization of coal-derived liquid and solid fuels. Sulfur and nitrogen, two trace species found in fossil fuels, contribute significantly to the total air emissions from stationary combustion sources. This paper is concerned with the influence of the combustion process upon the fate of fuel nitrogen in order to identify those combustor conditions which minimize nitrogen oxide emissions. The primary method of controlling the formation of fuel NO (NO produced from nitrogen bound in the fuel) is to operate the initial stages of combustion fuel rich. This can be achieved either by delaying the mixing of the fuel with all the air or by dividing the total air into two streams, one of which is injected downstream of the point of fuel injection. Although staged combustion techniques have been successfully used to reduce NO emissions from utility boilers there is considerable concern that these conditions might lead to enhanced fireside corrosion rates.

Table 1 lists the nitrogen content of various coal, petroleum and coal-derived fuels. It can be seen from these figures that any conversion from conventional fuels to coal-derived or shale oils will result in an increase in the production of fuel NO from stationary sources unless measures are taken to prevent it. Coal gasification is attractive because it allows the control of sulfur emissions and it produces a fuel suitable for combined cycle firing. From the viewpoint of overall efficiency it is desirable to utilize the gasifier product gas at its gasifier exhaust temperature; however, the ammonia

Table 1. Nitrogen Content of Various Fuels

Fuel	Nitrogen Content	Ref.
Low Btu Fuel Gas produced by Westinghouse Fluidized Bed/ In-Bed Desulfurization	0.0007 mole fraction NH_3	1
Low Btu Flue Gas, GE Advanced Fixed Bed Gasifier	0.04 % by weight NH_3	2
Coal-Derived Liquid Fuel (H Coal)	1.3 % by weight	1
Wilmington Crude Oil (exceptionally high nitrogen content)	0.6 % by weight	3
Shale Oil - Heavy Distillate	1.6 % by weight	3
Shale Oil - Residium	2.04 % by weight	
U.S. Coals	1 - 2.1 % by weight	3

content of the high temperature clean gas will be significant. Estimated NO_x emissions from four advanced power plant systems are presented in Table 2, together with the New Source Performance Standards and the EPA emission targets.

Table 2. Estimated NO_x Emissions from Advanced Power Plants

Advanced Plant Description	Estimated NO_x Emissions	Current Standard (Coal)	Advanced Target 1	Ref.
	g/MJ (lb/10^6 Btu Input)			
Open Cycle Gas Turbine - Coal-Derived Liquid Fuel	0.731 (1.7)	0.301 (0.70)	0.13 (0.3)	2
Open Cycle Gas Turbine - Integrated Gasifier Cold Product Gas	0.15 (0.35)	—	—	2
Open Cycle Gas Turbine - Coal-Derived Liquid Fuel	0.26 (0.6)	—	—	1
Open Cycle Gas Turbine - Integrated Gasifier Hot Product Gas	0.28 (0.65)	—	—	1

The fate of fuel nitrogen during combustion is linked to three processes:

1. Pyrolysis. The fuel droplet or particle is heated, vaporization occurs and volatile nitrogen compounds are released.
2. Heterogeneous Reactions. Char contains refractory nitrogen compounds which will be oxidized during char burnout. In fuel-rich mixtures char may reduce NO. Other heterogeneous reactions may involve reduction or absorption of NO by soot (solid carbon formed from the gas phase).
3. Homogeneous Reactions. Volatile nitrogen compounds are converted to NO or N_2 by gas phase reactions involving fuel fragments.

This paper will review the processes occurring in all of the above; however, the main emphasis will be placed upon gas phase fuel nitrogen reactions. Information on the kinetic mechanism of fuel nitrogen reactions will be used in an attempt to define those combustor conditions which lead to the minimum conversion of fuel nitrogen to NO. No claim can be made that the following discussion represents a complete set of references for what is a considerable body of data. Rather it was the intention to present the general consensus of opinion as well as any significant variations.

PYROLYSIS OF NITROGEN-CONTAINING FUELS

During the initial stages of combustion fuel droplets and particles undergo pyrolysis. Axworthy et al (4) carried out a series of pyrolysis experiments with several model compounds and real liquid fuels. Under inert conditions there was almost quantitative conversion of the nitrogen in the organic model compound to HCN at pyrolysis temperatures of 1373 K. Pyrole, pyridine, quinoline and benzonitile were used, and this conversion was found to be relatively insensitive to the nature of the model compound. In all instances the rate of decomposition of the nitrogen compound was accelerated by the presence of oxygen and the products were found to contain N_2, N_2O, NO and NO_2. These same investigators studied the pyrolysis of coal and liquid fuels and found that the HCN yield varied from 23 to 51 percent for liquid fuels at 1373 K. The highest yield of HCN was obtained from a coal-derived liquid fuel.

Axworthy et al do not give information on the nitrogen content of the solid residue remaining after pyrolysis. Sarofim and co-workers at MIT have studied the retention of nitrogen in char as a function of particle temperature (5,6). Relatively long times (of the order of one second) and high temperatures (in excess of 1500 K) were necessary to reduce the nitrogen content of the char to 40 percent of the value of the parent coal when the coal was heated in an inert gas stream. Peak temperatures in pulverized coal flames may exceed 1800 K although residence times at these temperatures are less than one second.

Blair, Wendt and Bartok (7) studied the devolatilization of single particles of coal under controlled pyrolysis conditions. The coal particle was heated rapidly by contact with a resistance heated platinum ribbon and the rates of mass evolution as functions of temperature heating rate and coal composition were determined. Species measured were CH_4, CO, CO_2, C_2H_2, C_2H_4, C_2H_6, HCN and NH_3. It was found that the total nitrogen volatilized was a more sensitive function of pyrolysis temperature than was the total mass pyrolyzed. The former was strongly coal composition dependent, while the latter was not. Blair reports that approximately half of the total volatiles and 85 percent of the nitrogeneous species do not evolve as the light gases listed above. The results of time resolved volatilization rates support the hypothesis that, during combustion, nitrogeneous species are evolved late in the volatilization sequence.

From a practical viewpoint pyrolysis studies indicate that:

- the speciation of volatile nitrogen compounds may not be important. It appears that the compounds that are evolved initially will be rapidly converted to HCN in the absence of oxygen;
- nitrogenous species may be evolved rather late in the volatilization sequence. Thus fuel/air mixing must be delayed beyond this time to prevent NO formation; and
- at high temperatures the major portion of the fuel nitrogen is evolved as a volatile compound.

HETEROGENEOUS REACTIONS INVOLVING NITROGEN

Comparatively little is known concerning the conversion of char nitrogen to NO. Sometime ago Heap et al (8) suggested that char nitrogen was not the most significant source of fuel NO in uncontrolled pulverized coal flames. Recent work by Sarofim et al (5) and Pershing and Wendt (9) have confirmed this hypothesis experimentally. Figure 1, taken from the work of Sarofim et al, shows the conversion of coal and char nitrogen to nitric oxide as a function of fuel/air equivalence ratio and furnace temperature. The char had been previously devolatilized at 1500 K for one second. The results indicate that the conversion of char nitrogen to NO occurs with a lower efficiency than coal nitrogen under comparable conditions.

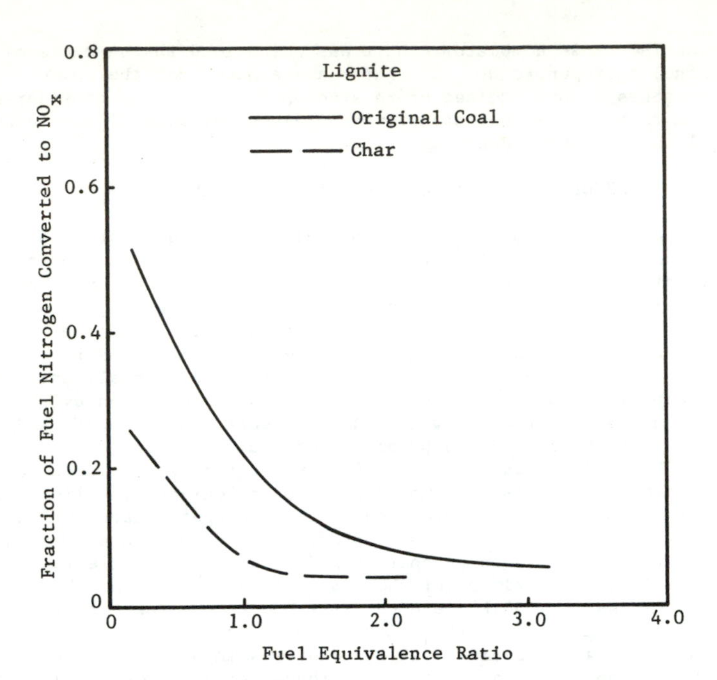

Fig. 1. Comparison of fuel nitrogen conversion to NO for a char and the parent coal (after Sarofim et al (5))

The results shown in Figure 2 were taken from the work of Pershing and Wendt who burned several coals in argon, oxygen mixtures to determine the fraction of total emissions attributable to fuel NO. The char NO formation was calculated based upon information obtained by Blair et al (7) and upon the measured conversion of NO and NH_3 injected with the fuel. The results of both Pershing and the MIT group are in agreement, the volatile nitrogen compounds account for the major fraction of the NO produced in coal flames. Furnace investigations involving the combustion of both coal and pulverized petroleum coke (a low volatile, high nitrogen solid) under comparable conditions give total emissions of around 800 ppm for coal and 200 ppm for coke at a stoichiometric ratio of 1.15 (8). The results of these three different studies appear to be internally consistent — under uncontrolled conditions NO formed from coal char is of the order of 25 percent of the total emission. However, under staged combustion conditions NO formed from char may represent greater than 75 percent of the total emission and may present the greatest barrier to the attainment of very low levels of NO.

Wendt and Schulze (10) carried out a series of theoretical calculations on the conversion of char nitrogen to NO during char combustion. These theoretical calculations indicated that the conversion was relatively insensitive to temperature, but quite strongly dependent upon particle size and the oxygen content of the free stream. In practical combustors the objective is to burn out the char as rapidly as possible, thus small particles, high oxygen levels and high temperatures are desirable. Unfortunately, these appear to be factors which tend to maximize char nitrogen conversion to NO. It would appear that minimum NO emissions from coal-fired systems are most likely to occur when all the nitrogen has been removed from the char under fuel-rich conditions.

Evidence exists suggesting that char and soot will reduce nitric oxide. Heterogeneous reduction reactions involving char particles were proposed by Gibbs, Pereira and Beér (11) to partially account for the fact that the concentration of NO within the bed was less than the concentration in the freeboard of a fluidized bed. An interesting series of experiments were carried out by Dobovisek and Cernej (12) involving soot collected from porous sphere

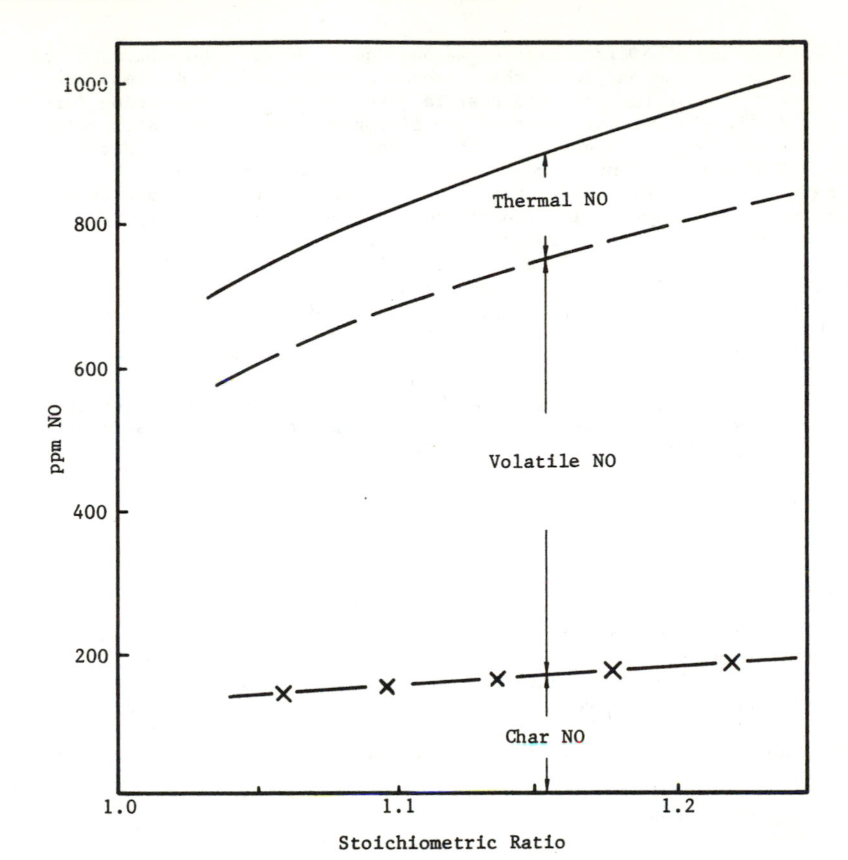

Fig. 2. Measured thermal, fuel and calculated char NO for a Western Kentucky coal (after Pershing(9))

supported distillate oil flames. The soot was collected on a glass wool plug from various heights above the porous sphere. In the first experiment a heated inert gas was passed over the soot and analyzed for NO. It appears that NO was absorbed by the soot in the flame and the amount absorbed increased with soot residence time. When a heated inert gas containing NO was passed over the soot the NO concentration was reduced, but it is not known whether this was due to absorption or to reduction of NO to N_2.

Staged combustion to control the formation of fuel NO involves operation with a rich first stage which will provide conditions which promote soot formation. If nitrogen compounds are absorbed by soot at temperatures typical of the rich first stage then any soot which is burned out in the second stage is a potential source of NO. Thus when burning oil or coal it appears that those conditions which minimize NO formation will minimize solid carry-over into the second stage.

HOMOGENEOUS REACTIONS

Many investigators have carried out experiments to determine both the mechanism of fuel nitrogen oxidation and the parameters which control conversion to NO. Equivalence ratio appears to be the most important parameter controlling the conversion of fuel nitrogen to NO in premixed gaseous flames. Typically conversion is of the order of 90 percent or greater for fuel-lean mixtures, but it drops off sharply as the mixture becomes more rich. At equivalence ratios around 1.5 the conversion is approximately 30 percent or less. The formation of NO from simple gaseous compounds does not appear to be strongly affected by the nature of the nitrogen compound, although compounds containing nitrogen-hydrogen bonds appear to give the highest conversion for lean mixtures. As the concentration of the nitrogen compound increases the total NO increases but the fraction of the nitrogen compound converted to NO decreases. This effect is most pronounced with rich mixtures.

Flame temperature does not appear to have a very strong influence upon fuel NO formation. De Soete (13) reports that for methane/oxygen/argon mixtures with an equivalence ratio of 1.61 the fraction of HCN converted to NO was 0.056, 0.179 and 0.363 for temperatures of 1800 K, 2000 K and 2200 K respectively, and of all the additives studied, HCN appears to show the strongest influence of temperature and the effect is accentuated for rich mixtures. The nature of the fuel does influence the formation of fuel NO. When hydrogen is the fuel there is almost twice the production of fuel NO than with ethylene (13).

It is generally recognized that there are two competing paths occurring in flames which account for the conversion of fuel nitrogen compounds to nitrogen or NO (13,14). These can be expressed as

$$NI + \text{Oxidant} \rightarrow NO + \ldots \quad (1)$$

and

$$NI + NO \rightarrow N_2 + \ldots \quad (2)$$

Fuel nitrogen fed as ammonia, hydrogen cyanide or pyridine to rich flames is normally present in the burnt gases as HCN, NH_i, NO and N_2. For high flame temperatures Morley (15) shows that regardless of their nature, nitrogen compounds are quantitatively converted to HCN in the reaction zone of rich hydrocarbon flames. It is most likely that the fuel nitrogen compound will be oxidized to NO prior to the formation of HCN. Guibert and Van Tiggelen (16) proposed

$$CH_3 + NO \rightarrow CH_3NO \rightarrow CH_2NOH \rightarrow HCN + H_2O \quad (3)$$

as the mechanism of HCN formation in methane/ammonia flames. An alternative route is suggested by Myerson (17) who conducted a series of experiments on the reduction of NO by hydrocarbons

$$CH + NO \rightarrow HCN + O \quad (4)$$

or

$$CH + NO \rightarrow CHO + N \quad (5)$$

The production of NH_3 in the combustion products of rich flames can be accounted for by the formation of NCO as an intermediate in the oxidation of HCN; thus

$$HCN + OH \rightarrow CN + H_2O \quad (6)$$

$$CN + OH \rightarrow NCO + H \quad (7)$$

$$NCO + H \rightarrow NH + CO \quad (8)$$

$$NH_x + H \rightleftarrows NH_{x-1} + H_2 \quad (9)$$

At lower temperatures Haynes (18) suggests that CN could be removed by CO_2

$$CN + CO_2 \rightarrow NCO + CO \quad (10)$$

Assuming that the NH_i system is internally equilibrated Haynes has identified the NO forming and removing reaction ((1) and (2)) as

$$N + OH \rightarrow NO + H \quad (11)$$

$$N + NO \rightarrow N_2 + O \quad (12)$$

whereas Fenimore (14) suggests

$$NO + NH_2 \rightarrow N_2 + H_2O \quad (13)$$

as the major NO removing reaction.

In the reaction zone of hydrocarbon flames molecular nitrogen will contribute to the formation of HCN and NH_i via such reactions as

$$CH + N_2 \rightarrow HCN + N \quad (14)$$

$$CH_2 + N_2 \rightarrow HCN + NH \quad (15)$$

These reactions were first proposed by Fenimore as contributing to the rapidly formed NO in rich primary reaction zones.

Fuels which contain bound nitrogen normally contain sulfur compounds which are themselves oxidized to sulfur oxides during combustion. Sternling and Wendt (19) suggested that since SO_2 is an effective catalyst in reducing super-equilibrium free radical concentrations, the presence of sulfur compounds in flames would inhibit the formation of thermal NO. Wendt and Ekmann (20) added large quantities of both H_2S and SO_2 (equivalent to 6,300 ppm SO_2 in the products) to premixed flat flames and showed that there was an inhibition effect. Working with much smaller concentrations De Soete (13) studied the influence of sulfur compounds on the formation of fuel and thermal NO in premixed flames. For rich mixtures there was an enhancement of NO production in the presence of sulfur compounds. At an equivalence ratio of 1.5 the presence of 190 ppm of H_2S in the inlet mixture increased the conversion of ammonia to NO from 50 to 70 percent.

Corley and Wendt (21) carried out a series of experiments at the University of Arizona to investigate the effects of fuel sulfur-fuel nitrogen interactions in turbulent diffusion flames. The fuels used were natural gas and distillate oil. The distillate oil was doped with pyridine and pyridine and thiophene. Natural gas was doped with ammonia and SO_2. The enhancement effects shown in Figure 3 were only observed with preheated air and under low combustion intensity conditions. These results, in conjunction with the flat flame studies of De Soete, appear to indicate that there are important interactions occurring between fuel sulfur and fuel nitrogen during combustion.

Reactions proposed to account for the sulfur-nitrogen interaction can be divided into two categories:

- Direct Effects. Sulfur species reacting with nitrogeneous intermediates

$$NX + SX \rightarrow NS, N2 + \ldots$$

or oxidation of nitrogenous species to

$$NX + SO \rightarrow NO + \ldots$$

- Indirect Effects. Regeneration or production of hydrocarbon radicals which are important to the production of nitrogen radicals

$$CX + SZ \rightarrow CS, CZ + \ldots$$

or the recombination of free radicals by SO_2.

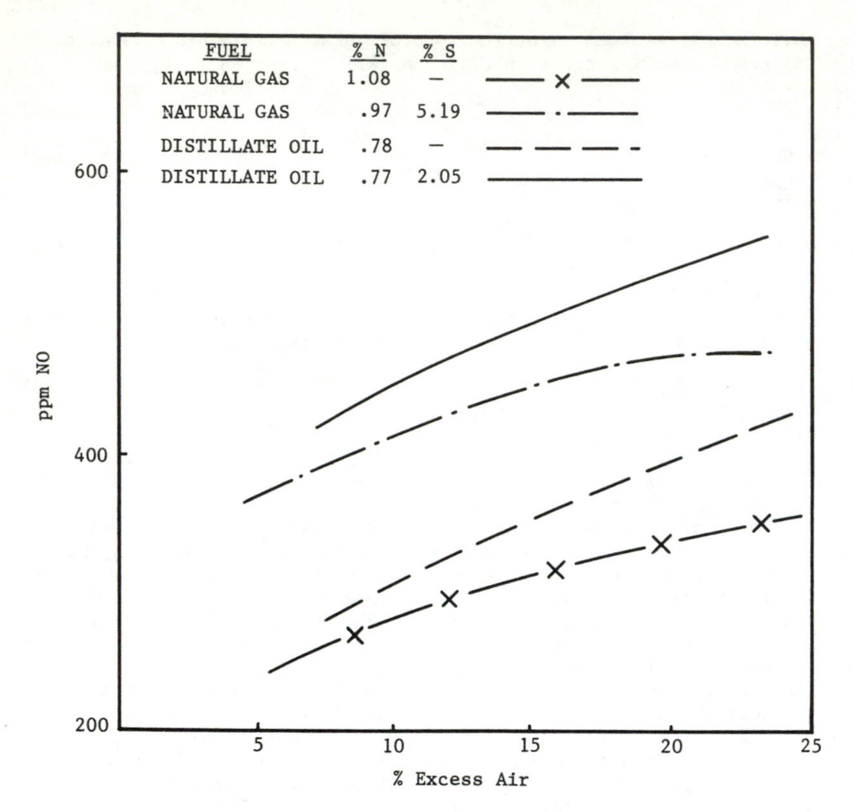

Fig. 3. Enhancement of fuel NO formation by the presence of fuel sulfur in gaseous and liquid fuels (Corley and Wendt (21))

LIMIT-CASE STUDIES FOR LOW NO_x COMBUSTORS

Limit-case situations are those in which some single aspect of the physics or chemistry dominates and the remaining phenomena can either be ignored or modeled by simple idealized processes. The term here refers to the methodology used to search for the lower bounds on NO_x emissions which are set by the chemistry. Implicit in this approach is the assumption that all physical processes associated with fuel-air-product contacting and heat transfer can be accomplished in an ideal and optimum manner dictated by the chemistry. Having established this chemical-lower-bound, the question of how close one can achieve the optimum physical transport behavior is then open to scrutiny.

Fundamental to the success of this approach is an authoritative description of the finite-rate chemistry pertinent to fuel nitrogen conversion to NO and thermal fixation of N_2. Of particular importance is the inclusion of the proper chemistry to describe the generation of stable intermediate nitrogen compounds (e.g., HCN, NH_3) and their subsequent equilibration path under rich conditions. The role of fuel breakdown and the effect of hydrocarbon radicals on the production and destruction of nitrogen compounds must adequately be described as must the influence of XN compounds on NO production and distribution.

In searching for the lower bound on NO_x emissions set by the chemistry it is assumed that the optimum chemical reactor system which achieves this lower bound can be described in terms of an interacting set of stirred and plug flow reactors which exchange matter and heat with one another and with the surroundings. The optimum coupling between these basic elements which yields minimum NO_x emissions is unknown at the outset and may indeed be very complex involving several parallel and feedback paths as illustrated in Figure 4. The simplest (computationally) of these systems is shown in Figure 4a which illustrates a "series" or sequentially staged set of coupled reactors. SR refers to stirred reactors and PFR to plug flow reactors. Q represents heat loss. The EER Modular Kinetics Analysis Program (MKAP) performs the nonequilibrium chemistry and element-to-element bookkeeping on such a coupled system and provides for an arbitrary distribution of mass and energy transfer along the PFR elements. Figure 4b illustrates hypothetical systems with feedback loops. In one case the feedback interaction is shown consisting of two well-stirred zones with different residence times and heat loss factors. The other illustration shows a plug flow zone with distributed feedback from both exhaust products and from a well-stirred zone. Figure 4c shows examples of "parallel" couplings in which no influence of downstream behavior is felt upstream. However, continuous exchange between reactor elements is allowed.

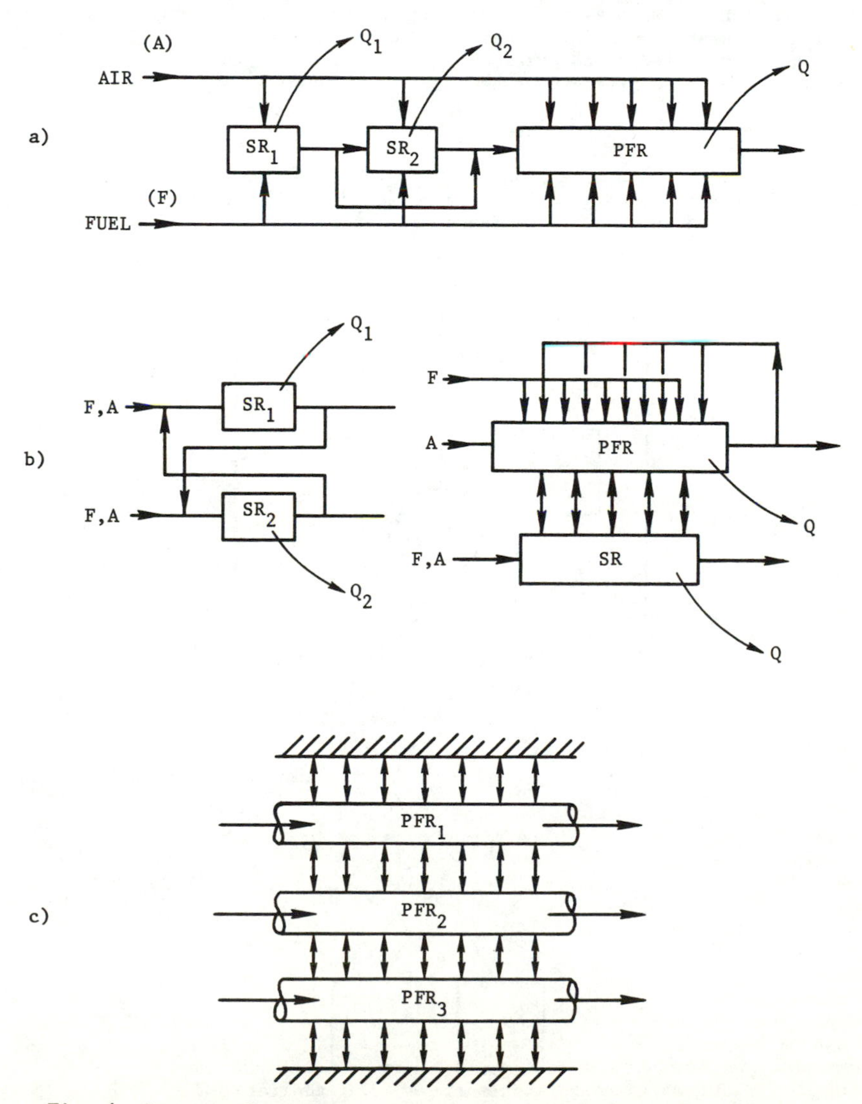

Fig. 4 Conceptual reactor sequences for staged combustion systems

Sarofim (22) has calculated the sum of the bound nitrogen specie concentration at equilibrium for C, H, O, N mixtures typical of coal and liquid fuels. Under fuel-rich conditions the sum of the bound nitrogen specie passes through a minimum as the equivalence ratio is increased. The minima in total bound nitrogen increases with increasing temperature from less than 1 ppm at 1600 K for a H/C fuel of 1.3 to 47 ppm at 2800 K. However, these values are considerably lower than experimental measurements of bound nitrogen species at these equivalence ratios indicating that kinetic constraints prevail. Sarofim suggests that these constraints can be limited if a combustor were to be operated with a rich first stage and a high degree of preheat. This would have two benefits for solid fuels; it would reduce the fraction of nitrogen retained in the char as well as speeding the reduction of total bound nitrogen concentration.

An alternative approach might well be to develop a parallel staged system which mixes products from different fuel/air systems together at a given time producing a mixture of species which would not normally exist, in the necessary concentrations, in any one single reactor. Used in conjunction with high preheat, this might well reduce the residence time in the rich sector necessary to equilibrate the bound nitrogen species.

Kinetic Mechanism

The general features of the kinetic mechanism used in the limit-case studies are illustrated in Figure 5 and have been described in detail elsewhere (23). Since there is ample evidence indicating that with solid and

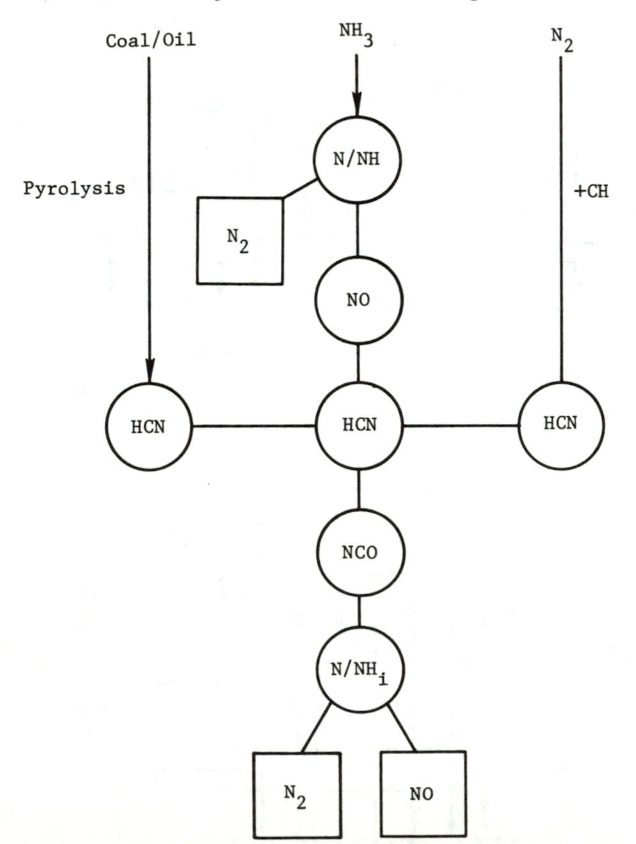

Fig. 5. General schematic showing the fate of fuel nitrogen during combustion

liquid fuels the pyrolysis reactions are rapid, it is assumed that the nitrogen in the fuel is converted to HCN or it remains with the char. For ammonia under rich conditions there are paths which allow breakdown to NH with subsequent production of N and NO. Reaction of N with OH provides the major path for NO production, whereas N_2 is almost exclusively produced by

$$N + NO \rightarrow N_2 + O \qquad (12)$$

HCN is produced by reaction of CH with either N_2 or NO. HCN oxidation proceeds by the path suggested by Morley (15).

The validation of a set of reactions to be used in the limit-case studies presents a major problem because of the lack of suitable data. Many of the reaction rate coefficients must be estimated and then adjusted to provide reasonable agreement with experimental data. Although the set of reactions have been used to calculate NO production in a stirred reactor fired with CH_4/air undoped and doped with NO or NH_3 no measurements of HCN were available and it is therefore not possible to ascertain whether the calculated total bound nitrogen concentration was reasonable or not. NO production in a stirred reactor (equivalence ratio 1.4 CH_4/air) appears to be most sensitive to the rate of three reactions

$$CH_4 \rightarrow CH_3 + H \qquad (16)$$

$$CH_3 + O \rightarrow CH_2O + H \qquad (17)$$

$$CH + N_2 \rightarrow CHN + N \qquad (19)$$

Doubling the rate of these reactions would result in a change of 40 percent or more in the NO produced.

LBG-Fired Combustors

Coal gas may contain 500 ppm of NH_3 if hydrogen sulfide has been removed at low temperatures (340 K) or 4000 ppm if high temperature (1008 K) clean up processes are used. Two different systems have been considered which utilize LBG in combined cycle plants:

- the lower temperature LBG was burned in a supercharged boiler exhaust gas turbine combined cycle;
- an advanced high temperature gas turbine was assumed for use with the high temperature LBG in a simple exhaust waste heat boiler combined cycle.

Limit-case calculations have been carried out to attempt to define combustor designs for minimum NO_x emissions from these two types of systems.

The supercharged boiler was assumed to operate at 10 atmospheres and feed a gas turbine at 1366 K inlet temperature. The boiler was fired with 588 K combustion air and 340 K LBG fuel gas containing 500 ppm of ammonia.

Figure 6 shows the calculated NO concentration for three reactor simulations of a supercharged boiler. Curve 'a' represents a system with a 1 ms stirred reactor for ignition followed by a plug flow section with a heat loss distribution as shown. The fuel and air are premixed. Curves 'b' amd 'c' show the effect of operating under staged combustion conditions. Curve 'b' represents conditions with a short residence time stirred reactor feeding a plug flow section with secondary air addition distributed over the first 500 ms. In curve 'c' the rich products are held at an equivalence ratio of 1.15 in a plug flow reactor (non-adiabatic) for 500 ms before secondary air addition. Heat extraction gives the same final temperature of 1366 K. The increase in NO level associated with second stage air is due to NO formation from the bound nitrogen species that exit the rich reactor. In curve 'c' it can be seen that the NO is slowly being reduced in the rich primary reactor.

As discussed earlier, the drive towards low bound nitrogen levels in the rich primary reactor may be accelerated by increasing the temperature of the primary reactor. However, unless the combustor is designed accordingly, this may result in excessive production of thermal NO. This is illustrated by the results presented in Figure 7 which simulate the conditions of a secondary

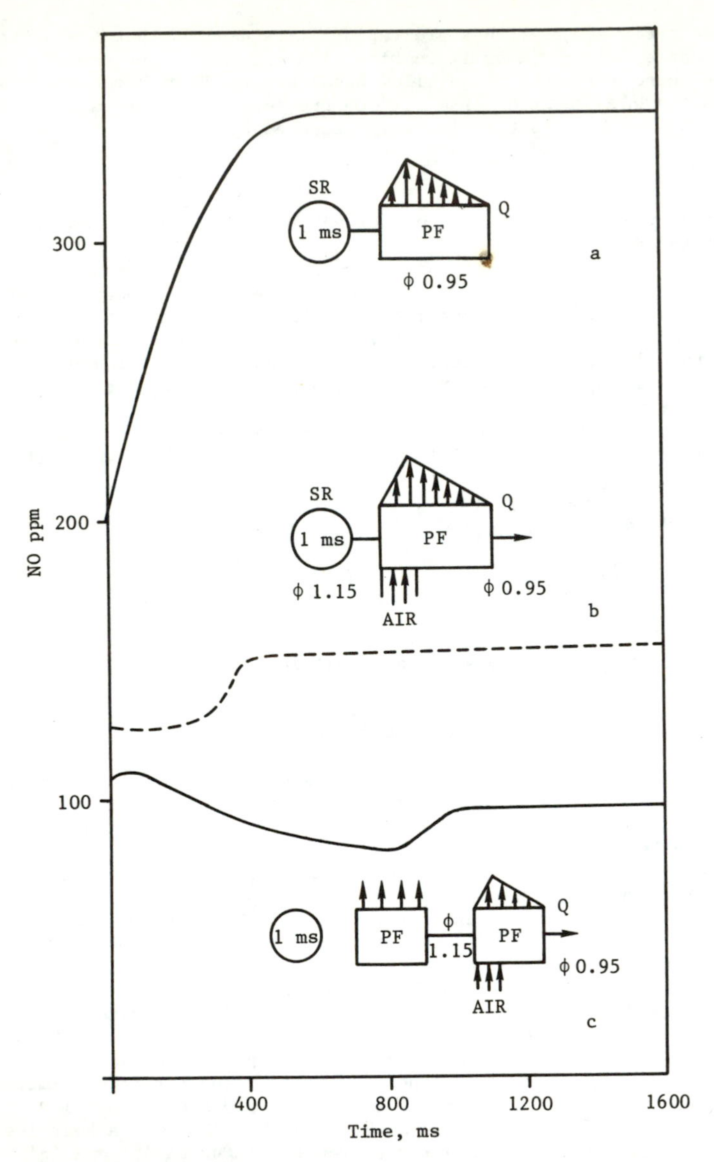

Fig. 6. NO concentration for various reactor simulations of a supercharged boiler

zone of an adiabatic combustor supplying products to a gas turbine at 1819 K. The fuel used was a high temperature coal gas containing 4000 ppm of ammonia. The long dash curves in Figure 7 present the instantaneously mixed situation. In this case NO rises rapidly in the first few ms as the primary zone bound nitrogen compounds are converted to NO. The NO level at the entry to the second stage is lower than the primary exit due to dilution as the second stage air is mixed with the primary products. Following the conversion of bound nitrogen the NO increases very slowly due to thermal fixation because the temperature does not exceed 1811 K. The solid curves represent second stage air addition over 50 ms. A temperature excursion to 2300 K occurs and the NO level goes over 1000 ppm. The dot-dash curves simulate conditions when the products of the rich primary section are added to the second stage air over a period of

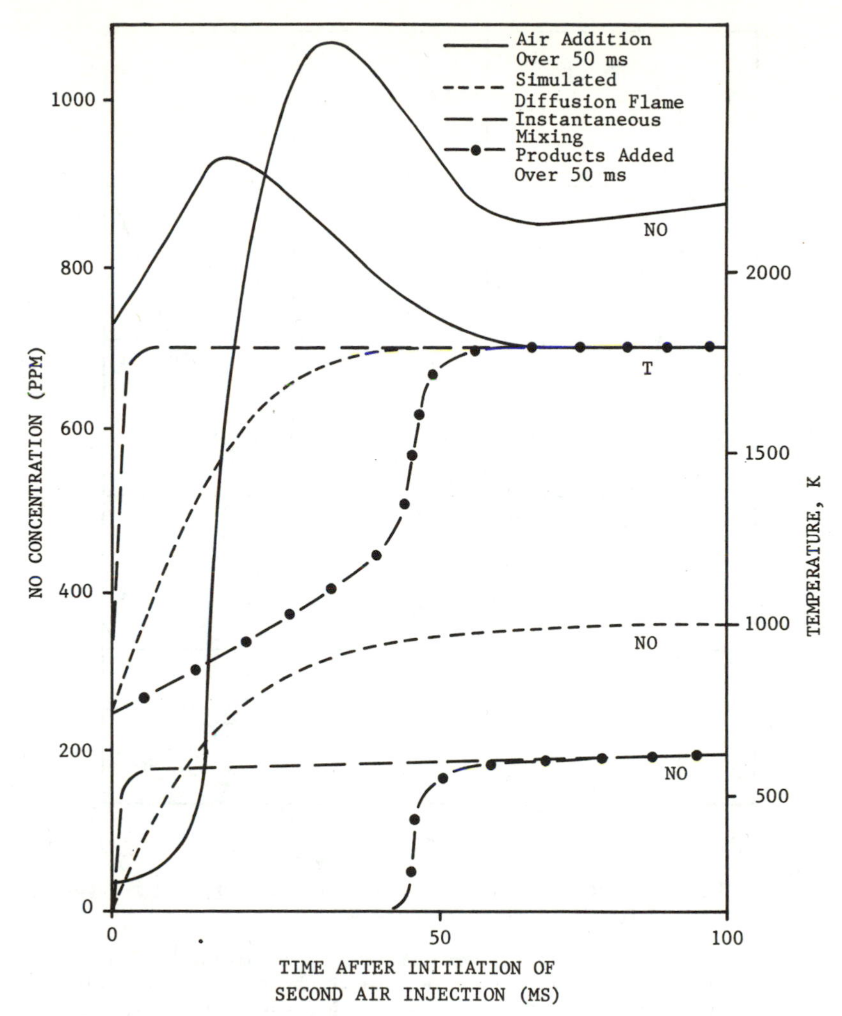

Fig. 7. NO concentration and temperature for alternative secondary stage configurations — adiabatic gas turbine combustor (at atmospheres)

50 ms. Since fuel is added to the secondary air the mixture increases in temperature until auto-ignition of the remaining fuel occurs at approximately 40 ms. There is no overshoot in temperature and the conversion of the bound nitrogen compounds is followed by a slow thermal fixation as in the instantaneously mixed case. The short dashed curves in Figure 7 correspond to a simulated diffusion flame. The temperature curve represents the mean axial temperature and does not reflect the local temperature excursion within the diffusion flames. Temperature and stoichiometry variations about the mean give rise to the increased NO.

One aspect of the strategy suggested by Sarofim (22) is that emissions become independent of the fuel nitrogen level. Figure 8 summarizes the results of an attempt to optimize the conditions in the rich primary reactor for an adiabatic combustor burning high temperature coal gas. An effective NH_3 conversion ratio is shown plotted as a function of equivalence ratio. The calculations were carried out with an adiabatic plug flow reactor ignited by a 1 ms stirred reactor. The effective NH_3 conversion ratio is the difference between the sum of the bound nitrogen specie at a given reactor residence

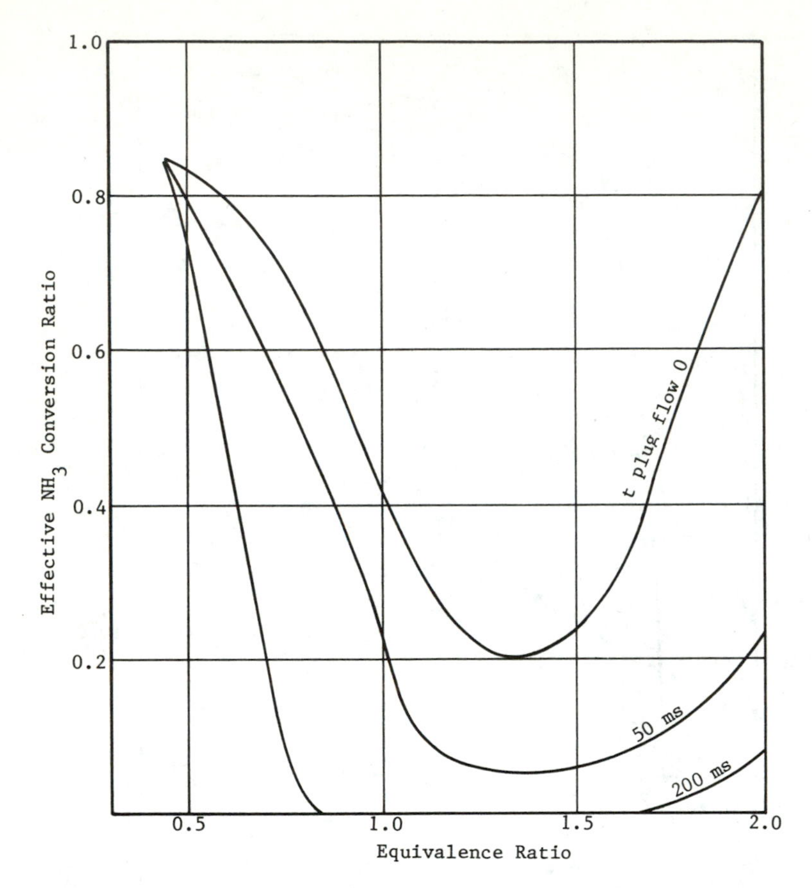

Fig. 8. Effective $(NH_3)_0$ conversion ratio premixed primary reactor — adiabatic gas turbine combustor

time for a fuel with and without NH_3 normalized by the initial NH_3 concentration of the doped fuel. When the ratio is zero the bound nitrogen specie which exit the rich reactor are independent of initial NH_3 concentration. These conditions occur for $t \geq 200$ ms and $0.8 < \phi < 1.7$. The minimum of the curve of bound nitrogen specie versus equivalence ratio occurs between 1.33 and 1.45. At 10 atmospheres the rich primary zone would require a residence time of 200 ms or more to achieve minimum levels. With instantaneous secondary air addition the final NO level at an equivalence ratio of 0.45 (temperature 1810 K) would be less than 70 ppm.

CONCLUSIONS

Conversion to coal or coal-derived fuels will increase the nitrogen content of the fuel being used in stationary combustors thus creating the potential for increased NO_x emissions. Fuel NO conversion is restricted by operating combustors with a fuel-rich primary section. Total bound nitrogen at the exit of the rich primary section may be minimized by using high preheat or by using combined parallel and sequential staged systems. Fuel sulfur appears to enhance the production of fuel NO. Reactions involving char, soot and nitrogen compounds require further study.

ACKNOWLEDGEMENTS

Much of the work described in this paper was carried out under EPA Contracts 68-02-1361 and 68-02-2196, and Mr. G. Blair Martin is the Project Officer for both contracts. One of us (T.L.C.) carried out the fuel sulfur experiments in part fulfillment of the requirements for a MS degree at the University of Arizona. We would like to thank Professor J. O. L. Wendt for helpful discussions on sulfur/nitrogen chemistry.

REFERENCES

1 Beecher, D.T., et al, "Energy Conversion Alternatives Study-ECAS-Westinghouse Phase II Final Report", Vol. 1, Oct. 1976, NASA Lewis Research Center, Cleveland, Ohio.

2 Corman, J.C. and Fox, G.R., "Energy Conversion Alternatives Study-ECAS-General Electric Phase II Final Report", Vol. 1, Dec. 1976, NASA Lewis Research Center, Cleveland, Ohio.

3 Mexey, D.J., Singh, Surjit, and Hissong, D.W., "Fuel Contaminants", Vol. 1, July 1976, Industrial Environmental Research Laboratory, Research Triangle Park, North Carolina.

4 Axworthy, A.E., Schneider, G.R., and Dayan, V.H., "Chemical Reactions in the Conversion of Fuel Nitrogen to NO_x", Proceedings of the Stationary Source Combustion Symposium, Vol. 1, Fundamental Research, EPA-600/2-76-152a, Sept. 1975.

5 Pohl, J.H. and Sarofim, A.F., "Fate of Coal Nitrogen During Pyrolysis and Oxidation", EPA Symposium on Stationary Source Combustion, Sept. 1975, Atlanta, Georgia.

6 Pohl, J.H. and Sarofim, A.F., "Devolatilization and Oxidation of Coal Nitrogen", Sixteenth Symposium (International) on Combustion, Aug. 1976, Cambridge, Mass.

7 Blair, D.W., Wendt, J.O.L. and Bartok, W., "Evolution of Nitrogen and Other Species During Controlled Pyrolysis of Coal", Sixteenth Symposium (International) on Combustion, Aug. 1976, Cambridge, Mass.

8 Heap, M.P., Lowes, T.M., Walmsley, R., Bartelds, H. and LeVaguerese, P., "Burner Criteria for NO_x Control: Vol. I, Influence of Burner Variables on NO_x in Pulverized Coal Flames", U.S. Environmental Protection Agency, Office of Research and Development, Environmental Protection Technology Series, Pub. No. EPA-600/2-76-061a (1976).

9 Pershing, D.W. and Wendt, J.O.L., "Pulverized Coal Combustion: The Influence of Flame Temperature and Coal Composition on Thermal and Fuel NO_x", Sixteenth Symposium (International) on Combustion, Aug. 1976, Cambridge, Mass.

10 Wendt, J.O.L. and Schulze, O.E., "On the Fate of Fuel Nitrogen During Coal Char Combustion", AIChE Journal, Vol. 22, No. 1, Jan. 1976, pp. 102-110.

11 Gibbs, B.M., Pereira, F.J. and Beér, J.M., "Coal Combustion and NO Formation in an Experimental Fluidized Bed", Sept. 1976, Institute of Fuel Symposium on Fluidized Bed Combustion, London, England.

12 Dobovisek, A. and Cernej, A., "Pollutant Emissions Produced in Fuel Droplet Burning", EPA Contract PR-2-517,7, Mar. 1977, University of Sarajevo, Yugoslavia.

13 De Soete, G., "La Formation Des Oxydes D'Azote Dans La Zone D'Oxydation Des Flammes D'Hydrocarbures", No. 23 309, June 1975, Institut Francais Du Petrole, France.

14 Fenimore, C.P., "Reactions of Fuel-Nitrogen in Rich Flame Gases", Combustion and Flame, Vol. 16, 1976, pp. 249-256.

15 Morley C., "The Formation and Destruction of Hydrogen Cyanide from Atmospheric and Fuel Nitrogen in Rich Atmospheric-Pressure Flames", Combustion and Flame, Vol. 27, 1976, pp. 189-204.

16 Guibet, J.C. and Van Tiggelen, A., Rev. Institute Fr. Petrol, Vol. 18, 1963, pp. 1284.

17 Myerson, A.L., "The Reduction of Nitric Oxide in Simulated Combustion Effluents by Hydrocarbon-Oxygen-Mixtures", Fifteenth Symposium (International) on Combustion, 1975, Pittsburgh, Pa.

18 Haynes B.S., "Reactions of Ammonia and Nitric Oxide in the Burnt Gases of Fuel-Rich Hydrocarbon-Air Flames", Combustion and Flame, Vol. 28, 1977, pp. 81-91.

19 Sternling, C.V. and Wendt, J.O.L., "Kinetic Mechanisms Governing the Fate of Chemically-Bound Sulfur and Nitrogen in Combustion", Final Report to Environmental Protection Agency, Contract EHS-D-71-45, 1972.

20 Wendt, H.O.L. and Ekmann, J.M., "Effect of Fuel Sulfur Species on Nitrogen Oxide Emissions from Premixed Flames", Combustion and Flame, Vol. 25, 1974, pp. 355-360.

21 Corley, T.L., "Fuel Sulfur Effects of NO_x Formation in Turbulent Diffusion Flames", Thesis, University of Arizona, 1976.

22 Sarofim, A.F., Pohl, J.H., Taylor, B.R., "Mechanisms and Kinetics of NO_x Formation Recent Developments", Paper Presented at 69th Annual Meeting AIChE, Chicago, Nov. 1976.

23 Folsom, B.A., Corley, T.L., Lobell, M.H., Kau, C.J., Heap, M.P. and Tyson, T.J. To be published.

HIGH TEMPERATURE REACTIONS INFLUENCING CORROSION

Session Chairman: **R. E. Barrett**
Battelle Columbus Laboratories

DEPOSITION OF INORGANIC MATERIAL IN OIL-FIRED BOILERS

P. J. JACKSON

Central Electricity Generating Board
Marchwood Engineering Laboratories
Marchwood, Southampton, Hampshire, England S04 4ZB

ABSTRACT

An experimental study has been made of deposit formation in superheaters of residual oil-fired boilers. Using a portable air-cooled probe, the mass transfer rates of ash constituents and magnesium compounds derived from anti-corrosion fuel additives on to surfaces at 510 to 730°C were determined. Data are presented showing the rates of deposition and the composition of the deposited material. Comparison with the long-term accumulations of deposits indicates the vapour diffusion inwards of sodium compounds and the in-situ sulphation of deposited magnesium oxide. It appeared that the deposition of magnesium compounds enhanced the deposition of sodium sulphate. Much higher deposition rates of ash constitutents were measured on probe surfaces below the acid dew-point.

INTRODUCTION

In the study of the formation of fireside boiler deposits, relatively little has been reported concerning the rate of deposition of fuel ash components onto surfaces in operating plants, particularly in the higher temperature sections. Most investigations of superheater deposits formation processes have been conducted using equipment in laboratories or confined to examinations of the products at the end of an operating campaign.

An experimetnal procedure for measuring directly the mass transfer rates of deposit constituents in boilers was described in 1961.[1] Results obtained in coal-fired boilers were reported at the Marchwood Conference in 1963[2] and some observations in an oil-fired boiler published later in the same year.[3] This paper presents more data obtained using this technique in oil-fired boilers with and without magnesium compounds added to the flue gases to neutralize sulphuric acid in the lower temperature regions of the plant. The objectives of the work were to determine the composition of the deposited material, the rates of mass transfer of its constituents and the influence on these of two methods of adding magnesium oxide or hydroxide.

PLANT CONDITIONS

The measurements were made in two types of water-tube boiler in CEGB power stations during commercial operation, under steady full-load conditions. On all occasions, the boilers had been operating with or without additive for some days previously, and as far as practicable, the fouled fireside surfaces were estimated to have reached a state of dynamic equilibrium.

Particulars of the boilers and the operating conditions at the time of this work are given in Table 1. Three types of magnesium-containing additive were under trial. X consisted of a suspension of magnesium oxide in gas oil injected into the residual fuel oil pipeline upstream of the burners; Y was a commercially available additive containing 78% magnesium oxide, 20% sodium bicarbonate and 2% urea, suspended in a stream of air and blown into the flue gas some 4 m (13 ft) upstream of the two regenerative air heaters; while Z was magnesium hydroxide added to the flue gases in a similar manner. At Station A, tests were made on one boiler, with and without additive X. At Station B, two boilers were tested, each with and without one of the additives, Y or Z: because these materials were added at the air heater inlet, only small proportions (ca. 1%) of them were available for deposition at the superheater entry, being transferred in the combustion air via the regenerative air heaters.

The probe was inserted from one side of each boiler into the gas space between the first and second pendent tube banks of the secondary superheater. Preceding the first pendent bank in Station A are two staggered rows of steam-generating tubes rising from the rear wall of the combustion chamber, and in Station B are superheater platens in line with the first pendent elements. The spacing of the first pendent elements is about 23 cm (9 in.) in Station A and about 33 cm (13 in.) in Station B. In considering the results, it is assumed that a negligible proportion of sodium was deposited from the flue gases before they reached the probe.

DEPOSITION PROBE TECHNIQUE

The same deposition probe[1] 4 m (13 ft) long, 2.54 cm (1 in.) external diameter, of AISI Type 347 steel, bearing four chromel-alumel thermocouples was used for all tests. The output of the thermocouples was recorded continuously during the exposure periods which were between 30 and 120 minutes.

Before each exposure the probe surface was washed with deionized water, rubbed vigorously with a cotton rag and rinsed thoroughly with deionized water. The surface was allowed to dry before the probe was moved into the flue gas, the probe temperature allowed to rise to near the desired value as quickly as possible, then the cooling air supply controlled manually to maintain temperatures within ± 5°C. The aperture through which the probe had been inserted was packed with refractory wadding to reduce air infiltration. Approximately 3 m (9 ft) of the probe was exposed to flue gas. Probe metal temperatures ranged from 510 to 630°C (950 to 1,160°F) at Station A, and from 530 to 730°C (990 to 1,350°F) at Station B.

After withdrawal from the boiler the probe was cooled to near ambient temperature and 15 cm (6 in.) lengths were washed with deionized water, the washings (80 to 100 ml) being collected in well-rinsed polyethylene bottles; generally, seven washings were taken from each probe exposure (see Figure 1). In the laboratory, the contents of each bottle were diluted to a standard volume, 120 ml. The pH of this solution was measured, hydrochloric acid added to yield a concentration of 0.02 M, and the solution was boiled, cooled and diluted to 200 ml. Sodium was measured by flame photometry, vanadium and magnesium by atomic absorption and sulphate gravimetrically. In the samples from Station A, nickel and iron were determined by atomic absorption. Samples of the deionized water used for washing the probe were also analyzed, as were 'blank' washings of the probe which had received the pre-exposure treatment only.

The superheater deposition measurements were made during trials arranged to determine the effectiveness of the additives for neutralizing sulphuric acid. The testing procedures included measurements of acid deposition rates[4] at the air heater inlet and outlet. For comparison with the exposures in the superheater, analyses were made of the material deposited on an acid deposition probe at the air heater inlet in Station A, operated with the probe surface temperature in the range 65 to 85°C (150 to 185°F), that is, well below the sulphuric acid dew-point.

TABLE 1

DETAILS OF BOILER PLANT AND OPERATING CONDITIONS

STATION		A - designed for coal firing converted to oil firing		B - designed for oil firing	
Alternator unit generating capacity	MW	60		500	
Evaporative capacity	tonne h^{-1}	249		1,600	
	klb h^{-1}	550		3,550	
Pressure	MN m^{-2}	6.55		16.55	
	psig	950		2,400	
Final steam temperature	°C	496		541	
	°F	925		1,005	
Excess oxygen at superheater outlet	%	0.5 to 0.6		0.1 to 0.3	
Oil Fuel consumption	tonne h^{-1}	16		100	
Number of oil burners		10		32	
Oil fuel		without additive	with additive		
sulphur	%	2.9	2.4	2.8	
sodium	µg g^{-1}	75	58	40	
vanadium	µg g^{-1}	109	90	82	
Additive		X in fuel		Y in air heater	Z into air inlet gas
Feed rate Mg	kg h^{-1}	12.5		47	54
equivalent	lb h^{-1}	5.7		21	25
to: ≡ µg g^{-1}	in oil	355		210	250
Na	kg h^{-1}	nil		5.5	nil
	lb h^{-1}	"		2.5	"
≡ µg g^{-1}	in oil	"		25	"
Temperature of gas at probe positions	°C	900 to 930		960 to 1,940	
	°F	1,650 to 1,700		1,800 to 1,900	

RESULTS

A typical temperature range along the probe at Station B is illustrated in Figure 1, together with the relative positions of the mid-points of the washed sections; this also shows estimates of the local flue gas temperatures in the two boilers at Station B, measured using an unshielded 2 mm (1/16 in.) diameter sheathed chromel-alumel thermocouple. From previous calibrations, it is estimated that the probe thermocouples under-read the 'surface' temperature by 5 to 10°C; a correction of + 8°C has been applied to the measured values.

From the quantities of various elements found in the washings, corrected for any in the 'blank' solutions, the rates of deposition were calculated, expressed in the units previously used in this work and for sulphuric acid deposition rates,

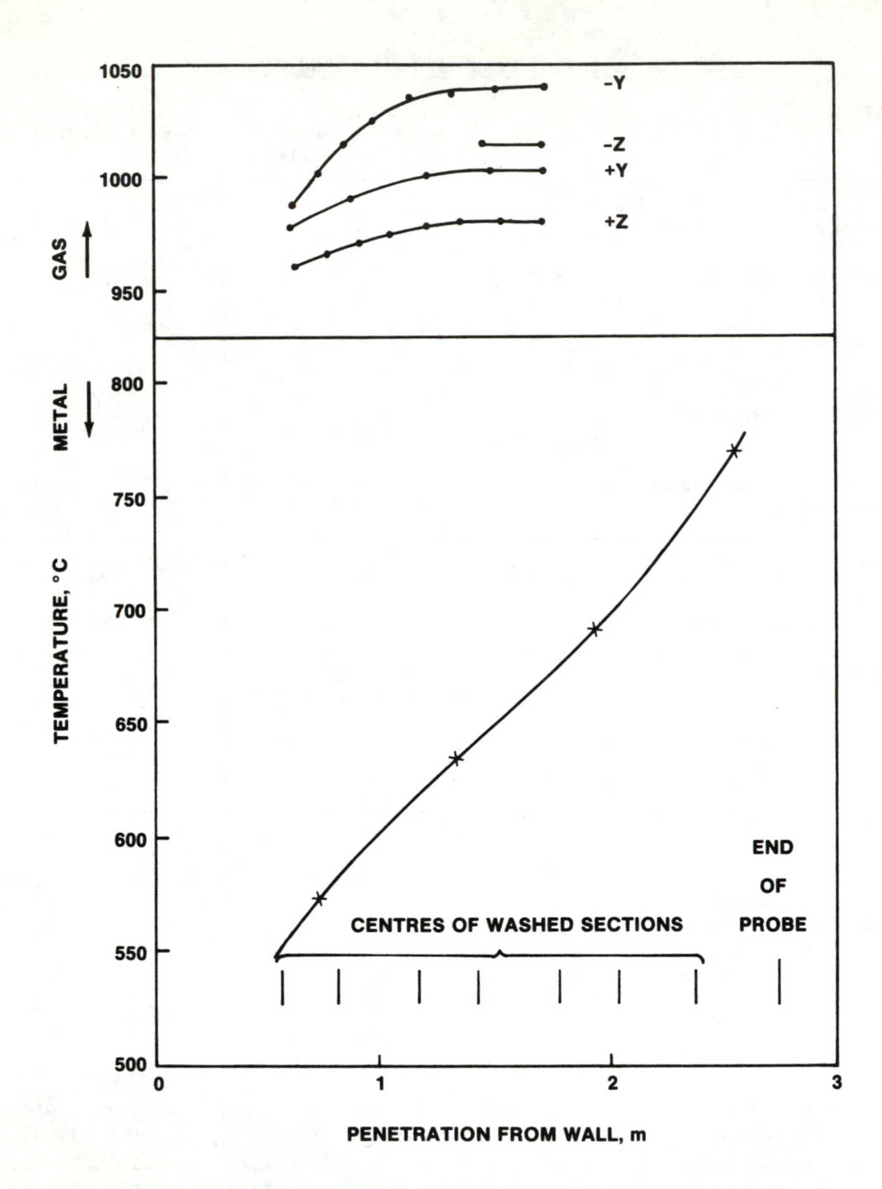

FIG. 1 GAS AND PROBE METAL TEMPERATURE DISTRIBUTIONS

that is, $\mu g\ cm^{-2}\ h^{-1}$. These values are plotted against the probe surface temperature in Figures 2 to 10; 'extra' deposit is that included material derived from upstream tubes (see "Mass Transfer Calculations").

Sulphate analyses for observations in the Station B boilers with additive Y and without additive Z are not available, but there is no reason to suppose that these results would be significantly different from those shown in Figures 9 and 10.

The deposition rates of the soluble material in the acid deposition probe washings are given in Table 2. Negligible quantities of nickel and iron were found in the superheater probe washings from Station A, so these results are not reported.

TABLE 2

SOLUBLE MATERIAL ON ACID DEPOSITION PROBE AT AIR HEATER INLET, STATION A (expressed as $\mu g\ cm^{-2}\ h^{-1}$)

Surface Temperature °C / °F — Constituent		65 150	75 167	85 185
Sodium	as Na	93	83	93
Magnesium	as Mg	41	41	36
Vanadium	as V	20	20	17
Iron	as Fe	187	185	184
Nickel	as Ni	48	48	44
Sulphate	as SO_4	594	511	458

DEPOSITION RATES

Boiler Fouling

To put the deposition units used here in broader perspective, 100 $\mu g\ cm^{-2}\ h^{-1}$ is equivalent to 10 kg m^{-2} $(10^4 h)^{-1}$ (2.05 lb-ft^{-2} per 10,000 h).

Considering sodium alone, if its measured deposition rate applied to the whole superheater, the total deposition rate of sodium sulphate (40 $\mu g\ cm^{-2}\ h^{-1}$ of Na) on the 2,847 m^2 (30,630 ft^2) of superheater surface in a Station A boiler would be 35 tonne per 10,000 h (the sodium in the fuel in this period is equivalent to 37 tonne sodium sulphate). Similarly, the sodium deposition rate of 17 $\mu g\ cm^{-2}\ h^{-1}$ in Station B (18,095 m^2; 194,700 ft^2) would be equivalent to 95 tonne sodium sulphate per 10,000 h deposited, out of a total of 120 tonne from the fuel. Since these deposition rates were measured at virtually the superheater entry, the mean rate of deposition through the whole superheater would be considerably less than this, due to the progressive depletion of the sodium from the gas -- say, a mean of one-third the observed rate. For the total deposited material, other components, e.g. vanadium and nickel compounds, say one-third of the total, must be added (see "Composition of Deposits"). However, the calculated figures are not unreasonable when compared with the total quantities of deposit found in a boiler at Station B, as follows. On the sloping floor under the pendent superheater and reheater elements, approximately 220 m^2 (2,400 ft^2), there were about 10 tonne of debris, after some 25,000 h operation. Deposits on the superheater tubes at the same time totalled between 20 and 30 tonne.

If it were assumed that all the sodium sulphate which was deposited remained on the tubes, the deposition rate of 40 $\mu g\ cm^{-2}\ h^{-1}$ at Station A would build up a layer about 0.5 cm thick in 10,000 H. This is certainly of the order of thickness observed in the boiler but the deposits contain only about 50% sodium sulphate and have a bulk density lower than that (2.6) used in this calculation. So, manifestly, all the sodium deposited does not remain on the tubes.

Time-lapse cine photography of superheater tubes in an oil-fired boiler, burning fuels similar to those in this work, some years ago[5] showed that periodic exfoliation of deposits occurred every few hundreds of hours. Apparently without being disturbed by external agency, the outer layers of the deposit, less than 1 mm thick, thus fell off, but some of the material deposited stayed on the tube to build up gradually. The proportion which dropped off or stayed behind is not known. Evidence of this phenonomen was obtained at both stations where these deposition probe exposures were made. At random times and positions on the probe, pieces of deposit, obviously derived from the upstream tubes, were attached to the probe: these varied in size from about 1 cm to less than 1 mm in diameter. Some of these pieces were analyzed separately and the results are considered below. It is apparent that a proportion, probably about one-quarter, of the sodium which is deposited remained on the secondary superheater in Station B. The exfoliated deposits not found in the boiler move on to later stages of the plant, some of the material being removed from the 'dust' collector hoppers under the cyclone separators.

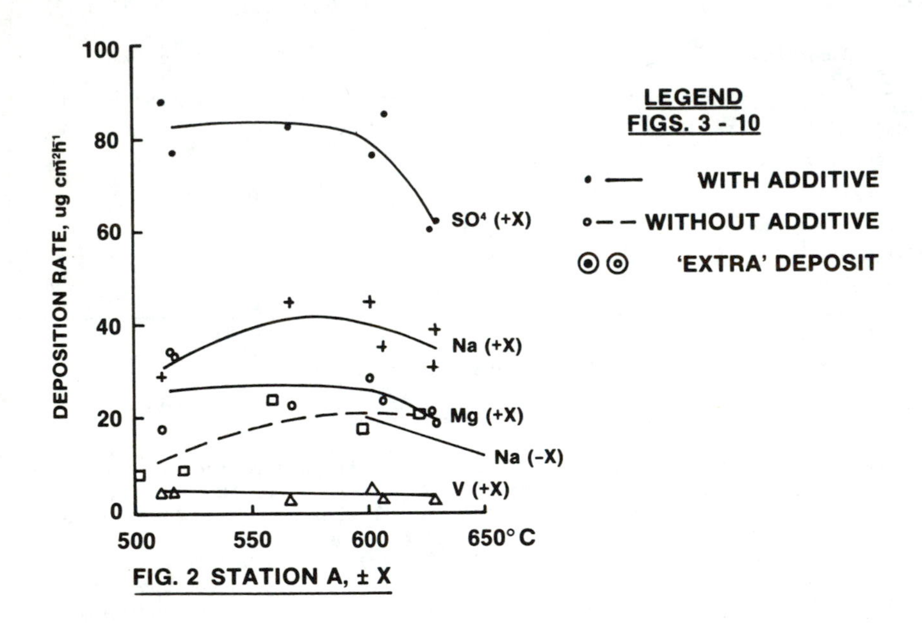

FIG. 2 STATION A, ± X

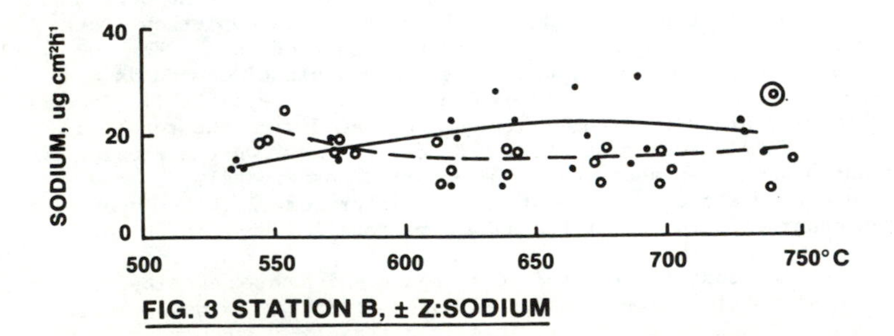

FIG. 3 STATION B, ± Z:SODIUM

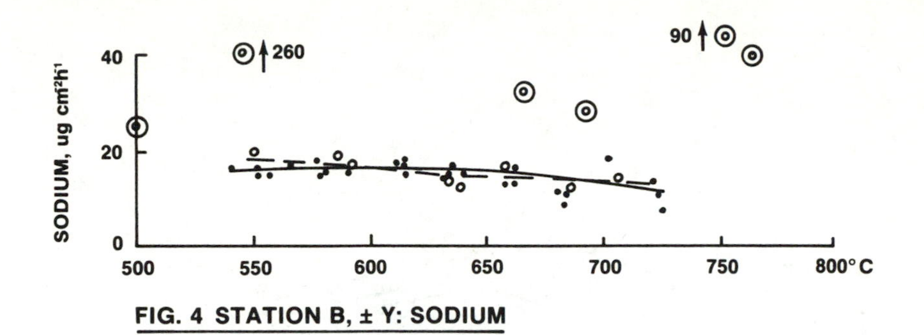

FIG. 4 STATION B, ± Y: SODIUM

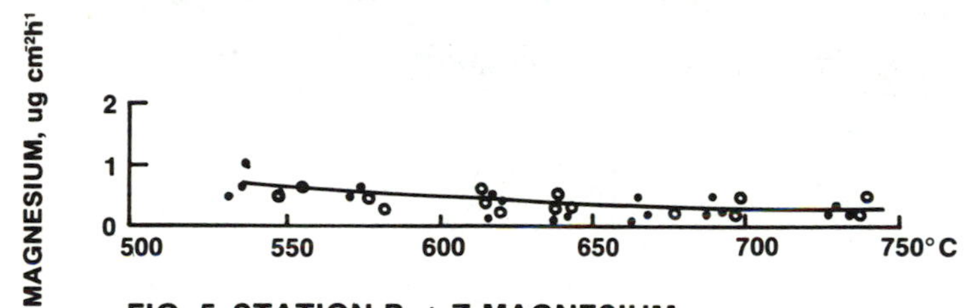

FIG. 5 STATION B, ± Z MAGNESIUM

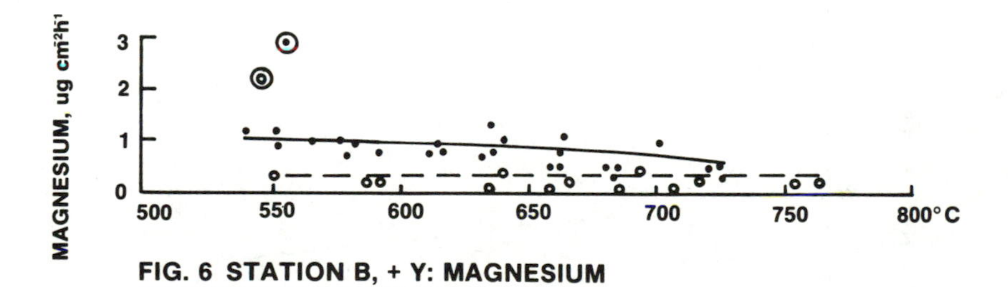

FIG. 6 STATION B, + Y: MAGNESIUM

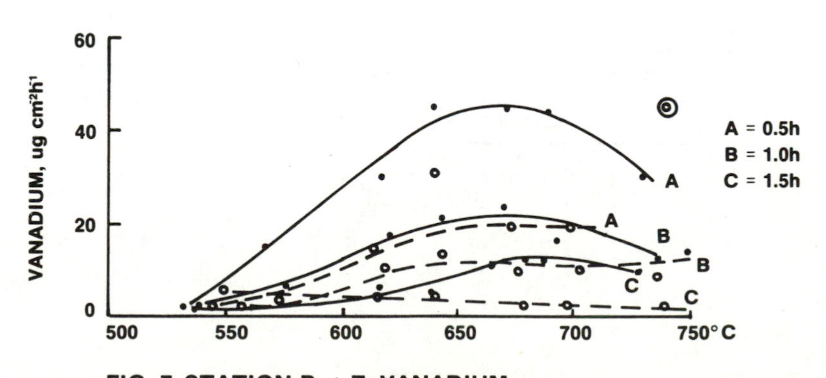

FIG. 7 STATION B, + Z: VANADIUM

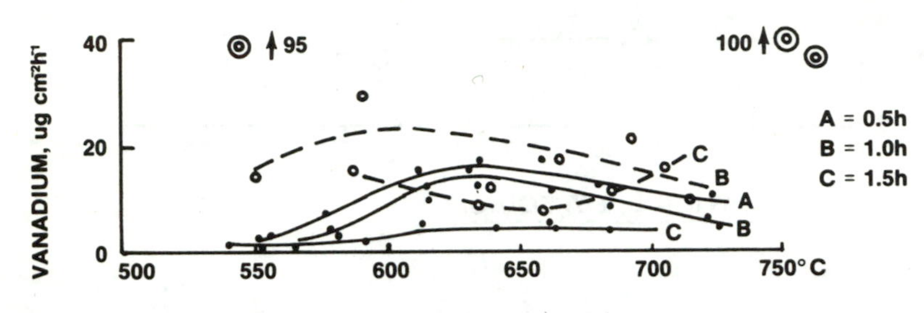

FIG. 8 STATION B, ± Y: VANADIUM

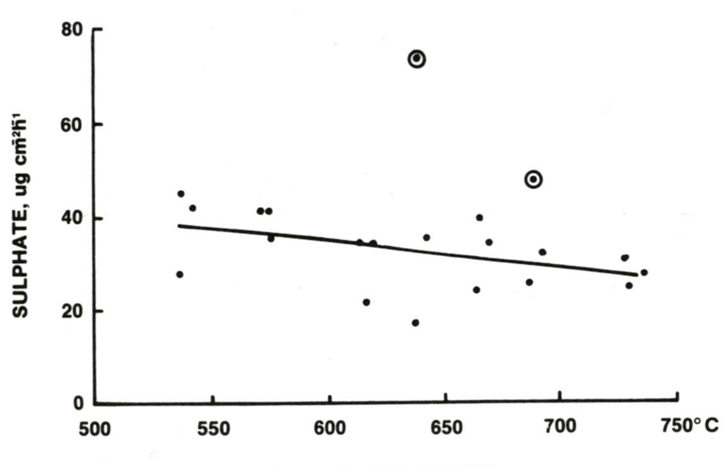

FIG. 9 STATION B, + Z: SULPHATE

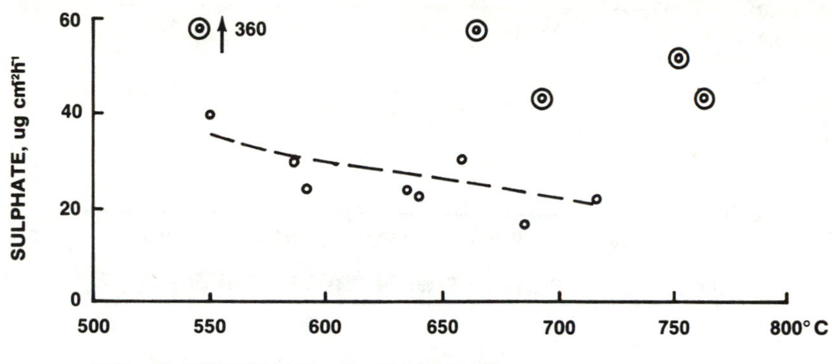

FIG. 10 STATION B, -Y: SULPHATE

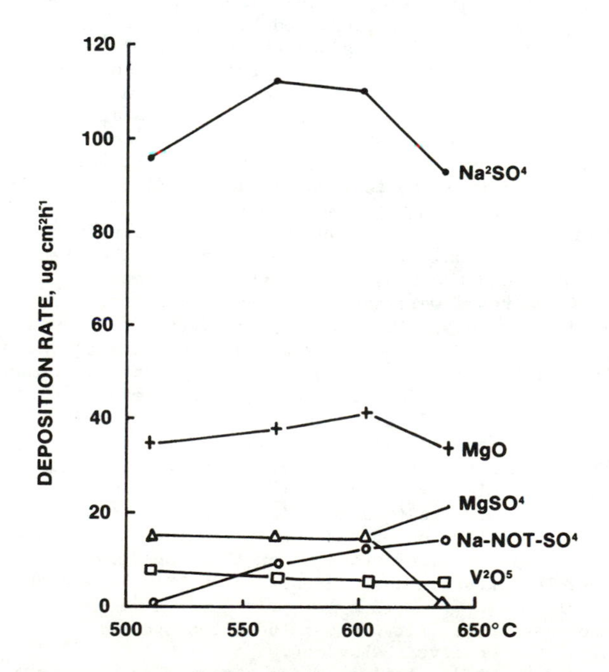

FIG. 11 HYPOTHETICAL COMBINATIONS, STATION A

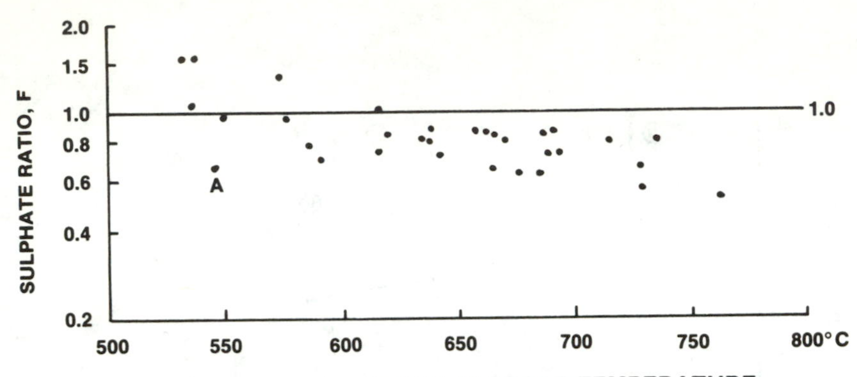

FIG. 12 VARIATION OF F WITH PROBE TEMPERATURE

Mass Transfer Calculations

A comparison was made between the observed rates of deposition of the major component, sodium sulphate, and those preducted by convective mass transfer theory. The basis for these calculations appears in Table 3.[7] It was fairly assumed that in the gas passing over the probe, the sodium compounds, probably NaOH or Na_2SO_4, were as vapour,[8] and that after deposition the principal product of deposition and post-deposition reactions, that is, sodium sulphate (see section under "Composition of Deposits"), had a relatively negligible vapour pressure: it is 1.3×10^{-2} Nm^{-2} (1×10^{-4} torr) at 750°C (1,380°F) equivalent to vaporization of 1 μg g^{-1} Na from the fuel, and 2.1 Nm^{-2} (1.6×10^{-2} torr) at 900°C (1,650°F) equivalent to 160 μg g^{-1} Na from the fuel. The calculations are subject to overall errors of about ± 30% due to contributions from the assumption of the Schmidt number as unity, and in the values for the viscosity and diffusivity of the gas[9] and the constants in the mass transfer correlation.

Standard significant tests were applied to the experimental results (values D in Table 3). The deposition at Station A with additive (6.5%) was significantly different from those at both stations without additive at the 99.99% level of probability; a 60% probability level applied to the difference between the two sets of results without additive: This is hardly significant. The agreement between the measured and calculated depletions in passing transversely over one circular cylinder (the probe) is remarkably good in the case of Station B and quite good for Station A without additive. Therefore, the assumption that vapor diffusion transfer dominates seems justified.

To determine whether the significantly higher rate at Station A with additive could be obtained by diffusion of particulate sodium, the mass transfer coefficient, and hence the proportion deposited on the transverse cylinder, was calculated from particle diffusion coefficients.[10] This showed that the percentage deposited would range from 8.2×10^{-7} for 1 μm diameter particles to 1.8 for 0.001 μm diameter particles. It was concluded that particle diffusion did not adequately explain the observed behaviour.

The higher proportion of depositing sodium measured at Station A could most probably be accounted for if some of it were transferred by inertial particle impaction. Comparison with the data of May & Clifford[11] for a similar geometrical situation showed that this higher deposition rate corresponds to an effective particle diameter of about 20 μm; this is typical of the solids normally encountered in boilers of this type. The major distinguishing feature of the conditions at Station A, not allowed for in the vapour diffusion calculations, is the presence in the fuel of some 350 μg g^{-1} magnesium as magnesium oxide, with a mean particle diameter of about 2 μm. To account for the apparent deposition of sodium here in 20 μm particles, one must seek an explanation of its relatively lower volatility. This might be caused by the formation of the compound $Na_2Mg_3(SO_4)_4$;[12] sodium sulphate can also absorb a considerable proportion of magnesium sulphate in solid solution.[13] Either of these might sufficiently reduce the vapour pressure of sodium compounds under the conditions at Station A so as to contribute enough particle borne sodium to attain the observed 6.5% deposition. This hypothesis should be checked experimentally.

TABLE 3

COMPARISON OF CALCULATED AND EXPERIMENTAL RATES OF DEPOSITION OF SODIUM SULPHATE

MASS TRANSFER CALCULATION		STATION A		STATION B
Gas temperature (T)	°C	900		1040
Gas density (ρ)	kg m^{-3}	0.31		0.29
Gas viscosity (μ)	kg m^{-1} s^{-1}	4.6×10^{-5}		5.0×10^{-5}
Gas velocity (u)	m s^{-1}	3.50		10.15
Diffusivity of Na_2SO_4 (D)	m^2 s^{-1}	7.7×10^{-5}		8.7×10^{-5}
Probe tube diameter (d)	m	0.0254		0.0254
(Schmidt No.)$^{0.3}$ = $Sc^{0.3}$		1		1
Reynolds No. (Re) = (ρdu)/μ		599		1495
Sherwood No. (Sh) = 0.26 $Re^{0.6}$ $Sc^{0.3}$ (reference 6)		12.06		20.9
Mass transfer coefficients (kg) = (Sh/D)/d	m s^{-1}	3.67×10^{-2}		7.2×10^{-2}
Proportion of Na SO deposited on one transverse cylinder (probe) = A	%	3.3		2.2
EXPERIMENTAL RESULTS		WITH ADDITIVE	WITHOUT ADDITIVE	WITHOUT ADDITIVE
Projected area of probe	m^2	3.81×10^{-3}	3.81×10^{-3}	3.81×10^{-3}
Concentration of Na in gas	kg m^{-3}	1.68×10^{-6}	1.30×10^{-6}	8.2×10^{-7}
Mass of Na in gas passing through probe area in one hour = B	kg	8.01×10^{-5}	6.2×10^{-6}	1.15×10^{-4}
Mean mass actually deposited in one hour = C	kg	5.2×10^{-6}	1.3×10^{-6}	2.2×10^{-6}
Proportion of Na deposited on probe (D = 100 C/B)	%	6.5	2.1	1.9
Standard deviation of D	%	0.71	0.51	0.20
Experimental/theoretical = D/A		1.97	0.64	0.86

It thus appears that most of the sodium was deposited by vapour diffusion, some by particle impaction, the latter constituting a greater propostion in the presence of the fuel-borne magnesium.

Comparison of the figures in Table 2 with those in Figure 2 shows the much greater proportional deposition of soluble sodium and vanadium on the surface of the deposition probe exposed below the acid dew-point than at the superheater. This is due most probably to the greater efficiency of adherence of the particulate matter by the sticky film of condensed sulphuric acid, and the greater extent of dissolution of the vanadium by the acid. However, the deposition rate of magnesium was not so much higher than at the superheater. The relatively low deposition rate of sodium at the superheater was probably due to its existence in the gases there as vapour, whereas at the air heater nearly all the sodium is present as aerosol or associated with the coarser particulate material, which deposits faster. The iron and nickel were derived from acid solution of the probe steel by the great excess of sulphuric acid.

COMPOSITION OF DEPOSITS

Effects of Additives

Direct assessment of the effect of additive X on the deposition of sodium at Station A was complicated by paucity of experimental data without the additive and the burning of fuel with slightly lower sodium (58 $\mu g\ g^{-1}$), vanadium (90 $\mu g\ g^{-1}$), and sulphur (2.4%) levels. Sodium deposition when additive was injected was nearly 50% higher than in proportion to the sodium content of the fuel (Figure 2).

There was little effect of either of the additives Y or Z on sodium deposition, though this might be expected because so little of them reached the superheater. However, results in Figures 5 and 6 show that more of additive Y than of Z was deposited on the probe.

Sulphate Deficiency

From the results at the superheater at Station A, hypothetical combinations of the metals sodium and magnesium with the only anion determined, sulphate, were calculated. There was a deficiency of sulphate from that needed to form both sulphates, Na_2SO_4 and $MgSO_4$, and the order of preference for combination with the sulphate found was assumed to be Na, Mg, with vanadium uncombined with sulphate. This led to the figures which appear in Table 4, yielding mean values represented in Figure 11. The deficiency of sulphate increased slightly with temperature of the collecting surface. Only a small proportion, of the order of 10%, of the magnesium was present as sulphate. The quantities of magnesium deposited in Station B were too low for a similar calculation to be at all valid, but the factor

$$F \equiv \frac{\text{sulphate found}}{\text{sulphate} \quad \text{sodium found } (Na_2SO_4)}$$

was calculated. These results, shown in Figure 12, confirm the 'sulphate deficiency' above 600°C and its increase with rising temperature. Those samples exhibiting the lower values of F also showed a significant alkalinity with pH values of the solutions up to 8.0. This is in evidence in support of the hypothesis that the 'sodium not as sulphate' was deposited as hydroxide,[8] or present as hydrolysable vanadate(s).

The figures in Table 4 for the 570°C sample (121 min. run) illustrate the case where a piece of deposit derived from tubes upstream was included in the analysis of the probe washings. In addition to the obviously inflated rates of deposition, the composition of the 'extra' deposit is of interest. This shows a higher proportion of magnesium sulphate, relative to the oxide, than in the other, freshly-deposited material and represents some progress towards more complete sulphation of the magnesium with time of residence on the tubes. However, analysis of a similar sample from a Station B run showed a higher-than-expected deficiency of sulphate (point A in Figure 12); Station B operated with a lower proportion of excess combustion air and it is possible that there was less reaction of the deposits with SO or $SO_2 + O_2$.

Sodium Vanadates

At Station B, with little magnesium present, there was evidence of reaction of sodium with the vanadium, to form soluble vanadates, although this was not so clear as that previously reported.[3] It is apparent from Figures 7 and 8 that the proportion of soluble vanadium on the probe generally increased with temperature above about 580°C (1,076°F) and increasing period of exposure. This latter phenomenon had been observed before,[13] when it had been shown that beyond 2 h exposure (to 24 h) there was no further change in the ratio soluble vanadium:total vanadium. Longer exposures, 2, 4, 12 and 24 h, also showed a decreasing mean rate of deposition of vanadium with increasing temperature. Perhaps the increasing quantity, therefore thickness, of deposit on the probe raised its surface temperature sufficiently to approach the 'dew-point' of the relatively volatile vanadium oxide.

A calculation was made of the atomic ratio of V:Na for the excesses of each over a datum value at 550 C, at which temperature the sodium could be taken as fully sulphated (see Figure 12). These ratios varied with temperature thus:

600°F	V:Na = 1.4
650°C	1.0
700°C	0.7
750°C	0.5

Comparison With Boiler Deposits

It is interesting to compare the composition of the short-term deposits on the probe with those of various layers of deposits taken from a boiler after an extended operating campaign. Such analyses were not available for Stations A and B, but data from a boiler at Station C, to which magnesium hydroxide had been added continuously, are given in Table 5. This boiler burned fuel containing about 80 $\mu g\ g^{-1}$ vanadium, 70 $\mu g\ g^{-1}$ sodium and 4.0% sulphur; the equivalent of about 180 $\mu g\ g^{-1}$ magnesium was injected by an air stream into the combustion chamber, in order to prevent corrosion of the superheater at a final steam temperature of 570 C (1,058 F). The samples of deposit were taken after about 15,000 h operation under these conditions; the superheater samples corresponded in position in the boiler to the tube bank immediately downstream of the deposition probe in Stations A and B, and the reheater was behind that, in cooler gas. The analyses show almost complete sulphation of magnesium on the superheater with a small proportion of the oxide in the outermost layer. There was relative enrichment of sodium and vanadium in the inner layers, probably by diffusion of their compounds as vapour, the more volatile species concentrating in the cooler position under the influence of the steep thermal gradient in the deposit.[8] The high proportion of magnesium carbonate in the reheater is unusual.

CONCLUSIONS

The rate of deposition of oil ash and additive components onto convection tubes is such as to progressively deplete the flue gases of inorganic material during passage through the boiler. Thus, the upstream tubes receive more material and though much of this exfoliates from at least the high temperature superheater during operation, experience shows that the net amount of deposit is greater at the gas inlet, where gas temperatures are higher. Of the order of one-third of the ash is retained in the boiler as deposits, the rest being discharged as fine dust in the flue gas or as coarse dust and deposit debris in the collecting hoppers, whether sootblowing is practised or not. The greater thickness of deposit on the superheater inlet tubes reduces heat transfer more and results in the outher portions of the deposits achieving temperatures close to those of the flue gas. This in turn accelerates differentiation of the deposit constituents, leading to high concentrations of the more volatile and corrosive sodium and vanadium compounds close to the tube metal. It also allows the outer parts of the deposit to act as a catalyst for the oxidation of SO_2 to SO_3; the greater bulk of the deposits allows better transfer of material between gas and deposits.

In the absence of significant proportions of magnesium, sodium appears to deposit on high temperature surfaces according to convective mass transfer theory theory. Diffusion of vapour of sodium hydroxide or sulphate through the boundary layers appears to dominate.

Addition of magnesium to the fuel, at a proportion of three to five times (by mass) the sodium increased the deposition rate of sodium, by a factor of two to three. This could be explained by depression of the vapour pressure of sodium compounds in the particulate solids arriving at the superheater.

In short-period deposition (30 to 120 min.), some of the soluble sodium was present not as sulphate, this proportion varying from zero at 550°C to 40% at 750°C; aqueous extracts were significantly alkaline. Over this range of temperature, the proportion of soluble vanadium increased up to about 650°C, thereafter falling slightly; the ratio of the 'excess' soluble vanadium to 'excess' sodium (datum at 550°C) decreased with increasing temperature.

ACKNOWLEDGEMENTS

This work was carried out at Marchwood Engineering Laboratories and is published by permission of the Central Electricity Generating Board.

TABLE 4

HYPOTHETICAL COMBINATIONS OF DEPOSITED CONSTITUENTS, STATION A
(all values expressed as μg cm^{-2} h^{-1})

Temperature of Probe Metal °C	°F	Compound: Na_2SO_4	Na not as Sulphate	$MgSO_4$	MgO	V_2O_5
78-Minute Exposure						
510	950	89	-	35	18	7
565	1049	123	14	-	38	5
600	1112	114	22	-	48	9
628	1162	92	25	-	32	5
121-Minute Exposure						
515	959	102	-	10	51	9
570	1058	241	-	141	139	39
605	1121	108	-	15	35	5
627	1161	90	5	-	36	5

REFERENCES

1 Jackson, P. J. & Raask, E., J. Inst. Fuel, 34, 275, (1961).

2 Raask, E., The Mechanism of Corrosion by Fuel Impurities, Paper 7, p. 145, Butterworths, London, (1963).

3 Jackson, P. J., Mitt. V.G.B., 85, 220, (1963).

4 Alexander, P. A., Fielder. R. S., Jackson, P. J. & Raask, E., J. Inst. Fuel, 33, 31, (1960).

5 Murray, G., private communication to the author.

6 Coulson, J. M. & Richardson, J. F., "Chemical Engineering", 1, 344 and 383, Pergamon Press, Oxford, (1966).

7 Hotchkiss, R., CEGB Laboratory Note R/M/N893.
8 Jackson, P. J., TheMechanism of Corrosion by Fuel Impurities, Paper 34, p. 484, Butterworths, London, (1963).
9 Perry, J. H., "The Chemical Engineers Handbook", 4th Ed., 3-197, McGraw Hill, Tokyo, (1963).
10 Dorman, R. G. "Dust Control and Air Cleaning", XVII, Pergamon Press, Oxford, (1974).
11 May, K. R. & Clifford, R., Ann. Occup. Hyg., 10, 83, (1967).
12 Janecke, E., Z. Anorg, Chem., 261, 213, (1950).
13 Rahmel, A., The Mechanism of Corrosion by Fuel Impurities, Paper 39, p. 556, Butterworths, London, (1963).
14 Raask, E., CEGB Laboratory Report RD/L/R1050, (1961).

IMPROVED THERMAL EFFICIENCY BY CONTROL OF FACTORS AFFECTING METAL WASTAGE IN PULVERIZED COAL-FIRED STEAM GENERATORS

RICHARD W. BORIO, ARTHUR L. PLUMLEY, and WILLIAM R. SYLVESTER

C-E Power Systems
Combustion Engineering, Inc.
Windsor, Connecticut, USA 06095

INTRODUCTION

The question of how best to meet the nation's future energy needs is one of great concern to industry, government, and the country as a whole. Of course, this is a highly complex question. There are many options and each one has a range of difficult technological, economic, and/or environmental questions that must be addressed.

One available option is to increase the thermal efficiency of conventional pulverized coal-fired plants by increasing steam temperature (and pressure) to the turbines. With this option, however, there is concern:

1. That the materials of construction will not give long-term, satisfactory performance at the higher temperatures.
2. Over increased use of alloying elements such as chromium and nickel, which are, in the long term, of critically short supply.
3. That higher steam temperatures will adversely affect unit reliability and availability.
4. For coal ash corrosion of the superheater and reheater materials operating at higher temperatures.

This paper will consider coal ash corrosion and reliability/availability as affected by corrosion. This shall be done by reviewing the extensive knowledge available on the subject, discussing current research, and examining those areas requiring additional research.

KNOWLEDGE REVIEW

There are two major and interacting sources of this knowledge. One is long-term operating experience with large numbers of coal-fired utility boilers operating at various temperatures. Of the over 500 coal-fired units supplied by Combustion Engineering, over 300 operate at 1000 F steam, 90 operate at 1050 F steam, one at 1100 F (Cleveland Electric's Avon Lake), and one at 1200 F (Philadelphia Electric's Eddystone).

The other important source of knowledge is research supplemented by extensive laboratory and field testing under actual firing conditions. The research done by Combustion Engineering has led to an understanding of the mechanism of molten phase coal ash corrosion and the significance of metal temperature in this phenomenon. C-E has also gained information as to the mineral constituents in bituminous coals that act to produce a corrosive molten phase. This knowledge is extremely important because it provides a basis for predicting the "corrosiveness" of any particular coal at any given temperature. It also indicates the changes -- either by additives or coal processing -- that would have to be made to "make a corrosive coal non-corrosive."

Experience over the steam temperature range of 1000 F to 1200 F has established the following:

1. All coals are not "corrosive." It is important to know that many coal-fired units operate at steam temperatures of 1050 F with essentially no corrosion or at acceptably low corrosion rates. Only a small percent of the coal-fired units has experienced serious rates of corrosion requiring major operating or maintenance corrections. While this percentage is relatively small, the actual number of units to experience serious corrosion is sufficient to warrant the concern and attention given to coal ash corrosion.

2. Metal temperature is an important corrosion rate variable for coals that are "corrosive." As surface metal temperatures increase from 1000 F to about 1200 F, corrosion rates increase significantly. A unit delivering steam at temperatures of 1050 F or higher, will have a greater corrosion problem than a 1000 F steam unit <u>if</u> the coal is corrosive. The corrosion rates will be higher, and the amount of material undergoing corrosion will be significantly increased.

   For the types of coal that can be classified as non-corrosive, however, temperature does not appear to be a variable of importance with regard to corrosion. If the coals are non-corrosive, there need not be a greater corrosion concern as steam temperatures are increased above 1000 F.

3. The molten ash produced by corrosive coals is a highly aggressive corrosive and corrosion is not easily prevented.

4. Commercially available pressure part materials corrode at unacceptable corrosion rates. Presently, there does not appear to be a tube material with adequate high temperature strength that will not corrode if the coal is corrosive. Research continues in this area. The Electric Power Research Institute (EPRI) has a program investigating improved material systems for coal-fired superheater tubes. The corrosion resistance of base metals and coatings are being investigated in this program.

5. Various types of tube coatings have been tried. Electroplated chromium has proven effective in preventing corrosion, but it is an expensive solution with an indeterminate adherence life. Bi-metal tubes with an outer corrosion-resistant clad high in chromium such as Inconel 671 (50% Cr-50% Ni) show promise of having adequate corrosion resistance, but once again the cost of such tubing is high. Other less expensive corrosion-resistant coatings have been tried with no success. Coatings research and development continue.

This service experience indicates the importance of having a fundamental knowledge of the factors that promote coal ash corrosion. Because of promising research work done in the past, a basic understanding of what makes a coal corrosive appears to be quite close, and more importantly, will soon be put to use.

High Temperature Corrosion Background and Experience

During normal operation of utility boilers, wastage has occurred primarily (Fig. 1):

1. In the high temperature superheater and reheater sections.
2. On waterwalls near the firing zone.
3. In the low temperature gas passes and air heaters.

To reduce failures in these areas, Combustion Engineering has continuing programs to develop methods for practical prevention of wastage. During the course of such programs, the underlying causes of major types of corrosion have been established (1). Except for this promising work in identifying the corrosive constituents in coals, little has been achieved in this field during the last 15 years. There has been little incentive to design an economic superheater or reheater using materials, coatings, or clad that would provide long-time resistance to coal ash corrosion at metal temperatures significantly above 1000 F, since most of the boilers in the U.S. were designed for and operate at 1000 F main steam.

Nevertheless, the capability is there. The technology of materials testing and evaluation in the field of coal combustion has been highly developed at Combustion Engineering. C-E collaborated with the U.S. Bureau of Mines, under the auspices of the Furnace Performance Factors Committee of the ASME, in the development of the mechanism and resolution of furnace wall tube wastage in wet or dry bottom coal-fired units during the 1940's. This investigation, conducted over a six-year period, established techniques in the identification of chemical characteristics of coal ash and operating conditions that identified the furnace wall tube attack as being the result of the reaction of alkali and metal pyrosulfates (originating from the coal ash) with the metal oxide to form a complex alkali iron trisulfate (2).

In general, waterwall corrosion (tube metal 500 to 800 F) is associated with external deposits and also with flame impingement. Adherence to proper operating techniques has reduced the concentration of pyrites and SO_3 near the walls, and waterwall wastage is infrequent in normal coal firing. These techniques include:

1. Maintenance of burners to avoid excessive flame impingement.
2. Close attention to coal fineness to prevent coarse coal from reaching the furnace.

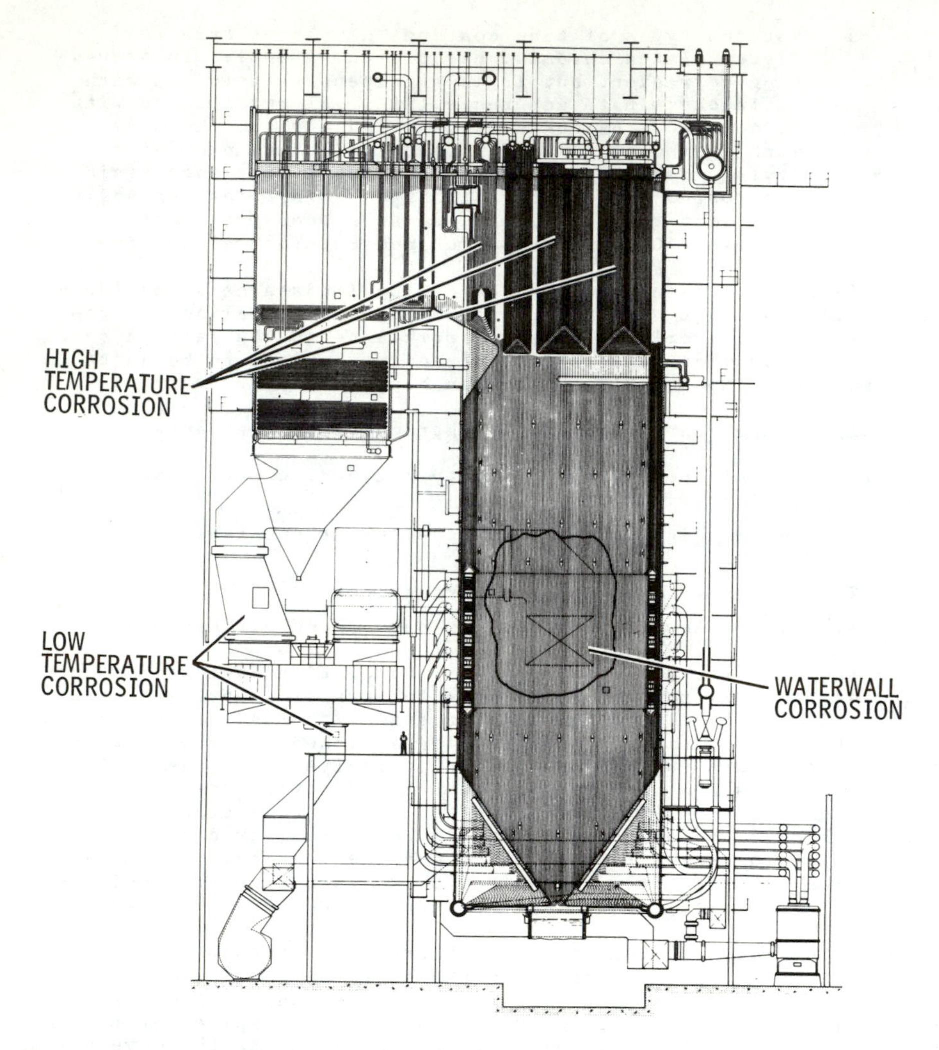

Fig.1 Steam generator side elevation showing areas where wastage has occurred

3. Maintenance of the distribution in coal feeding systems.
4. Good air distribution (3).

This paper will concentrate on control of high temperature superheater/reheater wastage.

During the early 1950's many utility steam generators were designed for 1050 F steam temperature. Some of these units experienced a severe attack on ferritic and austenitic surfaces where the surface metal temperature exceeded 1100 F. This liquid-phase coal ash corrosion associated with high-temperature superheater and reheater surfaces has been under intensive study by Combustion Engineering and other major steam generator manufacturers since 1955 when widespread observation of this type wastage was reported (1-11). During this time, C-E conducted extensive studies in the use of alloys and super alloys for the requirement of the most advanced cycle for the Eddystone plant. This system, designed for the highest thermal efficiency (42%), reached the highest steam pressure and operating temperature used for utility service.

Continuing research programs were undertaken to determine causes and suggest remedies for this type of wastage. Both austenitic (stainless) and ferritic steel were shown to be subject to this wastage, which was taking place under tightly bonded deposits. It was further shown that excessive deposit-type wastage of austenitic alloys was related to temperature. The "bell-shaped" curve shows that corrosion is limited to a temperature range of about 1000 F to 1300 F (Fig. 2). Visual observation revealed a thin white or yellow layer next to the tube surface. This layer, which was molten within the temperature range 1000 F to 1300 F, contained a high concentration of alkali and sulfur (Fig. 3).

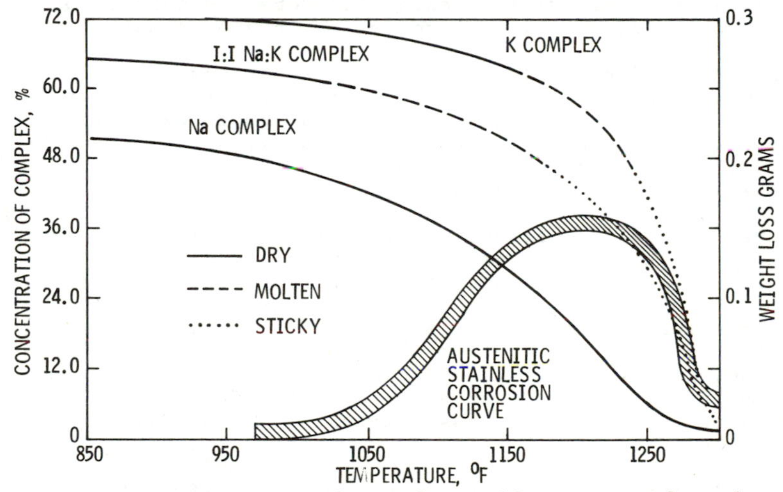

Fig.2 Physical state of complex sulfates as a function of temperature and the corrosion of austenitic alloys

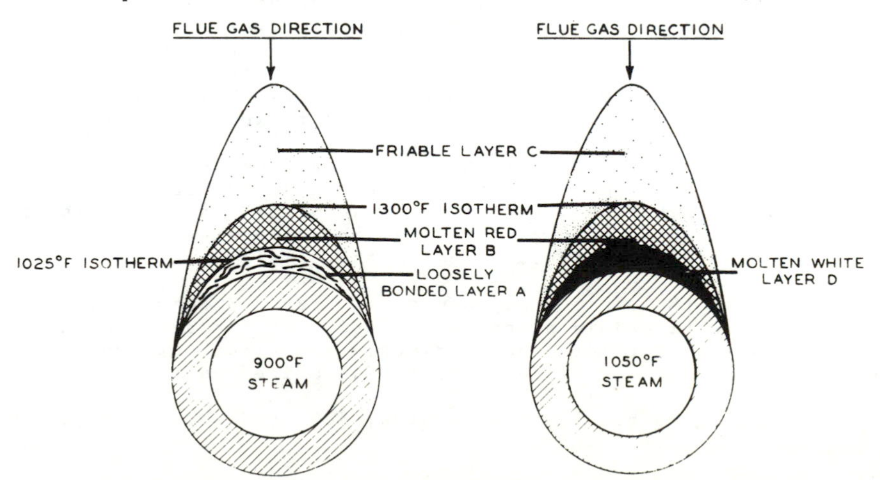

Fig.3 Effect of steam temperature on deposit structure

Simultaneous field and laboratory investigations were designed to determine the corrosion mechanism. Careful study, including chemical and X-ray diffraction analysis, showed that complex alkali iron trisulfates were major constituents of the white inner layers associated with wastage on both superheater and reheater tubes and on corrosion probes, as well as in controlled laboratory wastage experiments. By means of laboratory weight loss tests, the "bell-shaped" corrosion curve was reproduced under the same conditions needed for synthesis of the complex sulfate (6). As noted earlier, this same complex sulfate was formed as a by-product of waterwall wastage (2,11).

With the principal corrosive defined, efforts were turned toward establishing the mechanism of complex sulfate formation to assist in recommendation of adequate protective measures. It is significant that the content of compounds thought to be responsible for corrosion is considerably greater in probe and tube deposits than in the coal ash or fly ash from which they originate.

An important explanation advanced by early investigators is that the initial deposits may be a powder-like material containing alkalies and iron oxide which react with sulfur trioxide to form alkali iron trisulfates (12). Concentration of alkalies on the tube surface may also occur by thermal migration of molten material through the deposits to the tube surface (6).

On the basis of the selective deposition observed, it is felt that individual particles of fly ash vary in composition and, therefore, have different fusion temperatures. Some of the particles that are molten or semi-molten at relatively low temperatures continue to stick to the tubes. Sodium and potassium compounds released during the combustion process in a form capable of reaction with SO_3 in the flue gas, may condense or deposit on the tubes as the initial layer. This explains the formation of bonded deposits in regions where the gas temperature is significantly lower than the fusion temperature of the total coal ash.

Figure 2 also shows a comparison of the equilibrium curves and the "bell-shaped" wastage curve for austenitic stainless steel. It has been suggested that the ratio of potassium to sodium in the coal ash or tube deposit is significant. The molar ratio of alkalies determines the temperature range over which the complex sulfates are molten. The equilibrium concentration of sodium iron trisulfate was not molten from 1150 F to 1275 F. The greatest fluid range appears to be near the 1:1 molar ratio of sodium to potassium. This corresponds to the 1030 F to 1250 F region where corrosion was observed. This is to be expected since a molten corrosive is generally more reactive than the solid counterpart. Consequently, an alkaline molar ratio of about 1:1 in a deposit is corrosive over a wider temperature range than any other combination due to its effect on the fluid range of the complex sulfate.

The role of the various compounds necessary to form a corrosive deposit are placed in proper perspective by the equilibrium curve. The steps that influence formation and growth of deposits from combustion products of coal may be summarized as follows:

1. During combustion the pyrites, FeS_2 and organic sulfur react with oxygen:

$$2FeS_2 + 5\text{-}1/2O_2 \longrightarrow Fe_2O_3 + 4SO_2$$

$$RS \text{ (organic sulfide)} + O_2 \longrightarrow SO_2$$

$$SO_2 + 1/2O_2 \longrightarrow SO_3$$

2. In the high flame temperature, the Na and K in the clays, shales, etc., react to form Na_2O and K_2O.

3. The Na_2O and K_2O then react with SO_3 either in the gas stream or after deposition on the tube:

$$(Na_2 \text{ or } K_2) + SO_3 \longrightarrow (Na_2 \text{ or } K_2)\ SO_4$$

4. The alkali sulfates, iron oxide, and SO_3 then react to form the complex sulfate:

$$3(K_2 \text{ or } Na_2) + Fe_2O_3 + 3SO_3 \longrightarrow 2(K_3 \text{ or } Na_3)\ Fe(SO_4)_3$$

The melting point of a mixed alkali pyrosulfate ($K_{1.5}Na_{0.5}S_2O_7$) is 535 F when SO_3 is above 7 ppm. The low melting pyrosulfates, as well as the complex sulfates are considered primary agents in superheater corrosion as was previously noted.

5. The complex sulfate in the molten phase then reacts with the tube metal:

$$2(K_3 \text{ or } Na_3)\ Fe(SO_4)_3 + 6Fe \longrightarrow 3/2FeS + 3/2Fe_3O_4 + Fe_2O_3 + 3(K_2 \text{ or } Na_2)\ SO_4 + 3/2SO_2$$

The reaction products shown in Item 5 tend to slow down the corrosion rate, but if the deposit spalls, the reaction begins anew. The rate of spalling, no doubt, will affect the overall rate of corrosion.

Nomograph for Predicting Coal Corrosiveness

Based on a mechanistic understanding of the corrosion process, as outlined in the previous section, work was undertaken by C-E and West Virginia University to quantify the effects of specific coal ash constituents on high temperature corrosion. A nomograph was developed that quantitatively relates coal ash constituents to high temperature corrosion of the type that sometimes occurs in the superheater/reheater sections of utility boilers (1) (Fig. 4). The coals used in this study were obtained from six geographical areas that account for approximately 60 percent of the total bituminous tonnage mined in the United States per year. A total of eighteen tests were conducted in the laboratory Solid Fuel Burning Test Furnace to determine the relationship of coal ash constituents to high temperature corrosion. Although a total of six different raw coals were used, the combination of additive usage and coal beneficiation accounted for the total number of tests conducted. Flue gas temperatures of approximately 2000 F were generated as a standard condition and 321 stainless steel was consistently used in all cases as the simulated tube material.

Certain relationships were developed between coal ash constituents and the degree to which they affect corrosion rates. The study has shown that the alkalies, alkaline earths, iron, and sulfur were the most significant constituents relative to high temperature corrosion. It was significant that the alkali concentration had been measured by an acid leaching process. Such a procedure was found to give a better measure of the simpler forms of sodium and potassium than total alkali measurements. These simpler forms of the alkalies are thought to be more available

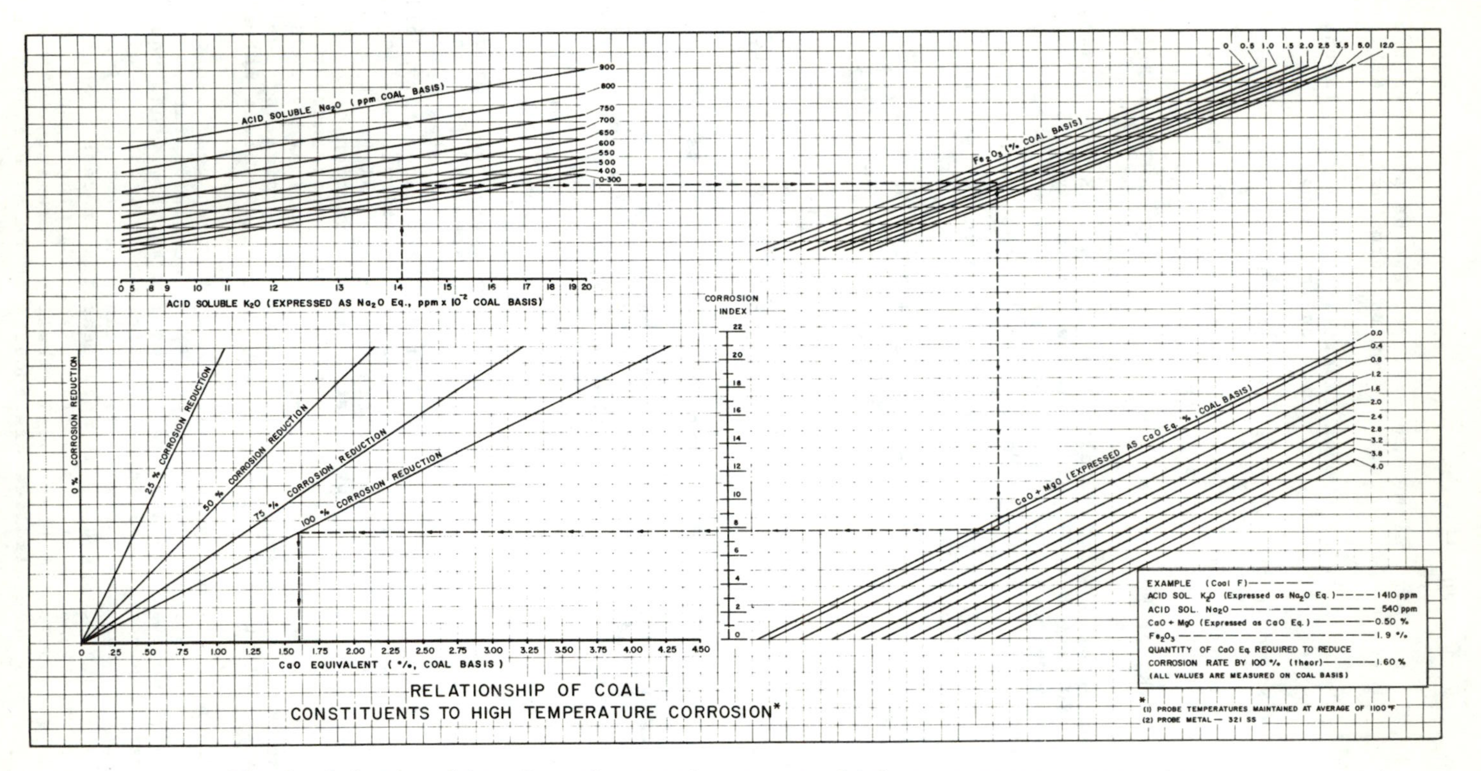

Fig.4 Relationship of coal constituents to high temperature corrosion

for reaction in forming alkali-iron-trisulfate, K_3 or $Na_3Fe(SO_4)_3$, the species directly responsible for tube wastage according to previous investigation by C-E and others. Since the alkalies are generally the least plentiful of the four constituents mentioned above, they are usually the most sensitive corrosion indicators.

The alkaline earth materials are important as they may preferentially retain the alkalies as sulfates or double salts of the type $K_2Ca_2(SO_4)_3$ thus preventing formation of the aggressive compound $K_3Fe(SO_4)_3$ (13).

Because of the similarity in concentrations of iron and pyritic sulfur, it was difficult to determine which of these variables had a greater bearing on coal corrosiveness. Since iron content (expressed as Fe_2O_3) showed a slightly better correlation when plotted against measured corrosion rates than did pyritic sulfur, iron was used in the overall relationship established.

By considering the combined effect of each of the constituents discussed, a nomograph was constructed to conveniently relate coal constituents to corrosion rates as shown in Figure 4. To use the nomograph, a coal ash must be analyzed in the conventional way for iron, calcium, and magnesium content. (These elements would be analyzed routinely as part of the normal coal ash analysis.) Sodium and potassium contents must be determined by an acid leaching procedure (1). By systematically reading the nomograph in terms of the above constituents (as shown in the example on the nomograph) a C.I. (corrosion index) is obtained (Fig. 5). When used in conjunction with Figure 4, the C.I. shows

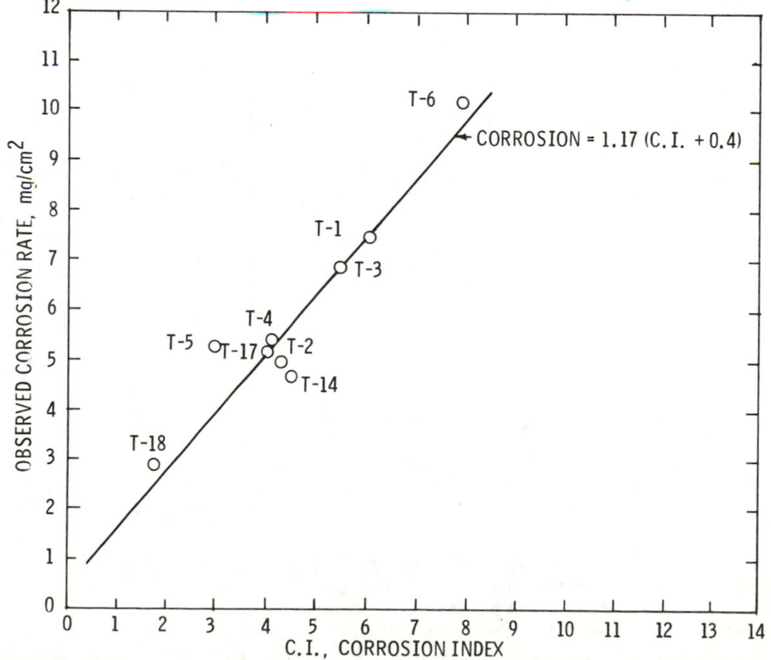

Fig.5 Corrosion index vs corrosion rate (Group I coals)

the corrosion rate in terms of test results units ($mg/cm^2/300$ hr.). The nomograph can also be used to determine the amount of CaO equivalent needed to theoretically reduce the corrosion rate by

100, 75, 50 or 25 percent. (See example, Figure 4.) Analyses of the prime coals used in constructing this nomograph are shown in Table I. Acid soluble sodium contents for these coals range

TABLE I

ANALYTICAL DATA (300-hr TESTS)

Test No.	1	2	3	4	5	6	7	8	9
Corrosion rate (mg/cm^2/300 hr)	7.46	4.90	6.82	5.27	5.12	10.18	4.62	5.13	2.94
Fuel identification	Coal C	Coal B	Coal A	Coal D	Coal E	Coal F	Coal C'	Coal F+ hydrated lime	Coal F prepared
Acid soluble (coal basis) (ppm)									
Na_2O	530	210	380	810	190	540	430	480	390
K_2O	2,020	1,580	1,870	660	1,180	2,170	1,400	2,390	1,320
CaO	1,700	1,600	3,140	16,940	3,110	2,170	1,480	7,320	2,270
MgO	700	870	1,470	2,700	820	1,550	340	2,740	1,360
Na_2O/K_2O acid sol. mol. ratio	.40	.20	.31	1.86	.25	.38	.47	.30	.44
Ash composition (coal basis) (%)									
SiO_2	17.1	7.5	6.5	4.6	6.3	9.4	8.7	11.0	7.1
Al_2O_3	7.5	3.3	2.6	2.5	2.7	4.5	5.4	5.3	3.5
Fe_2O_3	11.0	4.2	5.9	4.0	3.1	1.9	3.6	1.7	1.3
CaO	0.76	0.21	0.38	2.7	0.35	0.28	0.19	0.84	0.26
MgO	0.44	0.13	0.20	0.59	0.13	0.17	0.09	0.35	0.21
Na_2O	0.20	0.04	0.09	0.16	0.06	0.12	0.08	0.13	0.11
K_2O	1.00	0.36	0.40	0.19	0.28	0.72	0.32	0.79	0.56
TiO_2	0.28	0.13	0.10	0.09	0.09	0.15	0.20	0.18	0.13
P_2O_5	0.16	0.06	0.05	0.04	0.04	0.02	0.04	0.02	0.01
SO_3	0.88	0.41	0.46	3.6	0.54	0.46	0.07	0.48	.33
Cl	0.05	0.01	0.25	0.08	0.08	0.20			.33
Proximate Analysis (%)									
Moisture	0.2	5.5	2.0	1.3	3.5	1.5	0.5	1.7	2.1
Ash	40.0	16.3	16.6	18.9	13.8	17.5	19.4	21.1	13.9
Volatile Matter	17.2	39.2	35.9	36.3	34.8	33.4	20.7	31.9	33.9
Fixed Carbon	42.6	39.0	45.5	43.5	47.9	47.6	59.9	45.3	50.1
Total Sulfur	8.15	5.46	4.96	5.04	3.14	1.74	3.32	1.53	1.42
Sulfur Forms (%)									
S as Sulfate	0.22	0.52	0.47	0.12	0.29	0.02	0.11	0.14	0.23
S as Pyritic	6.62	2.75	2.98	3.04	1.68	0.94	2.19	0.61	0.52
S as Organic	1.31	2.19	1.51	1.88	1.17	0.78	1.02	0.78	0.67
H.H.V. (Btu/lb)	8,280	10,950	11,770	11,820	11,970	11,900	12,200	11,330	12,360
Ash fusibility									
I.T.	2,000	1,930	1,915	2,020	1,960	2,150	1,980	2,150	2,230
S.T.	2,080	2,070	2,025	2,060	2,080	2,380	2,330	2,240	2,360
F.T.	2,380	2,380	2,080	2,100	2,260	2,700+	2,700	2,620	2,700+

from about 200 ppm to 800 ppm as expressed on a coal basis; acid soluble potassium contents likewise range from about 650 to 2400 ppm. Combined alkaline earths range from 0.4 to 3.5 percent expressed as calcium equivalent on a coal basis. Iron concentration ranged from 1.3 to 11.0 percent as expressed on a coal basis. It is significant to note that within the range of coals tested the quantity of ash was not a factor.

If one were to contemplate using this nomograph for subbituminous or lignitic coals, the levels of alkalies and alkaline earths are likely to exceed the ranges outlined here. In addition to exceeding the given ranges, it may be erroneous to think that an extrapolation of existing data would accurately represent the corrosion potential of Western coals. Clearly, a program involving testing of Western coals would be the best way to determine the relationship of coal ash constituents to high temperature corrosion with Western coals. As discussed later, preliminary results from laboratory firing of such coals indicate low wastage as predicted by the extrapolated nomograph.

Presently the nomograph can be used for determining the relative corrosion potential of bituminous coals. The corrosion index as generated by the nomograph can also be expressed as metal weight loss in $mg/cm^2/300$ hr. These units, however, have not been expressed in mils/year, the conventional manner of expressing wastage, for several reasons. First and most important, the rate of weight loss due to high temperature corrosion is not linear with time. Usually, the rate of weight loss is highest in the beginning with a subsequent decrease in rate with time. Obviously, if long-term weight losses were calculated based on rates established in a 300 hr. period, premature failure would be indicated in most cases. Second, wastage is not uniform over the whole tube surface. For convenience of expression, however, the values given assumed uniform wastage over the whole tube surface.

It would be very valuable if long-term tests were conducted in field units burning coal that had been previously tested in the laboratory 300-hr tests. Laboratory generated wastage results could then be related to actual long-term wastage data gathered from the field. This would only need to be done in a few instances on bituminous coals to "calibrate" laboratory results. This would provide a frame of reference for coals evaluated with the nomograph in terms of conventional wastage expressions.

Short Term Probes as Corrosion Indicators

During the latter part of the referenced study probes were inserted for periods of five minutes for purposes of measuring the acid soluble sodium and potassium during each test firing (1). These probes were inserted during tests in which raw coals and prepared coals were burned as well as during tests in which additives were used. Acid soluble sodium and potassium values were expressed as a ratio and plotted against measured corrosion rate where they showed a remarkably good correlation (Fig. 6). Analysis of the coals used in constructing this figure are shown in Table II. It was noted that the acid soluble Na_2O/K_2O ratio measured from probe deposits is probably a closer approximation of the relative concentration of sodium and potassium-iron-trisulfate than that measured from the coal, for the following reason:

> The many physical and chemical phenomena that cause changes in the coal constituents during combustion have already taken place when the resultant probe deposits are measured for sodium and potassium content. Therefore, an acid soluble measure of the alkalies in the <u>deposit</u> rather than in the <u>coal</u> would automatically consider the effects of combustion on the mineral matter in the coal.

Further efforts on this short-term acid soluble alkalies measurement approach as a potentially simple and accurate way of assessing the corrosive potential of a coal is highly recommended.

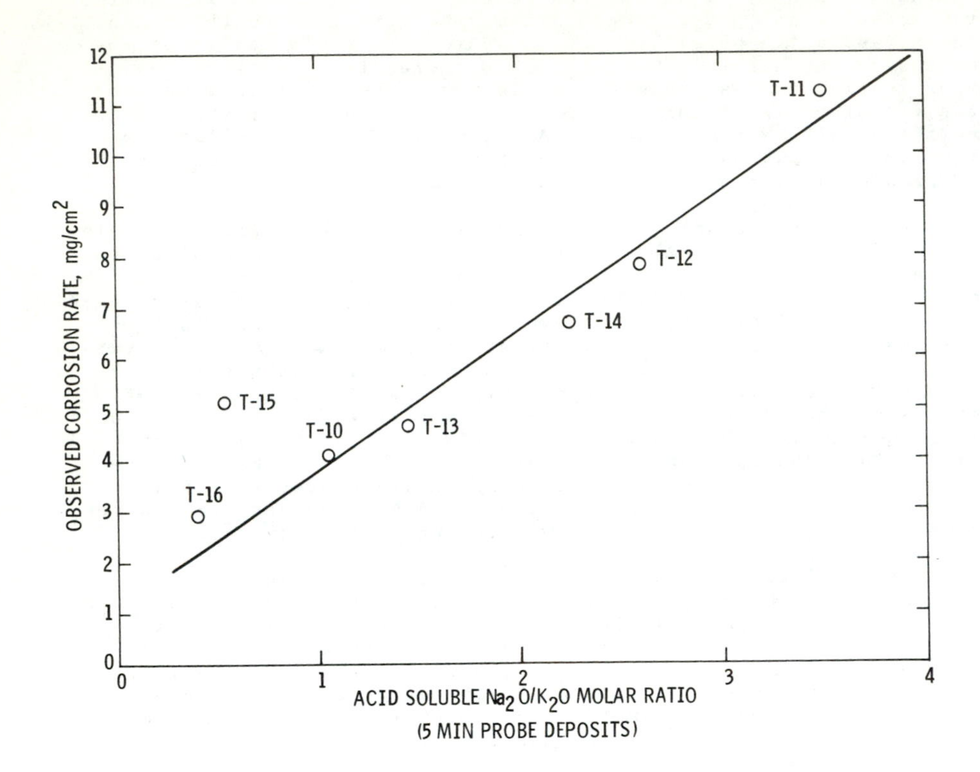

Fig.6 Acid soluble Na_2O/K_2O molar ratio vs observed corrosion rate

Methods of Minimizing a Coal's Corrosive Potential

The entire system of operations at the supplier facility, should be examined, from the seam face where mining is begun to the point of loading for shipment. These operations may be summarized as follows:

1. Mining: the coal is removed from the ground by one of several mining methods and transported from the mine.
2. Coal preparation: the run-of-mine (raw) coal may now be processed by any combination of many preparation procedures.
3. Coal additives and blending: additives or other coals may be blended with the original coal prior to shipment.

Mining

The coal is removed from a seam that is enclosed on two sides by "barren" material. Analysis of channel samples showed the corrosion-affecting minerals, as well as other mineral matter, originated primarily from the roof, floor, or partings in the seam (1). Therefore, the corrosiveness of a coal may be reduced by mining in such a way that the concentration of alkaline earth materials is increased with respect to the alkalies. This can be accomplished by inclusion or exclusion of: a) roof material, b) floor material, and c) material contained in partings.

TABLE II

ANALYTICAL DATA (300-hr TESTS)

Test No.	10	11	12	13	14	15	16
Corrosion rate ($mg/cm^2/300$ hr)	4.07	11.13	7.79	4.62	6.65	5.13	2.94
Fuel identification	Coal B' +NaOH	Coal B' +NaOH	Coal B'+ NaOH+ dolomite	Coal C'	Coal B'+ NaOH+ hydrated lime	Coal F+ hydrated lime	Coal prepared
Acid soluble (Coal Basis)(ppm)							
Na_2O	1,120	3,890	3,790	430	3,890	480	390
K_2O	3,130	3,220	3,380	1,400	2,990	2,390	1,320
CaO	3,270	3,730	13,210	1,480	26,220	7,320	2,270
MgO	1,310	1,550	6,981	340	1,450	2,740	1,360
Na_2O/K_2O acid sol. mol. ratio	.54	1.84	1.70	.47	1.98	.30	.44
Ash composition (Coal Basis) (%)							
SiO_2	12.7	13.3	12.7	8.7	12.9	11.0	7.1
Al_2O_3	4.8	5.1	4.8	5.4	5.1	5.3	3.5
Fe_2O_3	4.4	4.5	4.7	3.6	3.2	1.7	1.3
CaO	0.41	0.44	1.4	0.19	2.6	0.84	0.26
MgO	0.18	0.17	0.91	0.09	0.21	0.35	0.21
Na_2O	0.15	0.46	0.42	0.08	0.50	0.13	0.11
K_2O	0.55	0.59	0.57	0.32	0.58	0.79	0.56
TiO_2	0.19	0.21	0.20	0.20	0.20	0.18	0.13
P_2O_5	0.07	0.06	0.06	0.04	0.04	0.02	.01
SO_3	0.48	0.46	2.30	0.07	2.68	0.48	.38
Cl	0.02	0.02					.38
Proximate analysis (%)							
Moisture	1.4	1.5	2.1	0.5	2.2	1.7	2.1
Ash	24.1	25.7	28.3	19.4	28.4	21.1	13.9
Volatile Matter	34.8	34.1	32.8	20.7	33.0	31.9	33.9
Fixed Carbon	39.7	38.7	36.8	59.9	36.4	45.3	50.1
Total Sulfur	4.60	4.62	5.11	3.32	3.80	1.53	1.42
Sulfur forms (%)							
S as Sulfate	0.00	0.08	0.12	0.11	0.05	0.14	0.23
S as Pyritic	3.22	2.78	3.24	2.19	2.24	0.61	0.52
S as Organic	1.38	1.76	1.75	1.02	1.51	0.78	0.67
H.H.V. (Btu/lb)	10,080	9,810	9,740	12,200	9,950	11,330	12,380
Ash fusibility							
I.T.	2,050	1,980	2,030	1,980	2,060	2,150	2,230
S.T.	2,120	2,110	2,050	2,330	2,160	2,240	2,350
F.T.	2,400	2,400	2,190	2,700	2,280	2,620	2,700+

Coal Preparation

The procedures used to prepare coals for market should be examined. The results of bench-scale chemical and physical tests revealed that the individual chemical constituents of coal mineral matter often concentrate in specific size and/or weight fractions of the coal. Thus, preparation practices, which employ sizing and gravity concentrating techniques, can be used to alter the chemical composition of the coal product. Since high temperature corrosion can be related to coal mineral matter properties, it is possible to reduce corrosive potential of many coals through the judicious application of conventional coal preparation methods. In one particular case, the corrosiveness of a given coal was reduced from 10.2 $mg/cm^2/300$ hr to 2.9 $mg/cm^2/300$ hr because of the benefits derived from a beneficiation process (1).

Coal Additives and Blending

Materials such as dolomite and limestone may be added or two or more coals may be blended to counteract the corrosive nature of one of the coals. In one particular case, the corrosiveness of a given coal was reduced from 10.2 $mg/cm^2/300$ hr to 5.1 $mg/cm^2/300$ hr.

As previously mentioned, the use of additives results in the preferential formation of sulfates or double salts of the type $K_2Ca_2(SO_4)_3$ thus preventing formation of the aggressive compound $K_3Fe(SO_4)_3$. The use of additives can also have beneficial effects in reducing problems due to slagging and fouling. Test results showed that, in most cases, ash deposits became more friable when the additive was used and resulted in easier deposit removal (1). Although not showing a big difference, calcium compounds had a slight edge over magnesium compounds in terms of reducing corrosive potential.

The Role of Fuel Impurities in Tube Wastage

The extensive research performed by C-E in establishing the mechanism of the liquid phase deposit high temperature corrosion has not shown any significant corrosion effect at the chloride levels (0.1 to 0.2 percent) (14) normally encountered in coal firing. Several recent developments in the American utility fuels market, however, have placed a need for renewed investigation to better understand the role that chlorides and other fuel impurities play in tube metal corrosion and fouling. This research, important even for existing steam temperatures, but the concern for the adverse effect of these impurities is even more important for the higher temperatures presently contemplated.

Investigations at C-E showed that the probe corrosion rate for clean coal increased by a factor of 3.3 when firing raw and clean coals from the same mine source. In this case, the coal had been cleaned by a gravity separation in carbon tetrachloride. Therefore, the residual chloride in the cleaned coal was 1.3 percent as opposed to a 0.05 percent level in the raw coal. This corrosion rate was confirmed by subsequent testing (1).

Chloride concentrations in American coals range from 0 to 0.6 percent while coals from the United Kingdom peak at 0.8 percent. The 0.6 percent high for U.S. chloride levels is somewhat misleading in that the majority of U.S. coals have chloride levels less than 0.2 percent. Because of this, chlorine has not been implicated by U.S. investigators as a contributing factor to tube wastage (15). If the high chloride U.S. coals are to enter the market, however, a new look is justified.

Accelerated environmental cleanup, use of fuels with higher chloride content, and the firing of refuse containing increasing PVC separately or mixed with fossil fuels has resulted in an increasing concern about the effect of chloride on corrosion potential. The opening of new mines in the Illinois Basin (16) having high chloride coals (0.3 to 6.5 percent) and exploitation of lignite deposits in saline areas of Texas are examples.

Incinerators in both Europe and the U.S. have reported excessive corrosion rates (17,18), and combinations of sulfur, chlorine, lead, tin and zinc have been judged to be the corrosive agents. In these studies, where raw non-prepared refuse is fired, the concentration of lead, tin, and zinc in deposits has been thought to result in a lowering of the deposit melting point with an acceleration of liquid phase corrosion.

Battelle researchers have conducted investigations that show opposing effects of chlorine and sulfur in refuse (19). An increase in chlorine increases the corrosion rates of low alloy steels while an increase of sulfur decreases the corrosion rates of all the steels investigated. They reported that the major contribution to the corrosion reaction is in the type of compounds that are deposited on the corrosion probe. Their investigation confirms the importance of the chloride reaction as reported by other investigators.

Experiences to date in mixed refuse firing in the U.S., where prepared refuse is providing 10 percent of the heat input, indicate that the corrosion rates are not excessive. The responses with higher chlorine levels introduced with either the coal or the refuse or with higher metal temperatures are not known.

The role of chloride in increased corrosion may take several paths. Chloride is naturally occurring in coal as an inorganic material and may be carried over with the fly ash or released by strong sulfur acid as HCl (Fig. 7). It can also be volatilized as

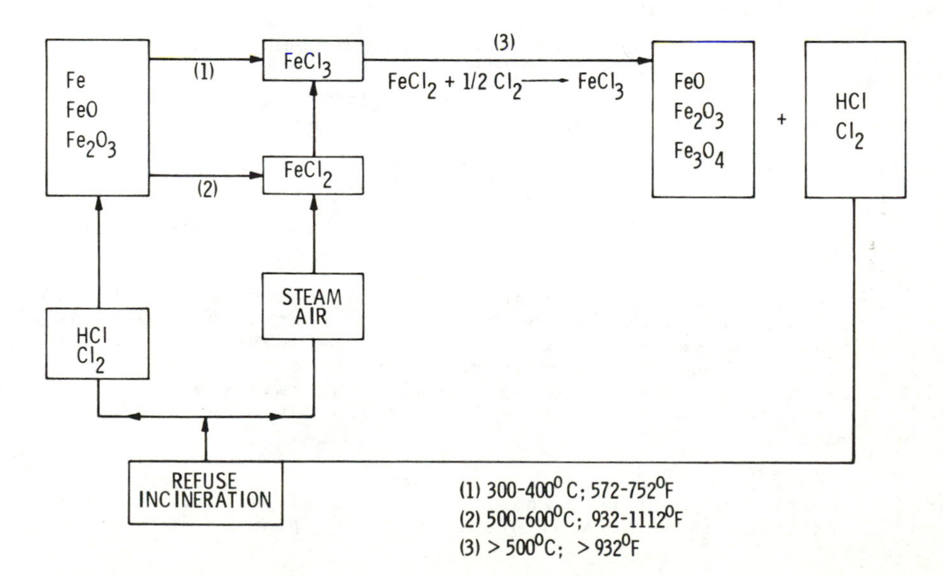

Fig.7 Role of chloride in corrosion

NaCl. In addition, chloride can enter with refuse in either an organic or inorganic form. The resulting chloride compounds usually are concentrated on cooled tube surfaces where additional reactions may occur.

Two major adverse mechanisms are thought to be involved in the potential increase in corrosion rate in incinerators and/or fossil fuel-fired boilers in the presence of chloride. One involves formation of eutectics or complexes resulting in a lowering of melting point of deposits. This phenomenon is of particular concern in waterwall wastage where the temperature range of any molten salts would be increased.

The second mechanism which is probably of greater concern in the superheater involves the reaction of gas phase sulfuric acid with deposit chloride to release HCl near the heated tube surface. Subsequent reaction (Fig. 8) may involve stepwise formation of

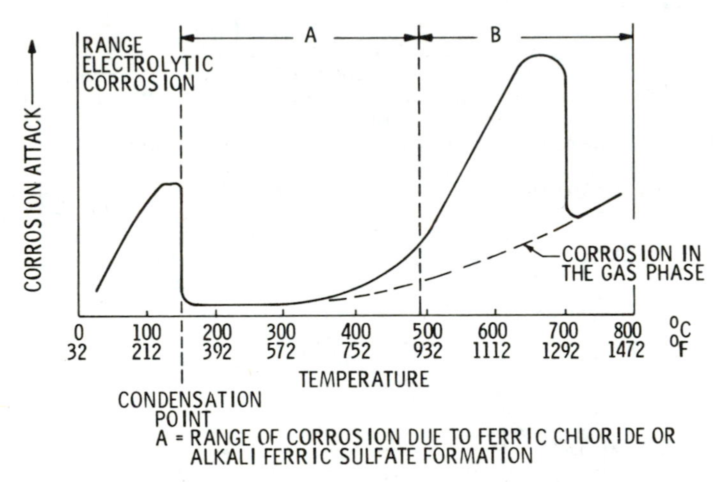

Fig.8 Rate of corrosion as a function of the tube wall temperature

volatile ferric chloride (20,21) and/or unstable chloride or oxychlorides of other alloy components.

In general, waterwall surfaces in steam generators (incinerators) with higher chloride or heavy metal concentrations operating in temperature ranges above 600 F appear to be subject to accelerated wastage. Superheater metals appear to be able to operate with tolerable corrosion rates up to 800 F in similar environments. However, maintenance may be markedly increased at metal temperatures above 900 F.

Development of new materials is continuing. In addition, work is continuing on firing modifications, fuel preparation, and use of additives as methods of preventing formation of corrosive deposits.

As previously indicated, new fuels and fuel firing procedures related to energy conservation and environmental cleanup have increased concern about the effect of various fuel "impurities" on corrosion.

Current R&D

A continuing C-E program involves laboratory studies to determine the influence of chloride, heavy metals, and other fuel impurities on deposit type corrosion in fossil fuel boilers. Results of this program will provide input for fuel preparation as well as a screening of alloys and materials for the practical prevention of boiler tube wastage.

The alkali-iron-trisulfate complex [$Na_3Fe(SO_4)_3$ and $K_3Fe(SO_4)_3$], the principal corrosive in high temperature coal ash corrosion, was made by treating synthetic corrosion mix, 1.5 Na_2SO_4 : 1.5 K_2SO_4 : 1.0 Fe_2O_3, with a typical boiler flue gas, 3 percent O_2, 0.25 percent SO_2, 13 percent CO_2, and 85+ percent N_2. The concentration of fuel impurity was varied to prepare sufficient samples to determine melting points and to determine impurity concentrations chemically. The same concentrating ratios were used for corrosion rate determinations in the tubular muffle furnaces.

Figure 9 is a schematic of the test equipment being used in this program. Pyrovane controllers are used to control temperatures in the Vycor tube at preset test conditions applicable to each type of corrosion medium. A sample is embedded in the corro-

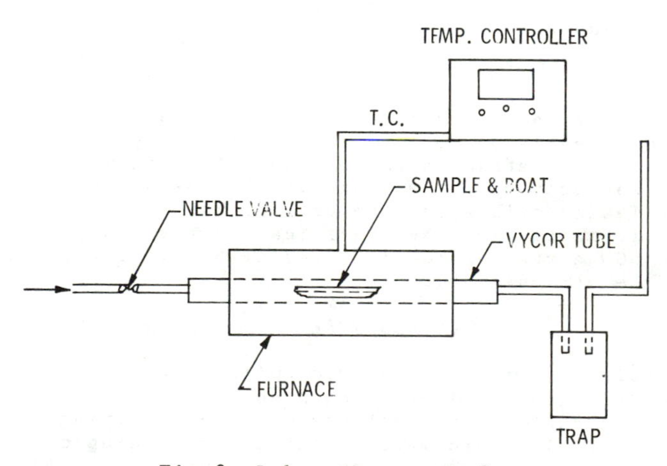

Fig.9 Laboratory test furnace

sion medium in a porcelain boat placed adjacent to the external controlling thermocouple in the Vycor tube. This controlling thermocouple was calibrated against a reference thermocouple in the Vycor tube. Gas is admitted into the Vycor tubes to establish the environment for each test. Both weight loss and penetration are determined during these tests.

Preliminary results indicate that combinations of chloride and some heavy metals seem to induce synergistic wastage on austenitic specimens at normal superheater metal temperatures.

Additional research at C-E includes an ERDA sponsored program for evaluation of commercially available alloys and coatings for possible use in advanced power cycles (metal temperatures 1200 to 1800 F) while burning coals of future economic significance in the

United States.

Preliminary evaluation of the selected materials and coatings was carried out exposing controlled temperature probes in the laboratory Solid Fuel Burning Test Furnace while burning the four selected coal feedstocks. The optimized materials are currently being evaluated in operating boilers using the same feedstocks. All four 300 hr tests while firing the selected coal feedstocks in the laboratory test facility have been completed. Probes containing the selected materials were exposed in the water-cooled "superheater" section.

Physical, chemical and metallurgical evaluation of the specimens did not reveal accelerated wastage of the type observed on conventional superheaters operating in a metal temperature range of 1000 to 1300 F. Such wastage is caused by molten phase complex sulfates, which are unstable at higher temperatures (>1300 F).

The liquid phase corrosion potential of the test coals in the above temperature range were derived from the nomograph (Fig.4) developed during previous testing of austenitic stainless steel while burning bituminous coals in the laboratory test facility used in these tests. The selected coals were rated:

Coals	Potential on a scale of 1-10	
Keystone	< 1	(Low)
Dave Johnston	0	(Very Low)
Baldwin	4.5	(Moderate)
Big Brown	0	(Very Low)

The validity of the corrosion potential formula for subbituminous and lignite coals or for new austenitic alloys has not been established. The attached table (Table III) shows that, although there was only slight wastage in all tests, the Dave Johnston and Baldwin coals appear to be somewhat more corrosive to the austenitic materials. No significant differences were noted when the 800 series or cobalt based alloys were exposed while burning the various coals.

FURTHER WORK

Conventional pulverized coal-fired boilers can be made up to 7 percent more efficient by increasing steam temperature. Should this option be exercised, efficient use of steel alloying elements could minimize costs if there were better understanding of the corrosive behavior of coals -- not only of bituminous coals but of lower rank Western coals, which are coming into wider use, but where corrosion potential has hardly been investigated.

For example, many lower rank Western coals have naturally occurring higher calcium and magnesium contents. This suggests they may be less corrosive than many bituminous coals. However, other Western coals, such as North Dakota Lignites, also have considerably higher alkali contents, which may offset the potential advantages of the higher calcium and magnesium.

The lack of understanding of how these constituents react and interact simply points up the necessity of expanding previous work to include the abundant, important Western coal resources so that the corrosiveness of all United States coals can be predicted better.

TABLE III

ERDA MATERIALS EVALUATION

AVERAGE WEIGHT LOSS DATA
mg/cm^2

		Keystone	Dave Johnston	Baldwin	Big Brown
316	Avg Low (1350)	2.97	5.6	3.7	2.0
	Avg High (1650)	----	---	---	---
310	Avg Low (1350)	13.3*	7.2	2.93	3.07
	Avg High (1650)	3.65	4.7	5.5	3.80
12R72	Avg Low (1350)	6.05	5.7	8.0	3.77
	Avg High (1650)	----	---	---	----
802	Avg Low (1350)	1.67	1.55	2.05	1.93
	Avg High (1650)	2.70	5.2	15.7*	1.23
617	Avg Low (1350)	1.3	1.45	0.22	1.42
	Avg High (1650)	0.67	12.0*	4.2	----
671	Avg Low (1350)	2.35	26.7*	3.0	1.3
	Avg High (1650)	8.0	4.7	7.8	1.96
188	Avg Low (1350)	----	----	---	----
	Avg High (1650)	1.9	1.2	10.5*	1.22

* Mechanical damage

The nomograph developed through previous work will provide an excellent basis for additional research on coal ash corrosion. The investigations necessary to complete and verify the nomograph are:

1. Undertake the necessary work to expand the present nomograph to include all U.S. coals, in particular subbituminous and lignitic coals.
2. Verify and calibrate corrosion indices generated with the nomograph by installing long term probes in utility boilers burning the same coals as were used to develop the nomograph.
3. Investigate credibility of short term probe (5 min.) exposure tests as a viable way of determining corrosive potential of a coal.

These additional investigations will permit the use of coal preparation or modification with additives or blending to nullify the effects of impurities in today's fuels.

REFERENCES

1 Borio, R. W., et. al., "The Control of High-Temperature Fireside Corrosion in Utility Coal-Fired Boilers," OCR R&D Report No. 41, (1969).

2 Corey, R. C., Grabowski, H. A., and Cross, B. J., "External Corrosion of Furnace-Wall Tubes III. Further Data on Sulfate Deposits and the Significance of Iron Sulfide Deposits," Trans. ASME, 67, 951-963 (1945).

3 Plumley, A. L., Jonakin, J., and Vuia, R. E., "A Review Study of Fireside Corrosion in Utility and Industrial Boilers," Canadian Chemical Processing, 51, 52 June; 70 July 1967.

4 Koopman, J. G., Marselli, E. M., Jonakin, J., and Ulmer, R. C., "Development and Use of a Probe for Studying Corrosion in Superheaters and Reheaters," Proc. Amer. Power Conf., 21, 236-245 (1959).

5 Jonakin, J., Rice, G. A., and Reese, J. T., "Fireside Corrosion of Superheater and Reheater Tubing," ASME Paper No. 59-FU-5 (1959).

6 Nelson, W., and Cain, C., Jr., "Corrosion of Superheaters and Reheaters of Pulverized-Coal-Fired Boilers," Trans. ASME, series A82, 192-204 (1960).

7 Reese, J. T., Jonakin, J., and Koopman, J. G., "How Coal Properties Relate to Corrosion of High Temperature Boiler Surfaces," Proc. Amer. Power Conf. 23, (1961).

8 Cain, C., Jr. and Nelson, W., "Corrosion of Superheaters and Reheaters of Pulverized-Coal-Fired Boilers," Trans. ASME, Series A83, 468 (1961).

9 Nelson, W. and Lisle, E. S., "A Laboratory Evaluation of Catalyst Poisons for Reducing High-Temperature Gas-Side Corrosion and Ash Bonding in Coal-Fired Boilers," J. Institute Fuels 37, 378 (1964).

10 Nelson, W. and Lisle, E. S., "High Temperature External Corrosion on Coal-Fired Boilers," Siliceous Inhibitors, J. Inst. Fuels 38, 179-186 (1965).

11 Corey, R. C., Cross, B. J. and Reid, W. T., "External Corrosion of Furnace-Wall Tubes II. Significance of Sulfate Deposits and Sulfur Trioxide in Corrosion Mechanism," Trans. ASME, 67, 289-302, (1945).

12 Anderson, C. H. and Diehl, E., "Bonded Fireside Deposits in Coal-Fired Boilers -- A Progress Report on Manner of Formation," ASME Paper 55 A 200 (1955), Abs. Mech. Eng. 78, 271 (1956).

13 Rahmel, A., "Influence of Calcium and Magnesium Sulfates on High Temperature Oxidation of Austenitic Chrome-Nickel Steels in the Presence of Alkali Sulfates and Sulfur Trioxides," Proc. Int. Conf. Marchwood, England (1963).

14 Plumley, A. L., "Incinerator Corrosion Potential," ASME Incinerator Division Corrosion Symposium, New York, (1970).

15 Cutler, A. J. B., Halstead, W. D., Laxton, J. W. and Stevens, C. G., "The Role of Chloride in Corrosion Caused by Flue Gases and Their Deposits," ASME Paper 70 Wa/CD-1, New York (1970).

16 Gluskoter, J. J. and Reese, V. W., "Chlorine in Illinois Coals," Illinois State Geology Survey, Circular No. 372 (1964).

17 Borio, R. W. and Plumley, A. L., "European Incinerator Inspections," Internal C-E Reports, April, 1970 and February, 1976.

18 Sommerlad, R. E., Bryers, R. W., and Shenker, J. D., "Systems Evaluation of Refuse as a Low Sulfur Fuel-Steam Generator Aspect," ASME Winter Annual Meeting, (1971).

19 Krause, H. H., Vaughan, D. A. and Boyd, W. K., "Corrosion and Deposits from Combustion of Solid Waste III," ASME Paper 74-WA/CD-5 (1974).

20 Fassler, K., Leib, H. and Spalm, H., "Corrosion in Refuse Incinerators," Mittelungen der VGB 48 2, 130 (1968).

21 Baum, B. and Parker, C. H., "Incinerator Corrosion in the Presence of Chloride and Other Acid Releasing Constituents," DeBell and Richardson, (1973).

THE INFLUENCE OF ELECTRODE POTENTIAL ON CORROSION AND CREEP BEHAVIOUR OF HEAT RESISTANT ALLOYS IN ALKALI SULFATE MELTS

ALFRED RAHMEL

Dechema-Institut
D-6000 Frankfurt/Main 97
Federal Republic of Germany

SUMMARY

The paper begins with a justification of electrochemical measurements and a brief explanation of the theory upon which they are based.

According to investigations of the influence of electrode potential on the anodic partial reaction of metal dissolution, the metals can be subdivided into two groups. The metals of the first group (Ag, Cu) form no protective surface films. Those of the second group form protective surface films within defined potential ranges. All heat-resistant materials belong to this second group. In their case, greater corrosion only occurs above a critical breakthrough potential that depends on the individual material.

The cathodic partial reaction is particularly strongly stimulated by acids such as SO_3, V_2O_5, $NaVO_3$ and MoO_3. These oxidants also stimulate free corrosion, because the free corrosion potential of heat-resistant materials is shifted above the breakthrough potential level. Neutral oxidants such as Na_3VO_4 or K_2CrO_4 stimulate scarcely any corrosion, as the free corrosion potential of heat-resistant materials generally remains within the passive range.

In the case of the nickel-base alloy IN 597, internal sulphidation that has a disadvantageous effect on creep behaviour only occurs within a certain potential range.

1. INTRODUCTION

It appears advisable to begin by justifying why electrochemical investigations of the problem of corrosion caused by deposits have been carried out and by stating the theory upon which these investigations are based. The theory of corrosion and the experimental methods employed are closely interrelated and therefore influence each other.

The author omits a description of the causes for the formation of deposits on material surface, their formation mechanism and the transport processes in the deposits themselves, referring only to summary descriptions 1 - 5 and the pertinent papers presented at this conference. The considerations begin with the existence of deposits on the surface of heat-resistant materials.

Deposits become particularly aggressive at temperatures above their melting point or above the eutectic melting point between the deposit and the scale of the material. This has been particularly impressively demonstrated by Meijering and Rathenau, using the example of MoO_3 deposits [6, 7]. The same however applies for alkali sulphate-rich deposits in the presence of SO_3, in which complex sulphates of the type (K,Na) [(Fe,Al) $(SO_4)_2$] and $(K,Na)_3$ [(Fe,Al) $(SO_4)_3$] then occur [8 - 10] . Rapid or catastrophic corrosion is therefore linked with the occurence of liquid phases on the material surface.

Molten salts have many similarities with concentrated aqueous salt solutions. For example, the electric current is conducted by ions in both systems. In aqueous solutions, metals corrode by electrochemical mechanisms. Because of the many similarities between electrolyte solutions and molten salts, it is today widely assumed that corrosion in and through molten salts is also due to electrochemical mechanisms [11] . Consequently, one should be able to use the same electrochemical methods as are employed for aqueous solutions in investigating corrosion in molten salts.

The electrochemical theory of corrosion says that corrosion of metals consists of the two partial reactions of anodic metal dissolution in accordance with

$$Me = Me^{z+} + ze^- \tag{1}$$

and cathodic reduction of an oxidant in accordance with

$$Ox + ze^- = Red^{z-}. \tag{2}$$

The rates of the two partial reactions are best measured as current density i and are functions of the electrode potential E. The functional relationship between i and E is called a current density-potential curve or i(E) curve, Figure 1. In all anodic partial reactions, the anodic partial current density i_a increases with the potential; in all cathodic partial reactions, the cathodic partial current density i_c decreases with rising potential. In an external circuit, one can only measure the sum of all partial currents as the total current i_{tot}:

$$i_{tot} = \sum i_a + \sum i_c \tag{3}$$

Cathodic currents are of negative sign.

When there is no external current, the anodic partial current must for reasons of mass and charge balance be equal to the cathodic partial current, i.e.,

$$i_a + i_c = 0 \quad (E = E_{cor}) \tag{4}$$

Figure 1. In view of this condition, the free corrosion potential E_{cor} and the corrosion current density i_{cor} result. i_{cor} is proportional to the rate of metal dissolution.

The corrosion current density i_{cor} and the corrosion potential E_{cor} depend on the course of the anodic $i_a(E)$ and the cathodic $i_c(E)$ curves. Figure 2 shows how different $i_a(E)$ curves of the metals Me(1) and Me(2) lead to differing values for i_{cor} and E_{cor} with the same $i_c(E)$ curve. Figure 3 demonstrates how different cathodic partial current density-potential curves $i_c(E)$ of the oxidants 1 and 2 lead to differing values for i_{cor} and E_{cor} with the same metal. If the $i_a(E)$ curve of the metal has a passive range, i.e., a potential range with good protective film formation and the resultant low dissolution rate, as was assumed in Figure 3, the location and course of the $i_c(E)$ curve will decide whether E_{cor} lies in the passive or transpassive range and thus whether i_{cor} is small or relatively large, Figure 3.

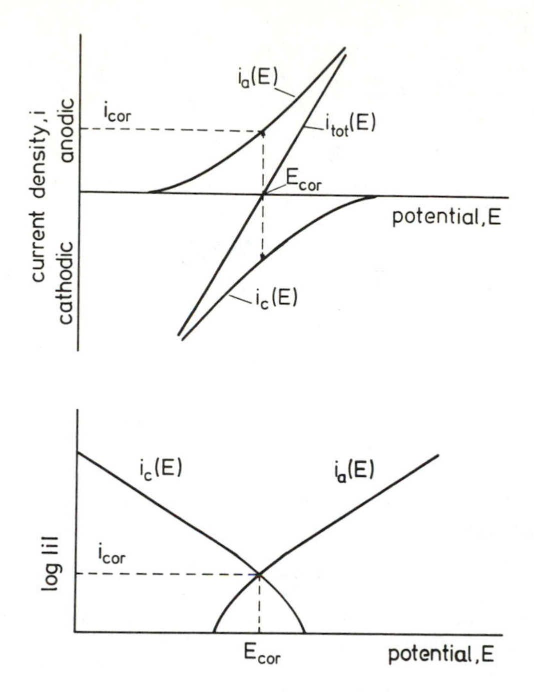

Figure 1. Potential dependence of partial and total current densities in the electrolytic corrosion of a homogeneous metal (schematic)

$I_a(E)$: partial current-potential curve of the anodic partial reaction $Me = Me^{z+} + ze^-$

$i_c(E)$: partial current-potential curve of the cathodic partial reaction $Ox + ze^- = Red$

$I_{tot}(E)$: total current-potential curve, $i_{tot} = i_a + i_c$

E_{cor} : free corrosion potential

i_{cor} : corrosion current density

From these brief explanations and Figures 1 - 3, it therefore follows that:

a. the free corrosion potential E_{cor} only provides information concerning the rate of corrosion when the $i_a(E)$ curve of the material in the medium in question is known;

b. the total current-potential curve $i_{tot}(E)$ approaches only at a greater distance from the free corrosion potential the anodic or cathodic partial current density-potential curves. Only in an electrolyte without redox system does one find a higher degree of coincidence between the $i_{tot}(E)$ curve and the $i_a(E)$ curve. Vice versa, the $i_{tot}(E)$ curve is practically identical with the $i_c(E)$ curve when the $i_{tot}(E)$ curve is measured on a metal that is resistant to attack. Here however, one must allow for the fact that the $i_c(E)$ curve can differ with the electrode material;

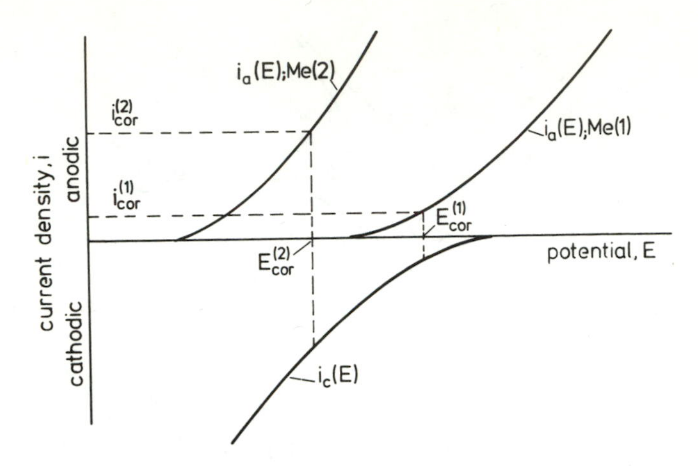

Figure 2. Partial current densities of the two different metals Me (1) and Me (2) in the same medium.

c. for the corrosion engineer, knowledge of the i_a(E) and i_c(E) curves - and particularly the i_a(E) curve - is desirable, as they describe the influence of electrode potential on corrosion. Allowances must be made for the fact that practically all local corrosion types (pitting corrosion, stress corrosion cracking) only occur above and/or below critical potentials. When dealing with these types of corrosion, it is particularly important to know the potential dependence of material behaviour.

In view of these facts, we carried out the following investigations:

α. Determination of the anodic partial current density-potential curves of various metals and heat-resistant materials in alkali sulphate melts.

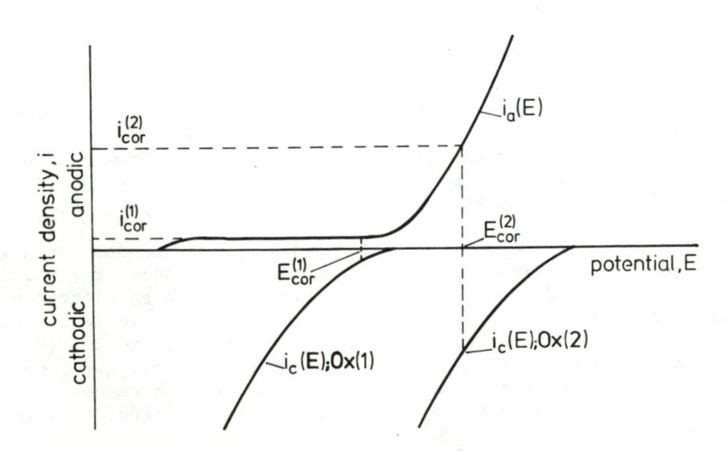

Figure 3. Effect of the stimulation of the cathodic partial reaction on corrosion current density and free corrosion potential.

β • The influence of various oxidants on the cathodic partial current density-potential curve, measured as a total current density-potential curve on platinum. Platinum does not dissolve in sulphate melts, so that the cathodic $i_{tot}(E)$ curves are practically identical with the $i_c(E)$ curves (cf. item b).

γ • Determination of the influence of the oxidants investigated under β on the free corrosion potentials and the corrosion rates of various materials, to check the applicability of the electrochemical model for corrosion in sulphate melts.

δ • Creep tests in sulphate melts at defined electrode potentials, as a means of determining the effects of corrosion on the creep behaviour of heat-resistant materials.

The main results of these investigations are described in the following.

2. EXPERIMENTAL PROCEDURE

Only a few general remarks can be made here; reference is made to the original papers for details [12 - 18].

The experiments were carried out between 900 and 1,075 K in melts of different composition. At 900 K, ternary-eutectic melts of (in mole %) 78 Li_2SO_4, 13.5 K_2SO_4 and 8.5 Na_2SO_4 were used; in the 975 to 1,075 K range, the melts were of 75 Na_2SO_4 + 25 Li_2SO_4 and 65 K_2SO_4 + 35 Li_2SO_4 (in mole %). The experiments with the nickel-base alloys Nimonic 105 and IN 597 were carried out in a melt of (in mole %) 53 Na_2SO_4, 7 $CaSO_4$ and 40 $MgSO_4$ at 1,075 K. This melt consists of the sulphates of the salts of seawater, which are frequently found in the deposits in gas turbines near the sea.

In determining mass loss-potential curves (anodic partial current density-potential curves $i_a(E)$), the specimens were working electrodes in an electrochemical cell and were kept at constant potential with the aid of a potentiostat. Adhering corrosion products were removed after the experiment to determine the mass loss [12, 15].

All potentials were measured against an Ag/Ag^+ reference electrode [13], whose potential is equated with zero. In the following, these potentials are designated E_{Ag}.

The electrochemical investigations were supplemented by metallographic, microprobe and X-ray diffraction analyses.

Table 1 shows the chemical compositions of the investigated materials.

Table 1: Chemical Composition of Alloys (WT %)

	C	Si	Mn	Fe	Cr	Ni	Al	Co	Mo	Ti	Nb/Ta
6 Cr Steel	0,05	1,04	0,77	Bal.	6,45	0,10	0,76	-	-	-	
18 Cr Steel	0,12	1,15	0,71	Bal.	17,70	0,18	0,91	-	-	-	
Nimonic 105	0,13	0,22	0,04	0,42	14,95	Bal.	4,59	19,60	5,13	1,28	
IN 597	0,05	0,23	0,09	0,64	24,39	Bal.	1,12	19,44	1,60	2,63	1,13

3. RESULTS

3.1 The Influence of Electrode Potential on Corrosion (anodic partial current-potential curve)

When one considers anodic dissolution, one can subdivide the metals into two groups:

a. metals that form only in the melt soluble corrosion products on dissolution, for example, Ag and Cu.

b. metals that form surface films. These surface films can be protective or non-protective, depending on the potential. All heat-resistant materials containing Cr belong to this group.

The metals of Group a. show a Tafel-like linear relationship between the logarithm of the anodic dissolution current i_a and the potential E, in accordance with

$$E = a + b \lg i_a. \tag{5}$$

a and b are constants. Thorough investigation revealed that not the charge transfer reaction is rate determining, but the transportation of the dissolved metal ions out of the boundary layer of convective diffusion [14].

More important from the technical point of view is the behaviour of heat-resistant materials. With these materials, two (or even three) potential ranges with differing corrosion behaviours were found in each case [15, 16, 18]. Figure 4 shows the example of the influence of electrode potential on the mass loss of iron and ferritic chrome steels with different chrome contents. The mass loss is proportional to the anodic dissolution current i_a in accordance with Faraday's Law. Below a critical potential, mass loss is small and practically independent of the potential. Oxide films are formed on the surface of the materials, protecting them against further attack. A terminological loan is made from the investigation of aqueous solutions, in calling this potential range "passive region". Above a critical potential called "breakthrough potential", E_B, that depends on the material mass loss increases greatly with rising potential. Porous scales with greater sulphide contents, particularly near the metal/scale interface, are formed. Internal sulphidation can occur in the marginal zone of the material. The range above E_B can be called "transpassive region".

An analogous relationship was found for all ferritic chrome steels, austenitic chrome-nickel steels and Ni- and Co- base alloys [15, 16, 18]. In exceptional cases, intercrystalline attack appears to occur at very negative potentials. Such attack was observed in the case of chrome steels at 1,075 K, Figure 5 [15]. The relatively large mass losses shown in Figure 5 for potentials $E_{Ag} \approx -1{,}600$ mV is due to such intercrystalline attack. The formation of protective oxide films can apparently be hindered when potentials are negative. Such negative potentials do not however occur in practice.

The level of the breakthrough potential is very greatly influenced by the chrome content of the alloy, as can be seen from Figure 6. Other components of the alloy will certainly also exert an influence, but this has not yet been investigated.

The potential's influence on the structure and composition of the scale was investigated in greater detail for the nickel-base alloys Nimonic 105 and In 597 [18]. Figure 7 shows the influence of the potential on the mass loss of the specimen and the quantity of corrosion products found in the melt and on the counter-electrode after 100 h of potentiostatic measurements. Nimonic 105 corrodes heavily at $E_{Ag} > -100$ mV. A porous scale with higher contents of Ni_3S_2 is formed, and the material suffers internal sulphidation, whereby four different

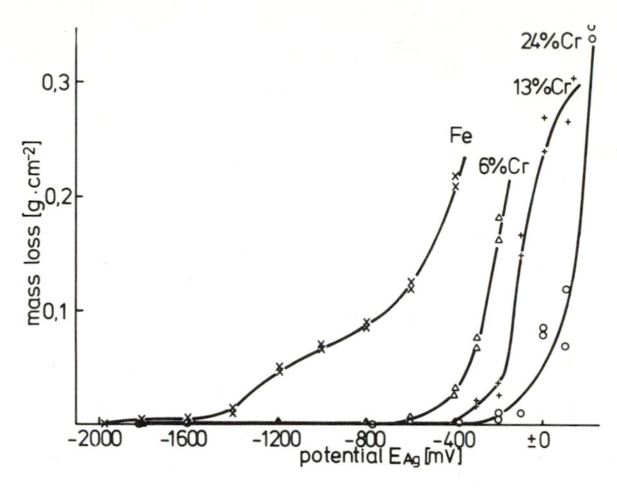

Figure 4. Potential dependence of the mass loss of iron and ferritic chrome steels after 15h potentiostatic conditions in a eutectic melt of (in mole %) 78 Li_2SO_4, 13.5 K_2SO_4 and 8.5 Na_2SO_4; 900 K.

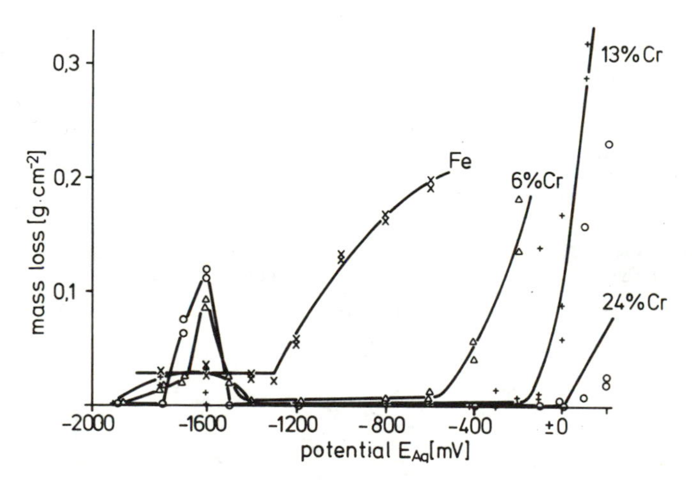

Figure 5. Potential dependence of the mass loss of iron and ferritic chrome steels after 15h potentiostatic conditions in a melt of (in mole %) 75 Na_2SO_4 and 25 Li_2SO_4; 1,075 K.

sulphides are formed [18]. At potentials of $E_{Ag} < -100$ mV, protective oxide scales with high contents of NiO and MgO are formed [18]. No internal sulphidation occurs in this case.

The behaviour of IN 597 is however particularly interesting, as it gives indications of the processes that play a role in the destruction of the scale. Figure 8 shows the influence of potential on the composition of scale that supplements the potential's influence on mass loss. At potentials of $E_{Ag} < -100$ mV, scales with high contents of MgO and NiO form again. Particular mention must be given to the potential range -100 mV $< E_{Ag} < +200$ mV. The specimen suffers

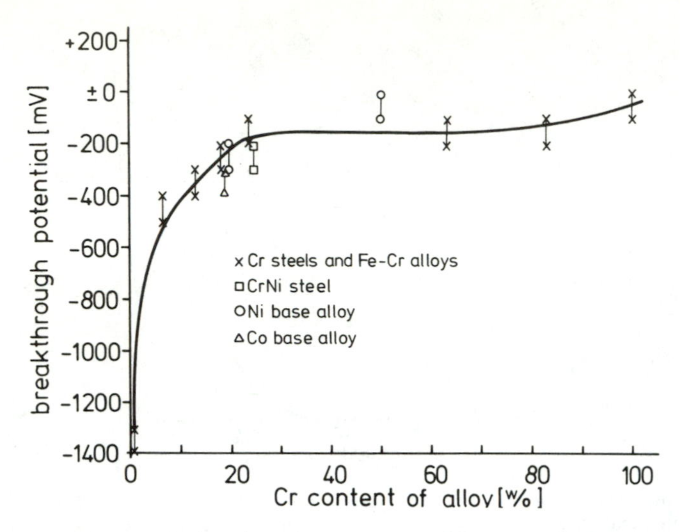

Figure 6. Influence of the chromium content of heat-resistant materials on the breakdown potential; 15h potentiostatic conditions in a eutectic melt of (in mole %) 78 Li_2SO_4, 13.5 K_2SO_4 and 8.5 Na_2SO_4; 900 K.

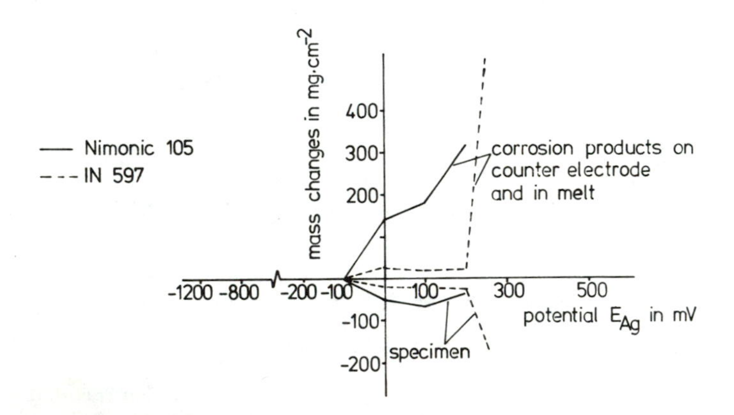

Figure 7. Influence of potential on the mass loss of Nimonic 105 and IN 597 and on the amount of corrosion products found in the melt and at the counter electrode after 100h potentiostatic measurements at 1,075 K in a melt of (in mole %) 53 Na_2SO_4, 7 $CaSO_4$ and 40 $MgSO_4$.

a slight mass loss, and NiO and MgO are precipitated at the counter-electrode. The external corrosion is slight, as a thin oxide scale covers the surface of the specimen. Figure 8 shows that the scale's contents of NiO, CoO and MgO decrease in the potential range around - 100 mV, whilst the contents of Cr_2O_3 and TiO_2 increase. The bivalent oxides are therefore selectively dissolved. The melt around the specimen increases in acidity with rising potential, because an

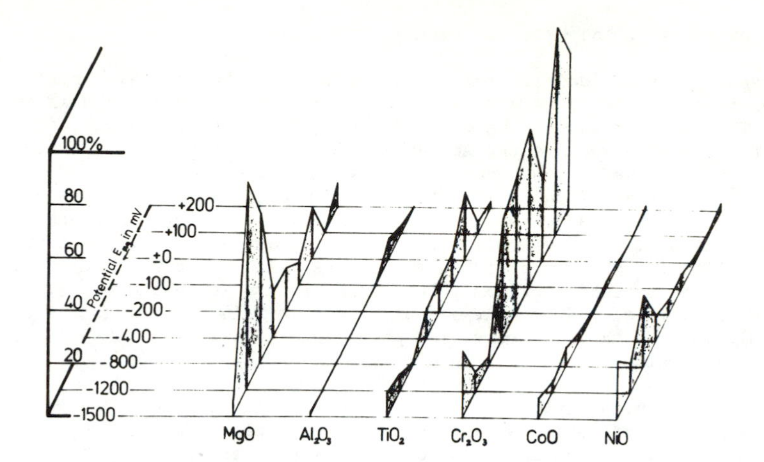

Figure 8. Influence of potential on the composition of the oxide scale formed on IN 597 during 100h potentiostatic conditions in a melt of (in mole %) 53 $Na_2S)_4$, 7 $CaSO_4$ and 40 $MgSO_4$; 1,075 K.

additional discharge of the SO_4^{-2} ions occurs at the anode in accordance with

$$SO_4^{2-} = SO_3 + 1/2\ O_2 + 2e^- \quad (6)$$

[12, 19]. The bivalent oxides are thus converted in accordance with

$$NiO + SO_3 = NiSO_4 \quad (7a)$$

$$MgO + SO_3 = MgSO_4 \quad (7b)$$

into sulphates that are soluble in the melt.

The counter-electrode is cathode. O^{2-} ions are formed there, in accordance with

$$1/2\ O_2 + 2e^- = O^{2-} \quad (8)$$

$$SO_4^{2-} + 2e^- = SO_3^{2-} + O^{2-} \quad (9)$$

The melt there therefore becomes basic, the oxides being precipitated in accordance with

$$NiSO_4 + O^{2-} = NiO + SO_4^{2-} \quad (10)$$

These experiments agree with earlier investigations [20 - 22] in showing that "acidic fluxing" plays an important role in the destruction of the scale. The first consequence of the changes in the scale is internal sulphidation, whereby in the case of IN 597 two sulphide phases are formed [18]. This has disadvantageous effects on creep behaviour, as will be shown in Section 3.4.

At potentials of $E_{Ag} > +200$ mV, severe corrosion occurs.

The effect of the differing breakthrough potentials of the materials Nimonic 105 and IN 597 on free corrosion are described in Section 3.3.

3.2 Stimulation of the Cathodic Partial Reaction

Oxygen plays no decisive role as an oxidant in the cathodic partial reaction Equation (2), as it is apparently only very sparingly soluble in sulphate melts. This can for example be concluded from the corrosion of Ag, in which the transportation of oxygen to the Ag surface determines the rate of corrosion [23]. At 900 K, the corrosion current density in the eutectic Li_2SO_4-K_2SO_4-Na_2SO_4 melt is only a few $\mu A\ cm^{-2}$, this corresponding with a corrosion rate of only a few tens of $\mu m\ a^{-1}$. Higher corrosion rates therefore only occur when additional oxidants that stimulate the cathodic partial reaction are present in the melt. The SO_3 that is almost always present in the combustion gases of fossilic fuels is such an oxidant. It is dissolved by sulphate melts, stimulates the cathodic partial reaction 12 and thus also stimulates free corrosion (cf. Figure 3), because the free corrosion potential is shifted in the direction of more positive values.

Also of technical interest is the influence of V_2O_5 and MoO_3, as V_2O_5 or V compounds can find their way into deposits when oil is burnt, and Mo compounds can be formed through the corrosion of materials containing Mo. The influence of increasing V_2O_5 contents of a sulphate melt on the cathodic parts of the total current density-potential curve measured on platinum is shown in Figure 9 in semi-logarithmic presentation [17]. Contents of as little as 2 mole% V_2O_5 increase the total current by some two orders of magnitude; they therefore stimulate the cathodic reaction very greatly. The steady-state potential is also shifted in the direction of more positive values. At E_{Ag} potentials approx. < -500 mV, diffusion limiting currents occur, because the transportation of V^{5+} ions to the electrode surface is the rate determining step [17]. This diffusion limiting current increases over-proportionally with rising V_2O_5 content of the melt, this probably being connected with the onset of electronic partial conductivity of the melt [17].

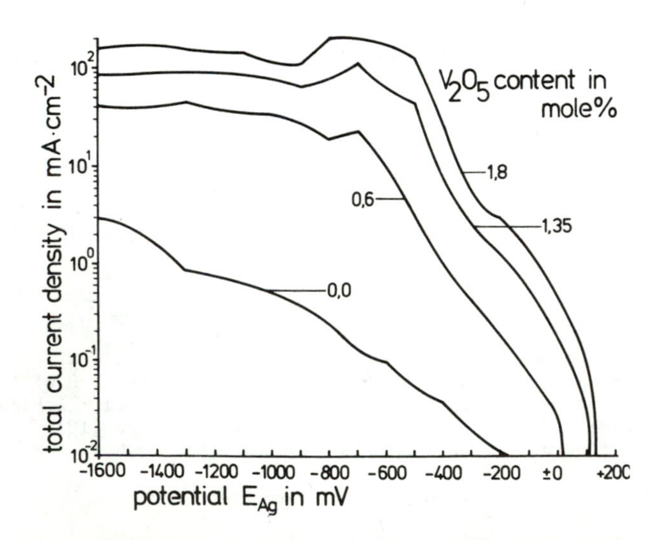

Figure 9. Influence of small V_2O_5 additions to a melt of (in mole %) 78 Li_2SO_4, 13.5 K_2SO_4 and 8.5 Na_2SO_4 on the cathodic total current-potential curve measured on a Pt electrode at 1,025 K.

Not only the V_2O_5 content but also the Na_2O/V_2O_5 ratio of the V compound plays a role. This is shown in Figure 10. V_2O_5, $NaVO_3$ and Na_3VO_4 were added to the sulphate melts in such quantities that the V_2O_5 content was equal in all cases. Figure 10 shows that all three curves take very similar courses, but that the location of the curve relative to the potential axis however differs. The $i_{tot}(E)$ curves after addition of V_2O_5 and $NaVO_3$ are shifted some 300 mV in the direction of more positive potentials than the Na_3VO_4 curve. According to Figure 3, this must lead to greater stimulation of corrosion.

Behind this potential shift lies a generally valid rule that cannot however be quantitatively formulated. Potentiometric acid-base titrations indicate that V_2O_5 and $NaVO_3$ are acids, whilst Na_3VO_4 is however an almost neutral salt [24]. These acid-base titrations [24] revealed that acidic melts always have a more positive and basic melts a more negative steady-state potential than neutral melts. In view of Figure 3, one must therefore expect acidic melts to be similar to acidic aqueous solutions in that they stimulate corrosion particularly greatly. According to the definition [25, 26]

$$\text{base} = \text{acid} + O^{2-}, \tag{11}$$

for example,

$$SO_4^{2-} = SO_3 + O^{2-}, \tag{12}$$

acidic melts have a low O^{2-} activity and thus a high pO value in accordance with

$$pO = -\lg c_{O^{2-}}. \tag{13}$$

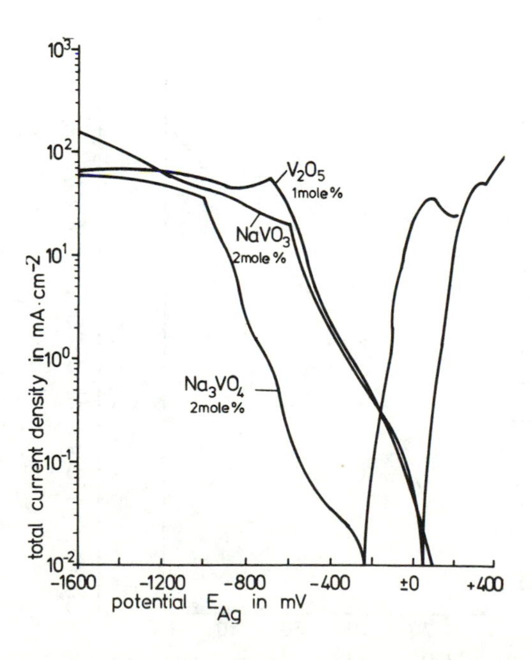

Figure 10. Influence of equivalent amounts of V_2O_5, $NaVO_3$, Na_3VO_4 added to a melt of (in mole %) 78 Li_2SO_4, 13.5 K_2SO_4 and 8.5 Na_2SO_4 on the cathodic total current density-potential curve measured on a Pt electrode at 1,025 K.

Very similar $i_{tot}(E)$ curves as are found after addition of V_2O_5, Figure [9], are also found when the acid MoO_3 is added to sulphate melts [17]. As MoO_3 however evaporates at a considerable rate because of its high vapor pressure, the stimulating effect decreases with time. Unlike MoO_3 and V_2O_5, WO_3 has practically no influence on the $i_c(E)$ curve when added to sulphate melts, as WO_3 is practically insoluble in sulphate melts [17].

3.3 The Effects of Stimulation of the Cathodic Partial Reaction on Free Corrosion Rate

Figure 3 shows that every stimulation of the $i_c(E)$ curve shifts the free corrosion potential E_{cor} towards more positive values, something that generally also causes an increase of the i_{cor} value. The presence of SO_3 in the gas phase or V_2O_5 or MoO_3 in the melt should therefore increase the corrosion rates of metals in sulphate melts. A prerequisite to this is however that the anodic $i_a(E)$ curve of the metal remains unchanged by the additions, something that must be verified where necessary.

Let us first consider stimulation of the corrosion of Ag by V_2O_5. The results are to be found in Figures 11 and 12 [17]. Figure 11 shows the mass loss rates or the proportional corrosion current densities of Ag in $(K, Li)_2SO_4$-V_2O_5 melts at 1,025 K in air. The corrosion current density rises from 5.10^{-6} A cm^{-2} in the pure $(K, Li)_2SO_4$ melt to approx. 3 A cm^{-2} in the pure V_2O_5 melt; that is, by some six orders of magnitude ! A corrosion current density of 3 A cm^{-2} completely dissolves an Ag wire of 2 mm diameter in approximately 2 minutes! Such high corrosion rates are apparently only possible because no surface layers hinder the anodic dissolution of the metal.

According to Figure 3, the corrosion potential E_{cor} should become more positive with rising i_{cor}. According to Figure 12, this is the case [17] : E_{cor} increases by approx. 1V!

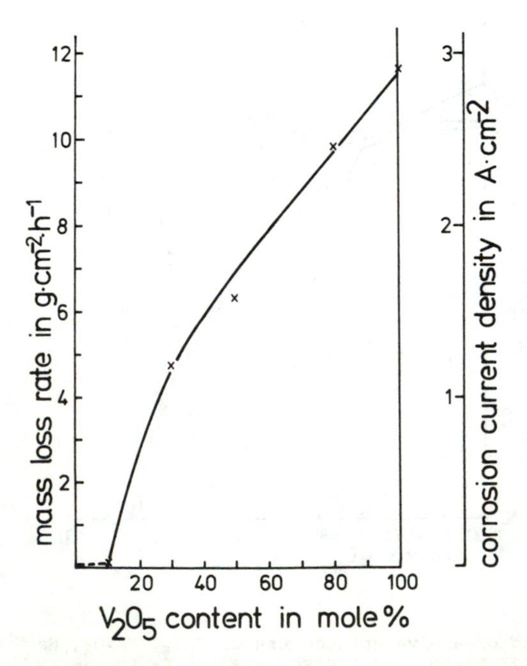

Figure 11. Mass loss rates and corrosion current densities of Ag in $(K,Li)_2SO_4$-V_2O_5 melts at 1,025 K in air. $(K,Li)_2SO_4$ = 65 K_2SO_4 + 35 Li_2SO_4 (in mole %).

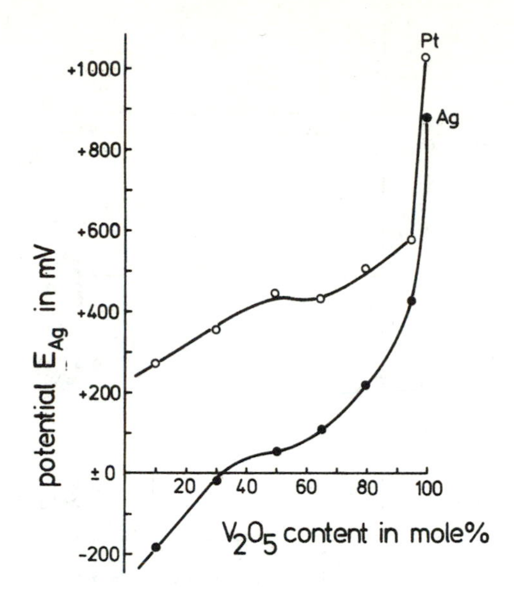

Figure 12. Free corrosion potential of Ag in $(K,Li)_2SO_4$-V_2O_5 melts at 1,025 K; air. (Pt = redox potential measured on Pt)

In materials with a passivity range (cf. Figure 4), a greater rate of corrosion first occurs when $E_{cor} > E_B$. Figures 13 and 14 show the examples of the influence of increasing V_2O_5, $NaVO_3$ and Na_3VO_4 contents of the 75 Na_2SO_4 - 25 Li_2SO_4 melts on mass loss and corrosion potential at the end of the experiments with ferritic steels containing 6 and 18% Cr (Nos. 1 and 2 in Table 1). Also stated is the breakthrough potential range E_B. From the experiments with the 6 Cr steel, Figure 13, follows that increasing melt contents of V_2O_5 and $NaVO_3$ shift the corrosion potential beyond the breakthrough potential E_B and thus increase the corrosion rate. V_2O_5 causes a greater potential shift and thus a greater stimulation of corrosion than $NaVO_3$. The addition of Na_3VO_4 lowers the potential; the corrosion rate should therefore stay low. The only way of explaining the unusually-high corrosion rate in melts with 20 mole% Na_3VO_4 is that the anodic $i_a(E)$ curve is also influenced by the addition of Na_3VO_4.

According to Figure 5, the breakthrough potential of the 18 Cr steel is approximately 400 mV more positive than that of the 6 Cr steel. The consequence according to Figure 14 is that $E_{cor} > E_B$ is only in melts containing V2O5 in the case of 18 Cr steel. The corrosion rate is however lower than with 6 Cr steel, Figure 13. The explanation for this is given by Figure 2. With additions of $NaVO_3$ or Na_3VO_4, $E_{cor} < E_B$ still applies; accordingly, mass loss is also slight.

The results described here can be understood on the basis of the initially-outlined electrochemical mechanisms, at least qualitatively.

In the case of materials which form protective surface films, E_{cor} is shifted towards more positive potentials with time. The reason for this is that the formation of scale causes the anodic partial current of metal dissolution to steadily decrease with time. According to Figure 2, E_{cor} must therefore become more positive when the $i_c(E)$ curve remains unchanged. This potential shift can however lead to $E_{cor} > E_B$ at a critical point in time. Something of this sort was for example observed with Nimonic 105 in the Na_2SO_4-$CaSO_4$-$MgSO_4$ melt at 1,075 K.

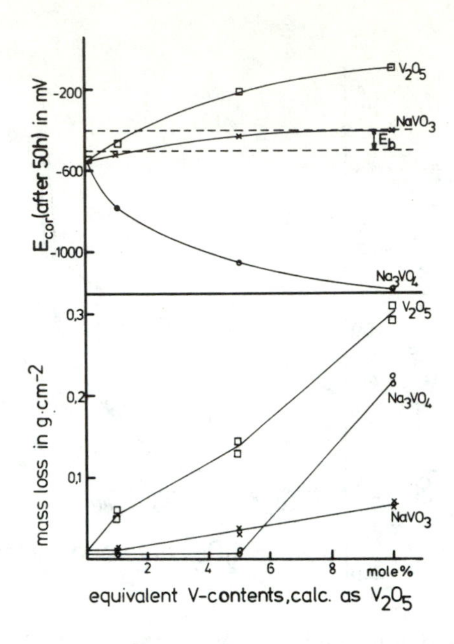

Figure 13. Influence of equivalent amounts of V_2O_5, $NaVO_3$ and Na_3VO_4 added to a melt of (in mole %) 76 Na_2SO_4 and 25 Li_2SO_4 on the mass loss and free corrosion potential of 6 Cr steel after 50h test duration; 1,075 k. E_B = break-through potential of 6 Cr steel in pure $(Na,Li)_2SO_4$ melt.

Figure 15 shows the corrosion potentials E_{cor} and the polarisation resistances R_p of the materials Nimonic 105 and IN 597 plotted against time. The polarisation resistance

$$R_p = \frac{dE}{di}_{(i \rightarrow o)} \qquad (14)$$

is largely a measure of the electric resistance of the scale in this case. Initially, E_{cor} and R_p increase with time. With the material Nimonic 105 however, $E_{cor} > E_B$ after about 200 h ($E_B \approx$ - 100 mV for Nimonic 105), this leading to local destruction of the scale, Figure 16. The beginning of destruction of the scale is marked by a drop in E_{cor} and R_p, Figure 15. As the break-through potential of IN 597 ($E_B \approx$ + 200 mV) is 300 mV more positive than of Nimonic 105, $E_{cor} < E_B$ remains for more than 2,500 h. This is why no destruction of the protective scale takes place in the case of IN 597. This example also clearly demonstrates the significance of having materials with break-through potentials as positive as possible.

The time that elapses until localised disruption of the protective scale on Nimonic 105 occurs depends on the "intensity" of the cathodic partial reaction. Every stimulation of the cathodic partial reaction, for example, by SO_3, shortens the time until disruption. Figures 2 and 3 make this understandable.

3.4 The Influence of Electrode Potential on the Creep Behaviour of IN 597 in Sulphate Melts

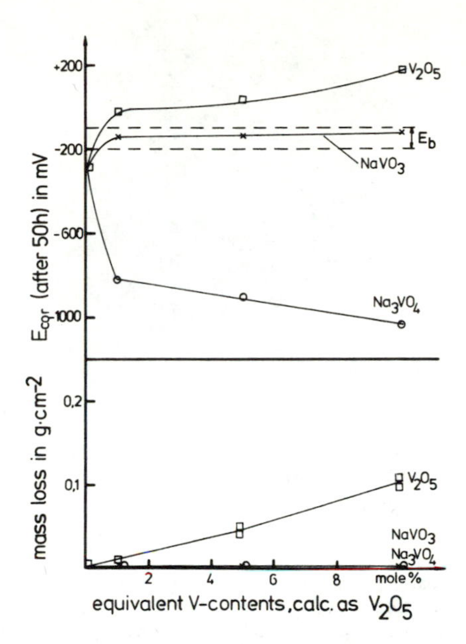

Figure 14. Influence of equivalent amounts of V_2O_5, $NaVO_3$ and Na_3VO_4 added to a melt of (in mole %) 75 Na_2SO_4 and 25 Li_2SO_4 on the mass loss and free corrosion potential of 18 Cr steel after 50th test duration; 1,075 K. E_B = breakthrough potential of 18 Cr steel in pure $(Na,Li)_2SO_4$ melt.

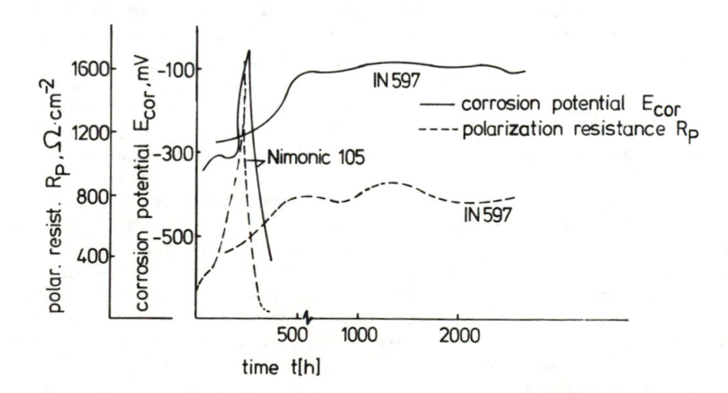

Figure 15. Time dependence of free corrosion potential E_{cor} and polarization resistance R_p of Nimonic 105 and IN 597 in a (in mole %) 53 Na_2SO_4 - 7 $CaSO_4$ - 40 $MgSO_4$ melt at 1,075 K.

⊢5mm⊣

Figure 16. Localized corrosion of Nimonic 105 after 370h free corrosion in a 53 Na_2SO_4 - 7 $CaSO_4$ - 7 $CaSO_4$ - 40 $MgSO_4$ melt at 1,075 K.

In Section 3.1, it was pointed out that one can differentiate between three potential regions for the material IN 597 at 1,075 K in the Na_2SO_4-$CaSO_4$-$MgSO_4$ melt; cf. Figure 7 [18]. At $E_{Ag} < -$ 100 mV, a thin protective scale forms and the material remains practically undamaged. Between - 100 $< E_{Ag} <$ + 200 mV, external corrosion remains very slight, but internal sulphidation however leads to damaging of the material. At $E_{Ag} >$ + 200 mV, severe external corrosion leading to very rapid destruction of the material occurs. To determine the influence of the differing corrosion behaviour in the three potential regions on the creep behaviour of material, creep tests in a Na_2SO_4-$CaSO_4$-$MgSO_4$ melt were carried out, using the experimental arrangement shown schematically in Figure 17. The specimen was the working electrode in an electrochemical cell and could therefore be kept at a potentiostatically controlled potential throughout the experiment. It was thus possible to obtain defined corrosion conditions.

The creep curves plotted for various potentials are shown in Figure 18. Only at $E_{Ag} < -$ 100 mV are the time values until failure and the elongations practically identical with the values obtained in air. At all $E_{Ag} > -$ 100 mV potentials, the consequence of material damage is always shorter times until rupture and reduced elongation. The results obtained up to the present date are summarised in Figure 19. When $E_{Ag} >$ + 200 mV, the stress has only a small influence on time until rupture, because the lifetime of the specimen is largely determined by external corrosion. When $E_{Ag} < -$ 100 mV, the time until rupture is practically identical with that in air, as no material damage occurs in this case. In the potential range - 100 mV $< E_{cor} <$ + 200 mV in which internal sulphidation takes place, one finds no influence at high stresses and thus shorter times until rupture, as the depth of damage due to internal sulphidation is still slight with a specimen diameter of 6 mm. One however finds a noticeable potential-dependent shortening of the time until rupture at lower stresses and consequently longer lifetime. This is shown in Figure 20 for a stress level of 230 N mm^{-2}. Figure 20 also shows the influence of potential on the elongation at fracture, which decreases considerably when the material is damaged. The electrochemical methods of investigation and testing discussed here therefore make it possible to examine the influence of defined types of corrosion on the creep behaviour of materials.

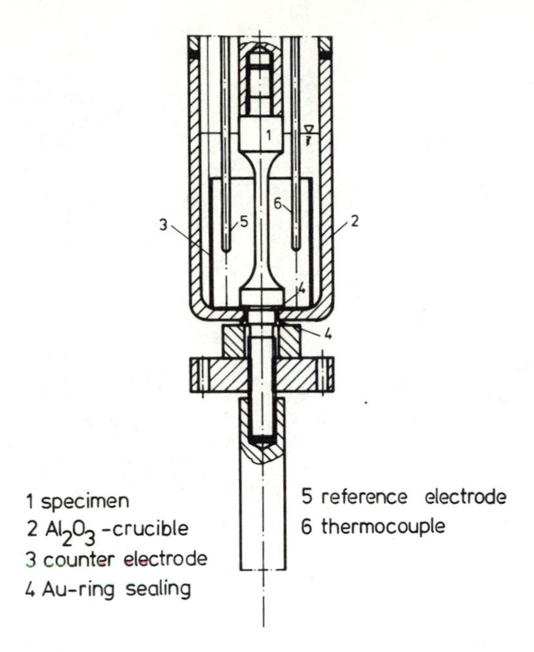

Figure 17. Cross-section through the electrochemical cell for creep test (schematic)
1 specimen (working electrode)
2 Al_2O_3 crucible
3 level of melt
4 Au ring sealing
5 reference electrode
6 thermocouple
7 counter electrode

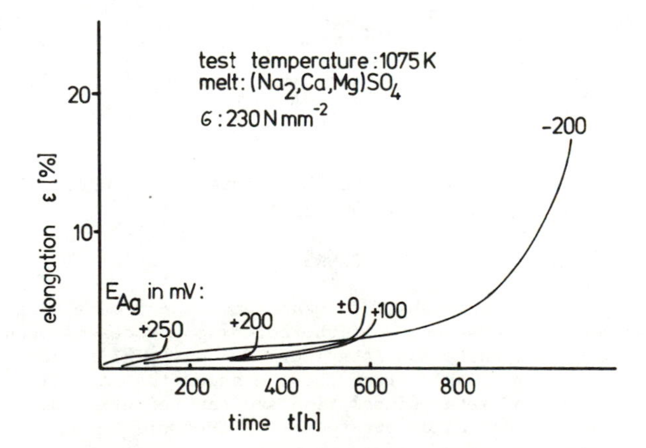

Figure 18. Elongation-time relationships of IN 597 at different potentials (potentiostatic conditions) in a (in mole %) 53 Na_2SO_4 - 7 $CaSO_4$ - 40 $MgSO_4$ melt at 1,075 K; δ = 230 N mm^{-2}.

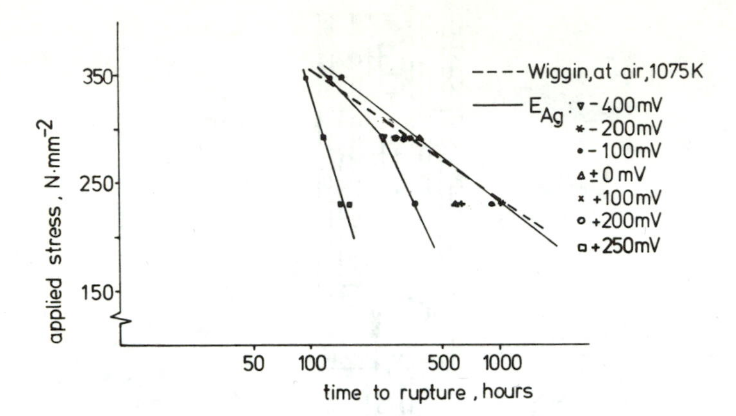

Figure 19. Results of creep tests with IN 597 in air and at controlled potentials in a $(Na_2,Ca,Mg)SO_4$ melt at 1,075 K.

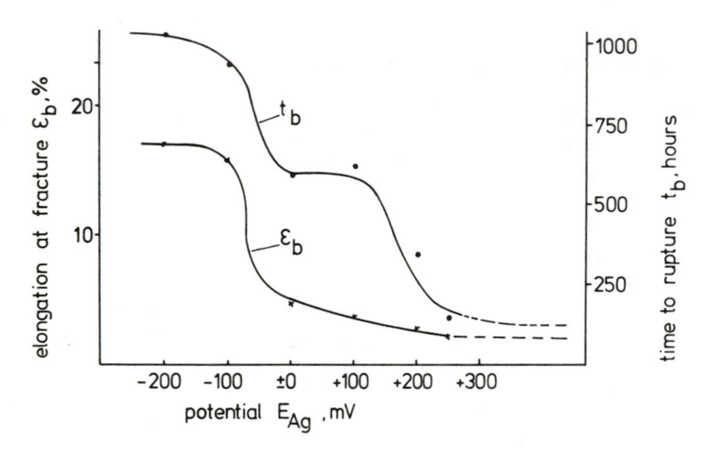

Figure 20. Influence of potential on time to rupture and elongation at fracture of IN 597 in $(Na_2,Ca,Mg)SO_4$ melt at 1,075 K; $\delta = 230$ N mm^{-2}.

CONCLUSIONS

The results show that electrochemical measurements can make considerable contributions towards an understanding of the corrosion processes occurring under liquid deposits. The corrosion behaviour of a material is determined by the corrosion potential that results from the course of the partial current density-potential curves for metal dissolution and the reduction of the oxidant. Electrochemical investigations allow one to determine the

a. influence of electrode potential on the nature and extent of corrosion of a material and

b. the influence of various components of the melt on the cathodic partial reaction and thus also on free corrosion.

This in turn allows one to estimate the

c. Corrosion potential range under practical conditions.

The results of investigations as described under a) provide data that are suitable for use in the preselection of materials. One characteristic value for heat-resistant materials seems to be for example the breakthrough potential. Materials should have as positive a breakthrough potential as possible. It would therefore be a sensible course to rank materials by their breakthrough potentials. The experiments that have been carried out until now allow this for only few materials. There is however qualitative agreement between conventional laboratory tests and behaviour in practice on the one hand and electrochemical measurements on the other hand in their evidence that increasing chrome contents of materials have favourable effects on their corrosion behaviours. Compared with conventional corrosion tests however, electrochemical measurements provide information concerning corrosion behaviour over a wider potential range, whilst conventional tests can only be carried out at one corrosion potential - the free corrosion potential - that can also change with time. In the purpose-specific development of alloys, the electrochemical testing method is clearly superior to the conventional method.

Electrochemical measurements allow recognition of whether impurities or additives of the melt have a stimulating or inhibitive effect. One can determine whether the anodic or the cathodic or even both partial reactions are stimulated or inhibited, or whether one is stimulated and the other is inhibited. This is only possible by means of electrochemical measurements. They could therefore make a considerable contribution towards the development of inhibitors.

Rapid electrochemical tests for the ranking of alloys have been proposed in recent years [27,28]. They supply interesting results, that are however difficult to understand and interpret without supplementary investigations similar to those described here having been carried out.

ACKNOWLEDGEMENTS

The investigations were sponsored by the Bundesministerium für Wirtschaft, via the Arbeitsgemeinschaft Industrieller Forschungsvereinigungen (AIF) and by the Bundesministerium für Forschung und Technologie. The author wishes to express his thanks for the financial support granted.

Hearty thanks are also due to Dr. E. Tatar-Moisescu, Mrs. H. Müller, Miss M. Schorr, Dr. U. Feld, Dipl.-Ing. M. Schmidt and Ing. (grad.) J. Mathy for the careful and painstaking way in which they assisted in the investigations.

REFERENCES

1. Johnson, H.R., and Littler, D.J., "Mechanism of Corrosion by Fuel Impurities", Butterworths, London 1963.

2. Reid, W.T., "External Corrosion and Deposits. Boilers and Gas Turbines". American Elsevier Publ. Co, New York 1971.

3. Stringer, J., Jaffee, R.J., and Kearns, T.F., "High Temperature Corrosion of Aerospace Alloys" AGARD Conference Proceedings No. 120, 1973.

4. Hart, A.B., and Cutler, A.J.B., "Deposition and Corrosion in Gas Turbines", Applied Science Publ. Ltd., London 1973.

5. Rahmel, A., and Schwenk, W., "Korrosion und Korrosions-schutz von Stählen", Kapitel 6, Verlag Chemie, Weinheim 1977.

6. Rathenau, G.W., and Meijering, J.L., Metallurgia 42, 169 (1950).

7. Meijering, J.L., and Rathenau, G.W., Philips' Techn. Rundschau 12, 217 (1951).

8. Rahmel, A., Arch. Eisenhüttenw. 31, 59 (1960).

9. Cain, C., and Nelson, W., ASME-Paper No 60-WA-180 (1960).

10. Rahmel, A., Mitt. VGB Heft 74, p. 332 (1961).

11. Inman, D., and Wrench, N.S., Brit. Corros. J. 1,246 (1966).

12. Rahmel, A., Werkstoffe und Korrosion 19, 750 (1968).

13. Rahmel, A., Electrochim. Acta 15, 1267 (1970).

14. Tatar-Moisescu, E., and Rahmel, A., Electrochim. Acta 20, 479 (1975).

15. Rahmel, A., and Tatar-Moisescu, E., Werkstoffe und Korrosion 26, 513 (1975).

16. Jäkel, U., and Schwenk, W., Werkstoffe und Korrosion 26, 521 (1975).

17. Rahmel, A., Werkstoffe und Korrosion 28, 299 (1977).

18. Feld, U., Rahmel, A., Schmidt, M., and Schorr, M., Werkstoffe und Korrosion, to be published.

19. Rahmel, A., Electrochim. Acta 21, 853 (1976).

20. Bornstein, N.S., and DeCrescente, M.A., Trans AIME, 245 1947 (1969).

21. Goebel, J.A., and Pettit, F.S., Met. Trans. 1, 1943 (1970).

22. Goebel, J.A., Pettit, F.S., and Goward, G.W., in Ref. 4), p. 96.

23. Rahmel, A., Corrosion Sci. 13, 833 (1973)

24. Rahmel, A., Electroanal. Chem. Interfacial Electrochem. 61, 333 (1975).

25. Lux, H., Elektrochem. angew.phys.Chemie 45, 303 (1939)

26. Flood, H., Førland, T., and Motzfeld, K., Acta chem. Skand. 6, 257 (1952).

27. Shores, D.A., Corrosion, 31, 434 (1975).

28. Erdos, E., private communication.

CORROSION IN REDUCING ATMOSPHERES—A DESIGNER'S APPROACH

P. B. PROBERT and L. KATZ

Babcock & Wilcox Company
Fossil Power Generation Division
Barberton, Ohio, USA 44203

ABSTRACT

The design of equipment for service in reducing atmospheres containing H S requires careful attention to metal temperatures. The effects of gas temperature and steam pressure on corrosion were studied for carbon steel and stainless steel for a typical coal gasifier. Areas of acceptable corrosion life and thermal stress were identified.

INTRODUCTION

Equipment designers strive to keep corrosion sufficiently under control so that products have acceptable service lives with reasonable expenditure for maintenance. This must be done at a reasonable price.

There are a vast number of corrosion processes with which the designer must be concerned. We will discuss only those corrosion processes associated with gaseous atmospheres that are reducing or that cycle between oxidizing and reducing conditions, and that may contain hydrogen and hydrogen sulfate.

Over the years, there have been failures in a short time under some conditions of equipment in service due to corrosion in reducing or cycling atmospheres. Extensive laboratory investigations have been carried on to to try to understand the mechanism of the corrosion processes.

In 1959 Eberle and Wiley[1] reported severe wastage in a pressurized reactor producing synthesis gas (CO and H_2) from partial oxidation of natural gas with oxygen. More than half the thickness of 3/4" thick (19mm) Type 347 and Type 310 stainless steel baffles had corroded away and several Type 310 sootblower elements were completely corroded through in 21 days. The gas temperatures were from 800F (427C) to 1300 (704C).

Laboratory investigations confirmed that at 1700F (927C) the corrosion was an oxidizing process, at temperatures of 1500F (815C) and below the corrosion process changed to carburizing with oxide penetration on surface decarburization, and at 700F (371C) corrosion was barely noticeable. This corrosion occurred in the absence of H_2S, but there was evidence that conditions had fluctuated from reducing to oxidizing. The problem was still being studied when the partial oxidation reactor was deactivated for process economic reasons.

The E. I. DuPont Company operated a pulverized coal fired slag tap gasifier in 1955-56. By its nature, this gasifier contained an atmosphere of CO and hydrogen with some H_2S. Ni-Cr-Fe Alloy 600 sootblower tubes were destroyed by corrosion in a few months and these were replaced with Type 310 stainless steel. Also, there was some low temperature corrosion of carbon steel tubes in the economizer that occurred at temperatures above the calculated dew point of the gases.

From time to time, there have been cases of tube wastage in boilers. These have occurred either in boilers operated with reducing conditions in the furnace such as paper mill recovery boilers, or in locally reducing zones in boiler furnaces that, on the average, are operated under oxidizing conditions. The furnace atmosphere contained significant amounts of H_2S. The method of coping with this corrosion when it occurred was to eliminate the local reducing zones by adjustment to burners and careful attention to the introduction of air and to the fuel/air ratio in every part of the furnace.

DESIGNING FOR ACCEPTABLE CORROSION LIFE

For operating pressures above 15 psi (0.1 MPA) coal gasification pressure vessels are designed to Section VIII-Div. 1 or 2 of the ASME Boiler and Pressure Vessel Code. The ASME Code establishes design procedures, materials properties, fabrication and non-destructive examination criteria. Environmental considerations, such as corrosion, are the responsibility of the designer. The designer of course must justify his analysis methods to the owner, his designated agent, his insurance carrier and government regulatory bodies.

Down time costs of a commercial gasification plant cannot be accurately estimated due to variations in plant size, etc. The costs will be very high, although probably below that of a typical utility power plant -- roughly $250,000 per day. This requires that the designer must approach the problem of corrosion from a conservative position.

The designer must consider both laboratory test data and relevant service experience. Laboratory results may be obtained in time to support design, but the schedule may require unrealistically severe conditions in an effort to get short-term data. Furthermore, short-term data may not be appropriate for long-term performance. Complex service conditions such as alternating variable oxidation and reduction environments may be difficult to model in a realistic manner. Because of such experimental limitations, the designer relies heavily on relevant service experience when available.

The total design responsibility includes consideration of the manufacturing function. Acceptable weldability, formability, commercial availability, costs, as well as ASME Code approval of the materials of construction are important factors in selecting materials and designing pressure vessels. Corrosion researchers developing new materials should appreciate these constraints and adjust their programs accordingly.

As an example of some of the corrosion concerns of a designer, let us review a study done recently on a coal gasifier. The coal gasification process produces a reducing gas containing CO and hydrogen with typically up to 1% H_2S. The designer of this equipment should expect corrosion and should consider means to control it in his design.

Figure 1 shows the corrosion rates of 0-5% Cr steels in H_2S.[3] Austenitic stainless steels corrosion rates are about one-tenth of those for ferritic steels. Independent longer time tests at the authors' company[4] have demonstrated that the published corrosion curves are not overly conservative.

Coal can be gasified with either air or oxygen and steam. In both cases the coal is reacted with a controlled deficiency of oxidant to produce a combustible gas. The sulfur in the coal appears as H_2S in the gas. In the case of air blown gasification, the nitrogen in the air dilutes the H_2S so it should be less corrosive to the gasifier; however, the nitrogen also dilutes the product gas and limits its transportability and usefulness to some degree. The gas composition for a typical oxygen blown gasifier is shown in Table 1. Figure 2 shows a portion of the Nelson curve for hydrogen damage. Hydrogen partial pressures corresponding to the composition shown in Table 1 are such that carbon steel is an acceptable material.

A design study was performed to examine the effect of gas temperature, heat flux and steam side pressure (saturation temperature) on the metal temperature and H_2S corrosion. The gasifier is a typical pressurized, oxygen blown, pulverized coal fired, slag tap gasifier used to produce industrial fuel gas having a

Higher Heating Value of approximately 300 Btu/scf (10.4 MJ/m^3) which could replace natural gas or oil in large industrial furnaces with minor changes to the burners.

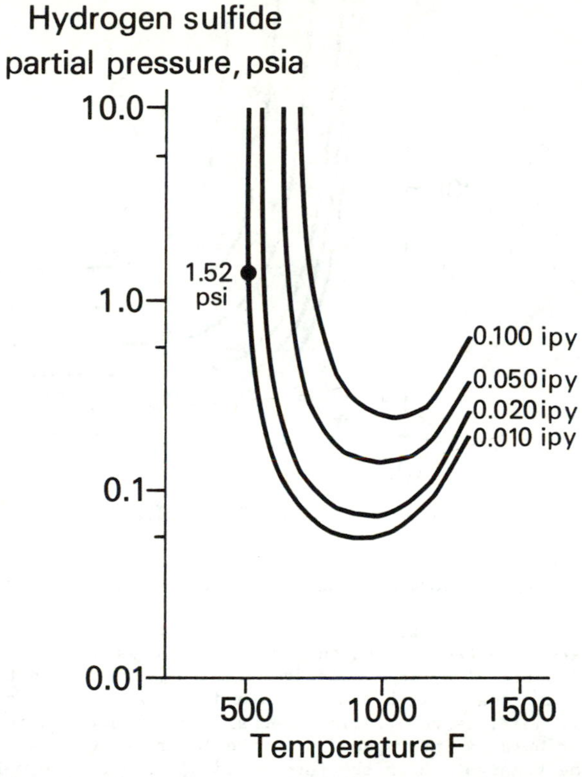

Fig. 1 Corrosion rate ferritic steel 0.5% chromium inches per year (ipy) in hydrogen sulfide.

TABLE 1

Composition of the Gasifier Atmosphere

Argon	0.6% by volume
CO	55.0
CO_2	6.8
H_2	28.1
H_2O	7.0
H_2S	0.8
N_2	1.7

Pressure in gasifier 190 psia (1.31 MPa)

Partial pressure of H_2S 1.52 psia (.36 MPa)

Partial pressure of H_2 53 psia (.36 MPa)

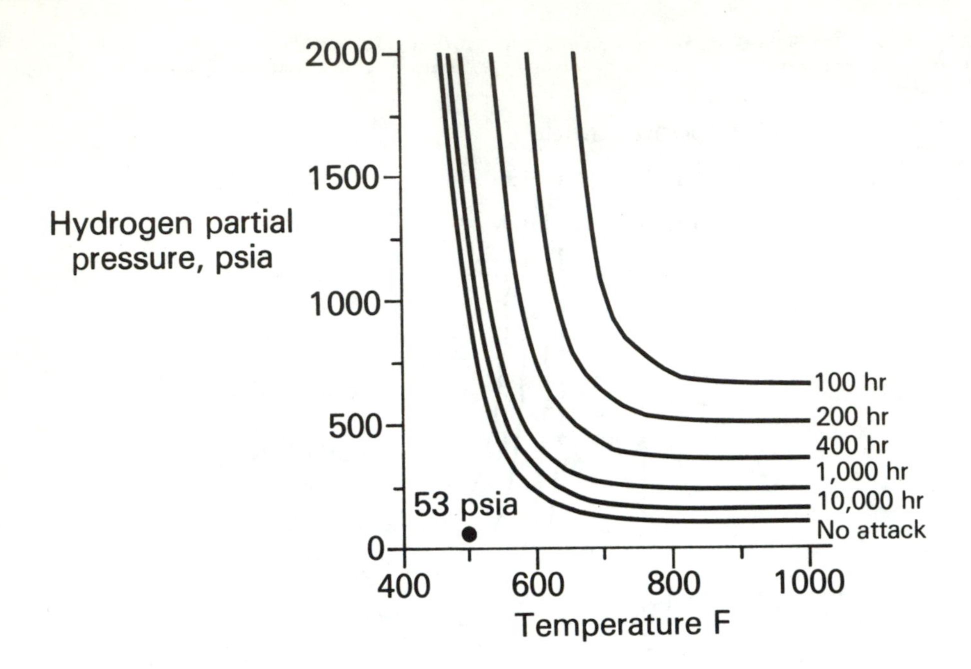

Fig. 2 Time for incipient attack of carbon steel in hydrogen service (Nelson).

A cut away drawing of the gasifier is shown in Figure 3. Pulverized coal, oxygen and steam are injected through the burners arranged around the lower portion of the gasifier. The finely divided coal is suspended in the upflowing gas, quickly heats and reacts with the oxygen and steam to form process gas. The unreacted material, char, is removed in external cyclone separators and reinjected through char burners interspaced with the lower row of coal burners in the lower furnace. The temperature in the furnace is high enough to melt the ash in the coal and char. Molten ash finds its way to the walls and runs down to the tap hole in the furnace floor. Below this tap is a water-filled ash pit where the molten ash is quenched to a stable granular material, depressurized and sluiced away to disposal.

The wall surrounding the high temperature gas is made up of vertical water-cooled tubes with welded steel membranes between. The lower portion of the wall has welded pin studs and thin slag resistant refractory as shown in Figure 4. In the upper portion of the furnace the tubes are left bare to cool the product gas below the ash softening temperature prior to leaving the gasifier.

The outer steel pressure shell is protected drom the high temperature corrosive furnace atmosphere by the gas tight wall of water-cooled tubes.

The highest tube metal temperature occurs at the zone of highest wall heat flux. This is at the location where the studded refractory covered construction ends and the bare tube wall construction begins.

Steam pressures from 200 to 2400 psig (1.4 to 16.6 MPa) gas temperatures from approximately 1000 to 3100F (593 to 1704C) having wall heat fluxes from 10,000 to 200,000 Btu/hr-ft^2 (31.5 to 631 KW/M^2) were studied for two materials of construction -- carbon steel and Type 304 stainless steel. An acceptable corrosion rate was assumed to be one that resulted in half the design minimum tube wall remaining after 20 years. Periodic inspection and local repairs will be made during the life of the gasifier.

For each material, steam pressure and heat flux, the temperature distribution and thermal stress throughout the tube wall and membrane web were calculated.

For the location of the highest metal temperature on the surface of the tube and web, the yearly corrosion rate was tabulated from the H_2S corrosion rate curves.

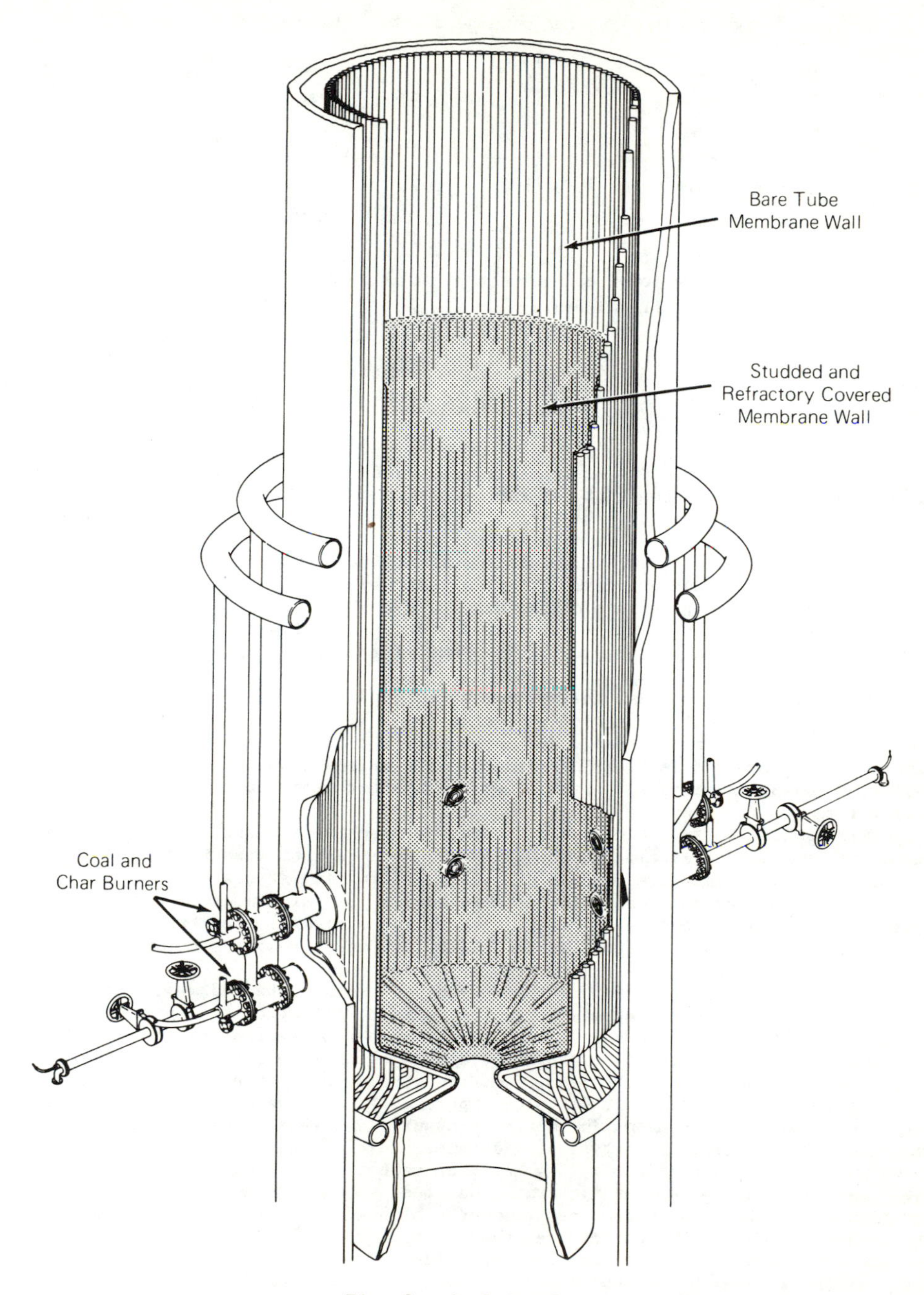

Fig. 3 Coal Gasifier

General design parameters that were held constant were:

Tube diameter	1.5" (38mm)
Tube thickness	0.188" (5mm)
Tube spacing	2" (50mm)
Membrane width	0.5" (12mm)
Membrane thickness	0.25" (6mm)
Partial pressure H_2S	1.0 psia (6.8 KPa)
Gas specific heat	0.43
Gas side convection conductance	8 Btu/hr-ft^2-F (45 W/m^{2}C)
Gas side overall emissivity	0.7
Tube metal conductivity	Carbon Steel K = 379-.114T
	Stainless Steel K = 100T + .057T
	where K = Btu/hr-ft^2-F/in.
	T = °F

In the case of Type 304 stainless steel, this corrosion rate was low enough that it was considered constant for the life of the gasifier. Since some repair of corroded areas during the 20-year life of a gasifier is acceptable, the corrosion rate was considered acceptable if half of the minimum initial primary design thickness (without corrosion allowance), was still remaining after 20 years.

In the case of carbon steel, analyses were made for each year to take into account that the outer surface of the tube gets cooler as the tube thins down due to corrosion (at a constant steam pressure and heat flux). The same criterion of half the primary design thickness remaining after 20 years was used to establish the acceptable design conditions.

For carbon steel tubes, the maximum calculated thermal stress in the tube and web was compared to an acceptance criterion of 1.5 times the ASME Code to the location of maximum stress. For carbon steel tubes, this very conservative stress criteria was acceptable because the life of the tube was controlled by corrosion considerations for steam pressures and gas temperatures of interest to the designer.

For stainless steel tubes, the thermal stresses were calculated in the same manner as for carbon steel, but the maximum stress was compared to the acceptance criterion of 1.5 times the maximum allowable alternating stress, S_a, at 10^6 cycles. This acceptance criterion results in a higher allowance thermal stress than 1.5 S_m used for the carbon steel tubes. For stainless steel tubes, the design is controlled by thermal stress limits at steam pressures below 1500 psi and by corrosion limits above 1500 psi.

From the average temperature on the surface of the tube and web, it was possible to calculate the effective gas temperature for each steam pressure and heat flux. This temperature is used by the designer in evaluating process performance and temperature to maintain the ash in the molten state.

Figures 5 and 6 show the results of the studies of carbon steel and stainless steel, respectively. It is quite apparent that for carbon steel H_2S corrosion seriously limits the maximum steam pressure (saturation temperature) available to the designer. At a typical gas temperature of 2400F (1300C), steam pressure is limited to 500 psi (3.4 MPa) for carbon steel or low alloy tubes or 1600 psi (11 MPa) for austenitic stainless steel tubes. Although an all stainless steel gasifier would produce steam at a higher pressure, which is desirable, the cost increases. Furthermore, another problem could be introduced -- stress corrosion of stainless steel on the water side.

If coatings with enhanced corrosion resistance are to be used, the process of preparing the coating must be sufficiently controlled so that critical holidays do not exist. The practical use of coatings is also limited by the design of the gasifier since all areas -- welds and tight bends as well as straight tube sections -- must be protected.

Scheduling of new material developments is of utmost importance to the designer. New materials which are proposed for Section VIII construction require long-time creep and stress rupture data. The total time from test initiation to final approval by the ASME Pressure Vessel Code could approach two years. Final design commitments must, however, be made on the basis of code approved materials.

Until someone invents a new alloy that is both corrosion resistant and inexpensive, the designer is still obligated to choose a conservative steam pressure at some sacrifice of thermodynamic efficiency to achieve a reasonable service life in the corrosive gasifier atmosphere.

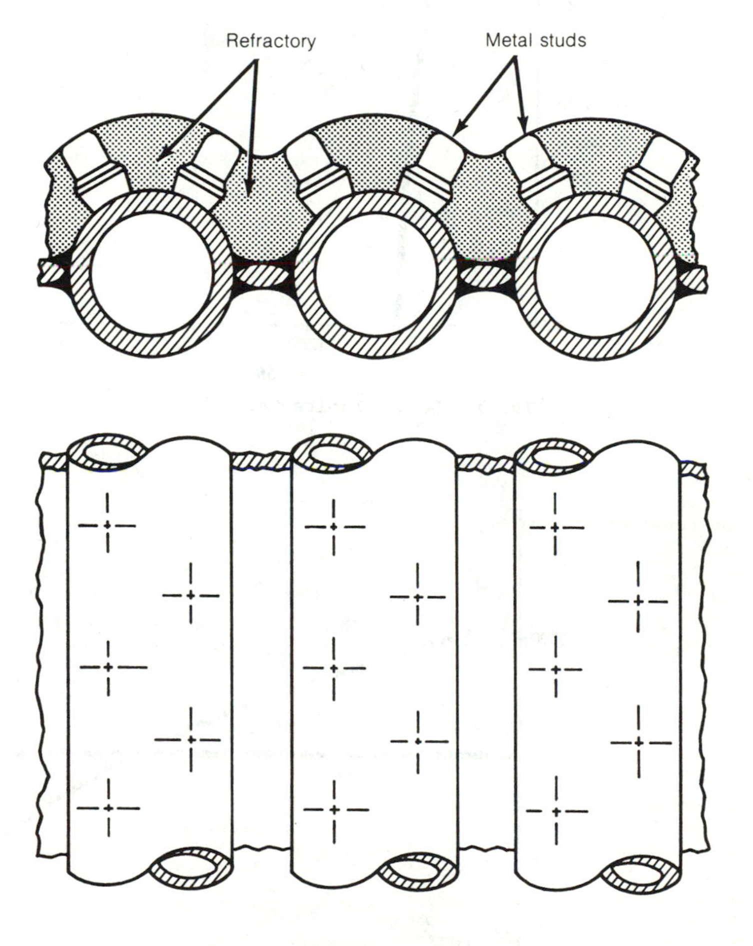

Fig. 4 Studded Membrane Wall

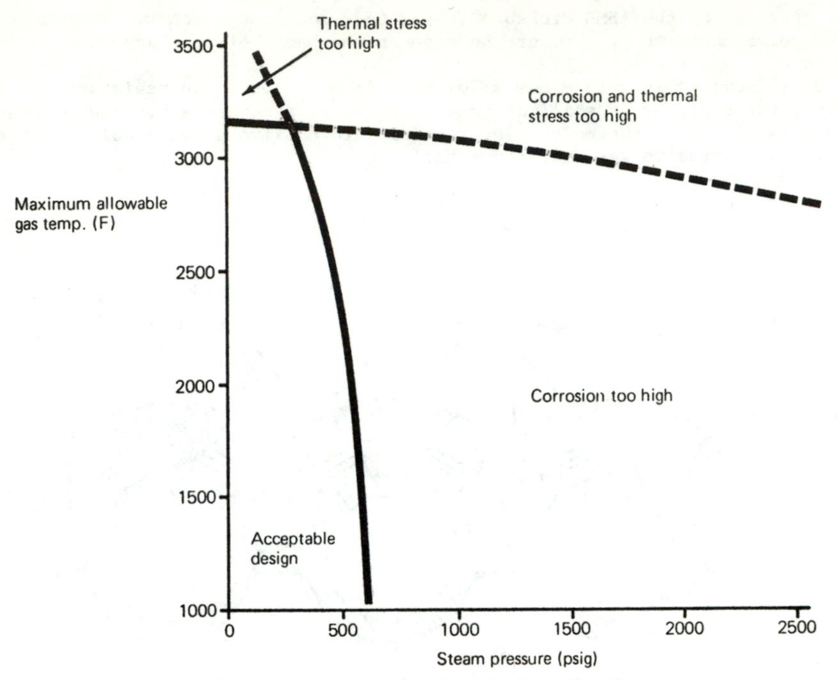

Fig. 5 Design Limits Carbon Steel

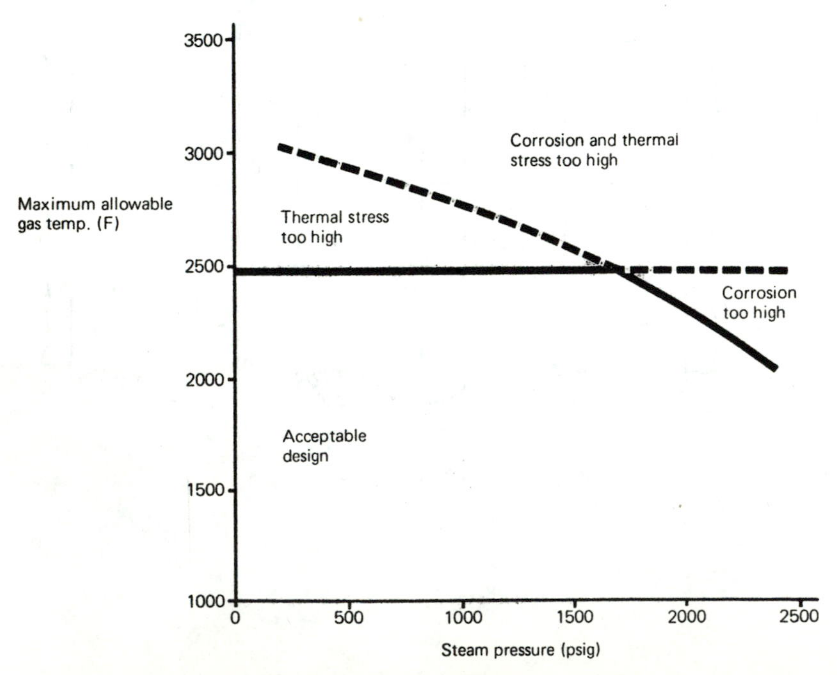

Fig. 6 Design Limits Stainless Steel

1 Eberle, F. and Wylie, R. D., "Attack on Metals by Syntheses Gas from Methane-Oxygen Combustion," Fifteenth Annual Conference, NACE, Chicago, Ill., March 16-20, 1959. Corrosion Vol. 15, December, 1959.

2 Stelling, O. and Vegeby, A., "Corrosion on Tubes in Black Liquor Recovery Boilers," Annual Meeting of the Technical Section, Canadian Pulp and Paper Association, Montreal, Quebec, January 28-31, 1969.

3 Metals Handbook, Vol. 1, Properties and Selection of Metals - 8th Edition, American Society of Metals, p. 268.

4 Paisley, M. A., "Material Study for High Temperature Gasifier Environment," June 2, 1976.

HIGH TEMPERATURE REACTIONS INFLUENCING DEPOSIT FORMATION

Session Chairman: **J. S. Wilson**
ERDA—Morgantown Energy Research Center
Co-Chairman: **R. K. Borio**
Combustion Engineering Inc.

DIRECT MASS SPECTROMETRIC SAMPLING OF CORROSION-RELATED GASEOUS SPECIES FROM LABORATORY PULVERIZED COAL-AIR FLAMES

THOMAS A. MILNE, JACOB E. BEACHEY, and FRANK T. GREENE

Midwest Research Institute
Kansas City, Missouri, USA 64110

ABSTRACT

An attempt to elucidate the nature and role of gaseous alkali and sulfur species in fireside corrosion and ash fouling is described. The approach chosen involves the direct molecular beam mass spectrometry of gaseous species sampled directly from small, aerodynamically simple, premixed, pulverized coal-air flames. Sampling by direct free-jet expansion through small orifices yields rapid quenching, high spatial resolution, and the ability to preserve such species as gaseous KOH and Na. The coal burner, sampling apparatus, and attempts to observe the initial rate of release and form of alkali and sulfur species in the primary flame zone are presented.

INTRODUCTION

In the study of the related problems of fireside corrosion, ash fouling and hot corrosion, much of the past work has focused on the condensed phases at heat transfer surfaces or turbine blades. However, it has often been speculated that gas phase species involving alkali metals, sulfur and perhaps silicon may play an important role (1-4). It has been established that the coal constituent alkalies, alkaline earths, iron and sulfur provide major correlation parameters with corrosion and fouling (5). Unfortunately, these correlations are far from satisfactory for predictive purposes and expensive trials must still be carried out with new power plants and coal sources. Knowledge of the kinetics and extent of the release of alkalies as the coal burns, the nature of the species formed with sulfur and their fate as they progress to various heat transfer or turbine surfaces has been largely inferential.

What is needed is direct observation of the nature and concentration of gaseous alkali and sulfur species from the earliest stage of coal volatiles combustion, through char burning, to flow past the various heat transfer surfaces. The coupling of controlled laboratory flames of pulverized coal and air with direct molecular beam sampling and mass spectrometric detection, which are capable of identifying and measuring very reactive and condensable transport species, appears to be a feasible research approach to these important practical problems. The present program has the twin goals of: (a) identifying the gaseous alkali metal-containing species that might play a role in corrosion and fouling, and (b) developing direct mass spectrometer probes to sample such species from both idealized laboratory flames and small-scale coal combustors. In this paper, we report on: (a) the successful stabilization of premixed, laminar, pulverized coal-air flames on small burners; (b) the sampling methods used to extract and detect stable and reactive species from one atmosphere coal flames; and (c) the preliminary results of a search for alkali and sulfur species in the combustion products of coal-O_2-N_2 flames.

STABILIZATION OF LABORATORY COAL-AIR FLAMES

In safety-related work for the U.S. Bureau of Mines (6-8), it has been demonstrated that both flat and conical premixed flames of coal dust and air can be stabilized on small burners. In studies with bituminous coals with volatile matter ranging from 30 to 35%, stable flames have been achieved on burners ranging from 1 to 12 cm in diameter. With air and dusts in the size range 10 to 20 μm, only flames with compositions richer than stoichiometric have been maintained without augmentation. The keys to achieving such burning, in our experience, appear to be: (a) smooth feeding of the coal dust from a fluidized bed feeder, (b) very turbulent mixing of the coal dust and air at the inlet of a conical burner, (c) a short transition to laminar flow at the mouth of the burner, and (d) vibration of all components contacting the coal dust. Details have been given elsewhere (7).

For the present studies the 6.3 cm diameter burner shown in Figure 1 is being used with 10 to 20 μm Freeport seam coal (29% VM, 21% ash). It has been found necessary to increase the O_2 content of the "air" to about 28% to permit stabilization of an unconfined flame at stoichiometric or lean compositions. Figure 2 shows the appearance of a rich coal-air flame, anchored closely to the honeycomb grid. To the eye, the flame has a striking similarity to a gaseous Meeker burner flame, showing a bright, narrow, primary reaction zone 2 to 3 mm wide in a waffled but merged flame front.

DIRECT MOLECULAR BEAM SAMPLING OF COAL-AIR FLAMES

Flat, premixed coal dust-air flames are being sampled for gaseous neutral species by extension of the well developed techniques of free-jet expansion to molecular flow with mass spectrometric detection of the resultant molecular beam (9). The system designed specifically for the present application is shown in Figure 3. Orifice sizes and pumping capability are matched to achieve a substantially shock free, free-jet expansion through the primary orifice (typically 0.025 to 0.075 cm diameter), reaching essentially collosionless flow before reaching the skimmer. The skimmer and subsequent orifices define a molecular beam which is modulated and passed through the mass spectrometer ion source.

Fig. 1 Photograph of 6.3 cm diameter burner, honeycomb and coal-air inlet being used for present studies.

Fig. 2 Photograph of coal-air flame on 6.3 cm diameter burner. Pittsburgh Seam coal, 10 to 20 μm, at about 200 mg/liter.

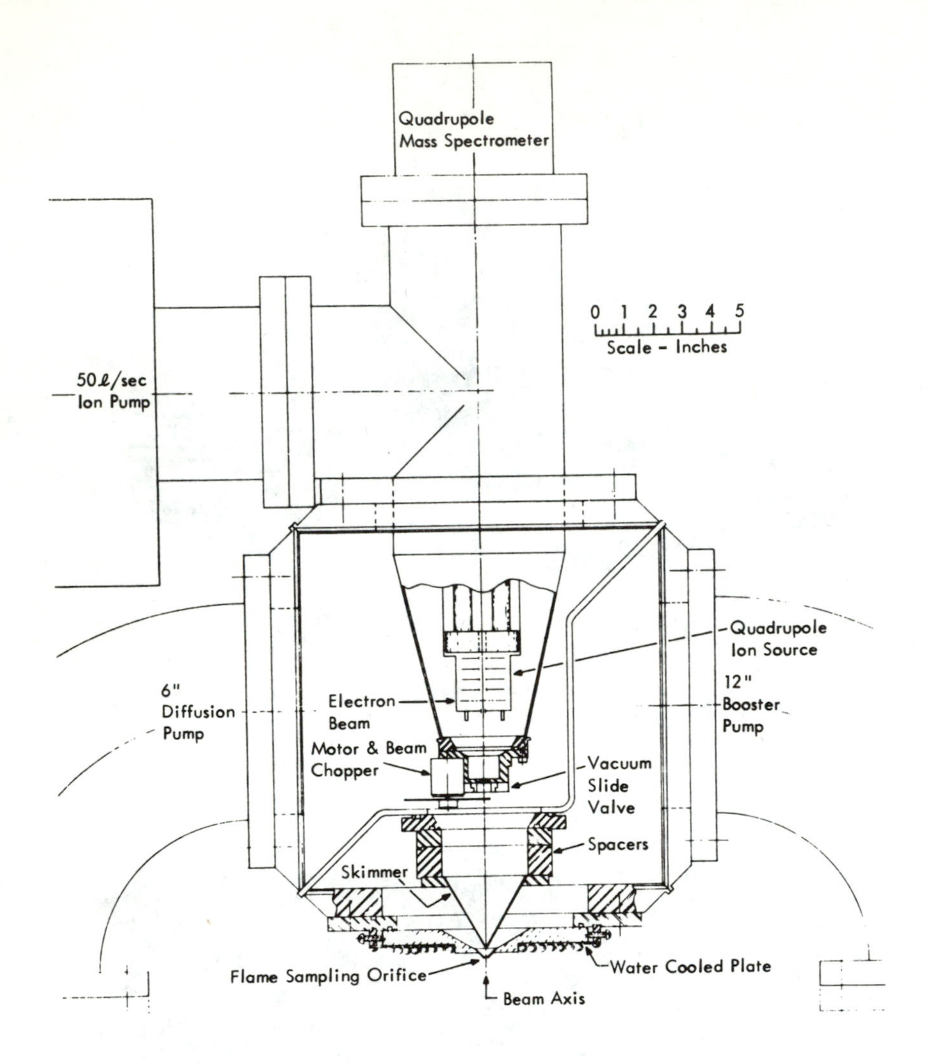

Fig. 3 Scale schematic of direct, free-jet, molecular beam, mass spectrometric sampling systems in the orientation for sampling upright burners. The system can be readily rotated 90 or 180 degrees.

The phenomena accompanying such sampling include mass separation, nucleation, background modulation and metastable noise effects. In the case of coal dust-air flames, the only unique factors would appear to be the problem of orifice plugging by the char and ash and the occurence of random, but phase-locked noise spikes due to passage of high-energy particles through the quadrupole mass spectrometer into the electron multiplier ion-detection region. The orifice plugging is postponed but not eliminated by means of a solenoid activated cleaning wire (7) that repetitively protrudes through the orifice from the inside of the sampling cone tip and by periodically blowing or brushing away soot, char and ash from the exterior of the sampling cone by a small air jet. The noise caused by particle impact can, in principle, be eliminated by structural changes in the multiplier region.

The kinds of results that a combination of aerodynamically simple flames and direct sampling of particulates and gases can provide are shown in the following figures. The mass spectrometric profiles were obtained with a system like that of Figure 3, but with a two-stage primary sampling-expansion system instead of the single-stage system shown. This system, a precursor to the system designed for this work, is shown in Figure 4. Such a system was expected to adequately quench and preserve relatively stable species like SO_2 and NO but would not likely preserve condensable species such as NaOH(g) or K_2SO_4(g). Figure 5 shows profiles of the major combustion species. In order to detect H_2, hydrocarbons and CO, without the troublesome interference problems in low resolution mass spectrometry, samples have been collected for gas chromatography with typical results shown in Figure 6. In Figure 7 data are shown on the production of a major pollutant species, NO, early in the primary reaction zone. Figure 8 shows the behavior of volatile matter and fixed carbon early in the reaction zone, as determined from standard prominate analysis of sampled particles. Figure 9 indicates the appearance of representative particles sampled from a rich flame. Figure 10 gives a typical uncorrected temperature profile through these flat flames. All these data were for rich Pittsburgh Seam coal flames using a 0.075 cm sampling orifice. Details of sampling and more extensive data are given in Reference 8.

The microstructure and behavior of laminar premixed pulverized coal-air flames can be summarized as follows: (a) small coal-air flames are easy to stabilize provided dust dispersion is uniform, (b) quenching diameters for 10 to 20 µm Pittsburgh Seam coal can be as small as 8 mm, (c) a narrow, bright initial reaction zone, 2 to 3 mm wide, is clearly visible, (d) the bright reaction zone involves mainly volatiles combustion, (e) the extent of pyrolysis at the start of O_2 consumption is small, (f) heating rates of the order 10^4°C/sec occur in the volatiles combustion zone, (g) small sonic-orifice probes permit sampling of particles and gases with spatial resolution of 1 to 2 mm and time resolution of a few milliseconds, and (h) coal and char particles sampled from the bright reaction zone show early softening and blow-holes.

SEARCH FOR ALKALI AND SULFUR SPECIES IN COAL-AIR FLAMES

In previous work, we have obtained two kinds of results. First, coal dust-air flames, sampled by a two-stage initial expansion, gave profiles for relatively stable species, as shown above (7,8). Second, CH_4-air flames, seeded with small amounts of dry-powder inhibiting agents such as $KHCO_3$, were sampled using quite small orifices (0.025 cm diameter) and a single stage free-jet expansion. In such a system, the species K(g) and KOH(g) were observed to be evaporating from the dust in or near the primary reaction zone. Typical electron impact (50 ev) positive ion mass spectral results are shown in Figures 11 and 12. The K^+ observed was presumably derived from K(g) or KOH(g). KOH^+ was also observed early in the reaction zone. With $NH_4H_2PO_4$ powder added to these flames, similar profiles for P^+ and PO^+ were observed. Details are given in Reference 12.

The system shown in Figure 3, by providing more pumping speed, was designed to permit the detection of such species during sampling of the more hostile environment of pulverized coal-air flames. The new system has been tested for performance, sensitivity and interferences, with cold gases, seeded and unseeded CH_4-air flames and coal dust-air flames. The preliminary results are summarized below.

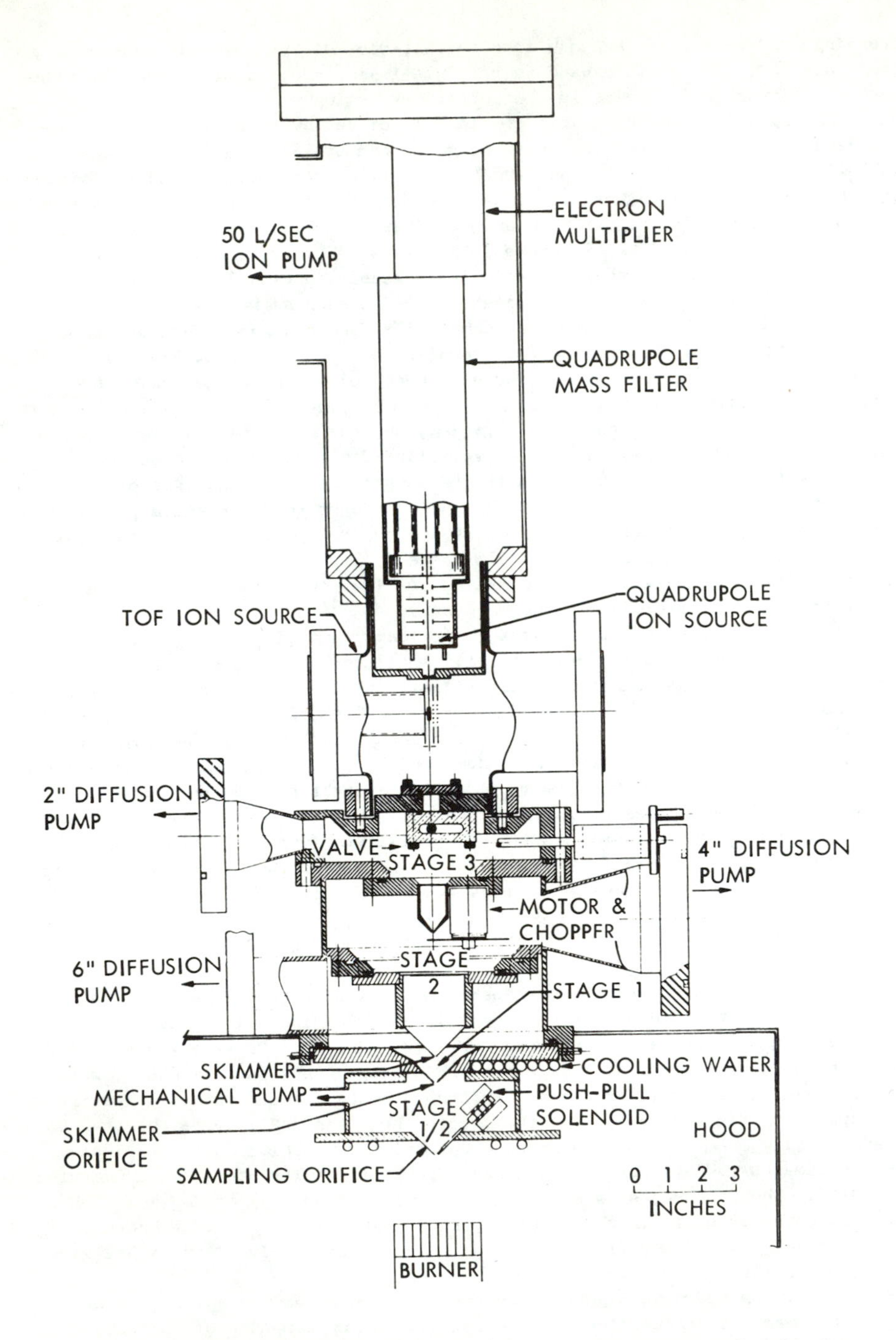

Fig. 4 Scale schematic of two-stage expansion, sonic orifice, molecular beam, mass spectrometric sampling system with mechanical orifice cleaner.

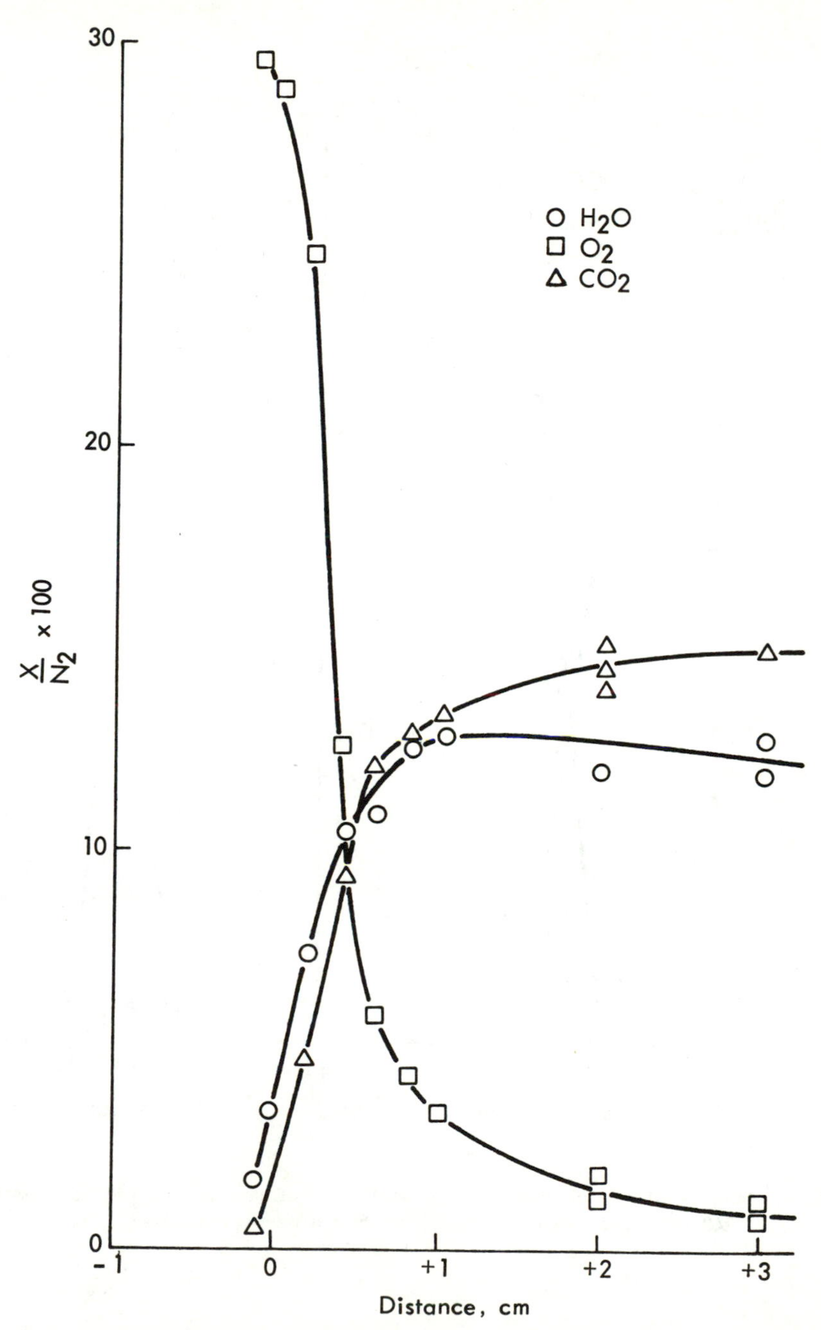

Fig. 5 Species profiles through a flat, 6.3 cm diameter flame of 10 to 20 µm Pittsburgh Seam coal at 146 mg/liter. Distance is measured from probe tip to burner grid. Two-stage expansion, through 0.075 cm diameter orifices, to molecular flow.

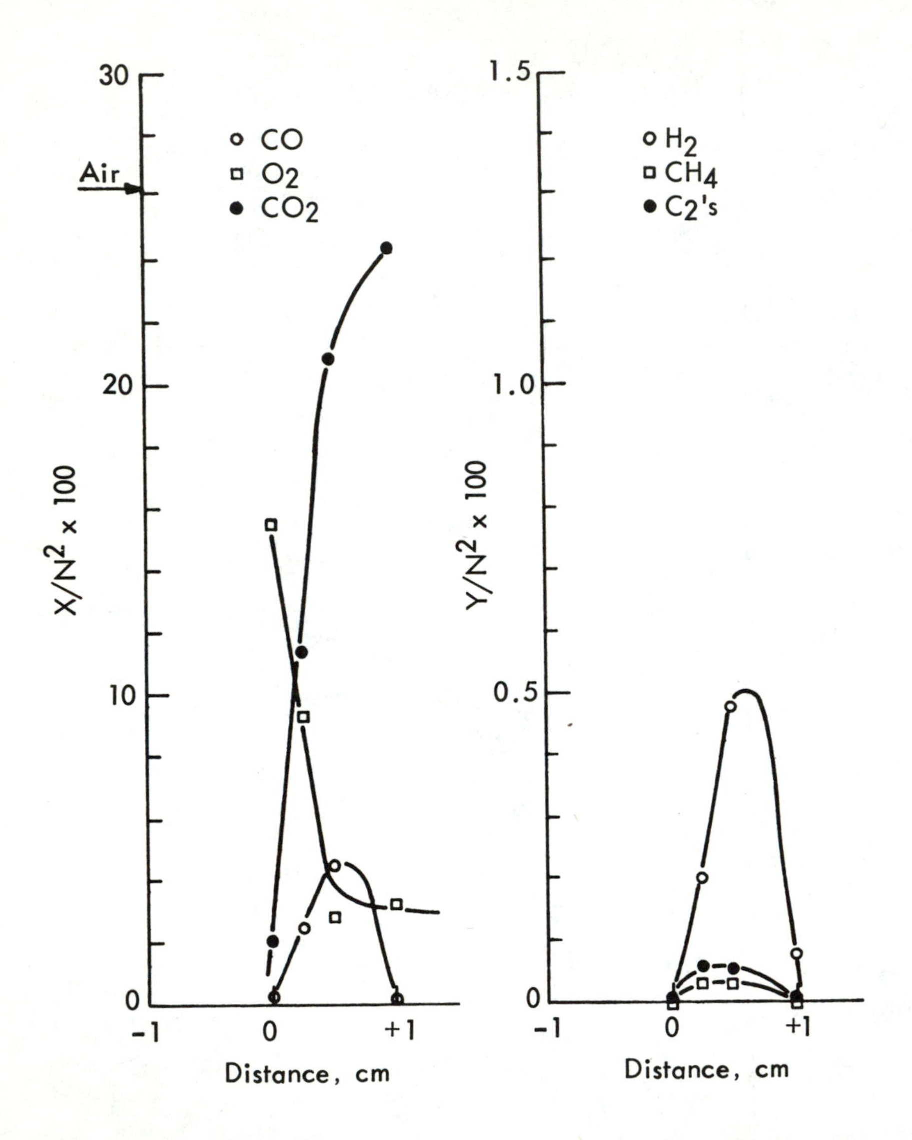

Fig. 6 Results of gas chromatographic analysis of samples collected from a 143 mg/liter, 10 to 20 μm, Pittsburgh Seam coal-air flame.

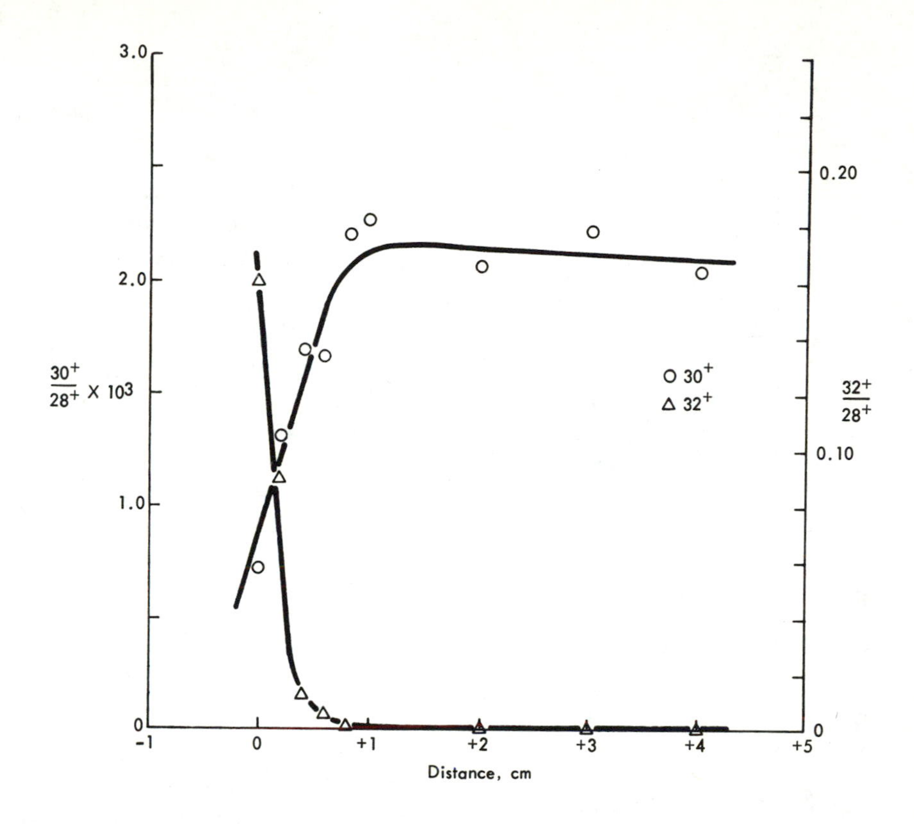

Fig. 7 Mass spectrometric profile of NO and O_2 through 208 mg/liter Pittsburgh Seam coal-air flame.

Permanent Gases and Air

Measurements of nucleation clusters (11) in sampling pure argon and air at one atmosphere indicate that a 0.050 cm diameter sampling orifice produces a relatively undisturbed free-jet, with stage one pressures up to 4 to 5 Pascales (30 to 40 μ). Normal constituents in air such as Kr^{84} can be detected in tens of seconds, indicating a sampling sensitivity of 1/2 ppm or less.

Gaseous Flames

Direct sampling of the burnt gases of CH_4-air flames indicates that in spite of lower beam intensities than for cold gas and a higher contribution of phase-locked noise at all masses from metastables, a detection capability of 2 to 4 ppm is readily achievable. For the hot gases, no evidence of clustering remains except perhaps a small contribution at 19^+ from $(H_2O)_2$.

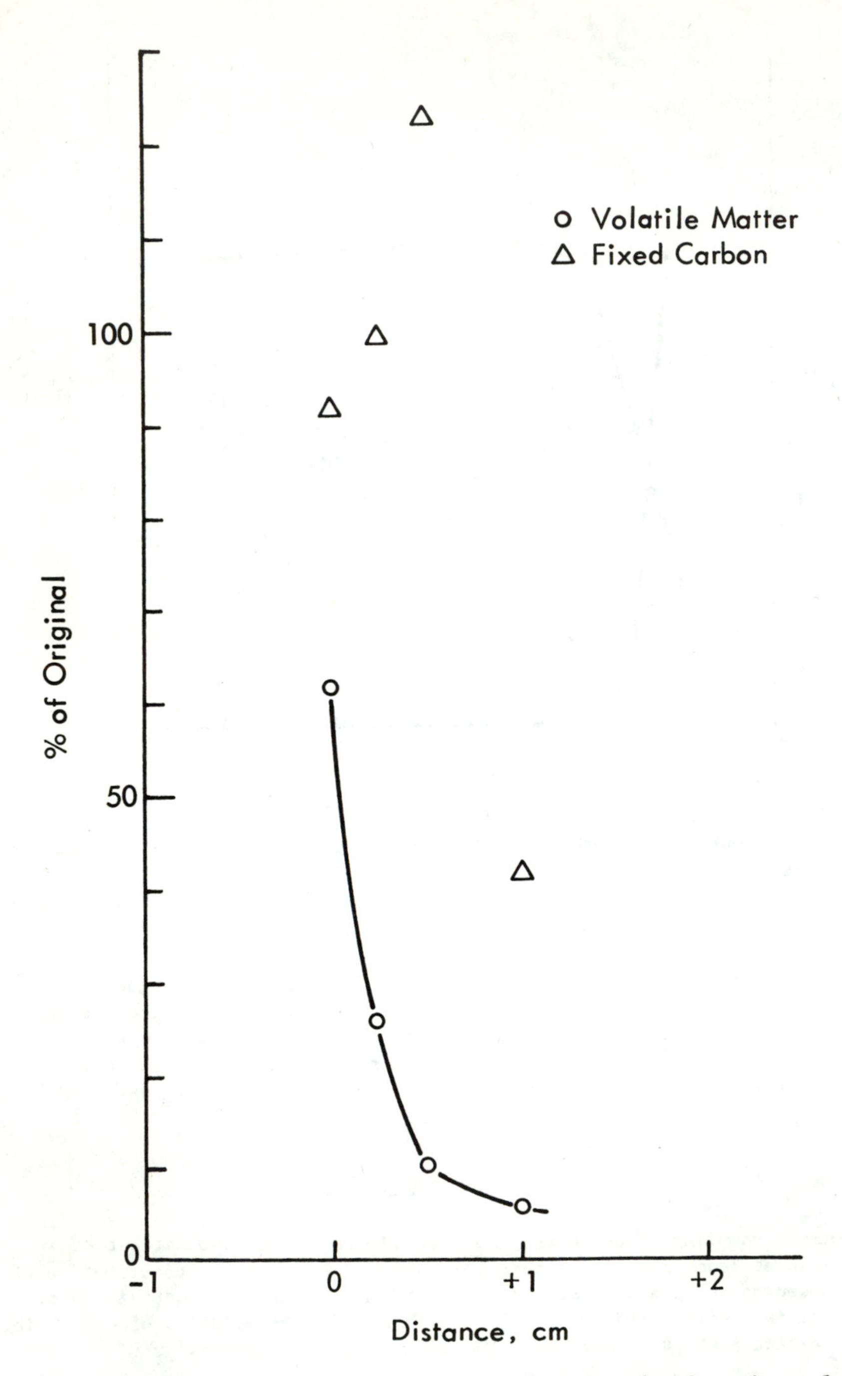

Fig. 8 Proximate analysis results for particulates samples from the early region of an unsieved Pittsburgh Seam coal-air flame, 214 mg/liter.

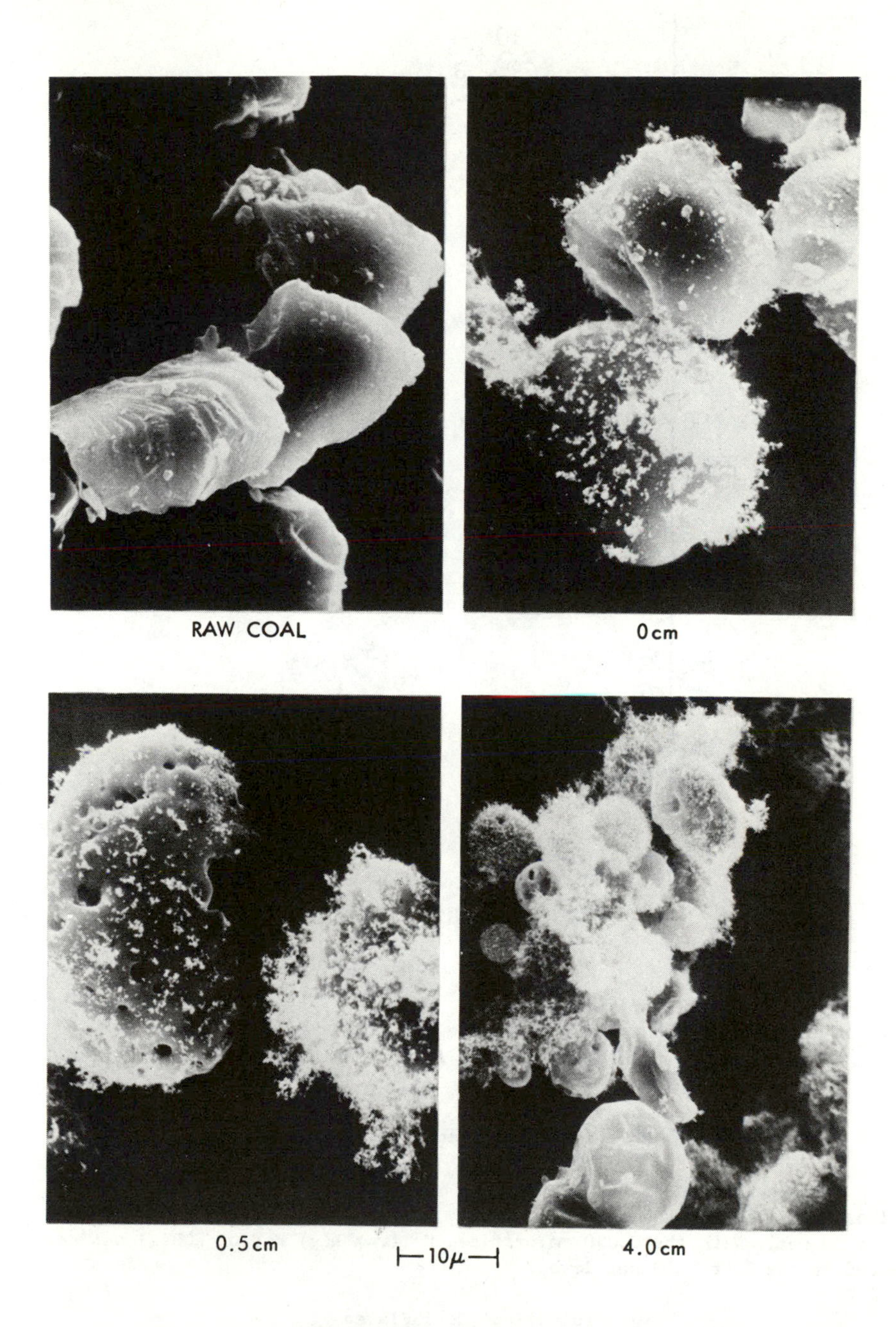

Fig. 9 Scanning electron microscope photographs of typical particles sampled from a 10 to 20 μm Pittsburgh coal-air flame. Coal concentration was 176 mg/liter.

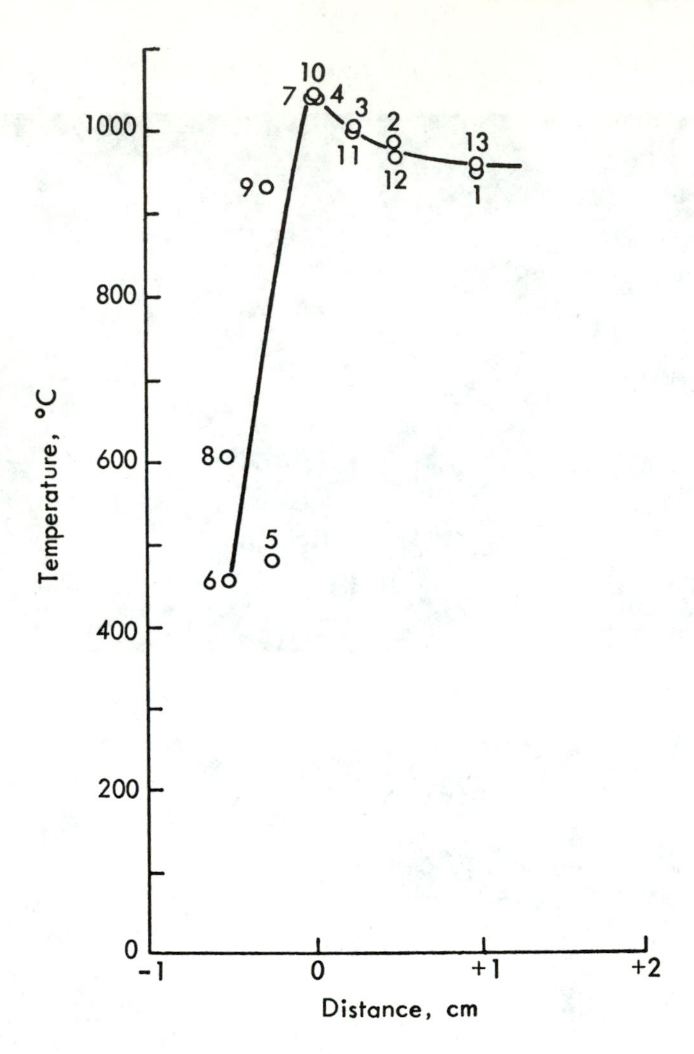

Fig. 10 Thermocouple temperature profile for a 210 mg/liter, 10 to 20 μm, Pittsburgh Seam coal dust-air flame. Numbers indicate order in which data points were taken.

Gaseous Flames Seeded with Powder

With powder loadings of about 3 to 4 mg/liter, for 37 μm $KHCO_3$, noise spikes appear across the entire spectrum, limiting detection sensitivity to about 10 ppm. With the 0.050 cm orifice, K^+ from K(g) and/or KOH(g) was detected at the 10 to 100 ppm level.

Pulverized Coal Flames

Molecular beam strengths are comparable to those in CH_4-air flames; however, noise spikes due to char or ash appear to be much worse than those observed in the seeded CH_4-air flames. In upright and inverted flames of 10 to 20 μm Freeport seam coal in "air" with 28% O_2, it was possible to burn at, or

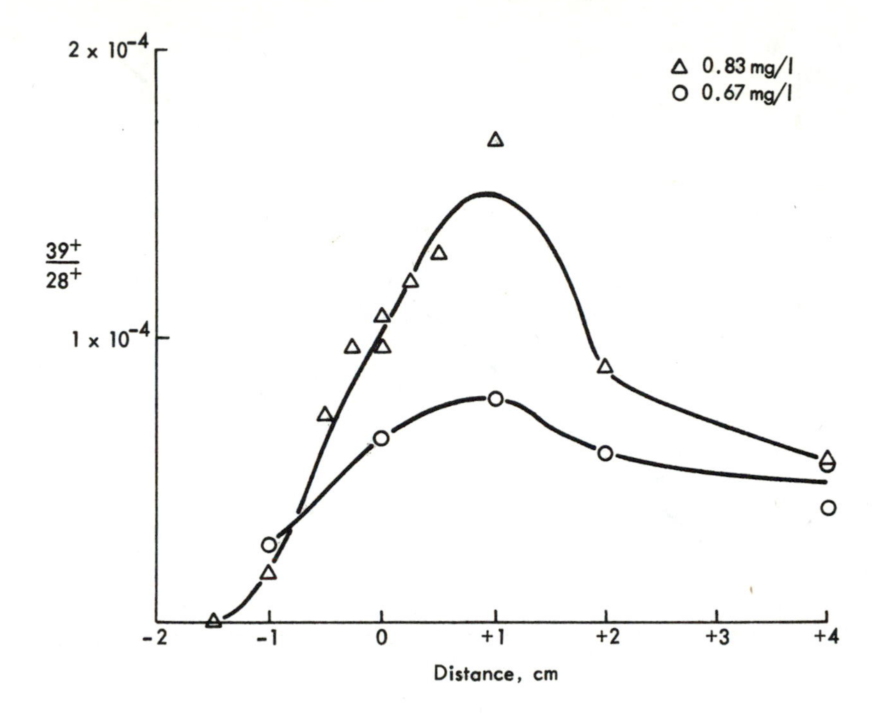

Fig. 11 Variation of $39^+/28^+$ ratio through a 0.9 equivalence ratio CH_4-air flame at two powder loadings of $KHCO_3$. Positive ion mass spectra at 50 ev. The reaction zone is at 0 probe distance.

slightly on the lean side of, stoichiometric. Evidence of a K^+ signal at the level of 10 to 100 ppm of neutrals was obtained in the first tests with such flames. These measurements were severely hampered by the large phase-locked signal at all masses which was presumably due to particle effects. Hence, attempts are being made to eliminate this source of indiscriminate phase-locked signal by multiplier modifications. Currently, studies designed to obtain definitive profiles of alkali, sulfur and stable combustion products species through a considerable region of the burnt gases of lean flames are underway using upright burners placed under the sampling system in the configuration of Figure 3.

DIRECTION OF FUTURE WORK

The primary emphasis of the present work is on analytical technique development rather than on a detailed survey of corrosion species behavior under different combustion conditions for different coals. Hence, as soon as satisfactory sampling, quenching and detection of alkali species is obtained from the coal-air flame, emphasis will shift to the feasibility of designing a geometrically restricted molecular beam sampling probe capable of insertion through thick firebrick walls of small-scale laboratory combustors such as the Morgantown Energy Research Center 10 lb/hr combustor.

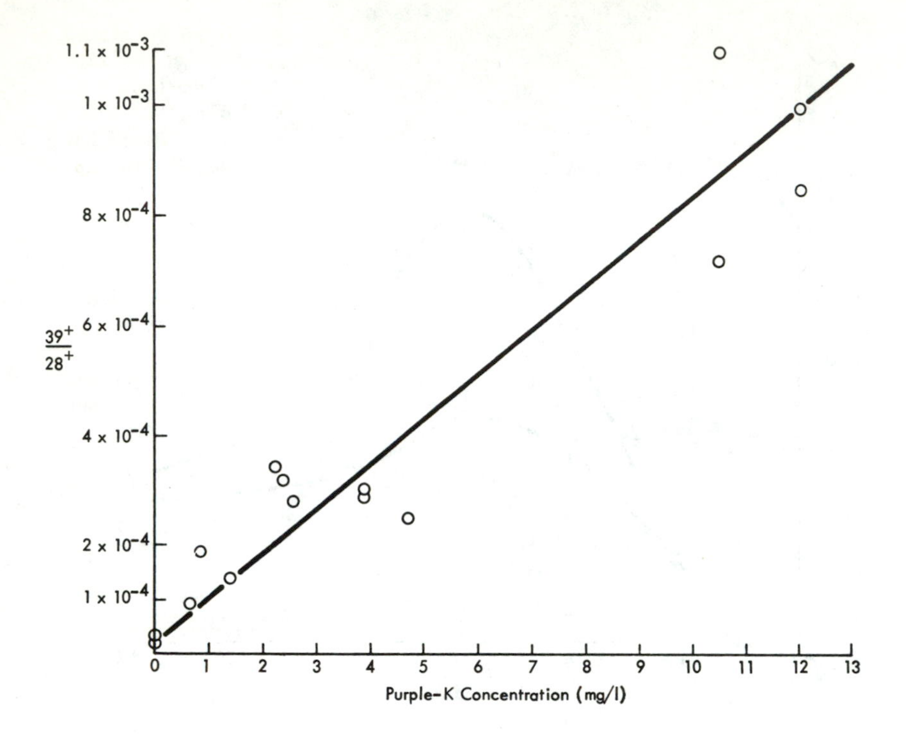

Fig. 12 Variation of the $39^{+}/28^{+}$ ratio observed 1 cm downstream from a 0.9 equivalence ratio CH_4-air flame to which $KHCO_3$ was added.

ACKNOWLEDGEMENTS

This work is supported by the Energy Research and Development Agency, under Contract No. E(49-18)-2288. Mr. George Cobb has helped with all phases of the experimental program and Mr. Walt Hodge designed many features of the sampling system. Consultation with Dr. John W. Wilson at Morgantown Energy Research Center and Dr. Andrej Macek of ERDA is gratefully acknowledged.

REFERENCES

1 Reid, T., "External Corrosion and Deposits, Boilers and Gas Turbines," American Elsevier, New York, 1971.

2 Wilson, J. S., and Redifer, M. W., Trans. ASME J. Eng. Power, April 1974, p. 145.

3 Kohl, F. J., Stearns, C. A., and Fryburg, G. C., "Sodium Sulfate: Vaporization Thermodynamics and Role in Corrosive Flames," International Symposium on Metal-Slag-Gas Reactions and Processes, Toronto, Canada, The Electrochemical Society, Princeton, New Jersey, 1975, p. 649.

4 Ulrich, G. D., "Investigation of the Mechanism of Fly-Ash Formation in Coal-Fired Utility Boilers," Interim Report, February to March 1976, FE-2205-1, May 28, 1976.

5 Borio, R. W., et al., "The Control of High-Temperature Fireside Corrosion in Utility Coal-Fired Boilers," OCR R&D Report No. 41, PB 183716, April 1969.

6 Milne, T. A., and Beachey, J. E., "Laboratory Studies of the Combustion, Inhibition, and Quenching of Coal Dust-Air Mixtures," Paper preprinted and presented at the Western States Section/The Combustion Institute, 1976 Spring Meeting, April 19 and 20, Salt Lake City, Utah.

7 Milne, T. A., and Beachey, J. E., "The Microstructure of Pulverized Coal-Air Flames, I. Stabilization on Small Burners and Direct Sampling Techniques," Paper to be published in Combustion Science and Technology, 1977.

8 Milne, T. A., and Beachey, J. E., "The Microstructure of Pulverized Coal-Air Flames, II. Gaseous Species, Particulate and Temperature Profiles," Paper to be published in Combustion Science and Technology, 1977.

9 Milne, T. A., and Greene, F. T., "Molecular Beams in High Temperature Chemistry," in Leroy Eyring (ed.), Advances in High Temperature Chemistry, New York: Academic Press, Vol. 2, 1969.

10 Milne, T. A., and Greene, F. T., "Exploratory Studies of Flame and Explosion Quenching," Summary Technical Progress Report, June 1976, U.S. Bureau of Mines Contract No. H0122127.

11 Milne, T. A., and Greene, F. T., J. Chem. Phys., Vol. 47, 1967, p. 4095.

CHEMICAL TRANSFORMATIONS OF THE MINERALS IN EASTERN BITUMINOUS COALS UNDER SIMULATED PULVERIZED COAL FIRING CONDITIONS

CHARTER D. STINESPRING, MICHAEL ZULKOSKI, and MARGARET H. MAZZA

Energy Research and Development Administration
Morgantown Energy Research Center
Morgantown, West Virginia, USA 26505

ABSTRACT

The transformation of coal minerals to ash under simulated pulverized coal firing conditions has been investigated for seven eastern bituminous coals. These coals were fired in a bench scale combustor and the resulting ash and deposits, as well as the original coals, were analyzed using x-ray diffraction and fluorescence. The transformations observed in this manner appear to be thermodynamically controlled. Evidence suggesting the existence of gas phase reactions leading to condensible sodium compounds has also been observed.

INTRODUCTION

Even in well beneficiated coals, the existence of inherent minerals as small inclusions in the coal matrix is well established. In coal conversion processes, the high temperature transformation of these minerals, extraneous as well as inherent, to ash is important due to the relation of ash composition to fouling, slagging, and corrosion problems, process stream cleanup procedures, and overall conversion efficiency. The objective of this study was to characterize the transformation of coal minerals to ash under simulated pulverized coal firing conditions for seven eastern bituminous coals. The basic experiment upon which this report is based consisted of firing these coals in a bench scale combustor and sampling the ash and deposits which were formed. These samples were then subjected to x-ray diffraction (XRD) and x-ray fluorescence (XRF) analysis to obtain their mineralogical and elemental composition.

The identification of coal minerals and their chemical transformations at high temperatures have received much attention (1-5). A list of the more common coal minerals is given in table 1. Typically the high temperature transformation work has consisted of thermogravimetric analysis (TGA) and differential thermal analysis (DTA) studies in which the coal minerals were heated slowly (6) or XRD-XRF studies in which the coal minerals were heated at a slow rate and then analyzed on a hot stage or quenched and then analyzed (7). The results of such a study are shown in table 2. The transformations observed in these TGA-DTA or XRD-XRF studies are apparently slow processes with respect to chemical kinetics, and it is not immediately clear that they should correspond to the more rapid transformations occurring in actual conversion processes. With regard to this question, a recent study by Padia, Sarofim and Howard (8) in which pulverized coal was heated to 1500 C at a rate of 10^5 C/sec, quenched, and then analyzed

Table 1

Characteristic Coal Minerals (9)

Classification	Mineral Species	Chemical Formula
Aluminosilicates	Kaolinite	$Al_2Si_2O_5(OH)_4$
	Illite	$K(Si_3{\bullet}Al)Al_3O_{10}(OH)_4$
Carbonates	Siderite	$FeCO_3$
	Calcite	$CaCO_3$
	Dolomite	$CaMg(CO_3)_2$
Sulfides	Pyrite and marcasite	FeS_2
Silica	Quartz	SiO_2
Accessory Minerals	Feldspar	$CaAl_2Si_2O_8$ $(Na,K)\ AlSi_3O_8$
	Gypsum	$CaSO_4{\bullet}2H_2O$

Table 2

Mineral Transformations due to Heating (7)

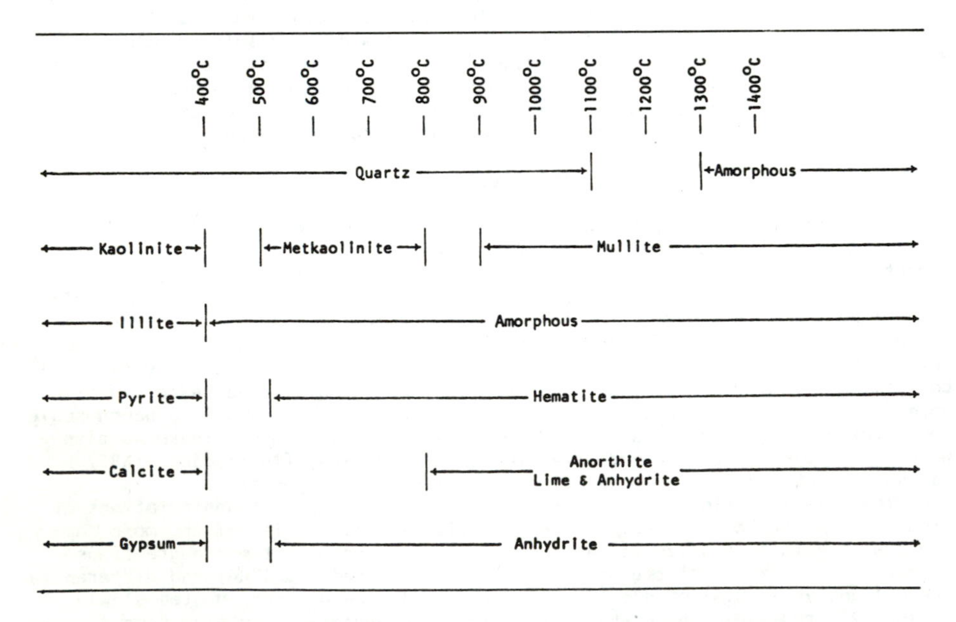

using XRD and XRF techniques has yielded results consistent with those described in table 2. Thus, the high temperature mineral transformations associated with pulverized coal combustion appear to be thermodynamically rather than kinetically controlled. With respect to this point, the work reported here may be considered a scale-up of that performed by Padia, Sarofim and Howard.

THE COMBUSTOR, RUN CONDITIONS AND AUXILIARY EQUIPMENT

The combustor used in this study was designed to be large enough to simulate conditions in a pulverized coal fired boiler but small enough to be easily controlled and monitored. In the combustor, coal ground to 70 percent through a 200 mesh sieve was fired at a rate of 1.25×10^{-3} kg/sec (10 lb/hr). The stoichiometry was maintained at 3 percent excess oxygen, and the combustion zone temperature was 1400 C. The combustor and sampling port locations are shown in figures 1 and 2.

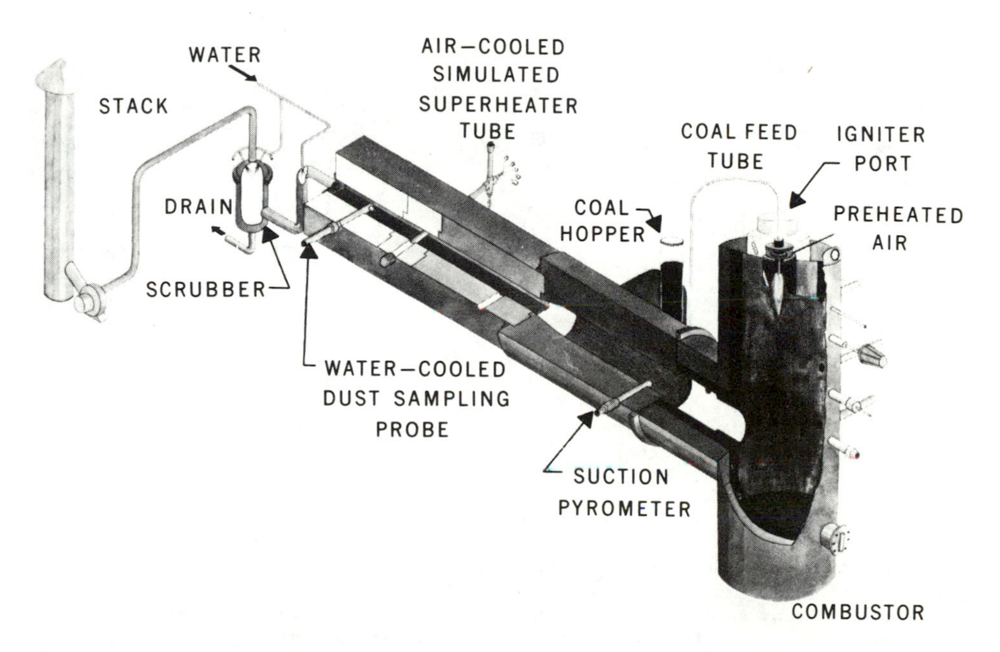

Fig. 1. The Coal-Fired Combustor Used in Transformation Study

The combustion gas was analyzed on line for CO and CO_2 using infra-red techniques, SO_2 using flame photometric techniques, and O_2 using paramagnetic techniques. Characteristic concentrations were 0.1 percent CO, 10 percent CO_2, 0.1 percent SO_2, and 3 percent O_2.

Temperature measurements were obtained using thermocouples in the refractory walls and supplemented by optical pyrometer measurements in the combustion zone. Typically temperatures were in the range of 1400 C in the combustion zone, 950 C near port V3 and 900 C near port H3 (see figure 2).

A water-cooled suction probe was used to sample the ash entrained in the gas stream. In this report ash samples taken at ports V3 and H2 are considered. Ash deposits were also allowed to build up on a simulated superheater tube located at port HP. The temperature of this probe was maintained at 650 C during four to six hour exposures to the gas stream. Usually this probe was covered with a fine white powder within a few minutes after insertion into the gas stream. This was followed by a gradual build up of the ash deposits. Frequently this deposit would drop off and another would develop in its place.

ANALYTICAL TECHNIQUES

The elemental and mineralogical composition of the suction probe ash, superheater tube deposits, and feed coals were determined using XRD and XRF

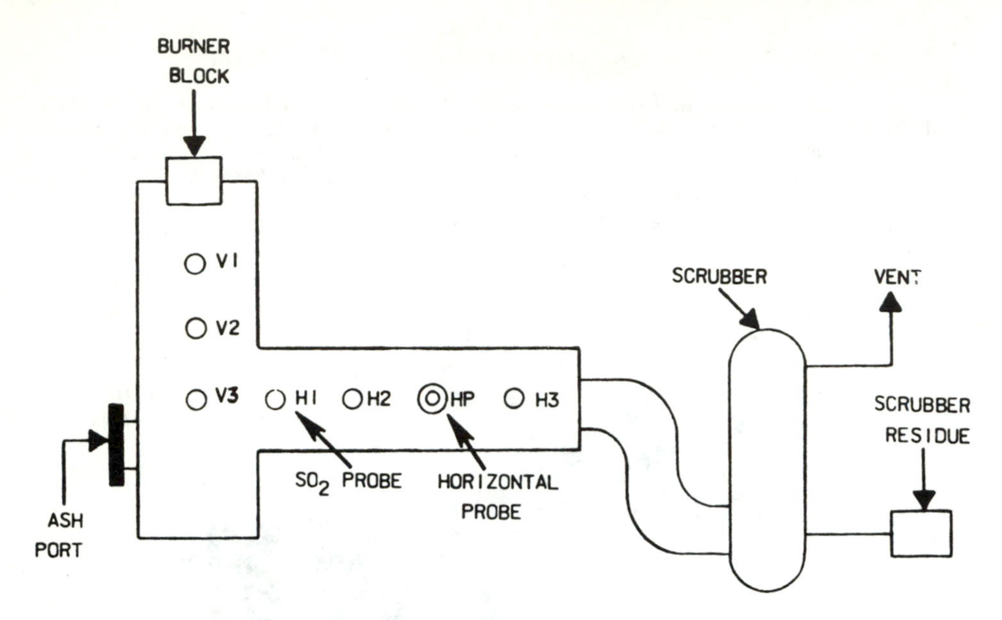

Fig. 2. Sampling Port Locations for the Coal-Fired Combustor

analysis. These techniques were developed as a part of the work reported here and have been published elsewhere (9).

The general sequence of events for the XRD analysis was as follows:

Grind → LTA → Grind → Slurry, Pack and Dry → XRD Analysis

The mineral species were separated from the organic matrix by low temperature ashing (LTA) each sample. This was accomplished by producing an oxygen plasma over the sample by exposing the oxygen to a 13.56 MHz RF field. In this manner the organic portion of the sample was oxidized while maintaining the sample temperature at 150 C. Prior to the LTA, samples were ground to increase the surface area exposed to the oxygen plasma. After the LTA, the samples were ground to pass through a 325 mesh sieve and packed into sample holders as a slurry with amyl acetate. After drying, the holders were loaded into the XRD unit, a Philips ADP-3500[1], and the data was collected. The uncertainty in this data was estimated at ±10 percent.

Sample preparation for the XRF analysis may be summarized as follows:

Form Solid Solution with $Li_2B_4O_7$ → Polish and Clean → XRF Analysis

The solid solution of sample with lithium tetraborate was formed by fusing a mixture of one part sample to thirteen parts lithium tetraborate in a Pt/Au crucible. The mixture was heated slowly during the devolatilization of the sample and then heated rapidly to 1100 C. (This procedure is later referred to as high temperature ashing--HTA.) The fused mixture was maintained at this temperature for 10 minutes and continuously mixed during this time. After cooling in air for four minutes, the sample, now in the form of a glass bead, was polished and ultrasonically cleaned. These samples were then analyzed on a Philips PW-1412[1] XRF unit. The uncertainty in the XRF data was estimated at ±2 percent.

[1] Equipment brand names are mentioned for information only and do not represent endorsement by the Federal Government.

COALS STUDIED

Seven eastern bituminous coals were studied in this investigation. Included among these were five coals mined in north central West Virginia: two Pittsburgh seam coals (denoted A and B in this report), a Waynesburg seam coal, a Redstone seam coal, and a Sewickley seam coal. A Freeport seam coal mined in central Pennsylvania and a coal blended from Harrisburg-Springfield No. 5 and Herrin No. 6 seams (denoted Illinois 5 and 6 in this report) were also studied. Analyses for these coals are given in tables 3, 4 and 5.

From the data in table 3, it may be seen that these eastern coals have a relatively high sulfur, ash and volatiles content. Although the data is incomplete, it seems generally true that pyritic sulfur represents half of the total sulfur. From table 4, it is seen that the minerals appearing in major quantities are the aluminosilicates, kaolinite and illite, followed by quartz and pyrite. The associated minerals, feldspar and gypsum, appear only in minor quantities, and the calcite content is significant only for the Pittsburgh A, Redstone and Illinois 5 and 6 seam coals.

Table 3

Analyses of Some Eastern Bituminous Feed Coals

	Moisture	Carbon	Hydrogen	Pyritic sulfur	Total sulfur	Volatile matter	Ash
Pittsburgh A	1.0	57.8	5.0	1.3	2.6	33.7	7.5
Pittsburgh B	0.9	50.3	5.3		1.0	30.1	10.7
Waynesburg	0.9	49.2	4.0	0.6	1.1	26.3	23.7
Redstone	0.8	56.0	5.2		1.4	30.3	13.0
Sewickley	0.8	50.6	4.9		2.3	30.3	18.3
Freeport	0.9	54.6	5.6	0.7	1.9	28.3	16.2
Illinois 5 & 6	3.3	49.7	5.1	1.2	3.5	35.0	12.5

Note: All numbers are percent by weight of total coal.

Table 4

Mineral Content of Some Eastern Bituminous Coals as Determined by X-Ray Diffraction of the Low Temperature Ash of Each Coal

	Illite	Kaolinite	Calcite	Pyrite	Quartz	Feldspar	Gypsum
Pittsburgh A	44	27	3	21	5	Trace	Trace
Pittsburgh B	31	48	Not Det.	10	11	Trace	Trace
Waynesburg	35	38	Not Det.	12	15	0.3	Trace
Redstone	40	30	5	10	15	Trace	Trace
Sewickley	36	30	Not Det.	19	15	Trace	Trace
Freeport	40	30	Trace	20	10	Trace	Trace
Illinois 5 & 6	31	27	5	27	10	Trace	Trace

Note: All numbers are percent by weight of LTA.

Table 5

Elemental Composition of Some Eastern Bituminous Coal Minerals (Reported as Elemental Oxides) as Determined by X-Ray Fluorescence of the High Temperature Ash of Each Coal

	SiO_2	Al_2O_3	Fe_2O_3	CaO	MgO	Na_2O	K_2O	P_2O_5	Ti_2O	SO_3
Pittsburgh A	42.0	18.1	20.3	7.1	3.5	3.3	1.5	0.7	0.9	2.6
Pittsburgh B	60.1	27.3	5.5	Trace	Trace	2.9	2.1	0.0	1.6	0.5
Waynesburg	56.2	27.3	7.8	Trace	0.9	2.9	2.9	0.1	1.3	0.6
Redstone	55.4	21.0	11.5	4.2	1.5	Trace	2.2	0.4	1.1	1.7
Sewickley	59.9	23.2	10.5	0.5	0.8	Trace	2.6	0.2	1.3	0.5
Freeport	53.2	21.6	12.2	1.5	6.5	0.3	2.2	0.1	1.4	1.0
Illinois 5 & 6	46.3	17.2	16.4	6.3	1.0	0.7	1.6	0.3	0.8	9.4

Note: All numbers are percent by weight of HTA.

The alkali bearing minerals are fledspar (Na^+, K^+) and illite (K^+). Based on the apparent relative abundance of feldspar versus illite, it appears that potassium should be the dominant alkali species. As seen in table 5 above, however, this need not be true. This discrepancy may be explained by the fact that feldspar, as it occurs in these eastern bituminous coals, may possess a structural phase with a relatively low degree of crystallinity. Moreover, since the elemental composition of ash and deposits should be similar to the elemental composition of the HTA of the parent coal, it appears that the ash and deposits of both the Redstone and Sewickley coals should be low in sodium. It will be seen later, however, that this is not true.

HIGH TEMPERATURE MINERAL TRANSFORMATIONS

Tables 6 through 12 give the mineralogical and elemental composition of the suction probe ash and simulated superheater tube deposit samples. In terms of minerals present and non-present, the composition of the ash and deposits is consistent with the 1400 C mineral transformations depicted in figure 2. The occurence of magnetite, in addition to the expected hematite, is thought to be due to the decomposition of pyrite in the reducing environment provided by the char particle. Moreover, quartz, mullite, lime, anhydrite, and the sum of the magnetite and hematite appear to remain relatively consistent and are uniformly distributed in the ash and deposits.

In the case of the iron bearing minerals, magnetite and hematite, the effects of residence time are observable. Specifically, ash samples taken at port H2 had a longer residence time than those taken at port V3. Likewise a deposit taken from the simulated superheater tube would have a much longer residence time than ash samples from H2 and V3. With this in mind, it is seen that generally the ratio of hematite to magnetite increases with residence time.

As noted above, based on the elemental analysis of the HTA of the coals, both Redstone and Sewickley coals should produce ash and deposits low in sodium. However, based on the elemental analysis of the actual ash and deposits (tables 9 and 10) this need not be true. Although no proof can be supplied at the present time, this suggests the possibility that volatilizable sodium may be reacting in the gas phase to produce compounds which then condense on the ash particles.

Table 6

Mineralogical and Elemental Composition of Ash and Deposits Extracted from Pulverized Fuel Combustor

Coal Seam: Pittsburgh A Additives: None

	Mineral Composition as determined by XRD (pct. by wt. of LTA)						Elemental Composition as determined by XRF (pct. by wt. of HTA)							
Probe type and location[1]	$2SiO_2 \cdot 3Al_2O_3$ Mullite	SiO_2 Quartz	Fe_3O_4 Magnetite	Fe_2O_3 Hematite	CaO Lime	$CaSO_4$ Anhydrite	SiO_2	Al_2O_3	Fe_2O_3	CaO	MgO	Na_2O	K_2O	SO_3
Suction-V3	3	12	12	2	Trace	0	38.8	18.2	29.3	7.5	1.7	1.4	1.4	1.7
Suction-H2	10	12	9	6	Trace	8	42.3	23.1	20.4	8.2	1.8	1.7	1.6	0.9
Front deposit	9	11	4	9	Trace	20	33.8	15.8	39.6	6.9	1.2	0.5	1.1	1.1
Back deposit	12	11	6	8	Trace	27	41.8	21.8	24.1	7.2	1.3	1.0	1.4	1.4

[1] Refer to figure 2 for location reference.

Table 7

Mineralogical and Elemental Composition of Ash and Deposits Extracted from Pulverized Fuel Combustor

Coal Seam: Pittsburgh B Additives: None

	Mineral Composition as determined by XRD (pct. by wt. of LTA)						Elemental Composition as determined by XRF (pct. by wt. of HTA)							
Probe type and location[1]	$2SiO_2 \cdot 3Al_2O_3$ Mullite	SiO_2 Quartz	Fe_3O_4 Magnetite	Fe_2O_3 Hematite	CaO Lime	$CaSO_4$ Anhydrite	SiO_2	Al_2O_3	Fe_2O_3	CaO	MgO	Na_2O	K_2O	SO_3
Suction-V3	29	51	trace	3	0	0	61.3	30.3	5.0	0.7	T	T	1.8	0.9
Suction-H2	31	50	0	4	0	0	62.2	29.6	3.8	T	0	1.6	2.0	0.8
Front deposit	35	45	0	6	0	0	60.9	26.8	8.6	T	0	1.4	1.7	0.6
Back deposit	39	45	0	5	0	0	62.6	29.3	5.4	T	0.1	0.1	1.9	0.6

[1] Refer to figure 2 for location reference.

Note: T referred to in table designates trace.

Table 8

Mineralogical and Elemental Composition of Ash and Deposits Extracted from Pulverized Fuel Combustor

Coal Seam: Waynesburg Additives: None

	Mineral Composition as determined by XRD (pct. by wt. of LTA)						Elemental Composition as determined by XRF (pct. by wt. of HTA)							
Probe type and location[1]	$2SiO_2 \cdot 3Al_2O_3$ Mullite	SiO_2 Quartz	Fe_3O_4 Magnetite	Fe_2O_3 Hematite	CaO Lime	$CaSO_4$ Anhydrite	SiO_2	Al_2O_3	Fe_2O_3	CaO	MgO	Na_2O	K_2O	SO_3
Suction-V3	26	26	3	2	trace	trace	61.1	27.7	5.2	0.6	1.5	0.4	2.9	0.6
Suction-H2	30	31	4	3	trace	trace	60.7	27.9	5.3	0.6	1.4	0.7	2.8	0.6
Horizontal probe composite	34	27	trace	4	trace	trace	59.9	27.4	6.6	0.6	1.7	0.6	2.7	0.5

[1] Refer to figure 2 for location reference.

Table 9

Mineralogical and Elemental Composition of Ash
and Deposits Extracted from Pulverized Fuel Combustor

Coal Seam: Redstone Additives: None

	Mineral Composition as determined by XRD (pct. by wt. of LTA)						Elemental Composition as determined by XRF (pct. by wt. of HTA)							
Probe type and location[1]	$2SiO_2 \cdot 3Al_2O_3$ Mullite	SiO_2 Quartz	Fe_3O_4 Magnetite	Fe_2O_3 Hematite	CaO Lime	$CaSO_4$ Anhydrite	SiO_2	Al_2O_3	Fe_2O_3	CaO	MgO	Na_2O	K_2O	SO_3
Suction-V3	21	30	7	1	trace	trace	57.3	22.6	9.5	4.0	1.4	1.7	2.4	1.1
Suction-H2	24	31	7	4	trace	trace	56.9	22.8	9.6	4.3	1.4	1.6	2.5	0.9
Front Deposit	14	31	2	9	trace	5	54.3	20.8	14.5	4.2	0.7	2.4	1.9	1.2
Back Deposit	24	29	2	5	trace	5	58.1	23.4	9.5	3.8	0.6	1.2	2.4	1.0

[1] Refer to figure 2 for location reference.

Table 10

Mineralogical and Elemental Composition of Ash
and Deposits Extracted from Pulverized Fuel Combustor

Coal Seam: Sewickley Additives: None

	Mineral Composition as determined by XRD (pct. by wt. of LTA)						Elemental Composition as determined by XRF (pct. by wt. of HTA)							
Probe type and location[1]	$2SiO_2 \cdot 3Al_2O_3$ Mullite	SiO_2 Quartz	Fe_3O_4 Magnetite	Fe_2O_3 Hematite	CaO Lime	$CaSO_4$ Anhydrite	SiO_2	Al_2O_3	Fe_2O_3	CaO	MgO	Na_2O	K_2O	SO_3
Suction-V3	23	32	6	4	trace	trace	58.6	23.5	10.4	1.3	0.5	2.2	2.6	0.9
Suction-H2	19	25	4	5	trace	trace	58.7	24.1	11.0	0.8	0.6	1.3	2.7	0.8
Horizontal probe composite	25	32	1	8	trace	trace	58.6	23.3	13.2	0.8	0.5	0.5	2.5	0.6

[1] Refer to figure 2 for location reference.

Table 11

Mineralogical and Elemental Composition of Ash
and Deposits Extracted from Pulverized Fuel Combustor

Coal Seam: Freeport Additives: None

	Mineral Composition as determined by XRD (pct. by wt. of LTA)						Elemental Composition as determined by XRF (pct. by wt. of HTA)							
Probe type and location[1]	$2SiO_2 \cdot 3Al_2O_3$ Mullite	SiO_2 Quartz	Fe_3O_4 Magnetite	Fe_2O_3 Hematite	CaO Lime	$CaSO_4$ Anhydrite	SiO_2	Al_2O_3	Fe_2O_3	CaO	MgO	Na_2O	K_2O	SO_3
Suction-V3	16	18	1	1	0	0	56.4	24.2	13.9	1.4	0.7	0.7	2.4	0.3
Suction-H2	20	22	1	1	0	0	54.6	27.2	11.6	1.7	1.0	0.8	2.6	0.5
Front Deposit	27	28	0	0	trace	3	52.7	22.5	17.9	1.5	1.1	1.0	2.1	1.2
Back Deposit	24	16	0	1	0	3	50.5	29.8	13.2	1.4	0.6	1.6	1.9	1.0

[1] Refer to figure 2 for location reference.

Table 12

Mineralogical and Elemental Composition of Ash
and Deposits Extracted from Pulverized Fuel Combustor

Coal Seam: Illinois 5 & 6 Additives: None

	Mineral Composition as determined by XRD (pct. by wt. of LTA)						Elemental Composition as determined by XRF (pct. by wt. of HTA)							
Probe type and location[1]	$2SiO_2 \cdot 3Al_2O_3$ Mullite	SiO_2 Quartz	Fe_3O_4 Magnetite	Fe_2O_3 Hematite	CaO Lime	$CaSO_4$ Anhydrite	SiO_2	Al_2O_3	Fe_2O_3	CaO	MgO	Na_2O	K_2O	SO_3
Suction-V3	11	17	10	2	trace	3	45.8	26.0	15.9	7.5	0.6	0.7	1.4	2.1
Suction-H2	10	18	5	1	1	3	50.6	18.7	18.4	6.4	0.7	0.4	1.6	3.2
Front Deposit	15	25	1	5	trace	12	45.8	16.5	22.3	5.5	0	1.8	1.3	6.7
Back Deposit	19	26	0	14	0	14	50.6	17.9	16.0	6.1	0	1.4	1.4	6.6

[1] Refer to figure 2 for location reference.

CONCLUSIONS

The seven eastern bituminous coals under consideration have been found to be high in sulfur, ash and volatiles. The mineral species occurring in major quantities and contributing to the ash are illite, kaolinite, pyrite, and quartz. The minerals occurring only in trace amounts include calcite, feldspar and gypsum. The corresponding transformation species are mullite, quartz, magnetite, hematite, lime and anhydrite. These transformations appear to be thermodynamically controlled. However, some residence time effects are apparent for the transformation species magnetite and hematite. The influence on ash formation of the reducing environment of the char particle has been inferred from the existence of magnetite in the ash and deposits.

The possibility of forming condensible sodium compounds by gas phase reactions involving volatilizable sodium has been suggested by the presence of ash and deposits having a high elemental sodium content but which were formed from coals having HTAs low in sodium. This emphasizes the need for techniques to establish the quantity and rate of release of volatilizable alkalis under process conditions.

BIBLIOGRAPHY

1 Selvig, W. A. and Gibson, F. H., "Analysis of Ash from United States Coals," Bulletin 567, U.S. BuMines, 33 pp (1956).

2 Gluskoter, H. J., "Electronic Low Temperature Ashing of Bituminous Coal," Fuel (London), 44, 285-291 (1965).

3 Littlejohn, R. F., "Mineral Matter and Ash Distribution in 'Ash Fired' Samples of Pulverized Fuels," J. Inst. Fuel, XXXIX, 59-67 (1966).

4 Gluskoter, H. J., "Clay Minerals in Illinois Coals," J. Sed. Pet., 37, 205-214 (1967).

5 O'Gorman, J. V. and Walker, P. L., "Mineral Matter Characteristics of Some American Coals," Fuel (London), 50, 136-151 (1971).

6 O'Gorman, J. V. and Walker, P. L., "Thermal Behavior of Mineral Fractions Separated from Selected American Coals," Fuel (London), 52, 71-91 (1973).

7 Gluskoter, H. J. and Mitchell, R. S., "Mineralogy of Ash of Some American Coals: Variations with Temperatures and Source," Fuel (London), 55, 2, (1976).

8 Padia, A. S., Sarofim, A. F., and Howard, T. B., "The Behavior of Ash in Pulverized Coal under Simulated Combustion Conditions," presented at Combustion Institute's Central States Meeting April 5-6, 1976.

9 Mazza, M. H. and Wilson, J. S., "X-Ray Diffraction Examination of Coal Combustion Products Related to Boiler Tube Fouling and Slagging," Adv. X-Ray Analy., 20, 85 (1976).

RATIONAL ANALYSIS OF FIRESIDE DEPOSITS FROM BROWN COAL FIRED BOILERS

A. R. DRUMMOND, P. L. BOAR, and R. G. DEED

Research and Development Department
State Electricity Commission of Victoria
Herman Research Laboratory – Howard Street
Richmond, Victoria, Australia 3121

ABSTRACT

Brown coals from the Latrobe Valley in Victoria, Australia, are low in ash content (1-3%) and also in occluded mineral matter of which the major minerals are α-quartz, kaolinite and marcasite/pyrite. During combustion, a range of mineral compounds are formed in deposits; these are difficult to quantify if only the conventional approach to chemical analysis is taken.

To assist in the assessment of deposit formation and the results of the use of removal techniques, an analytical scheme termed "rational analysis" has been developed which essentially combines the data obtained from x-ray diffraction studies or infra-red spectrometry with the chemical analysis of the deposits to give a pseudo composition for the compounds found in deposits. An integral part of the scheme is the quantitative determination of the α-quartz present in the deposit.

As well as providing an easy method for comparing various deposits, new insights can be gained into the mode of deposit formation in the boiler by using the scheme. At present it is primarily applied to sulfatic deposits originating from the superheater region, although it is equally applicable to deposits formed elsewhere in the boiler. A modified scheme has also been used to give preliminary information on deposits formed from the combustion of other types of coal.

INTRODUCTION

Brown coal from the Latrobe Valley in the State of Victoria, Australia, is won from open cut mines situated close to a complex of thermal power stations, and is the source of about 90% of the State's electrical energy. The brown coals have a high moisture content, 60-68%, and a low ash content, 1-3% (dry basis). Some of the inorganic constituents from which the ash is formed, namely sodium, calcium, magnesium, iron and sulfur, occur in direct combination with the coal organic substance. During combustion of the coal these elements react to form oxides and sulfates. Adventitious mineral matter may also be present in the coal and is included in the ash either in its existing state or modified during combustion. The minerals commonly present are halite (NaCl), marcasite or pyrite (FeS_2), α-quartz (SiO_2) and kaolinite ($Al_2Si_2O_5(OH)_4$).

The tendency of the coal to form deposits during its combustion in a furnace can be related to the relative concentrations of the inorganic constituents. The most significant compound in the formation of the deposits is thenardite-sodium sulfate (Na_2SO_4) which readily forms in the gas stream and on condensation cements together the solid ash particles, thus acting as both initiator and binder in the deposit formation.

In the past, studies related to the mechanism of ash deposit formation used the conventional mode of chemical analysis; expression of the major elemental constituents as their relative oxide form or, where appropriate, as the element. Qualitative x-ray diffraction was also employed to identify the mineral form of the compounds present in the deposits. Although many compounds have been readily identified in deposits, the quantitative measurement of these compounds is difficult, for, in many cases, wide variations in chemical analyses have been reported, but the mineralogical forms identified have been found to be remarkably similar. A scheme termed rational analysis, which utilizes the analytical data to provide a hypothetical compound analysis, was developed as an alternative to quantitative x-ray diffraction for a means of deposit classification.

ANALYTICAL APPROACH

An outline of the analytical approach is shown in Figure 1. The techniques for the chemical analysis (1) and the x-ray diffraction (2) were developed for brown coal ash, but can be applied to many other mineral species.

An outline of the rational analysis for deposits is presented in Figure 2, and the details of the steps involved are:

a Quantitative determination of the α-quartz (SiO_2) - this is the key to the rational analysis scheme. The α-quartz concentration is subtracted from the total silica concentration. Any residual silica is considered to form other silicon compounds.

b The concentration of free lime (if present) is expressed as elemental calcium and subtracted from the total calcium.

c Chloride, if less than 1%, is ignored; if greater than 1%, it is calculated as sodium chloride (NaCl).

d Sodium is totally accounted for as thenardite (Na_2SO_4) unless chloride is considered.

e The excess sulfur remaining after satisfying the requirement for Na_2SO_4 is then expressed as anhydrite ($CaSO_4$). This should leave an excess of calcium. If, however, the sulfur content is in excess of the stoichiometric balance of the calcium content, the anhydrite ($CaSO_4$) is then recalculated, based on the calcium content. The unassociated sulfur remaining is then expressed as a mixed sulfate compound ($Na_2SO_4.MgSO_4$). The thenardite (Na_2SO_4) concentration is then recalculated after allowing for the sodium required for the formation of the mixed sulfate ($Na_2SO_4.MgSO_4$).

f The silicate types commonly found in the combustion deposits from brown coal are:

Pyroxene	:	(Diopside	$CaMg(SiO_3)_2$
	:	(Monticellite	$CaMgSiO_4$
Olivine	:	Forsterite	Mg_2SiO_4
Melitite	:	Akermanite	$Ca_2Mg_2SiO_7$

In calculating the silicate types the excess calcium from previous calculations is assigned with the magnesium and silicon to form one of the abovementioned compounds. The magnesium and silicon remaining are then expressed as periclase (MgO) and silica glass (SiO_2) respectively.

g The remaining elements are then reported in the following form:

Iron is reported as hematite (Fe_2O_3) and/or magnetite (Fe_3O_4) depending upon the ratio indicated by their respective x-ray diffraction pattern intensities.

Any excess calcium is reported as the oxide (CaO).

Other elements present are reported as their respective oxides, e.g.

$Al \rightarrow Al_2O_3$
$Ti \rightarrow TiO_2$
$K \rightarrow K_2O$

h The bulk density is determined to indicate the physical form of the deposits.

To simplify the calculations involved in the analytical scheme, a computer program has been developed which used the data already outlined, namely:

- chemical analysis data;
- x-ray diffraction data;
- α-quartz and free lime concentrations;
- bulk density.

Examples of chemical analysis and x-ray diffraction data of deposits are shown in Table 1. Table 2 lists the calculated concentrations of compounds, density and porosity of these deposits by the application of the scheme.

DATA INTERPRETATION

Deposits formed in the boilers range from loosely bonded aggregates to formations which are hard, strongly bonded and partially fused (3). The compounds present and the density can be used to indicate the temperatures to which they have been subjected. Because of the above heterogenity of the structure of deposits, their density and porosity should also be considered when interpreting the analytical results, as deposits with similar compositions are often markedly different physically.

All deposits contain thenardite (Na_2SO_4) which is considered to act as both a cement and flux. It is readily formed in the gas stream as a vapour at a temperature of approximately 1050°C, condenses to a liquid at about 950°C and solidifies at 880-900°C. Thenardite condenses on boiler tubes providing a medium for the capture and binding of other ash particles. As the vapour it can enter the porous deposit structure formed, cool and condense in a temperature zone of 900-950°C, and while still molten, can migrate and solidify where the temperature is less than 900°C and form a strongly bonded and dense zone within the deposit.

Anhydrite ($CaSO_4$) is the other common sulfate compound found in deposits. This compound is predicted, by thermodynamic calculations, to be formed in the boiler in the temperature range of 850-1000°C, especially at the lower temperature. However, the rate of the reaction is also influenced by particle composition, size and residence time in the gas stream. Complete sulfation of calcium oxide would occur within the deposit.

Periclase (MgO) is a major ash constituent usually found in appreciable quantities in the deposits. It is present in the deposits and gas stream as small crystals of less than 10 μm size. Sulfation of periclase is not considered to occur as it has never been detected in the ash or deposits. This observation is reinforced by thermodynamic calculations which show that at the concentration of the oxides of sulfur (<500 ppm) in the gas stream, sulfation is only likely to occur at temperatures less than 800°C. Water soluble magnesium is present in ash and deposits presumably as a sulfate. Evidence of mixed sulfates of sodium and magnesium have been found during x-ray diffraction studies of some porous deposits and in magnesium-rich ash. However, the appropriate patterns were difficult to interpret as the lines were few and weak and did not completely match those in established patterns.

Silicates, when present in the deposits, are considered to have been formed in situ, and are indicative that temperatures in excess of 900°C had been obtained. Calcium magnesium or magnesium silicates may be formed from a reaction between α-quartz (SiO_2), anhydrite ($CaSO_4$) and/or periclase (MgO). The reactions are time and temperature dependent, and it is considered likely that molten Na_2SO_4, acting as a flux, dissolves SiO_2 which in turn reacts with $CaSO_4$ and/or MgO. It is considered that silicate formation begins at a temperature of 950°C, coinciding with the decomposition temperature of $CaSO_4$.

The tabulations below are the expected salient features of deposit formation with respect to temperature.

TEMPERATURE °C	RATE OF FLUX DEPOSITION	TOTAL DEPOSITION RATE
750	High	Low
800	Medium	Medium
900	Low	High
900-950	(Critical point)	
950	Flux melts and migrates to cooler region of the deposit	
1000	Extensive melting of the flux deposit, consolidation begins, anhydrite ($CaSO_4$) begins to decompose, silicates formed.	
>1000	Consolidation of deposit.	

The porosity of deposit formations are indicative of their history and also influences the ease with which they may be removed from a boiler by water washing. Table 2 lists a number of characteristic values.

EXTENSION OF RATIONAL ANALYSIS

As other brown coal fields are evaluated as possible future fuel sources, modification to the rational analysis scheme may be necessary if investigations indicate that major variations occur in mineral compounds formed in the deposits. To date, modifications have been relatively minor.

It should be noted that the scheme described is the culmination of experience and data obtained over an extensive period of investigation into many facets of formation and removal of deposits from brown coal. However, by using a slightly modified approach, the rational analysis scheme has been used for deposits originating from a sub-bituminous coal (Leigh Creek, South Australia). It is considered that the scheme could find application to investigations into deposits from other coal sources, provided that suitable adjustments are made for the compounds actually found in the deposits in question.

In conclusion, the rational analysis scheme has made the comparison of deposits easier and has provided additional information on the deposits formed during the combustion of brown coal.

REFERENCES

1 Boar, P. L., and Ingram, L. K., "The Comprehensive Analysis of Coal Ash and Silicate Rocks by Atomic Absorption Spectrophotometry by a Fusion Technique," Analyst, 1970, pp. 124.

2 Grinton, G. R., "Computer-aided Identification of Compounds by X-ray Diffraction," Victoria, State Electricity Commission, Research and Development Department, Report No. 282, 1973.

3 Bonafede, G., and Kiss, L. T., "Study of Ash Deposits from Brown Coal Fired Boilers with the Aid of Scanning and Transmission Electron Microscopy," Bonafede, G. and Kiss, L. T., New York, ASME, 1974, Paper 73-WA/CD-7.

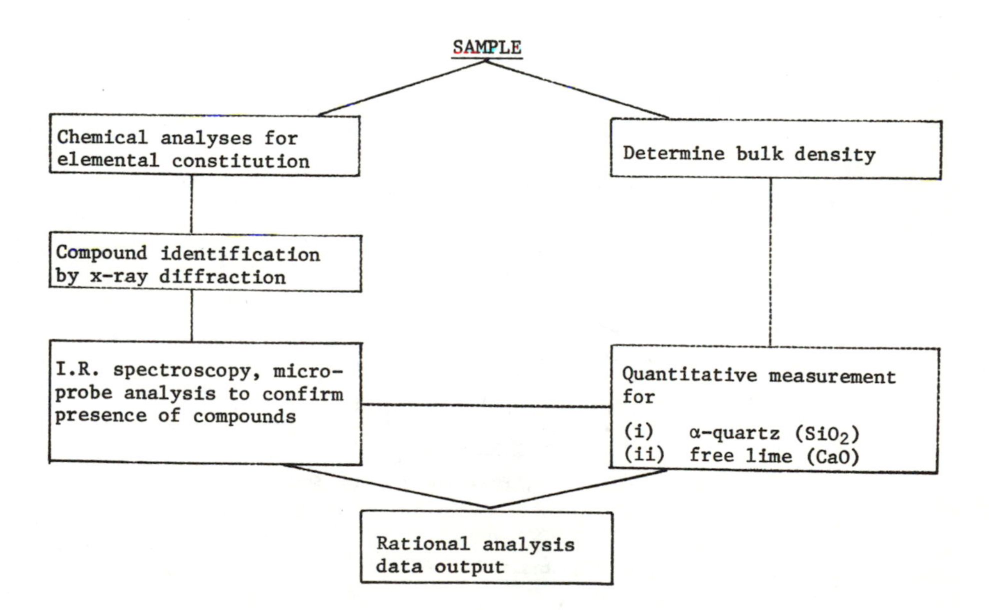

Fig.1 Analytical scheme

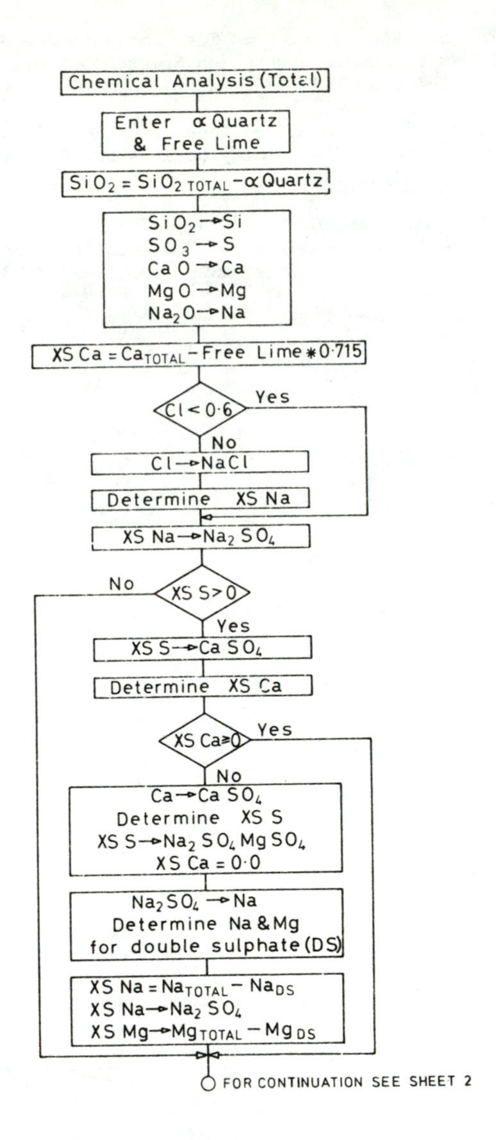

Fig.2 Outline of rational analysis
(Sheet 1)

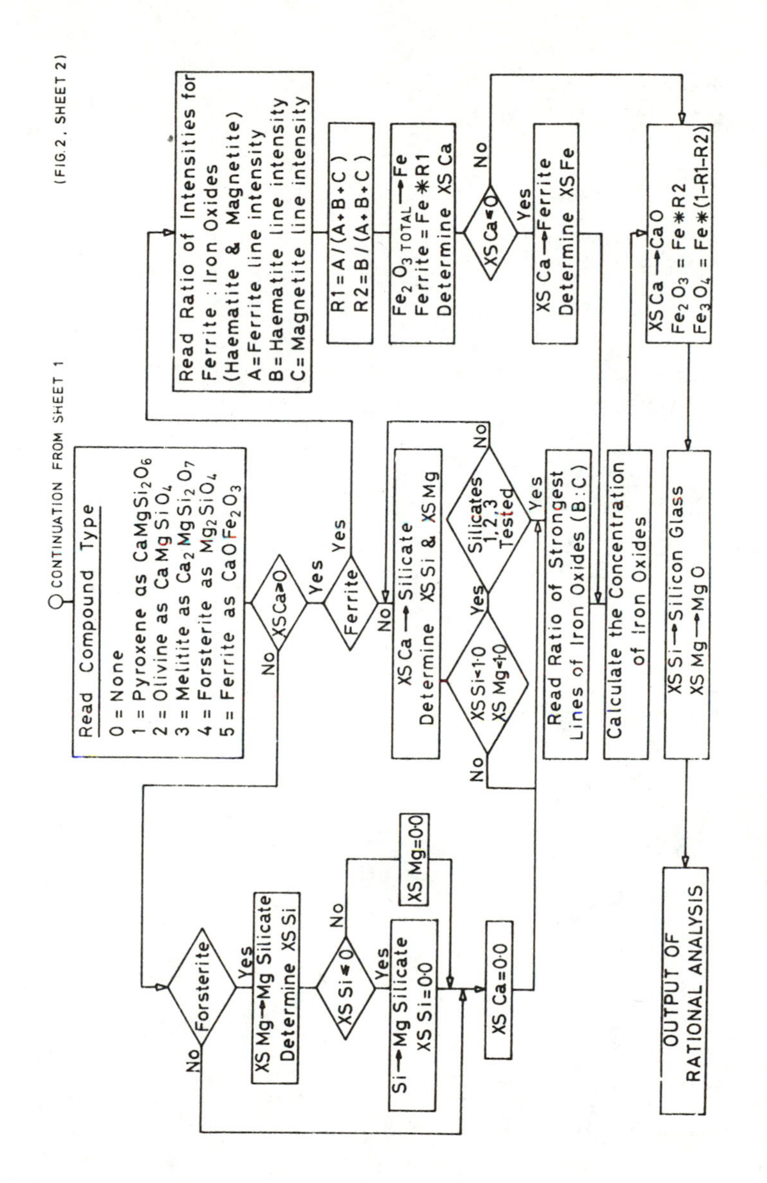

Fig.2 Outline of rational analysis (Sheet 2)

Table 1 Chemical analysis and x-ray diffraction data

CHEMICAL ANALYSIS		1	2	3	4	5
Silica	(as % SiO_2)	5.1	46.3	40.9	6.2	17.5
Alumina	(as % Al_2O_3)	1.4	7.7	5.1	2.5	11.8
Titanium Oxide	(as % TiO_2)	0.2	0.9	<0.1	<0.1	<0.1
Iron Oxide	(as % Fe_2O_3)	6.4	5.4	14.6	8.7	10.0
Calcium Oxide	(as % CaO)	18.4	15.4	12.4	13.2	2.0
Magnesium Oxide	(as % MgO)	12.7	12.2	7.4	8.7	12.0
Sulfur Trioxide	(as % SO_3)	40.9	6.7	10.3	43.5	10.8
Sodium Oxide	(as % Na_2O)	13.8	5.1	3.1	17.4	4.9
Potassium Oxide	(as % K_2O)	0.5	0.3	0.2	0.5	0.1
Chloride	(as % Cl)	<0.1	<0.1	<0.1	<0.1	<0.1
Loss on Ignition at 800°C		1.0	0.3	5.6	0.2	31.5
		100.4	100.3	99.6	100.9	100.6
COMPOUNDS IDENTIFIED BY X-RAY DIFFRACTION						
Thenardite	(Na_2SO_4)	√	√	√	√	-
Anhydrite	($CaSO_4$)	√	√	-	√	-
Sodium Magnesium Sulfate	($Na_2SO_4.MgSO_4$)	√	√	-	√	√
α-quartz	(α-SiO_2)	-	√	√	-	√
Periclase	(MgO)	√	√	√	√	√
Magnetite	(Fe_3O_4)	-	-	√	- -	√
Hematite	(Fe_2O_3)	√	-	-	-	-
Diopside	($CaMg(SiO_3)_2$)	-	-	-	-	-

√ Present
- No positive identification.

Table 2 Compound concentrations in deposits

SULFATES		1	2	3	4	5
Thenardite	as Na_2SO_4	31.6	11.7	7.1	38.0	9.8
Anhydrite	as $CaSO_4$	38.9	0.2	10.7	32.1	4.9
Mixed Sulfate	as $Na_2SO_4.MgSO_4$	-	-	-	3.5	2.6
OXIDES						
α-quartz	as α-SiO_2	-	5.0	20.0	-	14.0
Periclase	as MgO	11.1	1.2	1.7	8.2	11.6
Magnetite	as Fe_3O_4	3.1	-	14.1	8.4	9.7
Hematite	as Fe_2O_3	3.2	5.4	-	-	-
SILICATES						
Olivine	as $CaMg(SiO_4)$	-	-	-	-	-
Pyroxene	as $CaMg(SiO_3)_2$	8.8	59.1	30.9	-	-
Melitite	as $Ca_2MgSi_2O_7$	-	-	-	-	-
EXCESS NOT USED IN RATIONAL ANALYSIS						
Silica glass	as SiO_2	0.2	8.5	3.7	6.2	3.5
Calcium	as CaO	-	-	-	-	-
NOT DETECTED BY X-RAY DIFFRACTION						
Aluminium	as Al_2O_3	1.4	7.7	5.1	2.5	11.8
Titanium	as TiO_2	0.2	0.9	<0.1	<0.1	<0.1
Potassium	as K_2O	0.5	0.3	0.2	0.3	0.1
Chloride	as Cl	<0.1	<0.1	<0.1	<0.1	<0.1
Loss on Ignition		1.0	0.3	5.6	0.2	31.5
		100.0	100.3	99.1	99.4	99.5
Bulk density at 20°C		2.03	0.92	0.91	1.08	0.20
Calculated Particle Density		3.01	2.89	2.91	2.88	1.82
Porosity - % Voids		32.6	68.2	68.7	62.5	89.0

MECHANISM OF SUB-MICRON FLY-ASH FORMATION IN A CYCLONE, COAL-FIRED BOILER

GAIL D. ULRICH, JOHN W. RIEHL,[1] BRUCE R. FRENCH,[2] and RAYMOND DESROSIERS

Department of Chemical Engineering
University of New Hampshire
Durham, New Hampshire, USA 03824

ABSTRACT

Sub-micron fly-ash, because of its inordinate threat to public health, is receiving increasing attention. Through thermophoresis and selective enrichment, it may, as well, be more important in tube fouling and corrosion than the mass concentration suggests. Sub-micron ash is presumably formed by vaporization of certain metal-containing compounds at the burning coal surface followed by transport to the gas phase where nucleation and growth occur. The growth and morphology of sub-micron silica aggregates produced in a laboratory burner have been studied using laser light scattering and probe techniques. A growth model, based on collision and fusion rates, has been confirmed. Aggregates grow rapidly to a size of about 300 nm where a dramatic change in collision frequency occurs. The internal structure of the particle can vary from a chain-like aggregate of smaller primary units to a sphere, depending on the viscosity. Full-scale boiler studies and laboratory research are continuing. A method for computing the vapor composition at equilibrium has been refined. A comprehensive model for predicting the type of fly-ash generated from a given coal in a particular boiler is the goal.

NOMENCLATURE

E exponential viscosity parameter (deg-K)

SA specific surface area ($m^2\ g^{-1}$)

T temperature (deg-K)

α number of particles surrounding the central unit in a coalescing cell

β pre-exponential viscosity factor (Poise=g $cm^{-1}\ sec^{-1}$)

θ growth time (sec)

σ surface tension of particle (g sec^{-2})

ρ density of particle (g cm^{-3})

μ viscosity of particle (Poise=g $cm^{-1}\ sec^{-1}$)

---

1 Cabot Corporation, Billerica, Massachusetts 01821
2 Public Service Company of New Hampshire, Newington, N.H. 03801

INTRODUCTION

Coal-generated fly ash, as viewed in the optical microscope, is comprised of discrete, multicolored spherical particles, generally ranging from one to one hundred microns in diameter (1-3). Recent studies, employing the electron microscope, reveal that the size distribution is bimodal. The major mode, on a mass basis is larger than one micron (4,5), but on a number basis, the majority of particles is sub-micron.

The super-micron fraction (residual or entrainment ash) is, apparently, the molten residue from combustion of an entrained coal particle. Because of fracturing or bursting of cenospheres, there are approximately three to five such ash particles per coal particle (7).

The sub-micron fraction, so-called condensation ash, is allegedly formed by metal transport to the bulk gas phase followed by oxidation and precipitation. The migrating species may be reduced metals, sub-oxides, hydroxides, sulfides or, in the case of more volatile materials or higher temperatures, the stable oxide. The transport of silicon metal via the volatile silicon monoxide, for example, has been observed for many years in the electrothermal ferrosilicon process. A similar mechanism is suggested in the generation of silica-rich deposits in boilers (6). In a recent coal combustion study, up to four percent of the silica was vaporized by this process (7).

Environmentally, condensation ash is much more significant than its mass fraction suggests. Not only are sub-micron particles more difficult to remove from flue gases, but they penetrate deeply into the human respiratory system where their constituent elements are efficiently absorbed by the blood stream. These particles present large specific surface areas for adsorption and transport of toxic gaseous species. Also, recent observations reveal that the more volatile and more hazardous metallic elements are concentrated in the sub-micron ash fraction (8-10). Aside from physiological effects, other environmental factors such as visibility and weather are more strongly influenced by number concentration than mass concentration (11).

Because of their mobility in a thermal gradient, sub-micron particles may be similarly more significant in boiler tube fouling than their mass loading suggests. For example, in thermal precipitators, which are commonly employed to sample and segregate aerosols from suspensions, the small particles deposit nearest the intake (20). Generally, apparent migration velocities for particles larger than one micron in 1000 deg-K air at one atm pressure are lower by a factor of three to five than those for particles ten times smaller (19,21).

The theory of thermophoresis is still subject to controversy. In the free-molecule regime, where the Knudsen number (ratio of gas mean-free-path to particle radius) is considerably greater than unity, the theory is well established (12-14). For Knudsen numbers less than one, the continuum regime, some uncertainty remains (15-19). Perhaps surface deposition is influenced more by the adhesion/cohesion characteristics of particles than by their mobilities. Although the exact mechanism is not clear, fouling is a serious problem in the cooling of aerosol suspensions, and the severity is a strong inverse function of particle size (22).

In boilers, aerosol deposition is clearly a factor in tube fouling and corrosion, though its significance in these complex processes is unclear. Investigators typically find a dusty, white, alkali-rich, inner layer in slag deposits (23). Assuming it is formed by thermal deposition of submicron fly-ash particles, there are two reasons why it may serve a key function in irreversible slag build-up. First, the sub-micron particles, allegedly containing the more volatile ash components, will generally have a lower melting point than the bulk material and will fuse more easily to form a slag. Second, these particles tend to be bulky, low density aggregates having large void volumes and a thermal conductivity near that of the interstitial gas. (Carbon black, for example, is a similar aggregated sub-micron material. Its thermal conductivity is approximately ten thousand times smaller than the constituent graphite (24).) Thus, such a sub-layer will act as an insulator, allowing the ash surface-temperature to rise and impinging particles to stick and slag.

Tube deposition, of course, causes poor boiler efficiency, corrosion, and down-time. An intensive search for the underlying mechanism is certainly justified. Work in our laboratory is concerned with the formation and growth of sub-micron oxide particles and the role that they play in pollution and fouling. The research is divided into two categories, full-scale boiler studies and laboratory flame studies. Each will be considered separately below.

FULL-SCALE BOILER STUDIES

Most previous workers have analyzed either ash behavior during the coal combustion process or the characteristics of the final ash. Assuming condensation ash is precipitated somewhere between the combustion zone and the dust collector, intermediate samples are necessary for insight into the formation mechanism. Table 1, for example, contains the results of a preliminary study of an operating 340 megawatt cyclone-fired boiler (25). Originally, we had expected the surface area to decrease with time as growth occurred through coagulation. It does decrease between the combustion zone and the fifth floor, but probably because of porosity in the unburned char particles. No growth was evident between the fifth and seventh floor sampling ports. In fact, there was an increase in surface area in passing from the seventh floor, through the tube sections, to the electrostatic precipitator. This difference was not apparent from optical micrographs. At a magnification of four hundred, micrographs such as that shown in figure 1, for hopper ash, were indistinguishable from those for samples taken from the fifth and seventh floor sampling ports. This is

Table 1. Characteristics of Fly-Ash Samples Collected at Various Positions in an Operating Boiler (Number two unit, 340 megawatt, Merrimack Station, Public Service Co. of N.H.. Bow, N.H.)

Location	Residence Time (sec.)	Appearance in Optical Micrographs	Specific Surface Area (m^2/g)
Combustion zone	0.5	Single spheres 1 to 30 μm plus char	7.0
Fifth floor	1.1	Single spheres 1 to 30 μm	0.5
Seventh floor	1.5	"	0.5
Precipitator Hopper	3.0	"	1.2

illustrated in figure 2 by the similarity of cumulative particle-size distribution curves taken from the optical micrographs. A difference was detected, however, by election microscopy at a magnification of 26,700X. Although deposits were noted on the surfaces of the larger residual ash particles in all samples, only the hopper ash contained significant numbers of submicron aggregated particles in the dispersed field such as those shown in figure 3. The aggregates have a maximum length of about 200 nm, and the individual primary particle units are approximately 10 to 20 nm across. Based on the relationship to particle size, specific surface area differences noted in table 1 require only about 0.3 weight percent of the ash to be present as sub-micron aggregates. This small mass fraction, however, represents a number ratio of 400 to 1 in favor of the sub-micron particles. These observations are consistent with appearance in figure 3.

Fig. 1 Optical micrograph (400X) of the hopper ash described in table 1.

The lack of growth among residual ash is consistent with the low collision frequencies of super-micron particles at these concentrations. Although some have suggested that electrostatic repulsion is a factor, calculations reveal that the ash concentration must be larger by a factor of one hundred for significant coagulation to occur among the entrained ash particles in this boiler (25).

Formation of sub-micron condensation-ash particles seems to occur during rapid cooling in the superheating section of the boiler. Those studying enrichment phenomena in fly ash suggest surface deposition as an important mechanism whereby volatile oxides condense on the smaller entrainment-ash particles which exhibit the largest collision surface area (8-10). We propose a somewhat contrasting mechanism of homogeneous nucleation and growth producing sub-micron aggregates which then deposit preferentially on the smaller entrainment-ash particles. The final ash composition would be little different by either mechanism, but action taken to control deposition might be quite opposite in the two cases. To elucidate the real mechanism, more precise data must be obtained during the condensation-growth phase. Experiments with this aim are currently progressing in our laboratory.

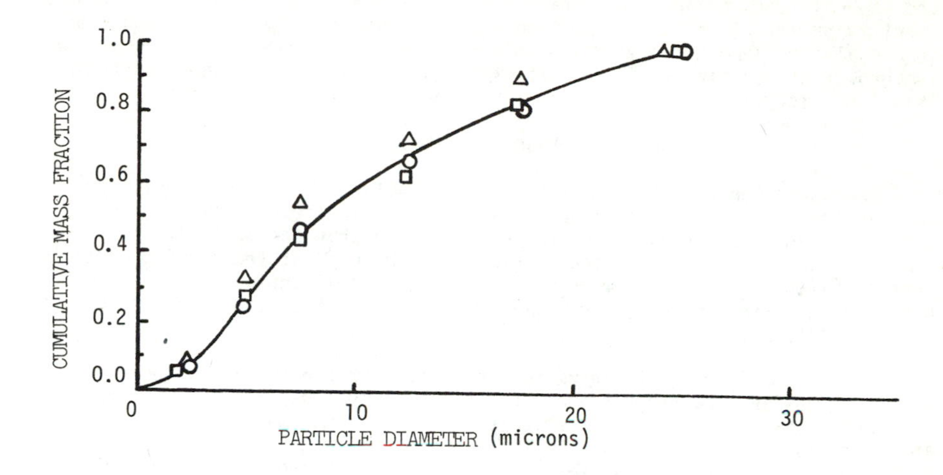

Fig. 2 Cummulative mass particle-size distribution curves for samples taken from the fifth floor ○, the seventh floor □, and the precipitator hopper △, of the operating boiler described in table 1.

Fig. 3 Transmission electron micrograph (26,700X) of the hopper ash described in table 1. The sample was dispersed by ultrasonic energy in dilute collodion using the technique developed by Medalia and Heckman (26).

LABORATORY STUDIES

To understand sub-micron ash formation and growth in a system as complex as coal combustion requires an understanding of the metal transport mechanism and the collision-coalescence behavior of small particles. As a first step in identifying important species, a technique for computing gas-liquid equilibria in a coal-air flame has been developed. This will be described briefly below. The growth of sub-micron silica particles has been studied at some length in this laboratory. It will be discussed in greater detail to follow.

Condensed-Phase Equilibria

The use of an equilibrium technique to elucidate a non-equilibrium process can, at the outset, be questioned. However, the kinetic restraints in coal particle combustion appear to be transport rather than chemical limitations. Thus, it seems that chemical equilibrium is approached at the burning particle surface (although the gas composition there is largely influenced by diffusion processes). Based on this, the escape of volatile, metal-containing compounds can be predicted.

During the 1950's, the drive for improved rocket propellents and higher specific impulse, spawned a number of computer techniques for determining chemical equilibria in high temperature combustion systems. One used profitably here is that developed at the Naval Ordinance Test Station, China Lake, California (27). It is accurate, flexible, converges rapidly, and employs a large, existing, open-ended data file (the JANAF Tables (28)). The program is designed to include condensed species as well as gases, making it suitable for coal-combustion calculations. However, as originally conceived, it did not allow for solid or liquid mixtures. This is not a limitation with rocket propellents which tend to produce pure, single-component, condensed phases. Even in coal combustion, the flame temperature and gas-stream enthalpies are little affected by this simplification. For insight into ash transport mechanisms, on the other hand, computer techniques must be refined.

Extension of the NOTS technique to include condensed mixtures would be simple if the condensed phase were an ideal liquid. Fortunately, even though these slag systems are not ideal, the liquidus temperatures are considerably below the burning particle temperature. Also, because of their importance in steelmaking, similar mineral slags have been studied extensively, and activity coefficients have been determined at the conditions of interest. The computer program has been modified accordingly and used to isolate significiant ash vaporization reactions. Results are shown in figure 4 for the medium-volatile bituminous coal (10% ash) used at the Merrimack Station. (The average ultimate analysis of the coal ash is shown in table 2.) Temperature and oxygen concentration are the parameters used in figure 4 since they are the major variables. Note that substantial amounts of silicon, aluminum, iron, calcium, magnesium, sodium, and potassium are volatilized under these highly reducing conditions. Silicon is evolved as the sulfide or monoxide, depending on the oxygen level. Aluminum, iron, calcium and magnesium evaporate as pure metals, whereas sodium and potassium appear as pure metals or cyanides depending on oxygen concentration. Based on figure 4, the molar fractions of each metal gasified at equilibrium are those shown in figure 5. Under these highly reducing conditions and at higher temperatures, significant fractions of the total sodium, potassium, silicon, magnesium, and iron appear in the gas phase. Substantial fractions of the aluminum and calcium are found at 6 percent stoichiometric oxygen but not at 35 percent. There are of course, numerous minor ash constituents which were not considered in this comptuation. These can be included readily if the thermodynamic data and activity coefficients are available or easily estimated.

Actual quantities of each metal found in the gas phase will, of course, depend on the temperature and oxygen concentration at the surface of the coal particle, as well as, the rate of diffusion. With this information, the compositions of submicron ash for various coals and under differing conditions can be predicted.

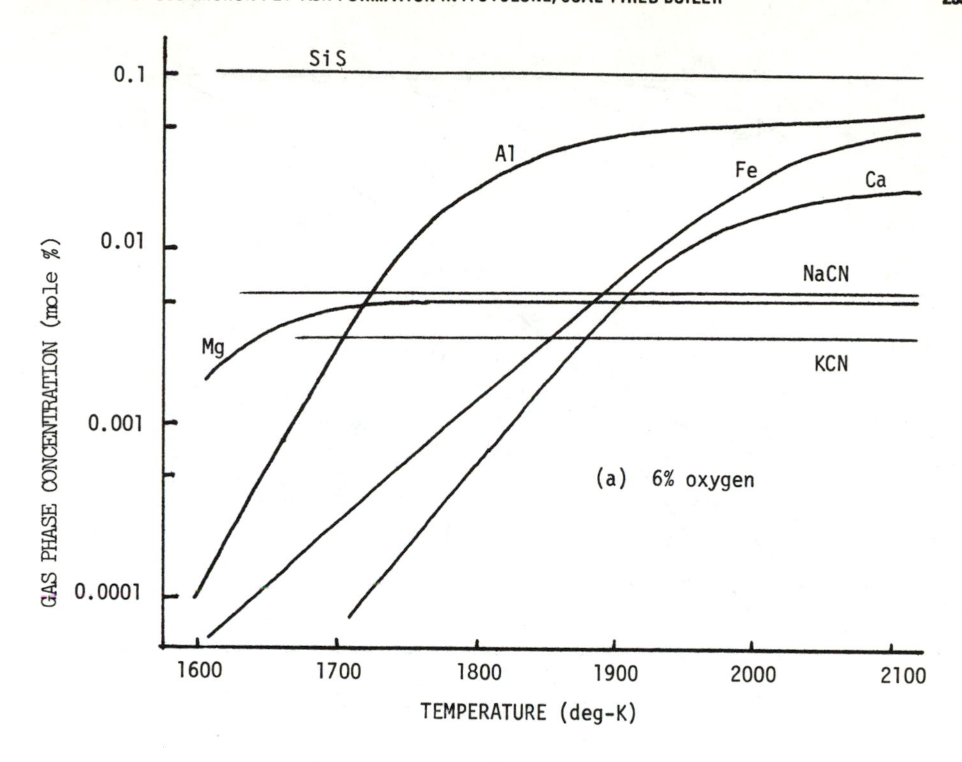

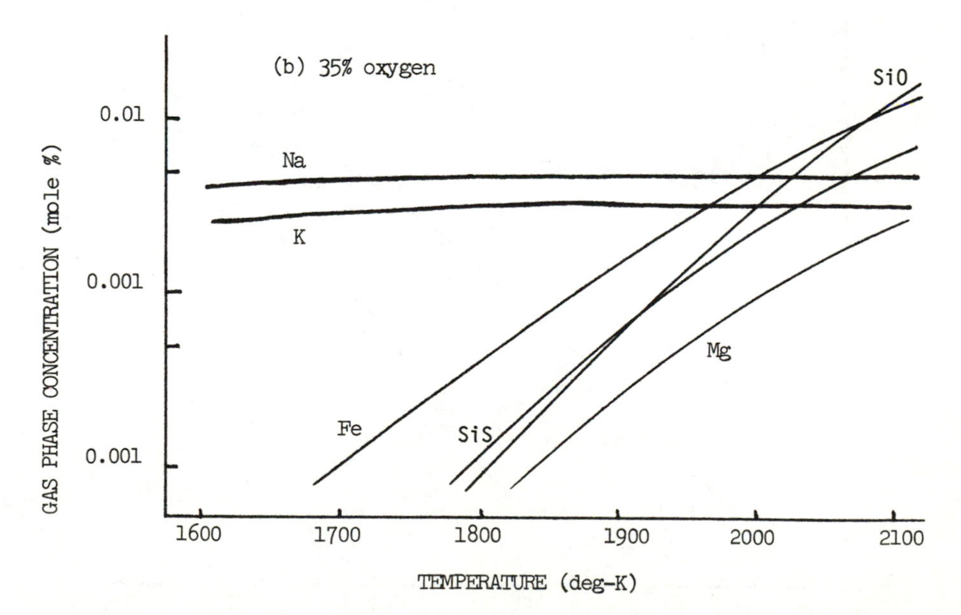

Fig. 4 Equilibrium compositions at the surface of a burning coal particle as a function of temperature. The oxygen concentration is a) 6 percent b) 35 percent of that theoretically required for complete combustion.

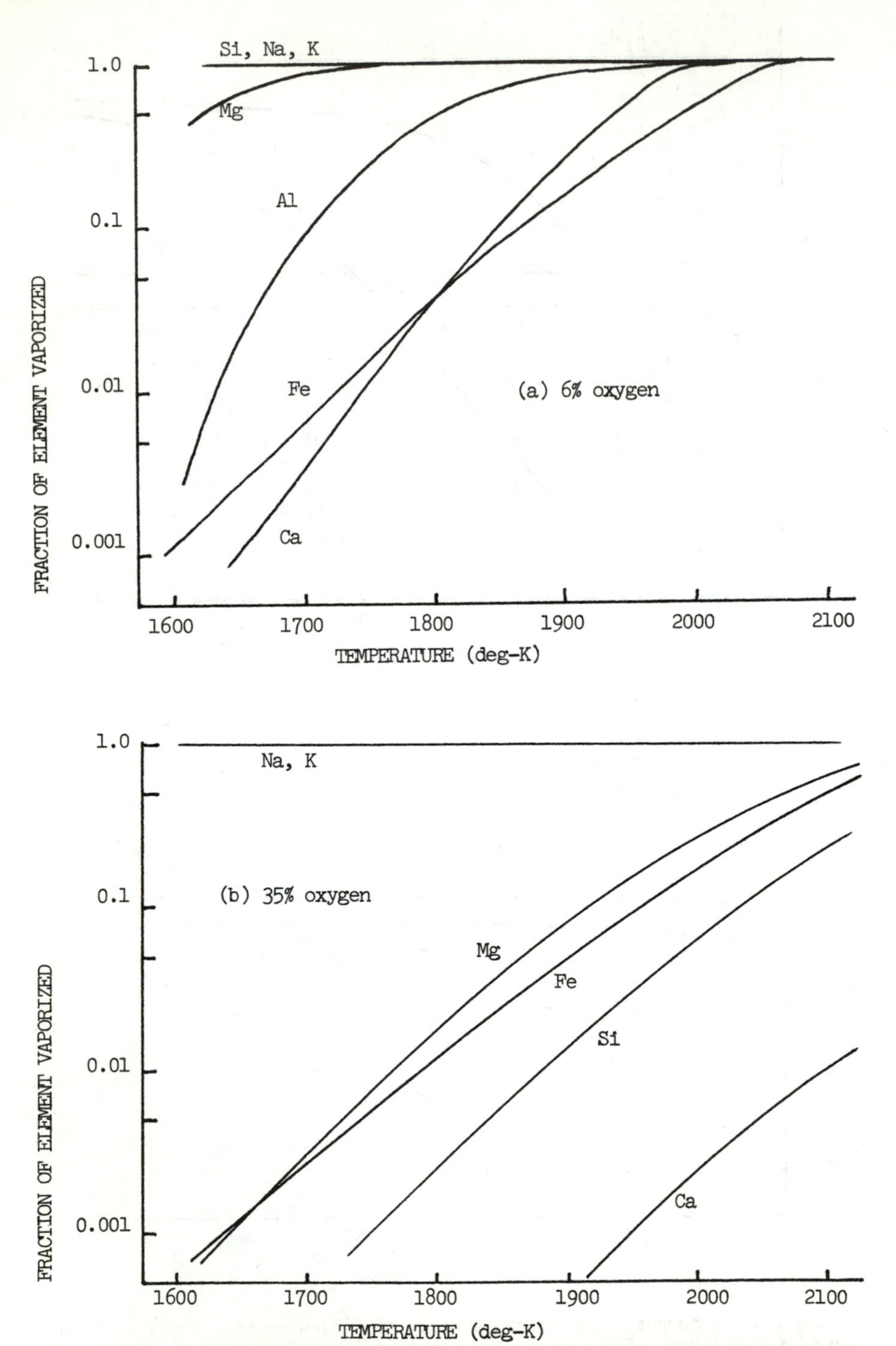

Fig. 5 Fraction of metal vaporized at equilibruim from coal ash. The oxygen concentration is a) 6 percent b) 35 percent of that theoretically required for complete combustion.

Table 2. Average ultimate analysis of ash in the coal used at the Merrimack Power Plant

Compound	Weight Percent	Compound	Weight Percent
SiO_2	37.02	MgO	1.26
Al_2O_3	20.52	K_2O	1.16
Fe_2O_3	19.62	Na_2O	1.12
CaO	6.77	SO_3	9.33

Ash softening temperature: 1188 C (2170 F)
Ash fluid temperature: 1260 C (2399 F)

Laboratory Flame Research

Because of experimental convenience, industrial prominence, and significance in fly-ash, silica has been studied extensively in this laboratory. Basic parameters such as sticking coefficient and fusion temperature have been sought in an effort to assemble a generalized growth theory.

Because of its pervasiveness and importance in industrial flame radiation, soot has been studied in depth by others. As a subject, however, it has two major disadvantages, the chemical formation processes are extremely complex, and oxidation competes with and obscures the growth mechanism. Prompted by the presence of aluminum in rocket propellents, there have been a few studies of flames containing this element. One other flame-produced oxide of commerical importance, titania, has also received attention as have some pure metals such as lead. (See (29-31) and the references cited therein for details on the various systems mentioned.)

For sub-micron particles, Brownian collision and coalescence appear to be the major growth phenomena. Nucleation generally occurs rapidly, forming a cloud of active particles which coagulate at a decreasing rate. Growth ceases, in the case of oxides and metals, when the temperature drops below the fusion point. In the case of soot, particles are thought to become inactive as they temper and harden through hydrogen evolution. Impaction, caused by turbulence and fluid motion, is generally insignificant for submicron particles (29).

Most growth correlations are based on the approach of Smoluchowski wherein the inventory of particles is related to the collision frequency (29). Different collision rates are used for the two regimes of Brownian motion. Theoretical growth rates coincide with experimental data for titania, molten metal droplets and carbon at high dilution.

However, for silica, an abnormally low sticking coefficent (.004) is needed for aggreement (30). Apparently, with its large viscosity, fusion rather than collision controls the rate of primary particle growth. A basic weakness in the Smoluchowski approach is the tacit assumption that coalescence is instantaneous. It may be valid for extreme conditions of low viscosity and dilute suspensions, but for highly viscous materials and high concentrations, growth is a balance between collision and fusion. In such cases, the colliding particles are actually multi-unit flocs or aggregates. The smallest identifiable (primary) particle, meanwhile, grows by coalescence with its neighbors. This combination collision-fusion model has been analyzed theoretically to yield expressions for both the number population of aggregates (N) and the inventory of primary or proto-particles (N_p) within an aggregate (31).

Experimentally, fumed silica or so-called "white smoke" is produced by the reaction between silicon tetrachloride, oxygen and a hydrocarbon fuel in a pre-mixed burner. (Experimental details are given in (30).) At various distances from the flame front, samples are extracted through a probe and collected on a filter. We measure the size of the non-porous primary particle by gas adsorption and that of the aggregate by light scattered from a laser beam passing through the combustion gases.

Surface area data for one set of runs are shown in figure 6. Note the rapid decrease (increase in primary particle size) near the flame front.

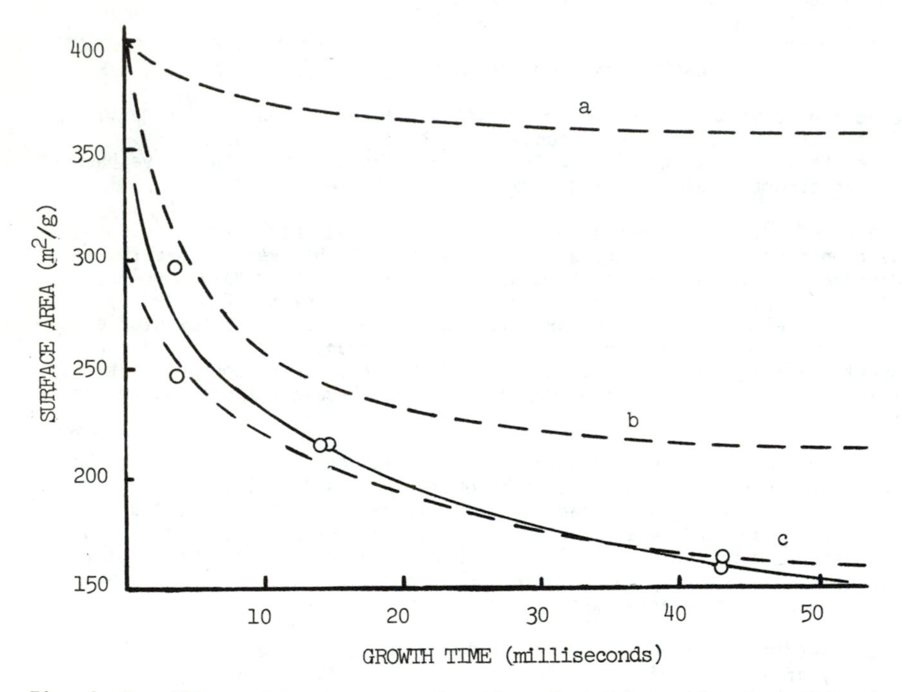

Fig. 6 Specific surface area as a function of residence time for silica in a laboratory burner. The circles and solid line represent experimental results. Curve a is calculated from equation (1) based on the estimated temperature profile and viscosity correlation of Hofmaier and Urbain (32). Curve b is calculated using the same viscosity correlation but with an increase of 150 deg-K in the temperature profile. Curve c is calculated based on the original temperature profile with viscosity parameters recommended by Rossin, Barsan and Urbain (34).

This stems from two factors, the inverse dependence of fusion rate on particle radius plus the exponential variation of viscosity with temperature which is, of course, higher at the flame front. The theoretical relationship (31) is expressed by

$$SA^{-1} = SA_o^{-1} + (2\rho\sigma\alpha/9\ [\alpha+2])\int_{\theta_o}^{\theta} \frac{d\theta}{\mu} \qquad (1)$$

where SA_o is the specific surface area at a reference time θ_o. All of the parameters except viscosity are essentially temperature independent and are removed from the integral. The temperature variation of viscosity can be expressed by the relationship (32):

$$\mu = \beta \exp(E/T) \qquad (2)$$

Based on election micrographs, α is assumed to be four. The surface tension is 307 (33) and the density is 2.2. The adiabatic flame temperature in figure 6 was 2000 K. Based on thermocouple measurements in a "clean" (silica-free) flame on a different burner, the estimated temperatures at 2, 4, 14 and 43 millisecond residence times are 1980, 1960, 1880 and 1710 K. Assuming various initial conditions and viscosity data, surface area values as calculated from equation (1) are those shown by the broken lines in figure 6. Curve a is based on the viscosity data of Hofmaier and Urbain (32), i.e. $\beta = 5.75 \times 10^{-7}$ and E = 62,000. Curve b is for the same viscosity correlation but with temperatures increased by 150 deg-K. Curve c is for the original assumed temperature profile using the viscosity data of Rossin, Bersan and Urbain (34), i.e. $\beta = 2.75 \times 10^{-7}$ and E = 60,000. Boundary conditions employed were surface areas of 400 and 300 respectively as indicated by the intercepts. We find qualitative agreement between theory and experiment, represented by the shapes of the curves, to be excellent. The sensitive relationshop between coalescence rate and viscosity is clearly evident in figure 6. Since the measured viscosity of silica has rather large uncentainty limits, depending markedly on impurity level, theoretical predictions cannot be more precise than is represented by the divergence between curves a and c. Another source of uncertainty is the possible variation of surface tension and viscosity with particle size. Despite this, theoretical surface area predictions, based on bulk properties, are, considering their accuracy, in reasonable agreement with our experimental results. Future experiments using other oxides and oxide mixtures may provide a firmer basis for theoretical prediction.

In contrast to surface area, the theoretical aggregate size is much less sensitive to physical properties and temperature. This is illustrated in figure 7 which shows experimental results for two flames compared with theoretical predictions. The aggregate size is the mass-average value (units of 10^6 atomic weight equivalents) determined by laser light scattering. Theoretical growth rates in the two Brownian regimes, free-molecule and continuum, are shown by solid lines (31). The highly-recommended extrapolation of Fuchs (35) is also shown. Details on the experimental procedure and theoretical predictions will be published later (36). Experimental results were obtained for different flow rates and temperatures in the same burner and at the same concentration. The influence of concentration is illustrated in figure 8 with data taken from two different burners at different flow rates and temperatures but four different concentrations.

Particularly noteworthy is the dramatic drop in growth rate at the point of transition from free-molecule to continuum behavior. Our results suggest a much narrower transition range than that proposed by Fuchs. Measured aggregate masses agree with predictions, and the theoretical dependence on silica concentration is confirmed. The sticking coefficient is near unity.

The collision cross-section of an aggregate, though slightly dependent on surface area, is primarily a function of the mass. Thus, growth of both highly-flocculated and spherical particles will occur at the rates represented in figures 7 and 8. This provides valuable insight into the behavior of submicron fly-ash particles in boilers.

One example where the physical mechanism of particle behavior may be important concerns the influence of a fire-side additive such as MgO. This material is effective in reducing boiler tube fouling. Used frequently as an oil additive, its application, as such, with coal, has generally required uneconomical amounts. Recent techniques which inject atomized liquid suspensions downstream of the coal burners but ahead of the superheater tubes have proven effective with much smaller addition rates (37).

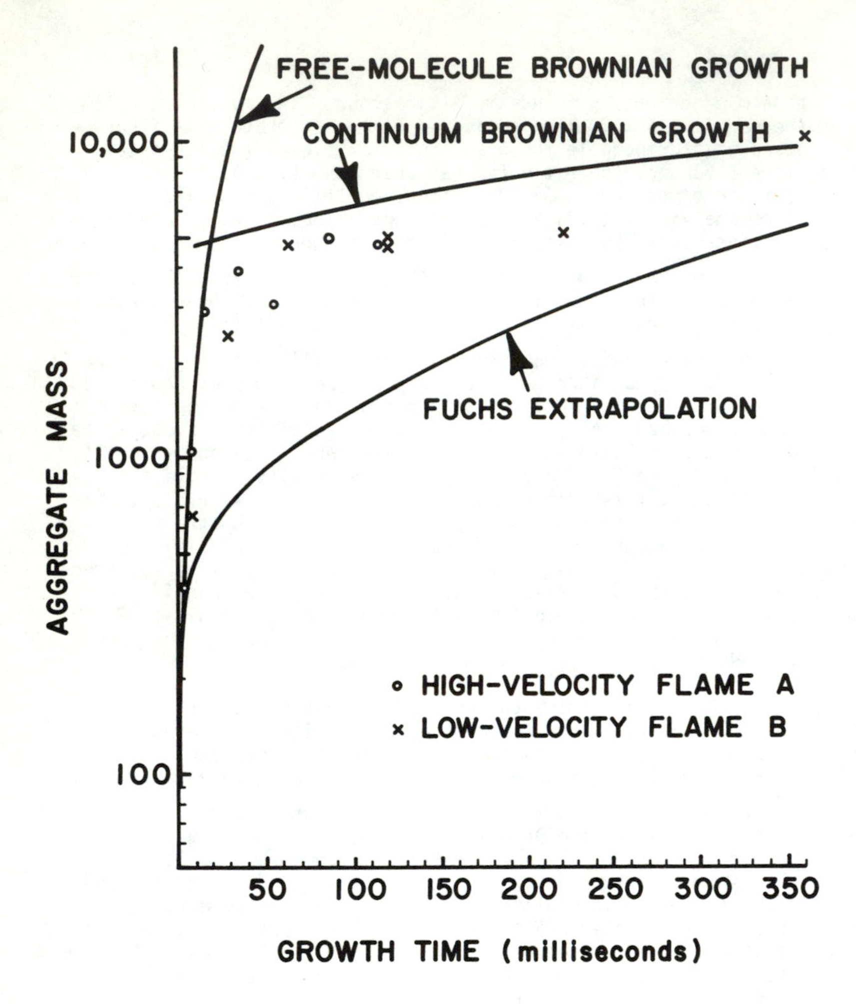

Fig. 7 Aggregate mass as a function of growth time in silica-yielding flames. Points represent data from the same burner at different flow rates and temperatures but the same silica concentration of 2.6 mole percent. Solid lines are theoretical predictions.

Of several possible mechanisms suggested, the generally preferred model views the magnesia as reacting preferentually with SO_3 to prevent its combination with alkalis to form low-melting complexes. One key to its effectiveness is the small size (sub-micron) of the well dispersed $Mg(OH)_2$ particles in the suspension. (Calcia is similarly effective but is not commonly used because of the possible formation of tough calcium sulfite coatings on cooler surfaces.) Even though acidity of deposits is reduced with the additive, it is also found effective with low-sulfur coals where sulfate formation is generally not as serious. Further, its effectiveness as a scavenger for SO_3 removal

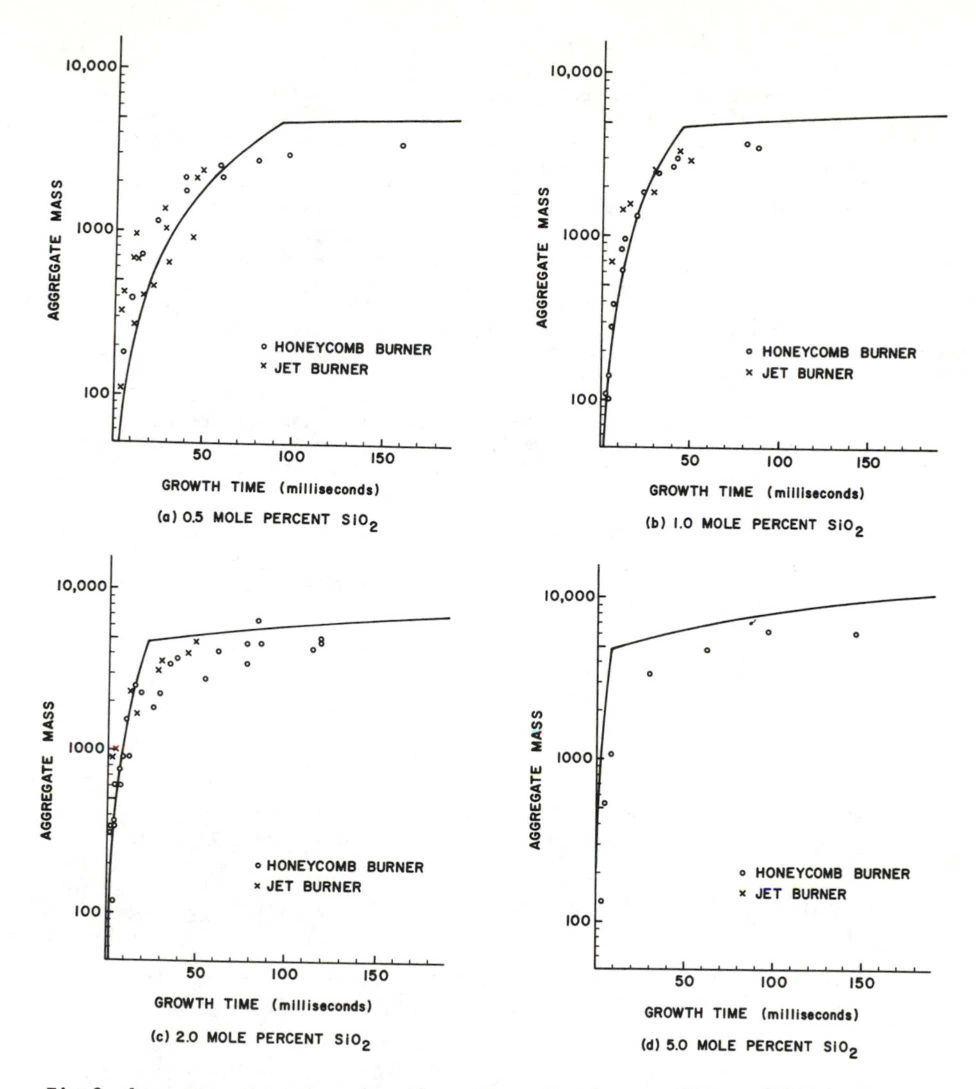

Fig 8 Aggregate mass as a function of growth time in silica-yielding flames. Points represent results from two different burners at varying flow rates and temperatures but the same silica concentration. Each frame, as indicated, is for a different concentration.

would not be dependant on particle size so long as enough adsorption area is available.

An alternate possible mechanism is that of physical inhibition. Not only are magnesia and calcia basic compounds, but they are unusually refractory. Unlike silica, they are probably present as reasonably unagglomerated single particles. The concentration, however, is so low that coagulation will not occur. The particles are small enough to be thermally mobile. Thus, as with the sub-micron ash particles, they will deposit preferentially on tube surfaces. Unlike sub-micron ash, they will not have a low melting point and, being unagglomerated, they will have high thermal conductivity. This may account for the observed friability and poor adherence of tube deposits obtained with these additives.

CONCLUSIONS

Submicron fly-ash particles, generated by vaporization and recondensation of volatile, metal-containing species, probably grow by a coagulation process similar to that of soot and oxide smokes in flames. Experiments in this laboratory reveal a strong relationship between aggregate morphology and viscosity. That is, if the viscosity is large, colliding particles are flocs comprised of large numbers of smaller primary units held together in a chain-like structure. If viscosity is small, the primary particles fuse rapidly resulting in, at the limit, a dense sphere. The flocs are too small to be influenced by turbulence and fluid motion but grow because of Brownian motion. These aggregates grow vigorously until a limiting diameter of approximately 0.3 μm is reached when the collison frequency drops dramatically. The ultimate size and structure of the stable particle is determined by its elemental composition and temperature history.

Further information on the growth of different oxides and an accurate model for metal transport are being sought. Once obtained, the nature of submicron ash from a given coal at various locations in a boiler can be predicted.

ACKNOWLEDGEMENTS

Financial support by the U.S. Energy Research and Development Administration (Contract number (E49-18)-2205), Cabot Corporation, The Public Service Company of New Hampshire, New England Electric System and the Babcock and Wilcox Company is gratefully acknowledged.

REFERENCES

1 McCrone, W.C., and Delly, J.G., Particle Atlas, 2nd ed., Vol. 2, Ann Arbor Sci. Pub., Ann Arbor, 1973, p. 547.

2 Ramsden, A.R., "Microscopic Investigation into the Formation of Fly Ash During the Combustion of a Pulverized Bituminous Coal," Fuel, Vol. 48, 1969, pp. 366-377.

3 Raask, E., "Fusion of Silicate Particles in Coal Flames," Fuel, Vol. 48, 1969, pp. 366-377.

4 Schulz, E.J., Engdahl, R.B., and Frankenburg, T.T., "Submicron Particles from a Pubverized Coal Fired Boiler," Atmospheric Environment, Vol. 9, 1975, pp. 111-119.

5 Flagan, R.C., Friedlander, S.K., "Particle Formation in Pulverized Coal Combustion," Symposium on Aerosol Science and Technology, 82nd National AIChE Meeting, Atlantic City, N.J., August 29-September 1, 1976.

6 MacKowsky, M.-Th., "The Mineral Constituents of Coal as the Causitive Agent of the Fouling of Heating Surfaces with Particular Reference to Pulverized Fuel Firing with Liquid Slag Removal," The Mechanism of Corrosion by Fuel Impurities, ed. by Johnson and Littler, Butterworths, London, 1963, pp. 80-89.

7 Padia, A.S., Sarofim, A.F. and Howard, J.B., "The Behavior of Ash in Pulverized Coal Under Simulated Combustion Conditions," Combustion Institute Central States Meeting, April 5 and 6, 1976.

8 Natusch, D.F.S., and Wallace, J.R., "Toxic Trace Elements: Preferential Concentration in Respriable Particles," Science, Vol. 183, Jan. 1974, pp. 202-204.

9 Davison, R.L., Natusch, D.F.S., Wallace, J.R., and Evans, C.A., "Trace Elements in Fly Ash, Dependence of Concentration on Particle Size," Environmental Science and Technology, Vol. 8, Dec. 1974, pp. 1107-1113.

10 Kaakinen, J.W., Jorden, R.M., Lawasani, M.H., and West, R.E., "Trace Element Behavior in a Coal-Fired Power Plant," Environmental Science and Technology, Vol. 9, Sept. 1975, pp. 862-870.

11 Friedlander, S.K., "Small Particles in Air Pose a Big Control Problem," Environmental Science and Technology, Vol. 7, Dec. 1973, pp. 1115-1118.

12 Waldmann, L., and Schmitt, K.H., "Thermophoresis and Diffusiophoresis of Aerosols," Aerosol Science, ed. by C.N. Davies, Academic Press, N.Y., 1966, pp. 137-162.

13 Goldsmith, P., and May, F.G., "Diffusiophoresis and Thermophoresis in Water Vapor Systems," Aerosol Science, ed. by C.N. Davies, Academic Press, N.Y., 1966, pp. 163-194.

14 Fuchs, N.A., The Mechanics of Aerosols, Pergamon, N.Y., 1964, pp. 56-69.

15 Derjagin, B.V., Storozhilova, A.I., and Rabinovich, Y.I., "Experimental Verification of the Theory of Thermophoresis of Particles," J. Colloid Science, Vol. 21, 1966, pp. 35-58.

16 Derjagin, B.V., and Yalamov, Y., "Theory of Thermophoresis of Large Aerosol Particles," J. Colloid Science, Vol. 20, 1965, pp. 555-570.

17 Jacobson, S., and Brock, J.R., "The Thermal Force on Spherical Sodium Chloride Aerosols," J. Colloid Science, Vol. 20, 1965, pp. 544-554.

18 Brock, J.R., "On the Theory of Thermal Forces Acting on Aerosol Particles," J. Colloid Science, Vol. 17, 1962, pp. 768-780.

19 Davies, C.N., Recent Advances in Aerosol Research, Pergamon, N.Y., 1964, pp. 61-64.

20 Fuchs, N.A., The Mechanics of Aerosols, Pergamon, N.Y., 1964, p. 65.

21 Davies, C.N., Recent Advances in Aerosol Research, Pergamon, N.Y., 1964, pp. 54-55.

22 Boothroyd, R.G., Flowing Gas-Solids Suspensions, Chapman and Hall, London, 1971, pp. 186-190.

23 Reid, W.T., External Corrosion and Deposits in Boilers and Gas Turbines, Elsevier, N.Y., 1971, p. 150.

24 Smith, W.R., and Wilkes, G.B., "Thermal Conductivity of Carbon Blacks," Ind. Eng. Chem., Vol. 36, 1944, pp. 1111-1112.

25 French, Bruce R., "Fly Ash Formation in a Coal-Fired Boiler," M.S. Thesis, Department of Chemical Engineering, University of New Hampshire (June 1976).

26 Medalia, A.I. and Heckman, F.A., "Morphology of Aggregates: Size and Shape Factors of Carbon Black Aggregates from Electron Microscopy," Carbon, Vol. 7, 1969, pp. 567-582.

27 Cruise, D.R., "Notes on the Rapid Computation of Chemical Equilibria," J. Phys. Chem., Vol. 68, 1964, pp. 3797-3803.

28 Stull, D.R., Project Director, JANAF Thermochemical Tables, The Dow Chemical Co., Midland, Michigan, 1965.

29 Ulrich, G.D., "Theory of Particle Formation and Growth in Oxide Synthesis Flames," Combustion Science and Techology, Vol. 4, 1971, pp. 47-57.

30 Ulrich. G.D., Milnes, B.A., and Subramanian, N.S., "Particle Growth in Flames II, Experimental Results for Silica Particles," Combustion Science and Technology, Vol 14, 1976, pp. 243-249.

31 Ulrich, G.D., and Subramanian, N.S., "Particle Growth in Flames III, Coalescence as a Rate-Controlling Process," Combustion Science and Technology (to appear).

32 Hofmaier, G., and Urbain G., "The Viscosity of Pure Silica," Science of Ceramics, Vol. 4, 1967, pp. 25-32.

33 Kingery, W.D., Introduction to Ceramics, Wiley, N.Y., 1960, p. 574.

34 Rossin, R., Bersan, J., and Urbain, G., "Viscosities of Molten Silica and Silica-Alumina Mixtures," Compt. Rend., Vol. 258, 1964, pp. 562-564.

35 Hidy, G.M., and Brock, J.R., "Some Remarks about the Coagulation of Aerosol Particles by Brownian Motion," J. Colloid Science, Vol. 20, 1965, pp. 477-491.

36 Riehl, J.W., and Ulrich, G.D., "Aggregation and Growth of Sub-micron Oxide Particles in Flames," (to appear).

37 Radway, J.E., and Boyce, T., "Reduction of Coal Ash Deposits with Magnesia Treatment," (this volume).

CORROSION IN GAS TURBINES

Session Chairman: **H. V. Doering**
General Electric Company
Co-Chairman: **R. L. McCarron**
General Electric Company

THE EFFECT OF NaCl(g) ON THE Na_2SO_4-INDUCED HOT CORROSION OF NiAl

J. G. SMEGGIL, N. S. BORNSTEIN, and M. A. DeCRESCENTE

United Technologies Research Center
East Hartford, Connecticut, USA 06108

ABSTRACT

Studies have been performed to examine the effect of NaCl vapor on the Na_2SO_4-induced hot corrosion of the alumina former NiAl. In the incubation period associated with such hot corrosion, NaCl(g) has been shown to be effective in removing aluminum from below the protective alumina scale and redepositing it as Al_2O_3 whiskers on the surface of the Na_2SO_4-coated sample. Similar effects seen in simple oxidation are associated with isothermal rupturing of the protective alumina scale.

INTRODUCTION

In gas turbines both nickel-based superalloys and the protective coatings, which include the simple aluminides and the many MCrAl(Y) compositions, develop aluminum-enriched protective oxide scales upon exposure to air at elevated temperatures (Ref. 1). However, simple oxidation is not the primary corrosion problem encountered in gas turbines operating in industrial and marine environments. Sea salt crystals composed in part of NaCl and Na_2SO_4, other species in the air and impurities present in fuel deposit onto turbine components increasing metal wastage rates orders of magnitude over those associated with simple oxidation. Based upon thermodynamic calculations, the environment within gas turbines operating under conditions favoring Na_2SO_4-induced hot corrosion contains low levels of NaCl vapor (Ref. 2). Extensive work by many investigators has shown that condensed NaCl or NaCl-Na_2SO_4 mixtures can be very corrosive (Ref. 3-19). Additionally, Hancock, using a vibrational technique, has shown that

*Work supported by NASA-Lewis Research Center, Contract No. NAS3-20039.

NaCl vapor under otherwise pristine oxidizing conditions adversely affects the isothermal oxidation behavior of a number of substrates (Ref. 6). The nature of these oxide scales is quite complex. Recently Fryburg, Kohl, and Stearns (Ref. 20) have examined oxide scales formed on a number of alloys in atmospheres containing gaseous NaCl. Some of these alloys approximate the compositions examined by Hancock (Ref. 6). Their results indicate heterogeneous oxide scales are formed during the oxidation of these alloys. Thus the conditions under which NaCl vapor adversely affects protective oxide scales is not well characterized.

Recently, Smeggil and Bornstein have shown that NaCl(g) substantially affects the high temperature oxidation behavior of the chemically simple alloy NiAl (or β), an alumina former (Ref. 21). This effect was observed at NaCl levels expected to be present in gas turbine hot-section atmospheres. In that work, aluminum was found to be selectively removed from below the normally protective Al_2O_3 scale and redeposited as Al_2O_3 whiskers on that scale. Concurrently the NaCl(g) caused spallation of the normally protective alumina scale at constant temperature.

In order for the accelerated attack associated with Na_2SO_4 to occur, the Na_2SO_4 must be present in the condensed state (Ref. 22). However, as stated earlier, low levels of NaCl(g) will also be present in such environments. Since only very small amounts, less than 0.01 mg/cm^2, of condensed NaCl are found on surfaces of first stage turbine hardware (Ref. 23) and since NaCl exhibits a relatively high vapor pressure (2.5 torr at 900°C (Ref. 24)), it is generally considered that NaCl does not normally play a significant role in Na_2SO_4-induced hot corrosion processes. However, since it has been shown that even low activities of NaCl(g) influence oxidation behavior, the objective of this study was to determine the extent of its effects in Na_2SO_4-induced hot corrosion processes. The substrate examined here was the alumina former NiAl, the phase principally responsible for the surface protection afforded by the simple aluminide and MCrAl(Y) compositions used to coat turbine hardware.

EXPERIMENTAL PROCEDURES

The preparation of the NiAl alloy samples and the experimental procedures involved in this work has been largely described elsewhere (Ref. 21). The NiAl used was single phase and contained 31.06 ± 0.09 wt.% Al. The specimens cut from the annealed ingot were ground to 600 grit SiC. The air used in this work was dried over anhydrous calcium sulfate (Drierite) and passed over the oxidizing samples at a flow rate of 300 scc/min with a velocity of 0.18 cm/sec. The NaCl and Na_2SO_4 used were ultrapure grades obtained from Alfa-Ventron. The chloride level of the Na_2SO_4 was determined to be 2.3 ± 0.4 ppm reported as NaCl.

The NaCl vapors were generated in the experimental apparatus by holding condensed NaCl in a platinum crucible on a movable pedestal inserted from the bottom of a quartz oxidation tube. The temperature of the NaCl-containing crucible was measured by an adjacent thermocouple fixtured into the tube. The empirical NaCl partial pressures were determined from the differences in weights of the NaCl-containing crucible before and after each experiment and the mass of air which flowed past the crucible in the course of the experiment. Since the NaCl crucible was held at temperatures well below those experienced by the samples undergoing oxidation, the NaCl vapors emanating from the crucible could not condense onto the oxidizing specimens.

Hot corrosion experiments were conducted with sodium sulfate-coated specimens. The amount of sodium sulfate deposited was determined by weighing the sample before and after the coating application. Although the deposited coatings visually appeared uniform, at temperature the salt melted and may have tended to flow to the lower part of the specimen.

Samples with and without Na_2SO_4 coatings were oxidized in pure air and in NaCl-bearing atmospheres at 900°C and 1050°C for periods of 24 hours. After exposure selected samples were examined via scanning electron microscopy, optical metallography and electron microprobe techniques. When necessary, sections of samples were polished under kerosine using SiC and diamond abrasive powders. Such samples were quickly inserted into the microprobe or scanning electron microscope after polishing to minimize oxidation or hydrolysis effects. X-ray diffraction studies were also used to identify oxidation products.

EXPERIMENTAL RESULTS AND DISCUSSION

NaCl Losses From Molten Na_2SO_4

Molten ultrapure Na_2SO_4 did not preferentially lose the chloride present as an impurity (low ppm levels) even after heating at 900°C for extended periods.

The level of sodium chloride observed in the ultra pure grade Na_2SO_4 used in this study is nominally larger than that needed in simple oxidation to yield both Al_2O_3 whisker growth and isothermal Al_2O_3 scale spallation (Ref. 21). Gram-size samples of the pure Na_2SO_4, contained in a platinum crucible in a box furnace, were held at 900°C for periods of 2, 6, 24 and 48 hours. The results of the chemical analyses, Fig. 1, show that the level of sodium chloride in the Na_2SO_4 melts increased over a period of 24 hours from less than 3 ppm to about 20 ppm. This increase in chloride content cannot be attributed to the preferential vaporization of Na_2SO_4 and concommitant enrichment of the residue in NaCl. At 900°C, Na_2SO_4 and NaCl exhibit partial pressures of 2.5×10^{-5} and 2.5 torr, respectively (Refs. 24 and 25). Thus the NaCl responsible for the chloride increase in the melt must have been derived from the furnace atmospheres. The furnace used for this work was an ordinary laboratory box furnace previously used to examine the effect of Na_2SO_4 corrosion effects on metal substrates and no special precautions were taken to assure NaCl was not present. Most importantly, these results indicate that NaCl at low ppm levels is not easily lost by preferential vaporization from

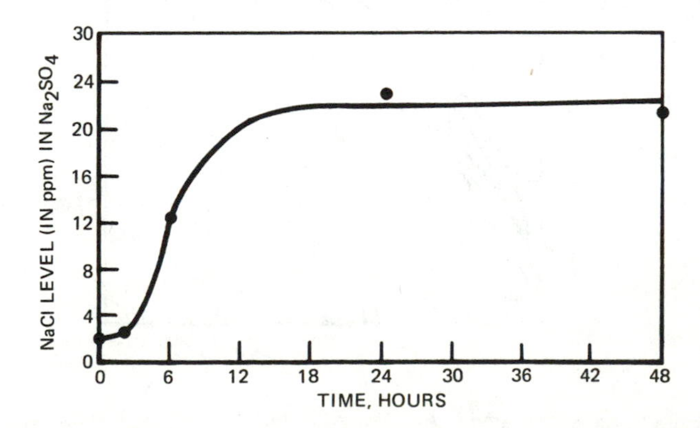

Figure 1. Change in NaCl concentration at 1050°C in air with NaCl vapors.

molten Na_2SO_4 deposits at 900°C. Wolters reported the NaCl-Na_2SO_4 phase diagram to be a simple eutectic with no significant solid solution of NaCl in the Na_2SO_4 (Ref. 26). Flood has remeasured the Na_2SO_4 branch of the liquidus and has shown that these two salts form a practically ideal solution (Ref. 27). Thus it is concluded that Na_2SO_4, normally used to study Na_2SO_4 induced corrosion effects contains NaCl in solution at impurity levels (ppm values) and, at these concentrations, the activity and behavior of the NaCl is similar to that observed in the gas phase.

The Effect of NaCl on the Sulfidation Incubation Period

The thermogravimetric data for specimens coated with Na_2SO_4 and oxidized at 900 and 1050°C are shown in Figs. 2 and 3, respectively. Based upon the experimental data for samples oxidized at 900°C, either (a) NaCl(g) at these levels has no effect on the Na_2SO_4-induced hot corrosion of NiAl or (b) the corrosion mechanism involving NaCl is already saturated by the lowest NaCl levels intrinsically present at impurity levels in the Na_2SO_4 layer.

On the other hand at 1050°C, the NaCl(g) added to the atmosphere appears to mitigate the corrosive effects of condensed Na_2SO_4 in the times investigated,

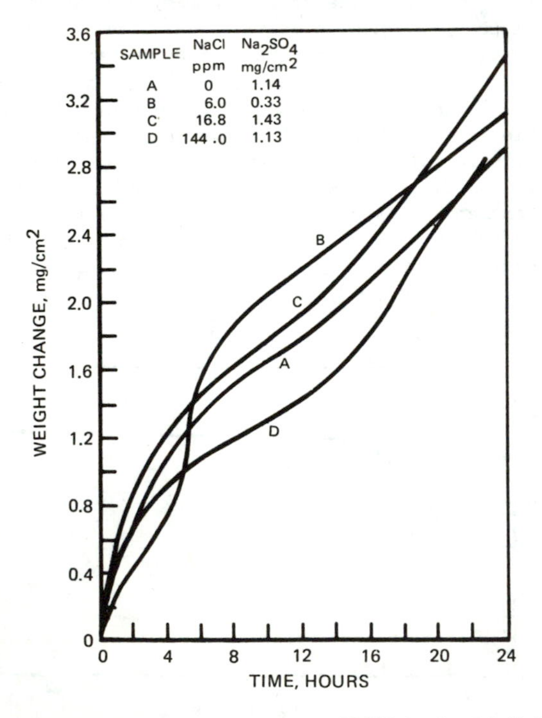

Figure 2. Na_2SO_4 coated NiAl oxidized at 900°C in air with NaCl vapors.

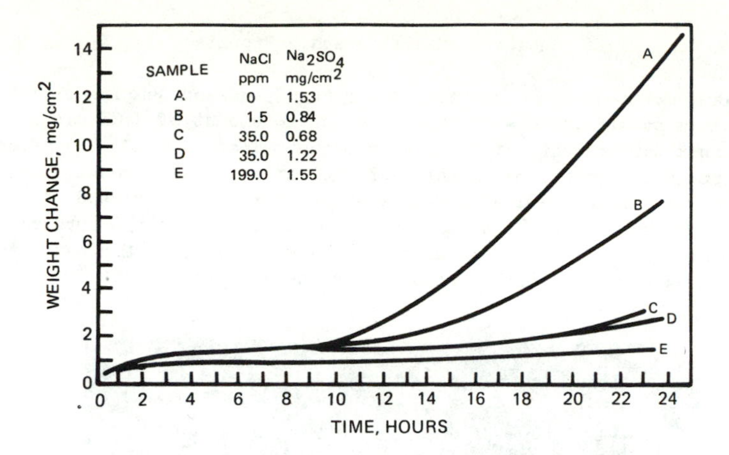

Figure 3. Na_2SO_4 coated NiAl oxidized at 1050°C in air with NaCl vapors.

Fig. 3. In fact, as the NaCl(g) partial pressure is increased, the duration of the incubation period is increased. Furthermore, the microstructure of the Na_2SO_4-coated NiAl sample oxidized at 1050°C for twenty-four hours in an atmosphere containing 199 ppm NaCl exhibits a structure unaffected by the Na_2SO_4 deposit, Fig. 4, in agreement with the thermogravimetric data, cf. Fig. 3.

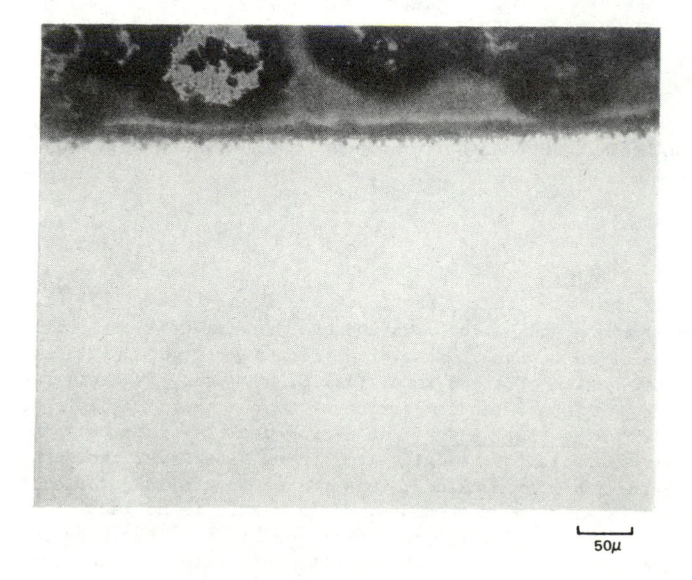

Figure 4. NiAl coated with Na_2SO_4 (1.55 mg/cm²) and then oxidized at 1050°C in air with 199 ppm NaCl(g) for 24 hours.

Substrate Metallographic Results

An examination of the substrate of the Na_2SO_4-coated samples oxidized at 900°C indicates no differences attributable to the amount of NaCl vapor introduced into the ambient oxidizing atmosphere. That is, NaCl has affected the microstructure to the same extent over the entire range investigated, 1 to 144 ppm. The role of NaCl at levels examined here is restricted to scale rupture and the development of an aluminum depletion zone in the substrate. Examination of the sample of NiAl oxidized at 900°C reveals a microstructure ostensibly in agreement with that reported in the literature, Fig. 5.

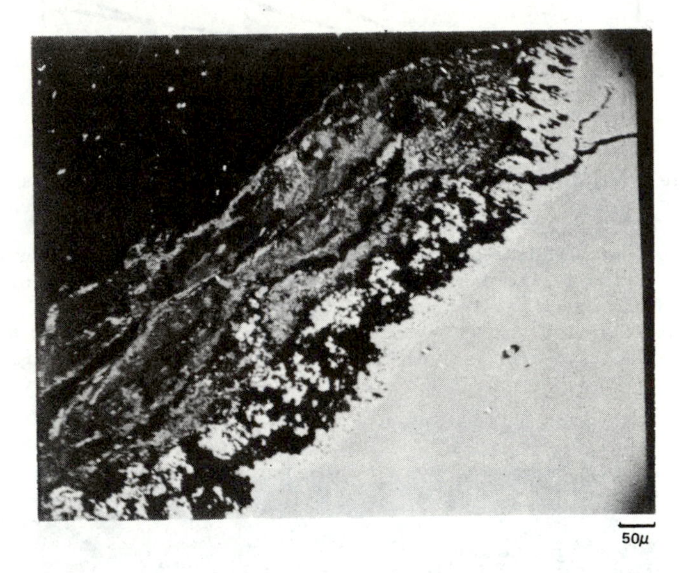

Figure 5. NiAl coated with Na_2SO_4 (1.14 mg/cm²) and oxidized in air at 900°C for 24 hours.

At the base of the oxide scale, particles are found in a γ' (Ni_3Al) layer that are largely aluminum and sulfur-rich and are presumably Al_2S_3. These particles are largely retained in the γ' zone. Occasionally larger particles extend into the Al-depleted region of the β zone, Fig. 6. However, in these larger particles, the portions extending further into the aluminum-depleted β area are enriched in <u>oxygen</u> <u>not</u> <u>sulfur</u>. In microprobe traverses of these particles from the NiAl substrate to the oxide-atmosphere interface, the composition of such particles changes from oxides, to sulfides to a mixture of sulfides and oxides and finally to only oxides. This result is not anticipated from a consideration of equilibrium thermodynamic conditions. Furthermore, sulfide particles have been reported within the NiAl substrate of specimens which have undergone Na_2SO_4-induced corrosion (Ref. 28). However in that report no mention is made of the presence of oxide particles as seen here. Furthermore, such oxygen-rich areas are not found in particles totally retained in the γ' zone. Such oxygen-rich particles are anticipated if the protective alumina scale ruptured and suddenly exposed the aluminum-depleted substrate to a high oxygen potential, i.e., the sulfate melt. The rapid depletion of oxygen by oxide formation would then result in a sudden increase in sulfur potential

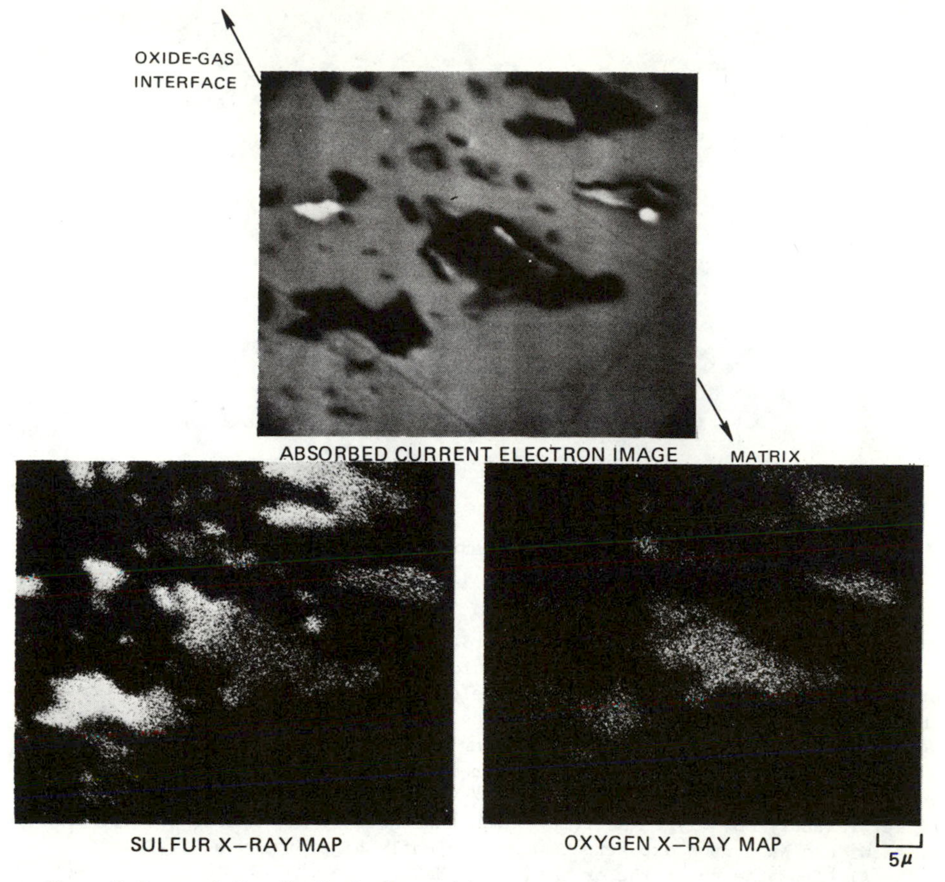

Figure 6. Oxygen-Sulfur distribution in particles extending into the Al-Deficient NiAl layer.

leading to aluminum sulfide formation behind the oxide front (Refs. 22, 29, 30).

Microstructural Effects Due to NaCl(g)* in Laboratory
Pure Na_2SO_4 Induced Sulfidation

An examination of the surface of the Na_2SO_4-coated NiAl sample exposed at 900°C to 144 ppm NaCl(g) reveals the presence of profuse quantities of Al_2O_3 whiskers, Fig. 7. Similar whiskers have been observed on NiAl specimens oxidized in NaCl vapor-containing atmospheres (Ref. 21). It has been hypothesized that the whisker growth involves stresses in the growing protective alumina oxide scale (Ref. 31). However the Al_2O_3 whiskers observed here have clearly not developed on a highly stressed oxide substrate. Moreover, the whiskers appear to have formed from a reaction occurring near the oxide scale-gas interface.

In light of the possibly enhanced vaporization of Na_2SO_4(c) effected by

*In this report the bracketed c implies the phase is present in the condensed state, either liquid or solid. The bracketed g similarly refers to the gaseous state.

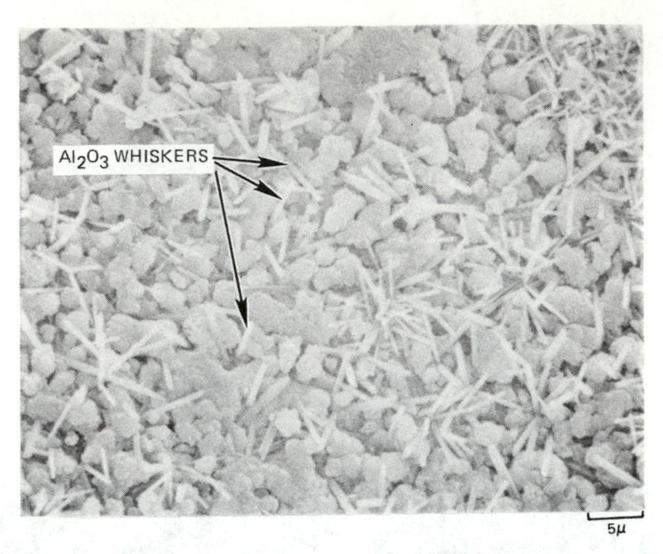

Figure 7. NiAl coated with Na_2SO_4 (1.13 mg/cm²) and then oxidized at 900°C in air with 144 ppm NaCl(g) for 24 hours.

NaCl vapor alone (Ref. 15), a spot test to confirm the presence of Na_2SO_4(c) on the surface of the sample after oxidation for 22 hours at 900°C in an atmosphere containing 144 ppm NaCl(g) yielded a positive result. Therefore, these whiskers did not grow on the surface of the NiAl sample after all the Na_2SO_4 had been removed from the oxidizing surface.

A proposed mechanism for the growth of such crystals involves chemical vapor transport processes by an "Al-NaCl" vapor species stable under conditions of sufficiently low oxygen activity which subsequently decomposes to some form of Al_2O_3 and NaCl upon exposure to a sufficiently high oxygen activity:

$$\text{Al}\begin{bmatrix}\text{from the substrate or the}\\ \text{scale at the metal-oxide}\\ \text{interface}\end{bmatrix} + \text{NaCl(g)} \rightarrow \text{"Al-NaCl" (g).} \qquad (1)$$

The "Al-NaCl" species, forming at the metal-oxide interface is stable under the low partial pressures of oxygen in this region, (Ref. 21). At the oxide-gas interface, the diffusing species encounters a high oxygen partial pressure in which it is thermodynamically unstable and decomposes to yield α-Al_2O_3 whiskers. Prior work has shown that the aluminum needed to form these whiskers is supplied at the oxide-metal interface and not the oxide-gas interface (Ref. 21).

In the presence of condensed Na_2SO_4, the "Al-NaCl" species experiences an increased oxygen activity in the form of a layer of molten Na_2SO_4. The presence of Na_2SO_4 deposit can be described in terms of the following equilibrium:

$$Na_2SO_4(c) = Na_2O(c) + SO_2(g) + 1/2\ O_2(g). \qquad (2)$$

The "Al-NaCl" species will remove oxygen from Na_2SO_4(c) or the phases in equilibrium with it (cf. Eq. 2) to yield a form of alumina "Al_2O_3" and release NaCl(g) to regenerate the cycle. If this "Al_2O_3" and not the alumina comprising the protective scale itself is available in sufficient amounts to react with the sodium oxide yielding $NaAlO_2$, the protective scale breakdown process will be delayed as will the sulfidation-initiation process.

Schematically these effects can be represented by the following schematic reaction:

$$Na_2SO_4(c) + "Al\text{-}NaCl(g)" \rightarrow NaAlO_2(c) + SO_2(g) + O_2(g) + NaCl(c,g). \quad (3)$$

This reaction will also enhance the loss of condensed Na_2SO_4 from substrate surfaces. If the flux of "Al-NaCl" species into the Na_2SO_4 deposit is sufficiently large such that the Na_2SO_4 deposits are removed prior to the scale rupturing, then metal wastage rates will approximate those seen in oxidation in atmospheres containing NaCl(g).

It is suggested, based on the results presented here, that the "Al-NaCl" flux is large enough at 1050°C but not at 900°C to supply the necessary amount of aluminum as "Al_2O_3" to delay the protective scale breakdown process for the times examined here, Figs. 3 and 4. Furthermore, the transport of "Al-NaCl" across the protective scale may be sufficiently high so that the production rate of Al_2O_3 is high compared with its removal rate by interaction with the Na_2SO_4 and phases in equilibrium with it, in particular sodium oxide. Under such circumstances, then alumina whiskers are anticipated to be found on the oxide surface. This situation will be particularly possible in atmospheres with high NaCl(g) levels, cf. Fig. 7.

The observation in the results reported here that kinetically no differences are observed in the oxidation behavior of Na_2SO_4-coated NiAl at 900°C in atmospheres with NaCl(g) concentrations ranging from 1 to 144 ppm suggests that similar effects are occurring in the Na_2SO_4-coated sample oxidized in nominally pure air. It suggests that in the case of no intentional additions of NaCl(g) to the oxidizing atmospheres, the NaCl needed for the transport derives from the impurity levels present in the Na_2SO_4 which is not lost upon heating at 900°C. In the case of Na_2SO_4-induced hot corrosion of NiAl, oxide scales which form are porous. Accordingly, the "Al-NaCl" vapor species, in the case of no intentional additions of NaCl(g) to the atmosphere of an oxidizing Na_2SO_4-coated NiAl specimen, are few in number since the NaCl content of the laboratory grade Na_2SO_4 is low. The "Al-NaCl" species will largely decompose within the porous oxide scale. Few "Al-NaCl" molecules will survive and escape to the gas-oxide interface. Therefore, large concentrations of Al_2O_3 whiskers will not subsequently be detected there.

Al_2O_3 Scale Breakdown Mechanism

The effect of a transport mechanism involving an "Al-NaCl" moiety on the sulfidation process is two-fold. Aluminum is removed from below the oxide scale thereby locally depleting the aluminum level in the substrate alloy, Fig. 8. Secondly, the mechanical adherence of the protective alumina oxide scale to the substrate is weakened as more and more aluminum is withdrawn. Eventually, when the alumina scale ruptures, the concentration of aluminum is locally insufficient to reheal the break. Moreover at these temperatures (900 and 1050°C), the diffusion rates of Al are insufficient to migrate to the broken oxide area. In the presence of molten Na_2SO_4, sulfidation attack will occur at the site of the break. Furthermore, if a protective oxide scale tries to reform, the continued presence of trace NaCl levels in the Na_2SO_4 deposit would negate the effective formation of such a protective scale. Moreover, even if the Na_2SO_4 were locally totally consumed in attacking the unprotected aluminum depleted substrate, the situation involving the presence of an extended aluminum depletion zone and insufficient aluminum diffusion rates at 900°C would result in continued attack at accelerated rates for extended periods of time. The local absence of a protective Al_2O_3 layer and the presence of an extended aluminum-depletion layer have been shown by Goebel and Pettit to be sufficient to yield accelerated

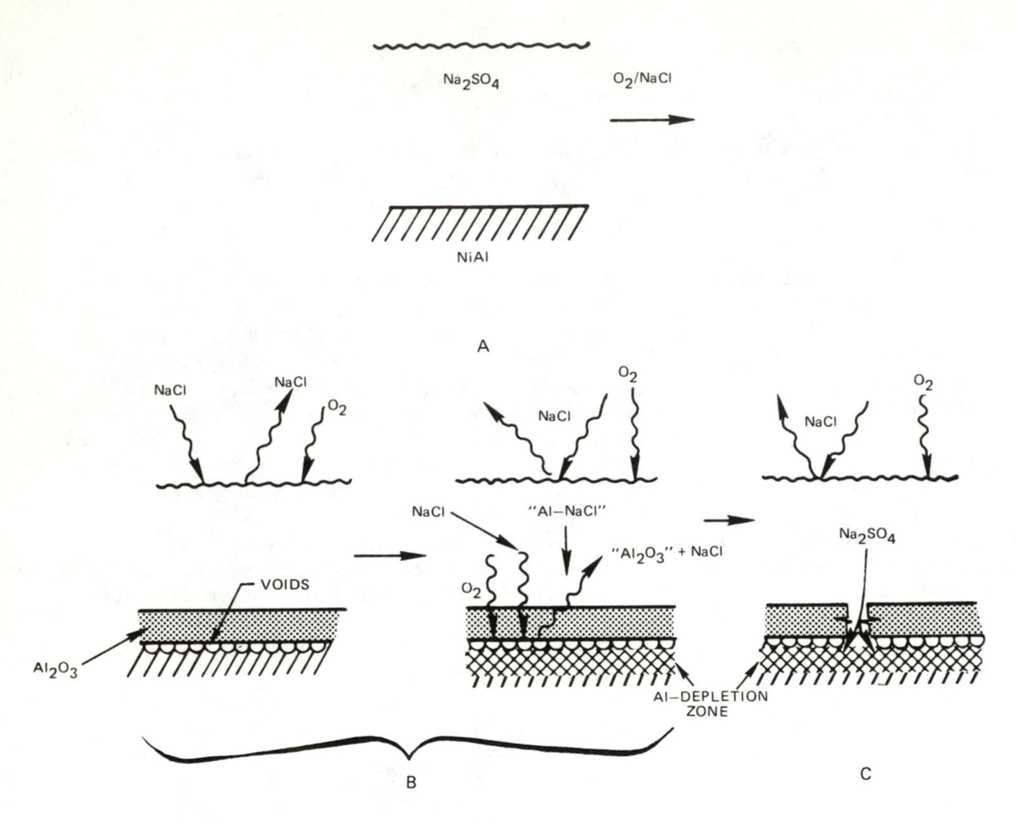

Figure 8. Effect of NaCl(g) on the Na_2SO_4 induced corrosion of NiAl.
A. Prior to oxidation
B. Incubation phase
C. Propagation phase

rates of attack of presulfidized Ni-31 wt.% Al subsequently oxidized at 1000°C in 1.0 atm oxygen (cf. Fig. 12 in Ref. 32).

This interpretation is not contradictory to the results in the literature in which pre-sulfidized B-1900 Waspalloy and U-700 were found to subsequently oxidize at rates comparable to the oxidation rates for unsulfidized samples (Ref. 33). That presulfidation treatment yielded substrate microstructures with sulfide precipitates in sulfidation affected zones grossly similar to those found after exposure of the base materials to Na_2SO_4-induced corrosion effects. Based on the composition of those alloys, the sulfides formed were likely Cr and refractory metal-rich. However, the aluminum content of the sulfide affected zone would have remained the same and probably even increased relative to the substrate because of the loss of the elements which are more prone to sulfide formation by precipitation from the substrate metal. The depletion zone formed in the work presented here and in the work presented by Goebel and Pettit involves aluminum depletion (Ref. 32).

It is tempting to describe the beneficial behavior of NaCl(g) toward the Na_2SO_4-induced hot corrosion of NiAl at 1050°C by the term "inhibitor." However, as is obvious from the discussion, NaCl(g) does not function as an inhibitor in the usual sense of the word. Cr_2O_3, an example of such an inhibitor, reacts directly with the Na_2SO_4 deposits on metallic substrates. The metallic substrate there does not directly interact with the Cr_2O_3. On the other hand, the beneficial effect of NaCl(g) is only obtained by direct interaction with the substrate. The mode of this interaction results in the direct sacrificial loss of aluminum from the substrate.

This work suggests the use of an aluminum compound as an inhibitor (in the usual sense of the word) injected into the combustion chambers may or may not be effective. The success however will depend on whether, in the gas or condensed state, suitable reducing conditions exist under which an "Al-NaCl" moiety can be preferentially formed with the aluminum contributed by the inhibitor.

SUMMARY AND CONCLUSIONS

1 NaCl(g) at 900°C behaves with respect to Na_2SO_4-coated NiAl, as it does toward NiAl in the absence of condensed Na_2SO_4. Specifically, Al_2O_3 whiskers are observed on the surface of the oxide. The source of the aluminum originates from below the outer surface of the protective alumina scale.

2 The removal of aluminum from below the surface of the oxide layer locally depletes the substrate of aluminum and leads to a progressive weakening of the protective scale-substrate bond.

3 Upon rupture of the protective Al_2O_3 scale, the substrate locally has an insufficient aluminum activity to reform the protective scale. At low temperatures diffusion rates are insufficient to allow aluminum to reform the protective alumina scale. This condition has been shown in oxidation studies to be sufficient to cause accelerated oxidation rates in the absence of a molten salt.

4 One role of the molten Na_2SO_4 in Na_2SO_4-induced corrosion is to maintain relatively low levels, i.e., ppm amounts, of species such as NaCl in immediate contact with oxide scales.

5 NaCl at the impurity levels, i.e., low ppm values, intrinsically present in laboratory grade Na_2SO_4 deposits has been shown <u>not</u> to be preferentially removed from condensed Na_2SO_4 at 900°C.

6 NaCl vapor mitigates the effects of Na_2SO_4 deposits in the hot corrosion of NiAl at 1050°C in the time periods examined here. However, in effecting this result, NaCl vapors function differently than do "normal" inhibitors, e.g., Cr_2O_3, in the Na_2SO_4-induced hot corrosion process.

ACKNOWLEDGEMENTS

The authors acknowledge helpful discussions with Mr. C. A. Stearns, Dr. F. J. Kohl and Dr. G. C. Fryburg of NASA-Lewis Research Center, Cleveland, Ohio and with Dr. R. A. Pike of UTRC and the efforts of Ms. J. Whitehead, Messrs. R. Brown, L. Jackman, J. Knecht and G. McCarthy for assistance in conducting the experiments presented here.

REFERENCES

1 Grisaffe, S. J.: Coatings and Protection. The Superalloys, p. 341, ed. by C. T. Sims and W. C. Hagel, John Wiley and Sons, New York (1972).

2 Kohl, F. J., C. A. Stearns, and G. C. Fryburg: Metal-Slag-Gas Reactions and Processes, p. 649, ed. by Z. A. Foroulis and W. W. Smeltzer, Electrochemical Society (1975); also NASA TMX-71641 (1975).

3 Hancock, P.: Corrosion of Alloys in High Temperatures in Atmosphere Consisting of Fuel Combustion Products and Associated Impurities - A Critical Review, Her Majesty's Stationary Office, London (1968).

4 Hurst, R. C., J. B. Johnson, M. Davies and P. Hancock: Deposition and Corrosion in Gas Turbines, p. 143, ed. by A. B. Hart and A. J. B. Cutler, John Wiley and Sons, New York (1973).

5 Hancock, P., R. C. Hurst, and A. R. Sollars: Published in Chemical Metallurgy of Iron and Steel, Special Publication by Ir. & Steel Inst. p143 (1973).

6 Condé, J. F. G. and B. A. Wareham: Proceedings of the 1974 Gas Turbine Materials in the Marine Environment Conference, p. 73, ed. by J. W. Fairbanks and I. Machlin, MCIC 75-27, Castine, Maine, July, 1974 and Hancock, P.: ibid., p. 225.

7 Shirley, H. T.: J. of Iron and Steel Inst. 182, 144 (1956).

8 Pickering, H. W., F. H. Beck, and M. G. Fontana: Trans. ASM 53, 792 (1961).

9 Seybolt, A. U.: Oxid. of Metals 2, 119 (1970).

10 Seybolt, A. U.: Oxid. of Metals 2, 161 (1970).

11 Alexander, P. A.: Laboratory Studies of the Effects of Sulphates and Chlorides on the Oxidation of Superheated Alloys. The Mechanism of Hot Corrosion by Fuel Impurities, p. 571, ed. by H. R. Johnson and D. J. Littler, Butterworths, London (1971).

12 Mansfeld, F., N. E. Paton, and W. M. Robertson: Met. Trans. 4, 321 (1973).

13 Johnson, D. M., D. P. Whittle, and J. Stringer: Corr. Sci. 15, 721 (1975).

14 Lewis, H. and R. A. Smith: Corrosion of High-Temperature Nickel-Base Alloys by Sulphate-Chloride Mixtures. First Int. Congress on Met. Corr., p. 202, Butterworths, London (1961).

15 Felten, E. J. and F. S. Pettit: Degradation of Coating Alloys in Simulated Marine Environments. Contract No. N00173-76-C-0146. First Quarterly Report, June 15, 1976.

16 Radzavich, T. J. and F. S. Pettit: Degradation of Coating Alloys in Simulated Marine Environments. Contract No. N00173-76-C-0146. Second Quarterly Report, September 15, 1976.

17 Barkalow, R. H. and F. S. Pettit: Degradation of Coating Alloys in Simulated Marine Environments. Contract No. N00173-76-C-0146, Third Quarterly Report, December 15, 1976.

18 Jones, R. L.: The Sulfidation of CoCrAlY Turbine Blade Coatings by Na_2SO_4-NaCl. Naval Research Laboratory LTR Report 6170859, December 1, 1975.

19 Jones, R. L., K. H. Stern, and S. T. Gadowski: Interactions of Sodium Chloride and Sodium Sulfate. To be published in Proceedings of the 1976 Gas Turbine Material in the Marine Environment Conference, Bath, England, July 1976.

20 Fryburg, G. D., F. J. Kohl, and C. A. Stearns: Oxidation in Oxygen at 900°C and 1000°C of Four Nickel Base Cast Superalloys: NASA-TRW VIA, B-1900, Alloy 713C and IN-738, NASA TN D-8388 (1977).

21 Smeggil, J. G. and N. S. Bornstein: Accepted for publication by the Journal of the Electrochemical Society.

22 DeCrescente, M. A. and N. S. Bornstein: Corrosion 24, 127 (1968).

23 Bessen, I. I. and R. E. Fryxell: Proceedings of the 1974 Gas Turbine Materials in the Marine Environment Conference, p. 219, ed. by J. W. Fairbanks and I. Machlin, MCIC Report 75-27, Castine, Maine, July, 1974.

24 Ewing, C. T. and K. H. Stern: J. Phys. Chem. 78, 20, 1998 (1974).

25 Fryxell, R. E., C. A. Trythall, and R. J. Perkins: Corrosion 29, 423 (1968).

26 Wolters, A. Neues: Jahrb. Mineral Geol. Beilage 30, 55 (1910); cf. Flood, H., T. Forland and A. Nesland. Acta Chem. Scand. 5, 1193 (1951).

27 Flood, H., T. Forland and A. Nesland: Acta Chem. Scand. 5, 1193 (1951).

28 Goebel, J. A., F. S. Pettit, and G. W. Goward: Met. Trans. 4, 261 (1973).

29 Bornstein, N. S. and M. A. DeCrescente: TMS-AIME 245, 1947 (1969).

30 Bornstein, N. S. and M. A. DeCrescente: Investigation of Sulfidation Mechanism in Nickel-Base Superalloys. Final Report conducted for U.S. Naval Ship Research and Development Laboratory, Contract N00600-68-C-0639, April 1969, Annapolis, Maryland.

31 Kuenzly, J. D. and D. L. Douglass: Oxid. of Met. 8, 139 (1974).

32 Geobel, J. A. and F. S. Pettit: Met. Trans. 1, 3421 (1970).

33 Bornstein, N. S. and M. A. DeCrescente: TMS-AIME 245, 1947 (1969).

CHARACTERIZATION OF CORROSION AND DEPOSITS IN THE LABORATORY GAS TURBINE SIMULATOR BURNING HEAVY FUELS WITH ADDITIVES

S. Y. LEE and C. J. SPENGLER

Westinghouse R&D Center
Pittsburgh, Pennsylvania, USA 15235

ABSTRACT

Gas turbine oil soluble additives were evaluated at various turbine blade metal temperatures in a laboratory gas turbine simulator burning fuels which contain potentially corrosive trace elements. Deposits that form when using magnesium and silicon based additives consist of spherical particles of MgO and cristobalite cemented together by magnesium orthovanadate and magnesium orthophosphate. In the additive testing, the fuel used always contained trace amounts of Na, Ba, P, K (usually less than 1 ppm) in addition to vanadium. For a high vanadium test ($\sim$100 ppm V) the small amount of these elements did not influence the performance of additives. But for a low vanadium test (< 10 ppm V), these trace elements were found to influence the effectiveness of additives significantly. Electron microprobe analysis scanning near the metal surface indicated that Na, Ba, etc., will segregate within the deposit and tend to concentrate near the deposit and scale interface where they can melt and cause hot corrosion of superalloys. The molten material was found to be a mixture of Na_2SO_4, $BaSO_4$ and Ni_2SiO_4 some of which will melt below 927°C. It is also possible that these compounds will form an eutectic mixture. It is concluded that trace elements other than vanadium in the fuel form significant phases when there is a low additive to trace element ratio and that segregation and reaction with oxide scales by certain elements produces localized regions of accelerated attack.

INTRODUCTION

In the present era of increasing cost of refined fuels, gas turbines used in either base-load or peaking service may have to burn available alternate fuels. However, industry-wide experience in burning crude or residual oils which contain levels of trace elements known to cause corrosion of the gas turbine hot parts is limited because most gas turbine operating experience is with clean distillates and natural gas. In addition, the modern gas turbines are designed to operate at higher gas and metal temperatures where the mechanical properties of the alloys become the limiting factor.

Increased temperature and increased level of trace metal content in the fuel are factors which promote corrosion attack of gas turbine alloys, and consequently reduce service life of the hot components. Recently, both ASTM and ADME revised the gas turbine fuel specification to limit the trace metal entering the turbine combustor as shown in Table 1 (1)[1]. Any fuel with a trace metal concentration exceeding the limit shown has to be treated in order to be used in the gas turbine. Vanadium concentrations in the oil range from a few ppm in a typical crude oil to several hundred ppm in some residual oil. Sodium levels in the crude or residual oil, can vary from a few ppm up to in excess of 50 ppm.

Vanadium, when introduced to a gas turbine combustor, reacts with excess oxygen to form oxides of vanadium which may deposit on the turbine blades and vanes and cause accelerated oxidation of the superalloys. Sodium and potassium will react with sulfur which is always present in sufficient quantities in the fuel and form alkali sulfate. When sufficient liquid condensate is on the surface of the alloys, it causes hot corrosion attack leading to a severe degradation of the metal. When both vanadium and sodium are present in the combustion gas, they form sodium vanadyl compounds which in general have very low melting points.[2]

Lead is found in gas turbine fuels probably due to contamination during transportation and is known to cause corrosion especially when both V and Na are already present (2). Calcium is not thought to be a very corrosive element, but it could lead to formation of hard to remove deposits.

Sodium, potassium, and calcium are present in the fuel as water soluble compounds and can be removed from the fuel by washing it with water. Current fuel treatment technology is capable of reducing the level of sodium below 1 ppm with several stages of washing (3). However, there are no practical means of removing vanadium and lead from the fuel, and some other means of reducing corrosion attack of the gas turbine hot parts due to these elements must be used.

---

[1] Numbers in parentheses designate References at end of the paper.

[2] The most catastrophic corrosion of a superalloy can be found under a molten deposit.

Table 1 - Trace Metal Limits of Fuel Entering Turbine Combustor(s).

Fuel Designation	Trace Metal Limits (Max.) PPM by Weight			
	Vanadium (V)	Sodium Plus Potassium (Na + K)	Calcium (Ca)	Lead (Pb)
No. 0-GT	0.5	0.5	0.5	0.5
No. 1-GT	0.5	0.5	0.5	0.5
No. 2-GT	0.5	0.5	0.5	0.5
No. 3-GT	0.5	0.5	0.5	0.5
No. 4-GT	*	*	*	*

*Consult turbine manufacturers

An option leading to the control of corrosion attack is to use fuel additives which during the combustion process combine with the corrosive elements producing a solid inert reaction product which can be washed out of the turbine periodically. Fuel additives have been used extensively in boilers fired with residual oil, and in few instances have been used in gas turbines burning residual oil (4,5,6). However, there have been no documented experiences on the use of fuel additives in modern gas turbines operating at high pressure and temperature with fuels containing significant amount of impurities.

Since chemical and physical processes taking place in the combustion gas between the combustion chamber and the vanes and blades are strongly dependent on pressure and temperature of the system (7,8), it is difficult to apply experience gained from the boilers and even low temperature turbines, without investigating the basic mechanisms that govern the chemical reactions between the additives and trace elements. This also extends to the formation of deposits in the gas turbine.

The experiments and subsequent analyses described below are an attempt to obtain a qualitative understanding of the processes using controlled laboratory simulation. It is assumed that the mechanism of the inhibiting action of additives can be elucidated and the results can be used in choosing additives for various turbine operating conditions by making a relatively detailed study of reaction products formed on superalloy surfaces with different additives and at various temperature levels.

EXPERIMENTAL APPROACH

Turbine Simulation

In order to reproduce some of the essential conditions of a gas turbine, the experiments were conducted in a pressurized rig called the turbine simulator. The turbine simulator approximates the operating conditions of low pressure single shaft, simple cycle industrial gas turbines. A detailed description of the gas turbine simulator can be found elsewhere (9). Briefly, compressed air supplied by a bank of 300-hp rotary compressors is heated 343°C (650°F) and fed into a 15.24 cm diameter film cooled combustor which can be operated up to 1093°C (2000°F) exit temperature. Fuel is injected through an air atomizing nozzle. At the combustor exit, there are thermocouple arrays for measuring stream temperature profiles. Beyond the thermocouple arrays a transition section is used to reduce the flow area to a 5 cm x 9.5 cm rectangular section which increases the velocity to that characteristic of gas turbine inlets, i.e., 150 m/sec. It is at this location that the superalloy specimens are exposed to the gas stream.

The test section holds a set of eight 0.3175 cm diameter pins or two internally cooled cylindrical sleeves 2.54 cm diameter and 5.08 cm long. The cylinder has 12 thermocouple wells spaced systematically to measure the specimen wall temperatures. Beyond the test section, a valve located in the passage maintains a pressure, up to 4 atm. (absolute), and finally the combustion products exhaust to a muffler.

Fuel Simulation

A distillate oil (Gulf No. 2 Diesel) which is relatively free of vanadium and sodium was used as the base fuel for this series of tests. The fuel contained traces of barium and phosphorus which appeared in the deposits in some of the tests. Chemical analysis of the fuel made both by Gulf and Westinghouse is listed in Table 2. The sulfur level of the fuel was adjusted up to 0.5 w/o with an addition of ditertiary butyl disulfide. Desired vanadium and sodium levels were achieved by adding vanadium carboxylate and sodium napthenate respectively - both oil soluble organometallic compounds.

Table 2 - Reported Chemical Analysis of Fuel Used For Corrosion Tests. Fuel Identified as Gulf No. 2 Oil.

Element	Westinghouse	Gulf
S % by wt	<0.1	0.07
P ppm		<2
V ppm	<0.1	0.07
Na ppm	0.04 ± 10%	<0.1
K ppm		<0.1
Pb ppm		0.16
Mg ppm		0.13
Ca ppm		0.21
Ba ppm	5 ± 10%	4.4

Fuel Additive Simulation

The most commonly used commercial fuel additives contain elements such as magnesium, silicon, aliminum, manganese, etc. Some proprietary formulations contain combinations of two or more of these elements. Physically, the additives are available in three general categories: (1) oil soluble organometallic compounds; (2) ultrafine metal oxide power dispersed in light oil; and (3) water-soluble compounds usually emulsified with oil.

The evaluation of the relative merit of these various types and forms of additives is beyond the scope of this paper. For the studies described in this paper the additives used were the organometallic forms of the Mg or Mg-Si combinations.

Turbine Operation Simulation, Test Variables

Tests were made at two different metal temperatures[1], 815°C and 927°C. Vanadium levels of 3 ppm and 100 ppm, respectively, in the fuel were compared with a constant additive dosage of Mg/V=3. When additives containing both magnesium and silicon were used, two different silicon levels Si/Mg=1 and Si/Mg=3 were compared. A cooled specimen test where the gas temperature is higher than metal temperature was compared with the isothermal tests with similar trace contaminants and additives.

Evaluation of Additive Deposits

The deposits and reaction products of the superalloy specimens were characterized with the use of the scanning electron microscope, electron-beam microanalyzer, and two x-ray diffraction techniques. The deposits formed by the combustion of the fuel with various contaminant concentrations and additives were removed carefully from the specimens and coated with carbon or Au-Pd alloy to insure electrical conductivity. The surfaces and fracture cross sections were then characterized morphologically with the scanning electron microscope.

[1]It has been shown that corrosion rate is dependent on metal temperature rather than gas temperature in air cooled gas turbine blades and vanes (Ref. 10).

Area and point qualitative chemical analyses were made with use of the energy dispersive analysis by x-rays system (EDAX).

Because of the porous and non-planar nature of the deposits, no attempt was made to do quantitative chemical analyses; however, some of the analyses are reported as the ratio of magnesium to vanadium, or magnesium to silicon. These ratios are based on simple ratio of the net Kα x-ray counts (corrected for background) of the respective elements. Since the operational characteristics of the SEM-EDAX were held constant between analyses it is assumed that the ratios have validity.

The deposits, still attached to detached oxide scale from the superalloys, were vacuum embedded with mounting material and ground and polished to a flat surface. These mounts were then analyzed with the electron-beam microanalyzer to characterize the elements present at the oxide scale-deposit interface.

The deposits and oxide scales were identified as to crystalline compounds with use of the 2°-radian Debye-Scherrer camera technique with use of Cr-Kα radiation. The patterns were indexed by comparison with ASTM standard x-ray data. In addition, the relative volume percent of the major phases in the deposits were estimated with use of the x-ray diffractometer.

The determination of the mode of corrosion attack of the superalloy specimen was done by standard metallographic sectioning of the speciments.

RESULTS AND DISCUSSION

Corrosion Attack Mechanisms - No Additives

Two experiments were run in the turbine simulator to determine the modes of corrosion attack produced without the presence of an additive. The first test with 0.2 ppm Na, 2 ppm V, 0.5 w/o S at a metal temperature of 816°C (1500°F) for 150 hours showed normal high temperature oxidation. The nickel based alloy (Figure 1) shows the types of internal oxides consistent with oxidation attack. Electron-beam microanalyses of compounds on the surface showed Na-Mg phosphate and Ni-Co-vanadate to be present. The cobalt based alloy (Figure 2) indicates that the mode of attack is normal high temperature oxidation.

The second test with 0.2 ppm Na, 10 ppm V, 0.5 w/o S, also at a metal temperature of 816°C (1500°F), for 150 hours was intended to produce V_2O_5 accelerated oxidation attack. The nickel-based alloy (Figure 1) formed a thick oxide scale that contained Ni, Co and Cr-rich phosphate and vanadate based phases. The type of attack is V_2O_5 accelerated oxidation. No Na or S was detected in the reaction products.

The cobalt-based alloy (Figure 2) had significant metal loss and oxide scale formation. The type of attack is also V_2O_5 accelerated oxidation. The tests were run in order to establish baseline corrosion microstructures for comparison with tests with additives in the contaminated fuels.

Corrosion Attack Mechanisms With Additives

Figure 3 shows the results of a pressurized test with a fuel that contained 0.5 ppm Na, 4 ppm Ba, 3 ppm V and, 0.5 w/o S. The additive contained Mg and Si and the exposure of the alloys was at 816°C (1500°F) for 100 hours. Although the overall corrosion rate was less than those tests with 2 ppm V or 10 ppm V without additives, some regions of the test specimen showed evidence of hot corrosion. The nickel-based alloy had a combined alkali sulfate-V_2O_5 accelerated type of attack. There are chromium sulfides in the depleted alloy zone and the oxide scales contain vanadium. Sodium is highly concentrated at the scale deposit interface.

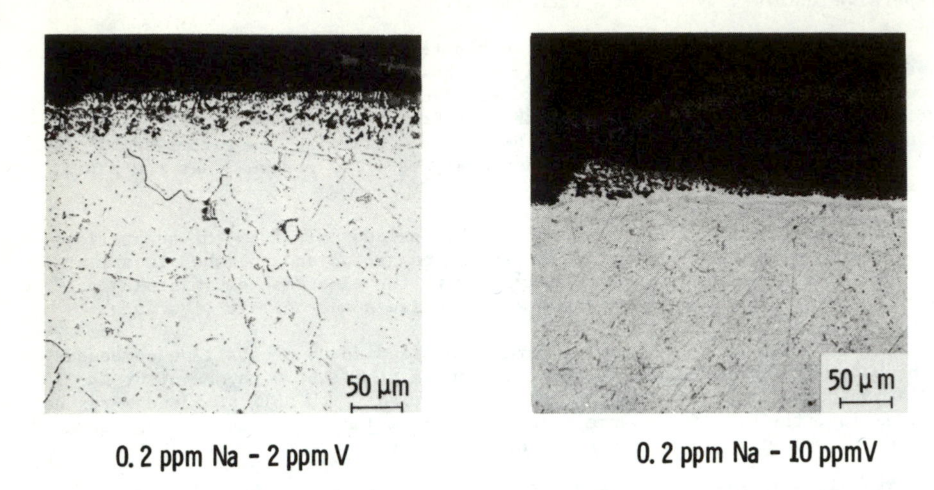

0.2 ppm Na - 2 ppm V   0.2 ppm Na - 10 ppm V

Fig. 1—Nickel base alloy at 1500°F (816°C) for 150 hours

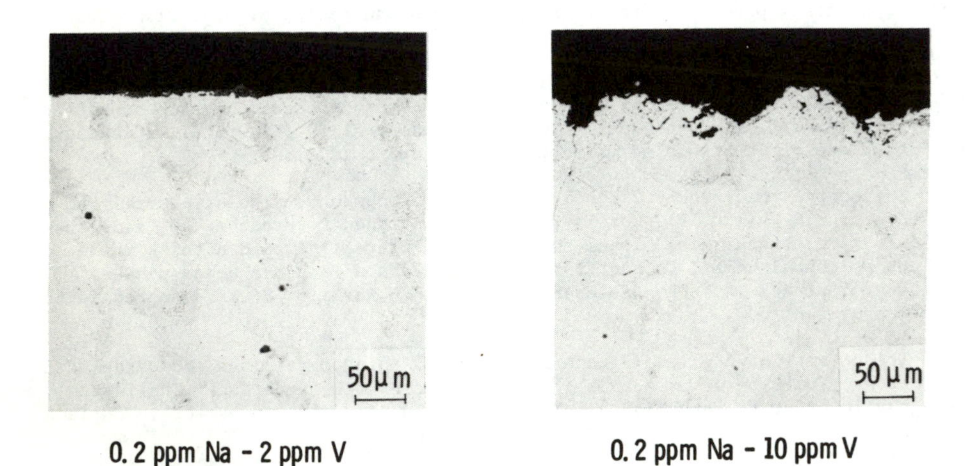

0.2 ppm Na - 2 ppm V   0.2 ppm Na - 10 ppm V

Fig. 2—Cobalt-base alloy at 1500°F (816°C) for 150 hours

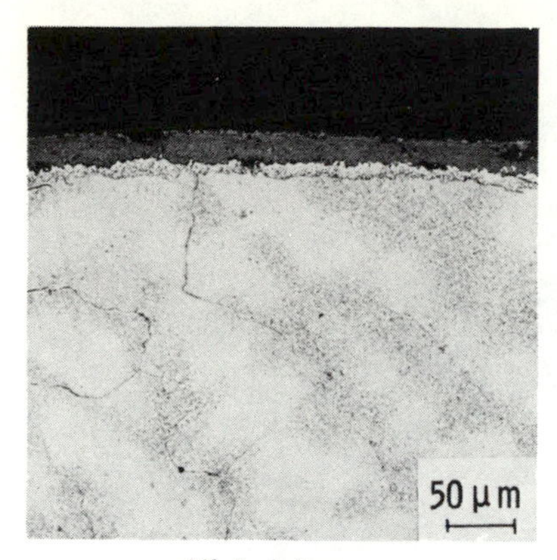

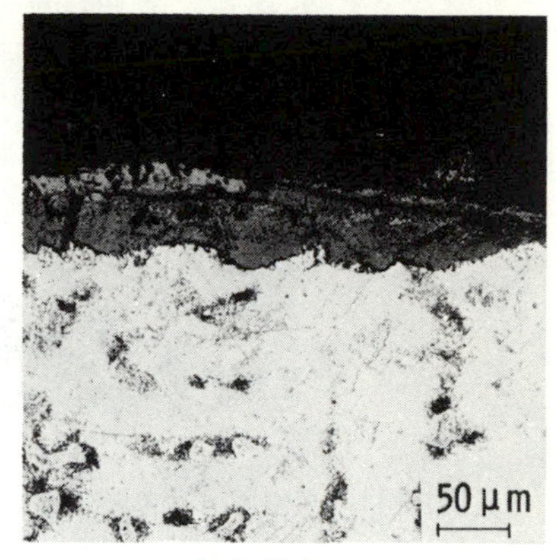

Nickel-base Cobalt-base

Fig. 3—Fuel: 3 ppm V, 0.5 ppm Na Additive: 3% Mg, 7% Si
Test at 1500° F (816° C) for 100 hours

The cobalt-based alloy (Figure 3) shows accelerated corrosion attack of the V_2O_5 type. Although Na is found to be concentrated at the scale deposit interface, there are no subscale sulfides present. X-ray diffraction analyses of the deposits show major $Mg_3V_2O_8$, MgO, and $Mg_3P_2O_8$ with minor SiO_2 (cristobalite), $BaSO_4$ and Na_2SO_4. The scales consist of $NiCr_2O_4$ and CO_3O_4.

Structure and Chemistry of Deposits

Tests were made in the turbine simulator at 927°C metal and gas temperature simulating a typical residual oil and a typical crude oil burning gas turbines. For the residual oil burning simulation the fuel contained 100 ppm V and 1 ppm Na, and for the crude oil burning simulation the fuel contained 3 ppm V and 1 ppm Na. Levels of other trace elements such as K, P, Ba, etc. were the same in both cases as originally contained in the base oil (Gulf distillate). The sulfur level in both cases was adjusted to 0.5% by weight. The simulated crude or residual fuels were treated with an oil-soluble additive containing magnesium and silicon. The additive treatment ratios were based on vanadium concentration, i.e., Mg/V = 3 in both cases.

Figure 4 shows the nature of a very thick deposit that adhered to a test specimen as a result of exposure in the test passage. The deposit may break off during the exposure and be rebuilt, or be stable during the length of the exposure (approximately 25 hours). If there is corrosion attack underneath the deposit there is a greater tendency for the deposit to flake off, especially during the cooling down period. This is illustrated in Figure 5 which is a multi-pin array of alloys.

Figure 6 is a schematic drawing of a wedge cross section of deposit that forms rapidly during exposure. The first mateial to adhere is on the direct leading edge. The deposit spreads laterally to the flow separation edges. Subsequently, the deposit develops quickly in thickness in the direction of the gas flow. The wedge cross section was then examined extensively in terms of structure and chemistry. Figure 6 lists the locations where samples were obtained for the characterizations.

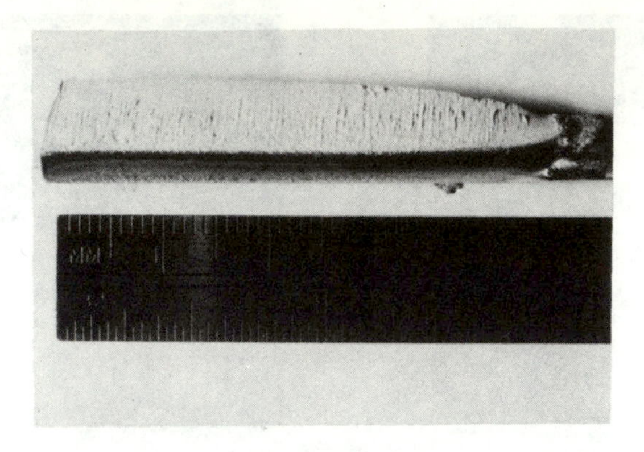

Fig. 4 — Typical buildup of deposit formed in tests in pressurized test passage

Deposits from Simulated Residual Oil Test

For the situation where the fuel contains a high concentration of vanadium, e.g., 100 ppm V, the deposits that form are over 1 cm in thickness. Table 3 shows the variation of the bulk composition of the deposit during the course of test. Where the Si/Mg ratio is 1:1 the relative volume percentages of major phases remained quite constant. This reflects the overriding effect that the higher concentrations of vanadium, silicon, and magnesium have on the steady accretion of trace elements such as sodium, barium and calcium.

Table 3 - Phase Composition of Deposit Formed at 927°C (1700°F) With Mg-Si Base Additives. Based on X-ray Diffractometer Analysis (Relative Volume Percent).

Fuel ppm	Si/Mg	Sample Hours	Mg_2SiO_4	$Mg_3(PO_4)_2$	$Mg_3V_2O_8$	MgO	$BaSO_4$
100 V <0.5 Na	3/1	25	40	20	25	15	-
100 V 1.0 Na	1/1	25	10	10	20	57	3
100 V 1.0 Na	1/1	77	10	10	10	67	3
100 V 1.0 Na	1/1	150	10	10	20	57	3
3 V 1.0 Na	1/1	23	5	5	30	30	30
3 V 1.0 Na	1/1	92	82	10	5	-	3
3 V 1.0 Na	1/1	92	80	10	-	-	10

Fig. 5 — Full array of eight test specimens showing some spalling of deposits as a result of scale loss

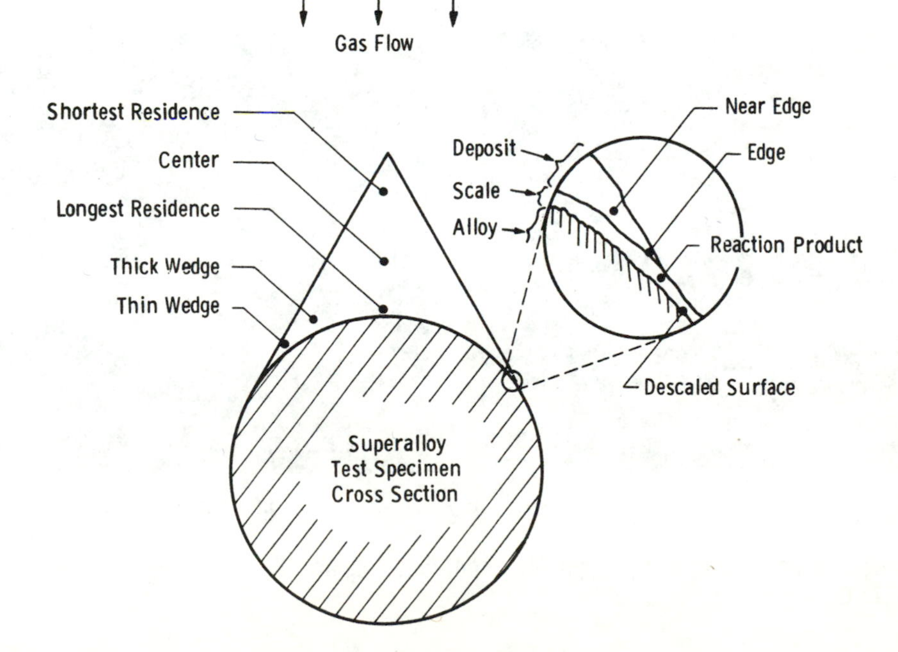

Fig. 6 — Locations where deposit samples were obtained

The increase of the Si/Mg ratio in the fuel to 3:1 results in a reduction of the MgO and an increase in Mg_2SiO_4. The $Mg_3V_2O_8$ content remains about the same from 1:1 to 3:1 Si/Mg but the increase in $Mg_3(PO_4)_2$ is unexpected.

The phosphorous is a trace element in the fuel oil but is actively precipitated as $Mg_3(PO_4)_2$ in all of the tests. The analytical sensitivity for the element phosphorous is about 2 ppm in the base fuel. We have to conclude that the variation in the relative volume percent of phosphorous compounds is due to variations in the concentration of phosphorous in the successive fuel shipments.

The structural studies of the deposits show that along the center of the wedge, the deposit consisted of dry, discrete spheroidal particles of approximately 1 to 4μ diameter. They are sometimes cemented together at the particle surfaces (Figure 7). However, the deposit with the longest residence has a less porous structure and shows more signs of interaction between the particles.

Table 4 lists the variation in elemental ratios in radial and circumferential directions of the deposit. X-ray diffraction analysis indicated that the major phases present in the deposit are MgO, $Mg_3V_2O_8$, Mg_2SiO_4, and $Mg_3(PO_4)_2$. The change in the Mg/Si ratio is probably due to the increasing rate of capture of Mg_2SiO_4 as the impact angle changes with the development of the wedge. The variation of the Mg/Si ratio in a direction along the circumference from the leading edge around to the flow separation edge tends to suggest a greater capture efficiency of the Si-rich phases with the lower impact angle.

The Mg/P ratio is very constant both in a radial as well as a circumferential direction, indicating $Mg_3(PO_4)_2$ is formed on the specimen surface. The Mg/V ratio is constant in the circumferential direction, but varies by about a factor of two in the radial direction. Since no free V_2O_5 is detected in the deposit, it appears that post deposition reaction has taken place between MgO already on the surface and gaseous V_2O_5 to form $Mg_3V_2O_8$, and/or that there is V diffusion toward the "long residence" region.

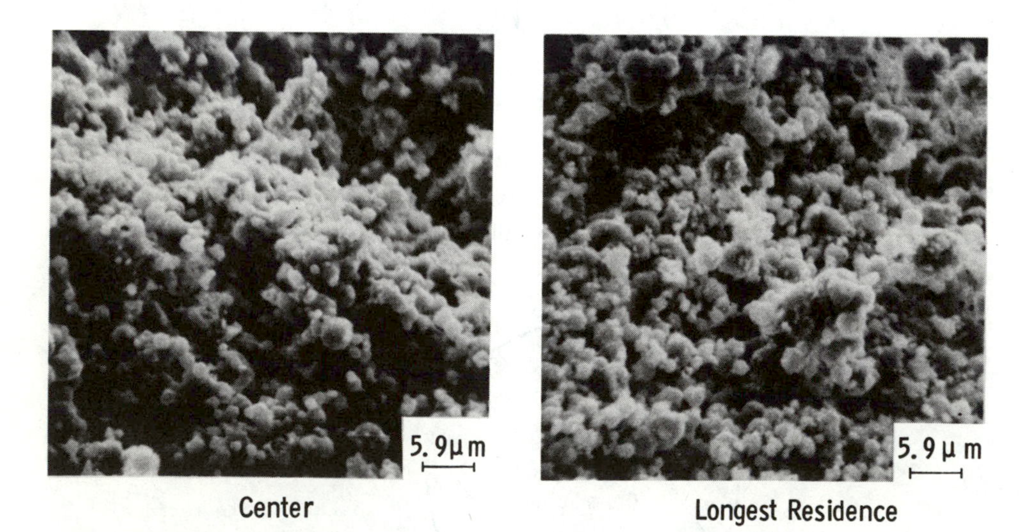

Fig. 7—Fuel: 100 ppm V, 1 ppm Na Additive: Mg and Si Temperature: 1700° F (927°C)

Table 4 - Variation of Elements From a Deposit Formed During the Residual Oil Simulation Test.

Deposit Radial Variation

Shortest Residence		Center	Longest Residence
Mg/V	4.0, 4.0	3.4, 3.4	2.2
Mg/Si	6.2, 7.5	17, 26	52
Mg/P	4.9, 5.5	5.4, 5.8	4.8

Deposit Circumferential Variation

Direct Leading Edge				Edge of Deposit	
Mg/V	2.2	3.4, 3.8	5.5	2.5	2.2
Mg/Si	52	15, 15	8.4	13	4.6
Mg/P	4.8	4.5, 4.8	5.3	4.1	3.4

The microstructure of the deposit becomes more dense with the decrease in the Mg/Si ratio as illustrated in Figures 8 and 9. In Figure 8 the thinner wedge section has the lowest Mg/Si ratio and is the most agglomerated. Figure 9 shows the very compact structure near the flow separation edge of the specimen. The reason for the greater degree of fusion where the deposit is silicon rich may be due to the lower melting points of cristobalite (1723°C) and forsterite (1890°C) compared to MgO (2800°C). These particulates are molten for a longer time during the transit from the primary combustion zone through the transition section until striking the specimen surface. This allows the opportunity for agglomeration of larger particulates and retention of heat that allows some degree of welding on subsequent impact.

When the deposits are scraped from the superalloy specimens it is evident that there has been some accelerated corrosion attack of the metal in the regions of the thin edges of the deposits. These regions are near the flow separation edges where the SEM Analyses have shown the highest silicon content in the deposit. Figure 10 shows such a region under the deposit where there is a glass-like reaction product. The primary constituent of the reaction product is a vanadate with secondary Mg, Si and P-V components. When the reaction product is removed by electrolytic descaling in molten salt there is indication (Figure 10) of pitting attack. The reaction product forms due to the reaction of V_2O_5 with the oxide scales in an area where there is no free MgO. The only available magnesium is present as relatively coarse fused masses of Mg_2SiO_4. Tables 5 and 6 list the ratios of elements of interest for the reaction products and alloy scales.

Deposits from Simulated Crude Oil Test

As discussed in the previous section, the deposit formed in the residual oil test contained principally compounds derived from the vanadium, silicon and magnesium with exceptions of magnesium orthophosphate and minor $BaSO_4$. In the deposits obtained from the crude oil tests, there are other elements such as barium and sulfur which show up as compounds of significant concentration. In addition, a considerable amount of sodium is concentrated in the deposit. Table 3 lists the relative volume percent of phases measured in the deposit formed from a fuel with 3 ppm V, 1 ppm Na. Whereas the deposit formed in 25 hours with 100 ppm V was close to the bulk composition of deposits formed at

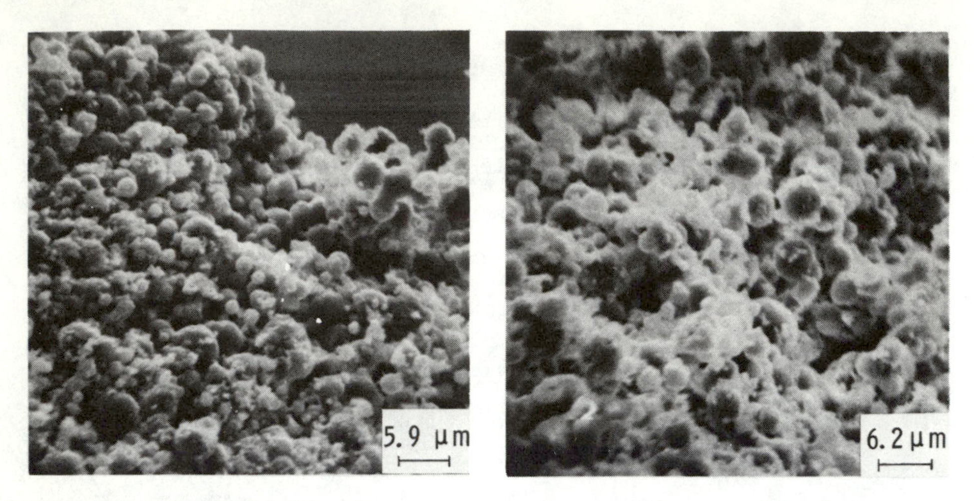

Thickest Wedge Section Thinner Wedge Section

Fig. 8–Fuel: 100 ppm V, 1 ppm Na Additive: Mg and Si
Temperature: 1700°F (927°C)

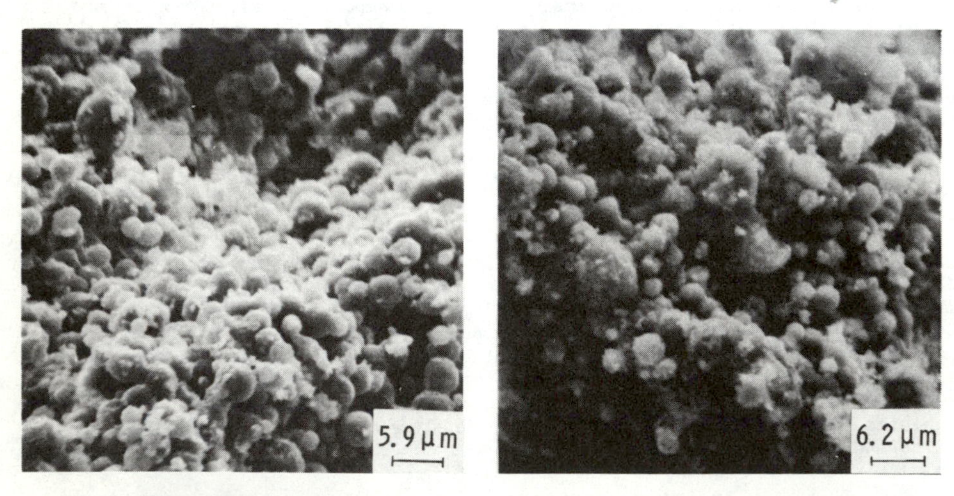

Near Edge of Deposit Edge of Deposit

Fig. 9—Fuel: 100 ppm V, 1 ppm Na Additive: Mg and Si
Temperature: 1700°F (927°C)

longer times, the deposit with the low vanadium fuel formed at 23 hours was markedly different than the long-time deposits (92 hours).

The deposit builds up very slowly and does not form the pronounced wedge shape illustrated in Figure 4. This may explain the much greater MgO content versus Mg_2SiO_4 at shorter times. The initial deposit is limited in volume and is dominated by $Mg_3V_2O_3$, a reaction product and $BaSO_4$ a crystalline condensate. With time the deposit becomes more voluminous and, as Table 3 shows, is principally Mg_2SiO_4. The $BaSO_4$ and $Mg_3V_2O_8$ are diluted by the Mg_2SiO_4. The volume percent of $Mg_3(PO_4)_2$ which, as discussed earlier, seems to be related to the magnesium content of the fuel, has increased. This also reflects a change in the absolute level of phosphorous in the fuel.

The deposits are characteristically hard and brittle. Figure 11 shows the compact nature of the deposit. There is a lack of spheroidal particles but instead there are crystals that are Mg-V and Mg-Si rich phases (presumably $Mg_3V_2O_8$ and Mg_2SiO_4, respectively). The fracture surface through the thin deposit shows that there is considerable phosphorous present presumably as the reaction product $Mg_3(PO_4)_2$.

Figure 12 shows the deposit-scale surface of a hard dense deposit. The base alloy (a nickel base superalloy) evidenced accelerated attack and the reaction products reflect this observation. The deposit contains a nickel-silicon rich reaction phase. The surface shown on the left in Figure 12 is primarily $BaSO_4$ but shows the characteristics of having been partially molten at 927°C (1170°F).

Electron-beam microanalyses of these thin hard deposits formed at 816°C (1500°F) metal temperature show significant concentrations of sodium at the scale-deposit interface. One such elemental profile is shown in Figure 13. The mode of corrosion attack beneath the scale of the nickel-based alloys is hot corrosion attack with subscale sulfide formation.

Reaction Product

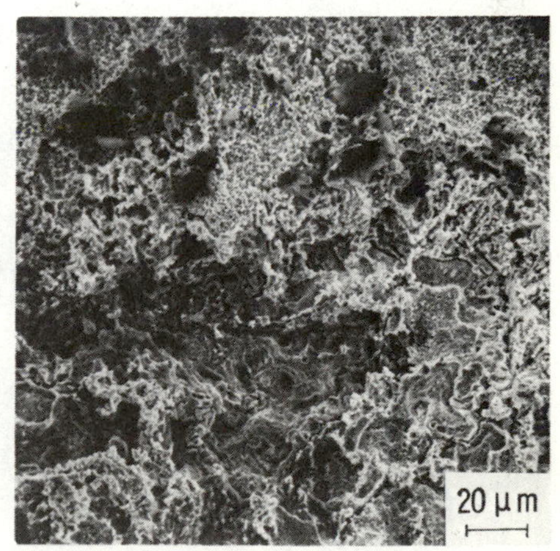

Descaled Surface

Fig. 10—Alloy: Cobalt-base Fuel: 100 ppm V, <0.5 ppm Na
Additive: Mg and Si Temperature: 1700°F (927°C)

Table 5 - Reaction Product Formed at 927°C (1700°F) in Simulated Residual Oil Test (Cobalt-Base Superalloy Substrate)

	Reaction Product		Alloy Scale
	Primary Phase	Secondary	
Mg/V	0.49, 0.47	8.3	10.0
Mg/Si	2.3, 3.6	1.6	8.0
Mg/P	6.1, 6.9	4.0	2.2
V/Si	4.8, 7.7	0.20	0.78
V/P	12, 15	0.48	0.21
V/Co	1.8, 2.0	0.62	3.0
Major	V Co Mg Si	Mg Si P Co V	Mg P Si
Minor	P	Ni	V Co Cr
Trace	Ca Fe	Ca Fe	Fe Ca

Table 6 - Reaction Product Formed at 927°C (1700°F) in Simulated Residual Oil Test (Nickel-Base Superalloy Substrate).

	Reaction Product	Alloy Scale
Mg/V	2.8	-
Mg/Si	1.0	<0.1
Mg/P	3.1	-
V/Si	0.36	-
V/P	1.1	-
V/Co	6.4	-
V/Ni	1.5	-
Major	Mg Si V P Ni	Si Ti Cr
Minor	Co Fe	Mg Ni Co

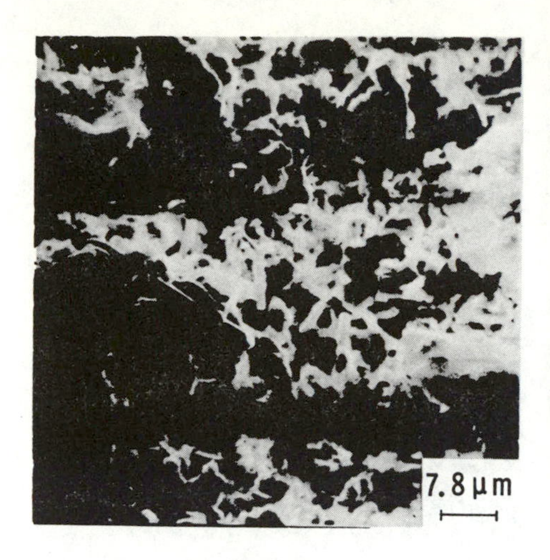

Surface

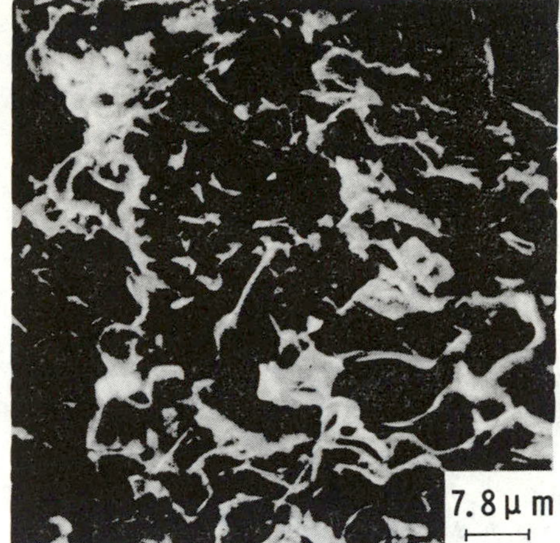

Fracture Surface

Fig. 11—Fuel: 3 ppm V, 1 ppm Na Additive: Mg and Si Temperature: 1700°F (927°C) Deposit: Hard

At the higher metal temperature of 927°C (1700°F) the attack of the metal is also of the hot corrosion type but Na_2SO_4 is not detected as a separate phase. On the basis of the molten characteristics of the $BaSO_4$ shown in Figure 12 it is concluded that sufficient sodium is present to form a liquid phase even though no Na was detected with the energy dispersive x-ray analyzer on the SEM.

Deposits - Phase Composition Variation

A considerable effort was made in the course of the tests in the pressurized test passage to determine the crystalline components of the deposits, reaction products, and corrosion products. Deposit-phases are those in the mass of the deposit sufficiently removed from the alloy substrate as not to contain elements derived from the substrate. Reaction products are those phases in the deposit that show interaction with the metallic substrate. The corrosion products are those phases that form primarily from elements derived from the alloys.

A number of compounds such as $Mg_3(PO_4)_2$ and $Mg_3V_2O_3$ have very similar x-ray diffraction patterns and are difficult to distinguish when present together. In addition, on cobalt-base alloys Co can substitute for Mg making it more difficult to identify the compounds. Therefore, many of the following tables list the compounds as $M_3*(PO_4)_2$ and $M_3*V_2O_8$ where there is uncertainty as to whether M* is primarily Mg or Co. Where there are SEM-EDAX of the deposits then the compounds are listed as either Mg or Co-based.

Table 7 lists the phases formed in the residual oil tests with an organometallic magnesium base additive. M is Mg for deposits on nickel-based alloys and mixed Mg and Co for cobalt-alloys. The phases $M_3(PO_4)_2$ and $M_3V_2O_8$ can also be reaction products. Except for $Mg_2P_2O_7$ the same phases form in the deposit at 816°C (1500°F) and 927°C (1700°F) gas and metal temperature.

Experiments with Mg and Si in an 1:1 ratio at 927°C (1700°F) gas and metal temperature formed the compounds listed in Table 8. For fuels with a low additive to fuel impurity (Na, K, P, Ba) ratio (crude oil) the greater degree of corrosion

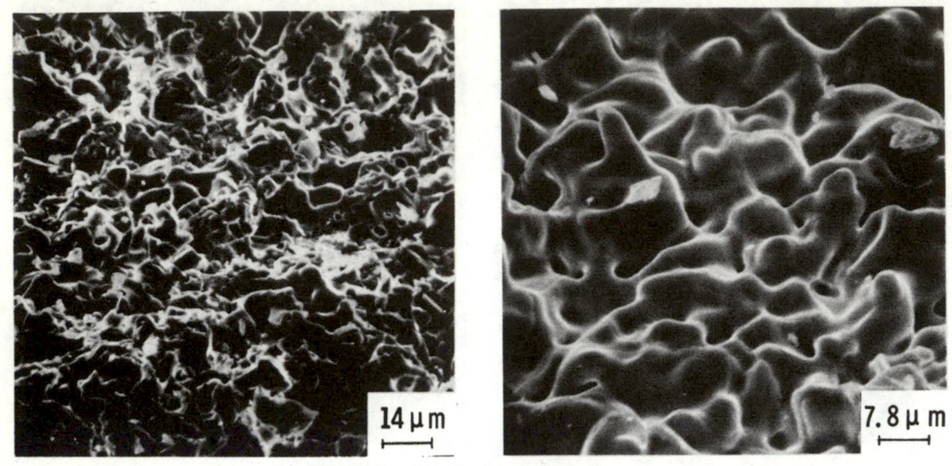

Deposit-Scale Surface

Fig. 12—Fuel: 3 ppm V, 1 ppm Na Additive: Mg, and Si
Temperature: 1700° F (927° C) Deposit: Hard

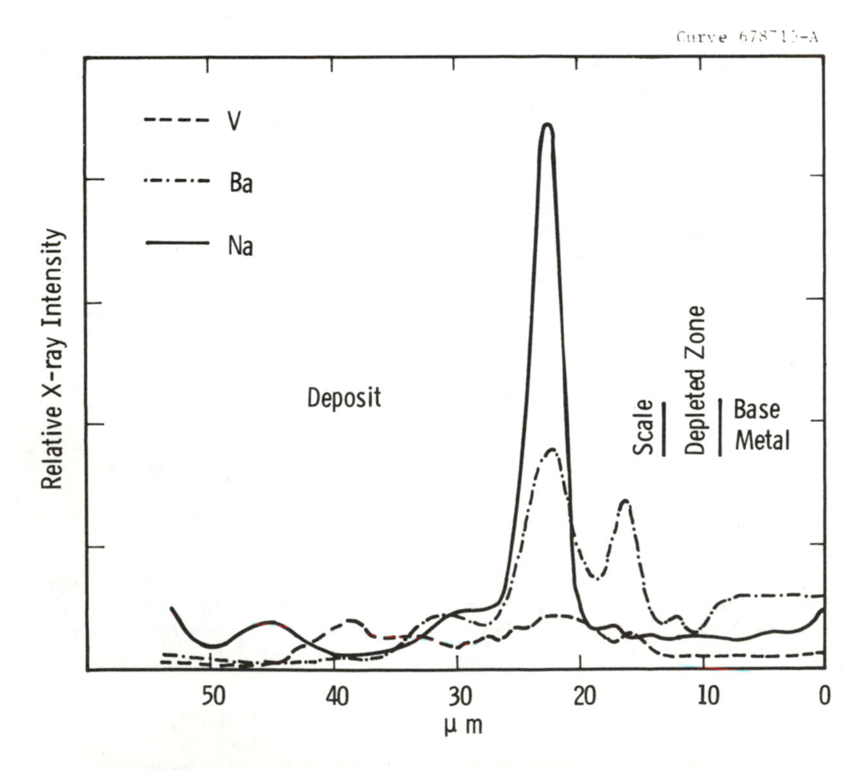

Fig. 13—Relative variation of sodium, barium, and vanadium across scale and deposit on edge of section

attack of the specimens is reflected in readily detected corrosion products. The phases $FeTa_2O_6$ and $Co_4Nb_2O_9$ form on the cobalt-based alloy. The phase Ni_2SiO_4 forms on the nickel-based alloy. The spinel $NiCr_2O_4$ forms on both alloys.

With fuels that contained a high additive to impurity (Na, K, P, Ba) ratio (residual oil) there was a considerable drop in the amount of corrosion attack thus the corrosion products were of the more usual oxidation mode of attack (Cr_2O_3, Co_3O_4). The Si-rich phase detected was the low temperature form of SiO_2.

Experiments with Si-Mg base additives with a 3:1 (Si/Mg) ratio and a high additive to fuel-impurity ratio produced deposits with much more Mg_2SiO_4 (see Table 3 and 9). No free SiO_2 was detected with the use of x-ray diffraction analysis. There was little attack of the base alloys hence the corrosion products reflect oxides associated with high temperature oxidation of the cobalt and nickel-based alloys.

An experiment was run to compare the composition of deposits formed on substrates of the same metal temperature (816°C) but at different gas temperatures. The fuel contained 3 ppm V, 1 ppm Na, 0.5 w/o S and the Mg-Si ratio was 1:3. In both the isothermal test (816°C gas and metal temperature) and cooled test (1038°C gas and 816°C metal temperature) there was accelerated attack of the substrate. Table 10 lists the compounds detected on the specimens.

The most striking difference between the two conditions of gas stream temperature is the complete lack of magnesium and vanadium containing phases that formed on the cooled substrate. The test was extended to 300 hours in order to examine a long time deposit but no Mg or V based compounds were detected. Cristobalite and $BaSO_4$ formed under both conditions. The corrosive condensate Na_2SO_4 was also found in the isothermal test.

Table 7 - X-ray Diffraction Results of Residual Oil Tests. M* Can Be Co or Mg, or Both.

Deposit	Reaction Products	Corrosion Products
	816°C (1500°F) Gas and Metal Temperature	
MgO (ASTM 4-829)	$Mg_2Al_2O_4$	$NiCr_2O_4$
$M^*_3(PO_4)_2$ (ASTM 22-1151)	Fe_3O_4	
α $Mg_2P_2O_7$		
$M^*_3V_2O_8$ (ASTM 16-832)		
	927°C (1700°F) Gas and Metal Temperature	
MgO		NiO (ASTM 4-835)
$M^*_3(PO_4)_2$		Cr_2O_3 (ASTM 6-504)
$M^*_3V_2O_8$		$NiCr_2O_4$

Table 8 - X-ray Diffraction Results of Mg-Si Base (1/1) Additive of Two Different Fuels at 927°C (1700°F) Gas and Metal Temperature.

Deposit	Reaction Products	Corrosion Products
	Crude Oil Test	
MgO	Ni_2SiO_4 (ASTM 15-388)	$NiCr_2O_4$
$Mg_3V_2O_8$	$Mg_3V_2O_8$	$FeTa_2O_6$ (ASTM 23-1124)
$BaSO_4$	Mg_2SiO_4	$Co_4Nb_2O_9$ (ASTM 13-464)
Mg_2SiO_4	$BaSO_4$	
$Mg_3(PO_4)_2$		
	Residual Oil Test	
MgO	$Mg_3(PO_4)_2$	$NiCr_2O_4$
$Mg_3V_2O_8$	α-SiO_2	Cr_2O_3
Mg_2SiO_4	Mg_2SiO_4	Co_3O_4
$BaSO_4$		

Table 9 - X-ray Diffraction Results of Mg-Si Base (1/3) Additive and High Ratio of Additive to Fuel-Impurity Concentration.

Deposit	Reaction Products	Corrosion Products
	927°C (1700°F) Gas and Metal Temperature	
Mg_2SiO_4	Mg_2SiO_4	$NiCr_2O_4$
$Mg_3V_2O_8$	MgO	αAl_2O_3
$Mg_3(PO_4)_2$		
MgO		

Table 10 - X-ray Diffraction Results of Si-Mg Base (3/1) Additive and Low Ratio of Additive to Fuel-Impurity Concentration. M* Can Be Co or Ni, or Both.

Deposit	Reaction Products	Corrosion Products
816°C (1500°F) Gas and Metal Temperature		
$Mg_3V_2O_8$ (ASTM-19-779)	SiO_2 (α-quartz)	$NiCr_2O_4$
MgO	SiO_2 (cristobalite)	Co_3O_4
$Mg_3P_2O_8$ (ASTM-19-767)	$BaSO_4$ (ASTM 5-448)	
	Na_2SO_4	
1038°C (1900°F) Gas and 816°C (1500°F) Metal Temperature		
	50 hrs	50 hrs
	$BaSO_4$	$NiCr_2O_4$
	SiO_2 (cristobalite)	
	300 hrs	300 hrs
	$BaSO_4$	$NiCr_2O_4$
	SiO_2 (cristobalite)	$M^*_2SiO_4$ (ASTM 15-865)
		CoO (ASTM 9-402)

Fracture Surface
Mg/V = 1.05

Surface of Deposit
Mg/V = 0.28

Fig. 14 — Structure of deposit from row one vane of gas turbine

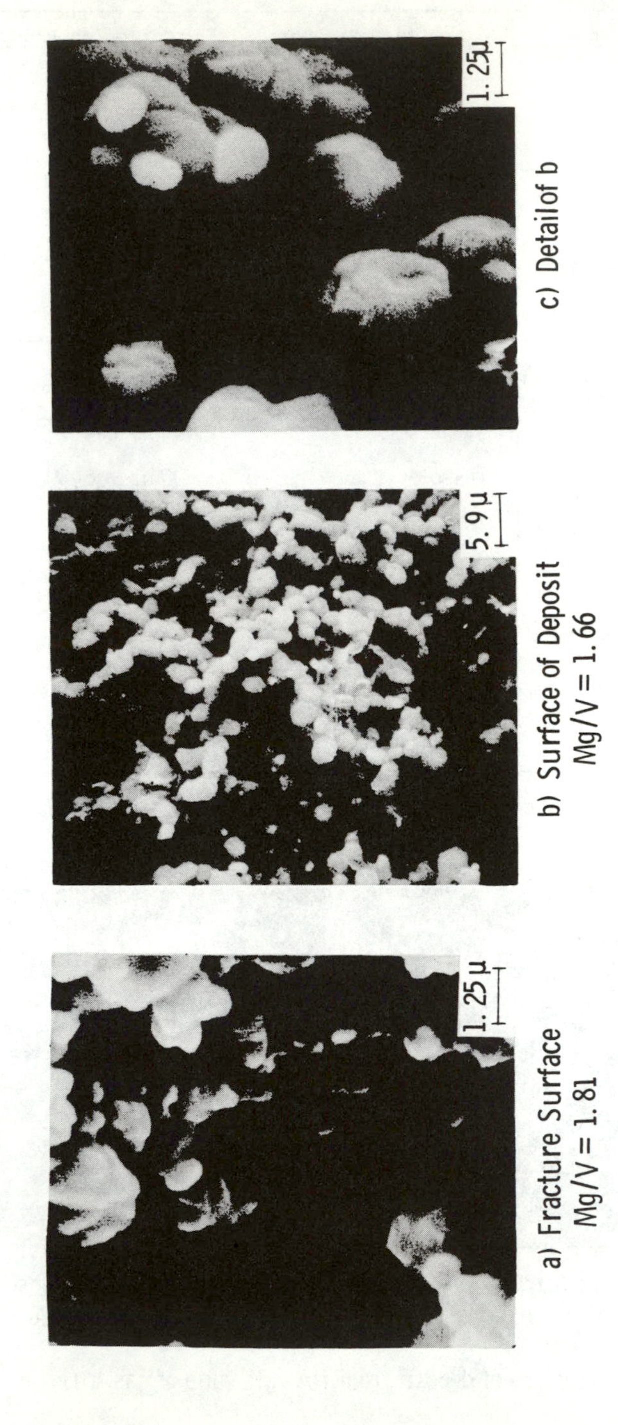

a) Fracture Surface Mg/V = 1.81

b) Surface of Deposit Mg/V = 1.66

c) Detail of b

Fig. 15 – Structure of deposit from row three blade of gas turbine

The x-ray diffraction analysis technique when applied to the evaluation of the deposits does not detect glassy or non-crystalline phases or phases present in less than about 5 w/o; therefore, the deposits at 50 and 300 hours were analyzed by emission spectrographic analyses. The deposit at 50 hours had >10 w/o each of Co, Cr, Ni and Si, and several w/o of Mg Ba and V. The deposit at 300 hours had >10 w/o each of Co, Cr, Ni, and Si; and several w/o each of Al, Ba, Fe, W, Mg and V. Some P and Na were detected. Electron-beam microanalysis showed that there was a phase present that is probably a Co-Ni vanadate. These compounds melt incongruently at 699°C (1290°F).

It is concluded that the vanadium present in the deposit is in the form of glassy or poorly crystalline compounds. The particles of MgO and Mg_2SiO_4 are hotter when they impact the cooled substrate than they are in the isothermal test. The deposit, due to its poor thermal conductivity compared to the metal substrate, must be elevated enough in temperature for vanadium-based liquids to be present (as evidenced by the Co-Ni-vanadium association). The magnesium, which can substitute for Co and Ni in the vanadates, has nevertheless, a limited effect on the solidus temperature of the compounds. The electron-beam microanalyzer results indicate that most of the magnesium is associated with silicon as a separate phase in the mass of $BaSO_4$. No vanadium is associated with the silicon-magnesium phase and little magnesium is present in the Co-Ni-V-rich phase.

Deposits from Gas Turbines

Deposit samples were collected from a gas turbine using a fuel containing 50 ppm V, 0.5 ppm Na and 4 w/o S treated with Mg-based additives. They were obtained from a row three rotating blade and a row one vane. These deposits had developed in over 500 hours of base load operation.

Analysis of the deposits with use of x-ray diffraction techniques show the following:

Row 1 vane - $MgSO_4$ - 6 H_2O
Row 3 blade - Major $MgSO_4$ - 6 H_2O, minor MgO

The deposits had crumbled in transit so that no attempt was made to do the systematic evaluation of the microstructure that had been applied to the tests in the turbine simulator. Instead, selected fragments of the deposits were examined with the scanning electron microscope. The deposit from the row one vane, in general, had a Mg/V ratios ranging from 1.05 to 1.18 for fracture surface and 0.28 for the surface of the deposit. The deposit from the row 3 blade has a range of Mg/V ratios of 1.4 to 1.8 for fracture surfaces with a minor structure that had ratios from 4.5 to 6. The analyses were done by area scanning on the fracture surfaces.

The microstructures shown in Figure 14 show the open structure of the interior of the deposit from the row one vane. The particles are not spherical but show the angularity associated with crystals; however, the deposit is $MgSO_4 \cdot 6 H_2O$ which indicates that extensive hydration of the original $MgSO_4$ took place.

Deposits from the row three blade are shown in Figure 15. The fracture surface has some porosity, but there is more densification of the particulates. There is less indication of angularity due to crystal formation compared to the row one deposits. The surface of the deposits (Figure 15) has a very dense crust of large spheroidal magnesium sulfate particles.

$MgSO_4 \cdot 6 H_2O$ is stable (10) only in the range of 48 to 65°C and is metastable at other temperatures. $MgSO_4 \cdot H_2O$ is the stable phase above 63°C but decomposes to $MgSO_4$ at 200°C. Anhydrous magnesium sulfate decomposes under a standard atmosphere at temperature of 1124°C to form MgO, O_2, SO_2 and SO_3.

It is interesting to note that the deposits on the row one vane contained only $MgSO_4 \cdot 6\ H_2O$ while the row three blade contained some MgO. Using a formula given by Lay (8), the equilibrium partial pressure of SO_3 at the row one vane is calculated to be approximately 1×10^{-3} atmosphere. At this pressure and turbine metal temperature of 871°C, the stable phase in the ash is predicted to be $MgSO_4$ and $Mg_2V_2O_7$[1]. However, deposits obtained from the third row blade, where partial pressure of SO_3 is expected to be lower, contained MgO as well as $MgSO_4$. Furthermore, none of the deposits from the turbine simulator tests contained $MgSO_4$ but only MgO and $Mg_3V_2O_8$. The equilibrium partial pressure of SO_3 in the turbine simulator test was calculated to be below 5×10^{-5} atmospheres which favors MgO formation. The partial pressure of SO_3 in the primary zone of the combustor is calculated to be very low ($\sim 1 \times 10^{-7}$ atm) so that MgO is the initial solid phase in the gas stream. MgO is converted to $MgSO_4$ in transit in the gas phase and subsequently to completion in the first stage if the equilibrium partial pressure of SO_3 favors $MgSO_4$ formation.

CONCLUSIONS

Vanadium, sodium and other trace elements (Ba, P, S, etc.) present in the fuel segregate with time within the deposit and concentrate near the deposit-alloy interface. This is particularly significant for the situation where the fuel contains only a few ppm V so that the additive to impurity (Na, K, P, Ba) ratio is relatively low. This segregation-time effect causes trace elements present in only minor amount in the fuel to play a major role in the corrosion attack of some regions of the superalloys even in the case where an adequate amount of additive is injected, e.g., Mg/V = 3. This is predicted to be the situation when the fuel contains only nominal amounts of vanadium as in the case of a crude oil.

In the case of residual oil where the additive to trace element (Na, K, P, Ba) ratio is large due to a large amount of vanadium present, the effect of trace elements on the accelerated corrosion attack is found to be relatively minor.

In the primary zone of the turbine simulator combustion chamber where the temperature is over 2000°C, magnesium and silicon react with oxygen to form oxides of Mg and Si. The partial pressure of SO_3 in the primary zone is theoretically very low due to the high temperature so that reaction between MgO and SO_3 is not favored. Although the melting point of pure MgO is 2800°C, the spherical nature of the particulates indicate that the compound existed for some short period in the liquid state. The molten particles of MgO agglomerate within the combustor to form spheroidal particles of approximately 2-5μ diameter. The molten particles traveling down the cooler regions of the combustor solidify and then impact on the target surfaces. The tendency of the particles to stick on impact is dependent on the final gas temperature and the residence time of particles in the gas stream to allow it to cool by radiative and convective heat transfer.

When silicon is added to magnesium, MgO, SiO_2 and Mg_2SiO_4 form in the primary combustion zone and subsequently deposit on the target surface. The latter two compounds form slightly larger particles than the MgO. It is noticed that the Si-rich phases are found more on the side surface of the cylindrical specimen than the front surface. This phenomenon may be due to one or both of the following explanations: (1) The melting temperatures of SiO_2 (1723°C) and Mg_2SiO_4 (1890°C) are lower than MgO (2800°C) and thus they may tend to agglomerate more than MgO while in transit in the exit gas of the combustor. The larger and hence heavier particles will deposit at lower impact angles while the smaller MgO particles will follow the stream lines and, (2) the lower melting phases would have a greater affinity to stick or weld, when they impact on the deposit than the MgO which can collide and rebound out into the gas stream.

[1]Reference 8, Figure 3.

After MgO particles are deposited on the specimen, they will react with gaseous oxides of vanadium to form magnesium orthovanadate ($Mg_3V_2O_8$). Other possible magnesium vanadate compounds such as $Mg_2V_2O_7$ and MgV_2O_6 were not detected in any of the deposits by x-ray diffraction techniques.

With the assumption that kinetic effects are not overriding, the relative concentration of MgO, $MgSO_4$, and $Mg_3V_2O_8$ are probably dependent on total pressure of the system, sulfur level in the fuel, and temperature of the gas as well as the metal. In deposits obtained from the turbine simulator where total pressure (4 atm) and sulfur level (0.5 w/o in fuel) are relatively low, the partial pressure of SO_3 is not sufficient for $MgSO_4$ to form so that only MgO and $Mg_3V_2O_8$ are identified as the bulk phases. In the gas turbine with a total pressure of 12 atm and 4 w/o S in the fuel, $MgSO_4$ constitutes almost 90% of the deposit.

A necessary condition for accelerated oxidation or hot corrosion attack to take place is for liquid to be immediately adjacent to the oxidized metal surface. This is a prerequisite regardless of the presence of fuel additives in the system. There can be a liquid phase present if one or more of the following conditions are satisfied: (a) high temperature of deposit (partial melting); (b) depression of solidus of deposit by accretion of Na, K, Ba, Pb, P, etc.; (c) segregation of liquid phase in the deposit and subsequent migration of liquid to the alloy surface; and (d) formation of liquid phase at alloy surface by reaction of trace elements with the scale. Catastrophic corrosion attack will not likely take place when the deposit adjacent to the metal remains solid. However, there still can be accelerated oxidation attack by gaseous V_2O_5 reacting with the scale if an adequate amount of magnesium compound is not available at the surface to neutralize it.

Silicon, introduced as a fuel additive in addition to magnesium, forms cristobalite (SiO_2) as well as Mg_2SiO_4 in the combustion chamber. X-ray diffraction analysis of the deposits detected no crystalline compounds that contain Si and V. Although Si reacts with some Mg so that it is not available to react with V, there is excess MgO present as long as a ratio of Mg/V = 3 is maintained. The reason for the beneficial action of Si (increases maximum effective temperature of V control about 28°C) is not clear. The presence of SiO_2 and Mg_2SiO_4 in the deposit may act as a diluent for trace metal build up, increase the solidus melting temperature of deposits, or increase the viscosity of resulting liquid phases.

ACKNOWLEDGMENTS

The authors are indebted to W. E. Young and R. M. Chamberlin for helpful discussions and suggestions and R. C. Kuznicki and C. W. Hughes for deposit analysis.

The continuous interest and support provided by the Westinghouse Generation Systems Division made this work possible.

REFERENCES

1. Standard Specification for Gas Turbine Fuel Oils, ASTM D2880-76, ASTM Standards 1976, Appendix XI, p. 761.

2. Zetlmeisl, M. J., May, W. R., and Annand, R. R., "High Tempterature Corrosion in Gas Turbines and Steam Boilers by Fuel Impurities, Part V - Lead Containing Slags," ASME Paper No. 74-WA/CD-4.

3. Krulls, G. E., "Gas Turbine Liquid Fuel Treatment and Analysis," ASME Paper No. 74-GT-44.

4. Nakao, K. et. al., "Two Years Experience of a Gas Turbine Firing Residual Fuel," ASME Paper 68-GT-11.

5. White, A. O., "20 Years Experience Burning Heavy Fuels in Heavy Duty Gas Turbines," ASME Paper 73-GT-22.

6. Lee, S. Y. and Young, W. E., "Gas Turbine Hot-Stage Parts in Aggressive Atmospheres," J. of Engineering for Power, Vol. 98, No. 4, Oct. 1976.

7. Halstead, W. D., "Calculations on the Effects of Pressure and Temperature on Gas Turbine Deposition," Deposition and Corrosion in Gas Turbines, John Wiley & Son, 1972.

8. Lay, D. W., "Ash in Gas Turbines Burning Magnesium-Treated Residual Fuel," ASME Paper No. 73-WA/CD-3.

9. Lee, S. Y., DeCorso, S. M., and Young, W. E., "Laboratory Procedures for Evaluating High Temperature Corrosion Resistance of Gas Turbine Alloys," J. of Engineering for Power, Trans. ASME Series A, Vol. 93, No. 3, July, 1971.

10. Lee, S. Y., Young, W. E., Hussey, C. E., "Environmental Effects of the High-Temperature Corrosion of Superalloys in Present and Future Gas Turbines," J of Engineering for Power, Trans, ASME, Series A, Vol. 94, No. 2, April, 1972.

11. Kirk-Othmer Encyclopedia of Chemical Technology, Second Edition, Vol. 12, Interscience, N.Y.

PART I: FORMATION AND REMOVAL OF RESIDUAL FUEL ASH DEPOSITS IN GAS TURBINES FORMED AT FIRING TEMPERATURES BELOW 982°C (1800°F)

T. A. URBAS and L. H. TOMLINSON

General Electric Company
Schenectady, New York, USA 12345

ABSTRACT

A series of combustion tests, nominally of 100 hours duration each, burning treated Bunker C fuel and synthetic Bunker C fuel were conducted in a simulated gas turbine combustor with a first stage nozzle vane segment. These tests have demonstrated that nozzle plugging rates, also called ash deposit rates, increase both as the firing temperature and sodium content in the fuel increase. Ash deposits result in a decrease in the effective area of the nozzle and, therefore, a similar reduction in power output. It has been shown that Ca in the fuel greatly enhances the deposit rate and also forms hard deposits. Ash deposit chemistry is highly temperature dependent. At a pressure of 5 atmospheres and with 1 to 2 wt % S in the fuel, the deposit changes from soft (primarily $MgSO_4$) to hard (primarily MgO) in the temperature range of 927 C (1700 F) to 982 C (1800 F).

Nutshells and thermal shock are effective on a variety of soft ash deposits formed at firing temperatures of 982 C (1800 F) and below, removing 20 to 90% of the deposits.

A synthetic Bunker C composed of an aqueous additive solution uniformly dispersed in a No. 2 fuel oil was shown to be feasible because both ash deposit rates and the chemical and physical nature of deposits formed were the same as those from real Bunker C fuels.

INTRODUCTION

Deposits formed from the combustion of ash-bearing fuels can adhere to hot gas path parts and restrict the flow of hot gas thereby reducing the output and efficiency of the gas turbine. A test facility (coined the "Turbine Simulator") was constructed at the General Electric Corporate Research and Development in order to study the parameters affecting the rate at which these deposits plug the first stage nozzle and to examine means for removing them.

Tests were conducted at a number of firing temperatures with a Bunker C fuel and a synthetic Bunker C fuel containing various levels of Na, Mg, V, Si, Pb, and Ca. Provisions for the injection of nutshells were provided to examine the effectiveness of removing ash deposited under many different conditions.

The construction of the facility and the conduct of the first six tests were sponsored by the Gas Turbine Division. The remainder of the tests were supported by MARAD contract (Contract No. C-0-35510, U.S. Maritime Administration, Dept. of Commerce).

TEST FACILITY

Two 800 hp Worthington three-stage compressors delivered air to the test apparatus (Figure 1) at a flow rate of up to 3.6 kg/sec (8 lb/sec) and a pressure ratio of 6 to 1. A gas-fired preheater brought the air up to 288 C (550 F) inlet temperature.

The combustor is a one-half size (in cross-sectional area) version (3.4 kg/sec or 7.5 lb/sec mass flow) from the MS 5000 series gas turbines. The nozzle sector is part of a MS 3002 series air-cooled nozzle sector, and the transition section has a cross-sectional flow area compatible with the above parts (Figure 1).

In order to remove deposits, crushed walnut shells could be injected into the test rig by compressed air at positions A, B, C, and D shown in Figure 1.

The nozzle area index number (NAIN), a measurement of the effective flow area of the nozzle, was computed by the formula NAIN = $W\sqrt{T}/P$. Temperature was measured by the control ring thermocouples downstream from the nozzle, and the pressure was measured upstream of the nozzle. Fuel consumption was measured by a flow meter with fuel tank level change as a check.

Airflow was measured by orifices in the main airline. From this information, fuel-air ratios and theoretical firing temperatures were calculated. Firing temperature was computer controlled by means of a motorized fuel valve.

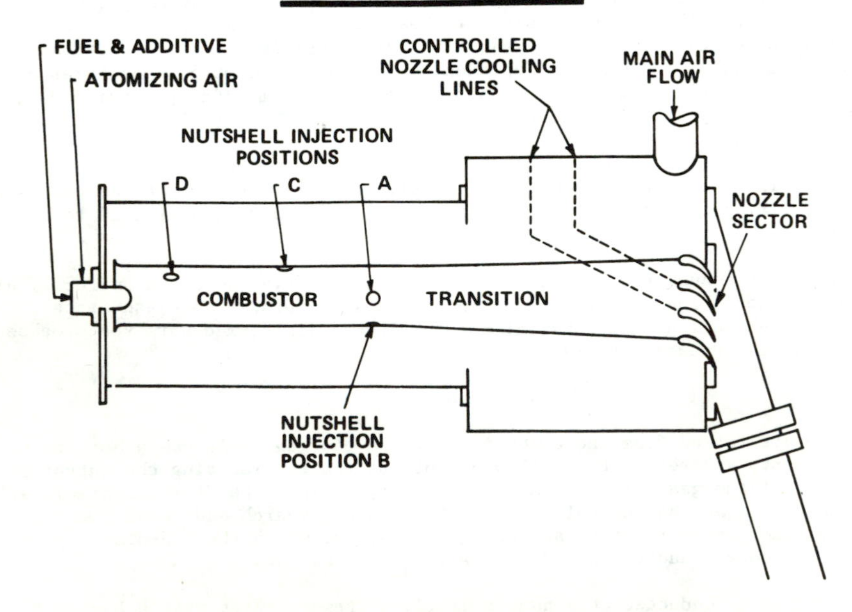

FIGURE 1

All test information was computer monitored and recorded. The more important data channels had set limits. The computer could signal if attention was needed or terminate the test if dangerous levels were reached.

The Bunker C used for the first fifteen (15) tests was a specially washed fuel from the Central Vermont Power Co. in Rutland, Vermont. See Table 1. It was considered to be a typical residual fuel.

The magnesium additive used for most of these tests, was a 12 1/2% magnesium sulfate water solution made from a commercial grade of epsom salts. This water solution was thoroughly mixed with the fuel and quickly burned before it could separate. A tap near the turbine simulator (which was used to take fuel samples) always released a milky brown colored sample showing the effectiveness of this mixing.

The supply of Bunker C from Rutland was no longer available after test No. 15. Therefore, tests No. 16 through No. 21 were used to calibrate the turbine simulator using a synthetic Bunker C fuel composed of an aqueous additive solution uniformly dispersed in a No. 2 fuel oil. The important mineral constituents such as sodium, vanadium, and magnesium were added as a water solution of sulfates, chlorides or nitrates. All other elements in the 0.1 and 3.0 μg/ml. (ppm) range were omitted since it was believed that at such low levels they would have no effect. In some tests additional elements were added by a second metering and pumping system. Since sulfur had to be used in relatively large amounts (2% sulfur level desired), carbon disulfide was chosen because it was inexpensive.

DEPOSIT RATE TESTING

The first twelve tests were devoted to an understanding of deposit rates as a function of sodium concentration and firing temperature. The firing temperature was varied from 760 C (1400 F) to 1010 C (1850 F) and the sodium from one to ten μg/ml except in one test where it jumped to forty μg/ml. (Sodium chloride was added along with the epsom salts.) Deposit rates varied from 1.2%/100 hours to 28%/100 hours. A prediction of nozzle plugging rates (ash deposit rates) as a function of sodium concentration and firing temperature was formulated from these results and is reported herein.

TABLE 1

Analysis of Rutland Bunker C Fuel

Element	Concentration	Element	Concentration	Element	Concentration
S	1.8 wt %	Al	7	Mo	0.4
Ash	665	Ni	75	Cr	0.4
Na	5-6	Si	3	Sn	0.2
V	215	Ti	0.3	Mn	0.07
Mg	41	Cu	1.3	Pb	0.14
Fe	31	Ca	1.3		

*μg/ml unless otherwise indicated

Since these were continuous tests, they were fully manned around the clock until test No. 6 when the computer system had become capable of controlling the test. These tests gave valuable insight into correcting equipment and unexpected control system difficulties. Nutshelling was done at various times to get an estimate on its effectiveness; cleaning data was also obtained from thermal shocks. These data are presented in a later section.

Test No. 6 was run in two parts; one with 2-3 μg/ml sodium and the latter part with 7-8 μg/ml sodium, both at a firing temperature of 927 C (1700 F). Figure 2 shows the change in NAIN as a function of time for this test. In the early part of the test, the deposit rate is only 1.5%/100 hours, but later increased to about 10.0%/100 hours reflecting the increased sodium level. This test gave the first positive evidence of a new form of deposit caused by high temperatures. Next to the cooled metal of the nozzle was the familiar yellow magnesium sulfate but on top of this in the much hotter outer layer was an area of white magnesium oxide. This can be seen in Figure 3.

A deposit rate of 9.6%/100 hours is shown in Figure 4. This test was run at 1010 C firing temperature and 1-2 μg/ml sodium. Deposit rate can be computed as follows: ΔNAIN is computed as the difference between the initial value of NAIN and the NAIN value at 100 hours. Then the deposit rate per 100 hours can be computed from (ΔNAIN/NAIN initial value) X 100%. Example, see Figure 4.

1. $\text{deposit rate} = \frac{\Delta \text{NAIN}}{\text{NAIN}_{\text{iv.}}} \times 100\% = \frac{3.54-3.20}{3.54} \times 100\% = \underline{9.6}$ (%/100 hours)
2. Or from the computer calculated equation $Y_t = 3.5473 - 3.40626\text{X}10^{-3}t$ (in hours) for 100 hours; $Y_t = 3.5473-0.3406 = 3.2067$ deposit rate $= \frac{Y_i - Y_t}{Y_i} \times 100\% = \frac{3.5473-3.2067}{3.5473} \times 100\% = \underline{9.6}$ (%/100 hours)

Test No. 9 had sodium levels in the fuel as high as 40 μg/ml due to salt water contamination of the Bunker C supply. Examination of the test rig showed very heavy deposits.

SYNTHETIC VS REAL BUNKER C

The feasibility of using a synthetic Bunker C fuel was established in tests No. 16 through No. 21. The constituents of this "synthetic Bunker C" are the same as shown in Table 1. Real commercial Bunker C fuels vary considerably in trace contaminant composition as well as physical properties. Therefore a distinct advantage in using a synthetic Bunker C is that the trace contaminant levels can be controlled. The major metallic constituents of the Rutland Bunker C were added in an aqueous solution along with the magnesium sulfate additive and were homogenized with the fuel oil before it was burned. Fuel contaminant levels were established by metering additive mixtures of known composition. A problem with this additive was corrosion of the Vickers pumps used to homogenize the fuel/water solution.

These tests were quite successful and will be discussed further in a later section.

Test No. 6 turned out to be quite unusual. No sodium had been put into the second barrel of additive, and therefore no sodium was added to the fuel from the 23rd to the 41st hour of the test. NAIN failed to drop during that period of "no sodium," but when the error was discovered and the sodium added, NAIN resumed its drop at about the same rate as before (Figure 5). This would indicate that sodium is one of the controlling factors in deposit rates.

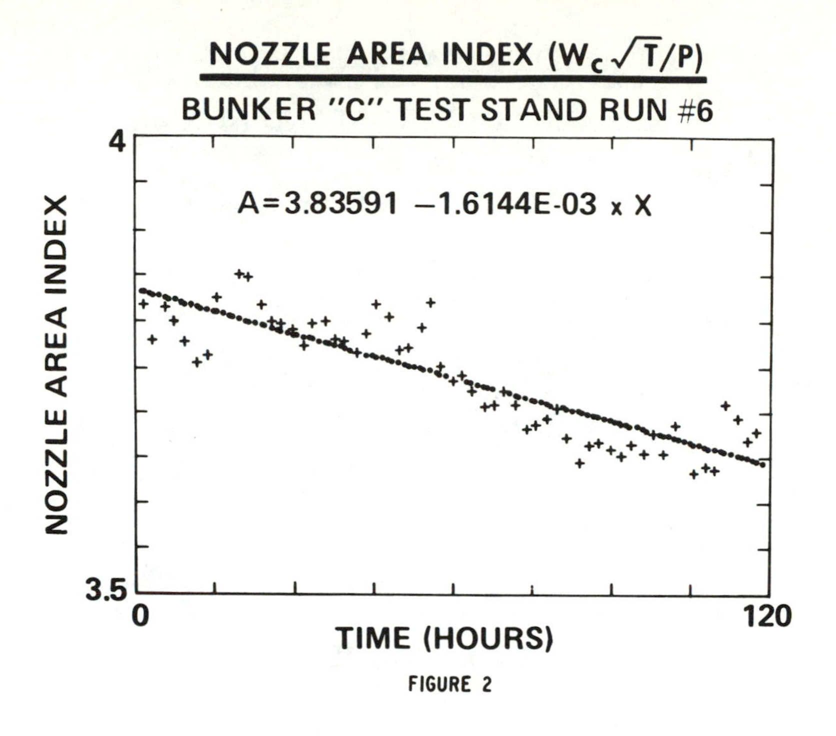

FIGURE 2

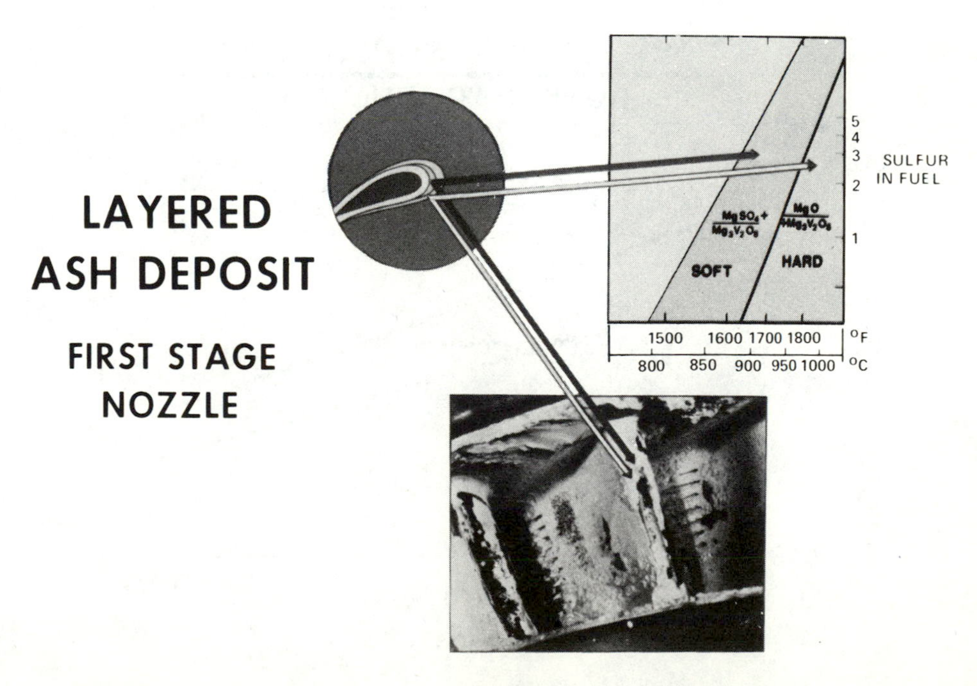

FIGURE 3

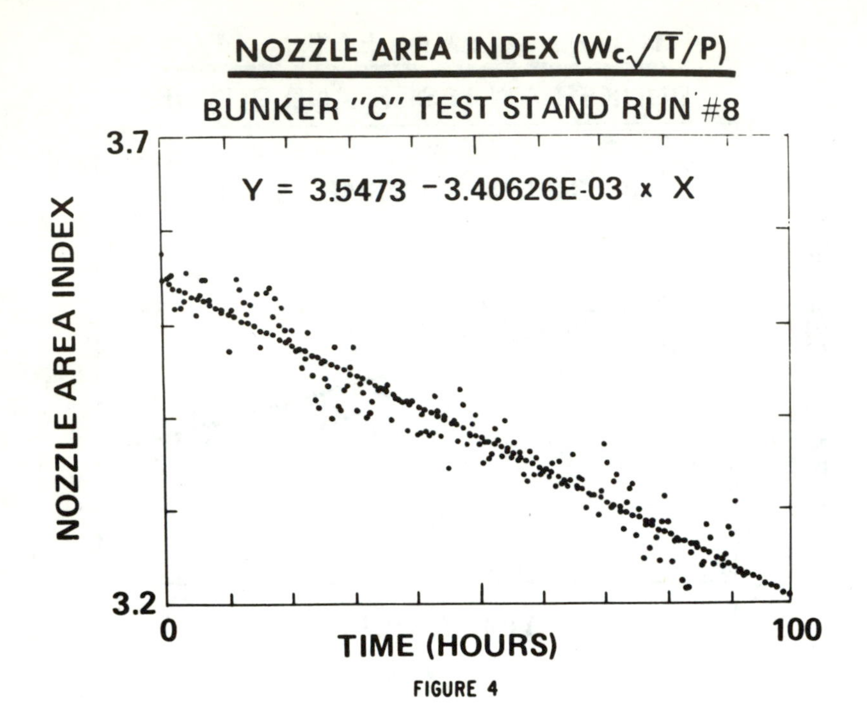

FIGURE 4

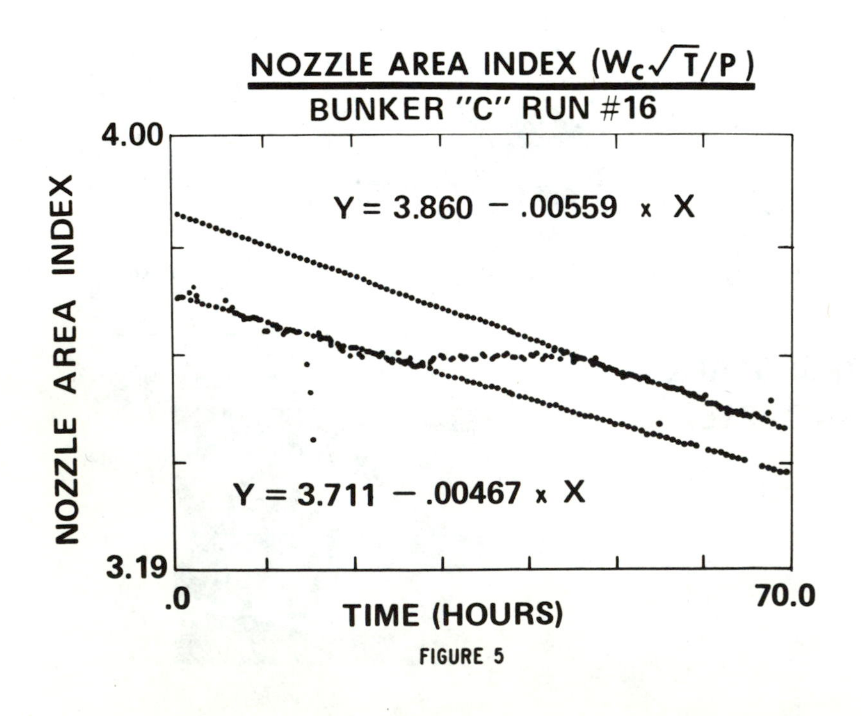

FIGURE 5

SPECIAL TOPICS

In test No. 13, calcium nitrate, at an effective concentration in the fuel of 1000 μg/ml was used to inhibit the vanadium rather than the usual magnesium additive. The nozzle sector showed a very heavy deposit of calcium sulfate which was removable by nutshelling or thermal shock. Deposit rates were three times higher than would be expected if magnesium had been used.

A complete breakdown of the fuel treatment system was simulated by test No. 14 at 538 C (1000 F) and was successful from a corrosion standpoint at least as far as it was run. Nutshelling was effective on this deposit.

An organic magnesium based additive rather than the normal epsom salts was used in test No. 15. This test showed that the organic magnesium was just as effective as the inorganic magnesium since no corrosion was noted on the test vehicle.

Three corrosion screening tests were run at 900 C (1650 F) firing temperature and 60 μg/ml sodium, to check the relative corrosion resistance of certain materials under actual turbine environments. This corrosion data was obtained under relatively short test times ($<$ 100 hours) and may not reflect incubation times, cycling, etc. These tests also showed that very heavy ash deposits can block combustor can cooling louvers creating local regions of buckling or burnouts.

DEPOSITS

Nature of Deposits

Examination of deposits were made after each run. The magnesium oxide deposits were found to be in an outer layer adhering to a layer of magnesium sulfate next to the nozzle partition surface. Since the nozzle is cooled, magnesium sulfate is stable at the metal surface, but as this layer reaches a given thickness the deposits more nearly reach the higher temperature of the gas at which point magnesium oxide is stable, as depicted in Figure 3. Magnesium oxide deposits are not water soluble and do not hydrate as does magnesium sulfate. An example of a magnesium sulfate deposit allowed to hydrate in air for several days is shown in Figure 6.

The similarity of ash deposits formed by synthetic and real Bunker C is summarized by the following observations:

1. Ash constituents fall within scatter of those formed from real Bunker C tests.

2. Structure of gas-borne particles are similar in both fuels.

3. Layered structures as discussed above were found in both cases.

4. The size of particles in the deposits themselves formed from synthetic fuel lie within the scatter of those formed from real Bunker C.

DEPENDENCE OF PLUGGING RATES ON SODIUM LEVEL AND TEMPERATURE

The results obtained in this program allow one to make a preliminary calculation of the nozzle plugging rates to be expected as a function of sodium concentration in the fuel and firing temperature. Sodium level and firing temperature were the main variables in these tests. Magnesium and vanadium were held constant because (1) it was fairly certain that a high level of vanadium would be present in the fuels requiring magnesium additives and (2) the sodium could be reduced to almost any level desired by the degree of washing employed. A model of the deposition mechanism was postulated; and its validity was tested by laboratory and field experience at hand.

Consider a nozzle in which ash builds up on all the walls at a uniform rate. Figure 7 is a model of how ash particles in the stream might be caused to stick to the sides of one of the walls of this nozzle. The "glue" in this model is assumed to be a sodium sulfate or perhaps one of the several low melting Na-V-O compounds.

EXAMPLES OF TREATED RESIDUAL FUEL ASH DEPOSITS ON NOZZLE SECTOR AFTER TEST

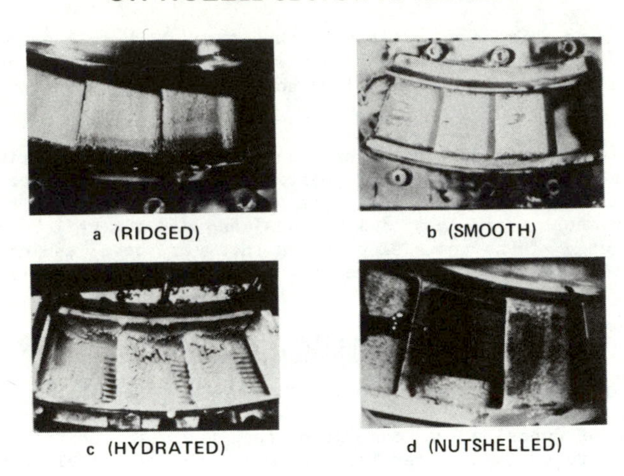

a (RIDGED) b (SMOOTH)

c (HYDRATED) d (NUTSHELLED)

FIGURE 6

MODEL FOR DEPOSITION OF ASH

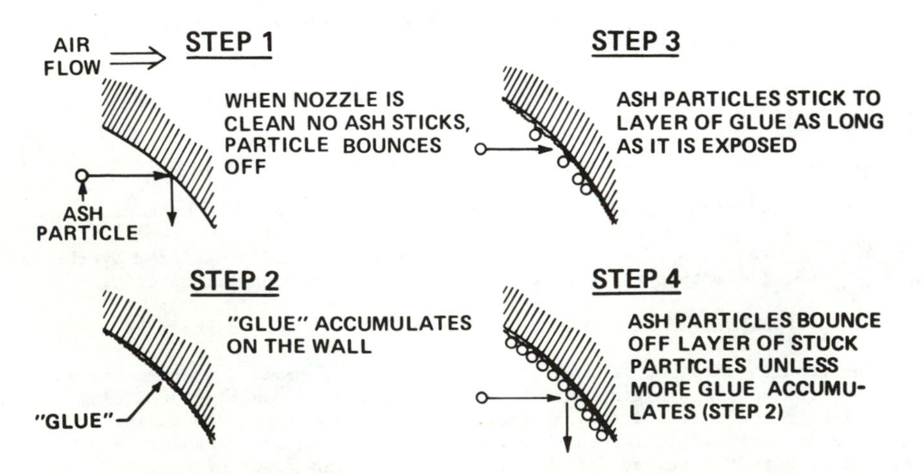

FIGURE 7

A corollary to this model is that the deposition of particles should be independent of the number of high-melting ash particles ($MgSO_4$, MgO or $3Mg{\bullet}V_2O_3$). Field experience bears this out in that deposition rates from fuels having as low as 12 µg/ml vanadium and 36 µg/ml magnesium are not substantially different from fuels having 200 µg/ml vanadium and 600 µg/ml magnesium.

The rate of change of thickness should be proportional to the mass of "glue" flowing through the nozzle and a thermally activated likelihood for a particle to stick.

An equation can be derived from these assumptions:

$$\frac{dN}{dt} = \frac{793}{L} [Na]^{0.7} \text{ EXP} \left(- \frac{11600}{273 + T_f} \right)$$

Since magnesium oxide forms at high temperatures and magnesium sulfate at low temperatures, the activation energy may not be the same for all temperatures. The density of the ash deposit, size of the ash particles, and boundary layer effects have been neglected here because they have not been experimentally defined well enough to indicate that their effect is significant.

Deposit rate data for both real and synthetic Bunker C are plotted in Figures 8 and 9, respectively. The deposit rates have been normalized with respect to the [Na] and are plotted versus the reciprocal of absolute temperature. The data are plotted as regions to reflect variations and uncertainties in laboratory and field test. Field experience is included from four types of turbines. An activation energy of 23000 kcal can be derived from these plots. The agreement, covering a wide range, between field and laboratory experience at temperatures between 760-1010 C gives some confidence that the use of the high pressure rig to predict plugging rates is valid.

The variability in the data leading to the equation suggests a one sigma error of ± 50% of the plugging rate calculated therefrom. The equation is evaluated for several sodium levels and firing temperatures in Table 2. It cannot be used to determine deposit rates at $T_F > 982$ C (1800 F) or for Na levels < 2 µg/ml due to uncertainties in the test data.

TABLE 2

Predicted Plugging Rates (%/100 hours)

Sodium Concentration in Fuel by Weight µg/ml	Firing Temperature (°C)			
	816	871	927	982
2	2	3	4	6
5	3	5	8	12
10	5	9	13	20

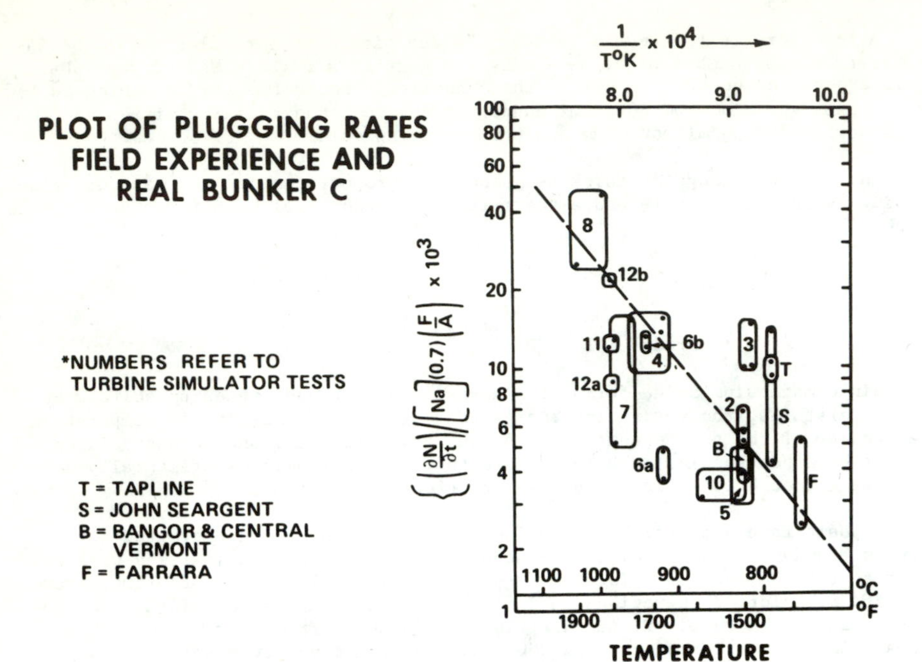

FIGURE 8

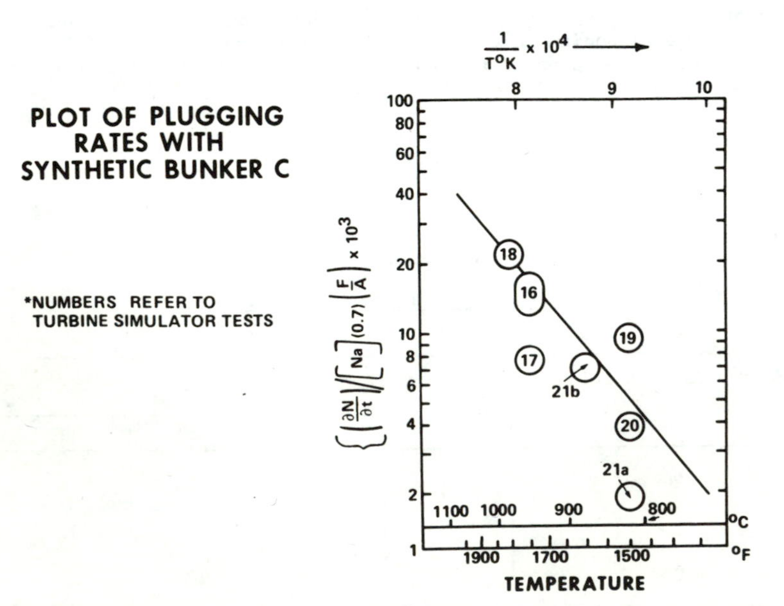

FIGURE 9

DEPOSIT REMOVAL

Nozzle Deposit Removal by Nutshelling and Thermal Shock

The effectiveness of thermal shocks and nutshell injections to remove soft ash deposits can be measured by the following: This is the change in NAIN during the cleaning sequence divided by the change in NAIN from the start of the test to just prior to the cleaning sequence times 100% or $(\Delta NAIN_{cs}/\Delta NAIN)X100\%$.

It would appear that position B (Figure 1) is the best overall injection point and that position D is the best for combustion and transition piece deposit removal. Nutshelling of soft ash deposits, those formed at or below 1010 C can remove an average of 50% of the deposit with the range being 25 to 75%. See Table 3.

Table 4 lists the nozzle area recoveries (for soft ash deposits) due to thermal shock to any of several temperatures. From this limited data it appears that a shutdown to 21 C (70 F, room temperature) is far more effective than shutdowns to 288 C.

Effect of Derating Before Nutshelling

It has been thought that by lowering the firing temperature before shutdown it would be possible to convert magnesium oxide to magnesium sulfate, which is easier to remove. However, the conversion at 816 C (1500 F) or 871 C (1600 F) in

TABLE 3

Percent Recovery of Nozzle Throat Area
by Nutshell Injection for Soft Ash Deposits

Na \ T_F	High 954 to 1010°C	Low 816 to 843°C
High (10-44 μg/ml)	55, 56	39
Low (1-5 μg/ml)	26, 49, 16	37, 61, 17

TABLE 4

Percent Recovery of Nozzle Throat Area
by Thermal Shocks for Soft Ash Deposits

	Shock to 288°C*	Shock to 21°C*	Shock to 1093°C*
% recovery	21, 28, 38	90	44

*From firing temperature to temperature indicated.

the presence of sulfur dioxide and sulfur trioxide is a rather sluggish reaction; very long times at reduced temperature are necessary because this conversion is diffusion limited.

SUMMARY

These tests have shown that:

1. The Turbine Simulator is a satisfactory device for these studies since (a) nozzle plugging rates are generally in agreement with those experienced in the field, (b) the chemical and physical nature of the deposits is the same, (c) results have a reasonable degree of repeatability.

2. A synthetic Bunker C made by homogenizing No. 2 fuel oil with contaminants in water solution can be used in a gas turbine type combustion system in the place of a real Bunker C to produce similar deposits and deposition rates.

3. Deposit rates (nozzle plugging rates) can be predicted as a function of sodium concentration and firing temperature.

ACKNOWLEDGEMENTS

The magnitude and complexity of the tests and the necessity for running an interrupted 100 hours required the involvement of a number of persons from both CR&D and Gas Turbine Division (some at very long and unusual working hours) to whom the authors are grateful and without whose enthusiasm and dedication little would have been accomplished; special mention should be given to Dan Smith (CR&D) and Harvey Doering (GTD).

NOMENCLATURE

NAIN = nozzle area index number = $W\sqrt{T}/P$

W = air mass flow in kg/sec

T = absolute temperature, °K

P = pressure in pascals

[Na] = concentration of sodium in fuel in μg/ml (ppm) by weight

L = nozzle throat width in cm

T_f = firing temperature in °C for $T_F \leq$ 982 C (1800 F)

$\frac{dN}{dt}$ = nozzle plugging rate in percent of original (clean) opening per hour = $\frac{\Delta NAIN}{NAIN_{i.v.}}$ x 100%

$NAIN_{iv}$ = initial value of NAIN, test start

ΔNAIN = change in NAIN during test

$\Delta NAIN_{cs}$ = change in NAIN during cleaning sequence

t = time in hours

PART II: FORMATION AND REMOVAL OF RESIDUAL FUEL ASH DEPOSITS IN GAS TURBINES FORMED AT FIRING TEMPERATURES ABOVE 982°C (1800°F)

T. A. URBAS and L. H. TOMLINSON

General Electric Company
Schenectady, New York, USA 12345

ABSTRACT

Ash deposition tests using synthetic Bunker C fuels (composed of an aqueous additive solution uniformly dispersed in a No. 2 fuel oil) were conducted in a simulated gas turbine combustor with a first stage nozzle vane segment. Silicon, when used in a ratio of Si/Mg/V of 7/3/1 in the fuel, has been shown to reduce nozzle plugging rates (ash deposit rates) by an order of magnitude with a fuel containing 2 to 10 μg/ml Na; use of less silicon results in a smaller reduction in deposit rates. An MgO slurry has the advantages of being less expensive than an oil soluble type and has much less bulk than epsom salt solutions. However, it has relatively much larger particles than those from oil soluble inhibitors or epsom salt solution, but was still shown to be an effective vanadium inhibitor.

A combination of washing and soaking for one half hour at a cold water flow rate of 0.14 dm^3/sec per kg/sec airflow (one gpm/one lb air per sec airflow) per combustion chamber followed by a refire to 538 C (1000 F) has been shown to be an effective means of removing hard ash deposits formed at 1066 C (1950 F) firing temperature.

INTRODUCTION

This paper details the second part of the work which was conducted on ash deposits and supported by the MARAD contract (Contract No. C-O-35510, U.S. Maritime Administration, Department of Commerce.) The tests performed primarily below 982 C (1800 F) are documented elsewhere in "Formation and Removal of Residual Fuel Ash Deposits in Gas Turbines Formed at Firing Temperatures below 982 C (1800 F)."

This series of tests was concerned with the evaluation of ash modifiers, a new MgO additive, and high temperature deposits including rates and in situ removal techniques. The techniques were later applied to low temperature deposits.

The test vehicle is the same as had been previously discussed in the aforementioned document.

EFFECT OF SILICON

Low Temperature and High Sodium Results

Five tests were run at various times during the program to examine the effects of silicon on ash deposit rates. The silicon was first added as an oil soluble compound and then as talc (a magnesium silicate). The organic additive was much more successful. These substances were added to the fuel in addition to the usual water solution of minerals.

The first test was run with an actual metal content in the fuel of 10 μg/ml (ppm) Na, 200 μg/ml V, 600 μg/ml Mg, and 1400 μg/ml Si; therefore the V/Mg/Si ratio was 1/3/7 as recommended by the vendor of the oil soluble silicon additive.

Figure 1 shows the history of NAIN (a measure of effective nozzle throat area) for this test. The first 40 hours showed that NAIN was not changing much; a decrease in NAIN indicates a deposit rate. During the first 20 hours of the test there were some problems with control thermocouples which in turn caused some problems with NAIN computation. In order to determine if the low deposit rate was being caused by the addition of silicon, the silicon additive was shut off for a period of time and then turned on again. The period from the 40th to the 70th hour shows a plugging rate of 11.7% per 100 hours with no silicon going in. The first 30 and last 20 hours of the 90 hours test show a plugging rate of only 1.29% per 100 hours with silicon additive turned on again.

Examination of the nozzle sector after the test (no nutshelling) showed deposits both yellow and white and mostly of a soft friable nature. Note that

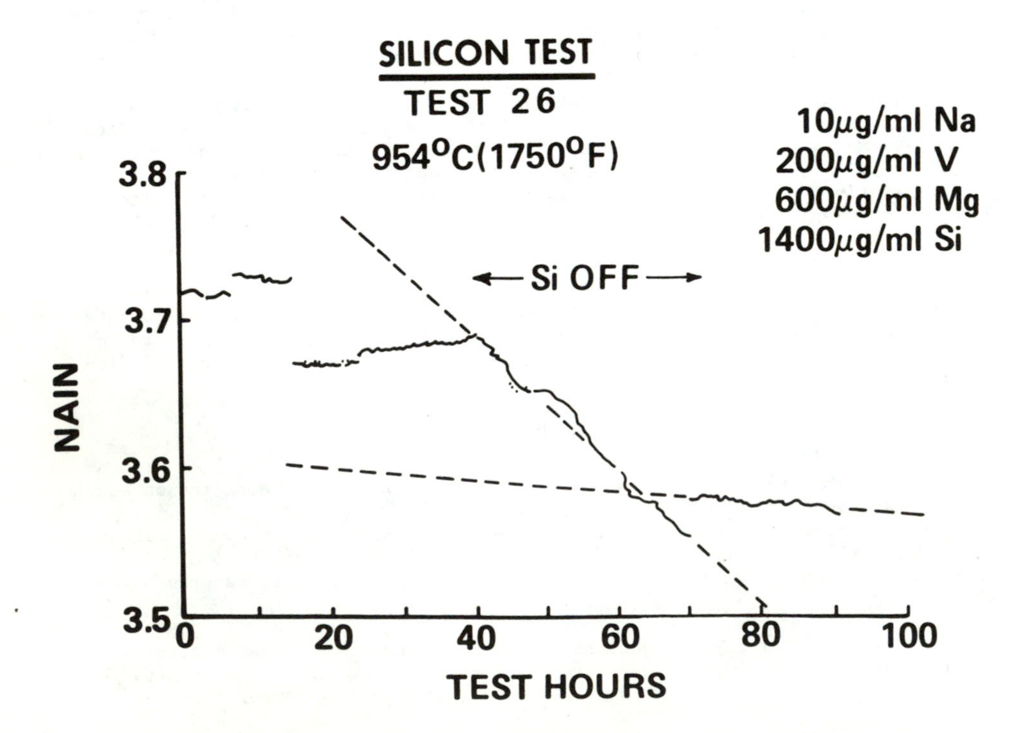

FIGURE 1

the low deposit rate of 1.29% per 100 hours is lower than that which would have been normally obtained even for well washed fuel (1 μg/ml Na). It appears that adding a high level of silicon to the fuel has a pronounced effect on deposit rates when burning fuels at low firing temperatures 954 C (1750 F) and high sodium values of 10 μg/ml.

High Temperature and Lower Sodium Results

Two tests were next run at a firing temperature of 1066 C and a fuel trace metal contaminant composition of 2 μg/ml Na, 200 μg/ml V, 600 μg/ml Mg, and 925 μg/ml Si. This was done in order to simulate the addition of talc (4 SiO_2•3 MgO•H_2O on a weight bases = 925 μg/ml Si + 600 μg/ml Mg) as an additive. Talc is an inexpensive, available product and contains silicon and magnesium in such a ratio that it could conceivably act both as an ash modifier (Si) and vanadium inhibitor (Mg). The deposit rate was reduced by a factor of five. The change in deposit rate was lower probably because less silicon was used. In the second of these tests, the deposit rate behaved erratically. It was negative (increase in area) for the first few hours, and then leveled off to zero for the remainder of the test. This can probably be accounted for by distortion of the nozzle partition.

Inspection of the nozzle sector after each of these tests showed a light white to yellow deposit corresponding to the observed deposit rate.

Use of Talc

The next step in this series was the introduction of talc directly into the combustion chamber in order to test the feasibility of eliminating other additive systems. It was thought that this would be the optimum additive system because it might eliminate the problems associated with slurry handling and pumping (which will be discussed in the next section), the high cost of organic additives, and the bulk storage problems associated with organic and water soluble additives. This airborne powder additive system proved to be very difficult to manage.

It was quite difficult to hold the powder injection rate constant and at times the flow rate (measured at 10 minute intervals by weight difference) would drop well below or rise well above the desired rate. Unfortunately, the deposit rate was ~ 52%/100 hours - very much above what was expected.

The last test was run at 1066 C, with a 2 μg/ml Na, 200 μg/ml V, and 600 μg/ml Mg + 925 μg/ml Si added as a talc slurry in No. 2 oil in order to answer these questions: (1) does talc added directly to the fuel create high deposit rates as had been previously observed in which talc was added as an airborne powder or (2) was the high deposit rate due to non-uniform powder additive feeding?

After six (6) hours of test time, it was readily apparent that there was a high deposit rate: ~ 42.5%/100 hours as calculated from computer data. The rig was run for a few more hours and then shut down. Inspection of the nozzle cascase after the test showed a very heavy deposit.

From these last two tests it is readily apparent that talc is not acceptable as an additive at these conditions due to the very high deposit rates.

MgO ADDITIVES

Two tests were conducted using an MgO slurry in diesel fuel in an effort to use an additive that would be less expensive than an organic type but would have much less bulk than the inorganic epsom salts solution. One concern, later proved not to be a problem, was the relative effectiveness of magnesium oxide as a vanadium inhibitor since the MgO particles are relatively much larger than those from the oil soluble inhibitor or epsom salt solution.

In the first test, the MgO pumping system had significant, intermittent flow interruptions forcing an early termination of the test.

Examination of the nozzle sector showed that some alloy corrosion samples which had been welded onto the nozzle trailing edge were badly corroded. This was probably due to interrupted MgO flow and resultant periods of non-inhibited operation.

Obviously, the problem of pumping and metering the MgO slurry is more difficult than with an oil soluble additive. Some of the problems were due to the fact that a No. 2 fuel was being used; this is very light and additive settles out readily. This is not expected with a heavy residual oil.

The second test in this series used an MgO slurry which was less difficult to handle because it had a much smaller particle size distribution. However, clogging of the fuel system in time forced a shutdown. Again, it was apparent that settling of the additive with a light fuel is a major concern.

A third test was run with MgO, but in this case the additive was added as an airborne powder added to the air stream. Some of the problems encountered in the airborne talc injection test had resulted in an improved powder feeder. The fine MgO powder, was agglomerated into larger particles in order to improve its flow characteristics. This was done by using an organic binder in a spray drying process.

The powder feeder worked quite well with only some minor fluctuations in the required feed rate. The deposit rate was somewhat higher than expected, but no major corrosion scaling of the nozzle sector was noted. This powder feeder would need much more work to determine its feasibility for sustained periods, but this test was the last that was run in support of the MARAD contract.

ASH REMOVAL BY WASHING AND SOAKING

Introduction

At firing temperatures above 927 C (1700 F) and depending on the sulfur content of the fuel, MgO deposits (insoluble in water) are more likely to form than $MgSO_4$ (water soluble), as indicated in Figure 2. Generally, $MgSO_4$ can be completely and quickly removed by water or wet steam. MgO deposits, however, are hard and have at times required mechanical removal.

It was discovered that if a nozzle segment coated with these deposits was allowed to sit for 12-14 hours in water, the hard MgO deposits disintegrated to a slurry and simply washed away. The postulated mechanism is that minute quantities of Na_2SO_4 and/or Na-V-O compounds formed during combustion from Na in the fuel acted as a "glue" to make the particles of MgO stick together. Since Na_2SO_4 is water soluble but resides between the particles of MgO, water does not dissolve it readily. It requires time for the water to reach the Na_2SO_4 and dissolve it. Once it is dissolved, however, there is nothing to keep the MgO particles bonded together and they separate to form a slurry.

Since high turbine firing temperatures are desirable for high output and high efficiency, it was necessary to learn more about deposits at high firing temperatures. It was found that deposits formed at these temperatures do not soak off readily compared to those formed at lower temperatures. Therefore, it was necessary to find ways to effectively clean the hot gas path, and a large number of tests were devoted to this. The wash/soak/refire procedures for removing ash deposits was the major development of this series of tests.

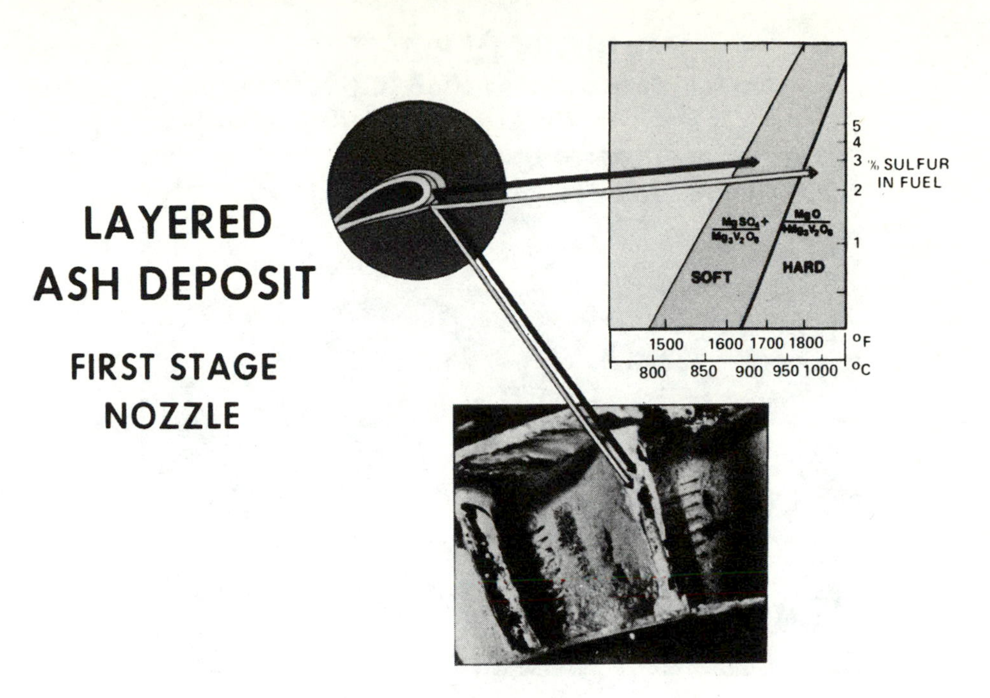

FIGURE 2

Development of the Wash/Soak/Refire Techniques

For the first in situ washing test of a high temperature deposit, the firing temperature was set at 1066 C with a 2 µg/ml Na, 200 µg/ml V, 600 µg/ml Mg fuel. This turned out to be a most challenging task. The deposit rate was very high, being greater than 20%/100 hours. Inspection of the nozzle sector showed a heavy white (MgO) deposit (Figure 3).

Removal of this deposit formed at 1066 C firing temperature was attempted by the following sequences:

1. Warm water from retractable wash probes (See Figure 4) for 30 minutes. This removed 60-70% of the deposits. Additional washing had no noticeable effect.

2. Both dry steam at 1034 k Pa (150 psi) and wet steam at 103 k Pa (15 psi) proved ineffective in attacking the residue.

3. Cold water at 1379 k Pa (200 psi) did not prove effective in removing residue.

The appearance of the deposit after this sequence is shown in Figure 5. About 30-40% of the original deposit remained and most of the trailing edge cooling holes were filled with deposit.

However, it was earlier noted that the deposit had "pulled away from" the metal surface after the first day's water wash, and it was believed that some sort of shock (thermal, acoustical, vibratory, etc.) might dislodge it. Therefore, the rig was refired to 1066 C for one hour; nutshelling was carried out just prior to shutdown. Soon after, visual inspection showed that the nozzle including the trailing edge cooling holes were essentially 100% clean (Figure 6).

VIRGIN DEPOSIT
AFTER 46 HOURS AT 1066°C (1950°F)
WITH 2 μg/ml Na, 200 μg/ml V, 600 μg/ml Mg.

FIGURE 3

TRANSITION PIECE WATER WASH PROBES

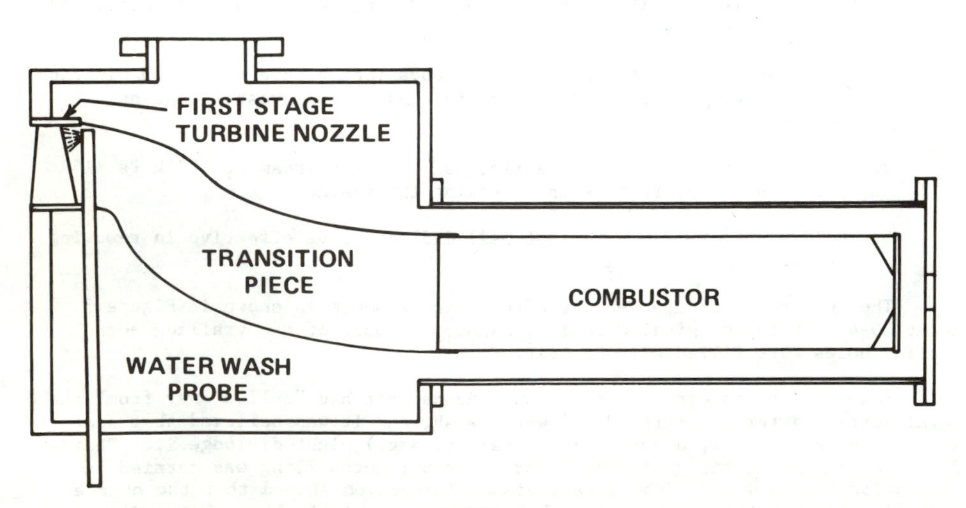

FIGURE 4

REMAINING DEPOSIT
AFTER REMOVAL ATTEMPTS BY STEAM AND WATER

FIGURE 5

NAIN history is shown in Figure 7. Apparently the shock of restarting and subsequent high temperature broke off the loosened deposits. One theory is that water penetrated through the deposit during the water wash; and as the rig was brought back up to 1066 C the water flashed into steam creating sufficient pressure to fracture the deposit. It has been reported that the deposit that forms has on the order of 50% porosity.

Simplification of Equipment

The use of retractable wash probes in front of each nozzle partition would require an enormous effort in redesign to implement such a system. A manual wash water lance (Figure 8) would be cumbersome to use and one could not be sure that he has the ideal location for washing.

A logical idea would be to hook the water up to an already available location such as the gas side of a dual fuel nozzle (Figure 9) if gas were not being used as a fuel. Another idea was to connect piping only to the outer combustion casing (Figure 9) such that there were no connections in high temperature or highly stressed areas.

To carry the water from the wash nozzle through the entire hot gas path required the use of crank airflow. That is, at ~ 20% speed, there is a ~ 10% airflow through the unit which is quite violent. Washing tests done with these latter two piping arrangements and with crank airflow were shown to be highly effective. At this time, water flow rate had been established at .946 dm^3/sec (15 gpm) for this test rig.

CLEAN NOZZLE

AFTER REFIRING TO 1066°C (1050°F) FOR ONE HOUR

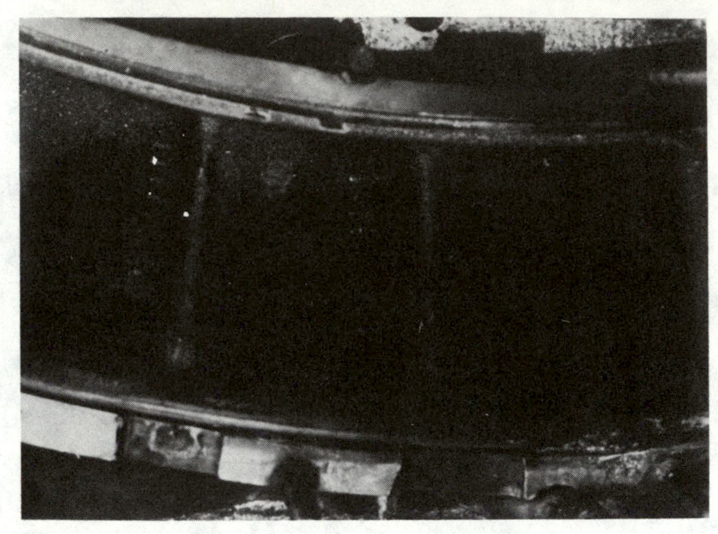

FIGURE 6

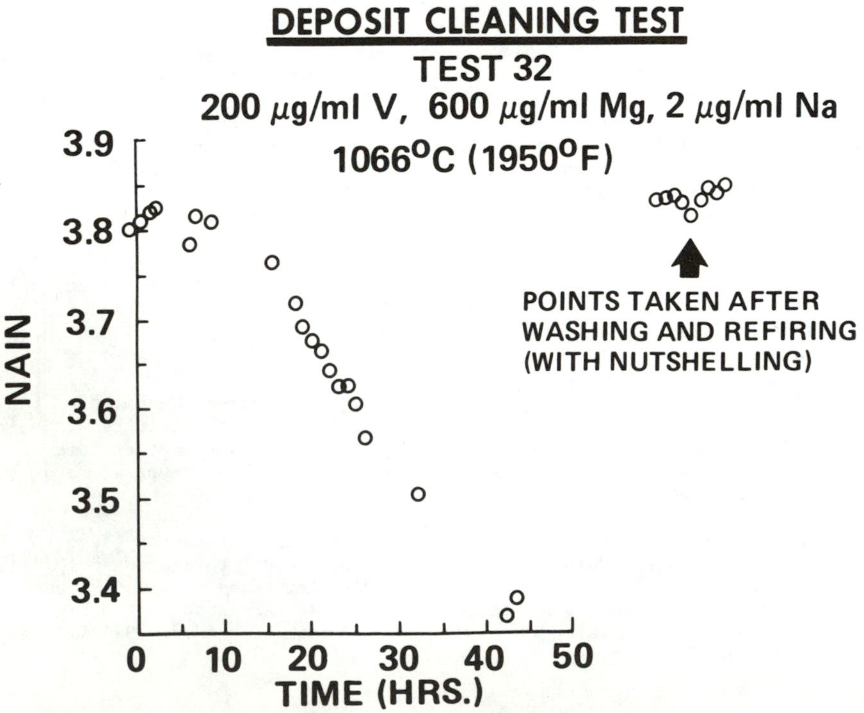

DEPOSIT CLEANING TEST

TEST 32

200 μg/ml V, 600 μg/ml Mg, 2 μg/ml Na

1066°C (1950°F)

NAIN

3.9
3.8
3.7
3.6
3.5
3.4

0 10 20 30 40 50

TIME (HRS.)

POINTS TAKEN AFTER WASHING AND REFIRING (WITH NUTSHELLING)

FIGURE 7

WASH WATER LANCE

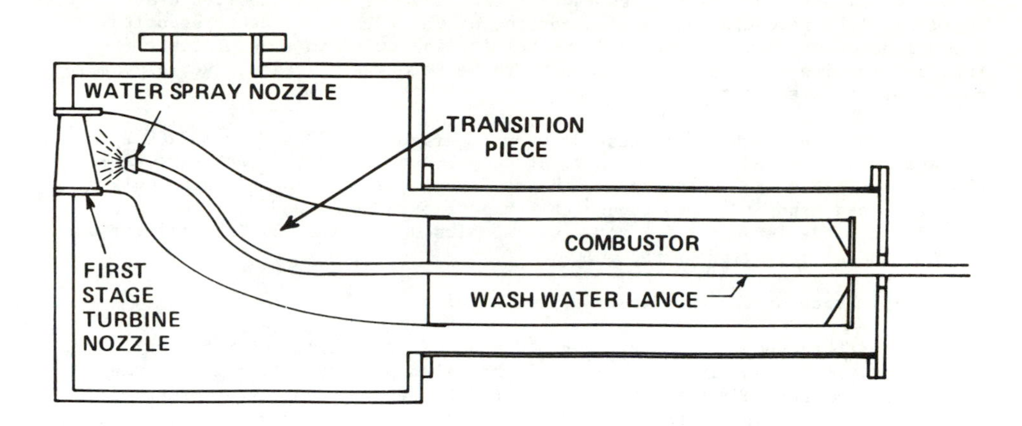

FIGURE 8

COMBUSTOR WASH WATER ARRANGEMENTS

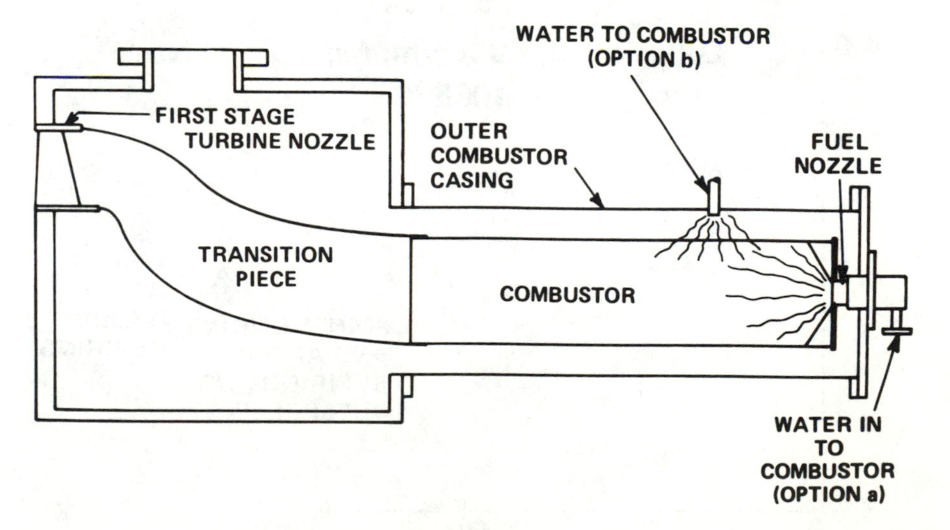

FIGURE 9

Effect of Thermal Shock and Nutshelling

Since a refire after washing had proved to be ~ 100% effective in removing high temperature deposits, it was thought that thermal shocks to 1066 C after shutdown might remove a substantial portion of the high temperature deposits. A combination of thermal shock and nutshelling was tried in several tests (before any washing was done) but was shown to be relatively ineffective; Figure 10 shows one of these tests.

However, in a subsequent test an overtemperature (> 1204 C or < 2200 F) of the system was caused by loss of part of the air supply to the test rig. Although the NAIN data had indicated a heavy deposit, inspection revealed only a light, burned deposit. Even though this had removed the deposit, it was not considered a viable option for removing deposits due to the possible effects on the mechanical integrity of the unit.

Optimization of the Wash/Soak/Refiré Technique

It has been shown that after 5 to 10 minutes of washing (no refiring), very little additional deposit is removed. It has also been established that the amount of time needed for water to penetrate the deposit was on the order of ten (10) seconds. Therefore, the limiting factor in the "soaking" effect would be the time it takes to get enough water to wet the surface and penetrate the deposit fully. A washing test was performed on a deposit formed at 1066 C firing by:

1. washing for five minutes and refiring - very little deposit removal.

2. washing for an additional fifteen minutes and then refiring - some deposit removal.

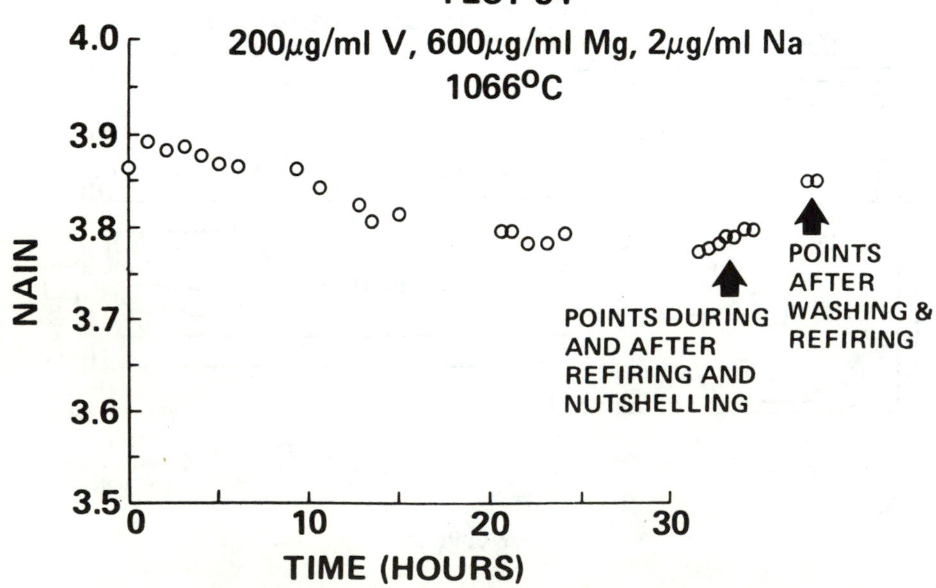

FIGURE 10

Since a previous washing test had a soak time of 35 minutes and a refire had removed essentially all the deposit, the wash/soak time was set at 30 minutes.

The use of .946 dm^3/sec for the test combustion chamber for one half hour washing requires a large amount of water. To lower the water usage, the following were tried:

1. a wash using 0.22 dm^3/sec (3.5 gpm) for one half hour - this required two washing periods and two refires to adequately clean the nozzle.

2. a wash using .47 dm^3/sec (7.5 gpm) for one half hour followed by a refire - very effective. This flow rate set the water usage at 0.14 dm^3/sec per kg/sec airflow (one gpm/one lb air per sec airflow).

Several washing tests which were visually observed showed that the combustor and transition piece are very readily washed in two minutes; the ease of washing is probably dependent on the low metal temperature of these parts (760°C or 1400°F). Furthermore, it showed that with hard deposits the water for washing can be recycled after the first five minutes to reduce total water requirements.

Since washing a complete turbine with one gpm water/one lb per sec airflow would require an enormous pump capacity, it was decided to determine if some combustion cans of a multicombustor turbine could be washed in sequence. To simulate this, a test deposit was purged with dry nitrogen for four days. Then, a refire was initiated and subsequent inspection showed the nozzle sector clean relieving any worries that the deposits might be dried out.

Up to this time, the refire after the soaking period has always been to the firing temperature of the test. In a following test, a refire was made to only 538 C (which is the lowest practical firing temperature for the test rig and close to the ~ 482 C (900 F) no load firing temperature for gas turbines) and was just as effective as refiring back to full temperature.

FUEL AND ADDITIVE ANALYSES

The analysis of Bunker C fuel from a utility in Rutland is shown in Table 1. For each test, several fuel and Mg additive solution samples were submitted for analysis. In the tests on Bunker C fuel, only Mg and Na were added to the fuel by use of the Mg additive solution.

Table 1

Analysis of Rutland Bunker C Fuel

S	1.8 Wt %	Al	7	Mo	0.4
Ash	655*	Ni	75	Cr	0.4
Na	5-6	Si	3	Sn	0.2
V	215	Ti	0.3	Mn	0.07
Mg	41	Cu	1.3	Pb	0.14
Fe	13	Ca	1.3		

Gross Heating Value 2120 kcal/kg (18,509 Btu/lb)

*μg/ml unless otherwise indicated.

Table 2

TRACE METAL CONTAMINANTS
(Partial Sample of Total Analyses)

TEST #	REQUIRED IN FUEL			CALCULATED IN FUEL #6 OIL			CALCULATED IN FUEL #2 OIL			MEASURED IN ** ADDITIVE	MEASURED IN * ADDITIVE	ADDITIVE FLOW RATE	ADDITIVE SP. GR.
	Na/V/Mg µg/ml (wt/wt)			Na µg/ml (wt/wt)	V	Mg	Na µg/ml (wt/wt)	V	Mg	Na/V/Mg µg/ml (wt/wt)	Na/V/Mg (µg/cc)	% of #2 OIL FUEL FLOW	
31	10	200	600	3.9	178	680	4.1	185	707	130/5806/22154	130/6630/25300	2.4	1.142
32	2	200	600	2.1-2.4	193-228	552-651	2.2-2.5	201-237	574-677	80/7570/21678	80/8660/24800	1.99-2.35	1.144
33	2	200	600	2.3-2.5	205-214	627-656	2.4-2.6	213-223	652-683	80/7089/21728	80/8124/24900	2.25-236	1.146
35	2	200	600	2.2-2.3	198-209	600-637	2.3-2.4	206-218	624-663	80/7295/22077	80/8360/25300	2.12-2.25	1.146
36	2	200	600	2.6	214	625	2.7	223	650	90/7351/21416	90/8410/24500	2.28	1.144
37	2	200	600	2.8-3.0	246-265	640-695	2.3-2.5	256-276	666-723	75/9290/21728	75/9500/24700	2.30-2.48	1.146
38	2	200	600	2.8-3.0	200-214	631-678	2.9-3.1	208-223	657-705	100/7030/22200	---	2.23-2.39	1.146

*Lab Analysis: Na measured in µg/ml (wt/wt); Mg, V measured in (µg/cc).

**V, Mg-µg/ml(wt/wt) calculated from V, Mg-(µg/cc).

For synthetic Bunker C fuel, the main elements Na, Mg, V were added to give a fuel closely representing the original Rutland Bunker C for critical metal contaminant level. Non-critical elements of only a few μg/ml or under were excluded since it was felt that they would have no influence on the test results. The synthetic Bunker C is a uniform, but unstable mixture of No. 2 fuel oil plus a water-based solution of elements which gave the same approximate analysis as for Rutland Bunker C. This made it difficult to analyze since the two phases separated rather easily and rapidly even though an emulsifier was added when the sample was taken. Later, only samples of the fuel oil and additive solution were taken separately. In general, the lab results agree quite well with what is expected as calculated from fuel additive flow rates, specific gravity and element concentration.

Table 2 is a compilation for a small portion of the tests of synthetic Bunker C of the required levels of contamination (column 2) in the fuel versus the actual levels (columns 3 and 4). Results of the additive analysis are shown in the fifth and sixth columns. The additive flow rate is shown in the seventh column and its specific gravity in the last column.

SUMMARY

The tests have shown that:

1. Silicon is an effective ash modifier at firing temperatures of 954 C to 1066 C and sodium levels in the fuel of 2 to 10 μg/ml because it greatly reduces ash deposit rates.

2. An MgO slurry is an acceptable vanadium inhibitor alternative to epsom salt solutions or organic additives. (MgO slurry fuel mixture must meet turbine fuel specification limits for Ca, Na, K, Pb).

3. High temperature ash deposits formed at 1066 C can be removed by a combination of washing, soaking and refiring. This technique can also be applied to low temperature deposits.

ACKNOWLEDGEMENTS

The authors are appreciative of the supporting efforts supplied by many individuals from the Gas Turbine Division and Corporate Research and Development; special mention should be given to H. von E. Doering (GTD) and D. P. Smith (CR&D).

WATER/OIL EMULSION COMBUSTION IN BOILERS AND GAS TURBINES

M. L. ZWILLENBERG, C. SENGUPTA, and C. R. GUERRA

Public Service Electric and Gas Company
Newark, New Jersey, USA 07101

ABSTRACT

The use of water-in-oil emulsions (wherein each fuel droplet leaving the atomizer contains a dispersion of microdroplets of water) is a promising technique which has been proposed for the reduction of smoke and NO_x emissions from boilers and gas turbines. This technique appears also beneficial in preventing gas turbine blade fouling from the deposition of fuel additives and residues from fuel combustion.

An evaluation of the effects of water/fuel emulsions if used in utility steam boiler and gas turbine operation has been performed to assess their relative importance. Factors considered include unit efficiency and operational factors, corrosion and deposits, environmental effects and costs.

A planned experimental study of water/fuel emulsion combustion in a 25 MWe PSE&G gas turbine/generator unit is outlined. This study includes emulsion characterization, monitoring of particulates, opacity, NO_x, CO and hydrocarbons and "hot-end" turbine inspection to detect corrosion effects.

INTRODUCTION

The use of water-in-oil emulsions (in which each fuel droplet leaving the atomizer contains a number of microdroplets of water) is a promising technique, which has been proposed for the reduction of smoke and NO_x emissions from boilers and gas turbines. Other possible benefits include: [1] the reduction in excess air with consequent reductions in SO_3 formation and cold end corrosion in boilers(4), [2] the elimination of metal-containing antismoke additives and the resulting deposits and fouling in gas turbines, [3] the ability to use heavier and cheaper turbine fuels and [4] reduced turbine inlet temperatures which reduce corrosion and extend blade life. Addition of water to oil in emulsion form promises to have greater NO_x reduction effectiveness than direct water injection. The improved effectiveness permits reducing both the amount of water needed and any efficiency losses incurred from water addition.

A multitude of fuel additives containing Mg, Ba, Mn, Fe, Zn, Co, Cr and other metals are in current use to improve combustion and alleviate corrosion and emissions problems in boilers and gas turbines[5-8]. Possible future restrictions on trace metal and fine particulate emissions [9] could limit use of such additives. Forthcoming gas turbine NO_x limitations[9] seem likely to require water use in the combustion zone. This water might be most efficiently introduced in emulsion form, since a smaller amount of water than for direct injection is required.

Historical Development

The possibility of improving combustion by use of water-in-oil emulsions was first suggested in 1957 by Ivanov et al[10-11]. They emphasized the improvement in atomization and fuel vaporization and the reduction in soot formation produced by the explosive vaporization (microexplosions) of the microdroplets of water contained within each fuel droplet. Dryer[2,12] and co-workers[1] have reviewed the literature in detail and analyzed the mechanisms of emulsion combustion. They list[14] a number of methods which have been employed to produce water-in-fuel emulsions [Table 1].

A wide range of "emulsion" systems is currently under study and reported in the literature. Figure 1 shows some of the systems discussed at a recent symposium[25] on "Water-in-Fuel Emulsions". The present paper will limit its scope to system 1a, emulsions of water in oil, with or without the use of a stabilizing surfactant.

Physical and Chemical Effects

Emulsions have both physical and chemical effects on combustion. The physical effect is based on the "microexplosion" phenomenon shown in Figure 2. Since the water has a lower boiling point than the fuel, the water microdroplets reach and surpass their boiling point and become superheated while the fuel droplet is still below its boiling point. At some point, at or below the homogeneous nucleation temperature (maximum superheat) of the water microdroplets, one or more of them boil explosively, shattering the fuel droplet, producing secondary atomization and improved fuel-air mixing. Since other water microdroplets may remain in the daughter fuel droplets, these in turn may cause subsequent microexplosions, subdividing the fuel even further. In addition to improved fuel atomization, vaporization and mixing, the dispersed water phase reduces the fuel droplet temperature (which otherwise would rise to the fuel high end boiling point) and droplet lifetime (by shattering into smaller droplets), thereby reducing the liquid-phase formation of soot by fuel coking. Moreover, any coke particles which should form would tend to be smaller and burn up more easily. Even in the absence of microexplosions, reduced droplet temperature due to the heat-sink effect of the water would reduce coking.

For microexplosions triggered by homogeneous nucleation, Law[27] has predicted theoretically that at 1 atmosphere pressure, only fuels with a high end boiling point above that of hexadecane, 288 degrees Celsius (550 degrees Farenheit) would experience microexplosions. All boiler and turbine liquid fuels (even No. 1 oil) have boiling points above this limit with the

Table 1

Methods Used to Produce Water-in-Fuel Emulsions

(After Dryer et al.[14])

Emulsor Principle of Operation	Trade Name	Patent Holder or Manufacturer
Pressure driven ultrasonic vibrating disc	TOTAL Emulsifier	Compagnie Francaise de Raffinage, Paris
Air/water/fuel co-injection	ELF Multi-fluid Burner	ELF Union, Paris
Piezoelectric driven ultrasonic	Cottell Reactor	E. C. Cottell
Gas/water/fuel venturi	MGD (Microgas Dispersion) Fuel Converter	SMS Associates, State College, Pennsylvania
Pressure drop/shear/ cavitation/jet impact	Gaulin homogenizer	Gaulin Corporation Everett, Massachusetts

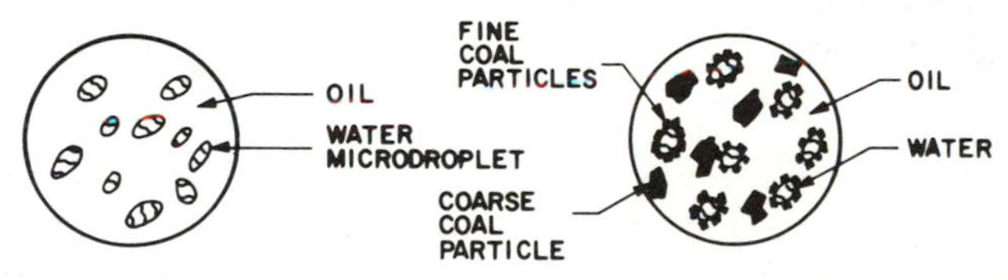

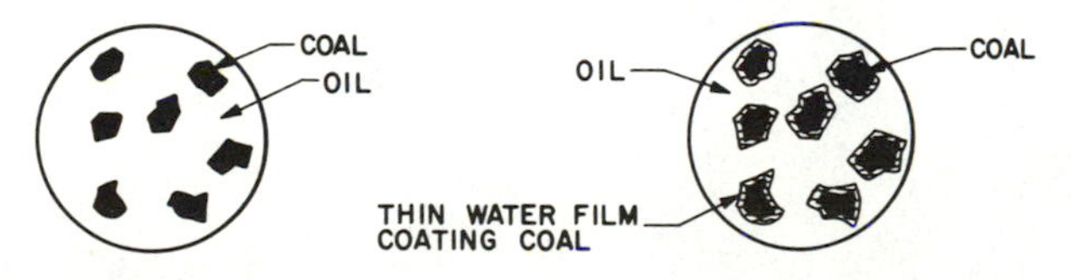

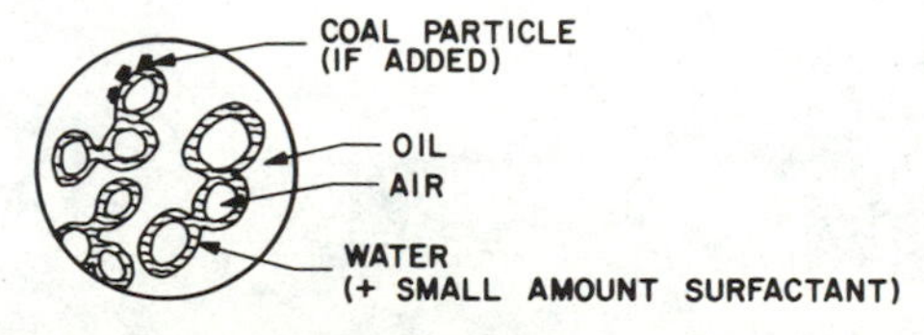

FIGURE 1. TYPES OF "EMULSIONS" DISCUSSED AT SYMPOSIUM ON THE USE OF WATER-IN-FUEL EMULSIONS IN COMBUSTION PROCESSES, U.S. DOT, CAMBRIDGE, MASS. APR. 20-21, 1977.

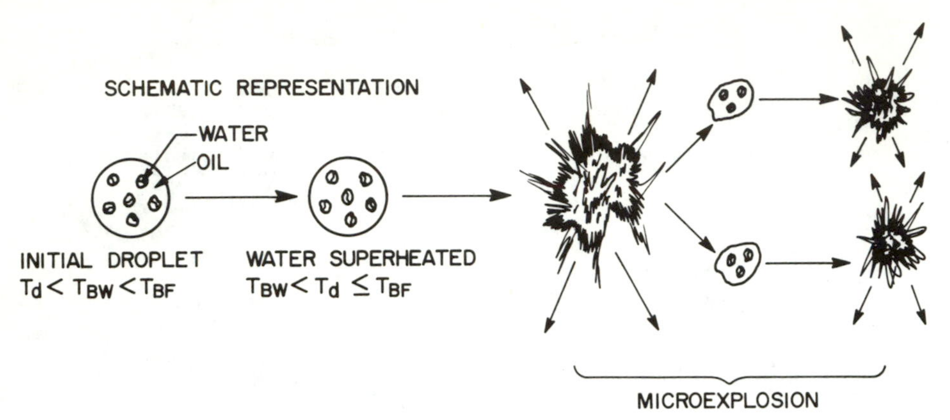

FIGURE 2a. MECHANISM OF EMULSION COMBUSTION ILLUSTRATING "MICROEXPLOSIONS"

FIGURE 2b. PHOTOGRAPHS OF ACTUAL BURNING DROPLETS (REFERENCE 1)

exception of JP-5 turbine fuel which is slightly below it. However, considerations of heterogeneous nucleation suggest conditions which lower the limit[28] to the point where microexplosions would be expected even with JP-5 at 1 atmosphere. Law's[27] analysis predicts that microexplosions occur more easily at higher pressures, conditions which are more typical of gas turbine combustors.

There are a multitude of chemical effects of emulsions on combustion, only some of which depend on the microexplosion phenomenon. Dryer[1] has noted that localized dilution within the fuel-rich region (both by improved fuel-air mixing and by dilution with water vapor) can reduce gas-phase soot formation, and that increased OH radical concentration produced by the reactions

$$H_2O + O \rightarrow 2OH \qquad \text{(Reaction 1)}$$

$$H_2O + H \rightarrow H_2 + OH \qquad \text{(Reaction 2)}$$

can react with soot precursors and further reduce soot formation. This is a more important mechanism than the direct reaction of H_2O with soot, once formed. The effect of added water is much greater than that of water produced from fuel hydrogen, since the latter is not formed until late in the combustion process, while added water is present from the beginning.

Water-in-fuel emulsions, like direct water injection, reduce thermal NO_x by reducing gas temperature. However, emulsions might be expected to be more efficient since all the water is added in the primary combustion zone where temperatures are highest. If the fuel burns in individual droplet flames, the fragmentation into smaller droplets by microexplosions would be expected to reduce total NO_x produced, as small droplets produce less NO_x than large droplets during their burning history[29-31]. However, in dense fuel sprays, individual droplet burning may not occur. Instead, an envelope flame may surround a cloud of droplets,[28] in which case this particular mechanism of NO_x reduction would not occur. However, microexplosions would still improve fuel vaporization and mixing, bringing the flame closer to a premixed rather than a diffusion flame. This reduces peak temperatures (for off-stoichiometric mixtures, as in staged combustion) and thus reduces NO_x production. Further, the use of emulsions would allow a much richer primary mixture in staged combustion without smoking. This would reduce both thermal and fuel NO_x. The removal of oxygen atoms by reaction(1) produces an additional, but smaller reduction in NO_x formation from fuel nitrogen and by the Zeldovitch (thermal) mechanism.

Applications

Table 2 shows some of the types of combustion processes to which emulsions have been applied. The present paper will be limited to boilers and gas turbines. The positive and negative effects of emulsions in these applications will be considered.

ADVANTAGES AND DISADVANTAGES OF EMULSIONS IN BOILERS AND GAS TURBINES

Tables 3 to 6 list the advantages and disadvantages of water-in-oil emulsions for boilers and gas turbines.

Table 2

Combustion Applications of Water-in-Oil Emulsions

Type of Combustor	Investigator (Date)	Benefits Claimed
Boilers	[13]Scherer and Trainie (1972) [14]Turner and Siegmund (1973) [15]Hall (1975) [16]Dooher (1975) [17]Laurendau et al. (1975) [18]Cato & Hall (1975)	Reductions in soot, NO_x, excess air
Gas Turbines	[3]Moses (1976) [19]Winkler (1977)	Smoke and NO_x reduction Allows use of heavier fuel
Diesel Engines	[20]Dryer et al. (1974) [21]Vichnievsky et al. (1972)	Smoke and NO_x reduction
Industrial Furnaces	[22,23]Essenhigh et al. (1976, 1977)	Smoke reduction, Heat flux profile adjustment to match that of natural gas
Residential and Commercial Space Heating	[13]Scherer and Trainie (1972) [24]Hall (1975)	Reduction in soot, excess air, NO_x

Table 3

Advantages of Emulsion Use in Utility Boilers

1. - Smoke and particulate emission reduction
2. - NO_x reduction, particularly in combination with staged firing
3. - Technique easily adoptable to existing boilers (vs. other means of adding H_2O)
4. - Reduction in carbon deposition and corrosion caused by adsorption and adherance to such deposits
5. - Moderate excess air reduction
6. - Reduction in SO_3 formation and cold corrosion (due to lower O_2)
7. - Reduction in need for SO_3 neutralization additives
8. - Increase in combustion efficiency (only for units with poor atomization)
9. - Elimination of need for "combustion catalysts"
10. - Better fuel atomization
    - easier, stable lightoff
    - improved cycling
11. - Improved combustion geometry

Table 4

Disadvantages of Emulsion Use in Utility Boilers

1. - Increased stack loss (not reduced, unless efficiency originally low and excess air > 20%)
2. - Emulsion generation, characterization and stability
3. - Emulsor and auxiliary power requirements
4. - Fuel line and injector scaling and corrosion (unless moderately purified water is used - at increase in cost)
5. - Increased fuel preheat required due to higher emulsion viscosity
6. - Need to regulate steam temperature - superheat and reheat due to change in gas and particulate radiation

Table 5

Advantages of Emulsion Use in Gas Turbines

1. - Reduce smoke and NO_x emissions
2. - Allow elimination of antismoke additives
    - cost saving
    - elimination of metal emissions (Ba, Mn, Cr, etc.)
    - Elimination of possible catalytic effect of metals on $SO_2 \rightarrow SO_3$ reaction
    - Elimination of Ba and Mn turbine deposits. These deposits:
        - reduce efficiency and output
        - require harder firing and higher gas temperature to maintain load, reducing blade life
        - could cause long-term corrosion effects
3. - Allow use of heavier and cheaper fuels, gas $\rightarrow$ kerosene $\rightarrow$ No. 2 $\rightarrow$ residual
4. - Reduce NO_x more efficiently than direct water injection, thus reducing water requirements and stack losses
5. - Lower turbine inlet temperature and possibly extend blade life
6. - Increased power output (if generator is designed for it)

Table 6

Disadvantages of Emulsion Use in Gas Turbines

1. - Efficiency loss (stack loss)
2. - Emulsion generation, characterization and stability
3. - Surfactant cost
4. - Emulsor and auxiliary power requirements
5. - Water purity and cost
6. - Turbine scale and corrosion due to water and surfactant mineral content
7. - Possible stack effects
    - plume buoyancy
    - plume visibility
    - ice fog
    - condensation and corrosion
8. - Degradation in primary atomization due to increased emulsion viscosity
9. - Increase in CO and HC emissions

Regarding utility boilers, it should be noted that with typical combustion efficiencies above 99% and excess air less than 20%, there is little scope for improvements in efficiency by emulsions. On the contrary, a moderate loss in efficiency must result. Use of emulsions in utility boilers must be justified on the basis of reductions in emissions, corrosion and deposits, and the elimination of the need for certain additives.

At a recent symposium[25], Seaworthy Engine Systems reported that water-in-oil emulsions (5-7% H_2O) allowed burning up to 75% residual oil in marine gas turbines (with anti-corrosion additives and frequent blade washing). Klapatch et al[32] reported no visible plume in their engine tests up to 0.85 weight ratio water/fuel but did not specify the ambient air temperature.

Because of the high temperature of gas turbine exhausts (~900 degrees Farenheit) plume visibility and stack corrosion will probably not be problems. The incidence of ice fog would not be appreciably increased but its intensity would, due to the greater quantity of water in the exhaust. EPA[9] has recognized ice fog danger as grounds for exemption to its proposed requirement for wet NO_x control. Atomization and hydrocarbon emission problems due to increased emulsion viscosity can be alleviated by injector redesign or fuel preheat.

EMULSION USE IN BOILERS

Anticipated effects of water/oil emulsions on boiler operating parameters are shown qualitatively in Figures 3 and 4. (In Figure 3, steam temperature rises since stack temperature is held constant. If fuel input is held constant and stack temperature allowed to fall, steam temperature would fall as well). As noted above, in utility boilers with high combustion efficiency and moderate excess air, the reduction in excess air permitted by use of emulsions is not sufficient to balance the increased stack loss. Table 7 shows values given by Bonne[33] for the excess air reduction necessary to maintain constant stack loss as a function of emulsion water content for an initial stack gas temperature of 149 degrees Celsius (300 degrees Farenheit). A 6% water emulsion would require a 6.5 percentage point decrease in excess air for no increase in stack loss. Table 8 shows results of calculations for a 30% water emulsion in a boiler with 19% excess air. A moderate (2-3%) increase in stack loss occurs and it is impossible to decrease excess air enough to balance it. Stack dew point is raised by 11 degrees Farenheit. Depending on the leeway between stack temperature and dew point, it may be necessary to raise stack temperature to avoid condensation and corrosion. It should be noted however, that if use of emulsions allows operation at very low excess air (Figure 5), an actual reduction in low temperature corrosion may occur, due to reduced SO_3 formation, and need for additives used to neutralize SO_3 may be reduced. Smoke reduction by use of emulsions will allow elimination of "combustion catalyst" additives. Efficient introduction of the water in emulsions into the primary combustion zone should reduce NO_x emissions by the mechanisms described earlier. No problems with plume visibility or buoyancy should be encountered, since the quantity of water introduced in a 25% water emulsion is comparable to the quantity of steam introduced in steam-atomizing burner nozzles[34]. Water does increase the viscosity of the emulsion over that of pure fuel at the same temperature, which may necessitate increased preheat and/or pressure, or nozzle modification. Moses[3] notes that a 30% water-in-JP-5 jet fuel emulsion has

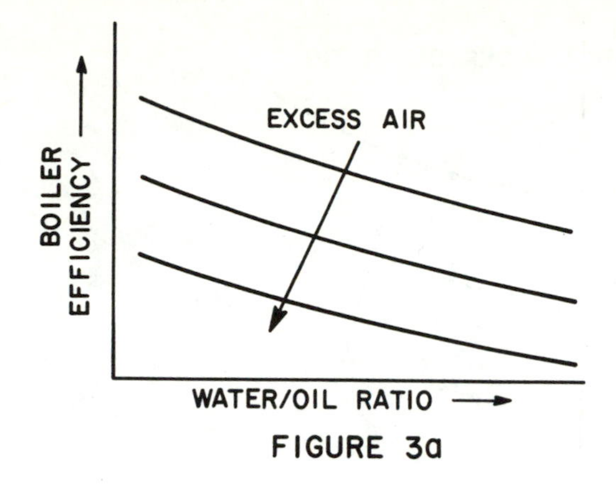

FIGURE 3a

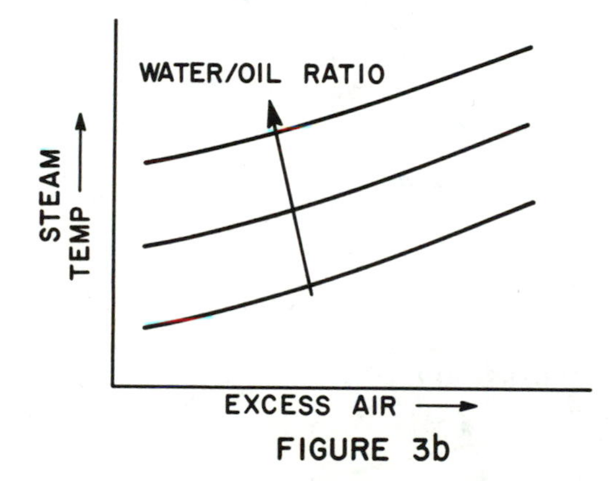

FIGURE 3b

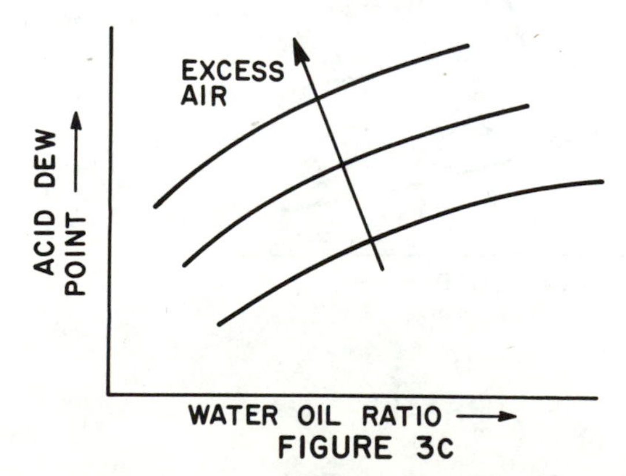

FIGURE 3c

FIGURE 3. EFFECTS OF WATER/OIL EMULSIONS ON FURNACE OPERATING PARAMETERS (CONSTANT STACK TEMPERATURE AND STEAM FLOW)

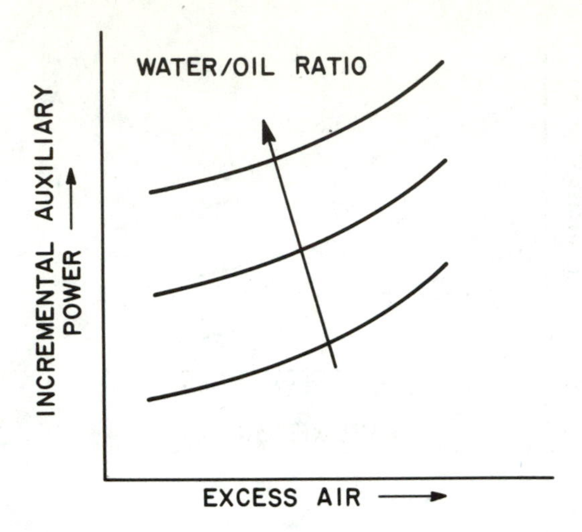

FIGURE 4a

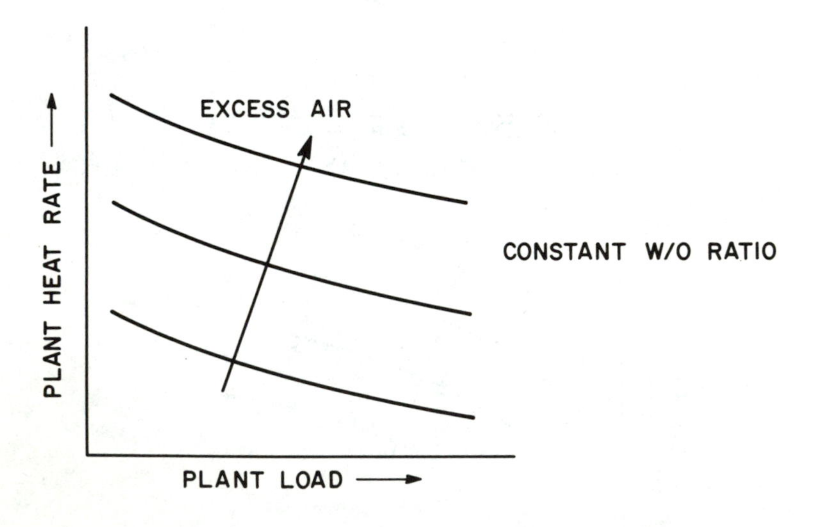

FIGURE 4b

Table 7

Excess Air Reduction Required to Maintain Constant Stack Loss

Combustion of water in No. 6 fuel oil emulsion
C/H = 1, Baseline: Stack O_2 = 3% (dry), T_s = 300°F
(From Reference 33)

Excess air reduction, % of theoretical air	Weight % water in emulsion, $\left(\frac{\text{water}}{\text{water + fuel}}\right)$ X 100%
1	1.1
2	2.1
4	4.2
6	6.5
10	11.
20	24.
40	70.
60	160.

Table 8

Calculated Effects of Water/Oil Emulsion on Boiler Performance

Given:

Fuel: Fuel Oil, Bunker "C"

Excess Air: 19%

Stack Temperature: 303°F

% water in emulsion = 30%

Calculated:		Required Decrease in Excess Air for Constant Stack Loss	Increase in Stack Loss
Stack loss without emulsion[34]	= 10.64%	-	-
Stack loss with emulsion (constant stack temperature)	= 13.30%	66.4%	2.66%
Increase in stack dew point	= 11°F	-	-
Stack loss with 11°F higher stack temperature	= 13.51%	-	2.87%

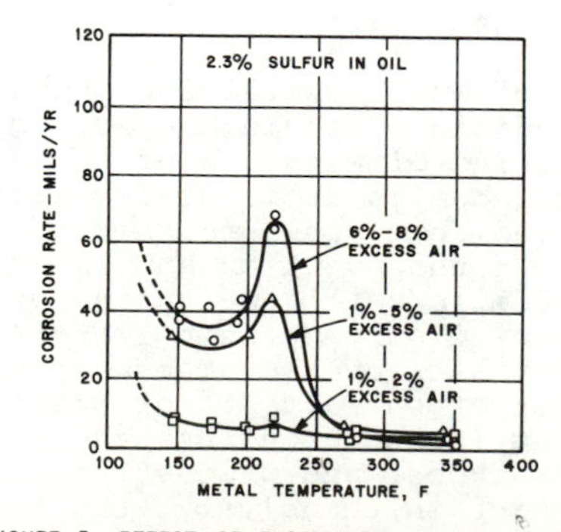

FIGURE 5. EFFECT OF EXCESS AIR ON LOW-TEMPERATURE CORROSION OF CARBON STEEL.

FROM BABCOCK & WILCOX, STEAM/ITS GENERATION & USE, 38TH ED., P.15-34 (REF. 4)

double the viscosity of pure fuel. Scherer and Trainie[13] presented results showing that at 66 degrees Celsius (150 degrees Farenheit) water-in-No. 6 fuel oil emulsions containing 20 and 40 weight percent water had viscosities 1.8 and 5 times, respectively, that of pure No. 6 oil. Correspondingly, increases of 15 degrees Celsius (27 degrees Farenheit) and 35 degrees Celsius (63 degrees Farenheit) respectively, in fuel preheat temperature, would reduce the viscosity of these 20% and 40% water emulsions to that of pure No. 6 oil.

Table 9 shows estimated capital costs and operating expenses for a 100 MWe utility boiler using 10% water-in-oil emulsion. The water cost given is for tap water, but use of semi-purified water or even demineralized water (to avoid possible fuel line and nozzle corrosion and deposits) would raise operating costs by less than 10%. The emulsifier is sized for maximum load.

EMULSION USE IN GAS TURBINES

Large reductions in smoke and NO_x have been demonstrated by Moses[3] in gas turbine combustor rig tests (Figures 6 and 7). At 30% water in the emulsion, about 70% reductions in both particulates and NO_x are seen. The latter is considerably higher than obtained with direct water injection. A study recently completed at PSE&G's Kearny Generating Station has shown that Mn and Mn/Ba turbine fuel additives produce a real decrease in exhaust particulate loading of up to 50%[35,37]. Dryer[2] has noted that the chemical effects of such additives and added water may have a common cause, an increase in OH concentration with subsequent reaction with soot precursors.

Tables 10A and 10B give capital and operating cost estimates for use of a 30% water in oil emulsion in a 25 MWe gas turbine unit. The addition to the cost of power by use of emulsions is 1.6 mils/kWh. Of this, 0.1 mils/kWh is due to the cost of demineralized water (to avoid turbine hot corrosion) and 0.4 mils/kWh is due to an estimated 1.5% increase in heat rate caused by water addition[9]. It should be noted that if NO_x control of gas turbines is required in the future, these two items will no longer count against emulsions, as water injection will be required anyway. However, since water in emulsions is a more efficient suppressant of NO_x than that injected separately, less water would be required, and these two "losses" would turn into credits for emulsions, depending on the reduction in water requirement realized.

The elimination of Mn/Ba containing antismoke additives would represent an additional savings of 0.2 mils/kWh credited to use of emulsions. The elimination of heat rate degradation by fouling caused by deposits from metal-containing additives has not been considered, but could reach 2-3% in heat rate (0.5-0.8 mils/kWh). In the event Mn and Ba emissions are banned, the 1¢/gal. (0.8 mils/kWh) lower price of No. 2 oil vs. kerosene could be enjoyed only be using water/oil emulsions to reduce smoke to acceptable levels. These results are summarized in Table 11. The cost of surfactant for stabilizing water/oil emulsions has not been included in this table since it is very sensitive to surfactant cost and prices range widely. For example, for 1% surfactant in the slurry, the addition to the cost of power is 3.1 mils/kWh for $1/kg surfactant and 0.06 mils/kWh for 2¢/kg surfactant. It should be noted that residual oil requires no added surfactant, and has been success-

Table 9A

Cost of Emulsion Combustion System - Boiler

The following capital and operating costs for an emulsion combustion system for a typical 100 MW generating station are based on the assumptions listed below:

1. Use of three emulsifiers, each with a capacity to supply 50% of the emulsion requirements for the steam generator.
2. The emulsifiers feed the burners directly.
3. Water flow rate is 10% (by weight) of the oil flow rate.
4. City water can be used for preparing the emulsion.
5. Any reduction in boiler efficiency due to the addition of water is offset by an increase in efficiency due to reduction in excess air requirements.
6. Neither the steam generator output nor the steam temperature changes significantly from normal design values.
7. The plant is assumed to run, on an equivalent basis, 50% of the time at full 100 MW load.
8. Costs are based on 1975 prices.

Table 9B

Cost of Emulsion Combustion System - Boiler

Capital Expenses	
Three (3) Emulsifiers @ $36,000/Unit	$108,000
One (1) Ratio Proportioner	5,000
Two (2) Recorder Controllers	6,000
Total Cost of Major Equipment	$119,000
Field Installation with Associated Piping, Instrumentation, etc., 40% of the above	48,000
Total Installed Cost	$167,000
Contingency, 10% of Total Installed Cost	17,000
Overhead, 40% of Total Installed Cost	67,000
Total Capital Cost	$251,000
Capital Cost in Dollars/kW = $2.51	
Operating Expenses	
For a Period of One Year:	
Cost of Water (2.5 million gallons/yr) @ 50¢/1000 gallons	$ 1,250
Auxiliary Power for 200 kW Emulsifier Motors @ 2.5¢/kWhr	21,840
Maintenance	5,000
Capital Amortization (20% of Capital Investment)	50,000
Total Operating Costs for 1 Year of Operation	$ 78,090
Power Generated in 1 Year	437,000 MWhr
Addition to Cost of Power	.02¢/kWhr (0.2 mils/kWhr)

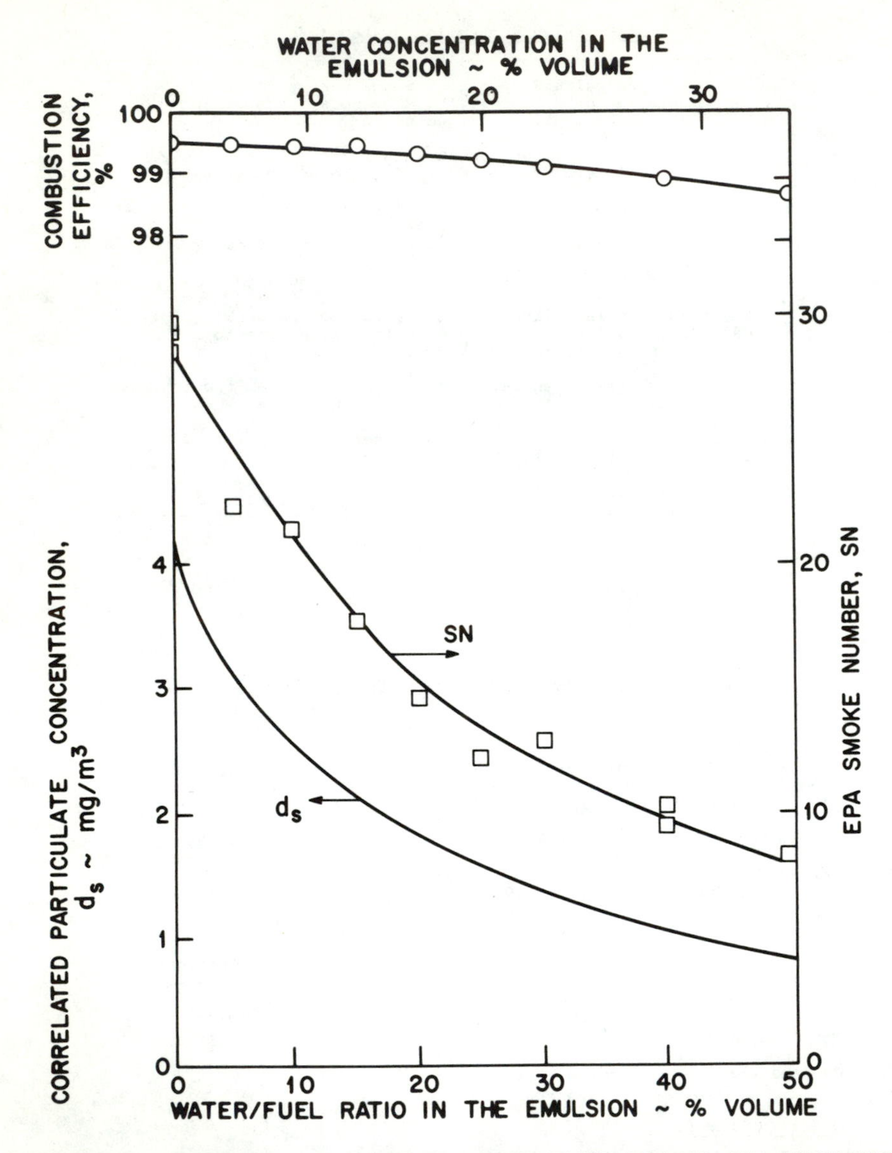

FIGURE 6. EFFECT OF WATER CONCENTRATION ON EXHAUST SMOKE AND COMBUSTION EFFICIENCY AT 100% ENGINE POWER (REF. 3)

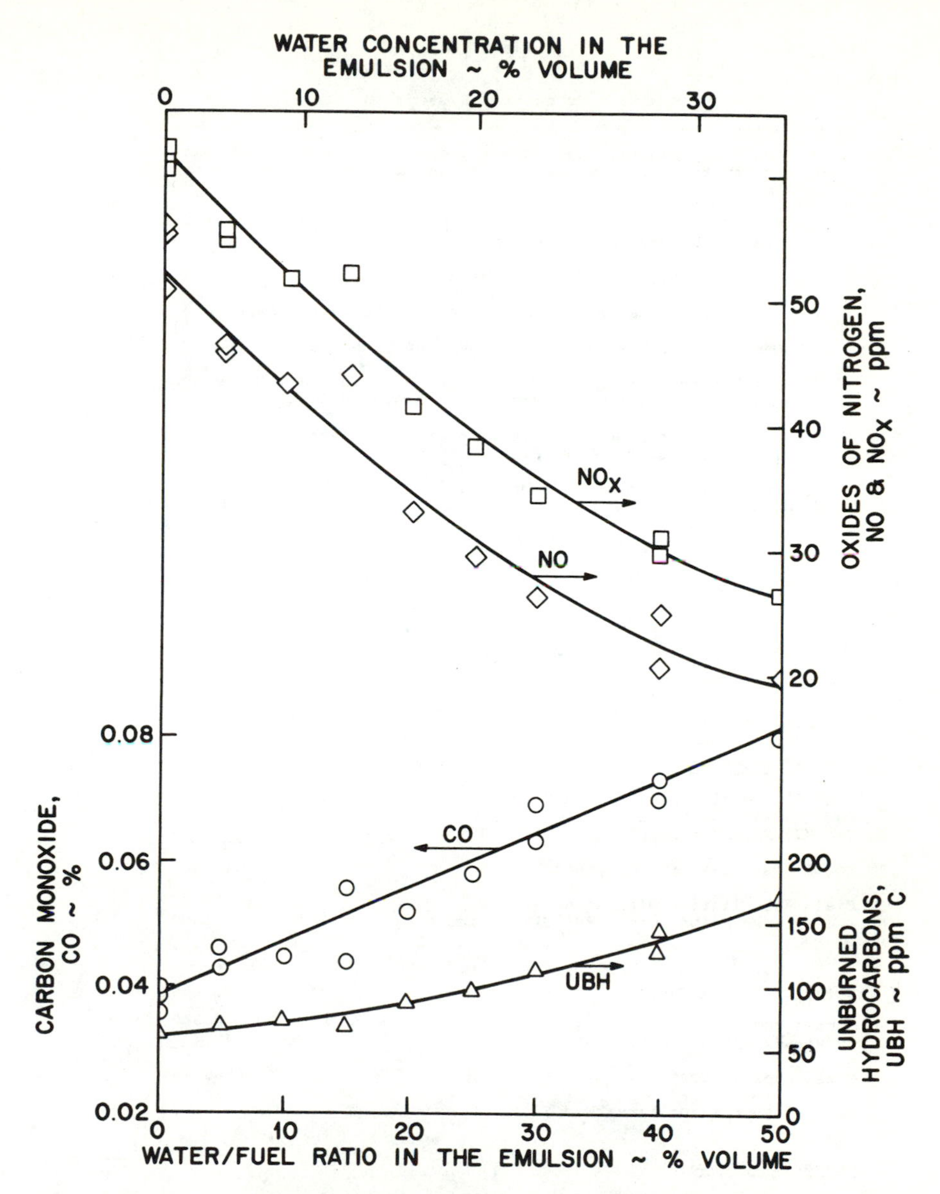

FIGURE. 7. EFFECT OF WATER CONCENTRATION ON THE GASEOUS EXHAUST AT 100% ENGINE POWER (REF. 3)

Table 10A

Cost of Emulsion Combustion System - Gas Turbine

The following capital and operating costs for an emulsion combustion system for a typical 25 MW gas turbine unit are based on the assumptions listed below:

1. Use of one emulsifier with a capacity to supply the emulsion requirements for the unit.
2. The emulsifier feeds the combustor directly.
3. Water flow rate is 30% (by weight) of the oil flow rate.
4. Demineralized water is used for preparing the emulsion.
5. Neither the generator output nor the stack temperature changes significantly from normal design values.
6. The plant is assumed to run, on an equivalent basis, 1000 hr/yr at full 25 MW load.
7. Costs are based on 1975 prices.

Table 10B

Cost of Emulsion Combustion System - Gas Turbine

Capital Expenses	
One (1) Emulsifier @ $36,000/Unit	$36,000
One (1) Ratio Proportioner	5,000
Two (2) Recorder Controllers	6,000
Total Cost of Major Equipment	$47,000
Field Installation with Associated Piping, Instrumentation, etc., 40% of the above	18,800
Total Installed Cost	$65,800
Contingency, 10% of Total Installed Cost	6,580
Overhead, 40% of Total Installed Cost	26,320
Total Capital Cost	$98,700
Capital Cost in Dollars/kW = $3.95	
Operating Expenses	
For a Period of One Year (1000 hr. operating time):	
Cost of Water (802,600 gallons/yr) @ $2.90/1000 gallons (Ref. 9, p. 7-20)	$ 2,330
Auxiliary Power for 100 kW Emulsifier Motors @ 2.5¢/kWhr	2,500
1.5% Heat Rate Penalty (Ref. 9)	9,300
Maintenance	5,000
Capital Amortization (20% of Capital Investment)	19,740
Total Operating Costs for 1 year of Operation	$38,870
Power Generated in 1 Year (1000 hr.)	25,000 MWhr
Addition to Cost of Power	0.16¢/kWhr (1.6 mils/kWhr)

Table 11

Additions to Cost of Power, Mils/kWh

	Antismoke Additive**	Water/Oil Emulsion
Present conditions, 1000 hr/yr	0.2	1.6
(A) Wet NO_x control required	0.7*	1.6
(B) Antismoke additive banned, requiring switch from No. 2 oil to kerosene without emulsion	0.8	1.6
Both (A) and (B)	1.3*	1.6

*The potential lower water consumption of emulsion vs. direct water injection for NO_x control has not been considered. Thirty percent water was considered for both.

**Does not include effect of heat rate degradation due to turbine fouling by metal deposits from antismoke additive. This could add 0.5 - 0.8 mils/kWh to figures in this column.

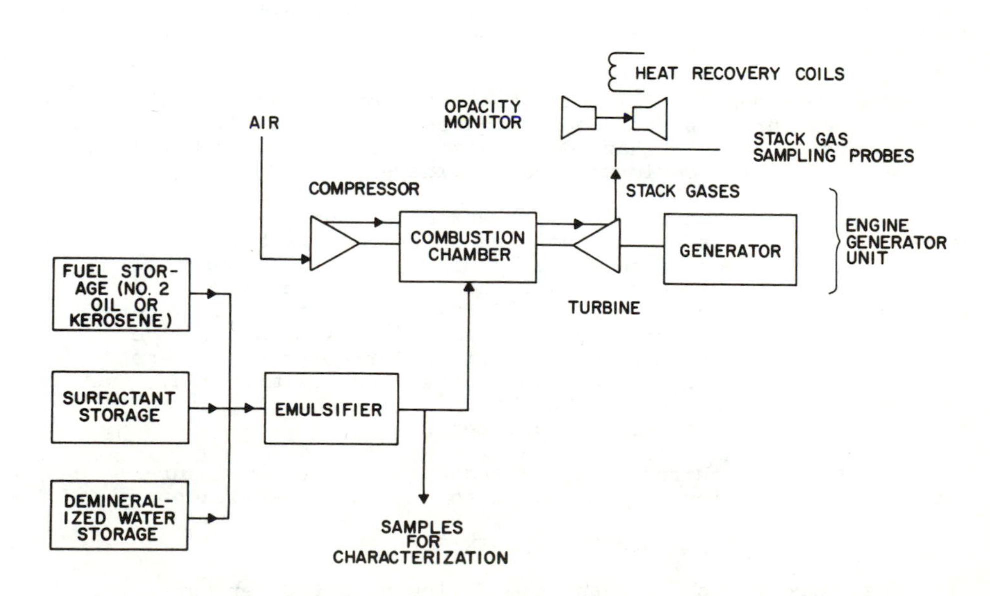

FIGURE 8. SCHEMATIC-TEST FACILITY FOR GAS TURBINE EMULSIFIED FUEL TEST PROGRAM

fully burned in water/oil emulsions in marine gas turbines with proper precautions [19]. It should also be noted that if surfactants are used, low Na and S content are essential. Luckily, there are many available surfactants which satisfy this requirement[36].

PSE&G RESEARCH PLANS

Figure 8 shows the outline diagram of a proposed test of the effects of water/oil emulsions on emissions and operation of a 25 MWe PSE&G gas turbine generator unit. Preliminary tests will be performed to optimize surfactant choice and concentration, water microdroplet size and emulsion water content relative to stack emissions and engine performance. The engine will then be operated on a regular schedule for two 500 hour tests, one under summer and one under winter conditions. Each test period will be followed by turbine hot end inspection for corrosion and deposits. Stack emissions and engine operating parameters will be monitored throughout the tests.

CONCLUSIONS

The major conclusions are listed below:

1. Emulsions have the potential to sharply reduce smoke and NO_x emissions from boilers and gas turbines.

2. Emulsions promise to reduce the amount of added water needed for wet NO_x control.

3. For units with good initial combustion efficiency, for gas turbines and for more than 10% water emulsion in boilers, increase in stack loss is unavoidable. However, use of emulsions may reduce this loss, when water must be added anyway for NO_x control.

4. Emulsions may eliminate the need for antismoke additives in gas turbines and "combustion catalysts" in boilers. Emission of potentially hazardous materials and turbine blade fouling by deposits could thus be eliminated.

5. The increase in cost of power due to use of emulsions ranges from 0.2 mils/kWh for boilers to 1.6 mils/kWh for gas turbines in those cases considered. These costs can both be significantly reduced when wet NO_x controls are required. (The cost of water purification and the heat loss due to water addition are then no longer chargeable to emulsion use). The costs of boiler combustion modifications for NO_x control were not analyzed here, but are probably more than 0.2 mils/kWh. For gas turbines, emulsions can show a slight economic advantage for one set of imposed requirements.

6. Low cost surfactants, or reduction of required surfactant concentration are essential to the use of emulsions.

ACKNOWLEDGEMENTS

The authors wish to thank the following PSE&G personnel for their assistance and advice on the application of emulsions to practical combustion systems: W. E. Somers (retired), V. S. Renton (retired), A. Prestifilippo, T. F. Gleichmann, Jr., H. B. Baranek, R. R. Mroczek.

REFERENCES

1. Dryer, F. L., Rambach, G. D. and I. Glassman. "Some Preliminary Observations on the Combustion of Heavy Fuels and Water-in-Fuel Emulsions". Aerospace and Mechanical Sciences Report No. 1271. Guggenheim Laboratories, Princeton University, April 1976.

2. Dryer, F. L. "Water Addition to Practical Combustion Systems - Concepts and Applications". Paper No. 22, 16th Symposium (International) on Combustion. The Combustion Institute. M.I.T. April 1976.

3. Moses, C. A. "Reduction of Exhaust Smoke from Gas Turbine Engines by Using Fuel Emulsions". Paper No. 76-34. Fall 1976 Meeting. Western States Section/The Combustion Institute. University of San Diego/La Jolla, California. October 1976.

4. Babcock and Wilcox Company, Steam/Its Generation and Use, 38th ed., New York (1972).

5. Finfer, E. Z. "Fuel Oil Additives for Controlling Air Contaminant Emissions". Jl.APCA, 17, 43-5 (1967).

6. Salooja, K. C. "Burner Fuel Additives". Combustion, 44, 21-7 (1973).

7. Lay, K. W. "Ash in Gas Turbines Burning Magnesium - Treated Residual Fuel". ASME Paper No. 73-WA/CD-3. Presented at Winter Annual Meeting, ASME, Detroit, Michigan. November 11-15, 1973.

8. Smock, R. W. "Heavy Fuels for Gas Turbines: Deliverance from the Oil Crisis". Electric Light and Power, E/G Edition. p. 10 (June 1974).

9. Durkee, K. R. and R. Jenkins. "Standards Support and Environmental Impact Statement: An Investigation of the Best Systems of Emission Reduction for Stationary Gas Turbines". (Preliminary Draft) U.S. Environmental Protection Agency, Research Triangle Park, North Carolina. July 1976.

10. Ivanov, V. M. et al. "Fuel Emulsions for Combustion and Gasification". J. Acad. Sci. USSR, May 1957, pp. 56-59.

11. Ivanov, V. M. and P. I. Nefedov. "Experimental Investigation of the Combustion Process of Natural and Emulsified Liquid Fuels". Trudy Instituta Goryachikh Iskopayemykh, 19, 35-45 (1962). NASA Technical Translation TT F-258 (January 1965).

12. Dryer, F. L. "Fundamental Concepts on the Use of Emulsions as Fuels". AMS Report No. 1224, Dept. of Aerospace and Mechanical Sciences, Princeton University. (1975). Presented at Joint Western/Central States Sections, Combustion Institute Meeting, San Antonio, Texas, April 21-22, 1975.

13. Scherer, G. and Trainie, L. A. "Pollution Reduction by Combustion of Fuel-Oil Water Emulsions. Pollutant Formation and Destruction in Flames and in Combustion Systems". Paper No. 83, Fourteenth International Symposium on Combustion. Penn State University. August 20-25, 1972. Not

published in the Proceedings).

14. Turner, D. W. and Siegmund, C. W. "Control of NO_x from Fuel Oil Combustion: Water in Oil Emulsions". Winter Symposium of the IEC Division of the American Chemical Society, Chicago, Illinois. January 1973.

15. Hall, R. E. "The Effect of Water/Residual Oil Emulsions on Air Pollutant Emissions and Efficiency of Commercial Boilers". Paper 75-WA/APC-1, ASME Winter Annual Meeting, Houston, Texas, December 1-5, 1976.

16. Dooher, J. et al. "Emulsions as Fuels". Paper 75-WA/FU-3. Fuels Division, ASME Winter Annual Meeting, Houston, Texas, December 1-5, 1976.

17. Laurendeau, R. et al. "The Reduction of Particulate Emissions from Industrial Boilers by Combustion Optimization". Paper 75-WA/APC-3, ASME Winter Annual Meeting, Houston, Texas, December 1-5, 1976.

18. Cato, G. A. and Hall, R. E. "Field Measurements of Pollutant Emissions from Industrial Boilers". Paper 75-WA/APC-7, ASME Winter Annual Meeting, Houston, Texas, December 1-5, 1975.

19. Winkler, M. F., Seaworthy Engine Systems. Discussion at Symposium on the Use of Water-in-Fuel Emulsions in Combustion Processes, U.S. DOT, Cambridge, Massachusetts, April 20, 1977.

20. Dryer, F. L. et al. "Fundamental Combustion Studies of Emulsified Fuels for Diesel Applications". Princeton University, Princeton, New Jersey. April 19, 1974. Initiating Proposal to Division of Applied Energy Research and Technology, NSF, Grant No. GI44215.

21. Vichnievsky, R. et al. "Employment of Fuel-Water Emulsions in Compression - Ignition Engines". Paper presented at CIMAC Conference, Barcelona, Spain, April 28-May 3, 1975.

22. Essenhigh, R. H. et al. "Smoke Point and Heat Transfer Characteristics of Oil/Water/Air Emulsions Without and With Coal Addition in a Hot-Wall Furnace". Paper presented at the Central States Section/Combustion Institute Meeting, Columbus, Ohio, April 1976.

23. Essenhigh, R. H. and A. Kokkinos. "Influence of Particulates on Thermal Efficiency in Emulsion Firing". Paper presented at Symposium on the Use of Water-in-Fuel Emulsions in Combustion Processes. U.S. DOT, Cambridge Massachusetts, April 20, 1977.

24. Hall, R. E. "The Effect of Water/Distillate Oil Emulsions on Pollutants and Efficiency of Residential and Commercial Heating Systems". APCA Paper No. 75-09.4. APCA Meeting, Boston, Massachusetts, June 16, 1975.

25. Symposium on the Use of Water-in-Fuel Emulsions in Combustion Processes. U.S. DOT, Cambridge, Massachusetts, April 20-21, 1977. Abstracts of papers and transcript of discussions to be available through NTIS.

26. Droplet photographs, Guggenheim Laboratories, Princeton University, reproduced in Catalog FE301.70, Gaulin Corporation (1976).

27. Law, C. K. "An Analysis for the Combustion of Water-in-Fuel Emulsion Droplets". Presented at Symposium on the Use of Water-in-Fuel Emulsions in Combustion Processes. U.S. DOT, Cambridge, Massachusetts, April 20-21, 1977. Combustion Science and Technology (To appear).

28. F. L. Dryer. Private communication. April 20, 1977.

29. Kollrack, R. and L. D. Aceto. "Nitric Oxide Formation in Gas Turbine Combustors". AIAA J., 11, 664-9 (1973).

30. Kesten, A. S. "Analysis of NO Formation in Single Droplet Combustion". Combustion Science and Technology, 6, 115-123 (1972).

31. Roberts, R., Aceto, L. D., Kollrack, R., Teixeira, D. P. and Bonnell, J. M. "Analytical Model for Nitric Oxide Formation in a Gas Turbine Combustor". AIAA J., 10, 820-6 (1972).

32. Klapath, R. D. and T. R. Koblish. "Nitrogen Oxide Control with Water Injection in Gas Turbines", ASME Paper 71-WA/GT-9. ASME Meeting, Washington, D.C., November 28 - December 2, 1971.

33. Bonne, U. "External Combustion of Fuel Oil-Water Emulsions. Technical Paper No. 970S. ASHRAE Session on Efficient Use of Fuels, II. 1976 Semi-Annual Meeting, Dallas, Texas, February 1-5, 1976. Quoted in Dryer et al.

34. W. Somers. Personal communication to C. Sengupta. July 8, 1975.

35. Hersh, S., KVB Inc. Presentation at PSE&G, Newark, New Jersey, December 8, 1976.

36. "Formulating with Hyonic[R] Surfactants", Bullentin PI-12. Diamond Shamrock Corporation, Morristown, New Jersey (1976).

37. Hurley, J. F. and Hersh, S. "Effect of Smoke and Corrosion Suppressant Additives on Particulate and Gaseous Emissions from a Utility Gas Turbine". EPRI FP-398, Final Report, KVB Inc., Scarsdale, New York, March 1977.

REMEDIAL EFFORTS TO COMBAT FIRESIDE PROBLEMS

Session Chairman: **J. E. Radway**
Basic Chemicals
Co-Chairman: **F. J. Moore**
Babcock & Wilcox Company

TECHNICAL EVALUATION OF MAGNESIA'S INFLUENCE ON FIRESIDE DEPOSITS, COMBUSTION, AND EMISSIONS ON OIL-FIRED BOILERS

ERIC BLAUENSTEIN

Pentol GmbH
D-7858 Weil a/Rhein, West Germany

ABSTRACT

In Europe the use of high sulfur, low quality fuels causes major combustion and environmental problems.

A controlled application of suspended magnesia in a Belgian 300 MW boiler clearly demonstrates that overall boiler operation is significantly improved.

New methods of liquid suspension injection enable short- and long-term improvements in combustion and thermal balance.

Environmental highlights of the test: unburned carbons reduced 80%, acid smut fallout eliminated, and particulate emissions greatly reduced.

INTRODUCTION

In Europe, the use of high sulfur, low quality fuels causes combustion and environmental problems. Fuels with 3% to 4% sulfur and high vanadium/sodium contents are commonly fired in most countries. In an effort to correct serious "acid smut" fallout, corrosion and air heater fouling problems at the Ruien Station of the Intercom utility system in Belgium, a long-term trial was initiated with a dispersed magnesia additive.

Three organizations cooperated to organize and run the comprehensive test program:

INTERCOM	Operators of the power station Centrale van Ruien, the largest power station in Belgium.
LABORELEC	The research and testing center of the Belgian electric utility industry (similar to EPRI in the United States)
PENTOL GMBH	Manufacturers of chemical additives.

Intercom had experienced at Centrale van Ruien, the usual problems of firing residual oil. To keep boilers 5 and 6 on line and maintain load, the DEKA air heaters had to be washed every three weeks. Boiler operability was gained at the cost of:

(a) extensive corrosion of the air heater, hoppers and ducts;
(b) pluggage and corrosion of the washing equipment; and
(c) problems with handling and treatment of acid wash liquors.

It became evident to Intercom management that the number of washings must be minimized and that evaluation of an additive which would reduce or eliminate these washings could be beneficial. A visit to the EDF power station in Loire-sur-Rhone (France) where a dispersed magnesia additive has been successfully used for two years to solve identical air heater problems, led to the trial on Boiler No. 6.

PROCEDURE

The trial was conducted on a 300 MW tangentially-fired Sulzer boiler burning a residual oil analyzing 3.4 to 3.5% sulfur, 80 to 100 ppm vanadium, 25 to 40 ppm sodium and 3.4% asphalt. Despite low excess air operation, typically 0.55 to 0.7% O_2, the operator had been forced to wash the air heater every three weeks for several years because the pressure drop would increase from the 75 to 100 mm range to 165 or more.

Instead of the traditional procedure of metering the dispersed magnesia into the fuel, a new technique was employed. The suspension was separately injected into each corner of the upper firebox using four air atomizers of a proprietary design. This avoided any complication of steam temperature control

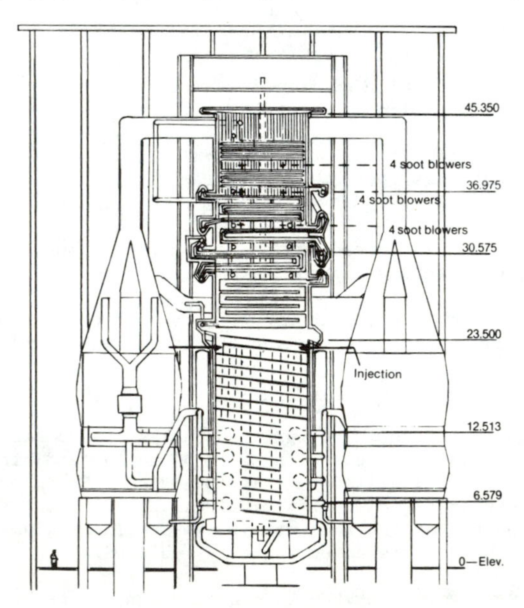

CENTRALE DE RUIEN Fig. 1
Trenchen's
COCKERILL-SULZER Steam generator
Steam Capacity: 860t/h
Pressure: 189 kg cm²
Superheat: 542 °C
Reheat: 542 °C

which might result from a reflective coating of magnesia on the waterwall tubes. It is, however, important to note that even when injecting the additive into the upper part of the furnace, some MgO is returned to the lower furnace due to recirculation of flue gas withdrawn ahead of the air heater. Figure 1 shows the boiler design, the soot blower locations, and the additive injection points.

Injection ahead of the superheater also insured realization of magnesia's inhibitory effect on the catalytic formation of SO_3. This benefit would be lost if the chemical treatment was applied on the cold end.

Ash samples collected from the air heater on a daily basis were checked for pH and analyzed chemically. Measurements were made of emissions and pressure drop across the boiler and air heater. The DEKA air heater (Figures 2 and 3) differs in design from the tubular air heaters commonly used in the United States and in the materials of construction (Pyrex and metal).

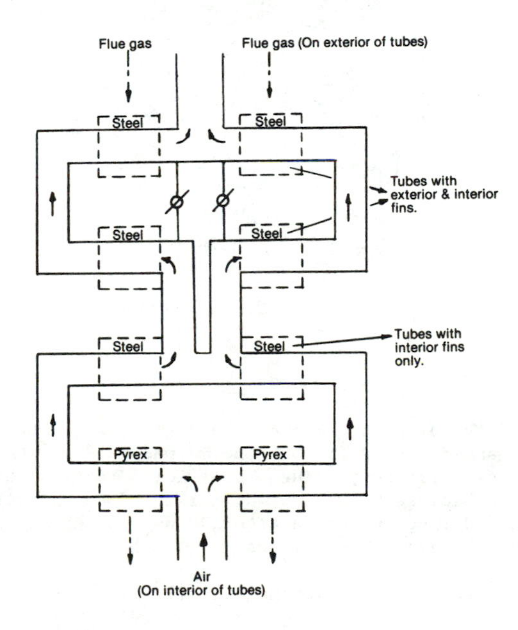

Fig. 2 **CENTRALE DE RUIEN**
DEKA Air heater
Two elements per boiler
Schematic showing upper side of one element

During an outage (after three months of treatment) deposit samples collected from various points within the boiler were analyzed chemically and fusion points determined.

An evaluation of the effect of the additive on overall fuel efficiency was also made.

The trial started on November 27, 1976, on a dirty boiler. Initially, the LiquiMag® feed was proportional to boiler load, but after January 17, 1977, the rate was held constant at 24 l/hr (about 24 kg dispersed MgO/hr). The boiler was cleaned during the February 28, 1977 outage.

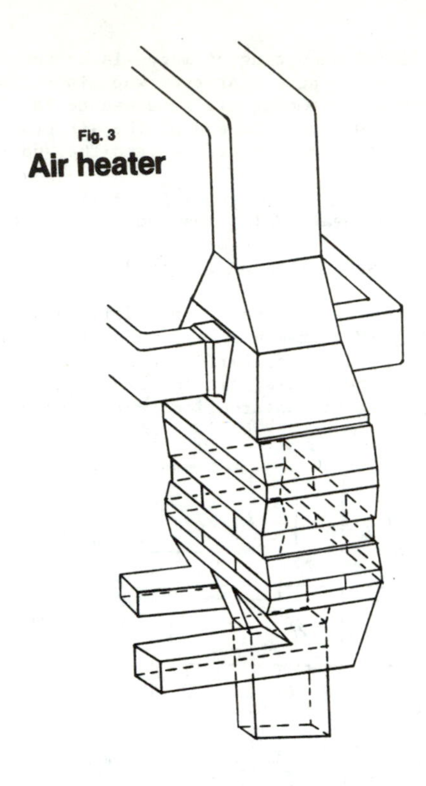
Fig. 3
Air heater

RESULTS

Injection of the dispersed magnesia additive, LiquiMag, favorably affected corrosion and deposits in both the furnace and air heater; stack emissions; and boiler efficiency. These beneficial results, however, were not realized immediately because the trial was initiated on a dirty unit. The boiler was inspected during an outage after 90 days of chemical treatment. The observations can be summarized as follows:

Spiral Evaporator	No fouling at all.
Vertical Evaporator	Small quantity of non-adherent deposits.
Superheater 1 and 2	Significant deposits, more adherent on S-1 than S-2. Less friable in the center of the boiler than at the edges.
Reheater 1 and 2	Deposits on elements non-adherent, friable, easily removed by blowing.
Economizer	Perfect condition (soot blowing had been maintained all the time).
DEKA Air Heater and Ducts	Clean and dry; the three weeks between washes without additive had been increased to ten.

The expected phenomenon of a white coating in the water walls was not observed.

High Temperature Deposits

The boiler inspection during the outage of February 28, 1977, generally showed a clean high temperature section where regular soot blowing had been maintained. However, Superheater 1 and to some extent superheater 2 had significant deposits. These deposits led to an increased delta P and need for constant attemperation. Once the unit was cleaned during the outage, the additive treatment prevented new superheater deposits. The delta P is stabilized and only minimum attemperation is now required.

Analyses of the many deposit samples taken from the high temperature section during the outage of February 25, 1977, showed the following compounds:

MgO	Magnesium Oxide
$MgSO_4$	Magnesium sulfate
$VOSO_4$	Vanadium oxide sulfate
$Na_2OV_2O_45V_2O_5$	Sodium vanadyl-vanadate. Fuel ash.
$Mg_3V_2O_8$	Magnesium vanadate
$FeSO_4$	Iron sulfate: indicating high temperature corrosion

The composition and fusion characteristics of samples taken from the center of the boiler differed from those obtained near the periphery. Figure 4 is a plot showing the variation in analytical results within the boiler. Low melting point-low MgO ash was confined to an inner core area which diminished in diameter with increasing elevation above the additive injectors. The interior of the core was primarily sodium vanadyl vanadate, $Na_2OV_2O_45V_2O_5$. These samples were dark and had a relatively low melting point.

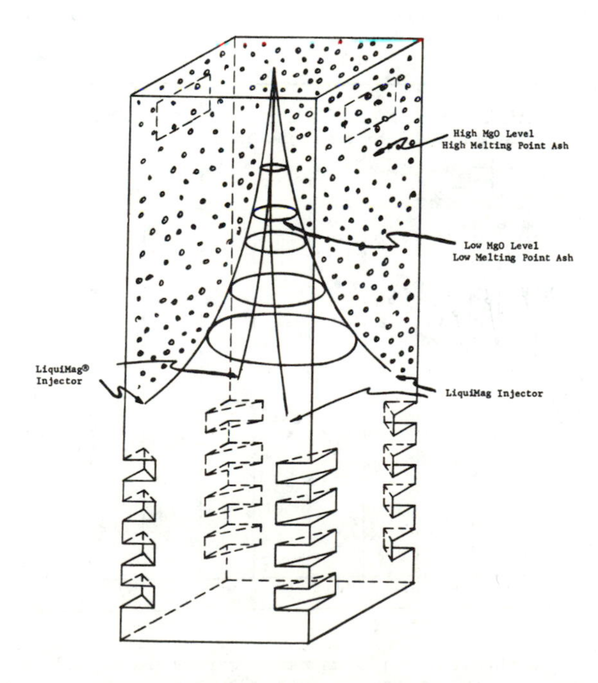

Fig. 4 BOILER CUT SHOWING ADDITIVE DISTRIBUTION

The exterior of the core was predominately high melting MgO and $MgSO_4$. The deposits were light colored, friable and easily removed.

It was obvious from these analyses that the additive increased the melting point and friability of the deposits while reducing their corrosiveness. However, distribution within the boiler was far from optimum. When treatment was subsequently initiated on the sister boiler, Unit 5, the injection point was changed to burner level in order to obtain better chemical distribution.

Air Heater Deposits

Since the trial was started on a dirty boiler, it took six weeks to both raise and stabilize the ash pH. It was necessary to wash the air heater twice during this period, but no additional washing was required during the boiler's usual outage two weeks later.

The initial pH readings were influenced by the spalling of old deposits. Soon after magnesia treatment started, periods of high ash content and low pH were observed when the combustion air preheater was in use (i.e. at low loads). This condition stabilized after two months of treatment but periodic operation of the air preheaters became standard operating practice to gain the cleaning action provided by higher inlet air temperature.

Inspection during the outage of February 25, 1977, showed that the air heaters and ducts were clean and dry after ten weeks without washing, i.e. a dramatic improvement over the usual three weeks between washings when operating without an additive treatment. The air heaters were mechanically cleaned during the outage and have not been washed since operation and additive feed were resumed 15 weeks ago.

Pressure drops across the air heaters were lower with dispersed magnesia injection even during the initial six-week boiler clean up period (Figure 5).

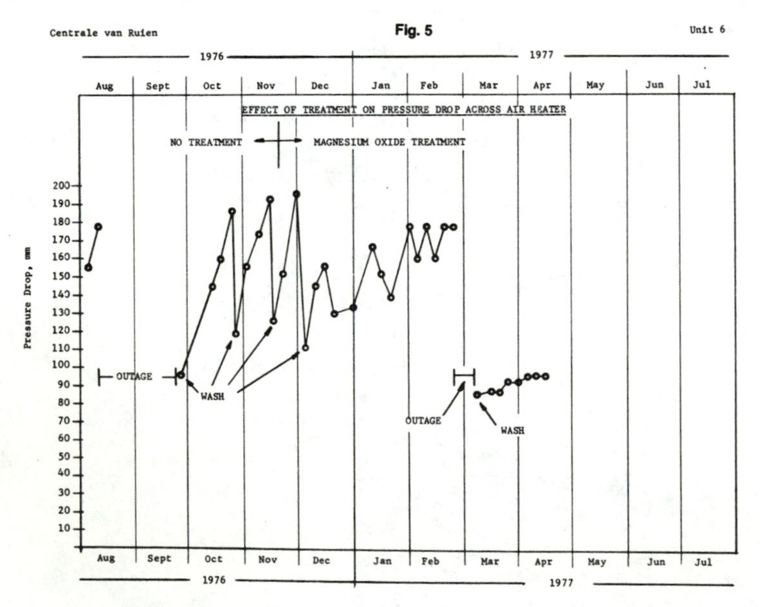

Maximum delta P's were in the 160 to 180 mm range versus nearly 200 without additive. Once the unit was thoroughly cleaned during the outage the additive has prevented deposits and held pressure drops in the 85 to 95 mm range for a 15-week period without washing the air heater.

It was possible to maintain the ash pH between 4 and 5 feeding additive at the rate of 24 l/hr. Upon reducing this to 18 l/hr, there was an immediate drop in pH. It was further noticed that on weekends when the boiler load was light, the pH generally dropped. Apparently the lower gas temperatures at low load increased H_2SO_4 condensation. Thus, when feeding additive in direct relation to boiler load, there was not sufficient MgO being added to react with the greater quantity of acid condensed.

SO_3 emissions with additive injection were reduced from a 30 to 40 ppm range down to 20 to 25 ppm. The apparent reduction of only one-third may be a result of the dirty boiler which numerous authors have shown to increase conversion of SO_2 to SO_3.

Much of the reaction between SO_3 and magnesia apparently occurred in the Pyrex section of the DEKA air heater. Table I below shows selected analytical data for ash entering and leaving that section. The pH of the sample entering, 9.7, was very nearly that of pure MgO, 10.45, and both the SO_3 and soluble magnesia content were low. The exit sample was quite different. The pH dropped to 4.65, the quantity of sulfate increased from 7.6 to 33.3%, and the soluble magnesium went from 1.8 to 8.24%. All of these changes indicate condensation of acid vapor and reaction with the magnesia.

TABLE I

SELECTED ANALYTICAL DATA - AIR HEATER INLET AND OUTLET ASH SAMPLES

	pH	Mg (insoluble)	Mg (soluble)	SO_3
Entering Pyrex	9.7	10.58%	1.80%	7.6%
Leaving Pyrex	4.65	4.14%	8.24%	33.3%

Emissions

Emissions were greatly influenced by the use of chemical treatment. Particulates which were three to four grams per kilo of oil prior to using the additive dropped more than 60% to 1.3 grams per kilo afterwards.

Analyses of ash samples taken on treated and untreated boilers firing the same oil and operating under the same conditions showed a reduction in combustibles of more than 80% (Table II below).

TABLE II

EFFECT OF LIQUIMAG TREATMENT ON TOTAL PARTICULATE AND COMBUSTIBLE EMISSIONS

	Total Particulate gms/kg oil	% Combustible	% Inert	gms Combustible kg oil	gms inert kg oil
Untreated	3.5	65.9	34.1	2.3	1.2
Treated	1.3	28.8	71.2	0.4	0.9

Note: Additive feed rate was approximately 0.4 gm/kg oil

Boiler Corrosion

The iron content of the ash entering the air heater was measured and compared with the reading when leaving, both with and without treatment. The reduction of iron content was from 4.5% down to 2.2%. This indicates a reduction in corrosion of approximately 50%.

Excess O_2

Within ten days after the trial started it was possible to reduce flue gas recirculation to a minimum and cut the excess oxygen from the normal 0.6% (at full load) to an average of 0.3 to 0.35% (approximately 50%) without notable production of CO or unburned solids.

DISCUSSION

LiquiMag treatment during January and February, 1977, resulted in a gain in overall boiler efficiency of 0.4% compared to that realized in the same months in 1976 without the additive (Table III). In March, 1977, after the unit had been cleaned, constant reheater attemperation became unnecessary, bringing the efficiency gain to the 0.5% range. The other operating benefits: elimination of acid smut fallout; reduction in corrosion; avoiding load restrictions caused by fouling; and the chemical and operating costs of frequent air heater washes, are more difficult to quantify, but nonetheless significant.

TABLE III

EFFECT OF LIQUIMAG TREATMENT ON BOILER HEAT BALANCE (AT FULL LOAD)

	Without Treatment	With Treatment
Time:	January-February, 1976	January-February, 1977
Stack Loss:		
t° gas	157.7°C	151.5°C
t° air	21.5°C	20.9°C
O_2	0.42%	0.3%
Loss	5.494%	5.227%
Loss due to Unburned Carbon:		
Incombustibles	3 g/kg fuel	0.4 g/kg fuel
Loss	0.261	0.0348
Reheater Attemperation		
Quantity	0	6.5 t/hr
Loss	0%	0.13%
Power for Gas Recirculation Fans		
Power	455 kw	210 kw
Loss	0.155%	0.07%
Air Preheating		
Loss	0	Once every two weeks for 24 hrs @60°C; Gcal/24 hr - 68 Gcal; Loss = 0.016%
Compressed air for injectors (30% of a 50 kw compressor)		
Loss	0	0.016
Total Loss	5.91%	5.506%

Difference = 0.404%

The increased efficiency is primarily the result of lower stack losses and lower combustibles, both related to improved combustion. The indicated improvement in combustion seems inconsistent with past reports that magnesia is a very poor combustion catalyst. However, it is more likely that higher furnace temperatures or other heat effects caused by reflective magnesia particles carried in the gas stream are responsible for the improved combustion. The improved combustion is particularly significant since it was obtained while lowering the excess O_2 from 0.6 to 0.3%. Normally, lowering excess O_2 results in an increase in unburned carbon and CO.

The improvement in air heater fouling and corrosion seems to be related to feeding sufficient magnesia to neutralize any acid condensed at the operating temperatures. As shown in Figure 6, a 4% excess of Mg (total - soluble) is needed to maintain a pH of 4.0

Fig. 6 INFLUENCE OF ADDITIVE ON ASH ACIDITY

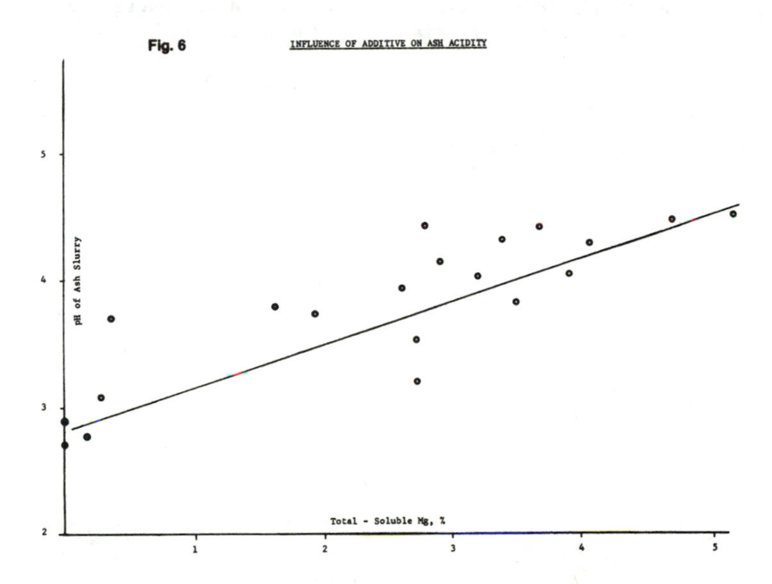

The 63% reduction in particulate emissions seems to be primarily the result of improved combustion, approximately 86% of the net decrease. However, the net reduction in non-combustibles becomes particularly significant when the fact that the LiquiMag addition was equivalent to 0.4 g inerts/kg oil (approximately 11% of untreated emissions) is considered. A significant part of this reduction can be attributed to less boiler corrosion as evidenced by the 50% reduction in iron content of the ash. Accurate assessment of emissions attributable to corrosion products would require determination of the kinds of iron compounds (oxides, sulfates, hydration state, etc.) present in the particulates. This determination was not made.

CONCLUSIONS

Injection of dispersed magnesia additive has yielded the following benefits on a 300 MW oil-fired boiler:

(1) Improved combustion as indicated by the ability to reduce excess O_2 by 50% while lowering combustibles emitted by approximately 80%.

(2) A 60% reduction in total particulate emissions.

(3) A 50% reduction in metallic corrosion as indicated by higher pH ash and confirmed by iron determinations on the ash.

(4) Improved boiler heat transfer - evidenced by a reduction in gas recirculation requirements and an overall improvement in heat balance of approximately 0.5%.

Realization of these benefits is best accomplished if:

(5) Trials are initiated on a clean boiler. Otherwise several months may be wasted in stabilizing conditions.

(6) Magnesia feed rate is not reduced with boiler load.

(7) The additive injectors are carefully located to insure good distribution of the magnesia within the boiler.

LiquiMag is a registered trademark of Basic Chemicals, a division of Basic Incorporated.

AN ANALYSIS OF THERMAL FATIGUE CRACK GROWTH IN 2.25% Cr 1% Mo STEEL SUPERHEATER TUBES CAUSED BY ON-LOAD WATER DESLAGGING

J. D. NEWTON and M. H. MELKSHAM

State Electricity Commission of Victoria
Herman Research Laboratory
Richmond, Victoria, Australia 3121

ABSTRACT

On-load water deslagging is an effective method of cleaning boiler heat transfer surfaces, but an attendant problem is the initiation and growth of thermal fatigue cracks. The growth of the cracks has been analysed and shown to be governed by the same basic relationship as that which holds for mechanical fatigue cracks. Fundamental to any analysis of fatigue crack growth is a knowledge of the alternating stress intensity. This was obtained by first determining, from a finite element stress analysis, the thermal shock stress, caused by on-load deslagging, and then calculating the stress intensity factor, using relationships available in the literature, and making a critical correction for the presence of multiple cracks.

The analysis has indicated that it is possible, within certain limits, to predict the rate of thermal fatigue crack growth.

NOMENCLATURE

a	active crack length (depth as measured from the quenched surface)
$\bar{a}$	maximum crack length
$2b$	surface length of a crack
C	constant (eq. 1)
C'	constant (eq. 6)
h	surface conductance
k	thermal conductivity
K	stress intensity factor
ΔK	alternating stress intensity
ΔK_o	threshold alternating stress intensity for fatigue crack initiation
m	constant (eq. 1)
M_B	stress intensity magnification factor for a semi-elliptical surface crack in a plate subjected to bending
n	number of stress cycles
r	outside tube radius
θ	temperature of surface to be quenched
θ_a	temperature of quench medium, adjacent to quenched surface
ϕ	elliptic integral
σ	stress normal to crack
σ_B	maximum bending stress at outer fibres of a plate

INTRODUCTION

A serious problem in the use of brown coal from the Morwell Open Cut has been the rapid deposition of ash on the surfaces of superheater tubes. The ability of on-load water deslaggers to remove ash deposits and maintain heat transfer efficiency has been conclusively demonstrated in plant trials (1). However, one of the results of the regular thermal shocks suffered by the tubes during the cleaning process has been the production of cracks on the tube surfaces (1,2,3). The extent and severity of the thermal fatigue cracks is the controlling factor in deciding if on-load water deslagging is a feasible cleaning method. The growth of cracks to a stage where they might seriously reduce the capacity of the steam-containing tubes to withstand the normal creep stresses, would make on-load water deslagging unacceptable despite its effectiveness. Thermal fatigue is caused by the repeated application of stresses that are thermal in origin; when stresses are generated by a sudden change in temperature the process is usually referred to as thermal shock. On-load water deslagging causes large non-uniform temperature gradients across the tube walls. In this situation the thermal contractions at the cooled surface are partially inhibited by the interior of the tube wall. Mechanical strains are therefore produced and, if they are repeated often enough, the associated mechanical stresses result in fatigue.

Experiments were performed in the laboratory to determine the extent of the thermal fatigue cracking that resulted from repeated quenching of tube surfaces from elevated temperatures (1,2). The experimental parameters were chosen so as to reproduce the transient metal temperature changes measured in superheater tubes during successful plant deslagging trials. Consequently, the severity of the thermal fatigue cracking produced in laboratory specimens could be correlated with that occurring in the superheater tubes. The following parameters were found to be of importance:

a metal temperature before quenching;

b specimen rigidity;

c duration of the quench cycle;

d specimen material.

Crack severity increased in a linear manner with increasing pre-quench metal temperatures up to the maximum service temperature. The rigidity of the specimen was found to influence the severity of the cracking. When strip specimens, cut from tube walls, were used for the sake of experimental convenience, it was found that the crack severity was less than in full tube-section specimens. However, a comparison of the two sets of results showed that there was a straightforward relationship between them. Crack severity increased with increasing quench duration up to one second, beyond which there was no further increase. Three typical superheater tube steels were used in the experiments, 1% Cr 0.5% Mo, 2.25% Cr 1% Mo, and AISI 347, each having a different level of cracking severity when tested under the same conditions.

A series of empirical relationships were produced relating crack depth and severity, over a range of pre-quench temperatures, with the number of quench cycles. The relationships are useful in predicting the extent of thermal fatigue cracking that will occur under the same conditions as those of the plant trials and the laboratory experiments. However, they cannot predict the extent of the cracking that might occur as a result of a different quenching regime with differing thermal shock stresses. To enable such predictions to be made, the crack growth process requires an analysis based on the important parameters mentioned earlier and a fatigue crack growth law.

There is evidence to suggest that a simple growth law for fatigue cracks (4):

$$\frac{da}{dn} = C(\Delta K)^m \tag{1}$$

where ΔK is the alternating stress intensity, and C and m are constants, is obeyed with little variation in C and m over a wide range of testing conditions (5). Most published work on the analysis of fatigue data has been carried out on the results of mechanical fatigue tests, where specimens of a standard geometry were used. However, attempts have been made to apply the simple growth law to the thermal fatigue process. In one case, the investigators studied the growth of isolated thermal fatigue cracks into the edges of the tapered disc specimens of the type that are favoured by workers in the gas turbine field (6). Changes in the alternating stress intensity range were achieved by using a number of discs, each having a different peripheral radius while keeping the heating and cooling cycle constant. The complicating factor created by the presence of multiple cracking was avoided by the introduction of two notches diametrically opposed into the disc edge, thus localising the thermal fatigue cracks. A second investigator (7) approached the problem by producing a network of thermal fatigue cracks 0.2 mm deep and then using them as crack initiators in mechanical fatigue tests. This approach allowed the very important effect of multiple cracks to be explored, without the necessity of becoming involved in the highly complex stress analysis required to convert thermal shocks to mechanical stresses. In both cases the simple law was shown to be a satisfactory method of describing the crack growth rates.

This paper describes the preliminary attempts that have been made to provide an explanation of the thermal fatigue crack growth caused by on-load water deslagging, in terms of the simple growth law.

THERMAL FATIGUE CRACK DATA

Thermal fatigue is unlike most fatigue processes in that multiple cracking occurs. Therefore two 18 mm long sections from each of the laboratory specimens were examined and the number and depth of all cracks present was recorded. After a large number of quench cycles it was common to find between 300-500 cracks present in each section (see Table 1). When this occurred an incipient crack was found at the intersection of almost every grain boundary with the quenched surface. However, only a small percentage of those cracks propagated as transgranular fatigue cracks. The cracks in excess of 0.025 mm in depth (greater than one or two grain diameters) were recorded separately (Table 1) since they were the cracks which had been extended by the cyclic thermal stresses. The extended cracks were subject to a competitive growth situation and the result was a positively skewed distribution of crack depths after each of the different numbers of quench cycles (see Figure 1).

NUMBER OF QUENCH CYCLES	AVERAGE NUMBER OF CRACKS	AVERAGE NUMBER CRACKS IN EXCESS OF 0.025 mm	INTEGRATED CRACK LENGTH mm	MAXIMUM CRACK LENGTH mm	DEEPEST CRACK mm
1 000	178	35	1.75	0.050	0.050
2 000	209	85	6.51	0.131	0.175
4 000	232	90	11.34	0.288	0.375
8 000	266	100	17.48	0.695	0.950
16 000	374	127	23.80	0.922	1.225
32 000	472	163	26.84	1.075	1.275
64 000	422	251	33.90	1.140	1.325

Table 1 Thermal fatigue cracks present after quenching a strip specimen from 550°C for various numbers of quench cycles

Two measures of the extent of the thermal fatigue damage have been used:

a A summation of the depth of the cracks in excess of 0.025 mm, designated the "integrated crack length";

b the "maximum crack length" defined as the average depth of the deepest 5% of all cracks in excess of 0.025 mm.

The first of the above measures proved to be of little use, since the majority of the cracks making up the totals were restricted in depth, and thus would not have a significant effect on the creep life of a superheater tube. Maximum crack length is a more useful indication of the damage, as it takes account of the relatively few deep cracks that are more deleterious to the creep life of a superheater tube. An alternative to maximum crack length is to use the deepest crack found in the 18 mm sections taken from each specimen. The depth of this crack was recorded for each experiment conducted, but it was considered an unrepresentative measure, as it took account of only one crack that emerged from the competitive growth situation. In the specimens which contained only very shallow cracks, the maximum crack length and the deepest crack were equivalent; this arose from the necessity of measuring crack depths within class limits (0.025 mm) and not individually. The maximum crack lengths had a mean value of 75% (standard deviation 9%) of the depth of the deepest cracks over the range of testing temperatures and numbers of quench cycles. Thus, it can be seen to represent both the depth and the number of the growing thermal fatigue cracks found over the whole range of testing temperatures and quench cycles.

The maximum crack length is dependent on both the pre-quench temperature and the number of quench cycles, for a given specimen configuration and quench regime (see Figure 2). In our results, however, a large increase in the maximum crack length values occured between the pre-quench temperatures of 575 and 600°C. This became particularly evident when the results were plotted in an alternative manner (see Figure 3). There was a simple relationship, over the whole range of quench cycles, between the maximum crack length data and the pre-quench temperatures between 500 and 575°C. This made it possible to reduce the four lower-temperature data sets into a single relationship between maximum crack length and the number of quench cycles, by the use of a series of simple factors (see Figure 4).

The departure of the maximum crack length versus pre-quench temperature relationship from linearity at temperatures above 575°C can be explained by a change in the oxidation behaviour of the steel, which occurs at 570°C (3).

Crack growth rates were obtained by differentiation of the data in Figure 2, and as would be expected from the shape of the graphs, the rate of growth first increases to a maximum and then begins to fall (see Figure 5). At pre-quench temperatures up to and including 575°C, the fall continued to the limit of the experimental observations; for the experiment conducted at 600°C, the growth rate after passing through a maximum attained a steady state as a result of the high rates of crack tip oxidation.

ANALYSIS

Preamble

To investigate the factors involved in crack growth caused by repeated thermal shocks - without the complications introduced by oxidation at temperatures above 575°C - the data from tests at 500 and 550°C have been used. The growth rates are in the range 1.15×10^{-3} - 1.0×10^{-1} μm/cycle; it had been established earlier (2) that the growth rate in tubes is greater than in strip specimens by a factor of 2.4. Therefore, the crack growth rates in tube walls under the same conditions of thermal shock would be 2.76×10^{-3} - 2.4×10^{-1} μm/cycle, which remains within the range where the simple growth law has been shown to be valid.

The simple fatigue crack growth equation (1) indicates that the crack growth rate/cycle is related to a single variable; the alternating stress intensity. If, therefore, the thermal fatigue cracking is following the normal pattern of fatigue crack growth, then the crack growth data from Figure 5 should, when plotted in logarithmic form versus the as yet unknown stress intensities, produce a straight line relationship. Thus, it is essential that the values of the alternating stress intensities are known.

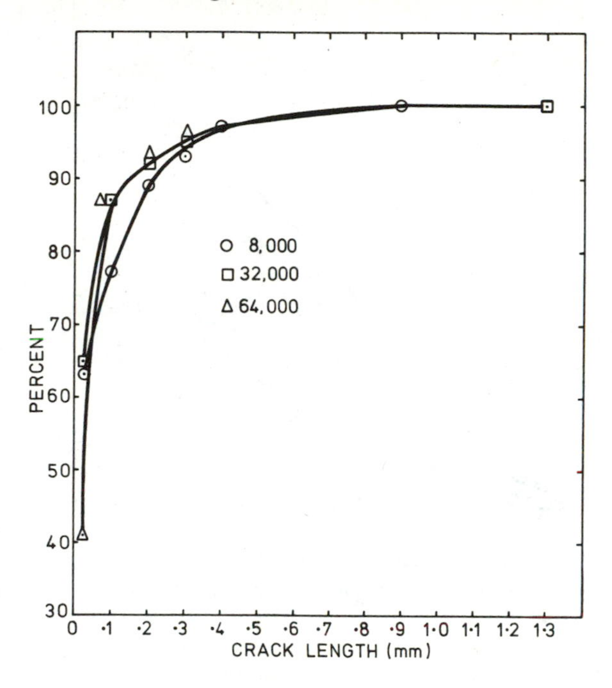

Fig.1 Percentage of those cracks present of less than the indicated depth for various numbers of quench cycles from 550 deg. C

Determination of ΔK (alternating stress intensity)

The stress intensity at a crack tip is related to the crack depth and shape as well as the applied stress. For a part through-thickness, semi-elliptical surface crack of depth (a) and length (2b), in an infinite plate under uniform applied stress, the maximum stress intensity factor is given by:

$$K = \frac{[1 + 0.12\ (1-a/b)]\ \sigma(\pi a)^{\frac{1}{2}}}{\phi} \tag{2}$$

where ϕ is an elliptic integral, which varies in value depending on the crack shape (8). For more complicated situations more complex analyses have been derived, for example (9), the stress intensity factor for a semi-elliptical surface crack in a plate subject to bending can be expressed in the following form:

$$K = \frac{M_B\ \sigma_B\ (\pi a)^{\frac{1}{2}}}{\phi} \tag{3}$$

where σ_B is the maximum bending stress at the outer fibres of the plate. M_B is the stress intensity magnification factor at the point of maximum crack depth, and is dependent on the ratio of both the crack depth to plate thickness and the crack depth (a) to half length (b). Thus, there are solutions available to determine K, providing that the applied stresses are known.

The relationships for the stress intensity factors (equations 2 and 3) are both calculated for the case of an isolated crack. It has already been pointed

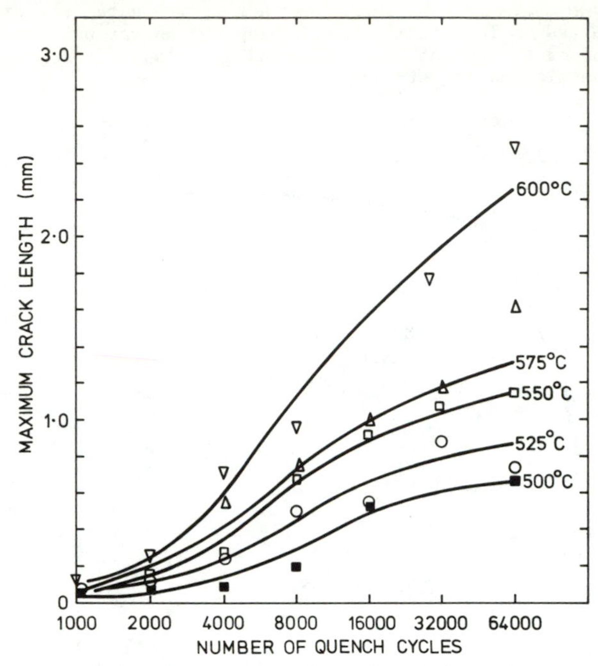

Fig.2 Maximum crack length versus number of quench cycles for a range of pre-quench temperatures

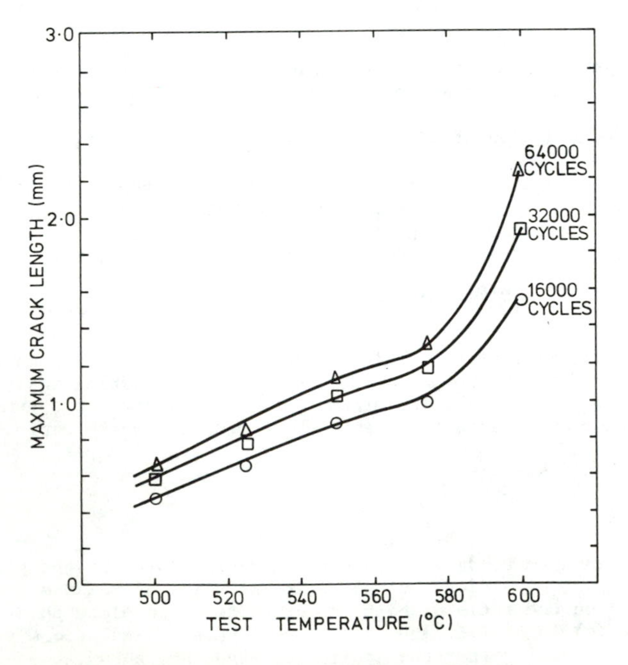

Fig.3 Maximum crack length plotted against pre-quench temperature for various numbers of quench cycles

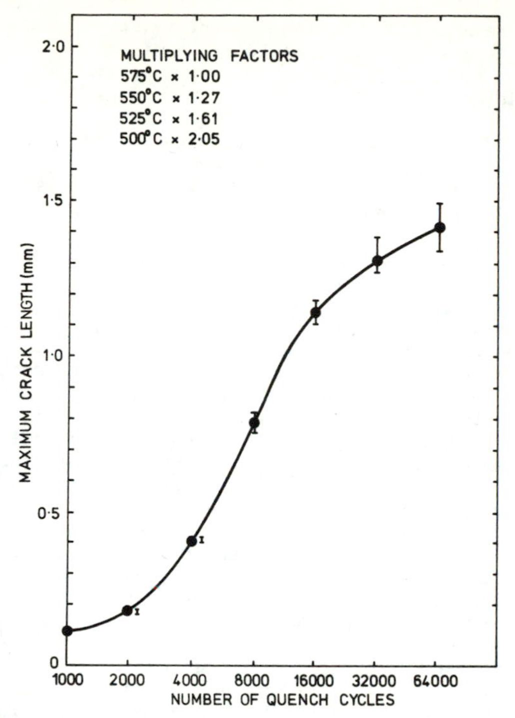

Fig.4 Maximum crack length versus pre-quench temperature : results normalised to 575 deg. C

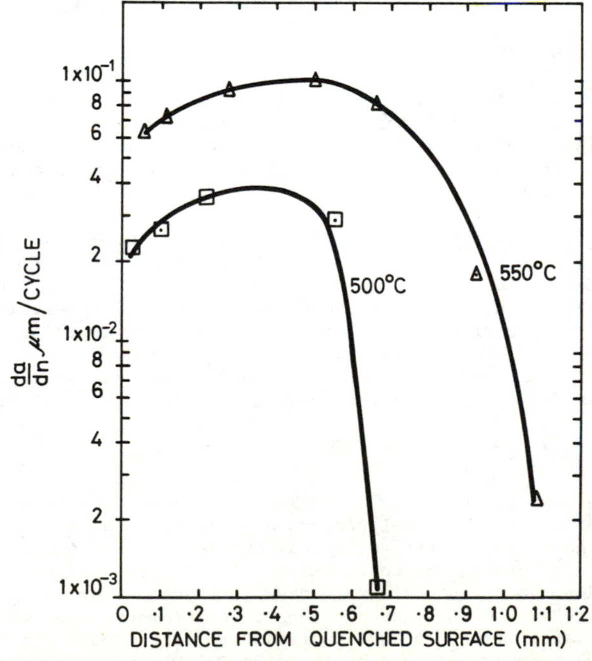

Fig.5 Crack growth rate compared with depth below the quenched surface

out that in the case of thermal fatigue, multiple cracking is the rule. The presence of multiple cracking reduces the stress intensity at each individual crack tip (7), the amount of the reduction being determinable from the crack depth spacing ratio.

For the case of fatigue with simple tension/tension loading, the alternating stress intensity, ΔK, is the difference between K_{max} and K_{min}. However, for the tension/compression case the determination of ΔK is not so simple, since compressive stresses cannot give rise to a stress intensity in opening mode loading. Crack growth rate measurements (5) from experiments where K_{max} was kept constant and differing values of K_{min}, between zero (no stress) and a 'negative' level, were used, showed that there was no significant increase in rate by cycling into compression. This indicated that compressive stresses can be neglected when calculating rates of crack propagation. Other results reviewed in the same paper (5) showed increased growth rates for fully reversed cyclic loading, when compared with zero/tension tests. To establish a method of calculating the alternating stress intensity for the stress regime caused by quenching followed by reheating, the stress-strain hysteresis loop must be examined. An idealised hysteresis loop for an elastic-plastic material was determined at an earlier stage (1). This showed that after the first quench and reheat cycle the tube walls were left in a state of residual compression. The numerical value of the residual stress was approximately equal to the maximum tensile stress that occurs during the quench. In on-load washing, the loading cycle is therefore one of steady state residual compression with superimposed fluctuating tensile stresses. The compressive portion of the fatigue cycle could be neglected entirely and ΔK equated with K_{max}. This would be justified by the results obtained from single edge notched pin-loaded specimens as detailed in (5). However, the other results reviewed in the same paper suggested that compressive loading could not be neglected when there was a departure from the plane strain condition. It is therefore proposed that in this analysis, where the cracks grow in a biaxial stress field and are subject to severe stress gradients, the alternating stress intensity is taken to be the difference between K_{max} and K_{min}, and is thus equal to $2K_{max}$.

Thermal Shock Stresses in Tube Walls

The stresses generated by the thermal shock of quenching result from the partial constraining of thermal contractions, thus the first stage in the stress analysis is to determine the temperature distributions at selected time intervals during quenching. This was subsequently used to find the corresponding stresses in the tube wall.

Heat conduction analysis. A three-dimensional finite element heat conduction analysis was carried out, the mesh used being shown in Figure 6, the quench region being indicated. A time-stepping scheme was used, initial temperatures being taken as uniform throughout the tube. The temperature (θ_a) adjacent to the quench surface was assumed to be constant at 100°C to simulate the quenching action of the water-washing operation. A value of 22 800 J/m^2 s deg K was used for the surface conductance h, which governs the heat flux across the quenched surface in accordance with the boundary condition,

$$K \frac{\partial \theta}{\partial r} = h(\theta - \theta_a) \quad (4)$$

where K is the thermal conductivity. The inner surface of the tube was prescribed to remain at the initial temperature for the time considered by the analysis, on the basis that water washing is carried out on-load with a continual flow of steam through the tube. Material properties of the tube steel are temperature dependent, and this was taken into account by using an iteration procedure at each time step. The finite elements used were the parabolic isoparametric type, and the analysis was carried out firstly for five steps of 0.02 seconds followed by ten steps of 0.1 seconds. Temperature

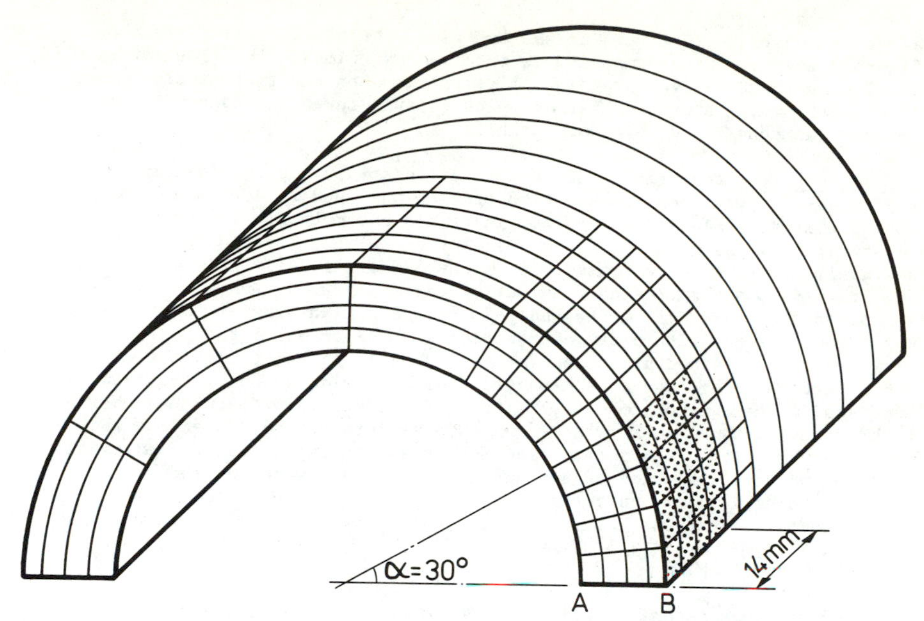

Fig.6 Finite element mesh for heat transfer analysis

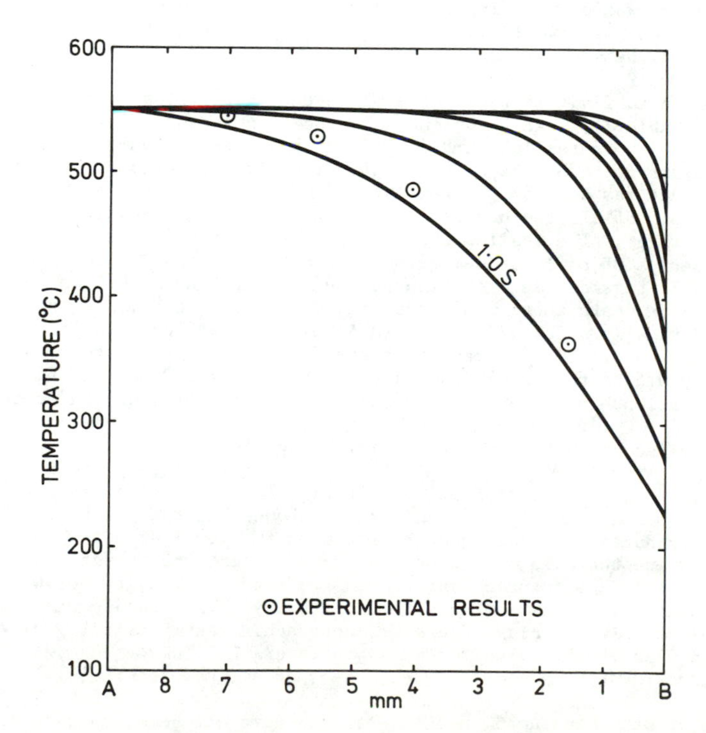

Fig.7 Temperature gradients through a tube wall, at centre of quenched zone, from finite element analysis, with experimental results at 1 sec. included (.02, .04, .06, 0.1, 0.2, 0.5 and 1.0 sec)

gradients through the tube wall at the centre of the quench region are shown in Figure 7 together with temperatures obtained experimentally. The latter values lagged those derived by this analysis, but when the inevitable thermal inertia of the metal sheathed thermocouples (1) was taken into account the difference was considered to be acceptable.

Elastic stress analysis. The temperature distribution obtained in the form of nodal temperatures at each time station was used to determine the resulting elastic stresses, using the finite element method. In order to keep the computer analysis down to an economically acceptable level, a harmonic stress analysis technique was used. This enabled a three-dimensional solution to be obtained by superimposing a number of harmonic two-dimensional solutions (11). Stresses obtained from the analysis showed a high biaxial tensile stress at the quench surface with a rapid reduction with depth below the surface. The axial stress obtained at the surface was slightly greater than the circumferential stress; this would be expected from a consideration of the relative stiffnesses in these directions. Variation of the axial stresses through the wall thickness is shown in Figure 8, and this pattern extended over the whole quench area, but with a sharp reduction in stress gradients near the edges of the quench. After initiation of the quench the surface stresses increased rapidly, reaching a maximum after about 0.1 seconds (Figure 9) and remained at a high level for the rest of the quench period.

Crack Shape Ratio (a/b)

A surface crack is characterised by its depth (a) and half surface length (b); it can vary between two extremes, the "long" crack where $a \ll b$, and the half circle where $a=b$. Thus, the ratio a/b varies over the range 0 to 1, causing a consequent variation in the value of the elliptic integral ϕ (equations 2 and 3) between 1.0 and 1.57.

It was common to find, in the specimens tested, 400 cracks in the 18 mm length sampled (Table 1), with the number of cracks in excess of 0.025 mm in depth increasing with increasing numbers of quench cycles. A sample was selected to measure the a/b ratios and to determine if the ratio changed with increasing a. This was done by progressive grinding back of sample and measuring the depth of all the cracks at many intervals. The results were not easy to interpret. The shallow cracks, those less than 0.1 mm in depth, appeared as a series of coplanar semicircular cracks close enough together to interact and be regarded as a continuous long crack, i.e. $a \ll b$. As the cracks increased in depth there was a tendency for some of the longer ones to coalesce and finally a few widely spaced deep cracks emerged from the large number of shallow ones. The a/b ratios of the deep cracks tended to a value of approximately 0.5 as the crack depth increased towards 1.0 mm. No firm mathematical relationship was evident relating depth and shape of the cracks.

Stress Intensity Reduction Factor for Multiple Cracking

The effect of the presence of multiple fatigue cracks on the crack growth rate has been examined recently (7), and it was shown that the presence of multiple cracks reduced the crack growth rate from the level expected for a single crack. Where there is more than one crack present, the elastic stress fields of adjacent cracks interact and reduce the stress intensity below that of a single crack. The reduction in stress intensity will increase with decrease in the relative spacing of cracks, approaching asymptotically a value between 0.3 and 0.35 of the value for an isolated crack. This value is reached when crack depth approaches or exceeds spacing (see Figure 10).

The crack growth rates used in this analysis have all been based on the maximum crack length (as defined earlier); this parameter averages the deeper cracks present after each experiment. Therefore, for the purposes of determining a stress intensity reduction factor, it was those cracks, growing in competition with each other, that were considered. The crack depth spacing

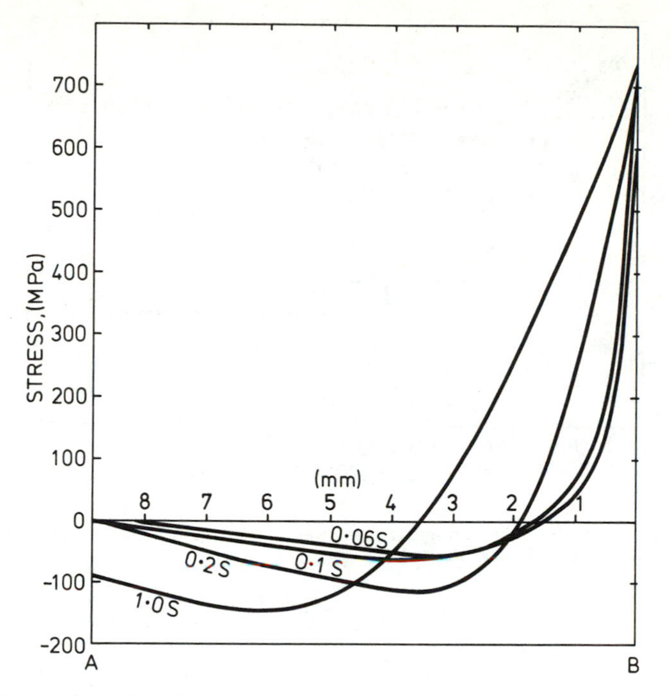

Fig.8 Variation of axial stress through wall thickness on quenching from 550 deg. C (quenched surface, B)

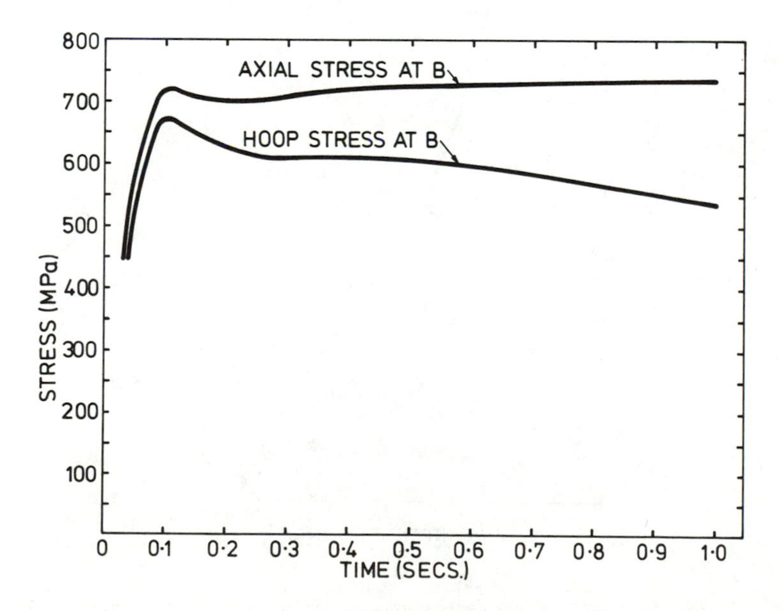

Fig.9 Change in surface stresses with time after quenching from 550 deg. C

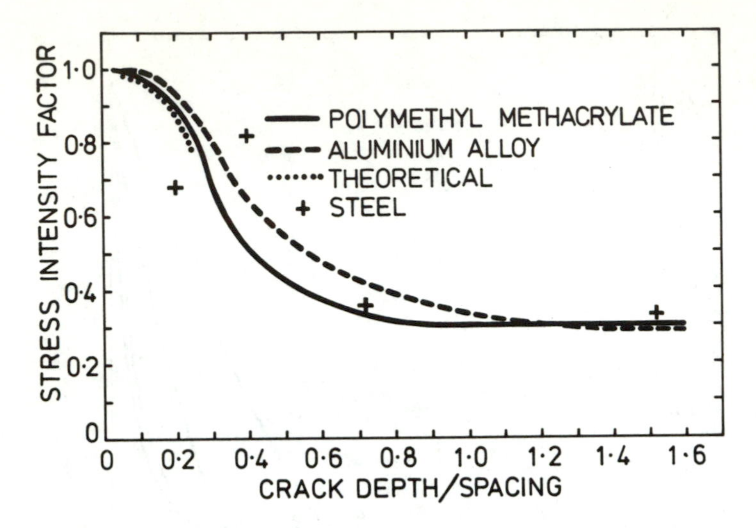

Fig.10 Change in stress intensity caused by interaction of parallel cracks, after Ritter (7)

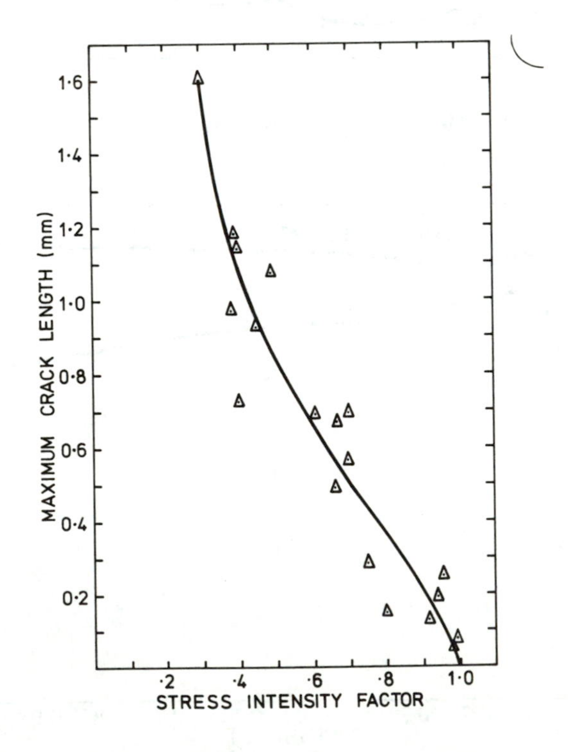

Fig.11 Stress intensity factors for cracks with different values of $\bar{a}$.

ratios were calculated for all the experiments carried out over the pre-quench temperature range 500 to 575°C. From these figures the stress intensity reduction factors were determined for cracks of various depths: the results are shown in Figure 11. It was found that the stress intensity reduction factor approached the minimum level of 0.3 when the maximum crack length exceeded 1.6 mm.

Results

The finite element stress analysis was used to calculate the stress gradients through the tube walls at different time intervals. Since the major thermal fatigue cracking intersects the tube surface circumferentially, it was axial stress in the uncracked tube at particular depths which was used as the basis for the calculation of stress intensity from equation (2). The ratio a/b was assumed to remain constant at zero, i.e. long cracks were always present. This was not strictly true: measurements made of crack shape profiles had shown that the value of the ratio tended towards 0.5 as the maximum crack length (a) approached 1 mm. The result was a loss of precision. However, the measurements made of the variation in the ratio did not permit a systematic adjustment to be made to the value of a/b (and thus ϕ) in equation 2. Therefore the nominal stress intensity for a single crack of any depth was to be calculated from a modified equation of the form:

$$K = 1.12\,\sigma(\pi a)^{\frac{1}{2}} \tag{5}$$

and the allowance for multiple crack interaction was made by introducing a stress intensity reduction factor. The final figure for the multiple crack stress intensity was doubled to produce the alternating stress intensity range. The results are presented in Table 2, where in addition, the crack growth rates obtained from experimental work are listed.

MAXIMUM CRACK LENGTH $\bar{a}$(mm)	STRESS σ (MPa)	$1.12\,\sigma(\pi)^{\frac{1}{2}}$ (MPa $m^{\frac{1}{2}}$)	STRESS INTENSITY FACTOR	ΔK (MPa $m^{\frac{1}{2}}$)	CRACK GROWTH* RATE IN A TUBE WALL $\frac{da}{dn}$, (μm/cycle)
0.050	720	10.11	0.98	19.82	1.49×10^{-1}
0.131	695	15.79	0.95	30.00	1.80×10^{-1}
0.288	650	21.90	0.86	37.67	2.26×10^{-1}
0.695	545	28.52	0.575	32.80	1.80×10^{-1}
0.922	495	29.84	0.465	27.75	6.24×10^{-2}
1.075	460	29.94	0.40	23.95	6.24×10^{-3}
1.14	445	29.83	0.38	22.67	$<2.5 \times 10^{-3}$
0.050	600	8.42	0.98	16.51	6.0×10^{-2}
0.065	570	9.12	0.975	17.79	6.48×10^{-2}
0.050	600	8.42	0.98	16.51	6.00×10^{-2}
0.200	535	15.02	0.91	27.34	8.40×10^{-2}
0.550	435	20.25	0.67	27.14	6.96×10^{-2}
0.670	410	21.07	0.59	24.86	2.4×10^{-3}

* The crack growth rate in a tube wall was obtained by multiplying the growth rates determined experimentally by a factor of 2.4 (2).

Table 2 Alternating stress intensities calculated at various crack depths for experiments conducted at 550°C and 500°C, with measured crack growth rates included

Figure 12 shows the relationship between the experimental crack growth rates and the calculated cyclic stress intensities for rates in excess of 1×10^{-2} μm/cycle. The results show that the thermal fatigue crack growth rate can be explained in terms of an equation of the type:

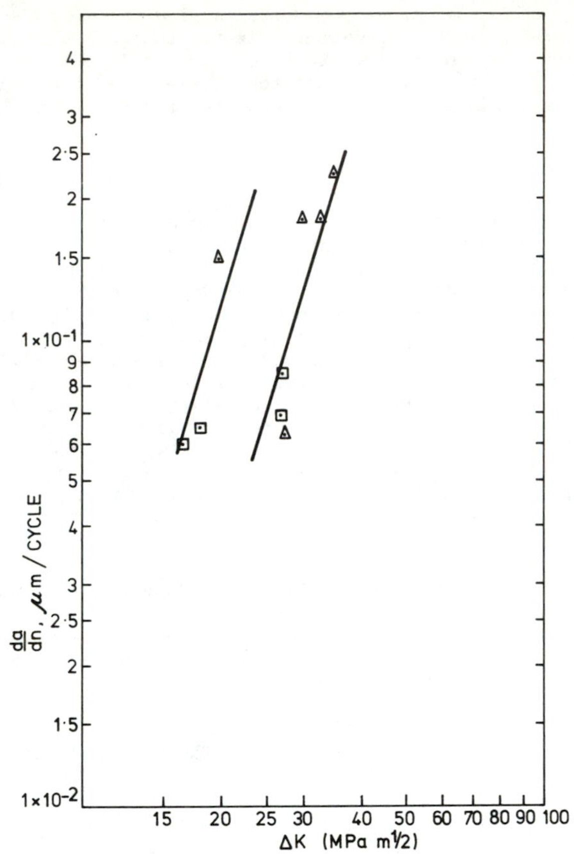

Fig.12 The relationship between the rate of crack propagation, da/dn and the cyclic stress intensity, ΔK, as a result of repeated quenching from 500 and 550 deg. C

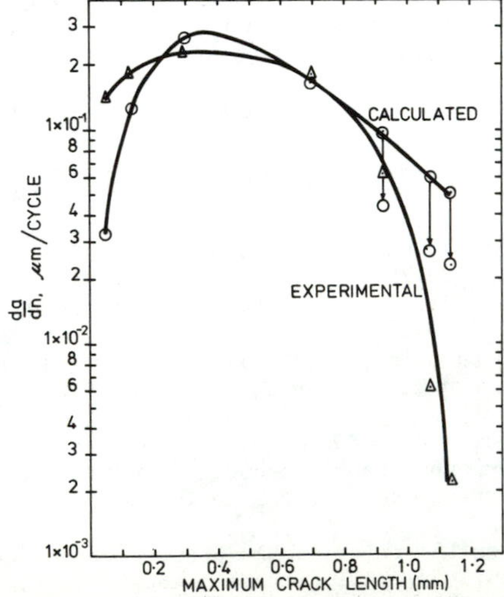

Fig.13 A comparison of the calculated and measured crack propagation rates with increasing crack depths after repeated quenching from 550 deg. C

$$\frac{da}{dn} = C(\Delta K)^m \qquad (1)$$

where m is 3.25 and C lies between 2.0×10^{-6} and 8.0×10^{-6}. This relationship is valid from the commencement of cracking until the maximum crack length reaches a depth of 1.0 mm, and in addition the values of the parameters C and m show a good agreement with those listed in Table 3 for several steels.

MATERIAL	C	m
Ducol W30B	3.5×10^{-5}	2.5
Spheroidised 1% carbon steel	1.3×10^{-5}	2.9
3 Cr Mo turbine steel	1×10^{-5}	2.9
H1 weld metal	3.5×10^{-5}	2.5
A2 weld metal	1.3×10^{-5}	2.9

Table 3 Values of the parameters C and m from equation (1) for several steels. Units of µm/c and MPa $m^{\frac{1}{2}}$, after Richards and Lindley (5)

DISCUSSION

The results show that crack growth rate in this thermal fatigue process is related to alternating stress intensity in the manner proposed by Paris and Erdogan for mechanical fatigue cracks. Furthermore the values of the parameters in the equation are very similar to those found for other steels. If the parameters, $m = 3.25$ and $C = 2 \times 10^{-6}$, are used to calculate crack growth rates over a range of depths, see Figure 13, then the calculated values tend to underestimate in the initial stages and overestimate as the cracks become deeper when compared with the measured rates.

Figure 12 shows the crack growth rate plotted versus the stress intensity range, with the scatter in the data enclosed by two lines having the same slope. The data appears to be made up of two sets, one at a higher growth rate for a given stress intensity range than the other. An examination of the data shows that the higher bound line passes through data points derived entirely from shallow cracks (<0.07 mm). In the early stages there is no doubt that corrosion plays a cosiderable part in the initiation and growth of cracks. Almost without exception the thermal fatigue cracks are initiated at grain boundaries, by a combination of thermal shocks and corrosion. This will have the effect of increasing the apparent growth rates in the early stages. It is therefore possible to look upon the lower bound line as the more representative of thermal fatigue crack growth.

After the crack propagation rate passed through a maximum and the crack depths increased, the calculated relationship in Figure 13 tended to overestimate the crack growth rates. This tendency became increasingly marked as the crack depth increased and the growth rate decreased. There is one factor which contributes to the overestimate that has already been mentioned, the change in value of the ratio of a to b as a (crack depth) increases. While it was possible to show that as a approached a depth of 1.0 mm, a/b tended towards 0.5, anything more systematic was not possible. The modification of the crack growth rates, for cracks in excess of 0.9 mm depth, by the introduction of new fixed a/b and ϕ terms would result in reducing the calculated values, see Figure 13. The insertion of the new terms changes the value of stress intensity equation (2) from:

$$K = 1.12\, \sigma(\pi a)^{\frac{1}{2}} \qquad (5)$$

to

$$K = 0.88\, \sigma(\pi a)^{\frac{1}{2}} \qquad (6)$$

Further efforts will have to be made to measure crack profiles so that the variation in the value of the a to b ratio can be systematically incorporated into the calculations. Thus, the effect of changes in the crack profile is to reduce the level of ΔK, which occurs at the same time that ΔK is falling for other reasons. This has an additional effect: the crack propagation rates at low levels of ΔK are described by another equation (5):

$$\frac{da}{dn} = C'(\Delta K - \Delta K_o)^m \qquad (7)$$

where ΔK_o is dependent on the mean stress and varies between 2 and 4 MPa $m^{\frac{1}{2}}$. The result is that changes in crack profile accentuate the reduction in alternating stress intensity and further decelerate crack growth.

Certain assumptions had to be made in order to carry out the analysis. The first was that the maximum stress intensity can be described by:

$$K = \frac{[1 + 0.12\ (1-a/b)]\ \sigma(\pi a)^{\frac{1}{2}}}{\phi} \qquad (2)$$

which had originally been obtained to derive K values for part through-thickness cracks in an infinite plate under uniform applied stress. The thermal shock stresses were known to vary through the tube wall, and therefore different values of stress were used at different depths from the quenched surface. There are other relationships available to calculate K, which take account of stress distributions normal to the crack face. Take for example equation (3):

$$K = \frac{M_B\ \sigma_B\ (\pi a)^{\frac{1}{2}}}{\phi} \qquad (3)$$

which is an expression for the stress intensity factor of a semi-elliptical surface crack in a plate subjected to bending. The axial stress distribution through a tube wall, during quenching, approximates to that of a plate in bending, see Figure 8. Consequently K values were calculated from equation (3), using the maximum surface stress σ_B. They proved to be very similar to those calculated from equation (2), indicating that the assumption on the use of a variable stress was reasonable. Further confirmation was obtained when a method of determining K due to Bueckner (12) was used. This method used a polynomial expression for the stress distribution normal to the crack face and yielded similar values of K to equations (2) and (3).

The second assumption was that the value of the alternating stress intensity, ΔK, was the difference between K_{max} and K_{min}. If on the other hand ΔK was equated with K_{max}, thus neglecting the compressive portion of the loading cycle, the only effect on the analysis would be to change the value of the constant C (equation 1). This assumption, which had also been made by others (6), appeared to be justified in that the results obtained using it compared well with other published data.

The stress analysis was limited to a determination of the linear elastic stresses, the magnitude of which would not be true stresses once the yield stress had been exceeded. However, it should be recognised that the thermal strains, from which the elastic stresses are derived in the finite element analysis, do provide a realistic measure of deformation. Consequently, the post yield stresses used to determine the stress intensity merely represent the level of strain involved. It is therefore implicit that it is strain range that is being used to carry out the calculations. This has been shown to be satisfactory for crack propagation in fatigue where cyclic plastic deformation occurs (13). Thus a logical extension of the stress analysis would be to consider elastic-plastic behaviour during thermal shock, and to further examine the stress-strain cycling produced by repeated quenching.

CONCLUSIONS

The growth of thermal fatigue cracks that result from on-load water deslagging can be interpreted using the same basic relationship as that used extensively for the more conventional fatigue processes. The values of the parameters C and m have been shown to be very similar to those reported for several other steels. It was necessary to reduce the stress intensity from the level of that of a single crack to make allowance for the multiple cracking that occurs as a result of thermal fatigue damage. This, together with the stress analysis, was the key to the problem.

In order to refine the analysis, further work is required on the stress analysis, and this should be directed towards an elastic-plastic solution. The stress analysis should also be extended to look further into the stress-strain hysteresis cycle after repeated quenching. Finally, the analysis would gain greater precision from further experimental work to determine the changes in crack shape that occur during crack growth.

ACKNOWLEDGEMENT

This paper is published with the permission of the State Electricity Commission of Victoria.

REFERENCES

1 Ellery, A. R., Johnson, T. R., and Newton, J. D., "Investigation Into the Likelihood of Thermal Fatigue Damage to Furnace and Superheater Tubes Caused by On-load Water Deslagging", Transactions of the ASME, Journal of Engineering for Power, Vol. 96, Series A, No. 2, April 1974, pp. 138-144.

2 Newton, J. D., "Thermal Fatigue and the On-load Water Jet Cleaning of Superheaters", Proceedings of the Metals Technology Conference, (Sydney) Australian Welding Institute, Australian Welding Research Association, Australasian Institute of Metals and the University of New South Wales, Vol. A, 1976, pp.2-4-1/2-4-17.

3 Newton, J. D., "Thermal Fatigue Cracking of 2.25% Cr 1% Mo Steel Caused by Repeated Quenching From Elevated Temperatures", Journal of Australasian Institute of Metals, Vol. 21, Nos. 2 and 3, June-September 1976, pp. 94-102.

4 Paris, P. C., Erdogan, F., "Critical Analysis of Crack Propagation Laws", Transactions of the ASME, Journal of Basic Engineering, Vol. 85, Series D, 1963, pp. 528-534.

5 Richards, C. E., Lindley, T. C., "The Influence of Stress Intensity and Microstructure on Fatigue Crack Propagation in Ferritic Materials", Engineering Fracture Mechanics, Vol. 4, 1972, pp. 951-978.

6 Mowbray, D. F., Woodford, D. A., and Brandt, D. E., "Thermal Fatigue Characterization of Cast Cobalt Alloys and Nickel-Base Superalloys", Proceedings of a Symposium on Fatigue at Elevated Temperatures, ASTM Special Technical Publication No 520, 1972, pp. 416-426.

7 Ritter, J. C., Private Communication.

8 Paris, P. C., Sih, G. C. M., "Stress Analysis of Cracks", Proceedings of a Symposium at the Sixty-Seventh Annual Meeting of Americal Society for Testing and Materials, Fracture Toughness Testing and its Applications, ASTM Special Technical Publication No. 381, 1964, pp.30-76.

9 Shah, R. C., Kobayashi, A. S., "Stress Intensity Factor for an Elliptical Crack Approaching the Surface of a Plate in Bending", Proceedings of the 1971 National Symposium on Fracture Mechanics, Part 1 : Stress Analysis and the Growth of Cracks, ASTM Special Technical Publication No. 513, pp. 3-21.

10 Glenny, R. J. E., "Thermal Fatigue", High-Temperature Materials in Gas Turbines, P. R. Sahm and M. O. Speidel, Ed., Elsevier, Amsterdam, 1974, pp. 257-276.

11 Melksham, M. H., Yeo, M. F., Cheung, Y. K., "The Initial Effects of Water Jet Cleaning on Superheater Tubes", Proceedings of an International Conference on Finite Element Methods in Engineering, Adelaide University, December 1976.

12 Bueckner, H. F., "Field Singularities and Related Integral Expressions", Mechanics of Fracture I, Methods of Analysis and Solutions of Crack Problems, George, C. Sih Ed, Noordhoff, Leyden, 1973.

13 Mummery, A., "Low-Cycle Fatigue Tests on Butt Welds Containing Slag Inclusions and Studies of Crack Propagation Under Low-Cycle Fatigue Conditions", The Welding Institute, Research Report 3322/8/72, issued October 1972.

STUDIES OF WATERWALL CORROSION WITH STAGED COMBUSTION OF COAL

E. H. MANNY, W. BARTOK, and A. R. CRAWFORD

Exxon Research and Engineering Company
Linden, New Jersey, USA 07016

R. E. HALL

NERC – Wing G
Research Triangle Park, North Carolina, USA 27711

J. VATSKY

Foster Wheeler Energy Corporation
Livingston, New Jersey USA 07039

ABSTRACT

The combination of low excess air and staged firing combustion modifications is being employed successfully to reduce NO_x emissions in coal fired public utility boilers. Reducing atmospheres produced in the first stage of the combustion process resulting from this type of operation potentially are conducive to external corrosion of the furnace tubes. Corrosion data obtained using corrosion probes failed to provide an unequivocal resolution to the corrosion question due to the inability to relate probe data to actual furnace tube life. Consequently, Exxon Research and Engineering Company under EPA sponsorship initiated a long term corrosion program using three methods of corrosion rate determination, i.e., (1) actual furnace tube panel test specimens; (2) a statistically designed pattern of ultrasonic furnace and panel tube measurements; and (3) corrosion probes exposed for varying periods. Corrosion probe data obtained in prior studies are reviewed and details of the long term corrosion probe test, now in progress, are discussed in the paper.

INTRODUCTION AND SUMMARY

Over the past seven years Exxon Research and Engineering Company has conducted research sponsored by EPA (and in part by EPRI) on the reduction of NO_x emissions in 32 large, pulverized coal fired, electric utility boilers throughout the United States. Concurrently with the Exxon Research field test programs, which were aimed at probing the effectiveness of combustion modification on NO_x emissions in various utility boiler designs fired with different coal types, boiler manufacturers and the electric utility industry also carried out programs in this area. Results of these investigations have shown that the most effective NO_x reduction combustion modification techniques are low excess air firing and staging of the combustion process, especially where these two methods are used in combination with one another. Staged combustion is now commercially practiced by most U.S. boiler manufacturers to control NO_x emissions in new boilers to comply with the New Source Performance Standard (NSPS) of 300 ng NO_x/J (0.7 lb. $NO_x/10^6$ BTU). In some cases existing boilers can be altered by installing overfire air ports or the top row of burners can be used as overfire air ports to reduce NO_x emissions, if required to meet local pollution regulations.

The reducing conditions generated in the furnace in the first stage of the combustion process potentially could lead to external corrosion and slagging of the furnace tubes. Such problems could be expected to be more pronounced when high sulfur, high iron coals are fired due to the possibility of iron sulfide attack on the furnace tubes under reducing conditions. The mechanism under which corrosion takes place is complicated and not fully understood but is generally agreed to be dependent on the formation of a liquid ash phase. Under reducing conditions in the first stage of the combustion process, ash fusion temperatures decrease, enhancing the possibility that the furnace tubes in the vicinity of the first stage combustion zone may tend to be covered with molten slag. Under these conditions the intimate contact of the molten slag with tube metal surfaces potentially could promote increased corrosion of furnace tubes.

The objective of an early research program conducted by Exxon Research in 1971 in an EPA-sponsored cooperative test program with TVA on the Widows Creek Boiler No. 7 and Shawnee Boiler No. 10 (1) was to develop the know-how and to demonstrate the technical feasibility of reducing NO_x emissions in pulverized coal fired electric utility boilers. Because of the potential of this approach to significantly reduce NO_x emissions (by about 50%) from coal fired installations, subsequent field studies by EPA under Contracts No. 68-02-0227 and 68-02-1415 not only concentrated on coal fired utility boilers (2) (3) but the programs were expanded to include investigations into potential adverse side effects resulting from "low NO_x" type operations.

During the course of these investigations (2) (3) it became apparent that the side effect of paramount concern would be that of potential increase in external furnace tube corrosion. To gain insight into this problem, initial approaches employed corrosion probes to obtain comparative data, but this approach was found to be inadequate due to the inability to effectively equate probe data to actual furnace tube life. Program modifications were made, accordingly, to obtain long term corrosion data using the boiler tubes themselves as test specimens under "low NO_x" operating conditions. This paper will review the extent of corrosion probe data obtained in prior programs (2) (3) and discuss details of the long term corrosion tests sponsored by EPA which have been initiated on the No. 7 boiler at the Crist Station, Pensacola, Florida in cooperation with the Gulf Power Company and the Foster Wheeler Energy Corporation.

FURNACE TUBE WASTAGE PROBLEMS

External corrosion of furnace tubes is generally a function of fuel, chemistry ash constituents and the composition of the furnace atmosphere along the sidewalls. Steam generators designed prior to the advent of NO_x emission regulations usually operated with all combustion air entering the furnace through active burners. Since excess air (usually 20-25%) is utilized, the lower furnace atmosphere is generally oxidizing. Thus, excessive slagging is usually prevented and tube wastage rates are minimal.

Existing Boilers

There does, however, exist an occasional, isolated, instance where a unit suffers severe sidewall wastage even though operating with burner air seemingly at least 20% greater than stoichiometric. Figure 1 shows a cross-section of a tube which was subjected to a reducing atmosphere along the sidewalls (4). The fuel was a high sulfur (>3.0%), high iron (Fe_2O_3 > 20%) bituminous coal with a low reducing ash fluid temperature. Tube wastage was observed after approximately 30 months of operation. It must be noted that similar units firing similar fuel do not have a tube wastage problem.

The sidewall reducing atmosphere was eliminated by redistributing the windbox air flow and adding sidewall slots to permit air entry. Tube wastage has since been significantly alleviated thus confirming that there is a strong link between this condition and oxygen level along the wall.

This experience [(4)] indicates that similar wastage rates are possible when overfire air is utilized as a NO_x control technique on a unit firing a similar coal. The sidewall reducing atmosphere on the unit (from which the sample in Figure 1 was taken), although caused by poor distribution, is essentially the same as that expected with overfire air operation, when the burners are operated substoichiometrically. Therefore, when firing substoichiometrically, other means may be required to retard corrosion, such as to raise the oxygen content along the walls, while maintaining the substoichiometry of the flames. Another means would be to use a different tube material.

Low NO_x Firing Effects

Nitrogen oxide emissions are limited most effectively in utility boilers by staging the combustion process and maintaining the overall excess air level at minimum values consistent with safe, efficient operation. Staging the firing pattern consists of first, burning the fuel in the lower regions of the furnace under fuel-rich or substoichiometric air supply conditions, followed by second stage air addition above the primary combustion zone to complete burnout of the remaining combustibles. As mentioned previously, several of the U. S. boiler manufacturers have included overfire air ports in new boiler designs so that the combustion process may be staged and New Source Performance Standards for NO_x may be achieved. Depending on individual boiler conditions, reductions in NO_x emissions in existing boilers may be reduced from about 20 to 50% in much the same way by using the top burners (or a combination of top burners) as overfire

Figure 1
Example of Fireside Tube Wastage*

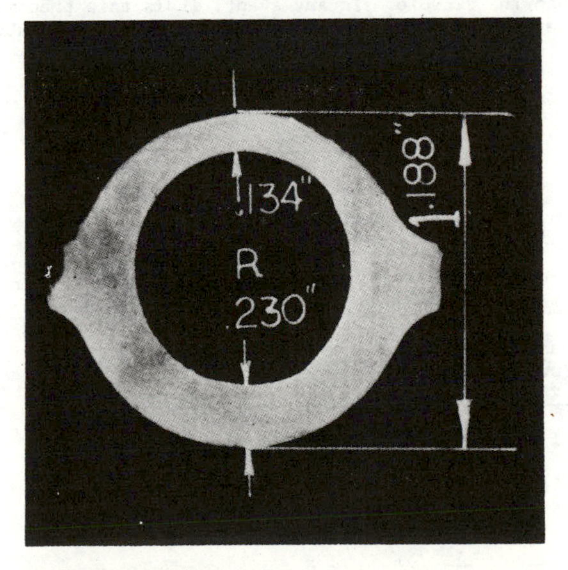

***Reference 4.**

air ports in a pseudo-staging pattern. Most of the NO_x emission reductions achieved by Exxon Research in field test programs have been accomplished in this manner for existing boilers.

As discussed below, furnace tube wall corrosion potentially could occur or be aggravated by operating a utility boiler at "low NO_x" firing conditions. Even though overall excess air is maintained at a reasonable level under "low NO_x" operation, conditions in the first stage combustion zone may approach levels as low as 80% of stoichiometric. Atmospheres at the furnace sidewalls where corrosion could take place under these conditions conceivably could be even lower. The simplified calculation of the estimated percent stoichiometric air supplied to each burner row shown in Table 1 for a front wall fired boiler with 3 elevations of burner rows illustrates how the atmosphere in the furnace changes under "low NO_x" staged firing conditions. In this illustration the top row of burners are used as overfire air ports. The calculations are based on the assumptions that:

- Overall excess air in the boiler = 20% (120% of stoichiometric) under both normal and "low NO_x" firing.
- Equal coal flow in all active burners.
- Equal air flow in each burner.
- Full load can be maintained by firing additional fuel through middle and bottom row burners.

Under the ideal conditions of the example it may be noted that for normal firing the atmosphere in the furnace is oxidizing. However, under "low NO_x" staged firing conditions using the top burner row as overfire air ports, only the top area of the furnace is under oxidizing conditions while the lower region (at the middle and bottom burners) is now in a reducing atmosphere (80% of stoichiometric) condition. Obviously, coal and air distribution between burners is not uniform and conditions in localized "pockets" in the furnace may be more or less severe than the example. In any event, it is this phenomenon that is anticipated to exert a potentially adverse effect on external tube corrosion.

Table 1

Calculation of Estimated Stoichiometric Air to Burner

	Normal Firing: All Burners Firing Coal			Low NO_x Firing: Top Row Burners on Air Only		
Burner Row	Coal Flow, %	Air Flow, %	% Stoich. Air	Coal Flow, %	Air Flow, %	% Stoich. Air
Top	33 1/3	40	120	0	40	∞
Middle	33 1/3	40	120	50	40	80
Bottom	33 1/3	40	120	50	40	80
Total	100	120	120	100	120	120

FURNACE FIRESIDE TUBE CORROSION PROBE MEASUREMENTS

As mentioned previously, corrosion probes were used by Exxon Research in two EPA sponsored programs, Contracts No. 68-02-0227 (2) and 68-02-1415 (3), respectively. Figures 2 and 3 show details of the corrosion probes used in these studies. The design of the corrosion probes was based on information supplied by Combustion Engineering, with appropriate modifications for this work. Essentially, the design consists of a "pipe within a pipe", where the cooling air from the plant air supply is admitted to the ring-shaped coupons exposed to furnace atmospheres at one end of the probe, through a 19 mm (3/4-inch) stainless steel tube roughly centered inside of the coupons. The amount of cooling air is automatically controlled to maintain the desired set-point temperature of the coupons. The cooling air supply tube is axially adjustable with respect to the corrosion coupons, so that the temperatures of coupons can be balanced. To simplify the pictorial presentation, the thermocouples mounted in each coupon are not shown in Figures 2 and 3. Normally, one thermocouple is used for controlling and the other one for recording temperatures. The cooling air returns along the 63.5 mm (2-1/2-inch) extension pipe and discharges outside of the furnace. Thus, the cooling air and the furnace atmosphere do not mix at the coupon location.

The approach used for measuring corrosion rates in the initial program was to expose corrosion coupons installed on the end of probes inserted into available openings located near "vulnerable" areas of the furnace under both baseline and low NO_x firing conditions. Coupons were fabricated of SA 192 carbon steel, the same material as that used for furnace wall tubes. Exposure of the coupons for 300 hours at elevated temperatures of 742 K (875°F) [higher than normal furnace tube wall temperature of about 489 K (600°F)] was chosen in order to deliberately accelerate corrosion so that measurable values could be obtained. Coupons were also mildly pickled to remove the existing oxide coating prior to exposure to eliminate differences potentially caused by surface conditions. The conclusion of these earlier corrosion probing tests was that no major differences in corrosion rates could be found between coupons exposed to low NO_x firing conditions compared to coupons exposed under normal boiler operating conditions. Coupon corrosion rates were, however, considerably higher under baseline and low NO_x conditions than those corresponding to normal furnace tube wastage rate because of the accelerated nature of these corrosion probing tests. Results illustrative of the type of corrosion rate data obtained in the initial program are shown in Table 2.

Significant changes were made in the conditions for measuring corrosion rates in the subsequent program to better relate rates obtained on corrosion probes to actual furnace waterwall corrosion. The approach was similar to the earlier work but with several important differences. First, corrosion coupons, which were all fabricated and machined in the same manner, were no longer mildly acid pickled but instead, were dipped in acetone and air dried prior to weighing to remove any oil deposited during machining. Second, coupon temperatures were controlled at temperatures approximating those of the furnace waterwall tubes, 603-658 K, (625-725°F) to more closely approximate actual furnace conditions. Third, three coupons were installed on each probe, to increase the amount of data obtained compared with only two coupons per probe in the prior program. Time of exposure (300 hours) was held the same so that the results of the corrosion probing runs could be compared to the earlier work. Other test conditions were also kept the same, e.g., probes were inserted through openings in the furnace wall as close as possible to vulnerable furnace areas, and the analytical procedures used were also the same in both programs. Each coupon was visually inspected after exposure and was photographed to record its appearance. Scale was then removed from the outside diameter surfaces by dry honing of the inside diameter surface scale after which the coupon was reweighed to determine the weight loss from the inside surfaces. Corrosion rates were then calculated as the loss in mils per year (m/yr) using the weight loss data, the combined exposure coupon areas, the metal and scale and densities, and the exposure time.

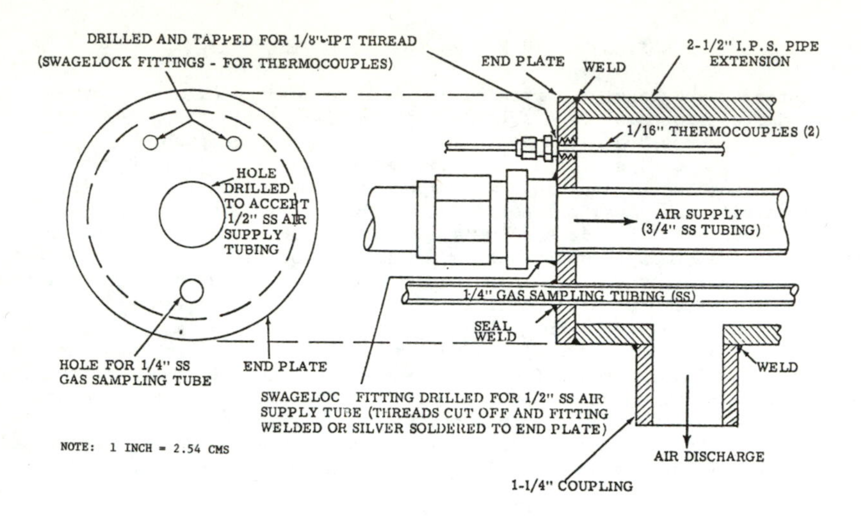

FIGURE 2

CORROSION PROBE
DETAIL OF 2-1/2" IPS EXTENSION PIPE AND END PLATE
(OUTSIDE OF FURNACE)

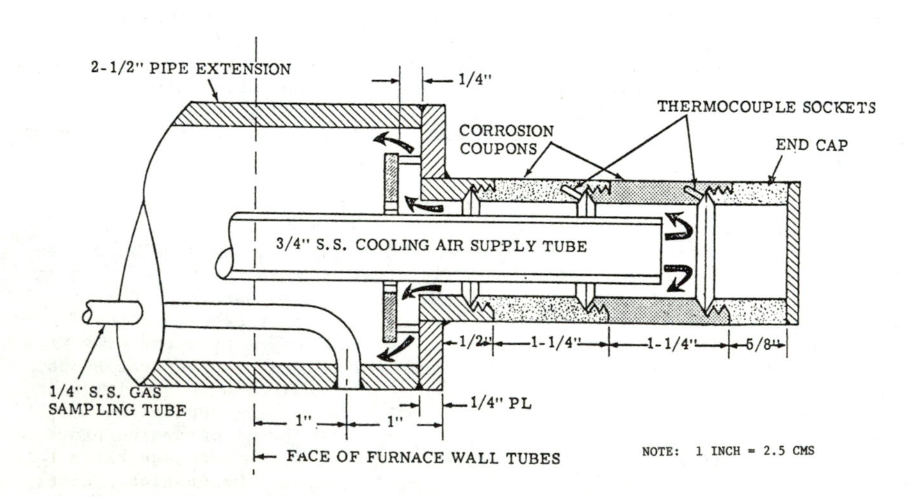

FIGURE 3

CORROSION PROBE
DETAIL OF CORROSION COUPON ASSEMBLY
(INSIDE OF FURNACE)

TABLE 2

ACCELERATED CORROSION RATE DATA

Boiler	Firing Conditions	Corrosion Rate, Mils/Yr. Paired Values*	Average
Georgia Power, Harllee Branch No. 4	Baseline	75 72	74
Georgia Power, Harllee Branch No. 4	Baseline	26 48	37
Georgia Power, Harllee Branch No. 3	Low NO_x	28 122	75
Georgia Power, Harllee Branch No. 3	Low NO_x	76 155	116
Arizona Public Service, Four Corners No. 4	Baseline	157 59	108
Arizona Public Service, Four Corners No. 4	Baseline	45 59	52
Arizona Public Service, Four Corners No. 5	Low NO_x	61 160	111
Arizona Public Service, Four Corners No. 5	Low NO_x	25 24	25

* Paired corrosion rate values obtained on two coupons exposed on the same probe.

Typical data obtained are tabulated in Tables 3 and 4. These rates are considerably lower than those measured under accelerated conditions in the initial program. The lower and more consistent coupon corrosion rates measured in the latter program reflect the changes made in test procedures to more closely approximate actual furnace wall tube conditions. However, the rates are still an order of magnitude greater than the 0.025 to .076 mm (1 to 3 ml/yr.) corrosion rates that are expected for the wastage of actual furnace tubes under normal firing conditions. Therefore, the corrosion probing results are viewed as only a relative measure of corrosion tendency under baseline and low NO_x firing conditions. It was concluded from these studies that only long term corrosion measurements of actual furnace tube wastage could answer the question of the magnitude of corrosion rate increase caused by staged firing of utility boilers with pulverized coal. However, corrosion probe data and techniques may still be valuable in the future after reliable correlations with actual tube wastage are established.

LONG TERM CORROSION TESTS

During the course of the corrosion probe investigations conducted by Exxon Research under EPA sponsorship, it became apparent that data obtained with probes could not provide an unequivocal resolution to the question whether external furnace tube corrosion might increase with "low NO_x" operation on utility boilers fired with pulverized coal. Even though "low NO_x" combustion conditions did not produce any major increases in corrosion rates as measured by corrosion probes, it was not only difficult to relate these data to actual tube wastage rates but one could not rely on such information for boiler design and operation. From the power generation industry standpoint it is imperative that this question be resolved and, if corrosion is indeed a problem, engineering solutions are required for the application of "low NO_x" operation to coal fired boilers. A comprehensive corrosion investigation program, therefore, was undertaken in an attempt to settle this issue conclusively. Long term tests sponsored by the EPA have been undertaken to a joint cooperative venture by Exxon Research and Engineering Company,

TABLE 3

CORROSION RATE DATA

WIDOWS CREEK STATION

BOILERS NO. 5 AND NO. 6

TENNESSEE VALLEY AUTHORITY

Boiler	Probe No.*	Firing Condition	Corrosion Rate, Mils/Yr. Coupon	Average
Boiler No. 6 (North Side)	4	Base	10.2** 12.0 14.6	12.3
Boiler No. 6 (South Side)	3	Base	9.7** 11.6 11.1	10.8
Boiler No. 5 (North Side)	2	Low NO_x	12.5** 13.4 18.8	14.9
Boiler No. 5 (South Side)	1	Low NO_x	11.7** 13.4 15.1	13.4

* Probes located between A & B burners, top & next to top rows, respectively.

** Coupons farthest into furnace.

Table 4

Corrosion Rate Data

Navajo Station

Salt River Project, Page, Arizona

Boiler*	Probe No.	Firing Condition	Corrosion Rate, Mils/Yr. Coupon	Average
No. 1 (North Side)	1***	Base	62.8 32.0 21.3**	38.7
No. 1 (South Side)	2	Base	15.1 10.6 13.2**	13.0
No. 2 (North Side)	3***	Low NO_x	43.8 29.2 19.7**	30.9
No. 2 (South Side)	4	Low NO_x	16.2 15.1 15.5**	15.6

* Probes inserted through observation doors in side walls between B & C burner rows.

** Coupons farthest into furnace.

*** Probes with damaged coupons.

Foster Wheeler Energy Corporation, and the Gulf Power Company. These tests have been underway since June of 1976 on the No. 7 boiler at the Crist Station of Gulf Power Company in Pensacola, Florida. Three types of corrosion rate determinations are being employed:

1. Measurements on specially installed furnace tube panel test specimens.

2. Ultrasonic mapping of the thickness of the furnace tubes and test panels.

3. Corrosion probes exposed for varying times.

Details of these methods will be discussed separately.

Test Panel Design, Installation and Measurements

Boiler manufacturers' experience indicates that external corrosion of furnace tubes occurs in areas in a largely unpredictable and random pattern, although the general problem areas are at the burner elevations of the furnace sidewalls. This presents a major problem in determining where to place the furnace corrosion panels to ensure that they are in areas where corrosion occurs. There is no ideal solution to this problem and from an economic standpoint the number of panels used must of necessity be limited. The best that can be accomplished is to use as many panels as practical and place them in areas where corrosion is most likely to occur with one or two panels located in "control" area.

Eight corrosion test panels were installed in the No. 7 boiler at the Crist power station during May, 1976 at strategic locations believed to be most helpful in defining potential corrosion effects. Windows were cut out of the furnace walls the size of the corrosion panels and the old panels were retained for future laboratory inspection. Panel and corrosion probe locations are shown in Figure 4. Referring to Figure 4 it can be seen that seven of the panels were installed in the left furnace side wall and one in the right wall. The reason for installing most of the panels in the one wall is to provide maximum areas of exposure to corrosion. Since the corrosion prone areas of the sidewalls are normally at the midpoint at the burner elevations, four of the panels (Nos. 3, 4, 5 and 6) were installed in this area. Three of these panels (Nos. 3, 4 and 5) are at the middle burner row elevation and panel No. 6 is at the top burner elevation. Note that panels No. 4 and 6 are located in the middle of the sidewall where the most severe corrosion can be anticipated. Panels No. 1 and 2 were located in the hopper area where corrosion on the boilers occasionally has been experienced. Panels 7 and 8 in the left and right sidewalls, respectively, installed in the upper reaches of the furnace (above the burners), are expected to experience lower corrosion rates (oxidizing atmosphere) and will serve as "control" panels for comparison purposes. This scheme of panel arrangement was conceived to provide the maximum amount of data within the constraints of reasonable cost.

Each test panel is five (5) tubes wide by 1.5 meters (5 ft.) in length. Tubes 1, 3 and 5 were made of the same low carbon steel material as the furnace tubes; SA-210 grade A-1. Tubes 2 and 4 are SA-213 grade T-2, a higher grade alloy carbon steel expected to have greater resistance to corrosion than normal furnace tube material. The use of two materials will provide additional useful information on rates of corrosion which would not be available if only one material was used.

Prior to installation, the panels were completely characterized in the laboratory after fabrication. Thickness measurements were made ultrasonically at 7.6 cm (3 in.) intervals on the side of the panel exposed to the furnace and at 15.2 cm (6 in.) intervals on the opposite side, for control purposes. In addition, points near the end of the tubes were measured independently by an accurate micrometer on both sides of the panel so that an independent measure of precision could be developed from paired mechanical vs. electronic measurements.

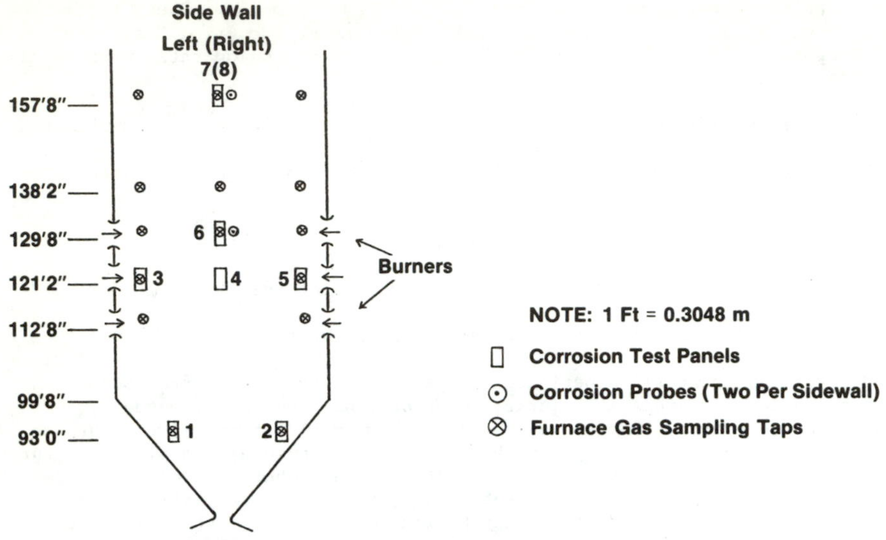

Figure 4 Gulf Power - Crist Station, No. 7 Corrosion Panel Location

Examination of the test panels while still in place in the furnace is almost identical to that for wall tubes. Samples of corrosion products will be removed periodically from the tube surfaces by chipping, and the extent of metal loss is being determined by ultrasonic measurement. The major advantage of test panels is that they can be removed from the furnace and sectioned to give precise indications of metal loss and of the composition of corrodents.

Metallographic examination of the panel tube metal will provide an important tool. Such methods, by examining the grain structure after exposure, give a good indication of the mode by which corrosion is occurring and can detect incipient metal attack along grain boundaries not detectable otherwise. Sulfidation, in particular, can be shown by metallographic examination. And if the metal should have been overheated at any stage, the extent of spheroidization will provide a rough measure of this change in metal structure.

Two steps have been taken to obtain information characterizing corrosion panel exposure conditions. These include (1) measurement of corrosion panel tube metal temperatures and (2) analysis of the gaseous atmospheres in the location of the panels. Three thermocouples were installed in each panel (in the 1st, 3rd and 5th tubes). The thermocouples are designed to measure the temperature of the tube as closely as possible to the surface exposed to the flame in the furnace. The objective is to determine whether any unusual temperature conditions may exist which would contribute to corrosion. Thirty gas taps were installed throughout the furnace area and in the front, rear and side walls including taps in the corrosion panels. Periodic surveys and analyses of the gases taken from these taps at different loads and combustion conditions will provide information to characterize the atmosphere in the furnace, especially under "low NO_x" operation.

Ultrasonic Tube Thickness Measurements

In order to determine the extent of potential furnace tube corrosion, the tubes must be measured before and after prescribed periods of time. Mechanical measurement of furance tube thickness is impractical for many reasons such as huge furnace sizes, the multiplicity of tubes and, mostly, the time consuming procedure in making the measurements. The most direct way to measure boiler tubes is through the use of an ultrasonic thickness measuring device. Sensors placed on the tubes at predetermined points (after the necessary cleaning operation) are easy to apply and a large number of thickness determinations can be made in a short period of time. The judicious procedures necessary to obtain definitive, reliable measurements, on the other hand, require a strictly executed program. The key to a reliable ultrasonic tube measurement program is the development of procedures which provide statistically reliable results. Such a program has been developed by Exxon Research and Engineering Company in consultation with Prof. John Tukey of Princeton University.

In the recent past, highly accurate ultrasonic thickness measuring equipment was not available. Accuracy within 0.13 mm (5 mils) only was possible (5). Today instruments are available capable of accuracy to one-tenth of a mil. In the program being conducted on the No. 7 boiler at the Crist Station, two Krautkramer-Bronson CL 204 ultrasonic instrument gages, the most accurate currently available, are being used in making the measurements. Accuracy of these instruments is 0.0001 of an inch, thus assuring that measurements to the nearest tenth of a mil are possible. This degree of accuracy is an essential feature for determining corrosion rates in a reliable fashion. Procedures and methods have been upgraded to achieve the highest degree of accuracy possible as experience with the instrument has been gained. Steps were taken to assure that the operators were thoroughly familiar with proper calibration procedures and a number of measurement demonstrations were conducted both in the laboratory and at the plant site to make sure that the required accuracy was achievable. The condition of the tubes is also a complicating factor requiring sand blasting or some other suitable form of slag, deposit, or scale removal before reliable measurements can be made. Then the question remains after the tubes have been cleaned bare, are they now more susceptible to corrosion and therefore no longer representative since oxide scales, coatings or protection have been removed? It is also difficult to relocate a spot previously measured for rechecking, and measurements at other places on the tube may be unreliable for comparison due to variations in tube manufacturing tolerances.

Unfortunately, sandblasting, wire brushing or other means of cleaning to remove ash or slag coatings from the tubes down to the base metal, necessary for accurate thickness measurement, undoubtedly will remove any protective coating making the tube more vulnerable to corrosion at this spot. This factor and the other considerations influencing tube measurements discussed above have been studied in detail in the design of a statistical measurement program providing reliable assurance that the tube cleaning and measurement process does not produce biased corrosion estimates or conclusions. In simple terms, patterns and frequencies of thickness measurement have been established to statistically eliminate bias in the results due to the greater corrosion vulnerability of tube areas previously cleaned and measured. This is accomplished by taking a large number of measurements at a given elevation at the beginning of the test in a random pattern of measurements on some tubes exactly at the elevation and on others slightly above and below the exact elevation. At the conclusion of the test many measurements are repeated and center measurements are made exactly at the elevation on those tubes previously measured above and below the elevation.

Ultrasonic tube thickness measurements were made at six different furnace elevations in the No. 7 boiler at the Crist station in May 1976 after the corrosion test panels were installed. A second set of measurements were taken at the same elevations in October of 1976 after the boiler had been operating for 5 months under baseline operating conditions. These data will provide corrosion rate information for future comparison with similar data presently being obtained under "low NO_x" firing conditions. Figure 4, showing corrosion panel locations, also shows the six furnace elevations where ultrasonic tube thickness measurements were made. The only elevation on Figure 4 at which measurements were not made is elevation 30.4 m (99'-8"), the junction of the furnace with the hopper. Elevations 28.4 m (93'), 34.4 m (112'-8"), 36.9 m (121'-2"), and 39.5 m (129'-8") include all elevations in the furnace area where corrosion may be anticipated. Measurements at elevations 42.1 m (138'-2") and 48.1 m (157'-8") are in the oxidizing zone where corrosion is least likely to occur. These measurements will provide "control" information.

Corrosion Probe Measurements

Corrosion probes (as described in Section 3) are again being employed on the No. 7 boiler at Crist Station to obtain corrosion rate information concurrently with corrosion rate data taken by corrosion panels and ultrasonic tube measurements. The objective of these investigations is to make a correlation of actual tube wastage experience with corrosion probe data so that confidence in the reliability of the less expensive corrosion probe methods may be realized. Several changes, however, have been incorporated in the corrosion probe procedures to improve the program and to provide a greater amount of useful information. First, probes are being exposed at 30, 300 and 1000 hrs. under baseline and "low NO_x" firing conditions. The 30 hour exposure will provide data on initial corrosion and the 300 and 1000 hour information will show the effect of corrosion with time. Also, correlation with previous investigations on other boilers at 300 hour exposure will be possible. Second, special openings were incorporated in the furnace corrosion panels to accommodate the corrosion probes so that for the first time corrosion probes could be located in areas of greatest anticipated corrosion and in "control" areas. In prior studies installations of necessity were made in available openings which were not always ideal from a corrosion standpoint.

Figure 4 shows the location of the corrosion probes at the Crist station. Two probes were located in the right sidewall and two in left sidewall at the locations of corrosion panels 6, 7 and 8. A special opening (inspection door) was installed in the right sidewall opposite panel No. 6 to accommodate the corrosion probe since no corrosion panel was placed at this point. Thus, two probes are located in the corrosion prone areas at the burner levels and two in the upper reaches of the furnace for "control" purposes where corrosion is expected to be considerably reduced due to the oxidizing atmosphere prevailing in this vicinity. All other corrosion probe procedures, as described in Section 3, remain the same.

STATUS OF LONG TERM CORROSION TEST AT CRIST STATION

As indicated above, eight corrosion panels were installed in the No. 7 pulverized coal fired boiler at the Gulf Power Company's Crist power generating station in Pensacola, Florida during May 1976. New openings were also provided in the furnace sidewalls during this outage for the more advantageous location of corrosion probes in areas vulnerable to corrosion, to better define the corrosion problem using this method of approach. At this beginning of the long term corrosion test (May 1976), over 1000 ultrasonic tube thickness measurements were made on the No. 7 unit (including measurements of corrosion panel tubes) at the six elevations mentioned earlier. The unit was then run over the summer during the peak load demand period under baseline operating conditions for 5 months until October 1976 when the boiler was taken out of service for a scheduled maintenance outage. Ultrasonic tube measurements were again made during the latter outage in accordance with the statistical plan developed for these tests. Comparison of the "before" and "after" measurements will afford an assessment of actual corrosion experienced on the furnace wall tubes under baseline operating conditions. Conclusions from these data have not yet been reached pending complete, detailed statistical analysis.

The No. 7 Crist boiler is now being operated under "low NO_x" emission firing conditions and will continue this type of operation until the next scheduled maintenance outage in September/October 1977. At that time the corrosion panels will be removed from the boiler (after about 1 year of low NO_x operation) and returned to the laboratory for re-measurement of the tubes and metallographic examination of the specimens. Ultrasonic tube thickness measurements will also be made at the previously prescribed elevations. These data, when compared to the measurements made in October 1976, should provide definitive information on external furnace tube corrosion experienced under "low NO_x" operation. A comparison can then be made to actual wastage experienced under baseline conditions which should prove whether or not "low NO_x" firing results in increased corrosion of the furnace tubes in this pulverized coal fired boiler.

Corrosion rate data have been taken under the baseline operating period and corrosion probe data are currently being obtained under "low NO_x" firing conditions at exposures of 30, 300 and 1000 hours for correlation with actual tube wastage rates determined on corrosion panels and by actual furnace tube ultrasonic measurements.

ACKNOWLEDGMENT

The research described in this paper was performed under the sponsorship of the U.S. Environmental Protection Agency, pursuant to Contract No. 68-02-1415. The Valuable comments of statistical design by Professor John Tukey are gratefully acknowledged. Many thanks are due to the Gulf Power Company for providing Crist Boiler No. 7 as the host site for this study.

REFERENCES

1. W. Bartok, A. R. Crawford and G. J. Piegari, "Systematic Field Study of NO_x Emission Control Methods for Utility Boilers", Esso Research and Engineering Company Report No. GRU.4GNOS.71, prepared under Contract No. CPA 70-90, EPA Report APTD1163, NTIS No. PB210739, December 31, 1971.

2. A. R. Crawford, E. H. Manny and W. Bartok, "Field Testing: Application of Combustion Modifications to Control NO_x Emissions from Utility Boilers", EPA-650/2-74-066, NTIS No. PB237344/AS, June 1974.

3. A. R. Crawford, E. H. Manny, M. W. Gregory and W. Bartok, "The Effect of Combustion Modification on Pollutants and Equipment Performance on Power Generation Equipment", Proceedings of the Stationary Source Combustion Symposium, Volume III, p. IV-3, EPA-600/2-76-152c, NTIS No. PB 257146/AS, June 1976.

4. J. Vatsky, R. P. Welden, "NO_x A Progress Report", Heat Engineering, July-September 1976, p. 125-129.

5. G. A. Hollinden, J. R. Crooks, N. D. Moore, R. L. Zielke and C. Gottschalk, "Control of NO_x Formation in Wall Coal-Fired Boilers", Proceedings of the Stationary Source Combustion Symposium, Volume II, p. III-31, EPA-600/2-76-152b, NTIS No. PB 256321/AS, June 1976.

REDUCTION OF COAL ASH DEPOSITS WITH MAGNESIA TREATMENT

J. E. RADWAY

Basic Chemicals
Cleveland, Ohio, USA 44114

T. R. BOYCE

The Rolfite Company
Stamford, Connecticut, USA 06901

ABSTRACT

Magnesium oxide was successfully used during the early 1960's to minimize fouling of heat-absorbing surfaces on coal-fired boilers, but the dosages required, 0.4% to 3% of the fuel burned, were uneconomic. Recent trials employing fine particle size MgO dispersions on operating boilers firing widely different kinds of coal have yielded reductions in deposits and/or slag at very low treatment rates, 0.015% or less. This study was designed to quantify the results and possibly identify the mechanism by which these low levels of MgO achieved such favorable results.

High and low temperature probes were used to obtain samples for examination by various chemical and physical techniques. The results suggest that the separately injected magnesia acts as a reactive "scavenger" for the flue gas SO_3 that is generally considered an important factor in coal ash deposits and corrosion.

INTRODUCTION

In past efforts to minimize the fouling of heat-absorbing surfaces in coal-fired boilers, one technique which proved effective was the addition of magnesium oxide powder, either with the fuel or by injection into the furnace. While effective, the quantity of oxide required, ranging from 0.4% to 3% of the fuel burned[1][2] made this approach uneconomical.

The success, in recent years, with magnesium-based additive dispersions on oil-fired boilers and the need to increase plant reliability and availability has created new interest in chemical techniques for controlling fouling on coal burning equipment. Recent trials employing fine particle size magnesium oxide dispersions on operating boilers firing widely different kinds of coal have yielded reductions in deposits and/or slag at very low treatment

---

Mr. Radway is Vice President, Technology, Basic Chemicals, division of Basic Incorporated, 2532 St. Clair Avenue, Cleveland, Ohio 44114.
Dr. Boyce is Combustion Consultant to The Rolfite Company, 300 Broad Street, Stamford, Connecticut 06901.

rates, 150 ppm (0.015%) or less.

One of the trials was on a pair of small PC boilers burning a low sulfur (0.5%) Saskatchewan lignite. Fouling of the superheater had required an outage every seven to ten days for mechanical cleaning. About five years ago, an MgO dispersion was tried and has been used ever since. Mechanical cleaning is required only every two to three months. The oxide addition rate is 50 ppm based on the coal fired.

Another recent application involves two 125 MW Cyclone boilers burning West Virginia coal(3). Magnesium oxide treatment was originally begun in an attempt to control air heater corrosion and plugging by injecting the additive into the furnace on one of the units. After three months of treatment, the convection surfaces were found to be much cleaner than they had been before. Similar results were achieved on the sister unit. In this case, treatment rate was about 130 ppm MgO.

Clearly sufficient indication has been shown by the limited field tests of separate magnesia injection to warrant more serious investigation of the potential for the amelioration of serious slagging problems in pulverized coal-fired boilers. The fact that beneficial results have been observed with a variety of coal types tends to suggest that more than one chemical/physical mechanism may be involved. Thus the authors have embarked upon a continuing series of rigorous field studies in an effort to clarify the picture. The present paper describes the results and theoretical implications of the first experiment in the series.

PROCEDURE

In an attempt to quantify the results and possibly identify the mechanism by which these low levels of MgO achieved such results, a series of tests were run on a pulverized coal-fired industrial boiler. The unit was a 1935 B&W slag tap boiler rated at 113 t/hour of steam at 48.2 bars and 370 deg C. The unit fired a 3.5% sulfur Ohio coal with a moderate to high fouling index. The superheater was located between the screen tubes and the inclined boiler tubes. The screen tubes and superheater fouled badly, requiring frequent soot blowing and hand water lancing.

The boiler was rear wall fired and the additive was injected into the furnace at two points on the front just above the burner level (Figures 1 and 2). The additive, in the form of an aqueous dispersion of magnesium hydroxide, was fed at a controlled rate through simple proprietary pneumatic atomizers. At the temperatures in the furnace, the water was immediately evaporated and the magnesium hydroxide was calcined to magnesium oxide.

The effectiveness of the magnesium oxide in modifying the deposits was evaluated by inserting a temperature-controlled deposition probe, similar to that of Jackson[4] into the furnace just above the screen tubes (Figures 1 and 2). Treatment rates ranged from 150 to 1,250 ppm MgO and probe temperatures were held in the 500 to 610 deg C range to better simulate conditions found in the modern utility boiler.

In addition to observation of the gross physical appearance and characteristics, the probe deposits collected, with and without magnesium treatment, were scrutinized by various analytical techniques in an effort to ascertain the mechanism by which such insignificant quantities of magnesia achieved such dramatic reductions in boiler deposits. Chemical analyses were supplemented by x-ray diffraction, differential thermal and optical microscopy procedures. Included in the latter was the optical refraction technique devised by Battelle Labs specifically for identifying alkali iron sulfates.

Since many references in the literature relate deposits problems in both the boiler convection pass and the air heater to the presence of SO_3 in the flue gases[2, 5 to 12], efforts were made to gauge changes in its concentration. Acid deposition rate measurements were made at the entrance to the air heater using an air-cooled probe similar to that of Alexander et al[13]. An effort was made to cross check the ADR data with a Land Dew Point Meter, but the test port design and location precluded meaningful measurements.

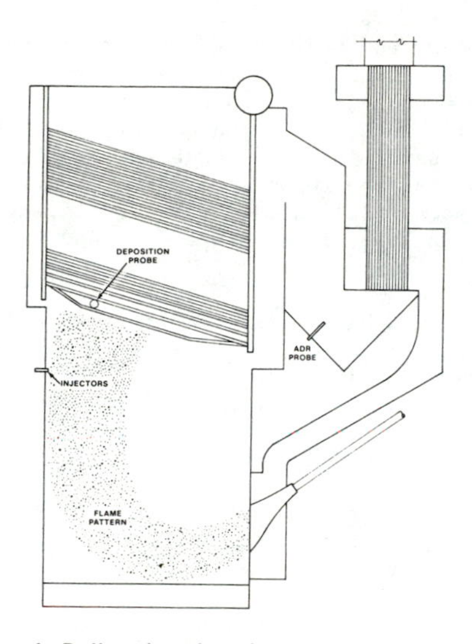

Figure 1. Boiler elevation showing location of sampling points, deposition probe and additive injectors.

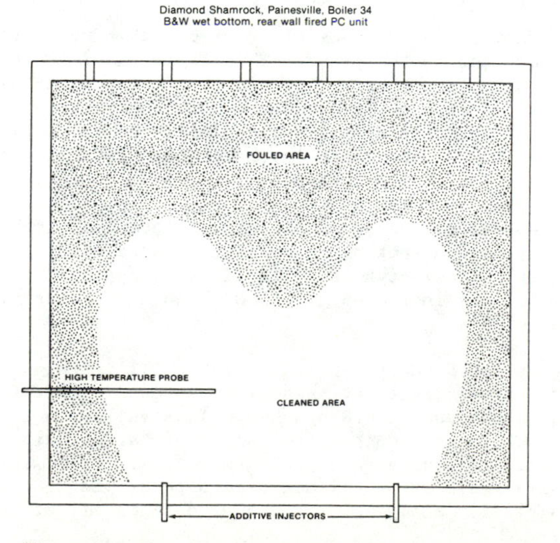

Figure 2. Boiler plan showing influence of MgO on slag screen deposits.

A minimum high temperature probe sampling period of three hours for significant deposit accumulation dictated the daily testing routine. Total test turn-around time required approximately four hours and this factor plus routine boiler maintenance operations mandated a maximum of two test runs per day. The boiler was slag-tapped, soot-blown, and hand-lanced at the start of each test day. These operations, coupled with boiler stabilization on test load conditions for at least one hour before the start of each test, required six to eight hours.

Additive injection when required was started at least one half hour before the start of each test run. The parameter measurement sequence began with the high temperature probe exposure in the boiler followed by acid deposition rate measurements (two runs, start and end of each test), oxygen level measurements, fly ash grab sample and coal grab samples from two of the three ball mills supplying the burners. Boiler operating data was recorded at regular intervals during each test run to insure consistent operation. Test conditions are summarized in Table I (see Appendix) and fuel characteristics in Table II (see Appendix).

High temperature probe graphics, sample collection and collation, and probe clean up occupied the remaining hour of each test period. Additive dosage rate changes and slag lancing when required, were conducted during this period to facilitate short test period turn around.

RESULTS

In all cases, the deposits formed on the probe were friable, more powdery and more easily removed when dispersed MgO was being used. Much more dramatic reduction in screen tube deposits were seen in the furnace itself.

Boiler design limited the injection sites causing the distribution of the additive within the furnace to be somewhat less than ideal. The screen tubes and wall tubes just above the injectors showed 85% to 90% less deposit accumulations than on those tubes not exposed to the additive (Figure 2). This difference was still evident 24 hours after additive injection had been discontinued and obviously affected gas flow distribution.

Probe deposits from the blank runs all showed a thin white to gray adherent layer on the flame side of the probe. This layer was in turn coated with a porous, sintered ash layer. When MgO was injected, the white bonding layer was also observed but the area covered receded progressively with increasing dosage to the higher temperature areas of the probe. At the maximum feed level of about 1,200 ppm MgO, the white layer was observed only on the highest temperature section of the probe.

The weight of deposit collected on the high temperature probe varied rather widely from run to run, but the amounts collected with magnesia injection tended to be greater, as shown in Table III (see Appendix). This could have been the result of the increased gas flow through the cleaner gas passes in the treated area.

Measurements of probe deposit densities (Table III in the Appendix) seem to be fairly similar with and without magnesia treatment except for the suggestion of a peak at about 300 to 600 ppm MgO. This variation is believed to be a sample effect. The measurements were made on total deposit and modifications caused by the additive were especially selective. At lower dosages the less sintered deposits away from the flame tended to be modified preferentially, leaving a denser sample for analysis. At higher dosages all of the deposits tended to be powdery.

Chemical composition of the deposits with and without magnesia injection (see Table IV in Appendix) were quite similar over the metal temperature range

TABLE I

BOILER TEST CONDITIONS - #34 BOILER, PAINESVILLE PLANT, DIAMOND SHAMROCK COMPANY, B&W WET BOTTOM REAR WALL FIRED P.F. UNIT

Date - 1976	8/24	8/26	8/26	8/27	8/27	8/29	8/29	8/30	8/30	8/31	8/31	9/1
Run Number	1	1	2	1	2	1	2	1	2	1	2	1
MgO Rate (ppm)	0	155	310	310	620	620	0	1240	1240	0	0	0
Boiler Load (t/hr)	113.4	108.9	108.9	108.9	106.6	108.9	106.6	113.4	108.9	108.9	108.9	106.6
Drum Press (bar)	46.2	44.8	44.1	44.8	44.8	46.2	44.5	39.3	40.6	41.4	43.4	42.7
Drum Temp (°C)	259	259	257	257	257	259	258	251	253	254	254	256
SH Press (bar)	44.8	43.1	42.7	43.4	42.7	45.5	42.7	37.2	39.3	39.6	42.1	41.4
SH Temp (°C)	349	343	343	349	349	349	352	349	343	349	340	343
Excess O_2 (%V)	5.5	5.2	5.4	4.8	5.2	4.6	4.4	5.6	4.4	5.7	4.7	4.4
Stack Temp (°C)	174	169	168	168	169	162	160	167	163	169	164	162
Δ P, SH (mm)	5.1	7.6	12.4	5.1	15.2	10.1	5.1	10.2	10.2	5.1	5.2	7.6
Δ P, Economizer (mm)	33	33	33	31	36	31	25	38	33	38	35	31
Δ P, A/H (mm)	81.3	76	74	74	76	69	66	81	71	86	71	64

TABLE II

ANALYTICAL DATA ON FUEL SAMPLES

Test Date	8/29	9/1	8/26	8/29	8/30
MgO Rate	0	0	310 ppm	620 ppm	1240 ppm
Composition	Coal Ash	Coal Ash	Coal Ash	Coal Ash	Coal Ash
Proximate Analysis					
Percent Volatiles	35.2	--	35.6	35.5	34.5
Percent Fixed Carbon	53.8	--	53.4	52.7	54.0
Percent Ash	11.1	--	11.2	11.9	11.5
Percent Sulfur	3.3	--	3.5	3.4	3.2
Ash Fusibility					
Initial Deformation (°C)	1288	1295	1291	1293	1285
1st Softening (°C)	1316	1313	1321	1321	1304
2nd Softening (°C)	1327	1324	1346	1341	1316
Fluidity (°C)	1352	1344	1388	1391	1338
Coal Indeces					
Base/Acid Ratio	0.49	--	0.48	0.43	0.44
Na_2O/K_2O Ratio	0.38	0.24	0.26	0.25	0.36
Na_2O Fouling Index	0.26	--	0.19	0.16	0.22
B&W Slagging Factor	1.6	--	1.7	1.5	1.4

TABLE III

PROBE DEPOSIT WEIGHTS AND SPECIFIC GRAVITIES

Test Date	Number	Treatment	Deposit Weight by Sections, grams 1	2	3	4	Total grams
8/24/76	1	none	4.64	9.43	3.76	3.58	21.40
8/29/76	2	none	2.38	3.34	1.37	3.14	10.23
8/31/76	1	none	1.90	2.42	1.45	1.77	7.54
8/31/76	2	none	2.70	3.71	3.01	3.01	12.43
9/1/76	1	none	2.37	6.29	2.50	2.76	13.92
8/26/76	1	150 ppm MgO	4.52	10.79	4.79	3.06	23.16
8/26/76	2	300 ppm MgO	4.42	9.83	4.78	3.09	22.12
8/27/76	1	300 ppm MgO	9.65	14.61	4.10	2.84	31.20
8/27/76	2	600 ppm MgO	4.41	3.60	2.52	2.56	13.09
8/29/76	1	600 ppm MgO	3.26	4.45	2.51	3.14	13.36
8/30/76	1	1200 ppm MgO	8.97	8.11	7.25	7.25	31.58
8/30/76	2	1200 ppm MgO	4.97	10.06	5.80	4.47	23.50
Temperature Range °C			590-623	572-619	528-582	499-549	

Test Date	Number	Treatment	Specific Gravities by Probe Section 1	2	3	4
8/24/76	1	none	3.14	3.31	--	--
8/29/76	2	none	2.97	3.05	--	--
9/1/76	1	none	--	3.09	--	--
8/26/76	1	150 ppm MgO	3.01	3.19	--	--
8/27/76	1	300 ppm MgO	3.25	3.68	--	--
8/27/76	2	600 ppm MgO	3.10	3.43	--	--
8/29/76	1	600 ppm MgO	3.01	3.03	--	--
8/30/76	1	1200 ppm MgO	3.01	3.16	--	--
8/30/76	2	1200 ppm MgO	2.96	3.10	--	--

TABLE IV

ANALYTICAL DATA ON FUEL ASH, PROBE DEPOSITS AND FLY ASH

Test Date	8/29 and 9/1			8/26			8/29			8/30		
MgO Rate	0			310 ppm			620 ppm			1240 ppm		
Composition	Coal Ash	H.T. Probe*	Fly Ash	Coal Ash	H.T. Probe*	Fly Ash	Coal Ash	H.T. Probe*	Fly Ash	Coal Ash	H.T. Probe*	Fly Ash
SiO_2 (%)	41.3 41.6	38.3/42.5 30.3/39.6	34.8 35.4	42.5	38.2/39.5	36.4	43.2	42.4/42.4	34.7	42.8	30.3/27.5	35.9
Fe_2O_3	21.8 22.1	27.3/24.9 42.9/27.8	19.0 18.7	23.7	33.8/22.7	24.6	21.5	27.3/27.3	19.0	19.4	32.9/45.7	17.2
Al_2O_3	19.3 20.9	17.0/18.8 14.8/18.5	16.5 17.8	24.8	20.3/20.2	18.7	21.5	20.7/20.3	17.5	20.5	14.1/12.9	16.5
TiO_2	0.62 0.77	1.5/ --	--	0.59	1.5/ --	--	0.65	1.5/1.5	--	0.62	--	--
P_2O_5	0.32 0.67	0.23/ -- -----	-- --	0.53	0.34/ --	--	0.39	0.23/0.23	--	0.29	---	--
Na_2O	0.53 0.44	0.30/0.92 0.50/0.74	0.65 0.54	0.39	0.30/0.84	0.50	0.37	0.41/0.41	0.46	0.50	0.58/0.53	0.74
K_2O	1.4 1.8	1.8/1.7 1.5/2.1	1.1 1.4	1.5	1.6/1.72	1.5	1.5	1.9/1.9	1.4	1.4	1.4/1.3	1.1
CaO	5.5 4.7	5.6/7.5 6.4/9.0	8.2 5.0	4.3	5.3/7.1	4.3	3.9	4.9/4.6	5.2	6.1	11.4/11.8	7.3
MgO	0.55 0.88	0.40/0.76 0.50/0.63	0.58 0.63	0.84	0.40/1.64	1.74	0.70	0.43/0.83	1.13	0.79	0.63/1.08	1.96
SO_3	6.8 5.4	1.8/ -- -----	1.3 --	5.3	1.5/ --	1.15	4.1	1.3/1.5	1.13	34.5	---	--

*Section I/Section IV

investigated, 500 deg C to 620 deg C. A preferential deposition of iron similar to that observed by Plumley et al.[5] was suggested by data on the probe samples.

X-ray diffraction scans with and without MgO were not significantly different. It was not possible to identify changes, if any, in the compounds present because of the high proportion of amorphous materials in the deposits. Optical diffraction studies failed to find alkali iron sulfates in either treated or untreated deposits. Microscopically, the deposit mineralogies were also similar.

Differential thermal analyses revealed some minor but apparently real differences in thermal characteristics of deposits with and without dispersed magnesia treatment. They were detectable only on high sensitivity equipment (10 µv/in.). Deposits analyzed on less sensitive equipment (100 µv/in.) yielded curves which appeared quite similar no matter what the treatment.

The only consistent differences in DTA measurements between deposits formed with and without magnesia treatment occurred on the cooling curve in the 1340-1380 deg C range. An exotherm was observed in all three treated samples and the spike height seemed to vary directly with magnesia dosage, i.e. 300, 600 and 1,200 ppm. This fact tends to suggest that the differences are not the result of an analytical fluke. Attempts at relating the observed "spikes" to possible changes in chemical composition have to date been unsuccessful.

The acid deposition rate measurements were found to be affected by the presence on the test probe of significant amounts of particulate fly ash. The alkaline material in this ash apparently becomes available for reaction with acid condensed on the probe once the deposit is wetted and diluted to volume for titration. This was evidenced (Table V) by the significantly different acid deposition rates found when aliquots of the samples were titrated immediately and after 18 hours at room temperature.

TABLE V

INFLUENCE OF FLY ASH LEACHING TIME ON APPARENT "ACID DEPOSITION RATE" MEASUREMENTS AT 84°C (190°F)

	CONTACT TIME	
	10 minutes	18 hours
	265 $\mu g/cm^2$-hr	190 $\mu g/cm^2$-hr
	330 "	90 "
	240 "	140 "
Average -	278 $\mu g/cm^2$-hr	140 $\mu g/cm^2$-hr

The naturally occurring base material is apparently not completely or readily available for reaction with the gas-phase SO_3 and appears to be tied up in the fly ash. Protracted immersion of the fly ash leaches out the "latent" alkali and partially neutralizes the acid solution.

In view of the wet phase reaction which could result in erroneously low estimates of the amount of free acid in the flue gas, immediate titration of the ADR sample became standard operating procedure. Thus, any biasing effect of fly ash alkali could be held constant, allowing assessment of the impact of varying the magnesia feed. In Figure 3, a least squares fit to the data shows a direct reduction in acid deposition rate with higher magnesia feed rates.

2. Reducing SO_3 formation by masking the catalytic iron oxide surfaces in the boiler;

3. Reaction with SO_3 either in the gas stream ahead of the tubes or on their surface;

4. Formation of more stable complexes with alkali sulfates, eg $Na_2Mg_4(SO_4)_4$[6].

The lack of a significant increase in MgO and SO_3 content of the high temperature probe deposits with additive injection would tend to refute both the protective layer theory and the proposed reaction with SO_3 on the tube surface. While the MgO content of the coal ash was low, circa 0.7%, the additive feed rates (from 17% to about 125% of the magnesia content of the ash) were sufficient to insure that any changes in deposit concentration would not be obscured by analytical difficulties.

Levy and Merryman[8] have suggested that the protective effect of MgO is " . . . through its ability to remove SO_3 from the gas stream and to a lesser extent through its ability to mask the iron surface physically". On a coal-fired boiler, one would expect that the gas phase reaction mechanism would be primary since Barrett has shown that fly ash which is present in large quantities also effectively masks the iron surfaces[14]. In the current trials fly ash levels were about 50 times the maximum magnesia injection rate and the boiler metal temperatures were generally well below the minimum 480 deg C that Barrett cites as necessary for oxidation catalysis by Fe_2O_3.

The current tests would lend some support to Levy and Merryman's theory of gas stream reaction between MgO and SO_3. There was a direct relationship between the amount of dispersed magnesia injected in the furnace and the reduction in condensable sulfuric acid at the air heater inlet. However, there was no obvious increase in sulfur level in the fly ash samples with chemical treatment. One would expect a difference from the baseline since the $MgSO_4$ reaction product would be a collectible solid at the high stack temperatures, circa 160 deg C, while unreacted SO_3 should pass through the boiler as a gas. Difficulties in obtaining a representative ash sample from the short U-turn duct between the economizer and air heater could account for the lack of such supporting evidence.

Of significance is the fact that injection of finely sized dispersed magnesia separately from the coal yields a "base" which seems to be more available chemically than the far larger quantities (ten to 30 times as much) of alkaline materials in the coal ash. This is not surprising in view of Raask's report[12] that " . . . all particles emerging from a typical boiler furnace flame are likely to be fused except for quartz particles with diameter greater than 50 microns". Separate injection of the magnesia inside the furnace but outside the flame envelope would tend to minimize fusion and/or chemical loss through reaction with ash materials.

Borio and Hensel have also observed that chemical additives react differently from the alkaline materials in the ash[15]:

> "It is significant to note that naturally occurring calcium or magnesium compounds in the coal have different inhibiting values than calcium compounds used for neutralizing additives. Some of the naturally occurring alkaline earth material is present as a silicate which would render it inactive for neutralizing purposes."

One might also speculate that conditions in the flame zone are suitable for converting additional alkaline earths to inactive silicates.

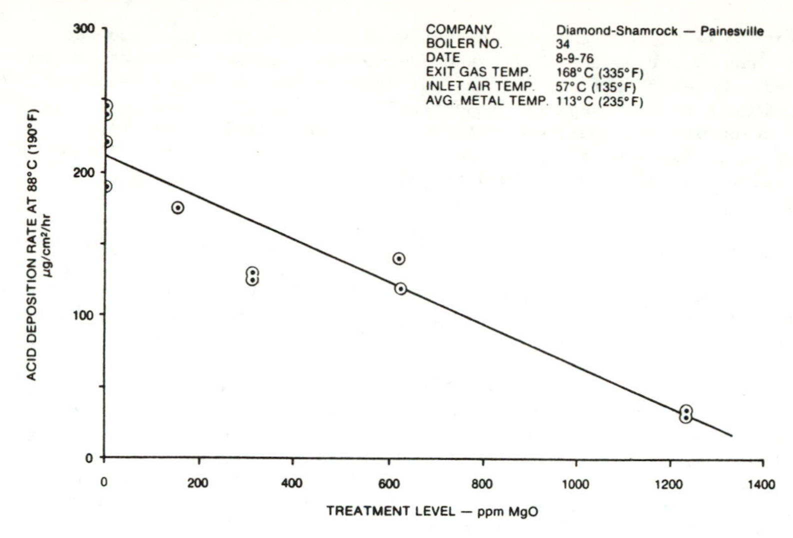

Figure 3. Acid deposition rate vs. MgO treatment level.

DISCUSSION

Borio et al[6] have postulated the mechanism for formation of corrosive bonded deposits as follows:

"1. Sodium and potassium are volatilized in part from the mineral matter in the high temperature flame forming Na_2O and K_2O.

2. During combustion, pyrites are disassociated thermally and with the organic sulfur in the coal, react with oxygen, forming mostly SO_2 and some SO_3.

3. The Na_2O and K_2O then react with SO_3 either in the gas stream or on the tube to form sodium or potassium sulfate, a low-melting material which deposits on the tube. This low-melting material attracts other ash particles eventually building up a moderately thick deposit on the tube.

4. Sulfur dioxide present in the flue gas is catalytically oxidized to SO_3 on iron oxide surfaces present in ash deposits on the tube and possibly on iron-oxide surfaces present as tube scale. Sulfur trioxide reaches near-equilibrium concentrations in localized areas near the tube ($\simeq$ 1,000 ppm).

5. The alkali sulfates, iron oxide and SO_3 then react to form the complex sulfates, $2(Na_3$ or $K_3)Fe(SO_4)_3$, which are molten at about 1100 deg F in a high-SO_3 atmosphere ($\simeq$ 1,000 ppm)."

The molten alkali iron sulfates are generally considered as providing a corrosive matrixing and bonding medium between the tube surfaces and the slag deposits.

Interferences by dispersed magnesia with this mechanism might take several avenues:

1. Preferential deposition of a high melting protective layer on the tubes;

Separate injection of the magnesia provides a more efficient chemical utilization than powder addition with the fuel because losses with bottom ash are minimized. However, this alone does not account for the recent success with low treatment rates. Separate injection was employed in some of the earlier studies which achieved only moderately satisfactory utilization rates. There is one other fundamental difference between the current work and that undertaken earlier: a much finer effective particle size MgO in today's dispersions.

Better performance with finer size would be consistent with Levy and Merryman's gas phase reaction hypothesis. The finer the size, the greater the surface area and number of MgO particles per pound fed. When effectively dispersed in the flue gas stream the probability of contact and reaction with an SO_3 molecule is increased accordingly.

A photomicrograph of the $Mg(OH)_2$ dispersion used in these tests is shown in Figure 4. For comparison purposes, an MgO produced by calcining and grinding the same $Mg(OH)_2$ is shown in Figure 5. It is apparent that the calcining process yields an agglomerated particle whose effective size is 10 times that of the dispersion. More important, relative to the dispersion, the effective surface area and number of MgO particles per unit weight of powder decrease by factors of about ten and one thousand respectively.

Figure 4. Photomicrograph of $Mg(OH)_2$ dispersion shows minimum particle agglomeration. 1 division = 5 microns.

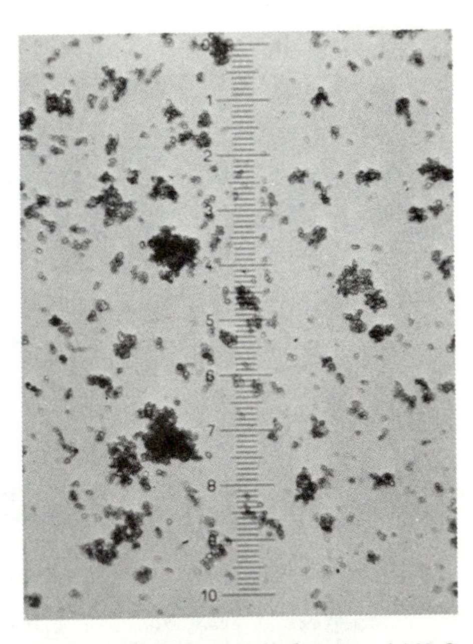

Figure 5. Photomicrograph of commercial MgO powder after grinding to minus 325 shows high degree of agglomeration. 1 division = 5 microns.

Highly agglomerated materials, powder and slurry, like that in Figure 5, were the only products commercially available to earlier experimenters. The material used in these tests was an exceptionally well dispersed $Mg(OH)_2$ which dehydrates and converts to an extremely reactive MgO in the furnace.

SUMMARY

1. The use of small dosages of finely sized and well dispersed magnesia to minimize coal ash fouling warrants further serious exploration.

2. Acid deposition rate studies show that reduction in concentrations of condensable sulfuric acid in the flue gas at the air heater entrance is a direct function of dispersed magnesia feed rate.

3. The injected magnesia is much more chemically available for reaction with SO_3 in the boiler than the far larger quantities of naturally occurring alkali in the fly ash. The latter become slowly available for reaction when wetted.

4. The superior chemical efficiency of the dispersed magnesia used in these studies appears to be related to:

   a. its extremely fine effective particle size

   b. injection separate from the fuel to minimize chemical loss with the bottom ash.

5. Chemical, x-ray diffraction, and petrographic analyses of high temperature probe deposits collected with and without dispersed magnesia treatment revealed no differences which might account for the exceptional results at low dosages.

6. Differential thermal analyses of the high temperature probe samples revealed minor differences between treated and untreated. They are apparently real but very difficult to detect and to relate to plausible chemical or physical changes.

7. The current test data lend some support to Levy and Merryman's hypothesis of gas stream reaction between MgO and SO_3. However, more than one mechanism may be involved in the reduction of coal ash fouling with dispersed magnesia, since benefits have been observed with both high and low sulfur fuels.

8. If the suggestion that reduction in fouling with separate injection of dispersed magnesia is independent of the kind of coal is confirmed by additional testing and field studies, the technique could have an impact on boiler design, operation, control and reliability analagous to that achieved with "gas tempering". The use of "chemical tempering" might provide both the designer and the operator with another useful control tool.

ACKNOWLEDGEMENTS

The authors thank the Diamond Shamrock Company for making the test boiler available and Messrs. Arthur Plumley and Robert Hensel of Combustion Engineering for their technical advice in collecting and interpreting the data. The authors also wish to express their thanks to the respective managements of Basic Chemicals and The Rolfite Company for permission to publish this paper.

REFERENCES

1. Michel, J.R. and Wilcoxson, L.S., "Ash Deposits on Boiler Surfaces from Burning Central Illinois Coal", ASME Paper No. 55-A-95, (1955).

2. Barnhart, D.H. and Williams, P.C., "The Sintering Test, An Index to Ash Fouling Tendency", ASME Paper No. 55-A-193, (1955).

3. Radway, J.E. and Rohrbach, R.R., "SO_3 Control and Steam Generator Emissions", ASME Paper No. 76-WA/APC-9 (1976).

4. Jackson, P.J., "A Probe for Studying the Deposit of Solid Material from Flue Gas at High Temperatures", COMBUSTION, October, 1961, pp. 22-27.

5. Plumley, A.L., Jonakin, J. and Vuia, R.E., "A Review Study of Fire Side Corrosion in Utility and Industrial Boilers", presented at Corrosion Seminar, McMaster University, Hamilton, Ontario, Canada, May, 1966.

6. Borio, R.W. and Hensel, R.P., "Coal-Ash Composition as Related to High Temperature Fireside Corrosion and Sulfur Oxides Emission Control", ASME Paper 71-WA/CD-4 (1971).

7. Babcock & Wilcox, Steam/its generation and use, 38 Ed., pp. 15-13, Babcock & Wilcox, New York (1975).

8. Levy, A. and Merryman, E.L., "Interactions in Sulphur Oxide-Iron Oxide Systems, ASME Paper 66-WA/CD-3 (1966).

9. Corey, R.C., Cross, B.J., and Reid, W.T., "External Corrosion of Furnace Wall Tubes - II - Significance of Sulphate Deposits and Sulphur Trioxide in Corrosion Mechanism", Trans. ASME, 67, 1945, pp. 289-302.

10. Nelson, W. and Cain, C. Jr., "Corrosion of Superheaters and Reheaters of Pulverized-Coal-Fired Boilers", Journal of Engineering for Power, Trans. ASME, Series A, 82, 1960, pp. 194-204.

11. Anderson, C.H. and Diehl, E.K., "Bonded Fireside Deposits in Coal-Fired Boilers -- A Progress Report on the Manner of Formation", ASME Paper 55-A-200 (1955).

12. Raask, E., "Boiler Fouling - The Mechanism of Slagging and Preventive Measures", Central Electricity Research Laboratories Report RD/L/N 97/72.

13. Alexander, P.A., Fielder, B.A., Jackson, P.J. and Raask, E., "An Air Cooled Probe for Measuring Acid Deposition in Boiler Flue Gases", Journal of the Institute of Fuel, January, 1960, pp. 31-37.

14. Barrett, R.E., "High-Temperature Corrosion Studies in an Oil-Fired Laboratory Combustor," ASME Paper 66-WA/CD-2 (1966).

15. Borio, Op.cit., p. 3.

FIELD EXPERIENCE WHILE FIRING REFUSE

Session Chairman: **C. O. Velzy**
C. R. Velzy Associates, Inc.
Co-Chairman: **R. S. Hekclinger**
C. R. Velzy Associates, Inc.

THREE YEARS OF OPERATING EXPERIENCE WITH A WATERWALL BOILER AT THE OCEANSIDE DISPOSAL PLANT

ROGER S. HECKLINGER[1] and CHARLES O. VELZY[1]

Charles O. Velzy Assoc., Inc.
Armonk, New York, USA 10504

JOHN C. KEMPISTY[2]

Zurn Industries, Inc.
Erie, Pennsylvania, USA 16503

ABSTRACT

Waste heat boilers commissioned in 1965 and 1966 at the Oceanside Disposal Plant in the Town of Hempstead, Long Island, New York, experienced serious tube failures commencing six months after initial operation. After a number of years of remedial effort, investigation, and observation, it was decided to replace the original forced circulation waste heat boilers with natural circulation waterwall boilers. The first replacement boiler was commissioned in June of 1974.

Tube thickness measurements were made periodically of selected areas of the furnace walls and in the convection bank. Operating parameters, particularly the use of overfire air, were monitored and controlled.

The operating results after 3 years have been encouraging. During this time, several modifications to operating procedures and construction details have been implemented.

The second replacement boiler will be commissioned in late 1977 or early 1978.

INTRODUCTION

In early 1962, the Town of Hempstead, Nassau County, Long Island, New York, authorized Charles R Velzy Associates, Inc. (CRVA) to begin the design of a power generating incinerator plant. This plant was to become one of the few incinerator plants in this country where the heat energy in municipal solid waste is used to generate steam and, in turn, electricity. A unique feature of this plant is that the steam is condensed in desalinization units so that fresh water for in-plant process use can be generated from adjacent salt water.

The original design included two 270 t (300 ton) per day refractory walled incinerator furnaces with horizontal tube, forced

---

[1] Charles R Velzy Associates, Inc.
[2] Zurn Energy Division, Zurn Industries, Inc.

circulation, waste heat boilers mounted on top and at the residue discharge end of the furnaces. Steaming conditions were 3.15 MPag (460 psig), saturated. A third furnace, rated at 135 t (150 tons) per day burning capacity, did not include a steam generator.

The plant commenced operation in late 1965 with great expectations. These expectations were not realized. Within the year, tube failures were experienced. Failures continued at an accelerated rate to the point where in the latter months of 1967 there was an average of 3.3 days between failures. The rate of failure was reduced from this point by limiting the burning rate and by reducing operating pressures.

Many reasons for these failures were advanced by those experienced in the mechanism of boiler tube wastage. Data from Oceanside was extensively cited in Battelle's report "Fireside Metal Wastage in Municipal Incinerators" prepared for the USEPA in 1972 (1). A summary of the various opinions and a description of the several remedial actions attempted are summarized in a paper by Mr. Charles O. Velzy (2).

During this same time period (1966 - 1967), a series of articles was published describing severe corrosion problems being experienced in several of the new European refuse fired boilers. Since none of the explanations offered for the metal wastage at Oceanside seemed to be totally satisfactory, members of this firm decided in 1968 to compare our observations first-hand with those of plant operators in Europe where they seemed to be experiencing similar problems. The extremely valuable and helpful information obtained during this trip was then evaluated in light of the improvement in tube life at the Town's other power generating incinerator plant, the Merrick Disposal Plant, where tube spacing was increased to reduce gas velocity.

Based on all of this information, CRVA was retained by the Town in 1970 to develop a program, including establishment of overall design parameters for waterwall boiler units, for rebuilding the furnaces so as to obtain satisfactory operating and maintenance experience. The purpose of this paper is to describe the design parameters incorporated into the replacement units and the operating experience to date.

DESIGN OF THE REPLACEMENT UNITS

In 1972, the Town contracted with Zurn Industries to furnish a natural circulation waterwall boiler for one of the two 270 t (300 ton) per day furnaces. In addition to combustion considerations, the new boiler was designed to fit within existing space conditions. Therefore, this particular boiler is unique and the product of a joint design effort between Zurn and CRVA.

The grate, charging chute, and the refractory furnace at the grate inlet were all that was retained from the original installation. Several basic parameters guided the overall design. The maximum gas velocity in the convection portion of the boiler was to be 10 m/s (30 fps). Average gas velocity up through the furnace was to be kept to a practical minimum, preferably below 5 m/s (15 fps). Overfire air was to be located for best effect, with respect to enhancing turbulence and mixing of combustion air with products of combustion. Horizontal tubes were to be avoided. The final furnace configuration and location of overfire air was to be determined by model testing. The design configuration was discussed with Mr. Franz Nowak from Stuttgart and Mr. Reinhold Mutke from Vereinigte Kesselwerke A.G. in Duesseldorf. Sootblower location was to be given careful consideration.

The result may be seen in Figure 1. A three-dimensional model test in Zurn's manufacturing plant was used to determine final furnace configuration and the location of overfire air nozzles. A

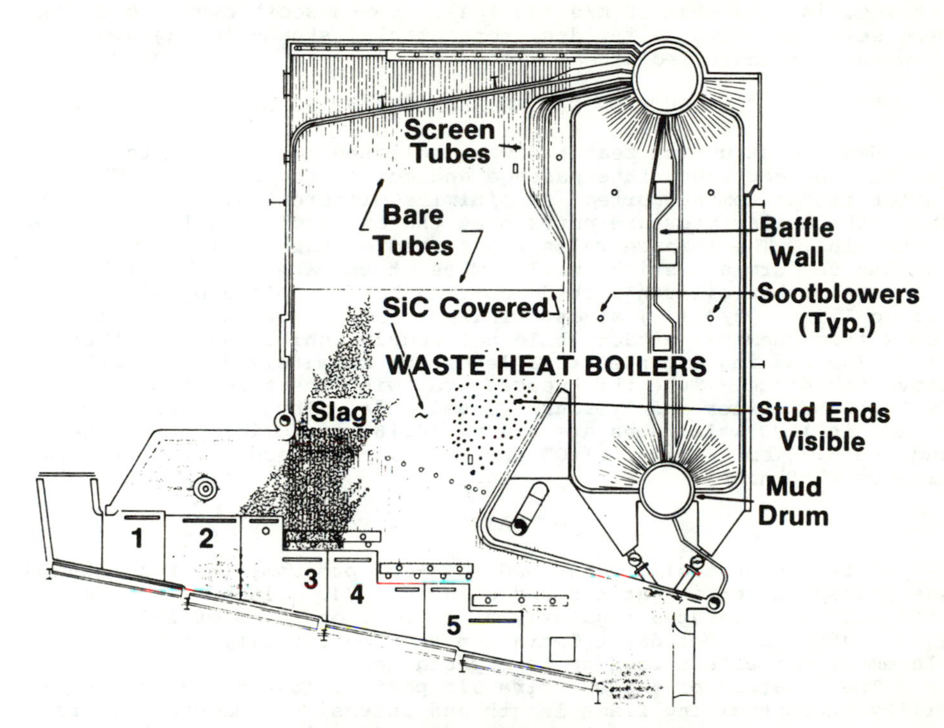

SECTION THRU FURNACE & BOILER
UNIT NO. 2
OCEANSIDE DISPOSAL PLANT
HEMPSTEAD, N.Y.
Figure 1

multiple pass convection bank, formed of essentially vertical tubes, was chosen which was of sufficient size to preclude the need for an economizer (Boiler exit gas temperature is 315 C (600F). The gas velocity in the initial downpass of the convection section, where the gas passes between 6.35 cm (2-1/2 in.) OD tubes on 15 cm (6 in.) spacing, is less than 10 m/s (30 fps). Seven sootblower locations were selected. Two half-width, retractable, steam-blowing soot-blowers were installed at each location.

CONSTRUCTION OF THE REPLACEMENT BOILER

Many construction features were selected to minimize the possibility of continued tube wastage and corrosion problems. The boiler is "bottom supported" to minimize differential expansion where the boiler pressure parts meet the air-cooled walls along the grate line. The furnace walls are formed of tubes welded one to another to form a gastight wall. These tubes are covered with a relatively thin silicon carbide coating held in place by closely spaced 1.9 cm (3/4 in.) studs. This provides protection of the tubes from chemical attack while not unduly inhibiting heat transfer. The coating extended to a level approximately 9 m (30 ft) above the grate. Tube thicknesses are twice the thickness required by the ASME Boiler and Pressure Vessel Code. Thus, the 8.25 cm (3-1/4 in.) OD wall tubes are 4.6 mm (0.180 in.) minimum thickness and the 6.35 cm (2-1/2 in.) OD convection tubes are 3.4 mm (0.135 in.) minimum thickness.

OPERATION

After an initial dry-out and boil-out program, the rebuilt unit was placed in service at the end of June 1974. Since that time, this unit has operated regularly at combustion rates of 270 - 360 t (300 - 400 tons) per day. Operation and availability of the replacement boiler has been up to expectations.

The location of the overfire air ports allows operator flexibility in controlling flame length and intensity. However, it requires an ongoing education program for effective use by the operating staff.

Slag buildup on the water-cooled furnace walls has never exceeded 15 - 20 cm (6 - 8 in.) in concentrated areas. Beyond this thickness, the slag is self-shedding. The silicon carbide coating has held up well. Slag buildup on the screen tubes at the furnace exit has never exceeded 2.5 - 5.0 cm (1 - 2 in.).

Slag and ash buildup in the convection bank has been controlled by sootblowing. The unit has never been cleaned in any way other than by normal sootblowing. At present, all blowers are operated once per day at a steam pressure of 515 kPag (75 psig).

The unit has experienced three extended periods (4 - 6 weeks) of down-time since it was placed in operation in June of 1974. Two of these periods were due to mechanical problems with the boiler feedwater pumps and the other was due to mechanical problems with the induced draft fan. The pumps and fan were originally installed in 1965. Regardless, no evidence has been found of low temperature corrosion.

The only outage attributable to the boiler was a tube failure in the convection bank which was followed, during a hydrostatic test, by a tube failure in a wall tube. Both tubes were in the vicinity of the sootblower in the cavity between the furnace screen and the boiler convection bank. A grooving pattern was observed on adjacent tubes which indicated that the problem was due to water-cutting as this sootblower, the first blower in the blowing sequence, entered the cavity.

The tubes were repaired and the problem alleviated by:

1. Installing stainless steel shields on seven tubes
2. Repitching the sootblower steam line
3. Maintaining pressure on the steam line at all times
4. Reducing blowing pressure from 1,200 kPag (175 psig) to 515 kPag (75 psig)

Total down-time was two weeks

TESTING

As part of the contract for the replacement boiler, the General Contractor was required to retain a testing company to periodically check the thickness of tubes in areas selected by the Engineer. Accordingly, tubes selected by us were ultrasonically tested on July 3, 1974, April 23, 1975, November 5, 1975, and January 12, 1976.

In addition, thicknesses were measured in the sootblower cavity by the same firm on March 12, 1977, following the tube failures described above. Finally, ultrasonic tests were made by Zurm Industries on April 12, 1977, in preparation for this paper. At the same time, tube deposits were removed for analysis.

RESULTS

When the ultrasonic test program began, we were particularly concerned about the potential for tube wastage of the screen tubes at the furnace outlet. The test results show that little or no wastage has occurred at this point, as may be observed in the following tabulation for a 8.25 cm (3-1/4 in.) OD screen tube in the center of the furnace outlet. The minimum thickness, as purchased, was 4.6 mm (0.180 in.).

Date	Thickness Measured
7/03/74	4.70 mm (0.185 in.)
4/23/75	4.83 mm (0.190 in.)
11/05/75	4.83 mm (0.190 in.)
1/12/76	Not Read
3/12/77	4.70 mm (0.185 in.)
4/12/77	4.69 mm (0.1846 in.) (Zurn)

The lowest reading recorded by Zurn in the screen area was 4.51 mm (0.1776 in.). In these tests, screen tube thickness was measured at two levels and selected tubes were scanned circumferentially.

No tube wastage has been detected in the boiler convection section with the exception of the grooving attributable to water from the sootblowers. Even there, no wastage was detected past the third tube from the sidewall.

No tube wastage has been detected under the silicon carbide coating.

However, tube wastage was detected during the November 1975 test program in the furnace side walls immediately above the area covered by silicon carbide. The January 1976 test program concentrated on that area in both side walls. Figure 2 shows the results. There was a definite wastage pattern extending up from the path of slag deposits on the side walls. Slag deposition appeared to follow the flame pattern. We learned that the unit was sometimes operated with a minimum of overfire air. This operating condition resulted in a long orange flame which, on occasion, extended to the screen tubes.

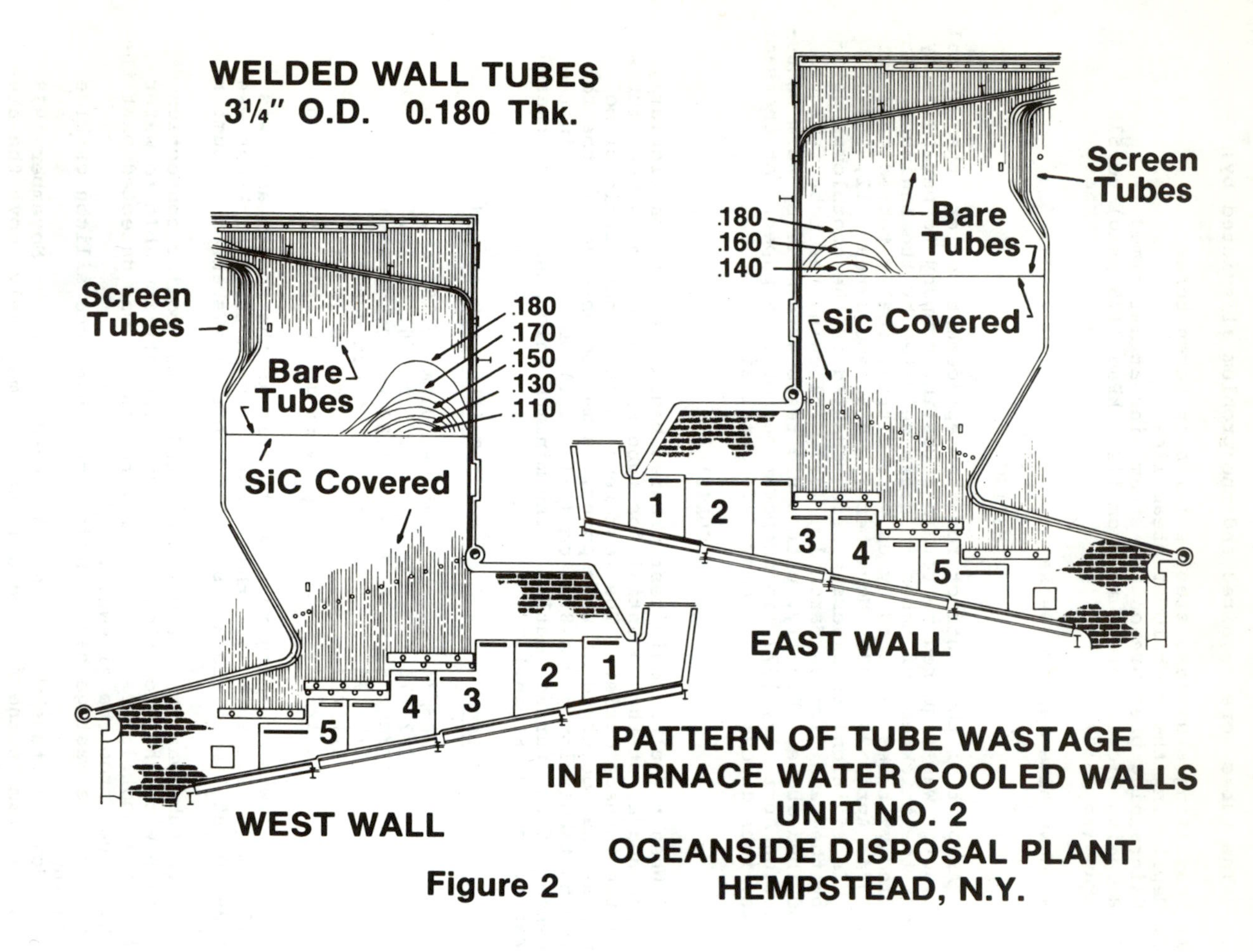
WELDED WALL TUBES
3¼" O.D. 0.180 Thk.
Screen Tubes
Bare Tubes
SiC Covered
.180
.170
.150
.130
.110
5
4
3
2
1
WEST WALL
Screen Tubes
Bare Tubes
Sic Covered
.180
.160
.140
1
2
3
4
5
EAST WALL
PATTERN OF TUBE WASTAGE
IN FURNACE WATER COOLED WALLS
UNIT NO. 2
OCEANSIDE DISPOSAL PLANT
HEMPSTEAD, N.Y.

Figure 2

An immediate program was instituted to assure that side wall overfire air was effectively used to insure complete combustion below the top of the silicon carbide coating. During the next extended outage (due to problems with the induced draft fan) the silicon carbide coating was extended upward, as shown on Figure 3, to provide additional insurance against continuing metal wastage. During the test program on April 12, 1977, Zurn measured the thickness of all side wall tubes on the west side of the furnace. One tube was found that gave erratic readings. This was probably due to pitting that was observed on the tube surface. This tube is located as shown on Figure 3. Tubes to the right and left showed no wastage. Readings taken on the erratic tube were:

Height above SiC	Thickness
15 cm (6 in.)	4.00 mm (0.1576 in.)
30 cm (12 in.)	4.47 mm (0.1763 in.) (erratic)
60 cm (24 in.)	4.51 mm (0.1776 in.)
90 cm (36 in.)	4.50 mm (0.1772 in.)
120 cm (48 in.)	4.51 mm (0.1777 in.)

This appears to be an isolated case. It will be watched closely.

At the time the Zurn ultrasonic test data was taken, sample tube deposits were removed from the furnace side wall just above the increased height of silicon carbide and from a screen tube in the middle of the furnace outlet. These samples were subjected to qualitative spectrographic analysis. The results are of interest, but because of the limited amount of data taken, one must be careful in drawing conclusions. Basic data are:

	Deposits	
	Waterwall Tube	Screen Tube
Most Prevalent Compounds	Lead	Calcium
Other Major Compounds	Sodium	Sodium
	Calcium	Silicon
	Silicon	Lead
pH	2.6	5.0

Note that lead, which has been suggested as a possible contributor to the corrosion mechanism by others investigating corrosion in refuse fired boilers, is detected in large quantities particularly on the side walls.

CONCLUSIONS

The basic design approach for the replacement units at Oceanside has resulted in significantly longer tube life than the original design.

Tube wastage other than that attributable to water in the sootblowing-steam has not been detected on heating surfaces beyond the point where the combustion process is complete.

The silicon carbide coating has been effective in protecting waterwall tubes from corrosion.

Convection surface formed of wide-spaced, essentially-vertical tubes can be kept free of plugging with normal sootblowing.

More definitive investigative work is required to relate tube deposits to metal wastage.

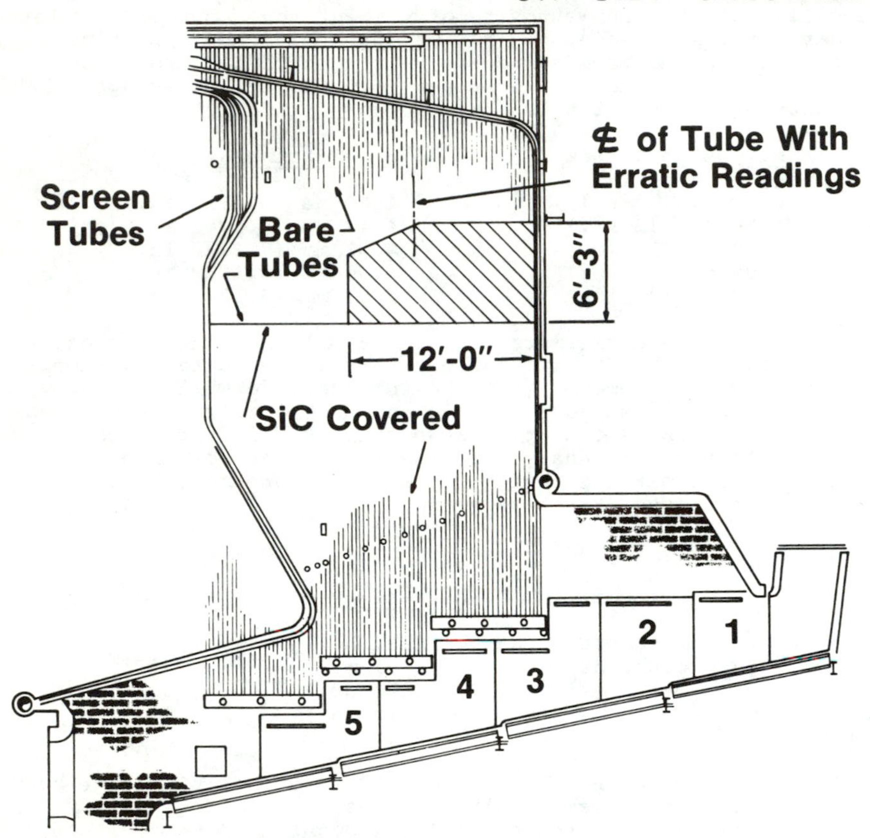

ADDITIONAL SILICON CARBIDE COVERING
IN FURNACE WATER COOLED WALLS
UNIT NO. 2
OCEANSIDE DISPOSAL PLANT
HEMPSTEAD, N.Y.
Figure 3

REFERENCES

1 Miller, P.D., et al, "Final Report on Fireside Metal Wastage in Municipal Incinerators (Research Grants EP-00325 and EP-00325-S1) to Solid Waste Research Division, National Environmental Research Center, Environmental Protection Agency," April 1972, Battelle Columbus Laboratories, Columbus, Ohio.

2 Velzy, C.O., "Corrosion Experiences in Incineration," Paper No. 133, Corrosion 74, March 1974, National Association of Corrosion Engineers, Houston, Texas.

CORROSION ON REFUSE INCINERATION BOILERS, PREVENTIVE MEASURES

FRANZ NOWAK

Techn. Werke Stuttgart
705 Waiblingen/BRD, West Germany

ABSTRACT

Over the years, extensive investigations and experiments at the Stuttgart refuse fired boiler plant have led to the development of a number of techniques, accomplished in the construction or operational control of these plants, that have resulted in control of the corrosion problem. The first technique involved the use of refractory coatings in the lower portion of the furnace. The second technique involved improvements in boiler configuration and design, including point of introduction of secondary air. The third technique involved regulation of the refuse feed slider, which controls the rate of feed of refuse onto the first two roller grates, by a mean furnace temperature measurement. The fourth technique for controlling corrosion is development of optimum use of refractory and tube coatings in the lower areas of the furnace.

* * *

Corrosion of heating surfaces, its extent and causes, and coating materials used in refuse incineration boilers have been treated in the lecture I read on the occasion of the National Incinerator Conference in Cincinnati in 1970 (1), and by other publications (2,3,4,5,6,7,8,9,10).

It has been demonstrated that the furnace wall surfaces are subject to essential material wastage, whereas the heating surfaces above the furnace, and particularly the superheater, are not corroded to the same extent. Therefore, the present paper has concentrated on furnace protection.

Intensive long-term research has determined that the main reason for tube metal wastage is the corrosive action of gaseous hydrochloric acid, set free particularly when burning PVC and other chlorinated plastics, as well as from heating sodium chloride and other chlorides always abundantly present in household refuse. Research has also shown, however, that certain conditions have to exist for the aforementioned metal wastage to take place. Thus, metal wastage only takes place in the simultaneous presence of a reducing atmosphere and a certain minimum temperature at the tube surface. If these conditions do not exist, the hydrochloric acid will not attack the tube metal. The latter criterion led to the conclusion that, for any refuse incineration plant to be built, it would be desirable to limit steam temperature to 450 C and steam pressure to 35 bar, taking into account the effect of the respective saturated steam temperature of water-cooled evaporator tubes being a function of the boiler operating pressure.

Different measures which have been taken from 1966 to date have cut down the corrosive effect on the tubing and its coating. These measures include the protection of the furnace tubes with resistant refractory coatings or injection moldings and by screening the most heavily stressed superheater sections with protecting caps at the tube inflow side. These means have been considerably improved lately.

Important factors in achieving control of corrosion include improvements in configuration of the furnace and operational control. In the above mentioned lecture, I have already suggested the way in which aerodynamic criteria should be taken into consideration in furnace design to obtain a speedy and complete burnout within the furnace chamber. Very valuable hints in this regard came from model tests. Furthermore, a reasonable introduction of high pressure secondary air in the zone situated just above the grates is necessary to achieve a perfect mixing of the air with the incompletely combusted gases present there. Flue gas analyses carried out on a boiler of a corresponding design have proven the effectiveness of these two measures.

Another success, at least as important as the aforementioned, has been scored with the achievement of an extremely uniform heat release by means of a recently installed refuse burnout regulation. This achievement was accompanied by an almost total elimination of CO content in the flue gases above the combustion chamber.

Finally, through temperature measurements at the cooled furnace walls, the conditions to be taken into account when selecting adequate ceramic coating materials have been found. These materials should not only provide a good insulating protection but also screen the tube walls against any corrosive gas components tending to diffuse through the coating.

Refuse incineration boilers which have been built in accordance with these criteria show a considerably reduced metal wastage so that the present state of technology can be called satisfactory.

PREVENTIVE MEASURES

Following is a more detailed view on the aforementioned development stages.

a) The furnace design of the first-generation type of refuse incineration boilers (Fig 1) was such that there were great variations in heat release along with constantly fluctuating CO and O_2 values (Fig 2). The constant variation in flue gas composition may be attributed to the constant variation in the composition of refuse supplied on one hand and to the impossibility of achieving uniform refuse feeding. Thus, refuse which has been stored for several days is so strongly entangled that it forms a mass which will not break apart when fed into the furnace. If this type of refuse is fed together with more recently received refuse or, even worse, with shredded bulky refuse, local heat release fluctuations are inevitable. Since adjustment of the required combustion air on a short-time basis is practically impossible, zones with incompletely burnt-out flue gases will exist, offering the basis for possible corrosive actions.

Figure 2 shows that the above condition is even more accentuated if larger quantities of refuse with a particularly high heating value, such as tires or plastics, are fed. Thus, the plant represented in the top diagram of Figure 2 was burning household refuse only, while the diagram in the center represents a plant burning 10% plastic material, and the diagram at the bottom represents a plant with about 20% plastic additions.

b) The third boiler (Fig 3) of the Stuttgart plant is equipped with a furnace of more advanced design (2nd generation refuse boiler). The operating behavior was found to be much better compared to the first generation boiler. Figure 4 shows the results obtained from measuring CO and O_2 values measured approximately 8 m (26 ft above the grate). Tests have been carried out with:

- household refuse only (see diagram at the bottom);
- a 50% addition of refuse with a high heating value (see diagram in the center); and
- household refuse with the addition of a high percentage of ground bulky refuse (see diagram at the top).

The improved combustion behavior (as compared to the diagrams of Fig 2) is clearly apparent and was reflected by a considerable extension of the service life of all tube material installed in the furnace. All tube sections of this boiler have already been operating 30,000 hours without the necessity for any replacement. The availability of the boiler has also been increased to 80%.

c) During the 5 years that this Stuttgart Unit 3 has been in operation, investigations have been conducted related to the problem of obtaining a more uniform flue gas atmosphere through a regulated heat release which would reduce or perhaps eliminate the corrosion problem.

Having in mind that refuse, as fuel, always differs in its composition, with average heating value variations of approximately $\pm$800 kJ/kg ($\pm$200 kcal/kg), the question is to regulate refuse feeding or combustion speed as a function of time, to assure a close to uniform heat release.

Measured taken to reach this goal are based on the following reasoning: Heat release in the combustion chamber depends on the amount and on the heating value of the refuse fed into the furnace. Both values are subject to constant fluctuation, depending on the proportion of household and industrial refuse burnt. The uniform thermal energy strived for corresponds to equal combustion air quantities which entail, on their part, a practically invariable flue gas production and more or less constant flue gas temperatures. To establish the basis for control, a definite combustion air volume was assumed, and the refuse quantity or the number of revolutions of the first and second rollers on a "Duesseldorf" grate system were controlled. This had to lead to a connection between the thermal energy released and the corresponding air volume required. Either the furnace temperature or the O_2 content could be chosen as output variable. Temperature measuring has the advantage in that it can be carried out by simple means and values are obtained quicker than O_2 contents.

The refuse feed slider drive, developed in the meantime, helped to realize our aim of control of refuse feed into the furnace. Refuse is fed through two sliders with hydraulic drives. Their control has been modified to the effect that the slider advance is slow and its return quick, with the result that refuse feeding has become almost uniform.

In the furnace, three thermo-couples have been arranged at the top of the first pass to produce a mean temperature value to compensate for the risk of different temperature levels due to flue gas stratification. The resulting difference between actual and desired temperature is sent to a transmitter which delivers the control signal for the speed adjustment of the grate roller gears. Up to now, only Rollers 1 and 2 have been controlled, their number of revolutions being added. In the future, Roller 3 will also be subject to regulation. The temperature value obtained also controls the action of the refuse feed sliders, the signal being transmitted through a time delay, due to reasons of regulating technique. Thus, first the combustion intensity is affected and then the rate of refuse feed. It would be too long and beyond the scope of the present paper to explain in detail the regulating mechanisms.

Figure 5 illustrates the positive results obtained with this method of control on a test day chosen at random. The flue gas temperature is shown on the left and the steam produced on the right, with and without regulation. Figure 6 shows the measured CO and O_2 contents for various refuse compositions as described in relation to the tests illustrated in Figure 4. A comparison between the diagrams in Figures 4 and 6 clearly illustrates that the combustion behavior has been further improved, which becomes evident from the reduction in the CO values. In the near future, the O_2 content may also be measured and used as a correcting factor in the regulating mechanism.

It should also be mentioned that the installation of the refuse combustion regulation system in the Stuttgart plant had a favorable secondary effect in that the refuse throughput was simultaneously increased. Due to a steady increase in refuse heating values and refuse quantities through the years 1965-1972 (Fig 7 and Fig 8), the refuse throughput capability of these boiler plants had been constantly decreasing, the steam generator having been designed for a maximum heat release of 126 GJ/h (30 G cal/h). The equalization of steam generation has resulted in an increase in the attainable mean value of plant throughput which meets local requirements when the plant is operating at full capacity.

d) Another success can be cited in the development of suitable coating materials for existing furnace designs. To check the thermal load on furnace walls, thermo-couples have been used to measure the temperatures in the most critical zones just below the surface turned towards the furnace. Figure 9 shows the location of points at which temperatures were measured, and Figure 10 shows the values found.

Based on these findings and in cooperation with a refractory material manufacturer that had carried out a research program,

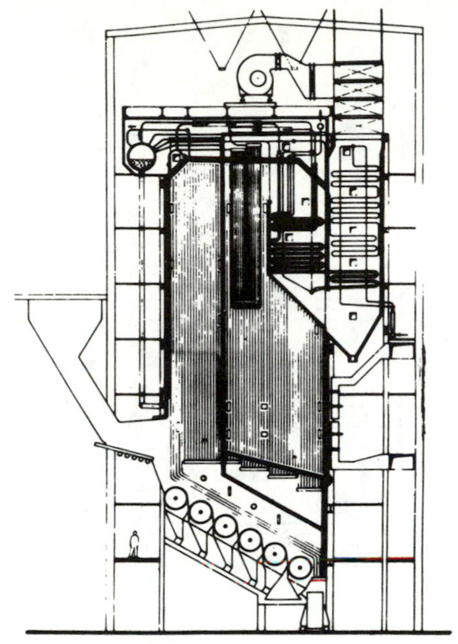

Figure 1. First generation refuse incinerator and boiler.

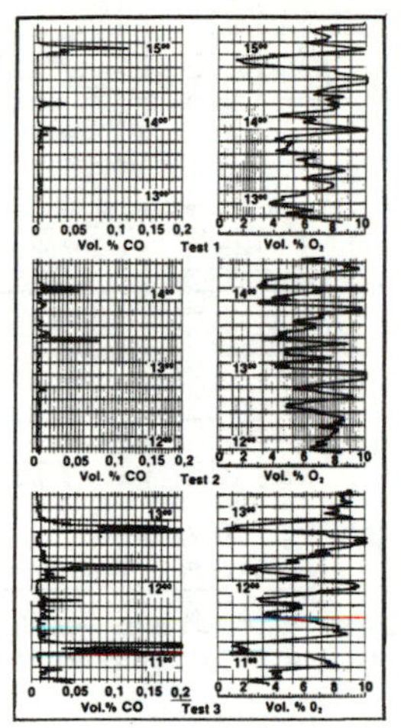

Figure 2. CO and O_2 at discharge from refuse furnace.

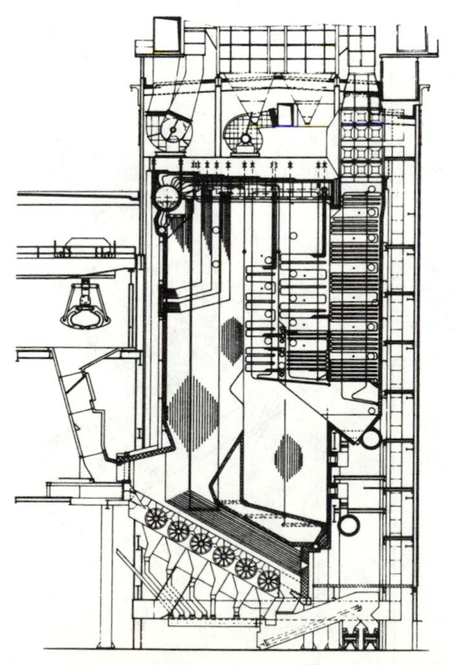

Figure 3. Second generation refuse incinerator and boiler.

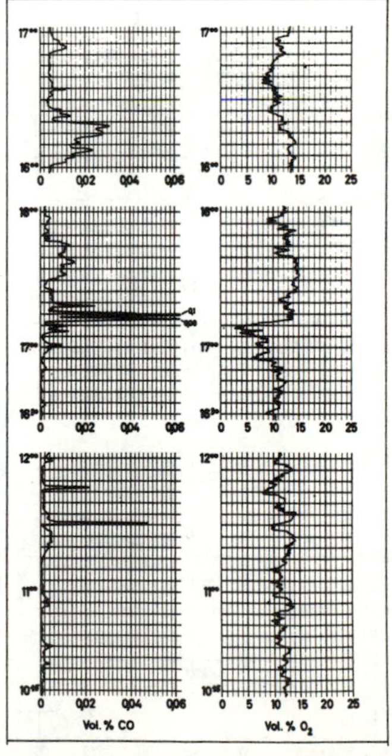

Figure 4. CO and O_2 at discharge from refuse furnace without regulation.

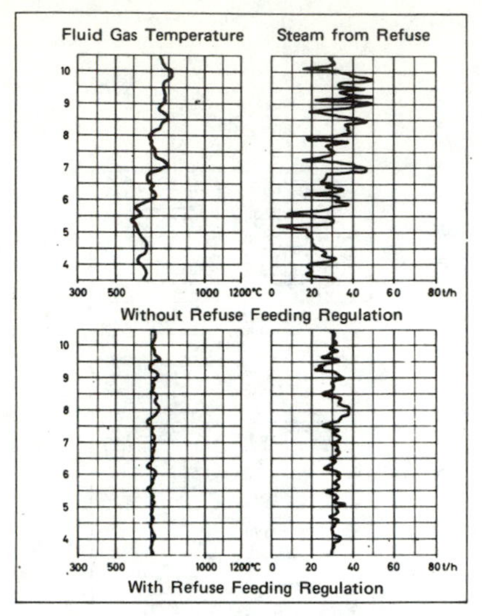

Figure 5. Refuse fired boiler flue gas temperature and steam produced from refuse.

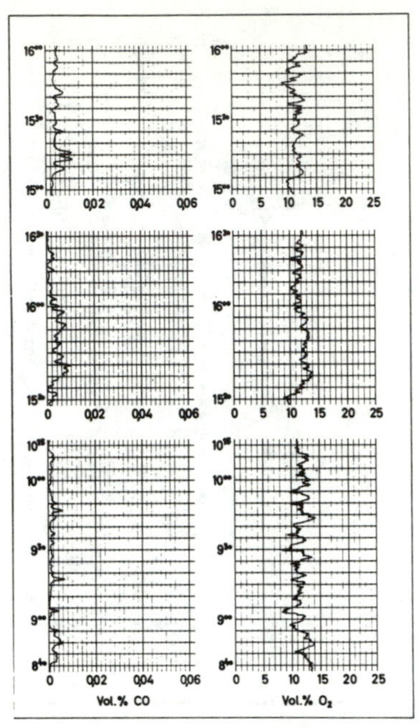

Figure 6. CO and O_2 at discharge from refuse furnace with regulation.

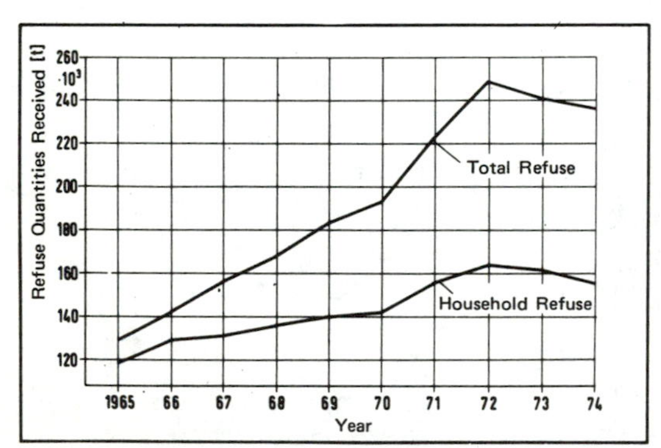

Figure 7. Refuse quantities in the years 1965-1974.

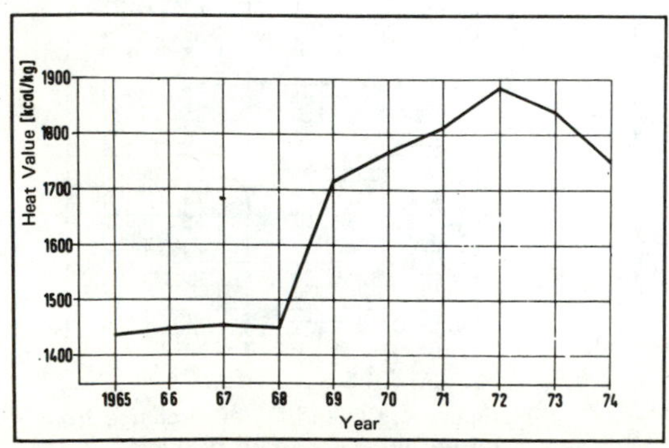

Figure 8. Refuse heat value in the years 1965-1974.

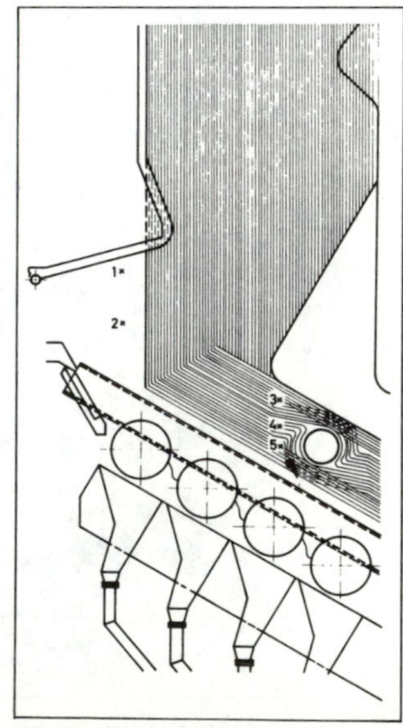

Figure 9. Location of thermocouples in furnace.

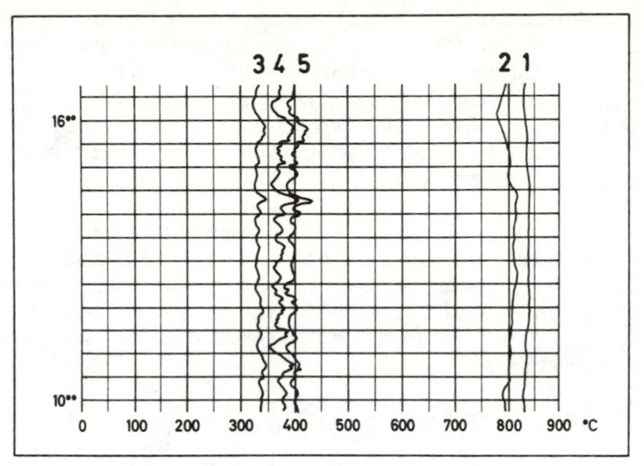

Figure 10. Temperature in the furnace walls.

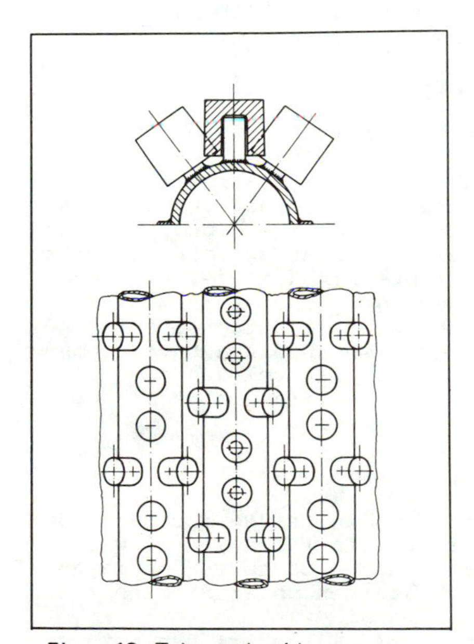
Figure 12. Tube studs with caps.

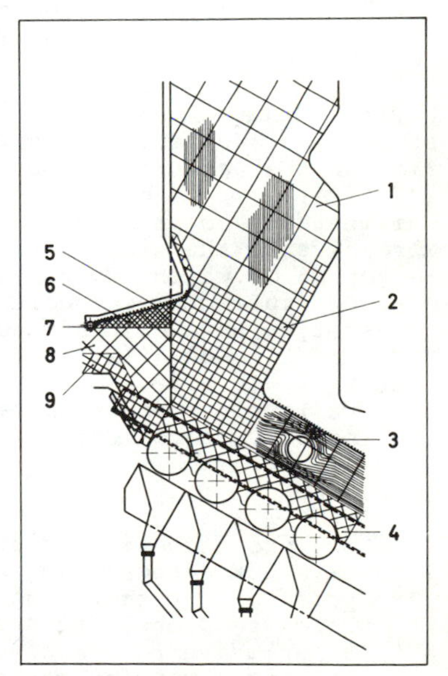

Figure 11. Application of lining materials to furnace.

suitable tube coating materials were developed. The criteria used to evaluate the suitability of the refractory materials included their resistance to corrosive gas constituents or corrosive refuse ash, and their resistance to the temperatures produced in the combustion chamber. Another criteria was the stability of these materials in the presence of variable quantities of water vapor set free in the furnace which, at the temperatures considered, tend to dissociate and to oxidize the refractory material used.

The materials selected provide sufficient protection to the boiler tubes against metal wastage. However, they have to be applied correctly. The material chosen exerts, therefore, a direct influence on the life of the tubing. Figure 11 illustrates the application of coating materials to a cooled combustion chamber, where:

Position 1 stands for SiC_{50} or SiC_{70} mass applied by spraying;
" 2 " " SiC_{50} or SiC_{70} mass with a protective coating of 69% SiC of slag rejecting effect;
" 3 " " mass of 52% Al_2O_3 and 38% SiC;
" 4 " " sintered SiC_{90} shapes;
" 5 " " mass of 52% Al_2O_3 and 38% SiC;
" 6 " " dry-formed hard-fireclay bricks with more than 42% Al_2O_3;
" 7 " " chrome brick mass with 33% Cr_2O_3;
" 8 " " high-alumina clay bricks with approximately 60% Al_2O_3;
" 9 " " sintered SiC_{50} bricks.

During these investigations, it was decided to protect the studs, indispensable for applying the monolithic coating material on the waterwall tubes, with caps of SiC_{90} material (Fig 12). Experience has shown that this method contributes considerably to increase the service life of the studs. This also extends the life of the monolithic coating material, the retention of which depends on well-preserved studs. Instead of 2200 studs/m^2 required up to now, only 530 studs/m^2 are now necessary. However, instead of the SiC_{50} coating material, SiC_{90} coating material should be used which has a better conductivity, in order to assure the same heat release to the tube walls situated underneath. From the standpoint of installed cost, both methods are about equal.

SUMMARY

The symptomatic aspects of corrosion in refuse incineration boilers have not changed basically in the last few years, with few exceptions. Our knowledge of the causes of corrosion is now, on the whole, complete, and studies have been undertaken to reduce this corrosion to a minimum.

As the metal wastage was concentrated in the refuse furnace, any preventive measures to be taken had to start there. The goal of reduced metal wastage has been reached by coating the furnace tube walls with appropriate monolithic coating material or bricks, by changes in the functional design of the furnace, and by a controlled introduction of secondary air in zones where the combustion remained incomplete.

Furthermore, a refuse furnace control system has been developed which assures a practically homogeneous atmosphere in order to reduce the zones where the flue gases still contain quantities of CO, because hydrochloric gas cannot have any corrosive effect without this gas component.

Finally, systematic investigations have been carried out which have resulted in the selection of the most suitable coating materials which have good stability and the best possible thermal endurance and which offer sufficient protection to the tubing against chemical action.

All these measures taken together have resulted in a remarkable extension of the lives of the tubing and of the coating materials, and at the same time have increased the availability of the refuse boilers and decreased repair costs.

Based on the present state of development and assuming that the above achievements and knowledge are applied, it can be concluded that refuse incineration boilers with heat recovery represent a mature technology.

REFERENCES

1 Nowak, F., "Considerations in the Construction of Large Refuse Incinerators," National Incinerator Conference, 1970, pp. 86-92.

2 Huch, R., "Chlorwasserstoffkorrosionen in Muellverbrennungsanlagen," - ("Hydrogen Chloride Corrosion in Refuse Incinerators"), Brennstoff-Waerme-Kraft, No. 18, 1966, pp. 76-79

3 Nowak, F., "Corrosion Problems in Refuse Incineration," Mitteilungen des Vereins der Grosskraftwerksbetreiber, No. 111, 1967.

4 Hirsch, M and Rasch, R., "Beitrag zur Bildung von Eisenchlorid durch Reaktion zwischen Chlorwasserstoff in Rauchgasen und Eisenoxid in Flugaschen und auf Waermeaustauscherflaechen," - ("The Contribution of the Reaction Between Hydrogen Chloride in the Combustion Gases and Ferric Oxide in Fly Ash and on Heat Exchange Surfaces in the Formation of Iron Chloride"), Aufbereitungstechnik, No. 9, 1968, pp. 614-623.

5 Hirsch, M., "Zur Rauchgasseitigen Heizflaechenkorrosion durch Bildung von Eisenchlorid bei Muellverbrennungsanlagen," - ("Fireside Corrosion of Heat Exchange Surfaces in Refuse Incinerators Due to the Formation of Iron Chloride"), Energie, No. 20, 1968, pp. 32-35.

6 Rasch, R., "Beitrag zur Bildung von Eisen (II) - und Eisen (III) - Chlorid bei Hochtemperaturkorrosionen in Feuerraeumen," - ("Formation of $FeCl_2$ and $FeCl_3$ and its Contribution to High Temperature Corrosion in Furnaces"), Battelle-Information, No. 5, 1969, pp. 18-22.

7 Wickert, K., "Versuche zur Bildung von $FeCl_2$ und $FeCl_3$ aus Fe, Fe_3O_4, Fe_2O_3 und HCl-haltigen Gasen," - ("Experiments to Produce $FeCl_2$ and $FeCl_3$ from Fe, Fe_3O_4, Fe_2O_3 and HCl Containing Gases"), Mitteilungen des Vereins der Grosskraftwerksbetreiber, 1969, pp. 449-452.

8 Nowak, F., "Korrosionserscheinungen an Muellkesseln und Abwehrmassnahmen," - ("Occurence of Corrosion in Refuse Fired Boilers and Preventative Measures"), "Corrosion in Refuse Incineration Plants" Vereinigung der Grosskraftwerksbetreiber, 1970, Sonderheft.

9 Wickert, K., "Zur Katalytischen Oxidation von Chlorwasserstoff in Feuerraeumen," - ("Catalytic Oxidation of Hydrogen Chloride in Furnaces), Waerme, No. 76, 1970, pp. 68-73.

10 Rasch, R., "Die Rolle von Natriumchlorid bei Feuerseitigen Hochtemperaturen," - ("The Effect of Sodium Chloride in Fireside High Temperature Corrosion"), Energie, No. 23, 1971, pp. 29-33.

THE CAUSES OF BOILER METAL WASTAGE IN THE STADTWERKE DUESSELDORF INCINERATION PLANT

K. M. KAUTZ

The Israel Electric Corporation Ltd.
Haifa, Israel

ABSTRACT

Since the first steam generator was put into service in the Stadtwerke Dusseldorf incinerator plant in 1965 high-temperature corrosion was observed on the evaporator, final superheater and primary superheater tubes. The cause of corrosion which varied throughout the boiler was due to reducing atmospheres, high temperature chloride attack and high temperature surface attack. A relationship was found between the rate of corrosion and the increase in the amount and the heating value of the burned refuse; a further intensifying influence on the rate of metal wastage of boiler tubes could be associated with the combustion in one boiler of precrushed bulky refuse with a high wood content.

INTRODUCTION

The first four units of the Stadtwerke Duesseldorf incineration plant were put into service during the years 1965/66. The first high-temperature metal wastage was observed after 400 h of operation on the gas side of the uppermost tubes of the primary superheater located at the top of the second pass. Further metal wastage in this region, caused by a common influence of corrosion and erosion (Kautz and Tichatschke, 1972 (1)), was avoided by attaching protecting strips of high-temperature corrosion resistant Cr-Si steel on the attacked tube surface.

After one year of operation, heavy metal wastages were found on evaporator tubes forming the walls of the first pass above the combustion chamber. Investigations proved that corrosion was due to direct reaction of the HCl in the flue gas with the tube material in the reducing atmosphere of some parts of the flue gas stream. By studding and coating the tubes with castable refractories and inserting a baffle wall in the combustion chamber, further corrosion of this type was avoided.

After more than four years of operation, the first tube failures occurred on the final superheaters located in the upper part of the first draft. The corroded tubes, with the heaviest metal wastages on the gas side of the lowest horizontal tubes, were replaced. After only two months of further operation, they had to be replaced again because of heavy corrosion. This time, the damaged tubes were replaced with austenitic steel. The corrosion rate was retarded and no new replacement has been necessary.

In 1972, a fifth unit of a different design was put into service. During the first two years, no tube damage occurred. Then severe metal wastage was observed on the lowest horizontal tubes of the platten-type final superheater on the tube side towards the flue gas stream. This corrosion continued despite a tube replacement by the same austenitic material, with which good corrosion resistance had been experienced in the other boilers. Detailed investigations lead to the conclusion that after the installation of a shredder during the second year of operation, nearly all the bulky refuse was burned in this unit. The combustion of crushed furniture, boards, and boxes caused changes in the composition of the flue gas and the deposits on the tube walls. These deposits, which favored the corrosion, (Thoemen, Tichatschke and Kautz, 1976 (2)) were of an unusual composition.

DESCRIPTION OF THE INCINERATION PLANT

The first four units erected in 1965/66 have the following data (Fig 1):

Steam output:	16/20t/H
Steam pressure:	80 bar (superheater outlet)
Steam temperature:	480/500 C
Furnace capacity:	70 GJ/h (16 G cal/h)

They are two-pass vertical tube boilers with natural circulation and roll grate stokers. Besides the foremost and the last part of the roof, the combustion chamber is uncooled. The heating surfaces of the first pass are radiant

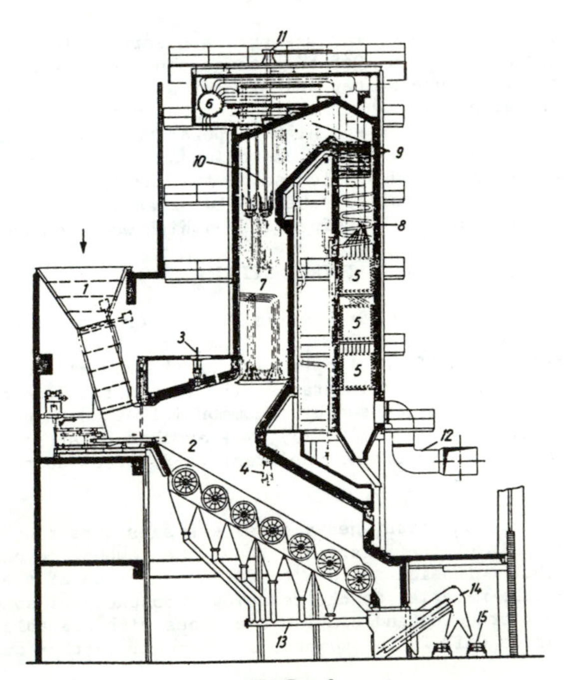

FIG. 1
FIRST GENERATION
STEAM GENERATOR

evaporators. In the upper part of the first pass, the double flow final superheater, designed as a platten-type superheater is found. The second step of the primary superheater is arranged in the gas pass between the first and the second draft, while the first step lies in the uppermost part of the second draft.

The tube materials are St 35.8 for evaporators and economizers, 15Mo3 for primary superheaters, and for the final superheaters 13CrMo44 for the inlet and middle part and 10CrMo9 10 for the outlet part. The steam temperatures within the final superheater are in the range 435-465 C at the inlet and 455-485 C at the outlet, depending on the grade of fouling of the boiler. The flue gas temperatures at the entrance to the final superheater lie in the range between 750 and 850 C.

The fifth unit has the following data (Fig 2):

Steam output:	30 t/h
Steam pressure:	80 bar (superheater outlet)
Steam temperature:	500 C
Furnace capacity:	109 GJ/h (25 G cal/h)

The boiler is designed as a four-pass vertical tube boiler with natural circulation. The construction of the combustion chamber immediately above the roll grate stoker is completely different than that of the first four units. To improve the burn-up rate, the furnace was designed so that the flue gas path is nearly twice as long as it was within the first four units before the first contact of the flue gases with tubes was made. Steam and flue gas temperatures are quite the same as in the first units.

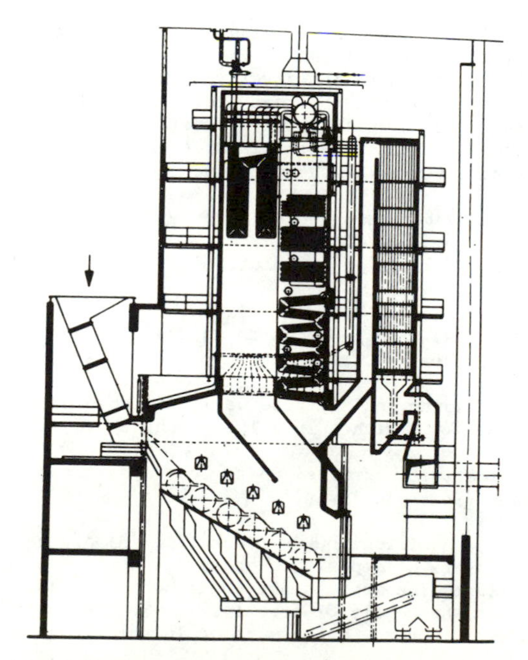

FIG. 2 SECOND GENERATION STEAM GENERATOR

TYPE OF CORROSION

1. Initial Metal Wastage of New Boiler Tubes

Investigation of superheater tubes during the first 400 hours of operation have clearly shown that an initial reaction takes place between the chloride content of the flue gas primarily as HCl and to some extent the alkali-chlorides and the steel surface. Chemical analyses of corrosion products and deposit layers from this initial period of operation show Cl contents of about 4% by weight. Deposit analyses from the same tubes after one year of operation gave Cl contents of less than 0.1% by weight. Following the initial period, the chloride containing corrosion products, as well as the precipitated chlorides, react continuously to form oxides and sulfates, i.e., a sort of protecting layer. This process can only take place as long as the presence of oxygen is guaranteed during the initial operation and the rate of fouling rich in chlorides, is not too high (Kautz, 1971 (3)). If a high fouling rate results in the chloride enrichment of the initial layer corrosion will continue.

2. Corrosion of Evaporator Tubes in Reducing Atmosphere

Corrosion of the evaporator tubes occurred in the lower parts of the front wall and the lower foremost regions of the side walls. The corroded parts of the tubes were nearly free of deposits. The deposit that formed consisted of a thin brownish layer of hygroscopic material. X-ray analyses identified the presence of Cl containing iron-oxidhydroxides believed to be products of the reaction between iron-chlorides and humid air after the sampling and a mixed chloride-sulfate $[(K,Na)_2\ (Zn,Pb,Ca)\ (So_4,Cl_2)_2]$. Chemical analyses revealed the presence of up to 11% by weight Cl6 in these thin layers. O_2 measurements of the flue gas showed that the O_2 contents in the region of the corrosion was less than <1-2% by volume. This was low compared to the values of >5% by volume O_2 normally expected.

All these results lead to the conclusion that in the absence of O_2, or at a low O_2 concentration in the flue gas, especially near the tube surfaces, the HCl content in the flue gas can react directly with the tube steel to form $FeCl_2$. This $FeCl_2$ evaporated from the tube surfaces because of the relatively low evaporation point of $FeCl_2$. Laboratory investigations of the reactions of different types of steel with HCl gas or mixtures of HCl with SO_2 and O_2 at operation temperatures proved this assumption to be true.

3. High-temperature Chloride Corrosion of Superheater Tubes

The gas-side metal wastages of the platten-type final superheater tubes of the first four units occurred.

a. Beneath the normally thick deposits on the upstream side of the lower horizontal superheater tubes;

b. Between the deposits and the tube wall, in those cases where a layer of hard scale was found with only a loose contact to the tube wall;

c. Directly on the steel surface, where the deposit consisted of a thin light brown hygroscopic layer.

Chemical and electron-microprobe analyses (Fig. 3) of the different layers showed that the thin layers of deposit on the steel surface contained 10-15% by weight Cl as chloride. The inner scale contained about 1% by weight Cl, while the outer deposit was nearly Cl-free (<1%). Neither sulfide nor sulfate could be detected directly on the steel surface in the first corrosion zone.

$[(K,Na)_2\ (Pb,Zn,CA)\ (SO_4,Cl_2)_2]$ was detected by X-ray examination. The deposit contained mainly sulfates:

$$(K,Na)_2\ (Ca,Pb,Zn)_2\ (SO_4)_3$$
$$(K,Na)_2\ (Pb,Ca,Zn)\ (SO_4)_2$$
$$(K,Na)\ (Al,Fe)\ (SO_4)_3$$
$$K_4Na_4Zn_2\ Fe_2\ (SO_4)_9$$

The K_2O concentration was about 11% by weight and Na_2O about 5% (total alkali-oxide. Concentration about 16% by weight).

The high concentration of chloride directly on the steel surface of corroded tubes originated from $FeCl_2$ formation by corrosion of the steel. The reactions leading to the $FeCl_2$ formation are as follows:

On tube surfaces on which the flue gas stream impinge directly, an intensified condensation of alkali-chlorides takes place. These chlorides react with sulfur trioxide releasing of chlorine in statu nascendi, which attacks the tube material:

$$2\ KCl + SO_3 + \tfrac{1}{2}O_2 \rightarrow K_2SO_4 + 2Cl$$

$$Fe + 2Cl \rightarrow FeCl_2$$

The rate of corrosion depends on:

a. the amount of chlorides reaching the tube surfaces which depends upon the total concentration of chlorides in the flue gas and the density of a protecting layer;

b. the rate of reaction of chlorine from the sulfatisation process with components of the deposit:

$$SiO_2 + 4Cl \longrightarrow SiCl_4 + O_2$$
$$Al_2O_3 + 6Cl \longrightarrow Al_2Cl_6 + 1.5\ O_2$$
$$Fe_2O_3 + 6Cl \longrightarrow FeCl_3 + 1.5\ O_2$$

4. High temperature sulfate-corrosion of the austenitic superheater tubes of the fifth unit.

While in Units 1-4, the austenitic tubes of the material X8CrNiNb 16 13 have shown a good corrosion resistance, in Unit 5 heavy corrosion occurred on tubes of this material after 6000 h of operation. As usually observed, the tubes were corroded on the side facing the flue gas stream, but the type of corrosion was different than that described before. Mainly white deposits formed firmly adherent to the tube wall and corrosion products covered the corroded parts (Fig 4). Locally, crystallized melt could be recognized directly on the tube surfaces. X-ray analyses (Fig 5) revealed alkali-sulfates (K_2SO_4, $K_3Na\ (SO_4)_2$) and minor quantities of $(K,Na)_3\ (Fe,Al)\ (SO_4)_3$ and $K_4Na_4Zn_2Fe_2\ (SO_4)_9$ were present as crystallization products of the melts. Chemical analyses showed that:

a. the deposits contained unusually high amounts of alkali ($\sim$20 weight-% $K_2O + Na_2O$ and only a negligible amount of chloride);

b. in the corroded regions, the chloride content was low (0.4% by weight) both in the deposits and in the corrosion products. The sulfate content was high in the corrosion products ($\sim$20 weight-%) due to the presence of alkali-iron-trisulfates beside metal oxides.

(a) BACK SCATTERED ELECTRON PICTURE

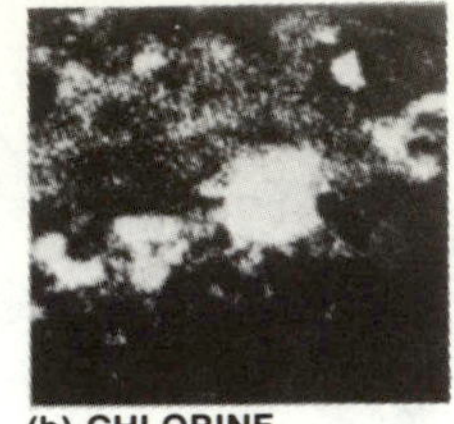

(b) CHLORINE

(c) IRON

FIG. 3
ELECTRON MICRO PROBE ANALYSIS OF A TYPICAL CHLORIDE CORROSION

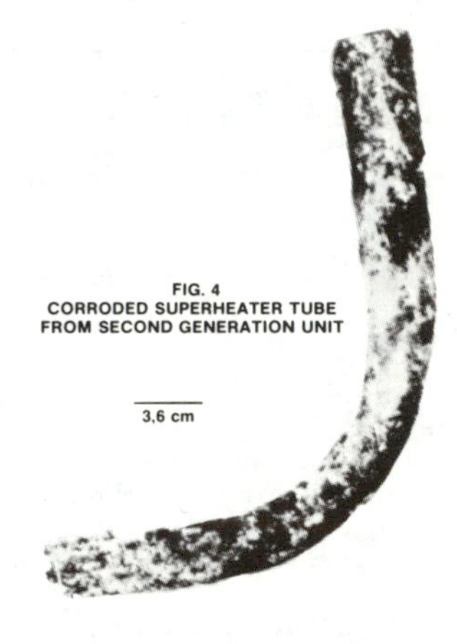

FIG. 4
CORRODED SUPERHEATER TUBE FROM SECOND GENERATION UNIT

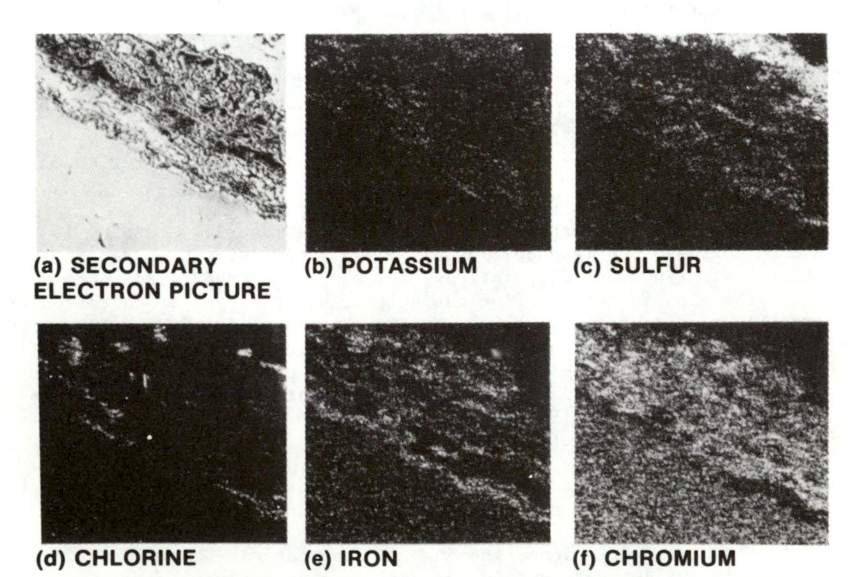

(a) SECONDARY ELECTRON PICTURE
(b) POTASSIUM
(c) SULFUR
(d) CHLORINE
(e) IRON
(f) CHROMIUM

FIG. 5
ELECTRON - MICRO PROBE ANALYSIS OF THE SULFATE CORROSION IN THE SECOND GENERATION UNIT

The lack of high amounts of chlorides in the corrosion products, the high alkali content of the deposits, the occurrence of melts, and the formation of alkali-iron-trisulfates on the steel surfaces lead to the conclusion that the corrosion was caused by:

a. attack of melts consisting mainly of sulfates as the influence of chloride may be small;

b. direct reaction of the protecting metal oxides on the steel surface with alkali-sulfates

$$Fe_2O_3 + 3K_2SO_4 + 3SO_3 \longrightarrow 2K_2Fe(SO_4)_3$$

These reactions were possible because the amount of precipitated alkali-sulfates and alkali-sulfates formed by the sulfatisation of chlorides exceeded the amount of precipitated alkaline-earth and metal oxides and sulfates. As long as there exists a balance between these phases, the alkali-phases react to form "harmless" mixed sulfates as mentioned in the previous Section 3.

The lack of chloride corrosion seems clearly a result of the good resistance of the steel X8CrNiNb 16 13 to chloride attack, even though the amount of chlorides in the flue gas and in the precipitations on the tube surfaces were higher than in the other units.

THE CAUSES OF CORROSION

The problems of initial metal wastage and of corrosion in reducing flue gas atmosphere are no longer of a major importance in incineration plants. Initial metal wastage can be kept low and stopped quickly by firing refuse of high ash content during the initial operation while reducing atmosphere can be avoided by a sufficient distribution of O_2 in the flue gas. The main problems are corrosion of different types on superheater tubes.

Early investigations during the first years of operation and the time of the superheater corrosion in Units 1-4 have clearly shown a relationship between the rate of corrosion and an increase in the heating value of the burned waste (from 1200 k cal/kg in 1966 to 1800 in 1972). Over the same period, there was a rise in chloride content in the fly ash (4.3 mg Cl in 1966 to 20 mg in 1971).

In February 1975, new investigations were started on the relation between chloride contents in fly ash and the combustion of the precrushed shredder waste. The results of chemical analyses on fly ash from Units 1+2 (without shredder waste) and Unit 5 (burning mainly shredder waste) revealed the following results (Thoemen, Tichatschke, Kautz, 1976 (2)).

Fly Ash Unit 5		Fly Ash Units 1+2	
K mg/g	Cl mg/g	K mg/g	Cl mg/g
21,9	40,2	14,2	16,9

In November 1975, the investigations were continued and extended and the result shows a clear dependence upon the combustion of (mainly) wooden products (shredder waste) and the chloride content in fly ash:

	Shredder in Operation		Shredder not in Operation	
	K mg/kg	Cl mg/kg	K mg/kg	Cl mg/kg
Fly Ash, Units 1+2	20,8	25,7	17,2	22,1
Fly Ash, Units 3+4	21,3	22,1	25,0	19,0
Fly Ash, Units 5	35,0	44,8	22,5	21,8

Chemical analyses of the flue gas showed that the alkali content of the wood reacted during combustion to form chlorides:

	Shredder in Operation			Shredder not in Operation		
	HCl	HF	SO_2(mg/m^3, 11%O_2)	HCl	HF	SO_2(mg/m^3, 11%O_2)
Unit 2	950	9,1	500	810	6,2	410
Unit 5	570	5,4	590	930	7,9	420

Thus, as the HCl content in the flue gas from the combustion of certain plastics is always above a certain level, the amount of formation of alkali-chlorides and sulfates and their precipitation on tube surfaces depends on the combustion of alkali containing materials, such as furniture and boards. The rise in precipitation of alkali-phases on tubes increase the danger of corrosion.

CONCLUSIONS

Incineration plants are normally designed for the combustion of a wide range of types of refuse under the name of municipal refuse. More than 10 years of operational experiences have shown that as long as the mixture of all refuse components can be controlled to contain a certain level of chemicals in the flue gas, no corrosion should occur. Although the combustion of only wooden products should never cause corrosion problems in incineration plants, the common firing of Cl-containing materials along with a larger amount of wood becomes dangerous in respect to corrosion. Therefore, a thorough mixing of all refuse components must be performed. If this is not possible, combustion of the wood products should be planned in a separate unit. If improvements of this type cannot be introduced, other local remedies might include using:

a. protecting strips of heat resisting high alloy Cr-Si-steel which has to be replaced every 3-4 months, fixed to the parts where corrosion is experienced;

b. and materials of higher corrosion resistance, such as an austenitic steel containing Columbium (nb).

The development of the latter possibility includes, besides laboratory work, a long period of testing of materials in the field. This is a task for the future.

ACKNOWLEDGEMENTS

I thank Mr. K. H. Thoemen and Mr. J. Tichatschke of Stadtwerke Duesseldorf for the year long cooperation which made this paper possible.

REFERENCES

1. Kautz, K. and Tichatschke, J., "Relationships between Flue Gas Ratios, Boiler Loading, and Corrosion in a Municipal Refuse Incineration Plant," Mitteilungen der Vereinigung der Grosskraftwerksbetreiber, No. 52, 1972, pp. 249-258.

2. Thoemen, K-H., Tichatschke, J., and Kautz, K., "Recent Findings about Heating Surface Corrosion in Refuse Incineration Plants," Mitteilungen der Vereinigung der Grosskraftwerksbetreiber, No. 56, 1976, pp. 715-721.

3. Kautz, K., "Causes of Corrosion in Municipal Refuse Incineration Plants," Mitteilungen der Vereinigung der Grosskraftwerksbetreiber, No. 51, 1971, pp. 398-402.

MECHANISMS OF CORROSION AND FOULING IN REFUSE-FIRED BOILERS

Session Chairman: **H. H. Krause**
Battelle Columbus Laboratories
Co-Chairman: **D. A. Vaughan**
Battelle Columbus Laboratories

CORROSION OF AN ELECTROSTATIC PRECIPITATOR IN A MUNICIPAL INCINERATOR IN SYDNEY, AUSTRALIA, AND SOME ASSOCIATED RESEARCH WORK

E. C. POTTER and P. C. NANCARROW

CSIRO Minerals Research Laboratories
North Ryde, Sydney, N.S.W., Australia 2113

ABSTRACT

Corrosion is described that, in a few months, put out of action the electrostatic precipitator attached to a municipal incinerator in Sydney, Australia. The problem has been successfully overcome by remedial measures requiring enlargement of a water spray tower, adjustment of sprays, improvement of gas distribution in the precipitator, and reliable temperature control. Laboratory studies are described of the oxidation of low-carbon steel at 300°C in mixtures of nitrogen, oxygen, water vapor, and hydrogen chloride. Protective films are formed even in dry hydrogen chloride, but their hygroscopicity and the occurrence of mixed films are considerable hazards to reliable protection of underlying metal.

INTRODUCTION

The City of Sydney, Australia, has customarily disposed of its refuse by tipping, but suitable sites are becoming more difficult to find. It was, therefore, a considerable departure when five of the city councils representing a combined population of 400,000 collaborated in building a new municipal incinerator to destroy 520 tonnes of refuse per 24-hour shift. The incinerator commenced operation in mid-April 1973 and in the following July corrosion of the electrostatic precipitator wires and plates was detected. The situation worsened in the ensuing months until the precipitator could no longer reduce the particulate emissions below the statutory limit. The councils were then obliged to close down the entire incinerator until repairs and vital alterations to the plant had been made. This paper describes the crippling corrosion that the incinerator plant encountered so soon after it had been commissioned, and goes on to consider the gaps in basic corrosion knowledge that the incident revealed.

DESCRIPTION OF THE PLANT

There being no exceptional features of the refuse, it was decided to burn it directly in a stepped-grate furnace without pre-sorting of any constituents. Since there was insufficient capital to incorporate a boiler/turbine combination and no market could be found for the output of a boiler alone, the heat of combustion of the refuse was wasted by cooling the combustion gases in a water spray tower. This produced a flue gas rich in water vapor (about 40%) at 350°C. The spray tower was immediately followed by a conventional wire-and-plate electrostatic precipitator consisting of two fields in series in the one casing. The combustor/spray tower/precipitator combination (see Fig. 1) was duplicated and all the cleaned flue gases were discharged through a single chimney.

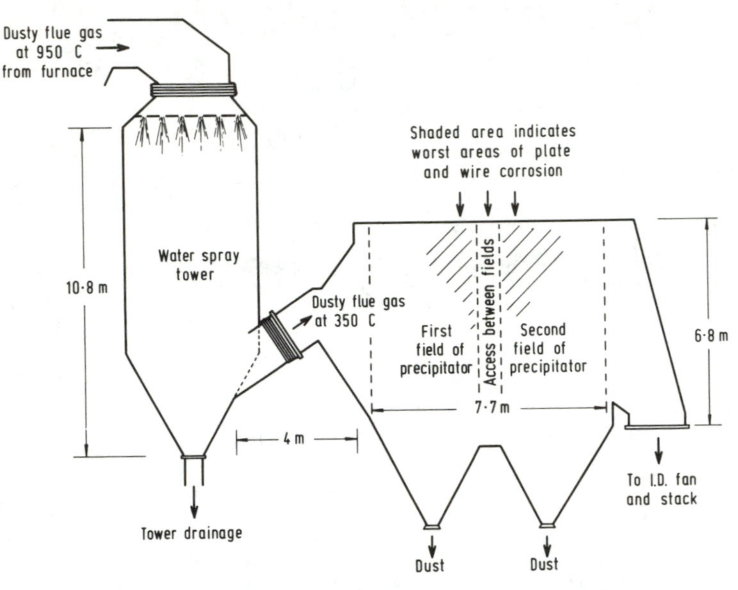

Fig. 1 Layout of spray tower and electrostatic precipitator at Waverley-Woollahra incinerator in Sydney, Australia (approximately to scale)

The conceptual design followed essentially the recommendations of a feasibility study, and the firm carrying out the construction work sub-contracted out the building of the spray tower and the precipitators. It is stressed that there was no fault to be found with either the combustion of the refuse or the cleaning of the flue gases, nor with the efficiency of either process. Two features of the design should be noted, however. First, the spray tower was rather close to the precipitator inlet, necessitating a steep entry upwards of dusty flue gases through the connecting duct into the precipitator; and second, the precipitator inlet was provided with gas-diffusing screens to distribute the flow, but no testing had been carried out to check their effectiveness.

FEATURES OF THE CORROSION

In all cases the corroding metal was low-carbon steel. The corrosion affected precipitator wires and plates, but was obvious first with the wires, a few of which gave way and caused short circuits. The plates also thinned and several that were extensively perforated were condemned. The corrosion product itself was not obvious. In terms of amount of oxidation the plates suffered most, but the rapping of the plates dislodged both collected dust and corrosion product, exposing the thinning and perforation. The wires thinned locally owing to corrosion, but the product merged with collected dust, forming a crust at the corroded areas that was almost impossible to dislodge by the normal wire-rapping mechanism. By the time any corroded metal was recovered for laboratory examination, nothing but iron oxide could be seen as corrosion product, although some early unconfirmed testing claimed to show a thin layer of ferric chloride next to the corroding metal surface.

The most significant feature of the corrosion was its distribution in the precipitator. The really serious corrosion was confined to the top area of the precipitator and was first evident at the end of the first field and the beginning of the second. Elsewhere the plates and wires were unharmed, being covered with the expected powdery coating that incombustibles from furnaces form in precipitators. Over the corroded areas, however, the deposits were a firm cake encrusted on to the metal, and on the wires especially the cake had segregated into toroidal masses. The deposit on the wires reached a diameter of 3 cm in places and could just be crumbled away by strong hand pressure.

FEATURES OF THE COLLECTED ASH

A sample of the encrustation from a corroded wire was 79% water soluble, a compatible analysis for this material being 50% potassium sulfate, 38% sodium sulfate, 7% potassium chloride, and 5% sodium chloride. The water-insoluble portion was a fine refractory silicate. Under the hot-stage microscope no molten phases were observed in the collected dust up to 600°C, the highest temperature reached, but a little water of crystallization was released at about 150°C. The flow temperature of ash collected from the precipitator hopper was found to be 1120°C, a reasonable value for incinerators. The water solubles from the ash gave an aqueous solution with a pH slightly above 6 (Sydney drinking water is a soft supply with pH 7.4 and averaging 30 ppm of chloride as sodium chloride).

FEATURES OF THE EARLY OPERATION OF THE PLANT

During the first weeks of operation after initial start-up in April 1973, it was extremely difficult to adjust the precipitator inlet temperature to the intended 350°C using the spray tower. Temperatures were generally in the range 300-320°C, but excursions to 450°C also occurred. The most frequent aberration was, however, the flooding of the spray tower, meaning that the evaporation was incomplete, and water had to be drained from it. When corrosion was first seen, the operators agreed with the contractors to reduce the precipitator inlet temperature to 250-270°C, but the additional water needed to achieve this could not evaporate in the tower in time, with the result that the inlet screens were clogged with damp dust and the hoppers and their dust conveyors likewise choked. Dust sampling equipment was also blocked, even when the temperature was raised to 310°C to keep the precipitator itself unobstructed. None of these measures checked the corrosion and within a year of start-up, major modifications and repair were necessary.

Two other observations made during this period appeared most relevant. First, a survey of gas velocities made close to the precipitator inlet showed large and unacceptable variations from top to bottom of the precipitator, although the mean velocity was acceptable at 1.2 ms^{-1}. The top of the precipitator at the inlet was encountering 2½ to 3 times the average gas velocity, while most of the middle and lower areas experienced less than half the average gas velocity. The extreme velocities at the top were 25 times the lowest velocities in the centre, and clearly the top corroded areas were subjected to a quite disproportionate burden of dust collection. The second observation concerned the nozzles in the spray tower, which were found to be incapable of spraying evenly owing to a maladjustment, probably in assembly.

DIAGNOSIS OF THE CORROSION

Several authorities were consulted about the corrosion, but it proved impossible to decide definitely which of two mechanisms was responsible. The *more obvious* mechanism was attack by hydrochloric acid released by the thermal decomposition of polyvinyl chloride (PVC) present in the refuse. One way to express the suggested sequence of events (see eqns. 1-3) is to suppose that the initial corrosion product, ferrous chloride, is oxidized to ferric chloride and ferric oxide, after which steam hydrolysis regenerates the hydrogen chloride.

Thus a minute amount of hydrogen chloride could do an enormous amount of damage to iron if the humid flue gas had good access to the corrosion site and the regenerated hydrogen chloride had difficulty in moving away (the hydrogen not surviving for long in the oxidizing conditions).

$$1 \qquad Fe + 2HCl \rightarrow FeCl_2 + H_2 \qquad (1)$$

$$6FeCl_2 + 3/2 O_2 \rightarrow 2Fe_2Cl_6 + Fe_2O_3 \qquad (2)$$

$$Fe_2Cl_6 + 3H_2O \rightarrow Fe_2O_3 + 6HCl \qquad (3)$$

However, it is difficult to visualize that, during the above sequence of reactions, the balance between the porosity of the deposits and the diffusion of gases through them could be so consistently favorable as to sustain such a rapid corrosive attack as occurred in the present instance (up to 7.6 mmy^{-1} (300 mpy)), particularly since the PVC content of the refuse (ca 0.1%) was subnormal by present-day standards.

The *more favored* mechanism of the corrosion supposed that for relatively long periods the upper areas of the precipitator were collecting particles that had had insufficient time to dry out, so that, in effect, the affected wires and plates were immersed in a boiling saturated solution of the mixed sulfates and chlorides of sodium and potassium in the presence of a plentiful supply of oxygen. Thus the metal at risk was subjected to an environment conducive to rapid rusting under near-neutral conditions.

It will be noted that the acid mechanism does not suggest a ready remedy to the corrosion, whereas the neutral mechanism suggests that the spray tower must be made to deliver a dry powder reasonably uniformly to all areas of the precipitator.

REMEDIAL MEASURES

The inadequacy of the spray tower came to be accepted and it was enlarged by increasing its height 40%. The faults in the spray nozzles were corrected, and vanes were installed at the precipitator inlet to improve gas distribution. Much better control of temperature (avoiding both non-uniformity and excursions) was instituted, this requirement overshadowing the actual value of temperature (250°C) selected for the precipitator. The two precipitators were repaired using low-carbon steel as before. The plant was returned to service in June 1976 and one year later had shown no sign of precipitator corrosion especially on wires, which may be considered the most vulnerable component, being subject to fatigue and stress. Most significantly, the spray tower does not now get wet internally, and the collected dust is powdery and dislodges properly from plates and wires.

RESEARCH AFTERMATH

The enquiries described above revealed a situation of great scientific interest, in that there seem to be no correlated scientific data available on the oxidation rates of iron or steel in air or comparable film-forming gases, at say 200-400°C in the presence of small concentrations of hydrogen chloride. Still less is known about the corrosion phenomena when flue gases (typically depleted air enriched in water vapor and carbon dioxide) are the carrier for the hydrogen chloride. The authors have thus set out to remove some of the more obvious gaps by a program of laboratory study.

Municipal refuse is rich in organic matter and burns well in excess air to form a gas containing, for example, 10% oxygen, 20% water vapor, 10% carbon dioxide, and minor but significant amounts of oxides of nitrogen and sulfur, hydrogen chloride and chlorine. This flue gas also contains particulate matter, mainly refractory siliceous material plus chlorides and sulfates of sodium, potassium, magnesium, calcium, aluminium, zinc, and iron. Corrosion of oxide-covered mild steel in such a mixture is complex, but a notable attempt to rationalize the thermodynamics has been made by Hirsch (1) publishing with Rasch (2). They do not include carbon dioxide fully in their analysis and the effects of fusible ash deposits are not considered in detail, but valuable indications come from their work. Incidentally, there is some appeal in thinking of oxygen, water vapor, and carbon dioxide as film-formers, ranged against hydrogen chloride, chlorine, and fusible salts as film-destroyers. In addition, condensed aqueous solutions forming at temperatures below the dew-point are also likely to be film-destroyers.

Hirsch considered a range of compositional variations, but a relevant choice is a dust-free flue gas containing 30% water vapor, no carbon monoxide, 9% oxygen, and hydrogen chloride and chlorine equivalent to 0.5% polyvinyl-chloride in the original refuse. He considered the hydrochlorination and chlorination of both ferric oxide and magnetite and found the equilibria heavily to the oxide side at relevant temperatures. Thus we expect that a sound oxide film on steel will survive attack by the dust-free flue gas, and there is even an opportunity to heal a faulty protective film if ferrous chloride has formed by some direct attack on exposed iron. Furthermore, except at the lowest relevant temperatures (say below 250°C), hydrogen chloride predominates over chlorine, and ferrous chloride predominates over ferric chloride. Accordingly, in none of the instances of mild steel corrosion seen in incinerator auxiliaries around 300°C have the present authors been able to detect ferric chloride - although ferric oxide has been common.

Notwithstanding the thermodynamics, there are possibilities of microenvironments depleted in oxygen and water vapor existing at metal surfaces beneath deposits and films, and consequently the kinetics of formation and healing of protective films may be slow or impeded under some conditions. The range of interesting variables is large (simultaneously 0-10% oxygen, 0-40% water vapor, 0-0.05% hydrogen chloride, and 0-10% carbon dioxide), not to mention time, temperature, metal-surface preparation, and effects of fusible or hydrolysable deposits.

Our oxidation experiments have been carried out with thin discs of low-carbon steel (3.5 cm diameter, exposed area 19 cm^2) in a slow stream (0.5 mms^{-1}) of mixed dust-free gases at 300°C.

The gas stream has been made up to contain various proportions of nitrogen, oxygen, water vapor, and hydrogen chloride (see Fig. 2), and weight gains have

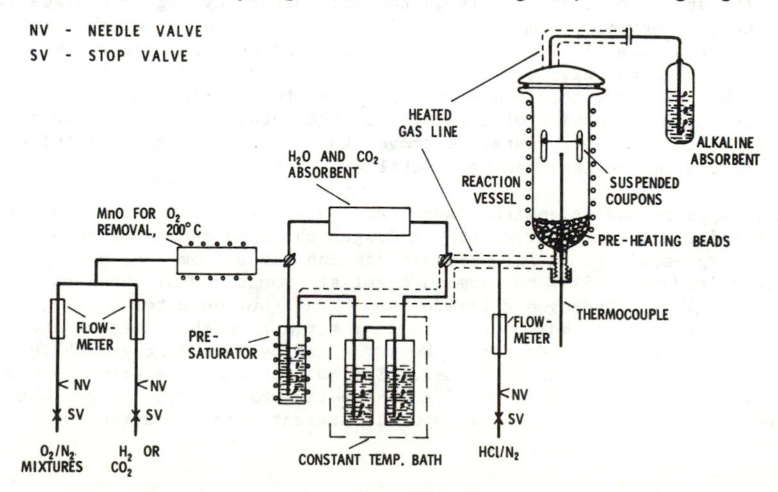

Fig. 2 Laboratory apparatus for corrosion tests in mixed gases

been traced with time and the appearance of the surface films assessed. Initially the metal surface carried the usual air-formed oxide known to be about 4nm thick, and oxidation then proceeded up to about 100 hours of exposure.

In all cases, the rate of corrosion fell with time, corresponding to the formation of protective films (see Fig. 3). Concentrations of hydrogen

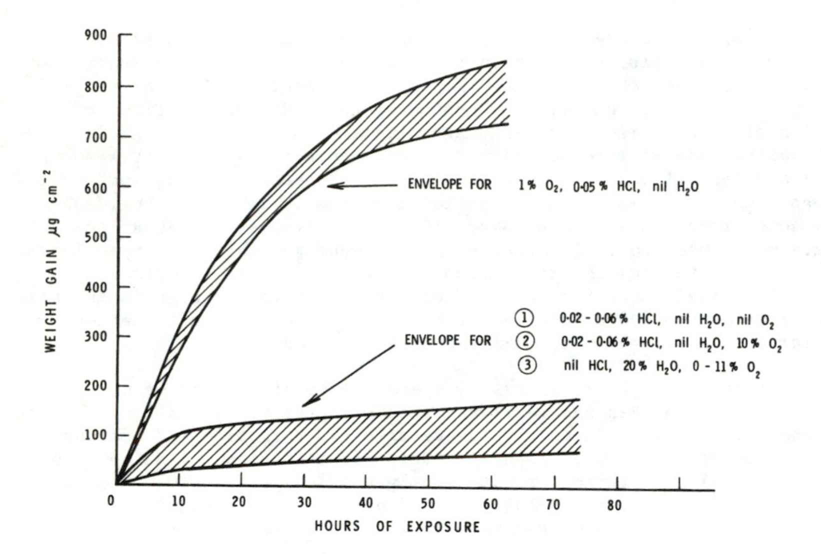

Fig. 3 Oxidation of low-carbon steel in various gases at 300°C

chloride up to 0.06% by volume have been studied, but it is difficult to discern any lasting kinetic effect of hydrogen chloride on film-formation because the initial rate of corrosion is higher, but film stifling is faster, leading to the same protective result in the end whether hydrogen chloride is present or not. However, an important difference in the nature of the protective films was that the film formed in gas containing hydrogen chloride (oxygen and water being very low or absent) was not stable in air at ordinary temperatures but was hygroscopic and caused quick atmospheric rusting to develop. Indeed, a proportion of this film on the metal was water-soluble, and leaching of the film with water destroyed the cohesion of the remaining insoluble portion (magnetite) and bare metal was exposed.

Thus it appears that plentiful oxygen and water vapor in flue gases may have a beneficial role in ensuring that hydrogen chloride and chlorine produced in large-scale waste incinerators are inhibited from causing hygroscopic protective oxide films to form on steel at around 300°C. When the oxygen level is 1% and hydrogen chloride is present (but no water vapor), there is up to eight times more corrosion in a given time than without the oxygen or when 2% or more is present. Nevertheless, a parabolic growth is observed and a protective film is formed. The magnitude of the effect can be seen in Fig. 4, which shows that an effect is noticeable at 0.1% oxygen, but there is no such effect if 20% water vapor is present with 1% oxygen

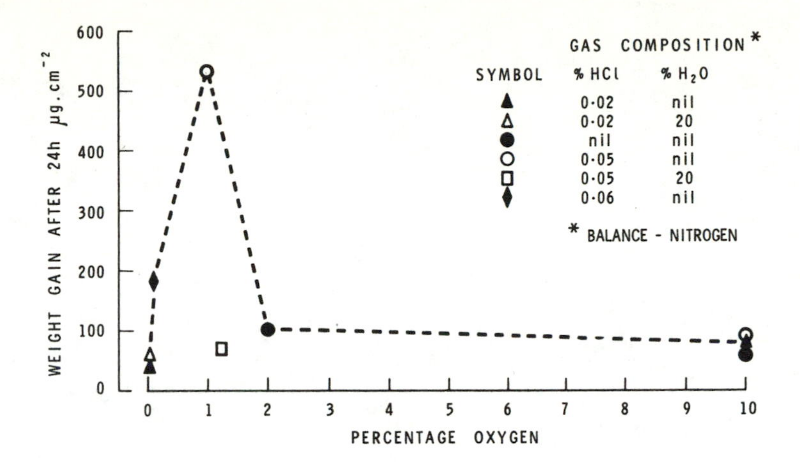

Fig. 4 Effect of oxygen concentration on oxidation of low-carbon steel in various gases at 300°C

Our explanation of this effect lies in the combined influence of oxygen and hydrogen chloride to produce mixed films of magnetite and metal chloride. These mixed films, we suggest, do not have the compaction and homogeneity that pure films of either magnetite or metal chloride possess, and consequently they allow more metal corrosion to occur before they reach a protective thickness. Apparently this mutual interference in protective ability is greatest when the oxygen level is near 1%, at least at the levels of hydrogen chloride we have examined. Above 2% oxygen it appears that the unimpeded protection of magnetite is guaranteed. Alternatively, this full protection may be assured by (for example) 20% water vapor, in which case the amount of oxygen present is irrelevant.

CONCLUSION

This work is continuing, but it will be some time before sufficient cover of the relevant variables is obtained for conclusions of applied importance to emerge. In particular, the modifying effects of ash and other salt-bearing deposits on the corrosion processes have yet to be assessed.

BIBLIOGRAPHY

1. Hirsch, M., Energie, 20(2), 32 (1968).

2. Hirsch, M., and R. Rasch, Aufbereit. Tech., 12, 614 (1968).

THE INFLUENCE OF HEAVY METALS Pb AND Zn ON CORROSION AND DEPOSITS IN REFUSE-FIRED STEAM GENERATORS

Z. E. KEREKES, R. W. BRYERS, and A. R. SAUER

Foster Wheeler Development Corporation
Livingston, New Jersey, USA 07039

INTRODUCTION

The burning of refuse in waterwall-cooled steam generators to reduce its bulk and provide an inexpensive source of energy from a low-sulfur fuel has introduced new corrosion and deposit problems. The cause for alarm that refuse might be a severe fouling and corrosive fuel originated from experience with the European units placed into service in the mid-1960s. Much of the original tube wastage in the European units has been minimized by improvements in design and operation. However, reports of corrosion and deposit problems continue as new units are commissioned. To gain a better understanding of the cause of fireside problems resulting from the combustion of refuse, Foster Wheeler initiated a program which included an extensive literature survey, visits to field installations and a comprehensive analysis of samples removed from tube surfaces at various installations. It was learned that the sulfates and chlorides of lead and zinc in addition to the alkalies were contributing to the deposit and corrosion problem. Recently the study was extended to include the analysis of deposits removed from six installations using thermal analysis, the scanning electron microscope and x-ray diffraction, to determine the roles of lead and zinc.

INITIAL EXPERIENCE IN EUROPE

The initial corrosion and deposit problems were brought out in a series of articles published in the mid-1960s describing severe corrosion problems in furnace and high-temperature gas passes shortly after start-up[1,2,3,4,5,6,7,8]. The corrosion in general occurred on the leading edge of superheater tubes in the first rows of the bundles underneath nonporcelainized deposits where tube-metal surface temperatures exceeded 825°F (441°C) and furnace wall tubes in the vicinity of 600°F (316°C) were subjected to reducing conditions. The corrosion product tenaciously clung to the tube surface in some instances while in other cases it could be peeled off like bark from a tree. In Munich there were few isolated cases in which corrosion occurred on the bare downstream side of the tube.

Corrosion in convection superheater banks occurred at a catastrophic rate at first, gradually subsiding as the ash accumulation increased. In numerous cases the deposits appeared to contain H_2S which was released upon heating or crushing. One investigator indicated that the corrosion rate decreased as the ash content in the winter refuse increased[8].

The most severe corrosion occurred on radiant superheater walls at tube metal temperatures of 950°F (510°C) and on waterwalls at tube metal temperatures of 600°F (316°C). The evaporator tubes were especially endangered in the corner of the furnace, while the material loss of the superheater tubes was distributed equally. Just as in the convection pass, thin oxide layers were found beneath the shell-like deposits.

The first furnace wall tube failures at Stuttgart were reported after 5500 operating hours, and the corrosion appeared to be linear with time[9]. Hilsheimer[10] reports similar problems at Mannheim after 3000 hours. The problem was corrected by installing studded tubes covered with SiC in the corrosion zone.

Nowak[11], Hilsheimer[10] and Maikranz[12] all report the existence of local reducing conditions on the front wall of the furnace despite 80% excess air supplied to the primary combustion chamber. This condition is believed to be the result of gases released by destructive distillation of the refuse in the first stage of combustion which remain unmixed with the combustion air or flue gas and pass as a laminar layer up the front wall. It is further believed that this reducing layer is one of the prime reasons for the high corrosion rates. This situation was remedied by a modification in design.

FIELD SURVEY

The first experimental program begun by Foster Wheeler attempted to identify the troublesome constituents in the ash and hopefully to establish the conditions under which corrosion and fouling were taking place[13,14,15]. The program included extensive sampling and analyzing ash deposits removed from about 30 different locations in each of three different boilers under study in Europe. Each boiler represented slightly different operating conditions. One boiler was fired by refuse alone, another boiler was fired with coal and refuse in separate furnaces and the third boiler was fired with oil and refuse in separate furnaces. A smaller number of samples was also taken from some other boilers operating in Europe as well as in the United States.

The sampling points include numerous locations on all four walls of the furnace, representing several elevations, upstream and downstream sides of the tubes located in the superheater bundle and numerous locations in the economizer. These sampling points provide good representative samples of the ash deposited at various temperature levels and gas-to-tube temperature differentials during its flight through the boiler.

The samples were chemically analyzed for the elements present using gravimetric, x-ray fluorescent and emission spectrographic techniques. They were also examined and tested for initial deformation temperature, softening temperature and fluid temperature. Laboratory analysis revealed zinc and lead in relatively large quantities. Neither of these two elements was anticipated. Zinc, like sodium and potassium, appeared rather consistently in most ash sampled, running between 5 and 10% and generally increasing in concentration with a decrease in gas temperature. Lead appeared somewhat sporadically, ranging from 0 to 11%. Presumably, these two elements became volatile during combustion and condensed on cold tube surfaces in the cooler gas zones similar to the alkalies.

The minor constituents, defined as sodium, potassium, zinc and lead, comprise about 18-25% of the ash sampled at gas temperatures below 1700°F (928°C). The concentrations decrease with an increase in temperature above this point, being approximately 15% at 1800°F (932°C) and 10% at 2000°F (1093°C), as shown in Figure 1. It was further noted that a rather constant relationship appears to exist between the percent zinc and the combined percentage of sodium and potassium, as shown in Figure 2. If lead is included with the zinc, it is found that a linear relationship exists, but in slightly different proportions. This could be interpreted to mean that zinc, sodium, potassium and lead are depositing as some discrete compound or that liquid solutions are solidifying with a particular composition. An examination of the ternary system of Na_2SO_4, K_2SO_4, and $ZnSO_4$, indicated that the percent zinc, sodium and potassium were reported in proportions that coincided with low melting temperature phases, 720-740°F (382-393°C), in this system (30-50% K_2SO_4, 40-60% $ZnSO_4$, and 10-30% Na_2SO_4), as shown in Figure 3.

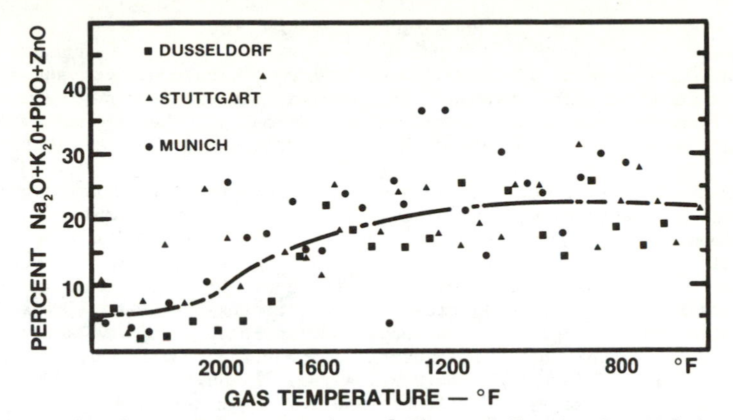

Figure 1 Percent Concentration of Elements Forming Low-Melting Compounds in the Ash Expressed as Oxides as a Function of Flue Gas Temperature.

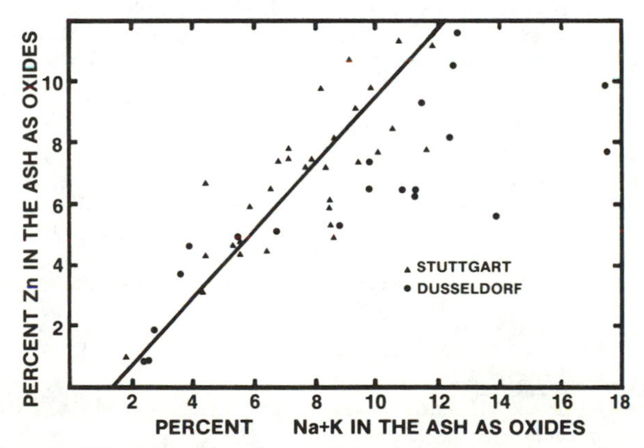

Figure 2 An Illustration of the Relationship Between the Percent Zn found in the Ash Expressed as an Oxide and the Alkalies K and Na also Expressed as Oxides.

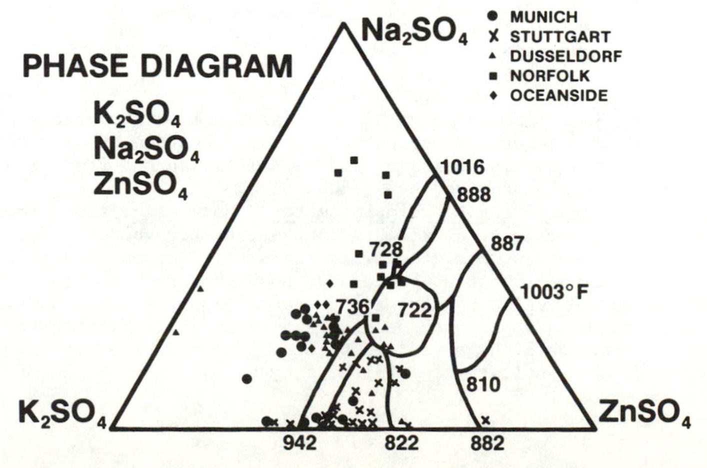

Figure 3 Super-Position of the Elements Na, K, Zn Expressed as Sulfates on the Na_2SO_4-K_2SO_4 $ZnSO_4$ Phase Diagram.

MOST RECENT PROGRAM

Recently the program was extended to explore the role of lead and zinc in deposit formation and tube wastage. Six samples were examined representing four different installations, two European and two domestic. The samples were selected on the basis of the zinc and lead concentration to reveal the contribution of each as well as their interrelationship.

All deposit samples were analyzed for the elements present using x-ray spectroscopy, atomic adsorption, optical emissions spectrography and volumetric analysis. The analyses were performed on composite analytical samples removed from each deposit. The corrosion product lining each deposit sample was then separated from the covering deposit, to permit examination of the tube-deposit interface (location designated as inside), cross section (center) and outer layer (outside), by means of scanning electron microscopy, nondispersive x-ray and x-ray diffraction techniques. A portion of the corrosion product was also subjected to thermal analysis including differential thermal analysis (DTA) from 80 to 1000°C, differential scanning calorimetry (DSC) from 80 to 600°C, thermal gravimetric analysis (TGA) from 80 to 900°C to determine the existence of low temperature melts and to verify the thermal behavior of various compounds responsible for the fireside problems. Thermal gravimetric analysis (TGA) or cycling of the DTA was used to distinguish melts from dissociation and dehydration of deposited material at low temperatures. Thermal analysis was performed on thirteen synthetically created binary systems to assist interpretation of the thermograms generated by the ash samples.

TESTING PROCEDURE

All deposit samples analyzed exhibited the same general characteristics. A dark-reddish dense corrosion product occurred at the tube-deposit interface partially surrounded by a white semi-dense formation throughout the thickness of the deposit. The outside layer of the deposit was generally smooth and very dense.

The following analytical procedures were used to examine the ash deposits:

A. Chemical Analysis

1. Analytical Techniques

Chemical analyses of the elements were conducted on all deposit samples using the following techniques:

X-Ray Spectroscopy: Si, Al, Ti, Fe, Ca, K, S, P
Atomic Absorption Spectrophotometry: Mg, Na, Mn, Ni, Cr, Zn, Pb, Sn, Cu, Ag, Sb
Optical Emission Spectroscopy: Mo, V
Volumetric Analysis: Cl

2. Instrumentation
   a) X-Ray Spectroscopy Apparatus: Model SRS-1 unit operates with 4000-watt generator sequential vacuum spectrometer, equipped with pulse-height discriminator and scintillation and flow-proportional counters manufactured by Siemens Corp.
   b) Atomic Absorption Spectrophotometer: Model AA-6 single-beam instrument featuring automatic gas control, background correction and digital readout operating with flame emission and flameless atomic absorption principles, manufactured by Varian-Techtron Corp.
   c) Optical Emission Spectrograph: Jarrel-Ash 1.5 meter Wadsworth Spectrograph equipped with Jaco arc-spark accessories operated by a Jaco vari-source of high voltage DC and AC arc-spark excitation. A 35-mm film is quantitatively evaluated by using a Jarrel-Ash microdensitometer.

B. Thermal Analysis (DTA, DSC, TGA)

1. Analytical Treatment

Thermal analyses were conducted on six deposit samples and thirteen synthetic binary systems of lead, zinc and iron. Three specimens were removed from the inside (tube-scale interface), center and outside surface of five of the ash deposits. The sixth deposit sample consisted of coarse and fine particle fly ash. All specimens were submitted for differential thermal analysis (DTA) using air and nitrogen; differential scanning calorimetry (DSC) by using air, nitrogen (N_2) and reducing gas (50% CO_2, 32% CH_4, 20% H_2) atmospheres; and thermal gravimetric analysis under air.

Thirteen synthetic binary systems were selected to study the thermal characteristics of the compounds by using DTA and TGA techniques. Twenty-five grams of each system were prepared by mixing each one of the thirteen for four hours in the Fisher-Kenthal mixer to obtain uniform distribution of the components. The thirteen systems were divided into two groups as follows:

TABLE 1

GROUP A			GROUP B		
Sample	Percent by Weight	Constituents	Sample	Percent by Weight	Constituents
1A	48%	KCl	1B	10%	KSO_4
	52%	$PbCl_2$		90%	$PbSO_4$
2A	20%	$PbCl_2$	2B	80%	PbO
	80%	$ZnCl_2$		20%	$PbCl_2$
3A	45%	KCl	3B	70%	NaCl
	55%	$ZnCl_2$		30%	$PbCl_2$
4A	58%	$ZnCl_2$	4B	55%	KCl
	42%	$FeCl_2 \cdot 4\ H_2O$		45%	$ZnCl_2$
5A	30%	NaCl	5B	21%	$ZnCl_2$
	70%	$PbCl_2$		79%	$FeCl_3 \cdot 6\ H_2O$
6A	20%	PbO	6B	4%	$ZnSO_4$
	80%	$PbCl_2$		96%	Na_2SO_4
7A	17%	NaCl			
	83%	$FeCl_2 \cdot 4\ H_2O$			

Three basic chlorides, namely zinc chloride ($ZnCl_2$), lead chloride ($PbCl_2$) and iron (II) chloride ($FeCl_2 \cdot 4\ H_2O$) were selected to make up the seven (Group A) synthetic binary systems. The combinations of the compounds grouped in the following order are tabulated below:

TABLE 2

BASIC COMPOUND	Group No.	$ZnCl_2$	Group No.	$PbCl_2$	Group No.	$FeCl_2 \cdot 4\ H_2O$
ADDITIVE COMPOUNDS	2A 3A 4A	$PbCl_2$ KCl $FeCl_2 \cdot 4\ H_2O$	1A 5A 6A	KCl NaCl PbO	7A	NaCl

Group B includes sulfates of Pb, Zn, Na and K and various combinations of chlorides appearing in Group A but in different proportions.

2. Analytical Equipment

A DuPont 990 thermal analyzer unit was used with various modules (DSC, DTA, TGA) operating in conjunction with a sensitive temperature programmer-control and X-Y recorder system.

C. SEM and Nondispersive X-Ray Analysis

1. Sample Preparation

The specimens were secured in place on aluminum sample mounts with conductive carbon paint. After the paint had dried, the specimens were put into a Model EMS-76 sputtering unit manufactured by Film-Vac, Inc. At a pressure of approximately 175 microns, gold was deposited on the specimens. The sputtering current and time were 10 ma and six minutes respectively. The film thickness for these conditions was approximately 600 angstroms.

2. Instrumentation

The Etec-Autoscan scanning electron microscope system, performing an electron optic and nondispersive x-ray pattern of the element spectrum, was used for the investigation. The electron-optical column lens system is capable of magnifying from 10 to 30,000 diameter, with a resolution limit of approximately 100 angstroms. The accelerating voltage was in the range of 5-30 KV.

D. X-Ray Diffraction Analysis

1. Sample Preparation

Samples were ground in the Model No. 8000 Spex Mixer/Mill and passed through a 200-mesh screen. The powder was then pressed onto an aluminum sample holder. When only a small sample is available, a glass slide with a thin film of petroleum jelly is used as the sample holder.

2. Instrumentation

The Siemens x-ray diffraction apparatus consists of a 3000-watt generator operating with a copper target x-ray tube. A diffraction pattern is obtained by a strip chart recorder from a diffractometer equipped with a diffracted beam monochromator and scintillation counter. The diffraction patterns were evaluated by using a computer-aided JCPDS telesearch program.

CHEMICAL ANALYSIS

The chemical analyses of the deposits sampled are summarized in Table 3. The results indicate the elements present in their oxidized state as a percent-by-weight of the sample. Samples 1 through 3 were taken from the same European boiler at different locations within the boiler. The first three samples were selected as the basis of the ash chemistry. Sample 1 is relatively low in zinc and medium in lead. Sample 2 was low in lead and high in zinc. The third sample is high in both lead and zinc. All three samples contained large quantities of sulfur reported as sulfur trioxide, calcium, potassium and sodium. The chlorine in all three was extremely low. Sample 4 is fly ash taken from a second European boiler reporting serious corrosion. The ash chemistry indicated large quantities of chlorine and alkalies and only moderate quantities of sulfates, lead and zinc. The calcium concentration is the highest in Sample 4.

Samples 5 and 6 removed from two domestic boilers contain large quantities of sulfates as in the case of Samples 1, 2 and 5. Sample 6 is rich in lead while Sample 5 may be considered nearly void of it. The zinc, potassium, and calcium concentrations in both samples are high while the chlorine is very low.

NONDISPERSIVE X-RAY AND SEM ANALYSES

The six deposit samples were scanned from the tube-side surface of the deposit to the outside surface of the inner layer as outlined by the white lines inscribed on the inner layer of the typical deposit shown in Figure 4. The x-ray and SEM photographs indicated that the tube-side interface of the corrosion product consists of a very thin layer of iron oxide (Fe_2O_3) crystal, characteristic of hexagonal growth habit, as shown in Figure 5. The bulk of the inner layers consisted of a crystalline matrix of K, Ca, Fe and SO_4 containing particulate inclusion of K, Fe, Zn and SO_4. Proceeding through the deposit, a molten phase appears to develop with the addition of Na to the matrix. As the covering deposit is approached, inclusions of spheroidal particles consisting of Al, Si, K, Fe and

TABLE 3

CHEMICAL COMPOSITION OF ASH DEPOSITS

ANALYSIS[1]		German			Spanish	Domestic	
		1	2	3	4	5	6
Al_2O_3		8.5	14.0	9.5	12.0	7.6	2.3
SiO_2		13.5	15.5	13.5	11.6	14.6	7.1
TiO_2		0.8	0.8	1.0	---	1.1	0.6
Fe_2O_3		3.5	4.0	2.5	2.8	3.2	1.6
CaO		9.4	11.0	7.2	12.0	8.3	5.5
MgO		3.3	2.9	3.3	1.0	5.0	1.2
Na_2O	Alkalies and Heavy Metals	4.9	5.1	4.3	10.7	10.8	5.3
K_2O		11.6	11.2	6.7	12.4	7.1	7.4
PbO		4.2	1.3	10.6	2.4	0.6	14.5
ZnO		0.5	7.2	7.5	5.6	4.8	5.5
P_2O_5		1.1	0.5	nil	nil	1.0	1.4
Cl		0.6	0.6	0.6	12.2	2.2	0.3
SO_3		34.8	30.8	37.0	15.9	29.0	33.5
ASTM - Ash Fusion Data							
Oxidizing Atm.							
I.D.[2]		2210	2180	2080	2360	2200	2040
S.T. (Sph)[3]		2300	2260	---	2440	2210	2100
S.T. (Hem)[4]		2400	2280	2320	2500	2300	2120
F.T.[5]		2440	2380	2480	2560	2340	2160
Reducing Atm.							
I.D.		2160	2000	---	2280	2140	2100
S.T. (Sph)		2200	2040	---	2300	2150	2240
S.T. (Hem)		2260	2080	---	2360	2180	2300
F.T.		2300	2120	---	2900	2200	2420

[1]Reported as % by weight as an oxide
[2]I.D. - Initial Deformation °F
[3]S.T. - Softening Temperature (spherical) °F
[4]S.T. - Softening Temperature (hemispherical) °F
[5]F.T. - Fusion Temperature °F

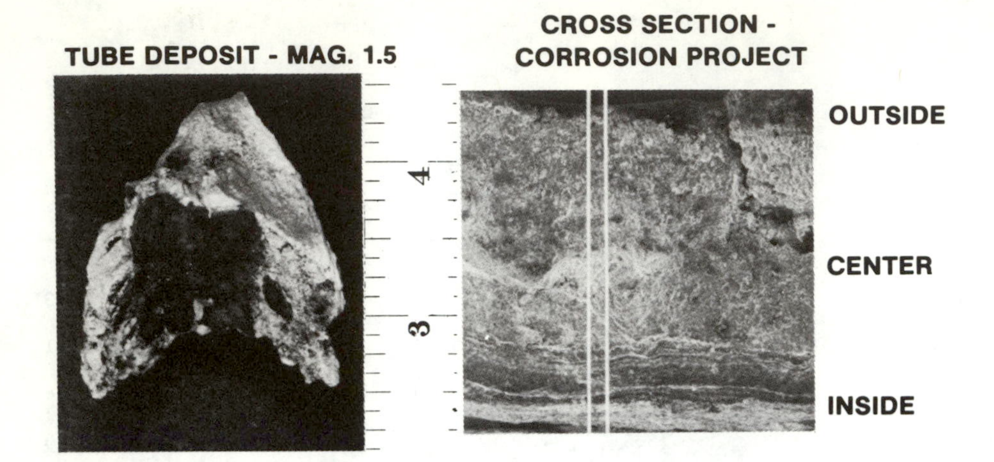

Figure 4 General Appearance of Tube Deposit Including Corrosion Product and a Scanning Electron Microscopic Photograph of the Cross Section of the Corrosion Product. White Lines Outline Area Scanned at 1200 to 6000X Magnification.

Zn become more frequent. Microphotographs and elemental distribution of additional representative samples appear in Figures 6 through 8. Several deposits revealed the presence of a crystalline substance consisting of sulfates of K, Fe, Zn, KCa and KZn. Kautz[16] was able to define new complex compounds consisting of complex sulfates of these elements. This experimental program has not proceeded far enough to positively identify these compounds. However, the data intimate their presence.

The appearance of a thin layer of iron oxide crystals lining the inside surface of the corrosion product and deposit was peculiar to Sample 5, which contained 2.2% chlorine identified as potassium and sodium chloride. The remaining deposit samples were lined with a matrix of Al, Si, S, Pb, K, Ca, Fe and Zn containing isolated crystals of S, K, Zn and S, K, Ca, Zn. Pure lead spheroids were found in a deposit from one of the smaller domestic boilers. Crystals of NaCl were found in the fly ash of one European boiler experiencing severe corrosion.

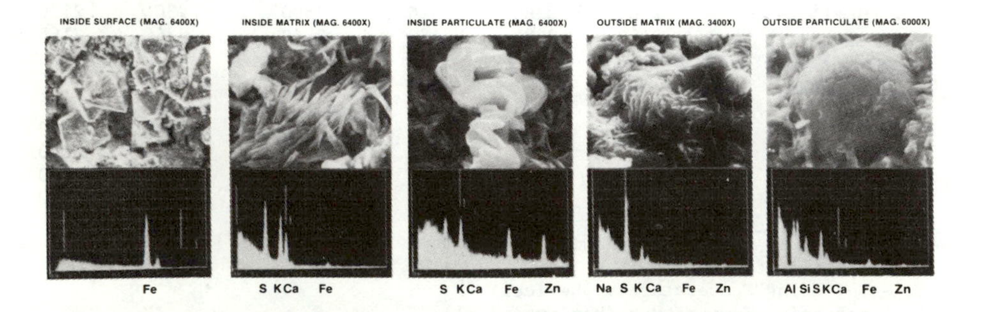

Figure 5 SEM Image and Non-Dispersive X-Ray Pattern of Particulate and Matrix Comprising the Corrosion Product.

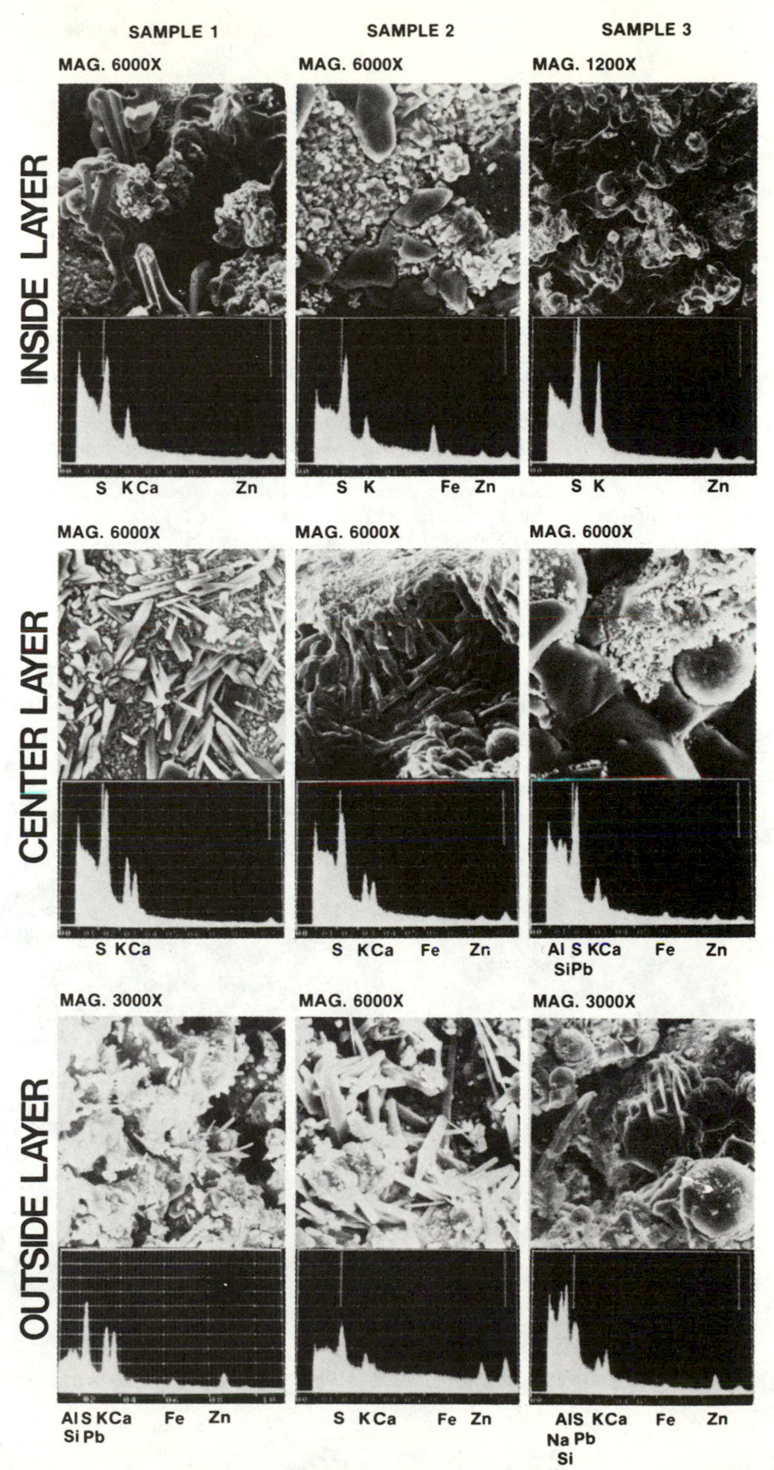

Figure 6 SEM Image and Non-Dispersive X-Ray Patterns of Samples 1 2 and 3.

THERMAL ANALYSIS

The thermal analysis for the synthetic compositions representing binary systems of key elements are summarized in Figure 9. Typical thermograms for deposit samples appear in Figures 10 through 13. They were selected to illustrate the influence of atmosphere and specimen sampling location within a given deposit on low and high temperature endotherms. The illustrations show the variation that can be expected between deposits. All the endotherms for the DSC and DTA are summarized in Tables 4 and 5.

The thermograms for the synthesized binary systems clearly indicate that lead chloride in combination with lead oxide or chlorides of such key elements as sodium, potassium and zinc all produce a melt in the vicinity of 400-450°C. Evaporation begins at 400°C and is completed by 1000°C. Sodium sulfate and zinc sulfate in the proportions presented in Table 1 also produce a melt within this temperature range. Melts of the K_2SO_4 and $PbSO_4$ systems do not occur until the temperature exceeds 600°C. The sulfates show no signs of weight loss below 1000°C.

Zinc chloride in combination with potassium, sodium, iron (II), and lead chlorides all produce melts that form at temperatures below the normal operating temperatures of evaporative tube surface. Binary systems including $ZnCl_2$ begin to lose weight at 200°C. Complete vaporization or dissociation takes place between 650 and 1000°C.

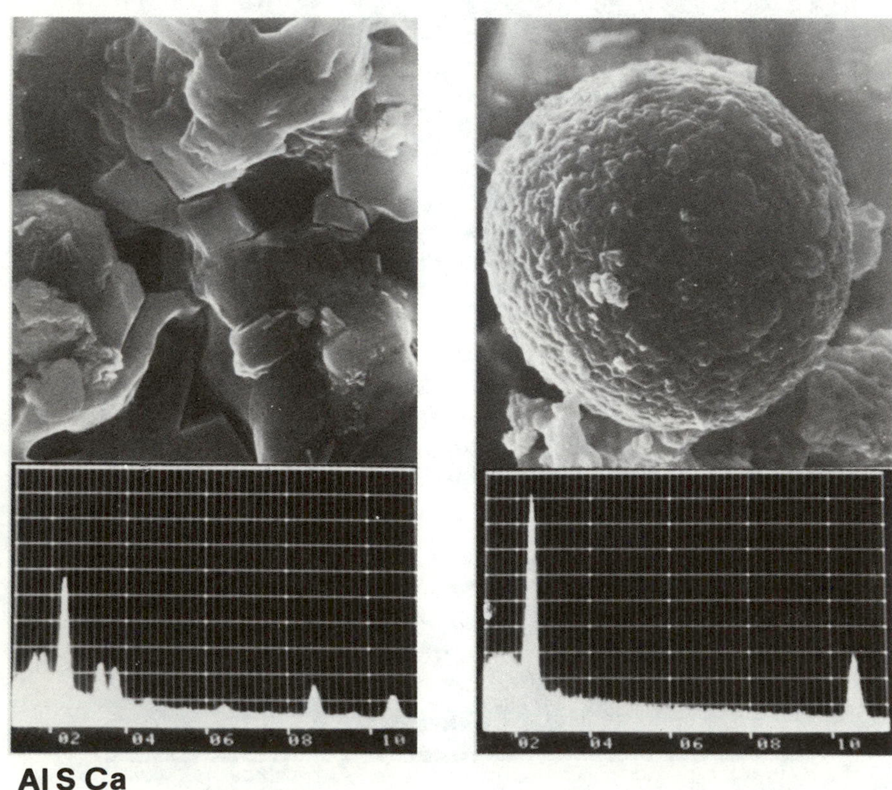

Figure 7 SEM Image and Non-Dispersive X-Ray Patterns of Fused Matrix and included Particle.

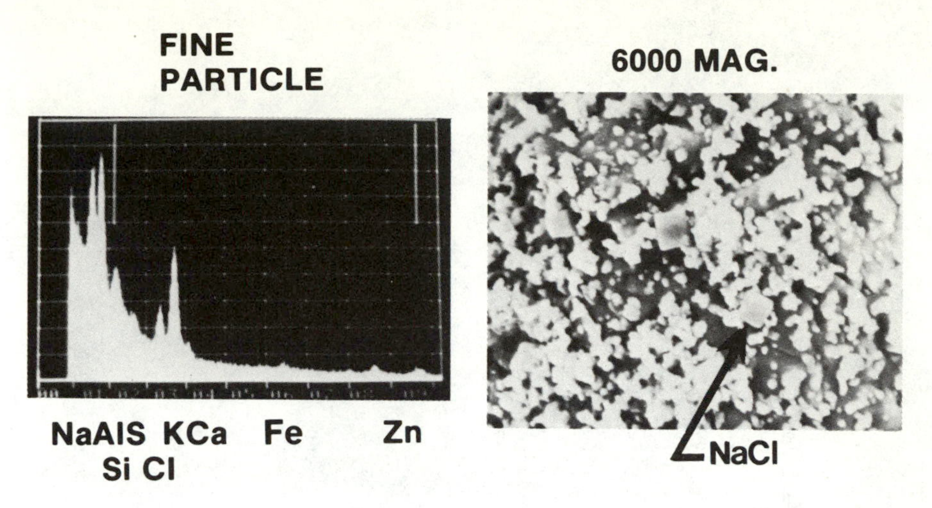

Figure 8 SEM Image and Non-Dispersive X-Ray Patterns of Fly Ash Containing Small Square Crystals of Sodium Chloride.

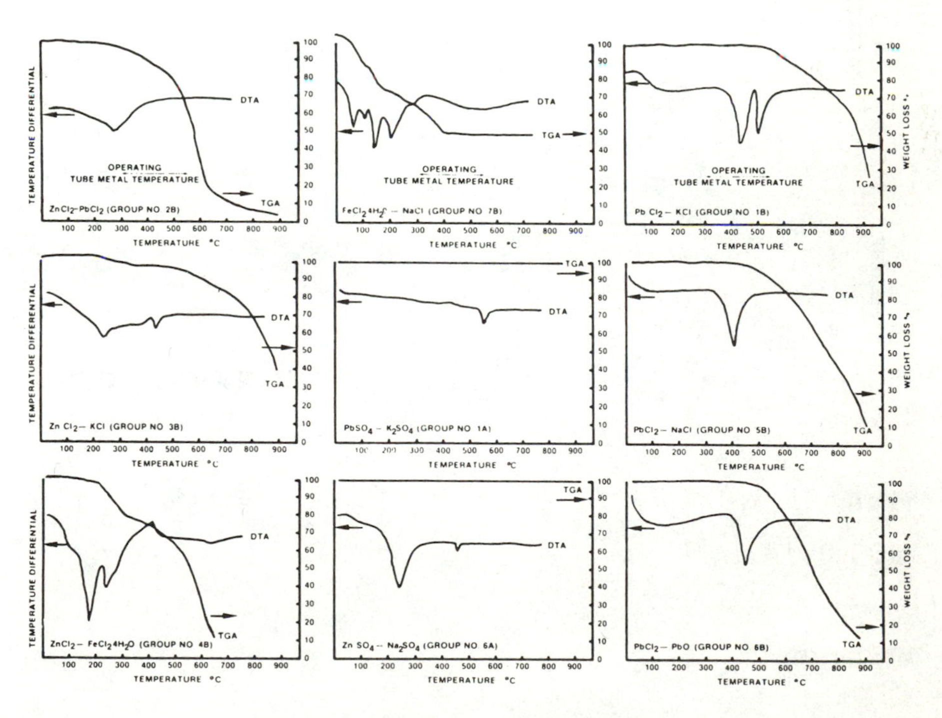

Figure 9 Thermographs of the Binary Synthetic Mixtures of Chlorides, Sulfates and Oxides.

TABLE 4

DIFFERENTIAL SCANNING CALORIMETRY - SUMMARY OF ENDOTHERMS
TEMPERATURES °C

SAMPLE	LOCATION	ATM	ENDOTHERM PEAK TEMPERATURE °C								
1	Inside	Air					180_{vs}	260_{ss}		400_{ss}	
		N_2					175_{ss}	260_{ss}		400_{ss}	
		Red.					180_{ss}	260_{ss}		400_{ss}	
2	Inside	Air					180_s		290_s		
		N_2				165_{ss}	190_s		280_s	400_s	450_s
		Red.	75_s			165_{ss}	190_s	260_s		400_s	
3	Inside	Air	60_s	90_{vs}	130_s	155_{ss}		260_s		400_s	480_{ss}
		N_2	60_{ss}			155_{ss}		260_s			480_{ss}
		Red.	65_{ss}			160_{ss}		260_s			480_{ss}
1	Outside	Air					190_{ss}		310_s	400_s	
		N_2					190_{ss}		340_s	400_s	
2	Outside	Air	60_s			165_s	190_s	260_s			450_s
		N_2					180_{vs}	260_s			
3	Outside	Air	60_s	100_{vs}	130_{vs}	155_{vs}		260_s			480_s
		N_2	60_{vs}		130_{vs}	155_{ss}		260_s			
5	Inside	Air				160_{ss}			300_s		480_s
		N_2				160_{ss}					440_s
5	Central	Air				160_s		250_s			
		N_2				160_s		280_s			540_s
5	Outside	Air				160_{vs}		250_{vs}			
		N_2				160_{vs}		280_{vs}	370_s		
6	Inside	Air	70_{ss}		140_s					400_s	440_s
		N_2	70_{ss}		140_s					390_s	440_s

TABLE 5

DIFFERENTIAL THERMAL ANALYSIS
SUMMARY OF ENDOTHERMS
TEMPERATURES °C

SAMPLE	LOCATION	ATM	ENDOTHERM PEAK TEMPERATURES °C									
1	Inside	Air						300_{ss}				
		N_2					200_{ss}				800_{s}	
2	Inside	Air	90_{vs}			180_{s}			420_{ss}			960_{s}
		N_2						300_{s}	420_{s}	720_{s}		
3	Inside	Air	100_{vs}	130_{vs}	155_{vs}				440_{vs}			
		N_2		130_{vs}	155_{vs}		260_{s}		440_{vs}			
1	Outside	Air					200_{ss}		400_{s}	740_{ss}	820_{ss}	
		N_2					200_{ss}		400_{s}	740_{ss}	820_{ss}	
2	Outside	Air					200_{ss}				820_{s}	920_{s}
		N_2					200_{ss}				820_{s}	970_{s}
3	Outside	Air	100_{vs}									
		N_2	100_{vs}									
6	Inside	Air				175_{vs}	260_{s}					950_{s}
		N_2				175_{vs}	260_{s}		440_{s}	700_{s}		950_{s}

vs = Very Strong
ss = Strong
s = Slight

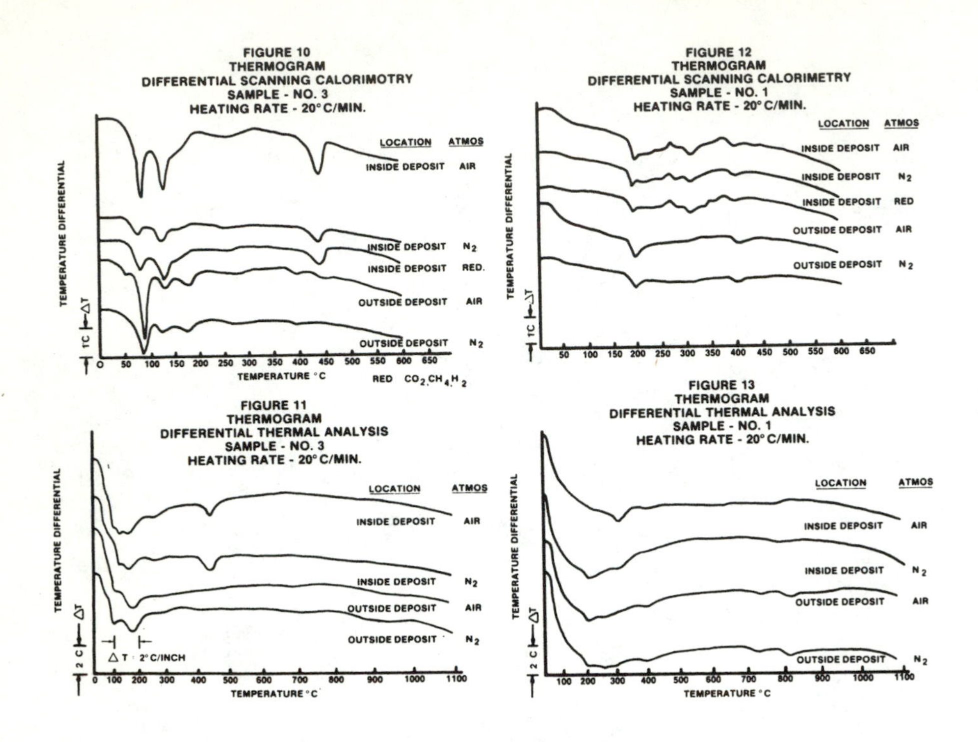

The thermograms for the ash deposit must be interpreted in terms of the physical chemistry of pure or synthesized compounds and the ash chemistry determined by traditional analytical techniques supplemented with x-ray diffraction and the SEM.

Thermograms of deposits containing moderate to high concentrations of zinc all contained strong endotherms at 60-70°C and 100°C. Sodium sulfide, sodium silicate and zinc sulfate all dehydrate at this temperature. Thermograms generated with the TGA, as well as thermal recycling of the DSC revealed a weight change confirming the suspected weight loss rather than a simple melt. X-ray diffraction patterns indicated minor quantities of zinc sulfate present in those samples displaying endotherms at 60°C. Therefore, the endotherm probably represents the dehydration of zinc sulfate.

A second endotherm occurred in the vicinity of 145 to 150°C which is in the temperature range at which various binary compounds of PbO and $PbCl_2$ dehydrate. Thermogravimetric analysis indicated a weight change of 1-4% between 100 and 200°C with no further weight loss until 800-950°C, suggesting the endotherm is indicative of a decomposition process such as dehydration. A second endotherm appears between 400 and 450°C which is coincident with the melting point of $PbO \cdot PbCl_2$ compounds. Although the x-ray diffraction did not identify the presence of PbO or $PbCl_2$, the traditional laboratory analysis which is more sensitive indicated lead was present in most samples from 1.3 to 14%. The DSC is also sensitive to minor constituents with substantial heats of fusion or vaporization and therefore can detect reactions or changes of state of constituents which might not be detected by x-ray diffraction. Chlorine was not detected during the quantitative analysis of the deposits. This is not alarming considering the quantity of chlorine present in the $PbO \cdot PbCl_2$ compound. Therefore, it appears that thermal analysis has uncovered small quantities of $PbO \cdot PbCl_2$ in the inner layers of the deposits.

A third endotherm occurs throughout most deposits at 260°C corresponding to the decomposition of 2 $NaHSO_4$ to $Na_2S_2O_7 + H_2O$.

X-ray diffraction patterns confirmed the melt of $K_2O_2O_7$ at 300°C on the inside and outside of deposit (Sample 1) and the inside of deposit (Sample 5).

Melts occurred in practically every sample between 400 and 450°C. The degree to which they occurred was less on the outside than on the inside. As indicated earlier, these endotherms correspond to melting temperatures of the PbCl binary systems and the binary and ternary systems of $(NaSO_4 \cdot ZnSO_4)(Na_2SO_4 \cdot ZnSO_4 \cdot K_2SO_4)$.

Above 450°C, the DTA showed only occasional endotherms of very small magnitude implying the absence of the complex potassium, sodium iron sulfates.

X-RAY DIFFRACTION

The results of the x-ray diffraction analysis indicate that calcium is found as an alumino-silicate or sulfate. The alkalies zinc and iron are found primarily as a sulfate. In several cases potassium is reported as a pyrosulfate; sodium and potassium appear as chlorides; zinc appears as a sulfide and sodium and zinc appear as a complex sulfate. Iron also appears as an oxide. Lead was found to occur as an oxide, a sulfate and a silicate. The x-ray diffraction analysis is summarized in Table 6.

The complex potassium or sodium iron sulfate is missing, which confirms the findings of the thermal analysis.

DISCUSSION

There appear to be several mechanisms for corrosive attack of tube metal surface in refuse-fired steam generators, depending on temperature level and the presence of corrosive agents (chlorides, alkalies and heavy metal sulfates) at the tube surface.

At temperature levels in the vicinity of 400-450°C, zinc combines with sodium sulfate and potassium sulfate to form a low melting substance which acts as a fluxing agent permeating and destroying the protective oxide layer, and resulting in severe corrosive attack. Sulfates of lead do not appear to contribute to the formation of this melt.

At the same temperature levels in the presence of small quantities of chlorine, lead chloride may form. When combined with lead oxide, the binary system forms a melt in the vicinity of 450°C. Chlorides of zinc, sodium and potassium arriving at the tube surface as particulate matter, somehow escaping destruction during combustion either as a result of poor combustion or short residence time, also combine with lead chloride to form corrosive melts at approximately between 400 and 450°C.

The role of chlorine in the corrosive attack is still uncertain. Cutler, et al[17], indicated that chlorides will not condense as an independent phase in the deposit and the HCl will have little effect on any metal surface it contacts. Furthermore, the absorption of gaseous chloride by the deposits will not enhance their long-term corrosiveness. They do recognize, however, the fact that chlorides may deposit on tube surfaces due to short residence time or poor combustion.

The presence of chlorides as an independent phase on a surface by whatever mechanism it is introduced, may be taken as an indication of enhanced rates of attack on the metal. If the general atmosphere permeating the deposit is oxidizing, then the chloride can play an active role in the corrosive process. If it is reducing, the atmosphere may cause enhanced rates of corrosion, via destruction of any protective oxide layer, sulfidation of the metal, etc. The chloride itself may play little or no direct part.

On the other hand, Brown, DeLong and Auld[18] show that elemental chlorine attacks carbon steel rapidly above 204°C, while the HCl attack increases rapidly above 480°C.

The x-ray diffraction patterns as well as scans made with the SEM have indicated the presence of potassium and chlorine in fly ash and deposits taken from the smaller domestic boilers and one European boiler in which metal wastage was severe. Ferric chloride and sodium chloride, as well as $FeCl_2$ and $ZnCl_2$ produce melts at 150 to 250°C which could be responsible for or contribute to corrosion of evaporating tubes.

TABLE 6

SUMMARY - X-RAY DIFFRACTION ANALYSIS

SAMPLE	1	2	3	4	5	6
INSIDE						
Major	$Ca_2Al_2SiO_7$	$KAlSi_3O_3$	K_2SO_4	$CaSO_4$ (1) NaCl KCl $Ca_2Al_2SiO_7$	Fe_2O_3	$PbSO_4$ αPbO_2
Minor	$CaSO_4$ K_2SO_4 $K_3Na(SO_4)_2$ $K_2S_2O_7$ Pb_4SiO_4 Ca_2SiO_4 $MgSiO_3$ SiO_2	$Ca_2Al_2SiO_7$ K_2SO_4 $CaSO_4$ Zn_2SiO_4 $K_3Na(SO_4)_2$ Ca_2SiO_4 $FeSO_4$	$Ca_2Al_2SiO_7$ $PbSO_4$ $ZnSO_4$ $CaSO_4$ SiO_2 Fe_2O_3	$K_{.7}Na_{.3}AlSiO_4$ Fe_2O_3 $CaCO_3$	Fe_3O_4 $CaSO_4$ $K_2S_2O_8$ NaCl K_2SO_4 $Na_2Zn(SO_4)_2$ $Ca_2Al_2SiO_7$ ZnS	βK_2SO_4 $CuSO_4$ ZnS $CaSiO_5$
OUTSIDE						
Major	$Ca_2Al_2SiO_7$	Zn_2SiO_4 $Ca_2Al_2SiO_2$	$PbSO_4$	$CaSO_4$ (2) NaCl KCl $Ca_2Al_2SiO_7$	$Ca_2Al_2SiO_7$	
Minor	$CaSO_4$ $K_3Na(SO_4)_2$ K_2SiO_7 Ca_2SiO_4 Pb_4SiO_6 Fe_2O_3 $FeSO_4$	$CaSO_4$ Ca_2SiO_4 $FeSO_4$ K_2SO_4 $K_3Na(SO_4)_2$ Pb_4SiO_4	$ZnSO_4$ K_2SO_4 $CaSO_4$ SiO_2 $Ca_2Al_2SiO_2$ Fe_2O_3	$K_{.7}Na_{.3}AlSiO_4$ Fe_2O_3 $CaCO_3$	SiO_2 $Na_2Zn(SO_4)_2$ $CaSO_4$ K_2SO_4 Fe_2O_3 NaCl ZnS	

(1)Coarse fly ash (2)Fine fly ash

CONCLUSIONS

Corrosion in refuse-fired steam generators is quite complex. Several mechanisms for attack seem to be occurring simultaneously under similar environmental conditions making it difficult to discern just what is the cause. The design and operation of the steam generator seem to have a bearing on the problem which adds to the difficulty of interpreting laboratory experimentation.

The heavy metals, lead and zinc, in the form of chlorides of lead sulfates and chlorides of zinc contribute to the corrosion in refuse-fired steam generators. It is doubtful that corrosion is due to pyrosulfate or complex alkali-iron sulfates.

Although refuse is a heterogeneous fuel, the problems appear to be similiar in many steam generators. Differences are probably due to variations in internal environmental conditions and operational parameters, and time-to-time deviation of presence of the various corrosion-causing elements.

Proper refuse preparation and mixing, with improved steam generator design and controlled operational procedures can be effectively exercised to minimize the catastrophic corrosion problems and reduce deposit phenomena in refuse-fired operations.

ACKNOWLEDGEMENTS

The authors wish to extend their gratitude to Gregory J. Stanko, metallurgist, for his excellent electron microscopic work, and Mrs. Linda Domino for the thermal analysis of the deposit and synthetic binary compounds. Their work is gratefully acknowledged.

REFERENCES

1 R. Huch, "Hydrochloric Acid Corrosion in Refuse Burning Installations," Brennstoff-Wärme-Kraft, Vol. 18, No.2, February 1969.

2 P. Stellar, "Experiments for Clarifying the Causes of Corrosion in Incinerating Plants," Energie, Vol. 18, August 1966.

3 F. Nowak, "Operating Experience at the Refuse Burning Plant at Stuttgart," Brennstoff-Wärme-Kraft, Vol. 19, No. 2, February 1967.

4 K. Perl, "Corrosion Damage to Steam Generators of Refuse Burning Installations," Energie, Vol. 18, No. 8, August 1966.

5 F. Nowak, "Corrosion Phenomena in Refuse Boilers," Mitt. der VGB, Vol. 102, No. 6, June 1966.

6 J. Angenend, "The Behavior of Boiler Tube Materials in Gases Containing HCl," Brennstoff-Wärme-Kraft, Vol. 18, No. 2, February 1966.

7 H. Kohle, "Fireside Deposits and Corrosion in Refuse Fired Boilers," Mitt. der VGB, Vol. 102, June 1966.

8 P. Stellar, "Corrosion Measurements in the Steam Generator of a Refuse Burning Installation," Energie, Vol. 19, No. 8, September 1967.

9 F. Nowak, "Corrosion Problems in Incinerators," Combustion, November 1968.

10 H. Hilsheimer, "Experience After 20,000 Operating Hours - The Mannheim Incinerator," Proceedings 1970 National Incinerator Conference, ASME.

11 F. Nowak, "Corrosion Phenomena in Refuse Firing Boilers and Preventive Measures," presented at VGB International Symposium on Corrosion in Refuse Incineration Plants, Düsseldorf, West Germany, 1970.

12 F. Maikranz, "Corrosion in Three Different Firing Installations," presented at VGB International Symposium on Corrosion in Refuse Incineration Plants, Düsseldorf, Germany, 1970.

13 R. W. Bryers and Z. Kerekes, "Recent Experience with Ash Deposits in Refuse-Fired Boilers," ASME Paper No. 68-WA/CD-4.

14 R. W. Bryers and Z. Kerekes, "A Survey of Ash Deposits and Corrosion in Refuse-Fired Boilers," presented at VGB International Symposium on Corrosion in Refuse Incinerator Plants, Düsseldorf, West Germany, April 1970.

15 R. W. Bryers and Z. Kerekes, "Corrosion and Fouling in Refuse-Fired Steam Generators," presented at the Symposium on Solid Waste Disposal at the Seventieth National Meeting of the AIChE, Atlantic City, New Jersey, August 29-September 1, 1971.

16 K. Kautz and J. Tichatschke, "Relationships Between Flue Gas Conditions, Boiler Load and Corrosion in a Municipal Refuse Burning Plant," Mitt. der VGB, Vol. 52, No. 3, June 1972.

17 A. J. B. Cutler, W. D. Halstead, J. W. Layton, and G. G. Stevens, "The Role of Chloride in the Corrosion Caused by Flue Gases and Their Deposits," Journal Engineering for Power, July 1971.

18 M. H. Brown, W. B. DeLong, and J. R. Auld, "Corrosion by Chlorine and by Hydrogen Chloride at High Temperatures," Journal Ind. Engineering Chem., No. 39, July 1947.

CHLORIDE CORROSION AND ITS INHIBITION IN REFUSE FIRING

D. A. VAUGHAN, H. H. KRAUSE, and W. K. BOYD

BATTELLE
Columbus Laboratories
Columbus, Ohio, USA 43201

ABSTRACT

Fireside corrosion from burning of municipal refuse alone and with high-sulfur coal was investigated in field studies by the use of corrosion probes. Corrosion rates of simulated boiler tubes were measured for periods up to 828 hours for carbon steel, low alloy steel, and stainless steels in an incinerator burning bulk refuse. Similar experiments were conducted in co-firing of shredded refuse with high-sulfur coal in a stoker-fired boiler. Detailed analysis of deposits, corrosion scale, and substrate metal showed that chlorine in the refuse is responsible for the most serious corrosion. Sulfur, sodium, potassium, lead, and zinc also take part in the corrosion reactions. The addition of sulfur to bulk refuse reduced the corrosion rates of carbon steel and stainless steels significantly. In the co-firing of refuse-coal mixtures the initial corrosion rates of carbon steel were about an order of magnitude lower than those for 100 percent refuse, while those for stainless steels were 10-50 times lower.

INTRODUCTION

There has been increasing incentive in recent years for disposal of municipal refuse by incineration. The need to recover energy from waste material, coupled with the difficulty in obtaining adequate landfill sites, has created a great deal of interest in disposal by incineration. In addition municipal refuse is being considered as a supplementary fuel for existing fossil-fuel power-generating stations and for industrial needs for process steam. However, the corrosion of heat recovery surfaces exposed to combustion products from the burning of industrial and municipal wastes has been a major problem in utilizing these materials as a source of heat energy. Research conducted by Battelle in recent years has demonstrated that chlorine in the refuse is the most important contributor to corrosion of boiler tubes.[1-8] The sulfur in the refuse definitely plays a part in promoting corrosion by chlorine, and other components which are not present in fossil fuels such as zinc and lead may contribute to the unusual amount of corrosion which can occur during incineration. Experiments have been conducted on this program to investigate the inhibition of chloride corrosion by the introduction of relatively large amounts of sulfur into the fuel. One of the most convenient ways of achieving this result has been to utilize high-sulfur coal in co-firing with municipal refuse in a stoker-fired boiler. The system employed is different from others being investigated elsewhere, in that refuse which has undergone only shredding and magnetic separation is burned on a traveling grate in conjunction with high-sulfur coal.

CORROSION IN FIRING OF BULK REFUSE

Corrosion rates from burning of bulk municipal refuse were obtained by exposing specially designed probes to the combustion gas environment at the Miami County, Ohio incinerator. The design of the corrosion probe, which has been used with either 18 or 34 specimens, is shown in Figure 1. The specimens were machined from 1-inch (2.54 cm) schedule 40 pipe or equivalent tubing and were nested together with lap joints, as shown in detail A. The probe was inserted into the incinerator through a side wall and was mounted at a 90° angle to the flue gas stream. The section of the probe extending through the wall was water cooled. The specimens exposed within the incinerator were cooled by air flowing inside the tubular specimens. The specimen temperatures were controlled by regulating the amount of cooling air admitted to the probe. The output from a control thermocouple was monitored by a proportional temperature controller. The controller maintained a preset temperature by varying the amount of cooling air bypassing the probe through a motorized butterfly valve located between the blower and the probe. A Roots blower delivering up to about 34 cfm (1000 ℓ/min) was used with a 5 hp (3.7 kw) motor as a drive.

Corrosion probe exposures up to 828 hours were carried out, and the corrosion rates of the various steels used as specimens were determined by the weight losses that resulted from the exposure. The data for the 828 hour exposure calculated to a basis of mils per month are presented in Figure 2. As seen from the figure, the weight losses increased with temperature and were especially high for the carbon steel and the low alloy steel. Over a metal temperature range of 250 to 1200°F (121-649°C) the high-chromium ferritic and austenitic stainless steel showed the best corrosion resistance. As a function of exposure time, the corrosion rate of carbon steel was found to be very high initially and then to drop rapidly, leveling off after about 800 hours, as shown in Figure 3. The effect of metal temperature is also shown by this data, which illustrate the increase in the corrosion rates as the temperature of the metal specimens was increased from 350 to 950°F (177-510°C).

DEPOSIT AND SCALE ANALYSIS

Both the bulk deposits and the corrosion product layers formed on the specimens were analyzed during this program to determine the reactions leading to metal wastage. The deposits were removed from all the probe specimens in numbered sequence. Thus the variations in composition could be determined as a function of temperature when the deposits were analyzed. Emission spectrography and wet chemical techniques were used to determine the concentration levels of the various elements in the deposits. Electron microprobe analysis and X-ray diffraction were used to obtain additional detailed information on the nature of the deposits. Analyses of the bulk deposits showed that chloride accumulated rapidly even at moderate HCl concentrations in the flue gases. Chloride concentrations ranging from 21 to 27 percent by weight were found on probes which were exposed for short periods of time ranging from 1.5 to 24 hours. Flue gases contained from 40 to 140 ppm HCl (corrected to 12 percent CO_2) during the time that these deposits were formed.

Eight analyses of typical refuse samples from the incinerator showed that the chloride content of the solid waste on a dry basis ranged from 0.32 to 0.79 weight percent, with an average of 0.49 percent. Based on an average 24 hour processing of 150 tons (136,080 kg) of waste at this incinerator, there would be about 1500 pounds (682 kg) of chlorine in the refuse. The sources of the chlorine are both inorganic, primarily as NaCl, and organic,

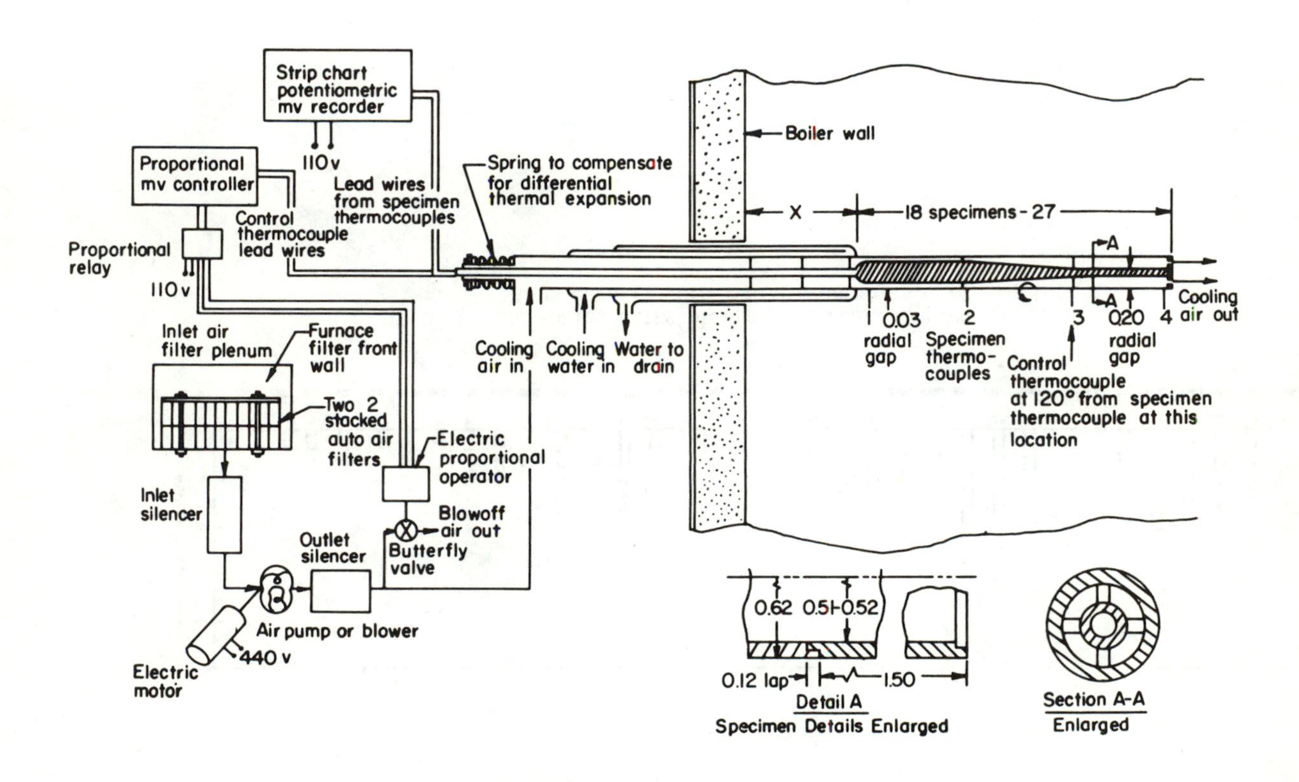

FIGURE 1. SCHEMATIC OF CORROSION PROBE APPARATUS

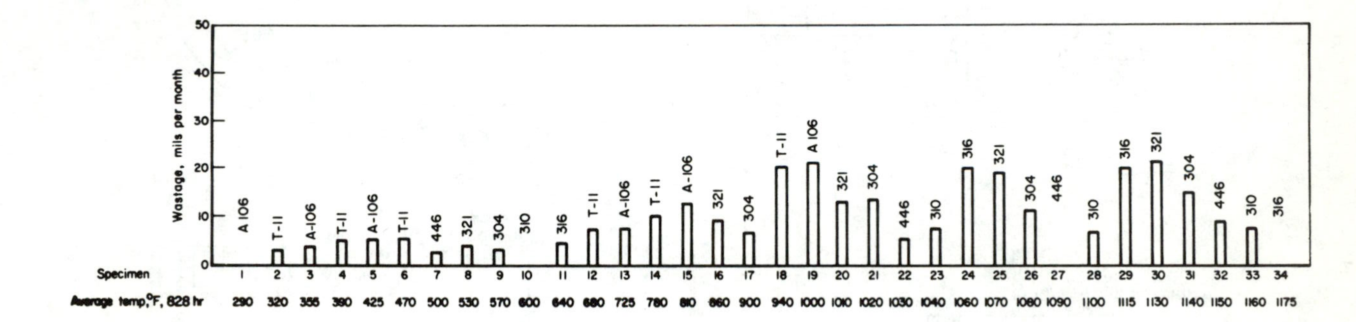

FIGURE 2. CORROSION RATES OF VARIOUS ALLOYS EXPOSED IN THE MIAMI COUNTY, OHIO INCINERATOR

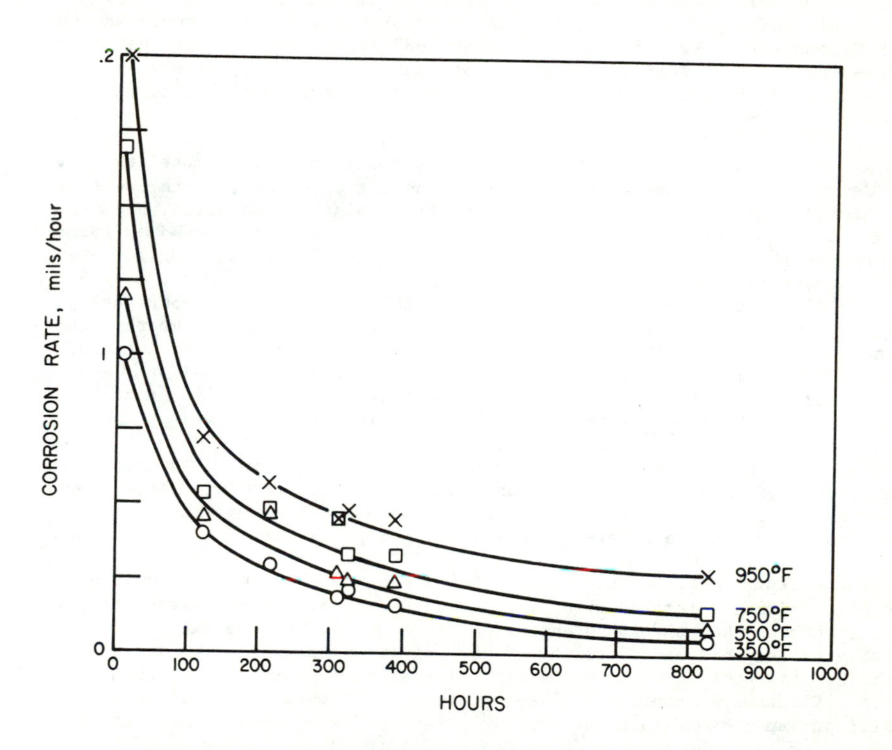

FIGURE 3. ISOTHERMAL CORROSION RATES FOR A106 STEEL VS EXPOSURE TIME IN BULK REFUSE INCINERATION

of which polyvinyl chloride plastic is the predominant item. Practically all of this chlorine is volatilized at combustion temperatures, as the ash contains only 0.01 to 0.02 weight percent chlorine. Refuse sortings made at four different times at the incinerator gave values for plastic content of 3 to 6 percent. Of this about 1/5 was estimated to be polyvinyl chloride. Using the lowest value of 3 percent total plastic, the polyvinyl chloride would account for 900 pounds (409 kg) of the 1500 pounds (682 kg) of chlorine in the refuse consumed daily. The remainder would be of inorganic origin. From these values it is obvious that large amounts of chlorine are available to cause corrosion.

As the exposure time increased and the deposit aged, the amount of chloride decreased. This change resulted from the action of SO_2, which converted the chlorides to sulfates. A temperature effect also was observed, because the amount of chloride generally was greatest in the deposits found on lower temperature specimens. Average chloride values in bulk deposits of long-term runs ranged from a few tenths of a percent at temperatures from 750 to 1250°F (399-677°C) up to 3.8 percent in the 250 to 500°F (121-260°C) temperature zone. However, the chloride was not equally distributed throughout the deposit and the bulk of it diffused into the layer at the metal-deposit interface. The deposits from the 828 hour probe exposure, which showed a uniform low average chloride content of 0.2 weight percent, were sectioned horizontally to separate the bulk of the deposit from the corrosion product scale at the interface with the metal. Analyses of the two sections of the deposit demonstrated that the chlorine was concentrated in the inner layer. The outer layer contained only 0.2 weight percent chlorine, while the inner layer showed values up to 4.2 weight percent. The other elements which were found to be concentrated in the inner layer were sulfur, potassium, lead, zinc, and iron.

The sulfur resembles chlorine in that its concentration built up rapidly in the deposits and leveled off after about 100 hours. On the higher temperature specimens the sulfur concentration was found to be somewhat lower, probably because of the volatility of some of its compounds. The inner layers of the deposit contained approximately twice as much sulfur as did the outer layers. Electron microprobe analyses of the corrosion scale layer showed that significant amounts of chlorine, sulfur, zinc, and lead were present at the interface of the tube metal and the scale (Figure 4).

Investigation by X-ray diffraction has demonstrated that a continuous layer of $FeCl_2$ exists at the interface with the metal. FeS occurs between the $FeCl_2$ and the overlaying Fe_2O_3 layer. Various other compounds, mostly sulfates, were identified in the deposits. About 20 different compounds have been identified as shown in Figure 5. On the higher temperature specimens, it was found that the $FeCl_2$ layer melted and collected into small pools. The electron microprobe showed that a zinc compound was also present in this area near the metal. It appears that this element is not in a form readily detected by X-ray techniques, but may be incorporated in the $FeCl_2$ or in the FeS as a replacement for some of the iron atoms in the host structures. These interface layers are of the order of 6 to 10 microns thick on the low temperature specimens. The thickness increased somewhat with temperature.

Between the interface scale and the bulk deposit was a multilayered scale of Fe_2O_3 and Fe_3O_4. It is interesting to note that a thin layer of alpha-Fe_2O_3 formed at the interface between the $FeCl_2$ and mixed oxides. At low temperatures FeS forms between the $FeCl_2$ and Fe_2O_3 layers. Part of the mixed oxide scale adhered to the substrate and part to the external deposit. The mixed oxide layer was a hard, brittle, magnetic, gray-black material which thickened with temperature and was made up of several layers with loose red Fe_2O_3 between layers. This suggests that reduction or reaction with other elements occurs in or below this layer. At the higher temperature where $FeCl_2$ melts and FeS forms on the substrate, the mixed oxide scale was less adherent, and in this area more of the mixed oxide was removed with the outer deposit.

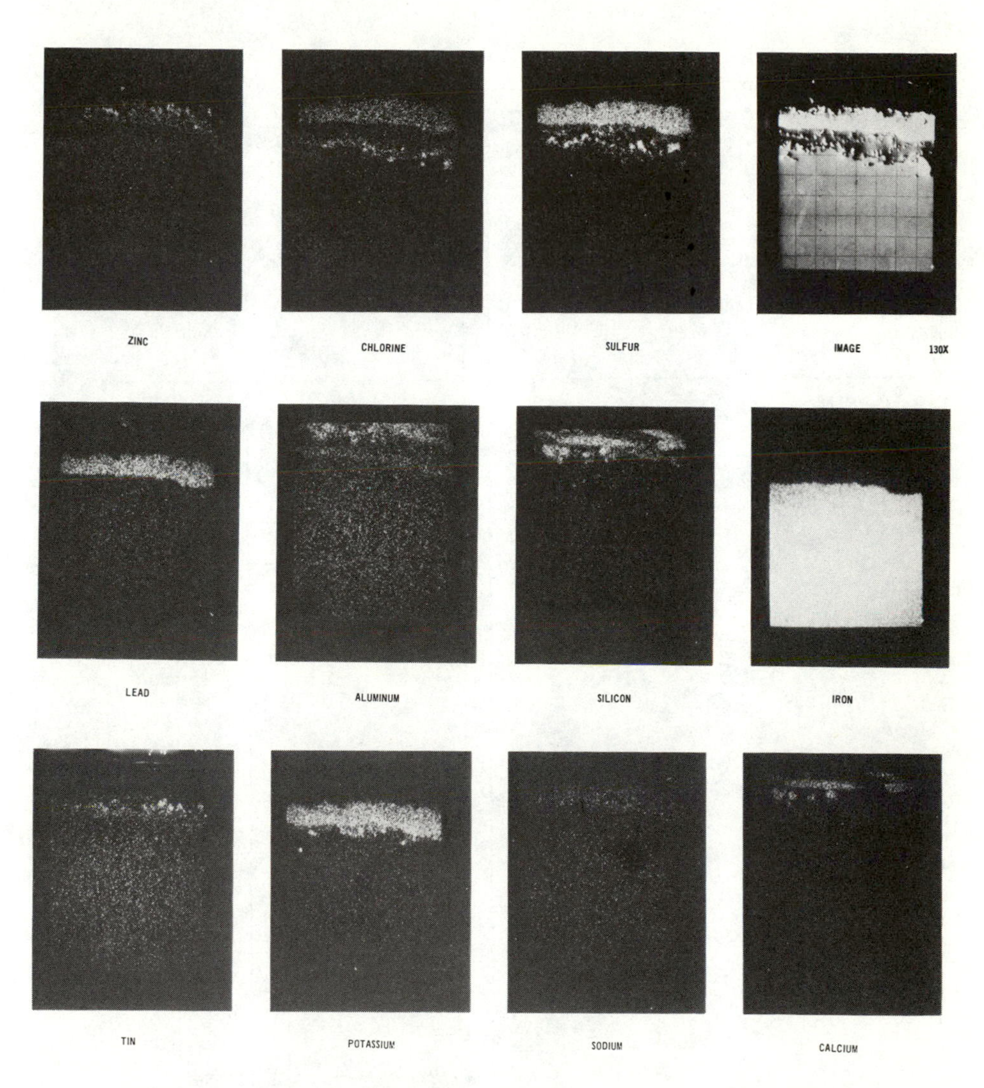

FIGURE 4. ELECTRON MICROPROBE PHOTOGRAPHS SHOWING DISTRIBUTION OF VARIOUS ELEMENTS IN THE CORROSION SCALE LAYER

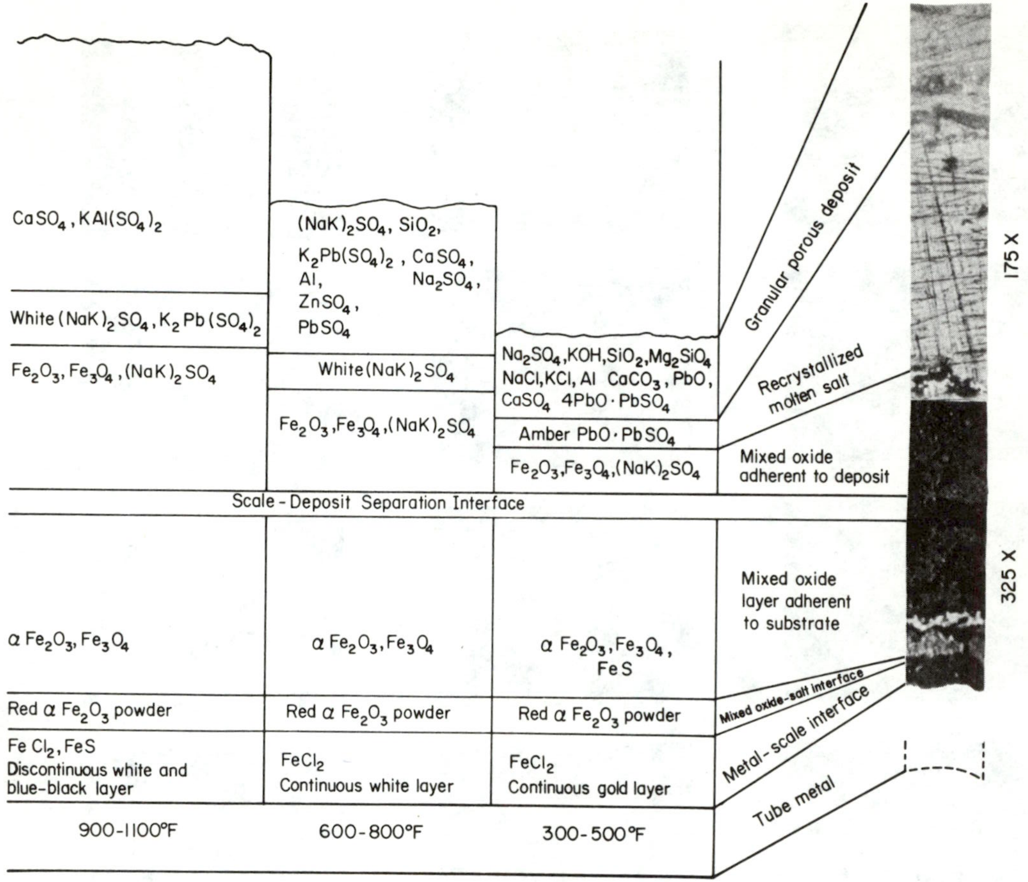

$CaSO_4$, $KAl(SO_4)_2$

White $(NaK)_2SO_4$, $K_2Pb(SO_4)_2$

Fe_2O_3, Fe_3O_4, $(NaK)_2SO_4$

$(NaK)_2SO_4$, SiO_2, $K_2Pb(SO_4)_2$, $CaSO_4$, Al, Na_2SO_4, $ZnSO_4$, $PbSO_4$

White $(NaK)_2SO_4$

Fe_2O_3, Fe_3O_4, $(NaK)_2SO_4$

Na_2SO_4, KOH, SiO_2, Mg_2SiO_4, NaCl, KCl, Al $CaCO_3$, PbO, $CaSO_4$ $4PbO \cdot PbSO_4$

Amber $PbO \cdot PbSO_4$

Fe_2O_3, Fe_3O_4, $(NaK)_2SO_4$

Granular porous deposit

Recrystallized molten salt

Mixed oxide adherent to deposit

175 X

Scale - Deposit Separation Interface

α Fe_2O_3, Fe_3O_4

α Fe_2O_3, Fe_3O_4

α Fe_2O_3, Fe_3O_4, FeS

Mixed oxide layer adherent to substrate

325 X

Red α Fe_2O_3 powder

Red α Fe_2O_3 powder

Red α Fe_2O_3 powder

Mixed oxide-salt interface

$FeCl_2$, FeS
Discontinuous white and blue-black layer

$FeCl_2$
Continuous white layer

$FeCl_2$
Continuous gold layer

Metal-scale interface

900-1100°F

600-800°F

300-500°F

Tube metal

FIGURE 5. SCHEMATIC DIAGRAM SHOWING LOCATION OF PHASES IDENTIFIED BY X-RAY DIFFRACTION

Examination with the optical microscope indicated that some melting had occurred in the deposit near the mixed oxide scale. These studies showed the presence of recrystallized continuous phases of the compounds $(Na, K)_2SO_4$ and $PbO \cdot PbSO_4$ on the external surface of the mixed iron oxide scale. The sodium-potassium sulfate phase was also dispersed throughout the iron oxide scale and adhered to the bulk deposit. It appears that separation in the iron oxide scale occurred at the depth to which this salt phase permeated in sufficient quantity to destroy the iron oxide. It is proposed that in the initial stage of incinerator operation, these salt phases as well as mixed alkali chloride salts and possibly $ZnCl_2$ and $PbCl_2$, permeate the iron oxide scale and destroy its protective characteristics.

The flue gas compositions in the vicinity of the corrosion probes were measured in order to determine the gaseous environment to which these specimens were subjected. The range of values found for the various components is presented in Table 1. HCl was by far the main component found in the flue gases, with the SO_2 concentration significantly lower. The other gaseous components found in the furnace were not considered to be making any significant contribution to corrosion.

CORROSION MECHANISMS

A number of chemical reactions are proposed to account for the formation of chlorides and sulfates in the deposits, as shown in Figure 6. Initial deposits are alkali metal oxides, which are then converted to chlorides by the action of HCl. Some chlorides such as NaCl may be volatilized in the flame and deposited as such. It is believed that corrosion is initiated by reaction 1 in which chlorides are converted to sulfates by the action of SO_2, oxygen, and moisture, thereby releasing HCl at the metal surface. The figure illustrates the reaction for potassium, and sodium acts in the same way. The hydrogen chloride formed can then react with the iron surface to form ferrous chloride as shown in reaction 2, or it may be oxidized to elemental chlorine as shown in reaction 3. As indicated by the laboratory studies at Battelle and supported by the work of Brown, DeLong, and Auld[9], the corrosive effects from HCl would not be expected to be severe below temperatures of about 600°F (316°C). Since $FeCl_2$ has been detected on corrosion probe samples exposed to temperatures well below 600°F (316°C), it is likely that corrosion by elemental chlorine is operating in the lower temperature range. Although elemental chlorine has not been identified in incinerator and furnace gases, it is believed that it may play a role in the corrosion reaction. Metal chlorides on the tube surfaces can catalyze reaction 3 as is done in the Deacon process. The chlorine, which is formed only near the catalytic surface, would then combine directly with the iron as shown in reaction 4.

The molten salt phase which was found as an interface layer between the scale and deposit would act as a barrier to the motion of gases in and out of the layer adjacent to the metal surfaces. Under oxidizing conditions, it would be expected that $FeCl_2$ would react with oxygen to form Fe_2O_3 as given by reaction 5. Once a scale formed on the metal tubes, such a cyclic reaction could occur. The mixed oxide in molten salt layers would limit the availability of oxygen, and retain the chlorine within the adherent scale. Similar corrosive conditions from deposited chloride were demonstrated by Cutler et al.[10], particularly as they applied to burning fossil fuels containing chlorine. The possibility of oxidizing $FeCl_2$ to release elemental chlorine also was shown by Fassler et al.[11]

The formation of FeS in the corrosion product layer indicates that sulfur is also playing a part in the corrosion reactions. The alkali pyrosulfates and bisulfates are known to be very corrosive materials, and the potassium compounds in particular have low melting points. Hence corrosion by these materials as shown in equations 8 and 9 is a very likely possibility for the formation of FeS. The corresponding sodium compounds have somewhat higher

TABLE 1. CONCENTRATIONS OF CHEMICAL SPECIES IN INCINERATOR FLUE GASES. (In dry gases, corrected to 12 percent CO_2)

Component	Range of Concentration, ppm
Sulfur dioxide, SO_2	59 - 195
Chloride, Cl^-	214 - 1250
Fluoride, F^-	3 - 34
Nitrogen oxides, as NO_2	53 - 115
Phosphate, $PO_4^{\equiv}$	0.5 - 6
Organic acids	35 - 178
Aldehydes and ketones	1 - 12

(1) $2KCl + SO_2 + 1/2\ O_2 + H_2O \rightarrow K_2SO_4 + 2HCl$

(2) $Fe + 2HCl \rightarrow FeCl_2 + H_2$

(3) $2HCl + 1/2\ O_2 \rightarrow H_2O + Cl_2$

(4) $Fe + Cl_2 \rightarrow FeCl_2$

(5) $4FeCl_2 + 3\ O_2 \rightarrow 2Fe_2O_3 + 4Cl_2$

(6) $K_2O + 2HCl \rightarrow 2KCl + H_2O$

(7) $Fe_2O_3 + 3K_2S_2O_7 \rightarrow 2K_3Fe(SO_4)_3$

(8) $2KHSO_4 + 3Fe \rightarrow Fe_2O_3 + FeS + K_2SO_4 + H_2O$

(9) $K_2S_2O_7 + 3Fe \rightarrow Fe_2O_3 + FeS + K_2SO_4$

FIGURE 6. CHEMICAL REACTIONS INVOLVED IN CORROSION

melting points, but it is possible that the melting points of these salts may be lowered in the deposits by the zinc and lead compounds which are known to be present on the corroded samples.

The presence of bisulfates, pyrosulfates, or alkali trisulfates in the incinerator deposits has not been proved by X-ray diffraction. This of course does not mean they are not present, since previous studies at Battelle and elsewhere have shown that these materials are difficult to detect in small amounts. These sulfate salts can react directly with the tube metal to form iron sulfide, which was identified in the corrosion probe scale by X-ray diffraction. At higher temperatures, where the bisulfates and pyrosulfates are not stable, the corrosion by sulfur can proceed through formation of the alkali iron trisulfates which can be formed as shown in equation 7. The trisulfates also attack the tube metal to form FeS. Both the chlorine and the sulfur in the tube deposits can be involved in cyclic reactions which continue the corrosion beneath the deposits as shown in Figure 7.

In order to make a more complete evaluation of chlorine and sulfur in the corrosion reactions, the concentration of these elements in the refuse was increased by the addition of either polyvinyl chloride (PVC) or elemental sulfur. As it was not feasible to conduct experiments with controlled additions with these materials over long periods of time, this program consisted of exposing corrosion probes for 10 hours. These short time experiments were intended to show differences in the corrosiveness of the combustion environment as the result of the increased concentration of chlorine or sulfur. With an increase of chlorine content of the refuse by PVC addition the corrosion rate increased sharply above the metal temperature of 900°F (482°C). The effect of adding polyvinyl chloride to the refuse is shown in Figure 8. At all temperatures the corrosion rate increased rapidly with the initial additions of PVC. After the amount added reached 0.5-1 percent the corrosion rates leveled off at metal temperatures up to 900°F (482°C). However, at a metal temperature of 1100°F (593°C) the corrosion rate continued to increase linearly with the amount of PVC added. During the combustion of the polyvinyl chloride, the chlorine content is converted to HCl, about 50 percent of this plastic being chlorine. The HCl which is formed reacts rapidly with the metals and particularly the alkali metal oxides in the combustion process and is rapidly deposited in the form of chloride compounds on the various boiler surfaces. Large amounts of chloride were found in the probe deposits as shown in Figure 9. The chloride content was the highest on the metal specimens which were at a temperature of 400°F (204°C), a concentration of over 12 percent being reached for the 2 percent addition of PVC although as much as 8 percent was found with normal refuse. As the specimen temperature increased, less chloride was found in the deposits, but even at 1000°F (538°C) significant amounts of chloride remained in the deposits.

Laboratory investigations conducted at Battelle on this program showed that a mixture of alkali metal sulfates and NaCl is slightly corrosive in a helium atmosphere. A synthetic flue gas atmosphere without SO_2 proved to be somewhat more corrosive. However, a large increase in corrosion rate occurred when SO_2 was included in the flue gas mixture. The increased corrosion was the result of the conversion of chlorides to sulfates, thereby releasing HCl as mentioned above. The potential also exists for conversion of some of this HCl to elemental chlorine by the catalytic action of metal chlorides on the tube surface. As shown in Figure 10, the work of Brown et al[9], demonstrated that elemental chlorine attacks carbon steel rapidly above 400°F (204°C) while the HCl attack increased rapidly above 900°F (482°C). Hence, the low temperature corrosion in refuse combustion can be attributed to chlorine formed by oxidation of HCl or by conversion of alkali chlorides to alkali sulfates.

When sulfur was added to the refuse, the chloride content of the deposits decreased significantly as shown in Table 2. The effect was particularly pronounced in the 700-750°F (371-399°C) temperature zone where chlorine reached a

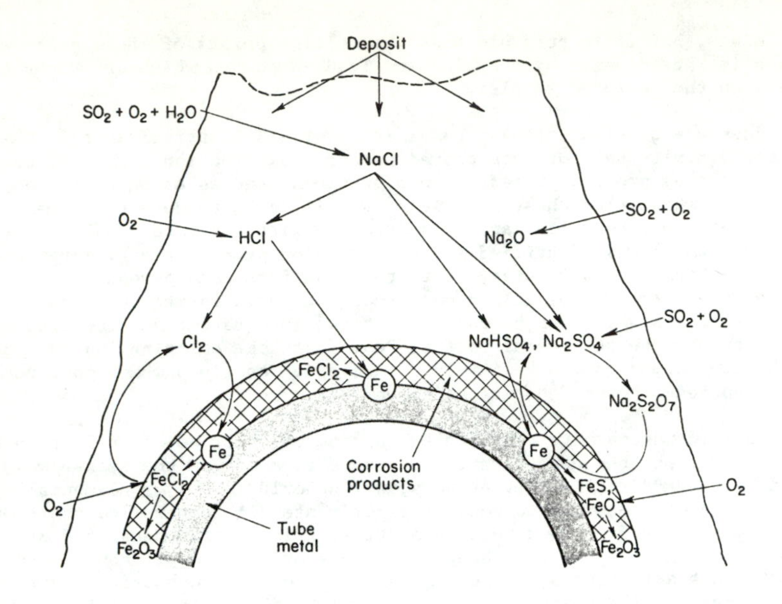

FIGURE 7. SEQUENCE OF REACTIONS OCCURRING IN THE INCINERATOR DEPOSITS

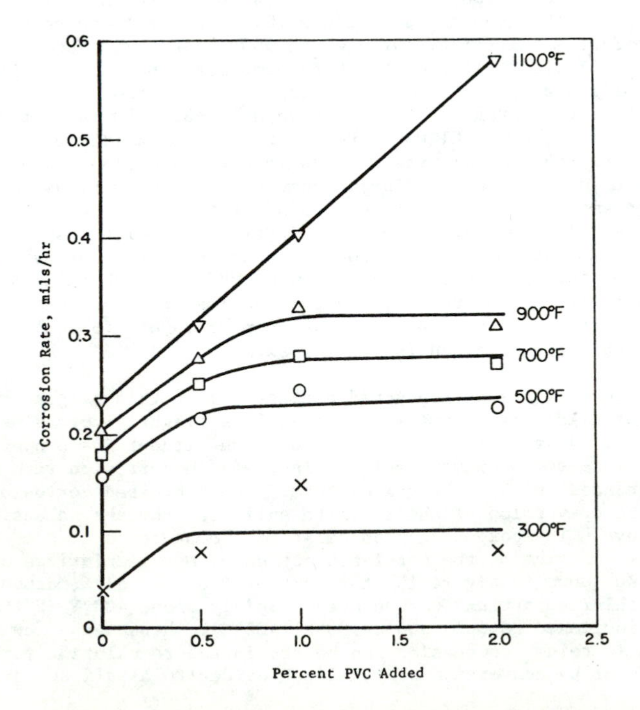

FIGURE 8. CORROSION RATE OF A106 STEEL AS A FUNCTION OF PVC ADDITION TO REFUSE

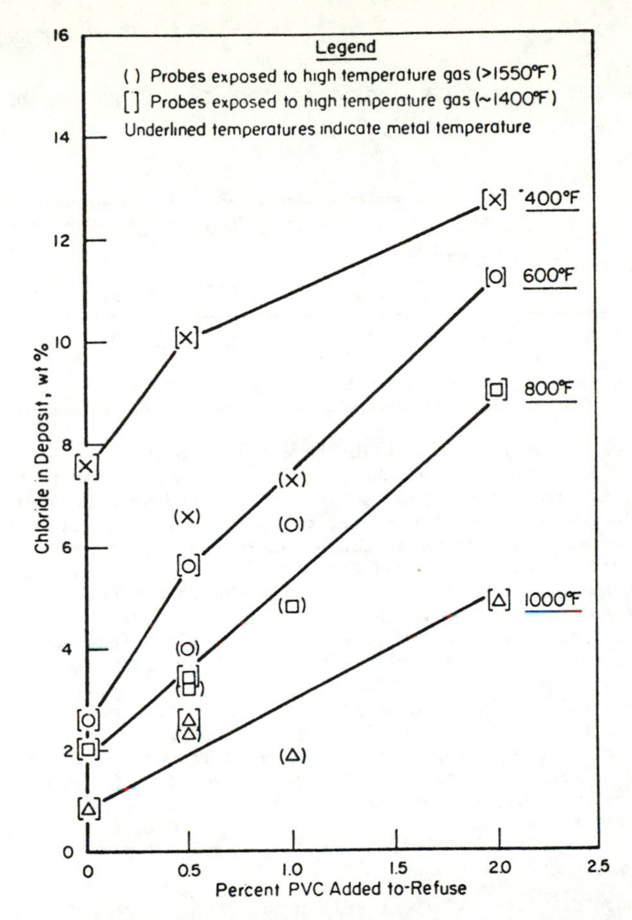

FIGURE 9. CHLORIDE CONTENT OF DEPOSITS AS A FUNCTION OF PVC ADDITION TO REFUSE

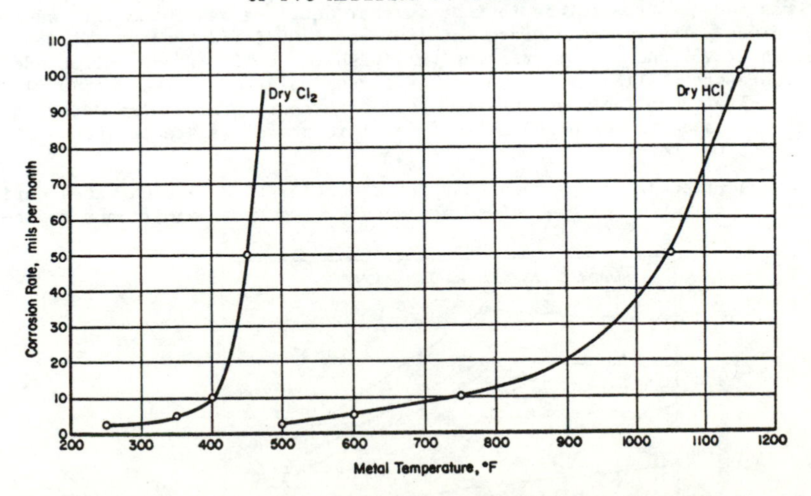

FIGURE 10. CORROSION RATES OF CARBON STEEL IN CHLORINE AND HCl AS A FUNCTION OF TEMPERATURE (REF. 9)

TABLE 2. SULFUR AND CHLORINE CONTENT OF DEPOSITS WITH SULFUR ADDITIONS TO REFUSE

	Concentration (Wt %) With Indicated Sulfur Addition (Wt %)					
	None		0.75% S		1.5% S	
Metal Temperature, °F (°C)	745 (395)	1120 (695)	730 (390)	1125 (605)	740 (343)	1175 (635)
Sulfur	4.6	2.8	3.5	6.0	5.0	3.8
Chlorine	7.5	2.5	1.9	0.1	0.1	<0.1

concentration of 7.5 weight percent in the deposits after 8 hours of operation with normal refuse. When 0.75 weight percent sulfur was added to the refuse the chloride concentration dropped to 1.9 percent, and when 1.5 weight percent sulfur was added the chloride concentration was further reduced to 0.1 weight percent. The effect was even more pronounced in the high temperature zone where the chloride concentration was reduced 25-fold by the addition of 0.75 weight percent sulfur, and below the detection limits for chloride when 1.5 weight percent sulfur was added. This decrease in the chloride concentration in the deposits indicates that the added sulfur in the refuse forms sulfur oxides rapidly, which then react with the metal chlorides while still in the flue-gas stream. This rapid formation of sulfates left very little particulate chloride to collect in the deposit. The effect of the sulfur additions on the corrosion rate of A106 steel is presented in Figure 11. At 300°F (149° C) metal temperature the corrosion rate decreased linearly with the amount of sulfur added. At higher temperatures the rapid decrease in corrosion rate began at the point of 0.75 weight percent sulfur addition. On the basis that the normal sulfur content of refuse is about 1.2 percent, these data indicate that when the total sulfur level in the refuse reaches 1 percent the corrosion rates can be expected to decrease.

An even more dramatic effect of sulfur addition to the refuse was found with Type 316 stainless steel, as indicated in Figure 12. At 800°F (427°C) the corrosion rate decreased by about an order of magnitude when 0.75 percent sulfur was added to the refuse, but no further decrease was noted with additional sulfur. The corrosion rates at 1000 and 1200°F (538 and 649°C) decreased about 25- and 30-fold respectively with the 0.75 percent sulfur addition. There was a slight increase in the corrosion rate at 1200°F (649°C) when the sulfur addition was increased to 1.5 percent. There also were significant decreases in corrosion rate of ferritic stainless steels when sulfur was added to the refuse.

The experimental data showed that the reduction in corrosion rates during the addition of sulfur to the refuse was accompanied by a significant reduc-

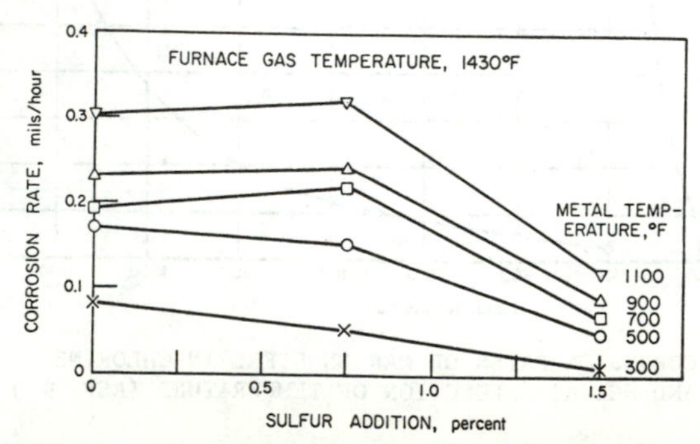

FIGURE 11. ISOTHERMAL CORROSION RATES OF A106 CARBON STEEL VS SULFUR ADDITIONS TO REFUSE

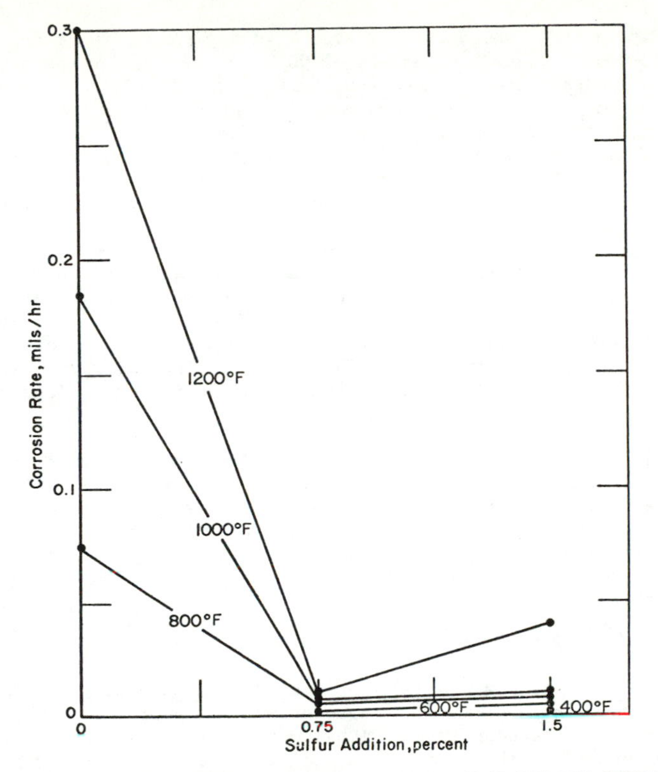

FIGURE 12. ISOTHERMAL CORROSION RATES OF STAINLESS STEELS AS A FUNCTION OF SULFUR ADDITIONS TO REFUSE

tion in the chloride content of the deposits and of the particulates extracted from the furnace or the stack. These observations indicate that the following chemical reaction is occurring in the flue gas stream rather than in the deposit:

$$\text{Metal chloride} + SO_2 + O_2 + H_2O = \text{metal sulfate} + HCl \text{ or } Cl_2 \quad .$$

Because the corrosive chlorine-containing agents are released in the gas stream they do not have the opportunity to act on the metal as they would by in situ formation beneath the deposit which adheres strongly to the tube metal. It also was noted that when sulfur was added to the refuse that the deposits formed on the corrosion probe were of a light powdery nature and much less adherent than in the case when normal refuse was burned. The conversion of the metal chlorides to metal sulfates before deposition on the tubes would account for this effect, and would result in the presence of compounds in the deposit which have much higher melting points than the corresponding chlorides.

Nevertheless, it is important to note that with these sulfur additions, sulfides did form near the metal oxide interface beneath the deposit on all of the corrosion probe specimens. This is undoubtedly the result of the reducing environment beneath the oxide layer, as it has been shown that FeS will form at a very low sulfur pressure (10^{-13} atmosphere) if the SO_2 or oxygen pressures are below 10^{-14} atmosphere.[12] At higher SO_2 or oxygen pressures the iron oxides form.

In order to establish the nature of the corrosive attack, metallographic examinations were made of the various alloys exposed to the incinerator environment. A106 low carbon steel exhibited general attack, with some evidence of pitting when exposed to normal refuse, as shown in the left and center specimens of Figure 13. This corrosive attack is typical of two metal temperature zones 400-800°F (204-427°C) and 900-1300°F (482-704°C), respectively. The higher metal temperature produced some surface decarburization as well as internal microstructural changes. When polyvinyl chloride was added to the refuse, similar attack was observed with more severe pitting. However, the addition of 1.5 percent sulfur to the refuse essentially eliminated the pitting attack as shown in the specimen on the right in Figure 13. The weight loss data showed that the general attack was reduced by a factor 3 in this case.

Type 316 and 302EZ stainless steels showed very similar behavior in that they underwent intergranular attack by the combustion products of normal refuse, and refuse to which PVC had been added. These steels were resistant and exhibited little or no grain boundary attack when exposed to combustion products of the refuse with greater sulfur content. Photomicrographs showing these differences for Type 316 stainless steel exposed to the various environments are shown in Figure 14.

The addition of sulfur to the refuse produced a sharp decrease in the corrosion rates of all types of steel investigated. In the case of low alloy steel, sulfur addition must be sufficient to reduce the chloride content of the initial deposit to the 0.1 percent level in order to achieve low corrosion rates. It is believed that the sulfur content of the refuse must be about 2 percent in order to achieve this low chloride level in the deposits and thus eliminate corrosion from this source.

REFUSE FIRING WITH HIGH SULFUR COAL

It was a logical step to proceed from the addition of sulfur to refuse to employing high sulfur coal as the source of the SO_2 needed to suppress chloride corrosion. Consequently additional work on the research program was directed to investigating the disposal of municipal refuse by co-firing with high-sulfur coal in a stoker-fired boiler, to demonstrate: (1) that corrosion of boiler tube materials in such a system would be minimal and (2) that the sulfur oxide emissions from the coal would be reduced to acceptable levels by dilution of the coal with refuse and by the interaction of the alkaline components of the solid wastes with the sulfur oxides. The coal-fired boiler at the Municipal Electric Plant, Columbus, Ohio has been used for this program. This site was selected to represent boiler furnaces having grates on which final burnout of the refuse could occur. The boiler used has a spreader stoker and is rated at 150 thousand pounds of steam per hour, but it normally operates up to 125 thousand pounds per hour with a maximum electrical output of 12.5 MW. This program is directed ultimately toward determining the limitations and benefits of burning the solid wastes with high-sulfur coal and limiting the processing of the solid wastes to shredding and magnetic separation. The corrosivity of the combustion products obtained from the burning of coal and the refuse-coal mixes was evaluated by inserting the corrosion probes in the superheater section of the boiler. At this location the gas temperature varied from 1100°F (593°C) under low load conditions to 1400°F (760°C) for the higher loads. The internal air cooling of the probe specimens maintained metal temperatures of 500-950°F (260-510°C). The initial corrosion rates of carbon steel and stainless steels were measured by 8 hour probe exposures during the combustion of coal containing up to 5 percent sulfur and in refuse-coal mixtures containing up to 70 weight percent refuse. With this maximum amount of refuse the heat input from the refuse constituted 60 percent of the energy supplied.

Initial corrosion rates of A106 carbon steel as a function of the metal temperature during the combustion of various coals, coals plus refuse, and 100 percent refuse are shown in Figure 15. The results presented in the upper part of the figure for burning of bulk refuse were obtained at the Miami County incinerator, Troy, Ohio. The corrosion rates with refuse alone proved to be

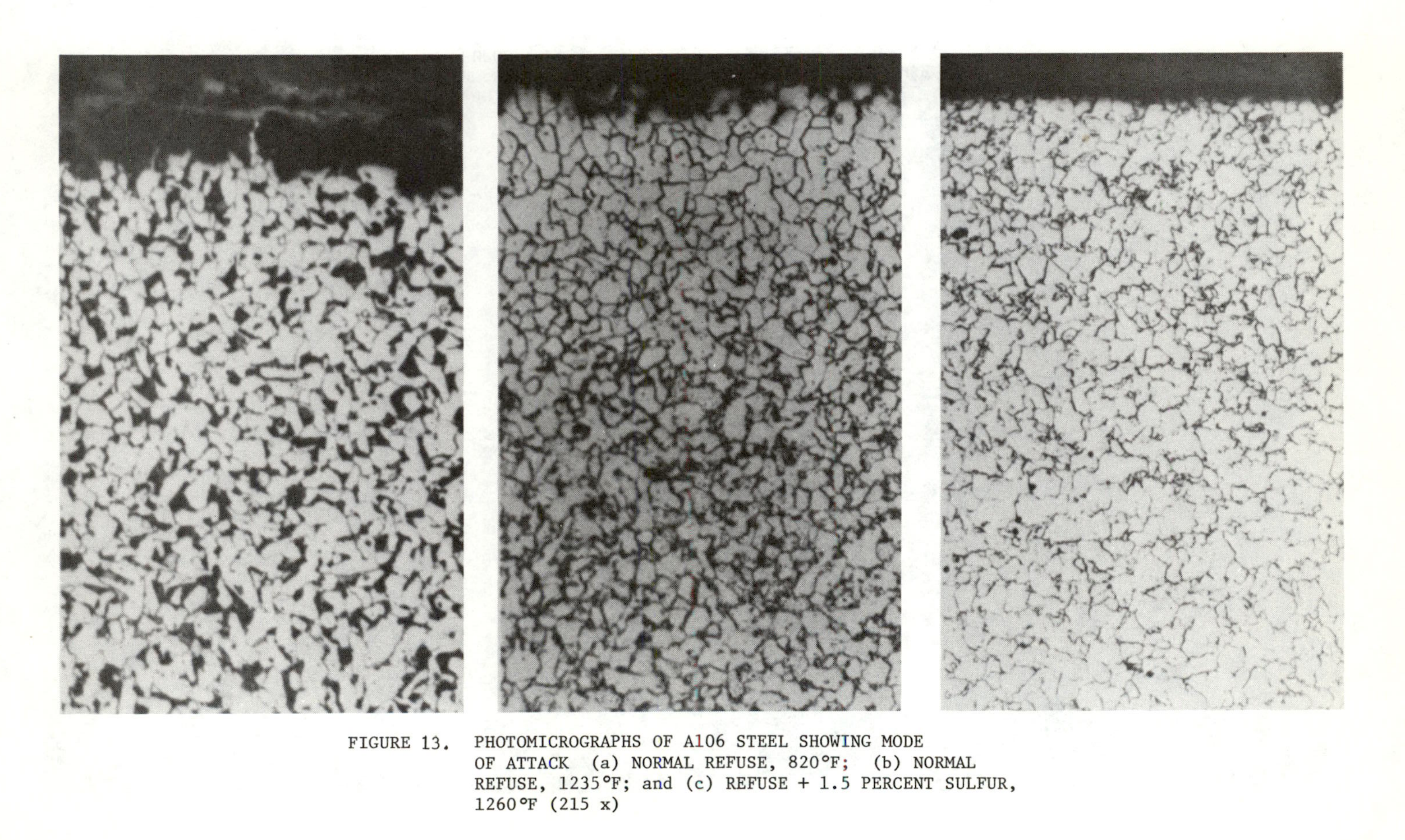

FIGURE 13. PHOTOMICROGRAPHS OF A106 STEEL SHOWING MODE OF ATTACK (a) NORMAL REFUSE, 820°F; (b) NORMAL REFUSE, 1235°F; and (c) REFUSE + 1.5 PERCENT SULFUR, 1260°F (215 x)

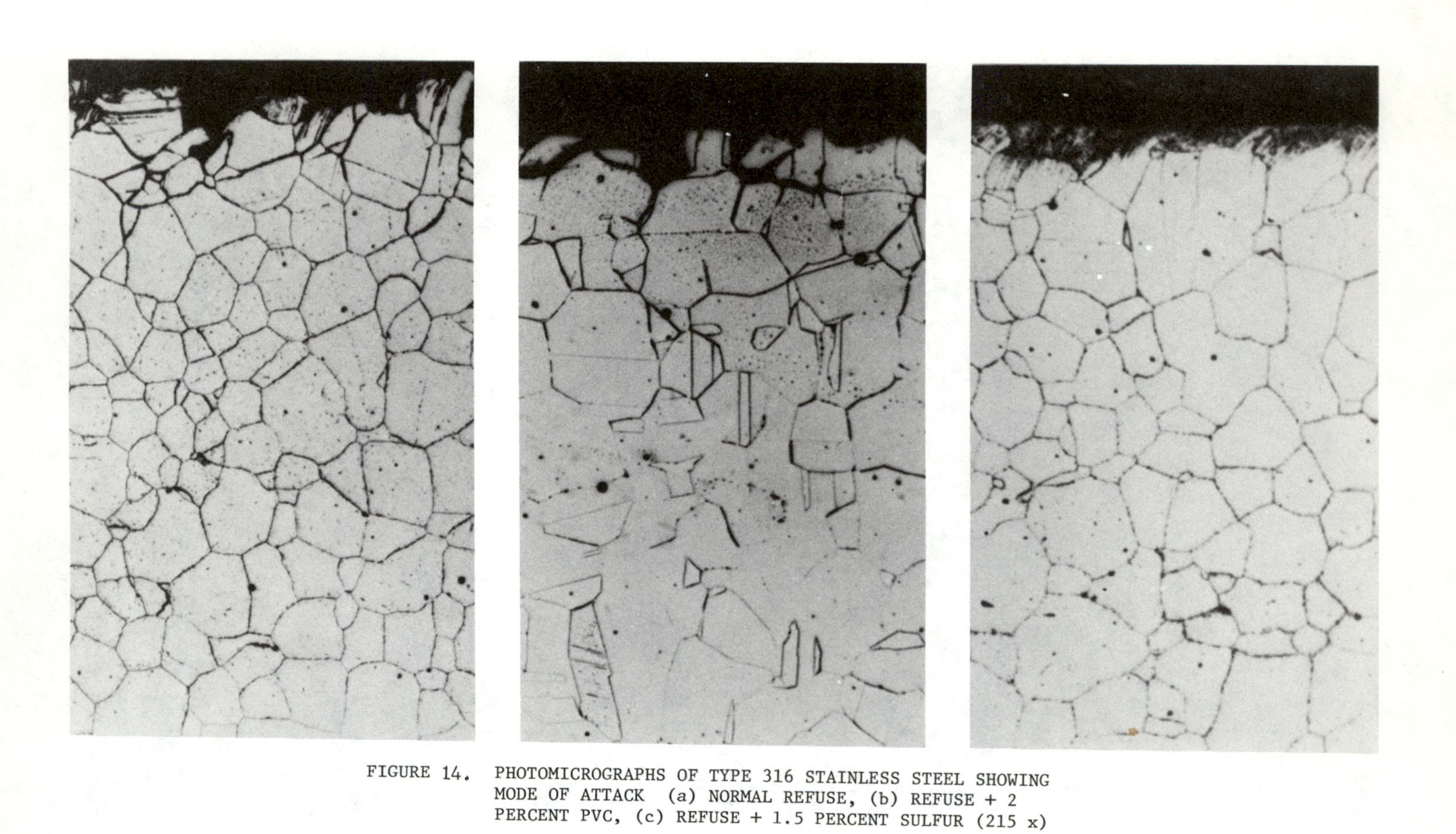

FIGURE 14. PHOTOMICROGRAPHS OF TYPE 316 STAINLESS STEEL SHOWING MODE OF ATTACK (a) NORMAL REFUSE, (b) REFUSE + 2 PERCENT PVC, (c) REFUSE + 1.5 PERCENT SULFUR (215 x)

an order of magnitude greater than those found with coal or coal plus up to 42 percent refuse. In the lower portion of the figure the two data points at about 750°F (399°C) which showed the highest corrosion rates, represent the coal with 5 percent sulfur and are the result of sulfide corrosion. It should be emphasized that these data represent corrosion rates obtained in 8-hour probe exposures and are useful for comparing materials and refuse-coal mixes but should not be extrapolated to represent long-term corrosion rates. These rates drop off rapidly with time, as protective oxide layers are built up on the metal surfaces, and then level off at a much lower wastage rate. It is significant that when the sulfur content of coal reaches 5 percent the corrosion rate of carbon steel is controlled by the sulfur and not by the chlorine content of the refuse, at least up to the amounts of refuse used in these experiments. The corrosion rates for the stainless steels are shown as a function of metal temperature in Figure 16. The lower plot presents the data for high alloy steel, the center plot for medium alloy steels, and the upper plot for both types exposed in the refuse incinerator. As was the case with carbon steel, stainless steels also are attacked much more rapidly by the combustion products of 100 percent refuse. With all of the fuel systems used to date the medium-alloy stainless steels have proved to be less resistant than the high-alloy stainless steels. The Type 316 stainless steel was affected most by the addition of refuse to the coal, particularly at the higher amounts of refuse. In general, the corrosion rates of the stainless steel with the various coals and coal-refuse mixtures were an order of magnitude lower than those found with the carbon steel.

CONCLUSIONS

The results of this experimental program have shown that the sulfur and chlorine contents of refuse have significant effects on the corrosion of metal surfaces exposed to hot combustion products. From the data collected it can be concluded that: (1) The conversion of chlorides to sulfates in the incinerator deposits by the action of SO_2 releases chlorine and HCl at the metal surface, thereby causing serious corrosion. (2) Chloride corrosion can be made negligible by increasing the total available sulfur in fuel to equal 2 weight percent of the refuse. (3) Co-firing of refuse with high sulfur coal up to a 60/40 mix on a Btu basis will not increase the initial corrosion rate beyond that of coal alone.

These results for the co-firing of refuse with coal have been obtained in short-term runs where the initial corrosion rates are high, and the actual wastage rates reported cannot be extrapolated to determine the long-term corrosion that might be expected under these conditions.

ACKNOWLEDGMENTS

This project has been financed in part with Federal funds from the Environmental Protection Agency under grants from the Municipal Environmental Research Laboratory and the Industrial Environmental Research Laboratory, Cincinnati, Ohio.

The cooperation of the administrative and operating staffs of the incinerator at Miami County, Ohio, and of the Electrical and Sanitation Divisions of the City of Columbus is greatly appreciated.

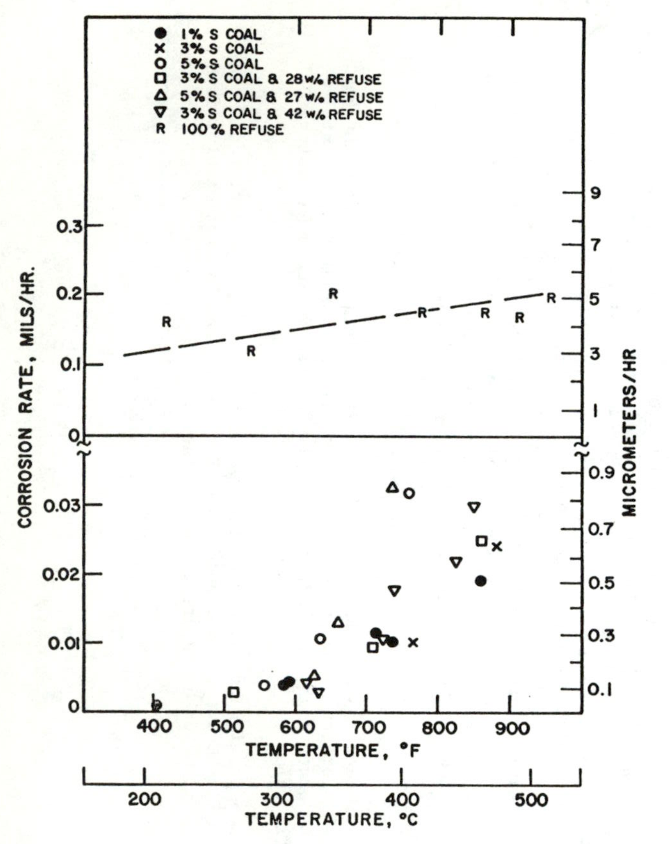

FIGURE 15. INITIAL CORROSION RATES OF A106 CARBON STEEL WITH VARIOUS FUELS

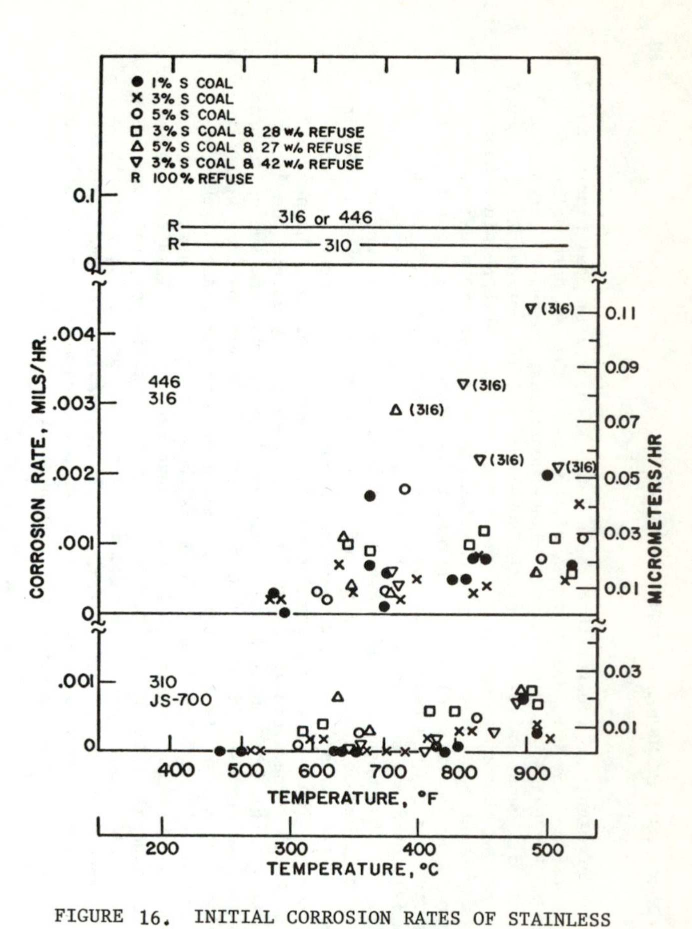

FIGURE 16. INITIAL CORROSION RATES OF STAINLESS STEELS WITH VARIOUS FUELS

REFERENCES

1. Miller, P. D., et al., "Corrosion Studies in Municipal Incicerators", U. S. Environmental Protection Agency Report SW-73-3-3, 1972.

2. Miller, P. D. and Krause, H. H., "Corrosion of Carbon and Stainless Steels in Flue Gases from Municipal Incinerators", Proceedings of the 1972 National Incinerator Conference, American Society of Mechanical Engineers, June, 1972, pp. 300-309.

3. Miller, P. D., Krause, H. H., Vaughan, D. A., and Boyd, W. K, "The Mechanism in High Temperature Corrosion in Municipal Incinerators", Corrosion, Vol. 28, July, 1972, pp. 274-281.

4. Krause, H. H., Vaughan, D. A., and Miller, P. D., "Corrosion and Deposits from Combustion of Solid Waste", Transactions of the ASME, Journal of Engineering for Power, Series A, Vol. 95, 1973, pp. 45-52.

5. Krause, H. H., Vaughan, D. A., Miller, P. D., "Corrosion and Deposits from Combustion of Solid Waste. Part 2 - Chloride Effects on Boiler Tube and Scrubber Metals", Transactions of the ASME, Journal of Engineering for Power, Series A, Vol. 96, 1974, pp. 216-222.

6. Krause, H. H., Vaughan, D. A., Boyd, W. K., "Corrosion and Deposits from Combustion of Solid Waste. Part 3 - Effects of Sulfur on Boiler Tube Metals", Transactions of the ASME, Journal of Engineering for Power, Vol. 97, 1975, pp. 448-452.

7. Krause, H. H., Vaughan, D. A., Boyd, W. K., "Corrosion and Deposits from Combustion of Solid Waste. Part 4 - Combined Firing of Refuse and Coal", Transactions of the ASME, Journal of Engineering for Power, Vol. 98, 1976, pp. 369-374.

8. Vaughan, D. A., Krause, H. H., Boyd, W. K., "Fireside Corrosion in Municipal Incinerators versus Refuse Composition", Materials Performance, Vol. 14, No. 5, 1975, pp. 16-24.

9. Brown, M. H., DeLong, W. B., and Auld, J. R., "Corrosion by Chlorine and by Hydrogen Chloride at High Temperatures", Journal of Industrial and Engineering Chemistry, Vol. 39, July, 1947, pp. 839-844.

10. Cutler, A.J.B., Halstead, W. D., Laxton, J. W., Stevens, C. G., "The Role of Chloride in the Corrosion Caused by Flue Gases and Their Deposits," American Society of Mechanical Engineers, Winter Annual Meeting Paper No. 70-WA/CD-1, December, 1970.

11. Fassler, K., Leib, H., and Spahn, H., "Corrosion in Refuse Incineration Plants", Mitt. V.G.B., Vol. 48, April, 1968, pp. 126-139.

12. Ross, T. K., "The Distribution of Sulfur in Corrosion Products Formed by Sulphur Dioxide on Mild Steel", Corrosion Science, Vol. 5, 1965, pp. 327-330.

FLUIDIZED-BED COMBUSTION

Session Chairman: **D. L. Keairns**
Westinghouse Electric Corporation

CORROSION/DEPOSITION IN FLUIDIZED BED COMBUSTION POWER PLANT SYSTEMS

D. L. KEAIRNS, M. A. ALVIN, E. P. O'NEILL, C. SPENGLER,
J. CLARK, J. R. HAMM, and E. F. SVERDRUP

Westinghouse R&D Center
Pittsburgh, Pennsylvania, USA 15235

ABSTRACT

Control of material corrosion, deposition, and erosion is required to demonstrate commercial fluidized bed combustion systems. The primary areas of concern are the heat transfer surface materials in atmospheric and pressurized systems and gas turbine expander materials used in pressurized systems. An understanding of the source of gaseous contaminants and particulates; the gaseous and particulate compositions throughout each fluidized bed combustion concept; the corrosion, deposition, and erosion potential under specified operating conditions; and the tolerance of alternative materials to the respective design and operating parameters is required in order to design a reliable and cost effective fluidized bed combustion process.

The scope of this assessment is limited to questions of material corrosion. Fluidized bed combustion power plant concepts are reviewed and reference operating parameters identified. Corrosive contaminants, corrosion potential, and materials selection are discussed, with emphasis on the gas turbine expander utilized in pressurized systems.

1. INTRODUCTION

Fluidized bed combustion processes are commercially used in over 300 installations in the U.S. for roasting pyrite ores, incinerating industrial wastes, incinerating sludge from sewage plants, and related applications. A high degree of reliability has been demonstrated with these systems. This experience is valuable for the development of coal-fired fluidized bed combustion systems.

The concepts considered in this document will be limited to those in which heat can be recovered for industrial or commercial use, steam production, or electric power generation. These fluidized bed combustion concepts can potentially utilize a wide range of fossil fuels, achieve lower costs compared with competitive systems, and meet environmental requirements. Numerous references are available to provide a historical perspective and an idea of the work that has been and is being performed.[1-8]

Control of material corrosion, deposition, and erosion is required to demonstrate commercial fluidized bed combustion systems. Of primary concern are the heat transfer surface materials used in atmospheric and pressurized systems and the gas turbine expander materials used in pressurized systems. An understanding of the source and composition of gaseous contaminants and particulates throughout each fluidized bed combustion concept; the corrosion, deposition, and erosion potential under specified operating conditions; and the tolerance of alternative materials to the respective design and operating parameters is required in order to design a reliable and cost effective fluidized bed combustion process.

The scope of this assessment is limited to questions of material corrosion. Fluidized bed combustion power plant concepts are reviewed and reference operating parameters identified. Corrosive contaminants, corrosion potential, and materials selection are discussed with emphasis on the gas turbine expander utilized in pressurized systems.

2. POWER PLANT CONCEPTS

2.1 Operating Conditions

Atmospheric and pressurized fluidized bed combustion systems are being developed. Four basic concepts - one atmospheric and three pressurized plant concepts - are shown in Figures 1a and b. Representative pressure, temperature, and gas composition profiles are also indicated. A summary of plant performance for these reference designs is given in Table 1.

A wide range of cycle options and operating conditions are being investigated for each concept. Thus, the cycles and conditions shown are selected to establish reference cases for evaluating materials corrosion in fluidized bed combustion systems. Several factors will determine the actual environment to which the materials will be exposed. For example:

- Feed materials. The trace elements in coals and sorbents vary over a wide range, and the release of contaminants will affect materials corrosion.
- Operating temperatures. Bed temperature and working fluid temperature will determine corrosion tolerance.
- Sulfur removal. The degree of sulfur removal achieved will affect corrosion.
- Concentration profiles. The local partial pressure of oxygen and other gaseous species in the fluidized bed will depend on the operating conditions (e.g., excess air, gas velocity, particle size) and design parameters (e.g., coal-feed location, heat transfer surface configuration).

TABLE 1

Fluidized Bed Combustion Reference Plant Performance

System	Steam Cycle	Plant Heat Rate
Atmospheric Pressure Boiler[1]	238 atm/811K/811K (3500 psia/1000°F/1000°F)	9535
Pressurized Boiler[2]	238 atm/811K/811K (3500 psia/1000°F/1000°F)	8750 Btu/kWh
Combined Cycle with Adiabatic Combustor[3]	48 atm/700K (700 psia/800°F)	9280 Btu/kWh
Combined Cycle with Partially Indirect Air-Cooled Combustor[4]	55 atm/716K (805 psia/830°F)	8915

1 t_{bed} = 1227K (1750°F)

2 t_{bed} = 1283K (1850°F)

3 t_{bed} = 1200K (1700°F)

4 t_{bed} = 1283K (1850°F)

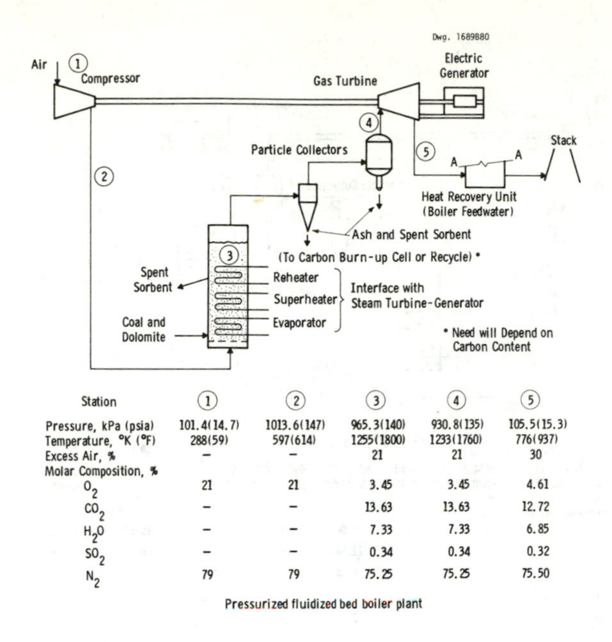

Station	1	2	3	4	5
Pressure, kPa (psia)	101.4(14.7)	1013.6(147)	965.3(140)	930.8(135)	105.5(15.3)
Temperature, °K (°F)	288(59)	597(614)	1255(1800)	1233(1760)	776(937)
Excess Air, %	–	–	21	21	30
Molar Composition, %					
O_2	21	21	3.45	3.45	4.61
CO_2	–	–	13.63	13.63	12.72
H_2O	–	–	7.33	7.33	6.85
SO_2	–	–	0.34	0.34	0.32
N_2	79	79	75.25	75.25	75.50

Pressurized fluidized bed boiler plant

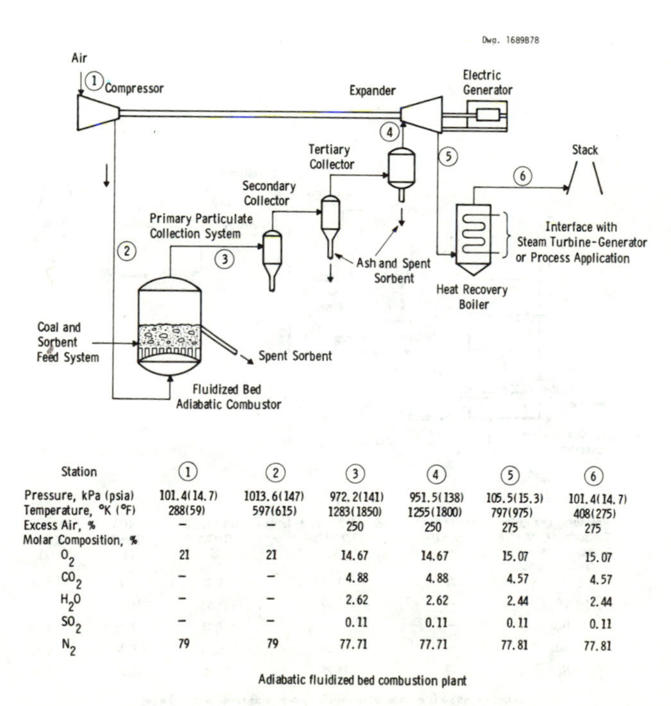

Station	1	2	3	4	5	6
Pressure, kPa (psia)	101.4(14.7)	1013.6(147)	972.2(141)	951.5(138)	105.5(15.3)	101.4(14.7)
Temperature, °K (°F)	288(59)	597(615)	1283(1850)	1255(1800)	797(975)	408(275)
Excess Air, %	–	–	250	250	275	275
Molar Composition, %						
O_2	21	21	14.67	14.67	15.07	15.07
CO_2	–	–	4.88	4.88	4.57	4.57
H_2O	–	–	2.62	2.62	2.44	2.44
SO_2	–	–	0.11	0.11	0.11	0.11
N_2	79	79	77.71	77.71	77.81	77.81

Adiabatic fluidized bed combustion plant

Fig. 1a - Fluidized bed combustion power plant concepts

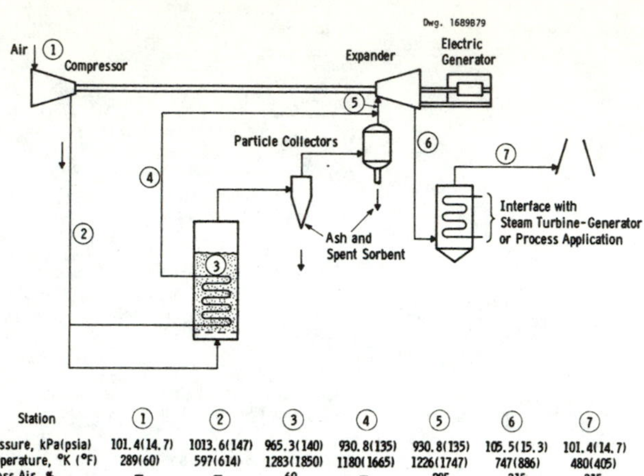

Station	①	②	③	④	⑤	⑥	⑦
Pressure, kPa(psia)	101.4(14.7)	1013.6(147)	965.3(140)	930.8(135)	930.8(135)	105.5(15.3)	101.4(14.7)
Temperature, °K (°F)	289(60)	597(614)	1283(1850)	1180(1665)	1226(1747)	747(886)	480(405)
Excess Air, %	–	–	60	–	285	315	315
Molar Composition, %							
O_2	21	21	6.08	21	15.24	15.63	15.63
CO_2	–	–	11.56	–	4.43	4.13	4.13
H_2O	–	–	6.64	–	2.38	2.22	2.22
SO_2	–	–	0.26	–	0.10	0.10	0.10
N_2	79	79	75.45	79	77.85	77.92	77.92

Pressurized fluidized bed indirect air cooled plant

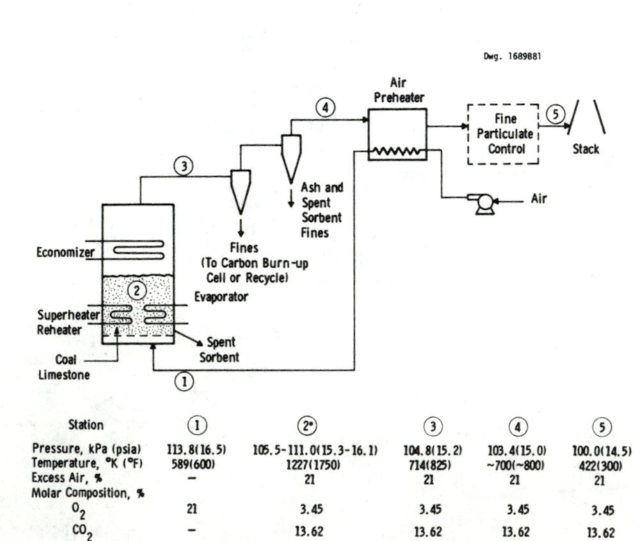

Station	①	②*	③	④	⑤
Pressure, kPa (psia)	113.8(16.5)	105.5-111.0(15.3-16.1)	104.8(15.2)	103.4(15.0)	100.0(14.5)
Temperature, °K (°F)	589(600)	1227(1750)	714(825)	~700(~800)	422(300)
Excess Air, %	–	21	21	21	21
Molar Composition, %					
O_2	21	3.45	3.45	3.45	3.45
CO_2	–	13.62	13.62	13.62	13.62
H_2O	–	7.33	7.33	7.33	7.33
SO_2	–	0.34	0.34	0.34	0.34
N_2	79	75.25	75.25	75.25	75.25

*Average values in bed; actual local values will vary over wide range within the bed

Atmospheric pressure fluidized bed boiler plant

Fig. 1b - Fluidized bed combustion power plant concepts

Operating procedure. Start-up, shutdown, and load-follow procedures will affect the concentration/temperature profiles which must be compatible with materials limitations.

These factors must be integrated with the reference steady-state conditions to evaluate the potential for materials corrosion and the options for control.

Three of the fluidized bed combustion concepts selected incorporate a heat transfer surface in the fluidized bed. The range of metal temperatures for each of the reference systems is shown in Table 2.

TABLE 2

Fluidized Bed Combustion
Heat Transfer Surface Temperature

System	Tube Material Temperature Range[a]
Atmospheric Pressure Boiler	650-908K (710-1175°F)
Pressurized Boiler	662-894K (732-1150°F)
Pressurized Partially Indirect Air-Cooled Combustor	1074-1272K (1473-1794°F)

[a] The temperature range will depend on heat transfer media conditions and on the bed operating temperature. Temperatures given are for reference designs.

2.2 System Analysis

Corrosion of heat transfer surfaces in the fluidized bed combustor, gas turbine expander materials, and heat recovery components must be controlled in order to achieve acceptable plant performance and to meet cost objectives. The process components and factors affecting corrosion must be understood in order to arrive at an optimum design. The plant components of primary interest include an in-bed heat transfer surface, an above-bed heat transfer surface, a gas turbine expander, and heat recovery elements. The corrosion of exposed metals in particulate control equipment is also a concern. The factors affecting corrosion of these components fall into four groups: the selection of feed materials (fuel and sulfur sorbent), the release of active species (elements and compounds important in corrosion reactions) from the feed, the dynamic "equilibrium" of the active species, and corrosion reactions. The matrix (Figure 2) of process components and corrosion vectors must then be understood within the context of a data base: thermochemical analyses, kinetic rate data, and experimental observations from actual plants. In this discussion attention is focused primarily on pressurized systems and on the difficulties of preventing corrosive conditions in the gas turbine expander.

2.3 System Design

The system designed to achieve acceptable materials performance will be based on the following factors (Figure 3):

- Selection of coal and sorbent (will affect the release of potential contaminants)
- Additive option (will affect the concentration of potential corrodants - e.g., the addition of alkali getters such as aluminosilicates)
- Selection of operating conditions (will affect release, concentration of potential corrodants, and corrosion reactions)
- Selection of design (will affect release - e.g., local reducing conditions may result in hydrogen sulfide release)
- Selection of materials (will affect corrosion reactions - e.g., the selection of advanced alloy materials or coatings).

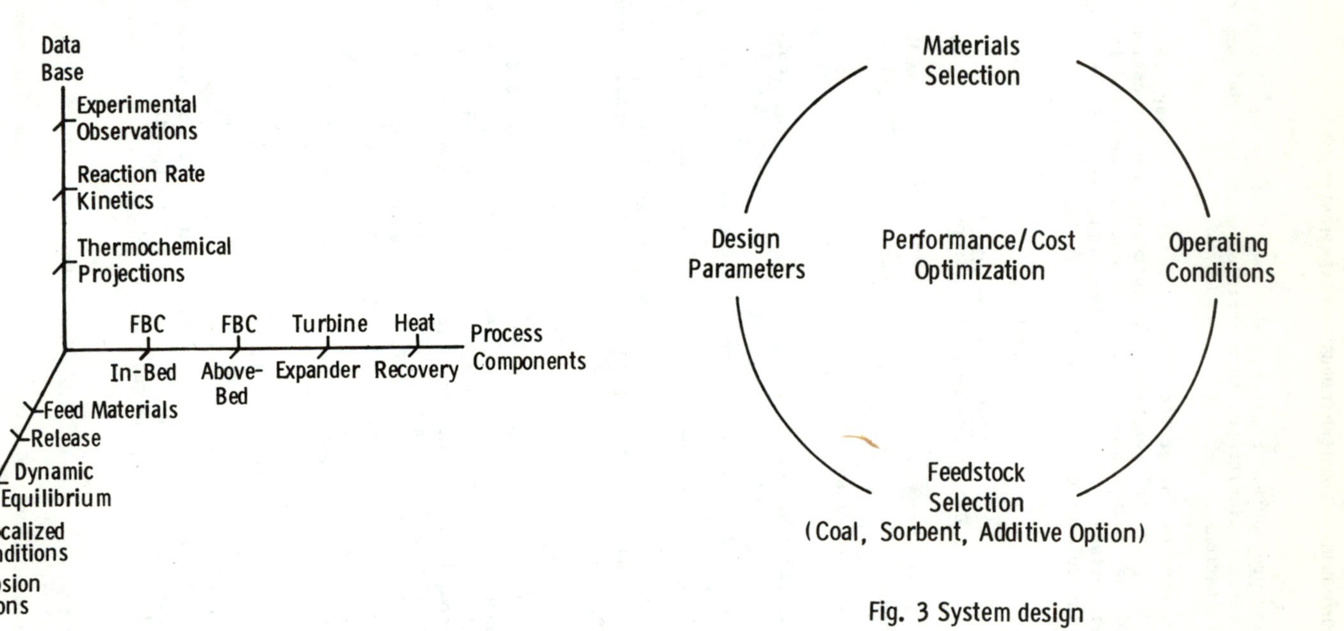

Fig. 2 System analysis

Fig. 3 System design

3. CORROSIVE CONTAMINANTS

3.1 Alkali Metal and Chlorine Components (Coal/Sorbent)

The feedstock materials - coal and dolomite - contain a variety of trace and minor elements. The elements of greatest technical concern are sulfur (S), sodium (Na), potassium (K), and chlorine (Cl). Corrosion problems may also arise from the presence in coal of lead (Pb) and vanadium (V).

The mean levels of alkali found in U.S. coals by Ruch, et al.[9] 0.05 wt % Na and 0.17 wt % K, are high in relation to the concentrations of concern (in the ppb range in the gas phase), as will be shown later. Similarly, dolomites have been shown to contain alkali in amounts ranging from 5 to 6000 ppm K and 5 to 1500 ppm Na. Wide variations in the alkali content of the same dolomite have been reported for samples obtained from different quarries over a seven-year interval. Some of the alkali is present as chlorides in both coal and dolomite; the remainder is bound up in such clay mineral matter as illites and the feldspars.

At 1400 ppm the halide content of coals is also high relative to that of fuel oils. A mean value of 782 ppm, with a range of 170 to 2290 ppm, has been reported[10] for nine samples of dolomite.

Essentially all the chlorine from coal will be liberated as hydrogen chloride [HCl (g)] or sodium chloride [NaCl (g)] in fluidized bed combustion. This chloride now serves as a gaseous carrier for the alkali metals. Thermodynamic equilibrium for the bed (assuming sodium is otherwise present as the sulfate) shows that between 20 and 100% of the sodium in coal could be released as gaseous sodium chloride, depending on operating conditions. This would result in a gas-phase composition similar to that presented in Figure 4, which shows the equilibrium distribution of products in the effluent gas from the fluid bed, assuming 10% of the sulfur, 100% of the chlorine, and 1% of the alkali metals are released to the gaseous phase at 1013.6 kPa (10 atm) total pressure.

For species which are liberated to the gas phase by volatilization of the compound actually present in coal, such as the alkali metal chlorides which have relatively high vapor pressures, a high efficiency of release is anticipated. Less efficient vaporization will occur where the trace-element species must be liberated by specific reaction of a solid with a constituent of the gas phase.

3.2 Factors Affecting Release

The factors that affect the degree of release of alkali and that might be adjusted to control release to acceptable levels are the feedstock composition, operating temperature, and addition of getter materials to the fluid bed, or to the gas cleanup train.

3.2.1 Feedstock Composition

The contribution of dolomites to the total alkali release may be lowered by using an extremely pure metamorphic dolomite such as Kaiser Dolowhite. Moreover, work has shown that alkali release from dolomites is not proportional to the total alkali content but is correlated with the chlorine content[9]. Pretreatment of the sorbent to drive off the volatile alkali is a possible method of control, particularly for regenerative sorbent systems. When coal is used, the presence of ash in a fluidized bed leads to competition for released alkali between the gaseous halide carrier and the solid aluminosilicates. The limit to this competitive process will be shown, but the kinetics will be governed to some extent by the location of chlorine and alkali release from the coal; if the bed height is relatively large (∿10 ft) significant retrapping of alkali may occur.

<u>Operating temperature</u>. Thermodynamic equilibria for the release of sodium to the gas phase as NaCl(g) show that higher temperatures should lead to greater release; in addition, more alkali will be released at lower pressures, as shown in Figure 5. These projections do not include the role of aluminosilicates in

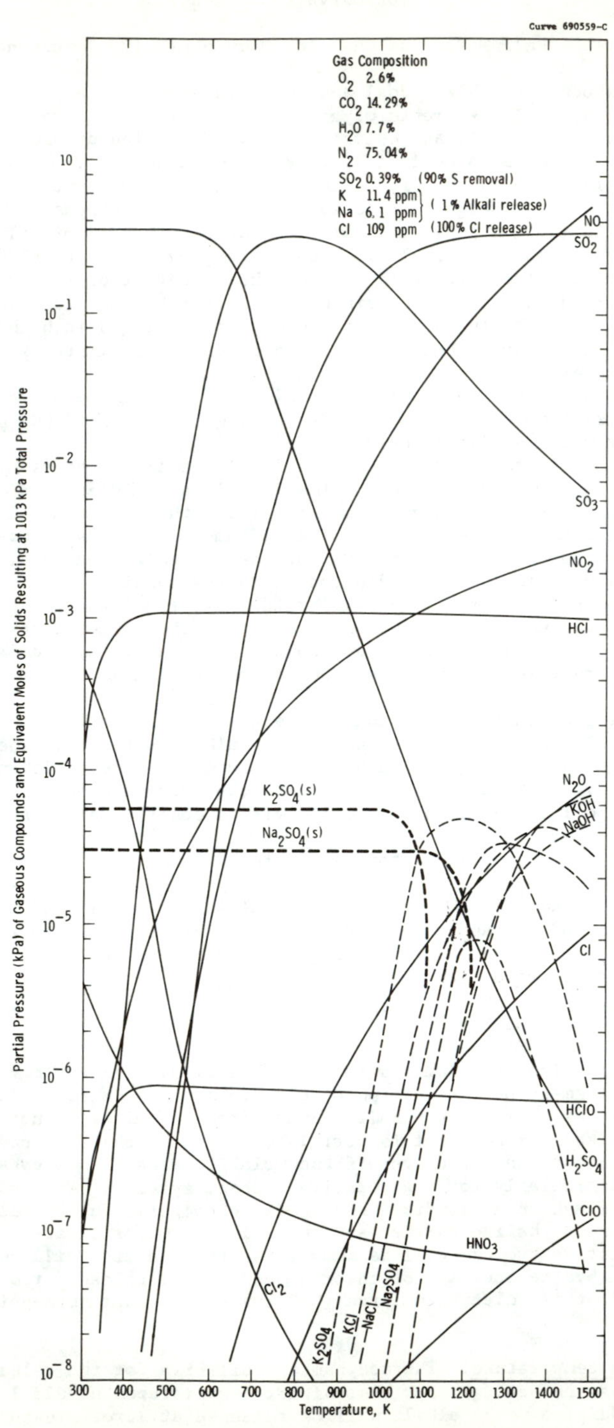

Fig. 4 – Equilibrium species concentration in fluidized bed combustion gaseous effluent

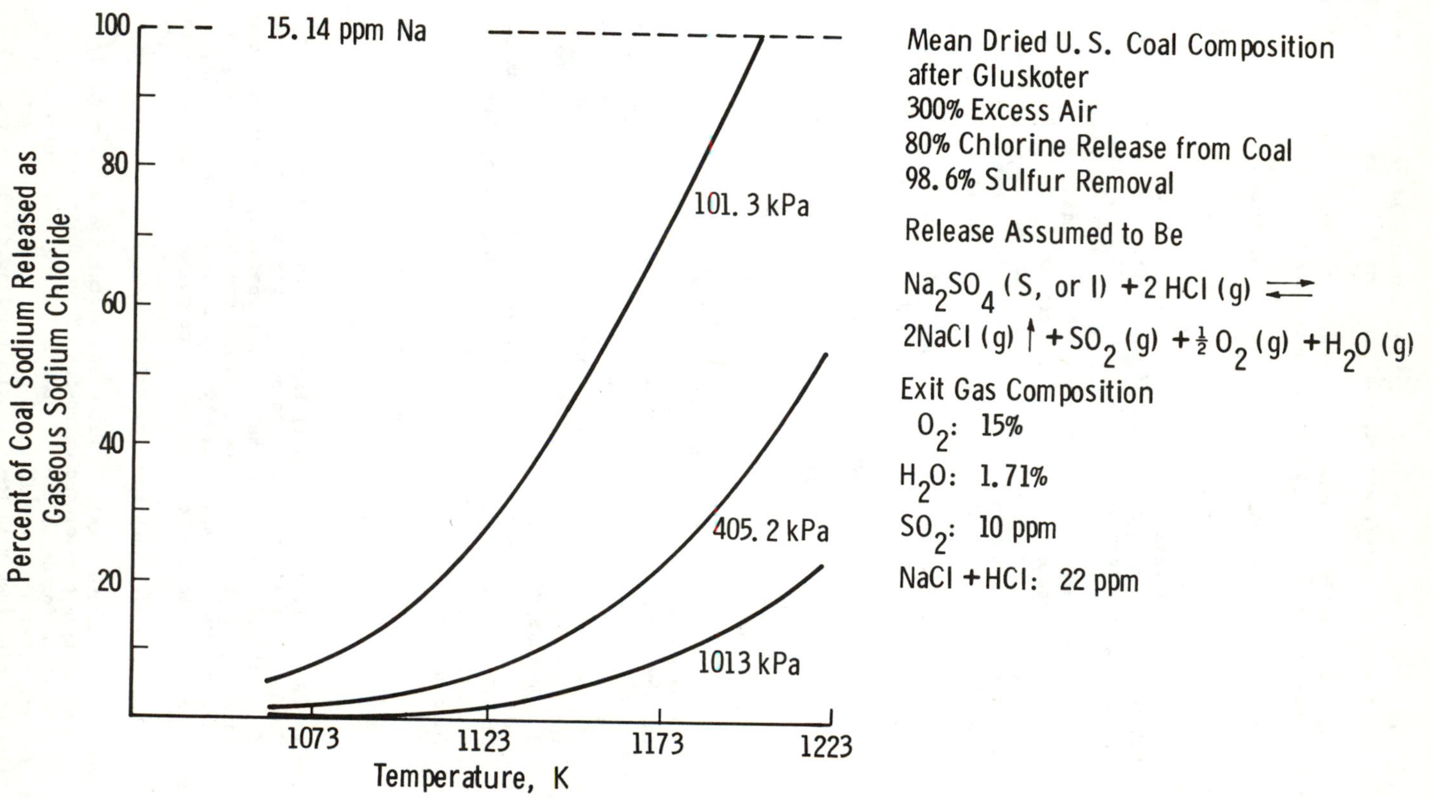

Fig. 5 — Equilibrium release of sodium from coal in the fluidized bed combustor

lowering alkali emissions. The major question to be determined is how effectively HCl(g) can release alkali from coal, coal ash, and dolomite in the fluidized bed.

Reducing the bed operating temperature should reduce the alkali emission rate, but for greater efficiency, higher bed temperatures are preferable.

3.3 Factors Affecting Removal - Gettering Action of Aluminosilicates

When aluminosilicates or other getters are present in sufficient concentration, they will reduce alkali emissions from the bed through the formation of feldspars. The potential reduction in alkali emissions is shown in Figure 6, which shows that potassium can be reduced to 2% of its previous concentration in the gas phase through the formation of adularia ($KAlSi_3O_8$). (The slight apparent decrease in gettering efficiency at 1013.6 kPa (10 atm) pressure should be read in the context of greatly reduced potassium emission as a function of increasing pressure in the absence of getter). Laboratory experiments with mixtures of dolomite and coal char, coal ash, and aluminosilicates have demonstrated the reduction in alkali release using both continuous vapor phase alkali concentration monitoring and solids analysis at the end of experiments.[10]

The addition of aluminosilicates as getter materials provides an attractive option since they are both cheap and effective over a wide temperature range. Neither their capacity nor required concentration is known, and in order to design effective alkali suppression stages, the kinetics of the gettering reaction must be studied.

3.4 Limitations

The limitations of thermodynamic projections must be considered when anticipating the level of contaminants and assessing their impact, as illustrated by the projected oxygen concentrations in the combustor. Within the fluid bed itself the presence of reducing zones has been indicated as a potential initiator of corrosion. Although oxygen profile measurements across a fluidized bed made by removal of gas samples show a continuous excess of oxygen, measurements made by a solid-state oxygen gauge have indicated regions where P_{O_2} is approximately equal to 1×10^{-5} Pa (10^{-10} atm).[11] Systems operating with incomplete combustion of carbon and high levels of carbon monoxide (CO) indicate that stagnant areas of the bed are vulnerable to the accumulation of deposits. The temperature of metal surfaces in the bed are so low that alkali sulfates rather than feldspars are stable, even with respect to the relatively high levels of HCl present in the gaseous phase. The circulation of solid bed material across the exposed metal surfaces is thought to be the most efficient method for preventing deposit accumulation.

4. CORROSION-DEPOSITION POTENTIAL

4.1 Concentration Profile

As shown in the equilibrium profile of Figure 4, the principal gaseous species likely to be released from the fluid bed are the chlorides, hydroxides, and sulfates of the alkali metals. On the basis of operating experience, current turbine operating practice is to specify a maximum contaminant level in the fuel. The alternative approach, attempting to calculate the contaminant levels at operating conditions which can lead to corrosion, is addressed here.

4.2 Turbine Metal Tolerance

The deposition of liquid alkali sulfates on hot turbine metal parts is a major factor in initiating accelerated corrosion of the superalloys. The thermodynamic conditions for deposition of sodium sulfate from fluidized bed combustion effluent gases are shown in the 3-D representation of Figure 7. Gas phase compositions which result in the formation of liquid sulfate can be avoided by lowering the metal temperature and system pressure, lowering the sodium concentration, or increasing the excess chlorine (as HCl) concentration.

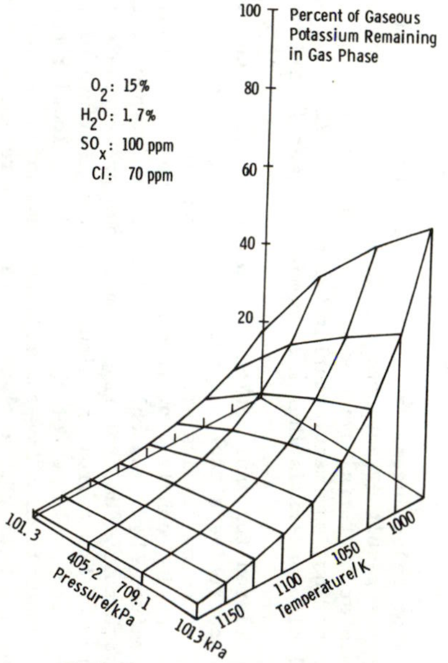

Curve 686622-A

Fig. 6—Thermodynamics of gettering action of aluminosilicates in reducing gaseous potassium emissions

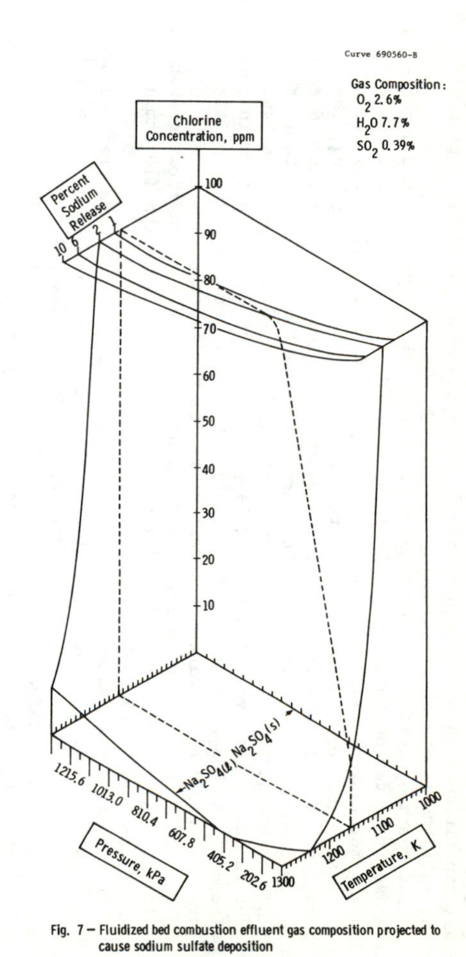

Fig. 7 — Fluidized bed combustion effluent gas composition projected to cause sodium sulfate deposition

A more exact formulation of the deposition conditions which considers the stability diagram for liquid eutectics of sodium (Na_2SO_4) and potassium sulfate (K_2SO_4) is under development. A complete analysis must ultimately include the exact deposition mechanism and kinetics from the gas phase, but the eutectic point estimate is probably adequate.

The thermodynamic theory that excess chlorine will prevent liquid deposition has to be demonstrated. The theory suffers from two limitations: excess chlorine may cause formation of the $NaCl/Na_2SO_4$ mixed melt, and excess HCl may cause direct gaseous attack which destroys the protective oxide coating, the superalloy. That the addition of excess chlorine reduces alkali sulfate deposition has been shown in isostatic laboratory scale reactors at 405.2 kPa (4 atm) pressure. Pressurized test passage tests at 405.2 kPa (4 atm) pressure have so far failed to verify that corrosion is prevented, but some reduction in corrosive deposits has been found. On balance, it is not yet clear that sufficient information has been obtained to accept or reject the hypothesis.

The upper limit to acceptable levels of HCl in the gas phase flowing over turbine superalloys is currently under investigation at 1013.6 kPa (10 atm) pressures in small-scale laboratory reactors.

In the three-dimensional projection of Figure 7, it is obvious that release to the gaseous phase of a small fraction of the sodium or potassium in coal is sufficient to exceed the turbine tolerance to alkalis. The potential of aluminosilicates to reduce alkali emissions is thus of critical interest. In Figure 8 the deposition conditions for potassium sulfate are shown as a function of temperature and pressure. Two points are important: potassium sulfate should only deposit as a solid below the sodium sulfate/potassium sulfate eutectic temperature. Secondly, the quantity available for deposition is greatly reduced. For conditions where a certain weight of solid potassium sulfate per unit area is required to accelerate oxidation the accumulation of this deposit may require a period of tens of thousands of hours.

5. MATERIALS SELECTION

5.1 Factors/Criteria

The complexity of selecting materials to use in fluid bed combustion process components is the result of advances in the process technology for producing new materials and material systems and advances in existing material combinations, fabrication, and manufacturing techniques. The cost and time required for developing materials that meet component specifications must also be evaluated.

Materials considered for application in fluid bed combustion processes must withstand hot corrosive attack and particulate erosion for extended in-service use. Hot corrosion results primarily from the accelerated oxidation of heat-resistant nickel-, cobalt-, or iron-based alloys in combustion gases containing sulfur oxides, hydrogen chloride, alkalis, and trace impurities from coal and sorbent materials. Maintaining a continuous protective oxide scale minimizes the degree of hot corrosion which can occur. Failure of the protective scale results from: solution of the scale surface; local reduction by carbon particulate deposition; alloy sulfidation; mechanical erosion; and faults in the as-grown surface scale, thermal and operating stresses, or contaminant reaction, with the underlying metal causing mechanical disruption of an overlying scale.

Additionally, alloy failure can result from intercrystalline cracking initiated either by stress rupture or creep, fatigue, thermal shock, and thermal fatigue or by grain boundary oxidation in addition to corrosion or erosion damage. Very small quantities of trace element impurities are detrimental in superalloys. Additions of aluminum and chromium to nickel-based alloys result, in general, in the reduction of high-temperature oxidation and an enhanced resistance to hot corrosion attack effect on strength and ductility. Considerations of the stability of secondary phases, and the attendant effect of strength impose a limit on chromium concentration such that a balance must be struck between mechanical properties and oxidation/corrosion resistance. Cobalt-based alloys are excellent in their resistance to corrosive attack but are usually inferior in strength.

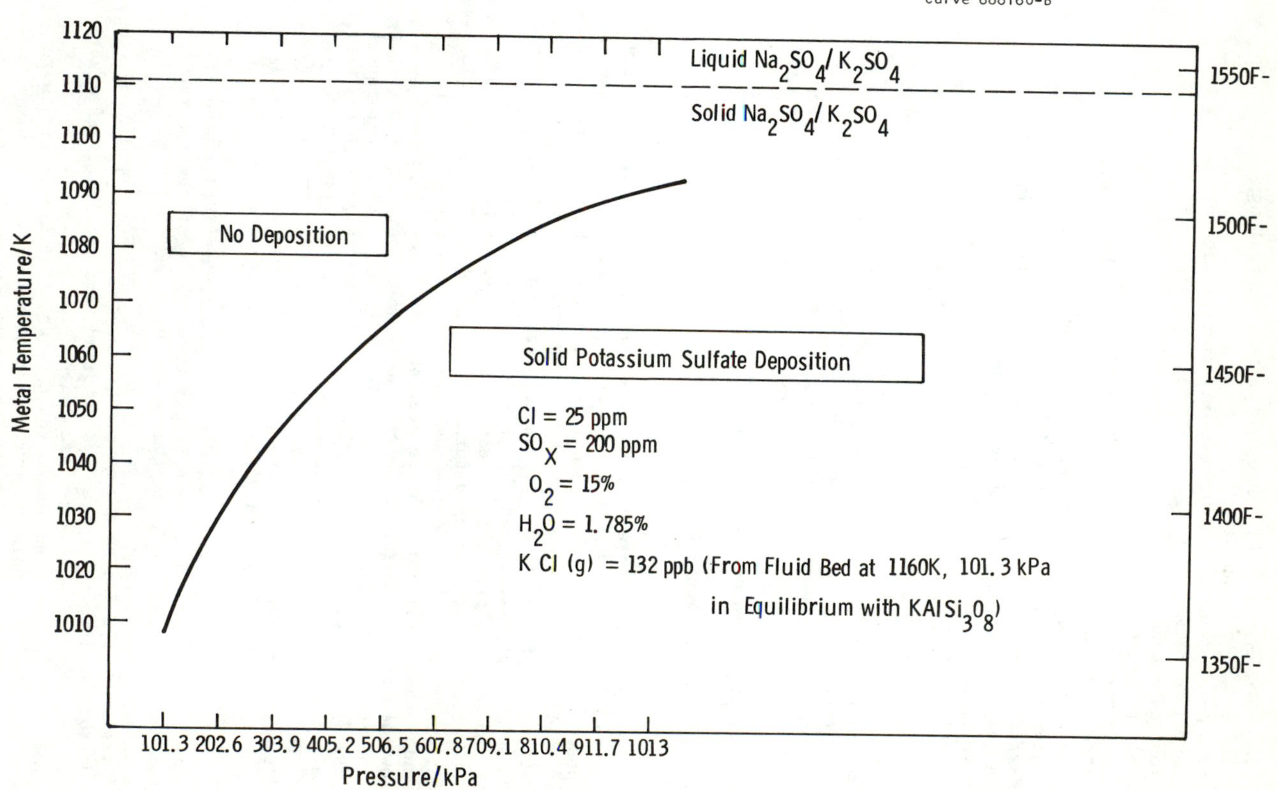

Fig. 8 — Deposition temperatures for potassium sulfate when aluminosilicate getter suppresses alkali release

To improve environmental resistance of the alloys, coatings can be applied. Protective coatings are to provide:

- Improved intergranular and surface oxidation resistance
- Hot corrosion resistance
- Erosion resistance
- Control of thermal fatigue.

The criteria for selecting coatings include:

- Ductility and spalling resistance
- Ability to bond metallurgically to the alloy substrate
- Diffusional stability with the substrate
- No degradation of the mechanical properties of the substrate
- Thinness, uniformity, and lightness of weight
- Reasonably low cost
- Easy application.

5.2 Experience

Experimental studies to characterize the resistance of superalloys to the fluidized bed combustion gases have continued in three phases:

- Isostatic alloy exposure to alkali sulfate-deposition conditions
- Isostatic alloy exposures to high levels of HCl gas in the combustion gases
- Alloy exposure in a pressurized combustion test passage.

5.2.1 Alkali Sulfate Isostatic Alloy Studies

A pressurized reactor was used to study the effect of the fluidized bed combustor effluent gas atmosphere on turbine alloy materials. The objectives were to generate a controlled atmosphere in which the corrosive constituents of the gas phase could be set to desired levels and deposition characteristics of the gases on different turbine alloy specimens could be determined.

The solution was forced from a pressurized reservoir containing a solution with a known sodium, potassium, and chloride ion concentration through a capillary at the top of a pressurized reactor into a quartz boiler. The boiler was set in the lower heated zone of a furnace so that the solution and dissolved salts rapidly evaporated. A simulated combustion gas was passed down the reactor to carry the alkali and chloride over two metal samples lashed to a support below the boiler. Thermocouples in the chamber monitored the temperature along the metal specimens. The furnace was controlled at a temperature of 1198K (1697°F) which produced a temperature of 1173K (1652°F) at the hotter end of the alloy sample. The temperature ranged from 1173K (1652°F) at the sample head, to 1158K (1625°F) at the center, and 1148K (1607°F) at the base. The reactor was protected from hot corrosion attack by use of a quartz sleeve and an air gap which was continuously flushed with nitrogen during the experiment.

The schedule of experiments carried out in the pressurized reactor is shown in Table 3. The first four experiments were trials of the apparatus; for subsequent experiments, the heating zone was extended to ensure that particulate chloride would evaporate before striking the alloy surface. Analysis of residual deposits on the quartz boiler showed that less than 1% of the alkali introduced into the system was retained, and the levels of alkali calculated to be present in the gas phase are, in fact, present.

The samples run in the pressurized reactors at 405.4 kPa (4 atm) were examined with use of the scanning electron-beam microanalyzer. Table 4 lists some of the observations made concerning the condensates on the specimens. The specimens were exposed in a manner which produced a longitudinal temperature gradient. The hotter end is defined as the end toward the boiler, the middle as the region of transition from crust-like scale formation to slight scale formation, and the cooler end as the end away from the boiler. The nominal metal temperatures are listed in Table 4.

X-ray diffraction analyses of some of the condensates indicate that the phases are Na_2SO_4 for tests with only sodium and mixtures ($KNaSO_4$ or $K_3Na(SO_4)_3$) for tests run with sodium and potassium.

TABLE 3

Experimental Conditions for the Fluid Bed Combustor Effluent Gas Simulator

Experiment	Time	Alloys	Pressure	Gas Concentration, ppmv K	Na	Cl
1	4m	U-710	405.4 kPa	3.0	3.0	9
2	80 hr cor.	U-710	"	3.0	3.0	9
3	90 hr cor.	U-500 X-45	"	3.0	3.0	3.0
4	30 hr ox. 66 hr cor.	" "	"	3.0	3.7	7.0
5	30 hr ox. 68 hr cor.	" "	"	3.0	3.7	30
6	26 hr ox. 69 hr cor.	" "	"	2.0	-	3.0
7	24 hr ox. 80 hr cor.	" "	"	1.8	-	30
8	24 hr ox. 70 hr cor.	" "	"	-	2.1	3.0
9	25 hr ox. 68 hr cor.	" "	"	-	1.8	30

TABLE 4

Condensate Characterization

Experiment	Alloy	Position of Specimen: Hotter End (1173K, 1652°F)	Middle (1158K, 1625°F)	Cooler End (1148K, 1607°F)
Low HCl Na+K	Ni-base	N.T.[a]	Molten K-Na sulfate	Crystalline K-sulfate
	Co-base	N.T.	Molten K-Na sulfate	Crystalline K-sulfate
High HCl Na+K	Ni-base	Molten K-Cl-chloride	Molten K-Na-Cl-sulfate	Crystalline K-sulfate
	Co-base	Molten K-Cl-sulfate plus crystalline K_2SO_4	Molten K-Cl-sulfate plus crystalline K-Na sulfate	Crystalline K-sulfate
Low HCl K	Ni-base	Crystalline K-sulfate	Crystalline K-sulfate	Crystalline K-sulfate
	Co-base	Crystalline K-sulfate	Crystalline K-sulfate	Crystalline K-sulfate
Low HCl Na	Ni-base	Crystalline Na-sulfate	Molten Na-Cl-sulfate	Molten Na-Cl-sulfate
	Co-base	Molten Na-sulfate	Molten Na-sulfate	N.A.[b]

[a]N.T. = not tested.
[b]N.A. = not analyzed.

Fig. 9 Liquid (at temperature) sodium sulfate (arrows) near hotter end of X-45 specimen

The test with potassium resulted in slight attack of the bare alloys, and only crystalline K_2SO_4 is present. This is expected because the test temperature is below the melting temperature of K_2SO_4 (1349K, 1969°F). The test with Na resulted in hot corrosion attack, also as expected, because the test temperature gradient crosses the melting point of Na_2SO_4 (1157K, 1623°F). In addition note that chlorine is still present in the quenched molten Na_2SO_4. At the test temperature the molten material is probably a mixture of $NaCl-Na_2SO_4$. Figure 9 is an illustration of the existence of such a mixture in the quenched state of the exposed alloy.

In some instances separate chloride compounds are present. Figure 10 shows a now-crystalline mass of K-Na chloride. This was probably molten at the test temperature. This area corresponds to the hotter end of the high HCl experiment with Na and K listed in Table 4.

Fig. 10 Potassium-sodium chloride condensed phase (arrow) on surface of Udimet-500 test specimen near hotter end

Figure 11 illustrates the case where the condensate is pure potassium sulfate. This phase always exhibits excellent crystalline morphology indicative of being grown from the gas phase at temperature. This phase is also detected in association with molten K-Na sulfate. Interestingly, K_2SO_4 is present on the cooler ends of specimens for tests which were run with both low and high HCl with sodium plus potassium, respectively. Below the liquidus temperature of K-Na sulfate only K_2SO_4 nucleates and no sodium-rich liquid condense.

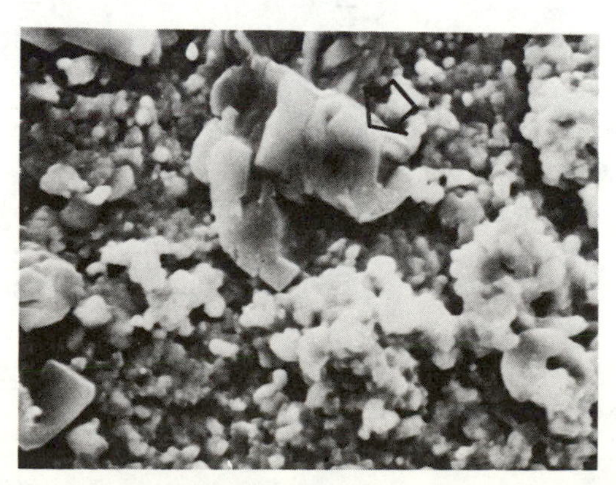

Fig.11 Crystalline potassium sulfate (arrow) near hotter end of X-45 test specimen

In the presence of HCl, mixtures of Na and K tend to form a potassium-rich sulfate. With higher pHCl even less Na is present in the sulfate condensate, although chlorine now can be found. Only in the middle portion of the specimen can sodium be found with the potassium sulfate condensate.

A comparison of tests made with just K or Na at high and low pHCl levels indicates that chlorine remains with condensed Na_2SO_4 but not with K_2SO_4. Interestingly, there is an indication that Na-sulfate was crystalline on the hotter end of the specimen but molten in the cooler regions. At the lower temperatures NaCl probably was also a condensed phase, and, therefore NaCl-Na_2SO_4 eutectic, which melts at temperatures below 1123K (1562°F), was formed.

In summary, the tests with equal molar potassium plus sodium result in condensates which are predominantly potassium-rich. Increasing pHCl further increases the K/Na ratio of the condensed phase. At the higher pHCl separate alkali-chloride phases form, as does alkali sulfate. As expected, hot corrosion attack is present only where molten condensates exist.

5.2.2 HCl Isostatic Alloy Studies

Since high HCl levels are projected to reduce or eliminate deposition of molten sulfates, experiments are in progress at a total pressure of 1013.6 kPa (10 atm) to determine the maximum HCl concentrations at which the superalloy oxide scales resist direct attack induced by the gaseous HCl itself.

The materials used are listed in Table 5 and test conditions in Table 6. The procedure adopted was to establish an oxide scale and then to introduce the simulated HCl and SO_x levels.

TABLE 5

Composition of Generic Turbine Alloy for HCl-Combustion Gas Studies

	Alloy					
Element	X-45	ECY-768	U-520	U-500	IN-738	B-1900
Al	--	--	2	2	3.4	6
B	0.01	--	0.005	0.005	0.01	0.015
C	0.25	0.56	0.05	0.08	0.17	0.1
Cb	--	--	--	--	--	0.1
Co	Balance	Balance	12	12	8.5	10
Cr	25.5	22.9	19	19	16	8
Fe	2	--	--	4	--	0.35
Li	--	--	--	--	1.75	--
Mn	1	--	--	--	--	0.2
Mo	--	--	6	6	--	6
Nb	--	--	--	--	0.9	--
Ni	10.5	10.5	Balance	Balance	Balance	Balance
Si	--	--	--	--	--	0.25
Ta	--	3.7	--	--	1.75	--
Ti	--	0.28	3	3	3.4	1
W	7	6.28	1	1	2.6	--
Zr	--	--	--	--	0.1	0.08

TABLE 6

Test Conditions for Isostatic Alloy Exposure to HCl Combustion Gases

Parameters	Conditions: Oxidation Exposures	Conditions: HCl Combustion Gas Exposure
Gas Composition	Oxygen	16.5% O_2, 4.4% CO_2, 2.7% H_2O 120 ppm SO_2/N_2 + 50, 100, 500 ppm HCl
Gas Flow	50 cc/min	2 l/min
Temperature	1173K (1652°F)	1173K (1652°F)
Pressure	101.3 kPa (1 atm)	1013.6 kPa (10 atm)
Time	20, 90, 200 hrs	48 hrs

The morphology of surface oxide scales is shown in Figures 12, 13, and 14. The physical and chemical characteristics of the oxide scales before and after exposure to HCl have been examined using scanning electron microscopy (SEM) and x-ray diffraction analysis.

SEM analysis indicates that accelerated oxidation of the X-45 material occurs when the alloy is subjected to combustion gases containing 500 ppm HCl. In reviewing the timed oxidation studies, however, a complete scale was not developed on the alloy surface prior to exposure of the combustion gas. In general, alloys which have formed uniform oxide scales prior to their exposure to combustion gases containing HCl show little or no evidence of accelerated oxidation when subjected to combustion conditions (Table 7).

The phase analysis by x-ray diffraction techniques indicates that chemical changes in the oxide or spinel form occur as the HCl level of the alloys X-45 and U-500 is increased (Table 8).

TABLE 7

Cross-Sectional Analysis by SEM of Oxide Thickness Formed on Alloys under Oxidizing and Combustion Gas Conditions

Alloy	Time: 24 Hr, μm	Time: 90 Hr, μm	Time: 200 Hr, μm
X-45	5	5-7	16
ECY-768			
U-520			
U-500	2-5	10	11
In-738			
B-1900	<1	1.5-2.5	4-5

Alloy	HCl: 0 ppm, μm	HCl: 50 ppm, μm	HCl: 100 ppm, μm	HCl: 500 ppm, μm
X-45	3	3	4	45-76
ECY-768	1	3	6	2-3
U-520		7	4-5	7-8
U-500	2-4	2-6	2-5	4-6
In-738	1.5-13.5	2-7	1.5-8	2-6
B-1900	2-3	4-5	1.5-2	2-3.5

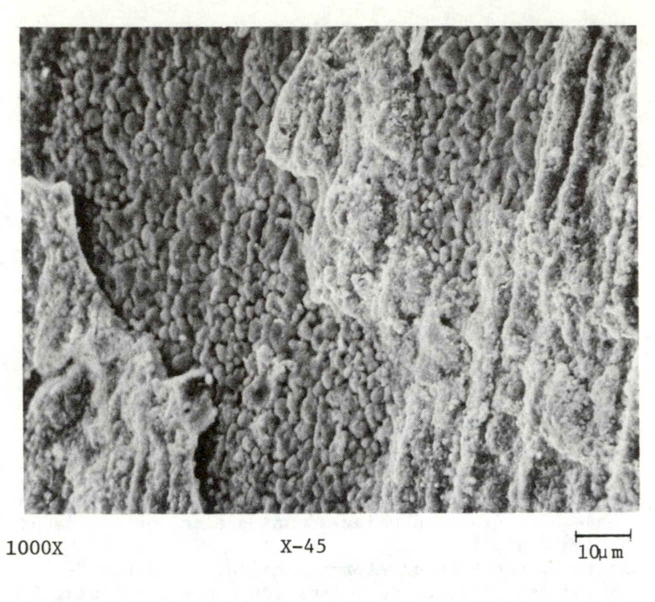

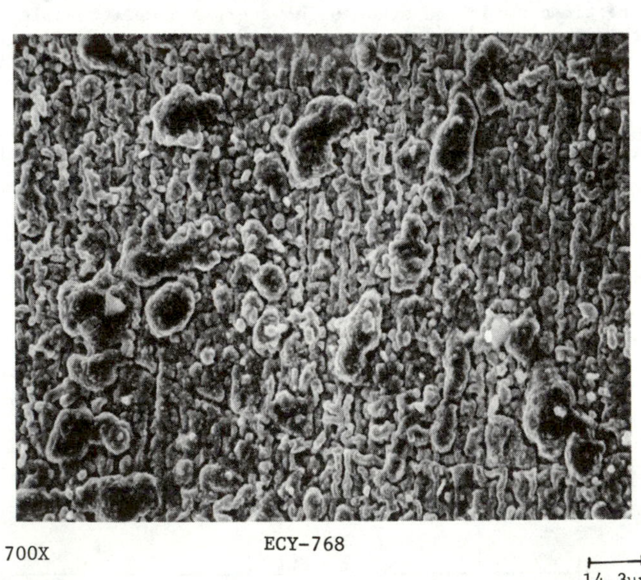

Fig. 12 Scanning electron micrographs of X-45 and ECY-768 surface oxides formed after 90 hours' oxidation at 101.3 kPa (1 atm), followed by 48 hours of combustion gas containing 100 ppm HCl at 1013.6 kPa(10 atm) total pressure and 1173K (900°C)

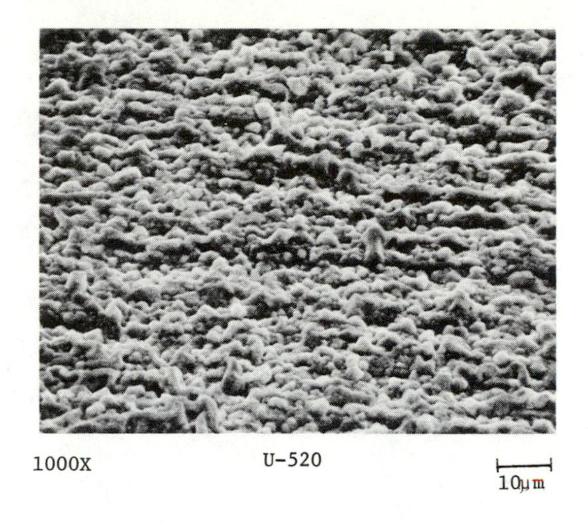

1000X U-520 10μm

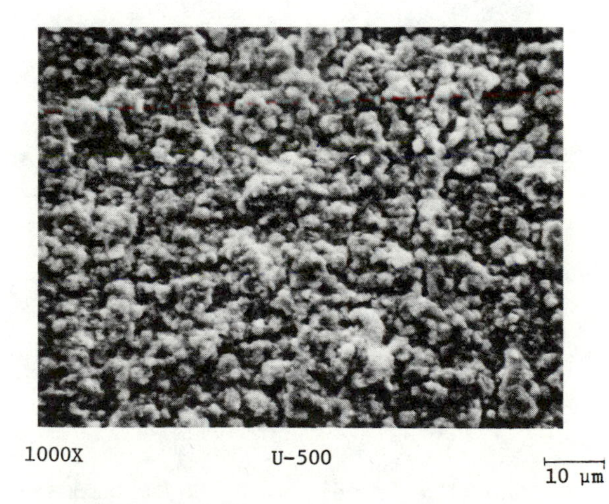

1000X U-500 10 μm

Fig. 13 Scanning electron micrographs of U-520 and U-500 oxidized and exposed to 100 ppm HCl in combustion gas atmosphere

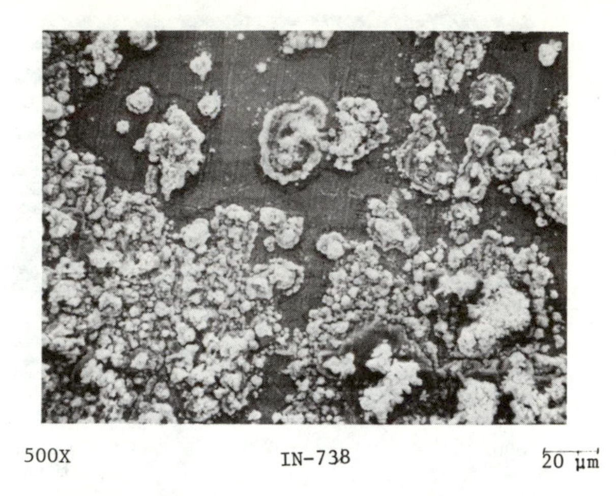

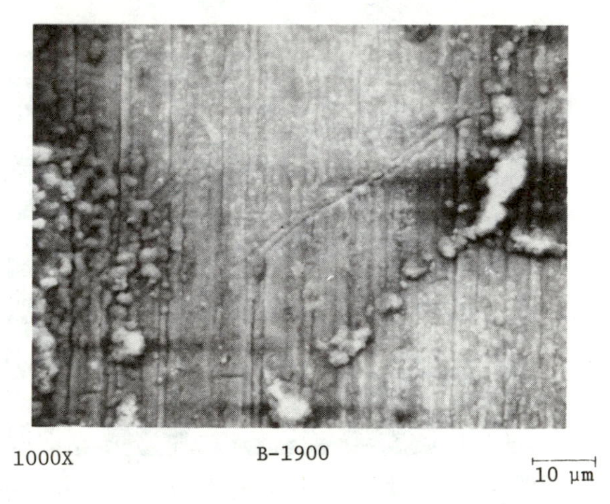

Fig.14 Scanning electron micrographs of IN-738 and B-1900 oxidized and exposed to 100 ppm HCl in combustion gas atmosphere

TABLE 8

Chemical Composition of the Oxide Scale Determinations by X-ray Diffraction Analysis

	Time					
	24 Hr		90 Hr		200 Hr	
Alloy	Minor	Major	Minor	Major	Minor	Major
X-45	$CoCr_2O_4$	Cr_2O_3	--	Cr_2O_3	$NiCr_2O_4$	$CaCr_2O_4$
ECY-768	$NiCr_2O_4$ and spinel	Cr_2O_3	$NiCr_2O_4$ type spinel	Cr_2O_3	Co_3O_4 type spinel and Cr_2O_3	$CrTaO_4$
U-520	TiO_2	Cr_2O_3	TiO_2	Cr_2O_3	TiO_2	Cr_2O_3
U-500	NA[a]		NA		NA	
In-738	Cr_2O_3	TiO_2	Cr_2O_3	TiO_2	Al_2O_3 Cr_2O_3	TiO_2
B-1900	NA		NA		NA	

	HCl							
	0 ppm		50 ppm		100 ppm		500 ppm	
Alloy	Minor	Major	Minor	Major	Minor	Major	Minor	Major
X-45	$CoCr_2O4$	Cr_2O_3	$CoCr_2O_4$	Cr_2O_3		Cr_2O_3	Cr_2O_3 $CoCr_2O_4$	CoO
ECY-768	CoO	Cr_2O_3	CoO	Cr_2O_3	CoO	Cr_2O_3	CoO	Cr_2O_3
U-520	Cr_2O_3	TiO_2	NA		NA		NA	
U-500	Cr_2O_3	TiO_2	TiO_2	Cr_2O_3	TiO_2 =	Cr_2O_3	Cr_2O_3	TiO_2
In-738	NA		NA		NA		NA	
B-1900	NA		NA		NA		NA	

[a]NA - not analyzed.

5.2.3 Materials Exposures in the Pressurized Test Passage

Two experiments were conducted in the pressurized passage to test whether hydrogen chloride from the coal gas would suppress the formation of condensed sulfates on heated superalloys. The results can be compared to results of base-line tests at levels of sodium which are known to cause, respectively, catastrophic and accelerated oxidation attack.

The long-duration corrosion test No. 1 was run in the pressurized passage at an average gas temperature of 1172K (1650°F) and 304 kPa (3 atm (gage)) pressure. A total of eight 0.64 cm diameter by 5.68 cm long test specimens were exposed to a combustion product gas containing 0.5 ppm Na, 25 ppm Cl, and 100 ppm SO_2-SO_3. The levels of sodium, chlorine, and sulfur in the gas were achieved by adding oil soluble compounds to a clean fuel oil. The test specimen pack contained three uncoated nickel-based alloys (IN 713C, IN 738, and Udimet 520) and a cobalt-based alloy (ECY-768).

The specimens were examined visually at 25-hour intervals, and the front surface of the top third of the specimens was coated with a crystalline deposit which an x-ray diffraction analysis indicated to be sodium sulfate. The lower two-thirds of the specimens were covered with a once-molten film of sodium sulfate. Due to a nonuniform temperature pattern in the combustor outlet gas, the top third of the specimens was exposed to approximately 1089K (1500°F) and the lower two-thirds to 1172K (1650°F) gas.

After 50 hours of exposure, corrosion had started on the top third of the surface of IN 713C (mostly under the solid Na_2SO_4 deposit.) The IN 738 and ECY-768 showed a wide area of corrosion on the surface of the lower two-thirds; the Udimet 520 showed no corrosion. Examination after 110 hours of testing indicated that the corrosion on IN 713C and IN 738 had spread to a wider area, and the corrosion can be classified as catastrophic. ECY-768 showed only moderate corrosion, and Udimet 520 showed no corrosion at all.

The test with 25 ppmv Cl and 0.5 ppmv Na was terminated after 153 hours. Uncoated specimens of IN 713C and IN 738 were heavily corroded on all surfaces. The nature of the corrosion is that known as a "catastrophic sulfidation." Uncoated alloys Udimet 520 and ECY-768 were moderately corroded.

The first test in the pressurized test passage lasted 153 hours. The specimens had deposits of Na_2SO_4 on them. The temperature distribution on the specimens was such that solid Na_2SO_4 was on the upper leading edges of the specimens. Liquid condensate was seen to be present on the edges of the cylindrical specimens after 52 hours.

Table 9 lists the total descaled metal losses for the superalloys. The specimens were also sectioned metallographically. The measurements are listed in Table 10. The specimens had been electrolytically descaled in molten salt to remove the surface corrosion products.

TABLE 9

Total Descaled Weight Losses

Alloy	Total Weight Loss, g	Duration, hrs
IN-713C	1.222	153
IN-738	0.647	153
U-520	0.181	153
U-520	0.232	153
ECY-768	0.367	153
ECY-768	0.261	109

TABLE 10

Pressurized Passage Test No. 1

Alloy	Radius Recession, in.	Duration, hrs
IN-713C	0.0164	153
Udimet-520	0.0021	153
IN-738	0.0149	153
ECY-768	0.0058	153

The IN 713C alloy exhibits the type of hot corrosion attack that has been found in aircraft and industrial gas turbines where Na_2SO_4 is present as a condensed phase. The IN 738 and Udimet-520 suffered the degree of attack more properly termed accelerated oxidation, but the corrosion microstructure indicates that longer exposure would lead to a metal loss comparable to that in hot corrosion attack.

On the basis of descaled weight loss measurements, it was concluded that alloys IN-713C and IN-738 experienced hot corrosion attack whereas ECY-768 and Udimet-500 had accelerated oxidation attack. The values of metal loss based on the total loss of useful base metal indicate that there is still a distinct difference between the two types of alloys. It is concluded that where high alkali-sulfate contaminant levels at a substrate temperature of 1172K (1650°F) occur Udimet-500 and ECY-768 would be more resistant to breakaway attack for short exposures of high contaminant levels; for long-term exposure to such severe contaminant conditions, however, coatings are necessary to resist the corrosion attack.

The next test run in the pressurized passage was selected to explore the limit of sodium sulfate condensation in the presence of HCl.

The passage operating conditions were:

N_2	78 v/o
O_2	16 v/o
CO_2	3 v/o
H_2O	3 v/o
SO_x	100 ppmv
HCl	25 ppmv
NaCl	0.1 ppmv
KCl	0
Particulates	nil

The second run was at conditions selected to ameliorate the massive corrosion which occurred during the first test. The turbine environment was selected to represent more effective gettering of alkali metal from the fluid bed combustor outlet gas, resulting in a sodium concentration in the turbine expansion gas one-fifth that of the first test.

There are no noticeable condensates and no significant spalling or blistering of the leading edge; but, interestingly, the trailing edge sides show significant flaking of the oxide scale on U-520 and IN-738. This downstream oxidation is affected by the presence of the HCl gas which may cause cracking of the normally protective scales. The absence of spalling on the leading edge is due to the scale on the front, which has a microstructure that forms in response to the thermally induced stresses.[12] The downstream side, which is beyond the flow separation of the gas stream, oxidizes more isothermally. Such scales are more liable to crack because they have cooled from the temperature of exposure and possibly as a result of HCl attack at test temperature.

The tested specimens were descaled electrolytically in molten salt in order to show the corrosion attack pattern on the leading edges, indicative of slightly accelerated oxidation, and not the type induced by alkali sulfate condensates. The U-520 shows significant penetration along the grain boundaries of the wrought microstructure because of high-temperature oxidation. The descaled weight losses are shown in Table 11 and indicate that the IN-713C has the lowest weight loss and ECY-768 the highest. The ranking in Table 11 is consistent with the expected oxidation of the respective superalloys.[13]

TABLE 11

Pressurized Passage Test No. 2

Alloy	Total Descaled Wt. Loss, g	Duration, hr
IN-713C	0.0234	48.5
Udimet-520	0.0792	48.5
IN-738	0.0733	48.5
ECY-768	0.0949	48.5

The metal loss measurements for each of the exposed specimens are listed in Table 12. The degree of attack is consistent with the descaled weight-loss values listed in Table 11. The corrosion based on weight loss and metal loss indicates the presence of the oxidation mode of attack.

Microscopic examination of metallographically prepared alloys from Test No. 2 shows incipient hot corrosion attack on the IN-713C, traces of subscale sulfidation in IN-738 and Udimet-500, but only internal oxidation for the ECY-768. This observation of subscale sulfidation indicates that there was some condensation of Na_2SO_4.

TABLE 12

Pressurized Passage Test No. 2

Alloy	Radius Recession, in.	Duration, hr
IN-713C	0.0002	48.5
Udimet-520	0.0004	48.5
IN-738	0.0003	48.5
ECY-768	0.0005	48.5

The amount of sodium in the combustion gas under conditions of lower chloride would normally be sufficient to form considerable liquid condensate and thus cause hot corrosion attack in 50 hours.[13] The reduction in attack in Test No. 2 is due to the HCl in the gas phase.

5.3 Assessment

Any materials considered for application to fluid bed combustion processes must withstand hot corrosive attack or accelerated oxidation and particulate erosion for extended in-service use. Corrosion of the combustor heat transfer surface may be due to a particular selection of operating and design parameters which produce stagnant regions in the bed or conditions which result in sulfide corrosion. Proper design and operation can be utilized to avoid these materials problems. Corrosion of materials as the result of trace contaminants in the coal and sorbent feed may not be simply controlled by design modifications.

The impact of release fractions of even 2% of the coal alkali is expected to cause liquid sodium sulfate deposition in the turbine expander unless the accompanying HCl level is high (>100 ppm). Aluminosilicate additions, however, can prevent deposition at temperatures above the K_2SO_4/Na_2SO_4 eutectic temperature. Deposition tests on superalloys for the turbine expander at 405.4 kPa (4 atm) pressure are incomplete but show reduced deposition with an increased HCl(g) level. The upper limits of the HCl(g) concentration that the alloy protective oxide scale can tolerate is under study. For certain alloys at 1013.6 kPa (10 atm) pressure, changes in the scale chemistry and extent of oxidation vary with the concentration of HCl(g). With concentrations of 500 ppm HCl in the gas phase, accelerated oxidation was observed on the partially oxidized alloy surface of X-45. Pressurized passage tests indicate that in conditions with high alkali sulfate contaminant levels and substrate temperatures of 1172K (1650°F), U-500 and ECY-768 are more resistant to scale spalling for short exposure than IN-713C or IN-738. Reduction of hot corrosion attack due to a relatively high concentration of HCl in the gas phase was also observed in pressurized passage tests.

Strategies for controlling corrosion and erosion factors of present materials for the turbine expander include choosing appropriate operating conditions and feedstock materials, pretreating the sorbent, adding getters, and using protective coatings on alloys.

Selection of feed materials with low alkali metal content may be possible in isolated cases; both coals and limestone or dolomites, however, normally contain alkali in excess of that required to initiate corrosion of most alloys. If chlorine acts as the major alkali carrier, then release of alkali to the gaseous phase can be extensive, ∿10 to 50% depending on temperature, pressure, gas composition, and original alkali species in the feedstock. Pretreatment of the sorbent may be used to drive off the volatile alkali. The gettering action of compounds such as aluminosilicates may be used in the combustor to lower the alkali release by two orders of magnitude. Experimental measurements under close simulation of the chemical environment are required to validate these projections. Protective coatings resistant to corrosive and erosive attack in the turbine expander are an alternative providing extended life to present component materials. However, continued research is required to develop suitable criteria for projecting the useful life of such materials in full-scale operations.

1. Archer, D. H., D. L. Keairns, J. R. Hamm, R. A. Newby, W.-C. Yang, L. M. Handman, L. Elikan, "Evaluation of the Fluidized Bed Combustion Process," Vols. I-III, Final Report, Office of Air Pollution, Westinghouse Research Laboratories, Pittsburgh, PA, November 1971, Contract 70-9, NTIS PB 211-494, PB 212-916, PB 213-152.

2. Skinner, D. G., The Fluidized Combustion of Coal, M&B Monograph CE/3, Mills & Boon Ltd., London, 1971.

3. Nack, H., K. D. Kiang, K. T. Liu, K. S. Murthy, G. R. Smithson, and J. H. Oxley, "Fluidized Bed Combustion Review," published in Fluidization Technology, Volume II, Dale L. Keairns, ed., Hemisphere Publishing Corp., Washington, DC, 1976.

4. Proceedings of the Second International Conference on Fluidized Bed Combustion, October 1970, EPA Publication No. AP-109, NTIS PB 214-750.

5. Proceedings of the Third International Conference on Fluidized Bed Combustion, October 1972, EPA Publication No. EPA 650/2-73-053, NTIS PB 231-977.

6. Proceedings of the Fourth International Conference on Fluidized Bed Combustion, December 1975, published by MITRE Corp., May 1976.

7. Proceedings Fluidized Combustion Conference, London, England, September 1975, Institute of Fuel Symposium Series No. 1, Volumes I and II.

8. Proceedings of the Fluidized Bed Combustion Technology Exchange Workshop, April 1977, Volumes I and II, NTIS CONF-770447-P-1 and 2.

9. Ruch, R. R., H. J. Gluskioter, and N. F. Shimp, "Occurrence and Distribution of Potentially Volatile Trace Elements in Coal," Illinois State Geological Survey report to EPA, EPA 650/2-74-054, July 1974, PB 238 091.

10. Chamberlin, R. M., D. L. Keairns, B. W. Lancaster, P. W. Pillsbury, L. A. Salvador, and E. F. Sverdrup, "Advanced Coal Gasification System for Electric Power Generation," Quarterly Progress Report, Second Quarter Fiscal Year 1976, Westinghouse Electric Corporation report to ERDA, FE-1514-45.

11. Cooke, M. J., A. J. B. Cutler, and E. Raask, "Oxygen Measurements in Flue Gases with a Solid Electrolyte Probe," J. Inst. of Fuel, March 1972, p. 153.

12. Dils, R. R., "Fatigue of Protective Metal Oxides in Combustion Chamber Exhaust Gases," Fatigue at Elevated Temperatures, ASTM STP 520, American Society for Testing and Materials, 1973, pp. 102-111.

13. Combustion Power Co., "An Investigation of Hot Corrosion and Erosion in a Fluid Bed Combustor - Gas Turbine Cycle Using Coal as Fuel," Final Report, May 5, 1977.

INVESTIGATION OF THE CORROSION PERFORMANCE OF BOILER, AIR HEATER, AND GAS TURBINE ALLOYS IN FLUIDIZED COMBUSTION SYSTEMS

J. C. HOLDER, R. D. LaNAUZE, E. A. ROGERS, and G. G. THURLOW

Coal Research Establishment
National Coal Board
Cheltenham, Gloucestershire, United Kingdom

ABSTRACT

The paper reports the behaviour of a range of airheater and turbine materials when subjected to long exposures to the bed environment and combustion gases typical of those proposed for advanced cycle power generation plants using fluidised bed coal combustion. A comparative assessment of the corrosion resistance of materials exposed in the fluidised bed and in the freeboard for two 1000 hour tests is given.

Materials exposed to the bed material, which was primarily calcium sulphate, underwent greater attack than materials placed in the freeboard. The austenitic steels generally showed least attack, while the nickel based alloys did not perform well.

The programme highlights the need for more detailed information on the micro-environment within the fluidised bed combustor. It would then be possible to determine the effect that operating variables might have on those conditions within the fluidised bed that lead to unacceptable metal loss.

INTRODUCTION

Problems of deposition and corrosion in conventional coal fired equipment are well documented and, although the mechanisms are not fully understood, the performance of alloys in different environments can be predicted. The choice of materials for use in new coal utilisation techniques, such as fluidised combustion, is less well defined. The purpose of this paper is to report tests on the behaviour of boiler, airheater and gas turbine alloys in a coal burning fluidised bed environment. The programme was sponsored by the Electric Power Research Institute (EPRI), the project manager being H.H. Gilman and more recently R.I. Jaffee. EPRI have been represented in the UK by their consultant Professor J. Stringer. The work was managed by Combustion Systems Ltd., UK. General Electric Co. (GE) and Foster Wheeler Development Corp. (FWDC) selected the materials for the tests which were carried out at the National Coal Board, Coal Research Establishment (CRE), UK. Examination was made independently on duplicate test specimens by GE, CRE and FWDC. This paper reports the results from specimens examined by CRE. Particular emphasis is placed on the results for materials in and above the fluidised bed. These apply to boiler and air heater applications. The gas turbine alloys are reported only briefly so as to complete the picture of the test undertaken. The results of the examination of the gas turbine alloys will be reported elsewhere.

SCOPE OF THE WORK

The objective of the work reported was to assess the behaviour of a range of turbine alloys and boiler/airheater tube and hanger materials when exposed to the bed environment and exhaust gases typical of those proposed for advanced power generation cycles. Bed temperature and metal temperatures were higher than used previously, (1), which together with a high sulphur American Coal, Illinois No.6, were chosen to provide aggressive conditions.

DESCRIPTION OF THE TEST FACILITY

The CRE facility, outlined diagramatically in Figure 1, consists of a 0.3 m square atmospheric fluidised bed combustor with external primary and secondary cyclones for gas clean up. The 2.4 m x 0.3 m x 0.3 m (internal dimensions) refractory lined combustor is shown in Figure 2.

Coal and limestone were fed pneumatically to the centre of the combustor just above the air distributor. Four specimen tubes passed horizontally through the bed and two were situated in the freeboard 0.6 m above the bed surface. Each tube consisted of a number of 50 mm o.d. x 22 mm rings of different alloys clamped together. The tube temperatures were maintained by passing air at room temperature through the tubes and controlling the flow rate on the mean of four thermocouples inserted into the wall of each tube. Variable cooling could be applied to four additional tubes for plant control.

Uncooled coupon plates were located within the bed and in the freeboard. A 3 m long test section was installed at the exit of the secondary cyclone in which turbine alloys could be placed radially as pins which were 6.4 mm dia x 70 mm long.

DESCRIPTION OF THE TESTS

Two 1000 hour tests were conducted. The conditions throughout each period were:-

Pressure	Atmospheric
Bed temperature	900 C
Bed height	0.6 m
Fluidising velocity	0.9 m/s
Excess air level	10 - 20%
Coal	9.1 kg/h, - 1.68 mm, 3.1% S, Illinois No.6
Limestone	Ca/S = 3, - 1.68 mm, Penrith UK
SO_2 emission level	Test 1 = 369 ppm Test 2 = 340 ppm

The mean tube wall temperatures, the free-board gas temperature and the mean gas temperature entering the turbine materials test section are given in Table 1. Mean temperatures can be misleading when considering the maximum corrosive penetration and maximum scale data on the outside of the tube specimens as quoted in this paper. It is important that radial variations in temperature that occur on the tubes in the bed are taken into account. The estimated maximum increase in temperature between the measured tube mean temperatures and the outside skin temperatures for the in-bed tubes are Tube 1, 86 C, Tube 2, 73 C, Tube 3, 60 C and Tube 4, 50 C.

To compensate for the high heat losses from the pilot plant it was found necessary to maintain the temperature in the free-board by the addition of 0.57 m^3/h of methane at a position just above the tubes situated in the free-board, Figure 2. Commissioning tests showed that the methane burnt diffusely. Gas analysis confirmed that complete combustion of the methane had occurred and there was no increase in the volatilised alkalis emitted. Table 2 shows typical analyses of the gas entering the turbine materials test section during the two tests.

The above-bed tubes, all coupons and turbine alloys, together with selected in-bed tube specimens were tested for the entire 2000 hours. The remaining in-bed specimens were replaced after the first 1000 hours. The metals tested are listed in Table 3 and typical compositions are given in Table 4.

SUMMARY OF RESULTS

Deposits

Deposits formed on the tubes in the bed were smooth and black to red brown in colour. These deposits varied in thickness from 0.02 mm on Tube 1 up to 0.5 mm on Tube 4. In both tests, a hard dark brown deposit up to 2 mm thick formed on the underside of a tube situated near the coal feed point.

The in-bed deposits consisted of particles of less than 1 micron. All had similar chemical compositions, as shown in Table 5. The compositions shown in Table 5 assume that all sulphur is present as $Ca\ SO_4$, and this is supported by x-ray diffraction analysis. On this basis the deposits contained from 60 to 70% $Ca\ SO_4$.

The deposits on the top of the tubes situated above the bed were similar to the bed ash and formed a loosely held ridge of from 10 to 20 mm thick. A finer ash deposit of up to 3 mm formed on the underside of these tubes, Table 5.

The material deposited in the turbine section, Table 5, consisted of fine particles which were loosely held to the turbine alloys pins.

Metallographic Examination

Turbine Materials

The turbine alloys tested are given in Table 3. Maximum corrosive penetrations in the hottest zone after 2000 hours were generally below 50 µm. The type of attack was oxidation although sulphidation was observed on a few specimens.

Freeboard - uncooled Coupons

The alloys tested are given in Table 3. Figure 3 illustrates the depth of penetration on these specimens after 2000 hours which varied from 60 µm on GE 2541 to 200 µm on U-500. Attack was mainly oxidation with traces of sulphidation on some specimens.

Freeboard - Above-bed Tubes

As shown in Figure 4, with the exception of 2¼ Cr 1 Mo, little attack on any of the specimens was found. Maximum attack, generally in the range 10 - 15 µm, occurred at the top of the tubes. At 650 C, 2¼ Cr 1 Mo suffered extensive oxidation, but this may just be confirmation that it was exposed at a temperature above its maximum service temperature of 600 C.

Fluidised Bed - uncooled Coupons

Figure 5 shows the results for the materials exposed directly to the bed environment. GE 2541 appeared reasonable, suffering only oxidation, while all the other specimens showed sulphidation. IN 617 and IN 671 suffered catastrophic corrosion.

Fluidised Bed - Tubes

Most of the specimens had grain boundary sulphide penetration. There was considerable variation in the depth of penetration occurring. Figure 6 and 7 present the results of maximum penetration and scale for the two 1000 hour tests. The austenitic steels had the least attack, while nickel based alloys showed severe corrosion.

DISCUSSION

The above-bed tubes exhibited little attack and there is a wide range of materials available to the designer for use in this area. Suitable hanger materials should also be available for this area. A subject which may require further investigation is the settling of coarse ash on the top of horizontal tubes which may lead to accelerated attack.

The turbine materials performed well under the conditions of these tests. Attack was primarily by oxidation although sulphidation was detected on a small number of specimens. Further longer term testing will be required under more representative conditions before positive conclusions can be reached in this area.

There was a significant difference in the performance of alloys at the same metal temperature when placed in the fluidised bed. Careful analysis of the data should allow a choice of materials for operation at superheater temperatures in the fluidised bed operating under conditions of these tests. Further testing is required to determine the effect of the operating variables which might lead to unacceptable metal loss at the higher metal temperatures used for tubes in the bed in advanced power generation cycles. CRE are performing further tests to investigate this aspect. It is clear that calcium sulphate can play a significant role in formulating a corrosive environment in the bed although the dependance on bed temperature, excess air level, coal and limestone properties has not been established.

ACKNOWLEDGEMENTS

While this paper concentrates on CRE's contribution to this project, the work could not have been completed without the contribution of GE and through them of FWDC. The efforts of Bob Frank of GE and his colleagues are gratefully acknowledged, as are those of EPRI's consultant Prof. J. Stringer. We are grateful to the EPRI and to CSL for permission to publish this paper. Any views expressed are the Author's own and not necessarily those of the National Coal Board.

REFERENCE

1 Cooke, M.J., Rogers, E.A., Investigations of fireside corrosion in fluidised combustion systems. Institute of Fuel Symposium Series No.1; Fluidised Combustion, Vol. 1, September 1975.

TABLE 1

MEAN TEMPERATURES (^{o}C) THROUGHOUT THE FACILITY DURING TESTS 1 AND 2

IN-BED TUBES					ABOVE-BED		FREEBOARD GAS TEMPERATURE	TURBINE TEST SECTION GAS TEMPERATURE
Tube No.	1	2	3	4	5	6		
Nominal	540	650	760	840	650	760	900	820
Test 1	542	648	762	843	653	760	909	821
Test 2	541	651	760	842	651	760	913	825

TABLE 2

TYPICAL OFF-GAS ANALYSES TESTS 1 AND 2
(Sampled at entry to Turbine Material Test Section)

COMPONENT	TEST 1	TEST 2
Mean O_2 vol % dry	2.7	2.7
Mean CO_2 vol % dry	15.6	15.5
Mean CO vol % dry	0.0	0.0
Mean CH_4 vol % dry	0.0	0.0
Mean SO_2 ppm v/v	369	340
Mean dust $g/m^3{}_n$	0.48	0.64
Mean Na ppm w/w	1.2	0.7
Mean K ppm w/w	0.5	2.0
Mean Ca ppm w/w	30	38
Mean Cl ppm v/v	33	17

TABLE 3

ALLOYS TESTED AND LOCATION

KEY

A = In-bed tube   C = Freeboard tube   E = Turbine Pin
B = In-bed coupon   D = Freeboard coupon

ALLOY	LOCATION				
FERRITICS					
Cor-Ten B	A				
2¼ Cr-1Mo	A		C		
9 Cr-1Mo	A		C		
Type 405 SS	A				
E-Brite 26-1	A		C		
GE 2541		B		D	E
AUSTENITIC Cr-Ni-Fe STEELS					
Type 310 SS	A		C		
Type 321 SS	A		C		
Type 329 SS	A		C		
Type 347 H SS	A		C		
21-6-9	A		C		
Incoloy 800	A		C		
Manaurite 36 X		B		D	
HK - 40		B		D	
Nitronic 50	A		C		
NICKEL BASE ALLOYS					
Inconel 690	A				
Inconel 601	A		C		
Inconel 617	A	B	C	D	E
Inconel 671	A	B		D	E
IN 713LC				D	E
IN 738				D	E
U - 500				D	E
U - 700					E
Hastelloy X	A		C	D	E
RA 333				D	
GTD - 111					E
COBALT BASE ALLOYS					
Haynes Alloy 188	A	B	C	D	E
X - 40					E
FSX - 414				D	E

TABLE 4

TYPICAL COMPOSITIONS OF ALLOYS TESTED

	C	Cr	Ni	Fe	Co	Mo	W	Al	Ti	Mn	Si	Others
Ferritic Steels												
Cor-Ten B	0.2	0.5	-	Bal	-	-	-	-	-	1	0.2	0.5 V 0.3 Cu
2¼ Cr-1 Mo	0.1	2.2	-	Bal	-	1	-	-	-	0.4	0.3	-
9 Cr-1 Mo	0.1	9	-	Bal	-	1	-	-	-	0.5	0.5	-
Type 405 SS	0.05	12	0.3	Bal	-	-	-	0.1	-	0.6	0.6	-
E-Brite 26-1	0.001	26	0.1	Bal	0.01	1	-	-	-	0.01	0.2	0.02 Cu
GE 2541	0.02	25	-	69.5	-	-	-	4.76	-	0.002	0.02	0.8 Y
Austenitic Steels (Cr-Ni-Fe)												
Nitronic 50	0.05	22	12	Bal	-	2	-	-	-	5	0.5	0.2 V
Type 321 SS	0.06	18	11.6	Bal	-	0.3	-	-	0.4	1.5	0.9	0.2 Cu
Type 310 SS	0.04	25	17.2	Bal	-	0.4	-	-	-	1.5	0.5	-
Type 347 H SS	0.06	18	9.5	Bal	0.24	0.4	-	-	-	1.5	0.6	0.8 Nb 0.3 Cu
Type 329 SS	0.05	28	4.2	Bal	0.2	1.5	-	-	-	0.4	0.4	0.2 Cu
21-6-9	0.03	20	7.2	Bal	-	0.1	-	-	-	9	0.7	0.1 Cu 0.3 N
Incoloy 800	0.04	19	31.5	46.5	-	-	-	0.4	0.4	0.8	0.4	0.6 Cu
Manaurite 36X	0.4	25	34	Bal	-	-	-	-	-	1.2	1.3	-
HK-40	0.4	25	20	Bal	-	-	-	-	-	1.2	1.1	-
Nickel Base Alloys												
Inconel 690	0.05	27	64.6	8	-	-	-	-	-	0.2	0.3	-
Inconel 601	0.04	22	61	15.1	-	-	-	1.5	0.4	0.2	0.1	0.01 Cu
Inconel 617	0.07	22	56	0.2	12.2	8.7	-	1	-	0.03	0.06	-
Inconel 671	0.05	48	51	0.4	-	-	-	-	0.2	0.04	0.3	-
Hastelloy X	0.1	21.5	Bal	18.5	2.1	9	0.7	-	-	0.6	0.4	0.002 B
RA 333	0.04	25	45	18	3	3	-	-	-	1.8	1.2	0.01 Sn 0.001 Pb
U-500	0.08	19	Bal	0.3	19	4	-	3.0	3.0	<0.1	<0.1	<0.1 Cu 0.004B
U-700	0.07	14.6	Bal	0.1	15	4.2	-	4.4	3.5	<0.1	<0.1	0.015 B <0.04 Zr
IN-738	0.1	15.7	Bal	0.2	8	1.7	2.7	3.5	3.5	0.01	0.05	2.5^{Nb}_{Ta} 0.05 Zr
GTD-111	0.1	14	Bal	<0.1	9.7	1.5	3.8	3.0	5	<0.1	<0.1	3^{Nb}_{Ta} 0.04 Zr
IN 713 LC	0.1	13.6	Bal	0.2	0.5	4.6	-	6.0	0.9	<0.1	<0.1	2.2^{Nb}_{Ta} 0.1 Zr
Cobalt Base Alloys												
Haynes Alloy 188	0.08	22.2	22.3	1.8	Bal	-	14	-	-	0.7	0.4	0.005 B 0.04 La
X-40	0.5	25.8	10.6	0.2	Bal	-	7.5	0.1	-	<0.1	0.3	0.16 Zr 0.002 B
FSX-414	0.2	30	10.8	1.2	Bal	-	7	-	-	0.8	1.0	0.008 B

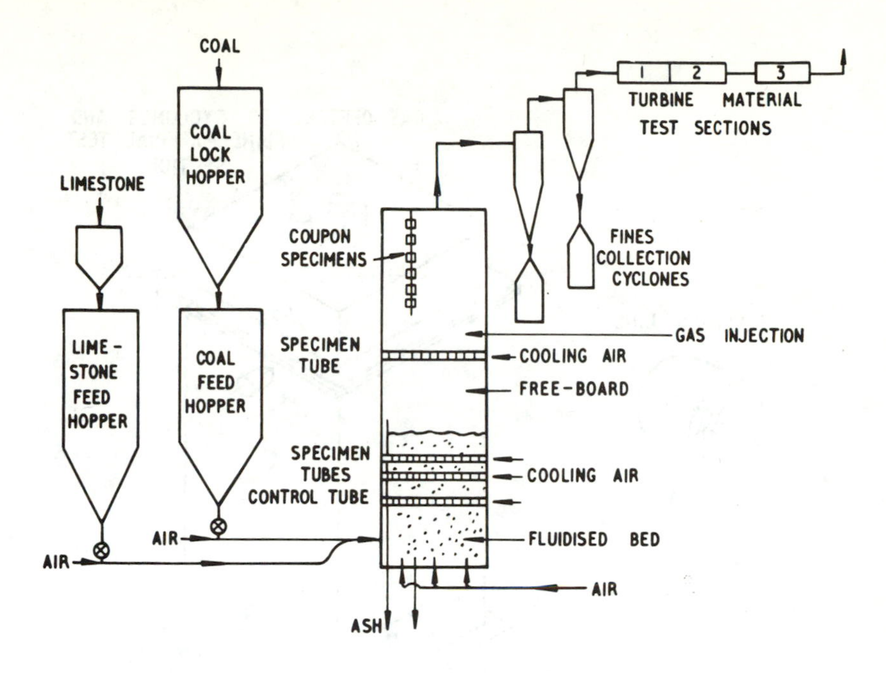

Fig.1 The 0.3m square fluidised bed combustor

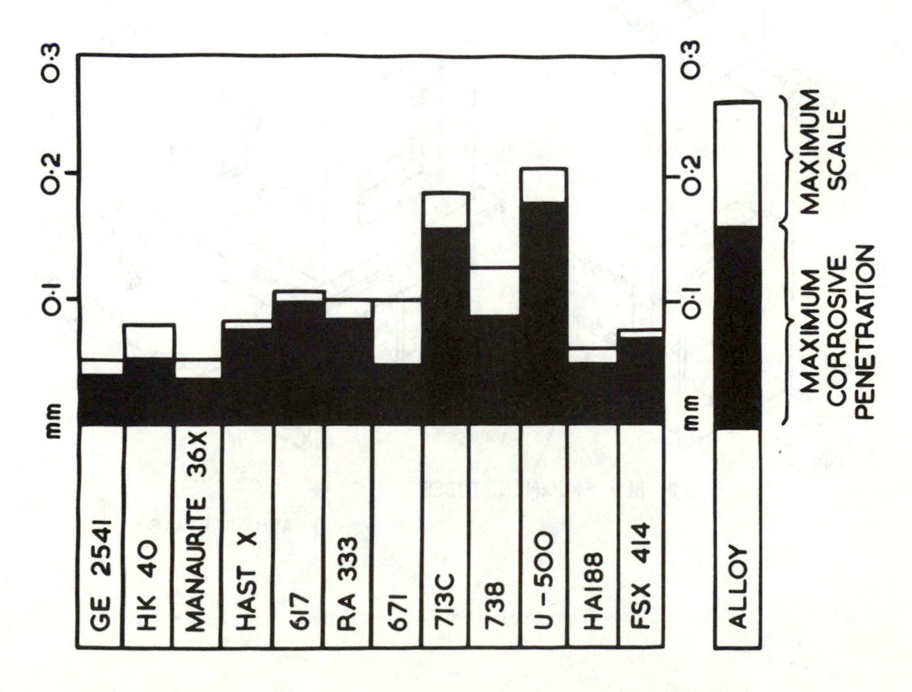

Fig.3 Maximum corrosive penetration and scale for freeboard coupons 2000 hours exposure at 913°C

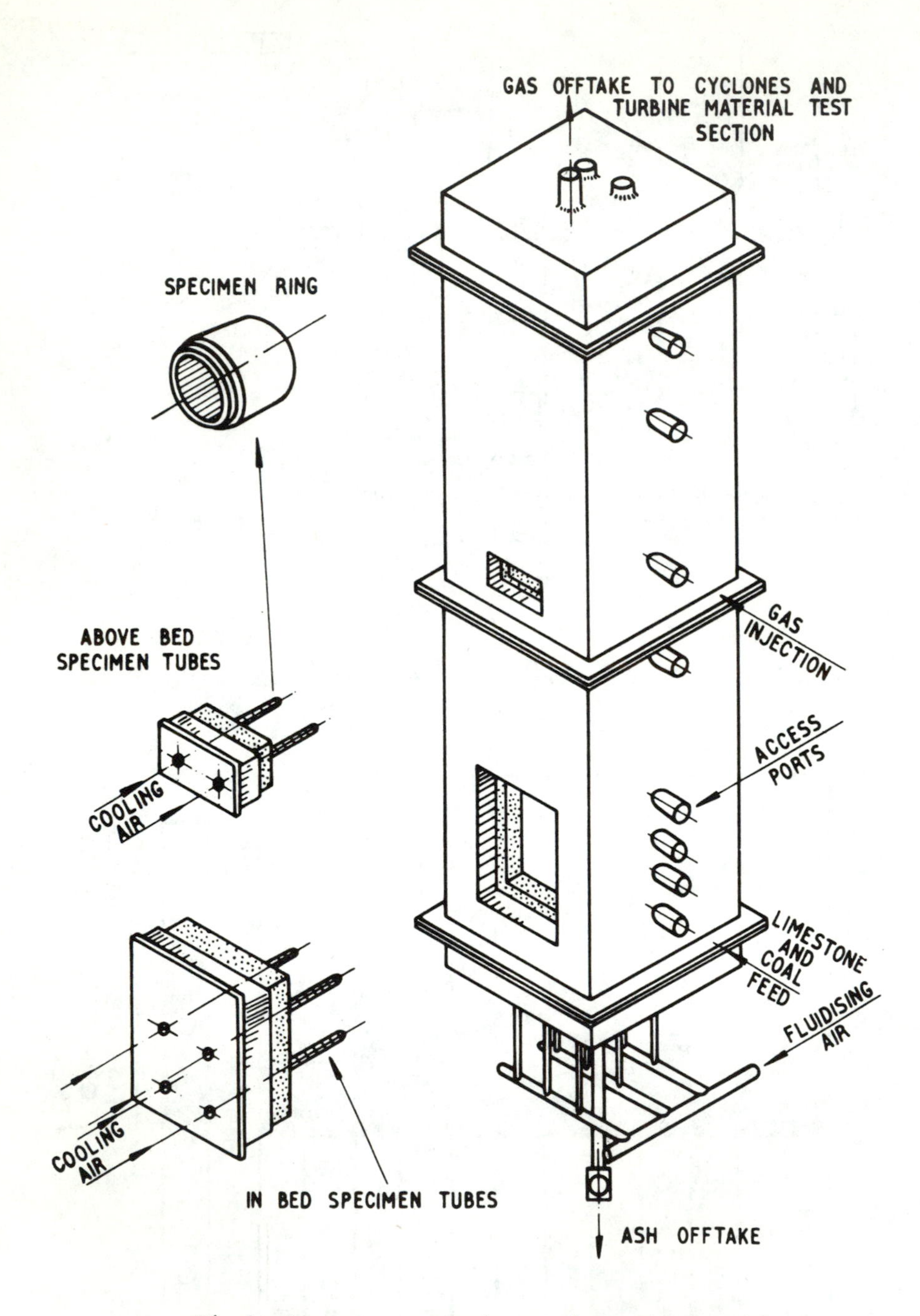

Fig.2 The combustor showing specimen tube location

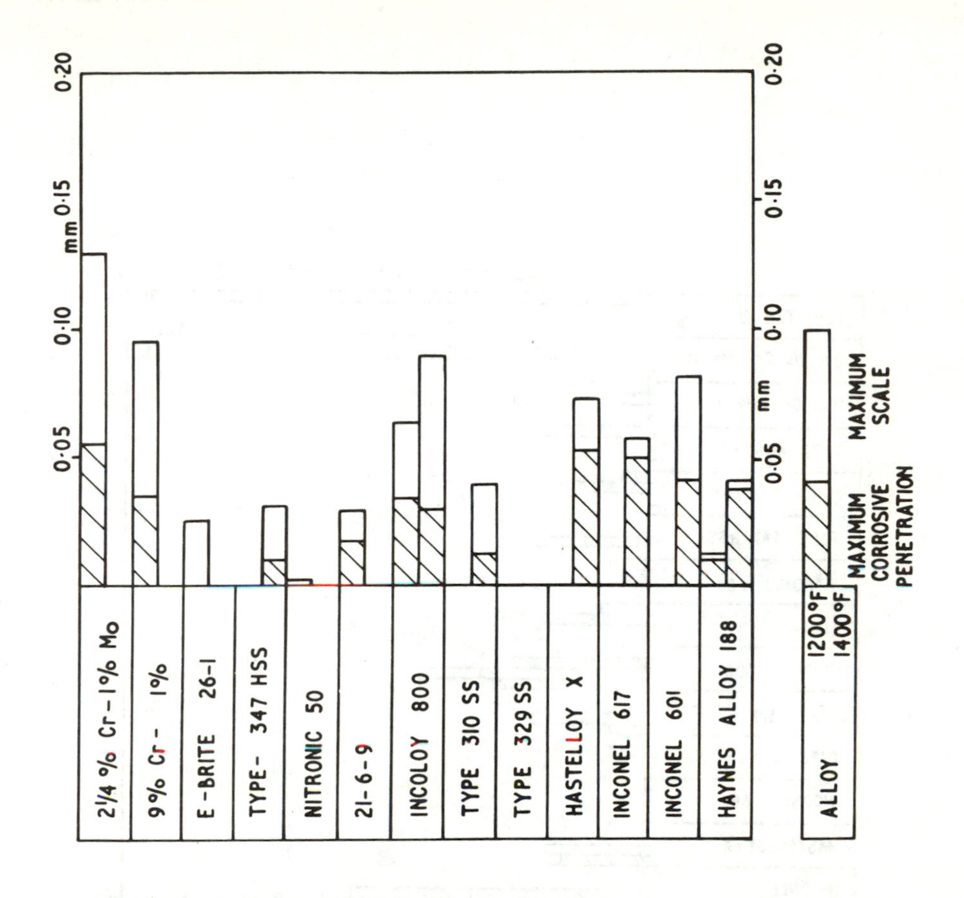

Fig.4 Maximum corrosive penetration and scale for above bed tubes 2000 hours exposure

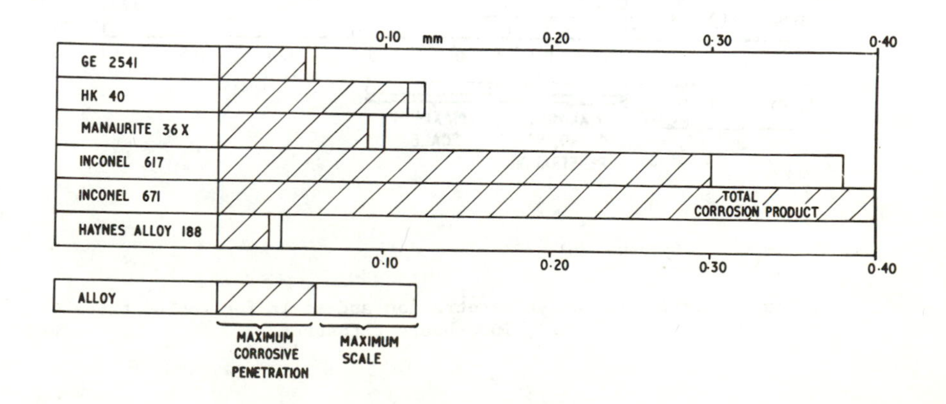

Fig.5 Maximum corrosive penetration and scale for in-bed coupons 2000 hours at 900°C

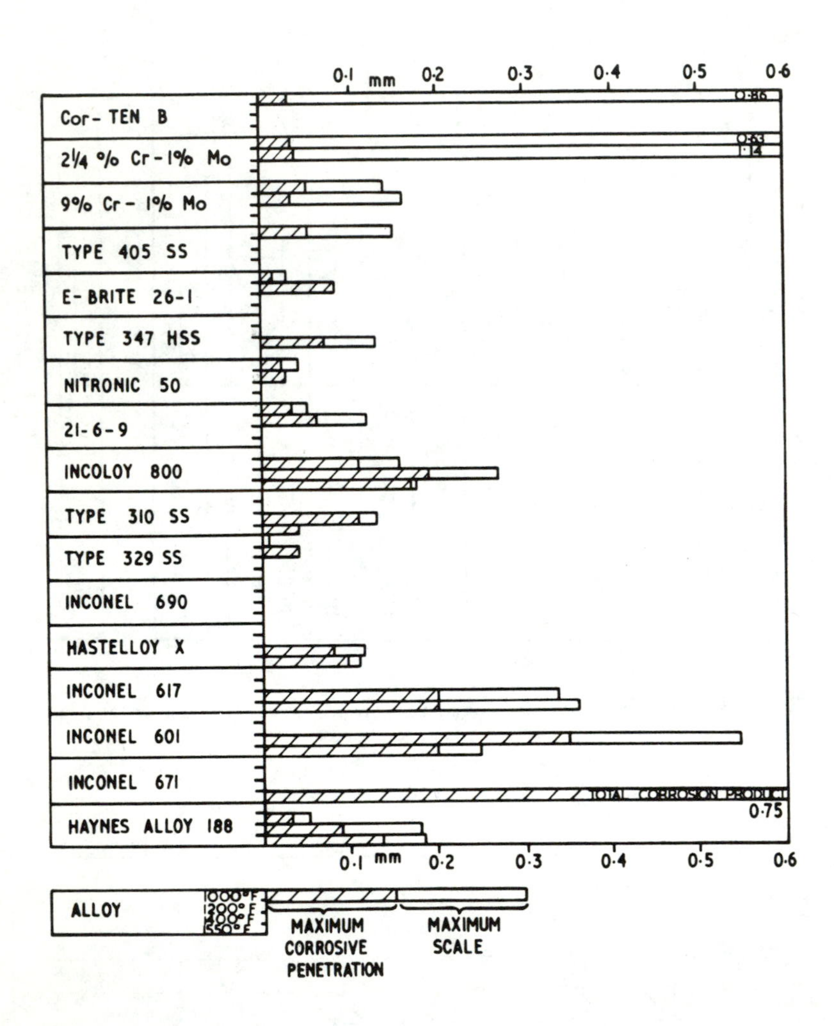

Fig.6 Maximum corrosive penetration and scale for in-bed tubes
Test 1 1000 hours exposure

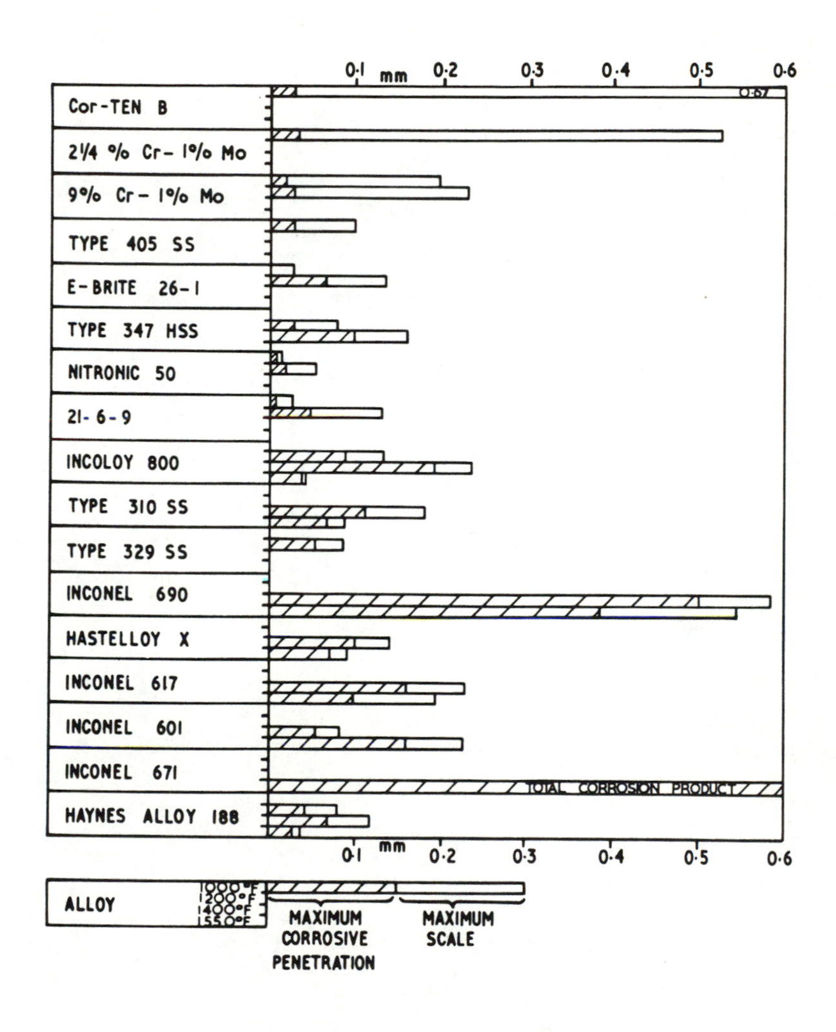

Fig.7 Maximum corrosive penetration and scale for in-bed tubes Test 2 1000 hours exposure

CORROSION OF SUPERALLOYS, INCONELS, AND STAINLESS STEELS BY THE PRODUCTS FROM FLUIDIZED-BED COAL COMBUSTION

HENRY F. WIGTON

Combustion Power Company, Inc.
Menlo Park, California, USA 94025

ABSTRACT

Metal specimens exposed to the products from the fluidized-bed combustion of high sulfur coal (Illinois No. 6) were examined for evidence of corrosion, particularly alkali-induced sulfidation.

The experimental results indicated that temperature level is the single most important factor in determining whether or not sulfidation will occur in a coal combustion system. Below 870°C no evidence of sulfidation was found in turbine blade materials. On the other hand, temperature excursions above 932°C for as little as one hour produced detectable sulfidation in some alloys. Above 954°C castastrophic sulfidation occurred in susceptible alloys. At a temperature of 910°C with clay and magnesium as additives no evidence of sulfidation was found on turbine blade material.

INTRODUCTION

The direct-fired gas turbine using a fluid-bed combustor is an attractions means of generating electrical power. However, as temperatures have increased, and as alloys have been developed to withstand the higher temperatures, problems have arisen with new forms of corrosion. One source of corrosion is vanadium. The vanadium compounds flux with and destroy the protective oxide coatings that would otherwise provide a barrier against destructive oxidation of the base metal. This kind of corrosion has been dealt with by using fuels that contain no vanadium or adding "inhibiting" compounds to tie up the vanadium in high-melting complexes, and by restricting temperatures to levels below that at which the fluxing will proceed.

Another form of attack is the phenomenon referred to either as sulfidation or as hot corrosion. This is an accelerated attack caused by the presence of liquid alkali sulfates on metal surfaces. It is characterized by the migration of sulfur into the metal ahead of the visible surface damage. The quantity of liquid needed to promote this type of attack is minuscule. Since turbine fuels all contain sufficient sulfur for the formation of the alkali sulfates, the amounts that can form are usually determined by the amount of alkalis present. For this reason major turbine manufacturers specify that oil fuels contain less than one part per million--and the air less than about ten parts per billion--of sodium or potassium.

Oils can be washed with water to eliminate alkali compounds. Coals, on the other hand, contain from two to three thousand parts per million sodium along with several thousand parts per million sulfur, and these cannot be easily washed from the coal. These facts would indicate that the hot-corrosion phenomenon poses a serious threat to the future development of coal-fired gas turbines.

Definitive information useful in assessing the magnitude of the threat has been lacking. To fill this void the Combustion Power Company has pursued a test program for three years in a smaller 2.2-sq-ft atmospheric model fluid-bed combustor. The atmospheric tests have been supported with additional testings at 4 atmospheres pressure in a full scale fluidized bed coupled to a gas turbine. This testing has been pointed toward the accumulation of definitive information with regard to hot corrosion in the coal-combustion process and the development of methods for dealing with it. Two major turbine manufacturers, Solar and Westinghouse, have participated in the program as subcontractors, supplementing the CPC work with theoretical studies of combustion thermochemistry, bench-scale corrosion tests, and testing in simulated turbine environments.

Because information at the outset was fragmentary and sometimes contradictory, the initial tests were exploratory in nature, and some of the tentative early conclusions were later found to be untenable. A fairly clear picture, however, is now emerging. Because it will be helpful to review the course that led to the present state of understanding, this report has to some extent the aspect of a chronology. Results reported here are from tests in the model combustor. Data from the full-scale pressurized combustor, while too limited to permit analysis on their own, are used as to support the conclusions drawn.

The overall effort reported here also includes concurrent work on the associated problems of erosion and deposition.

BACKGROUND

As late as 1974, the only known work in which the phenomenon of hot corrosion was studied under conditions of fluidized-bed coal combustion was that done by the British Coal Utilization Research Association.[1*] In the course of a program investigating fluidized-bed coal combustion, samples of five chrome steels were exposed in the bed for 98 hours. Although bed temperatures, ranged as high as 948°C, the samples were so positioned that none was exposed above 826°C, and most exposures were below 760°C. "Some" sulfidation was detected in one sample of 18%-chrome (AISI 321) steel, but in general the attack was minimal and mostly of the oxidation type. Incoloy 800 (21% chrome) showed only 2 microns of penetration.

Others found classic sulfidation occurred in nickel-based alloys and fine-grain sulfidation in cobalt-based superalloys[2] exposed to the same environment under similar conditions.

As part of preparing a test program, Combustion Power Company exposed some specimens in the model combustor during the course of other testing. After 288 hours (not at constant temperature) four types of state-of-the-art superalloys showed definite sulfidation. Figure 1 shows a photographic image and two element maps for M509. The sulfur map shows sulfur to be distributed within the metal in a manner characteristic of sulfidation, while the cobalt map shows a depletion of cobalt near the surface.

The extent of the damage, however, was no greater than what might be expected from burning oil with three parts per million sodium chloride and two parts per million vanadium at 871°C[3] and was clearly less than would occur if

---

*Superscript numbers refer to the list of references at the back of this report.

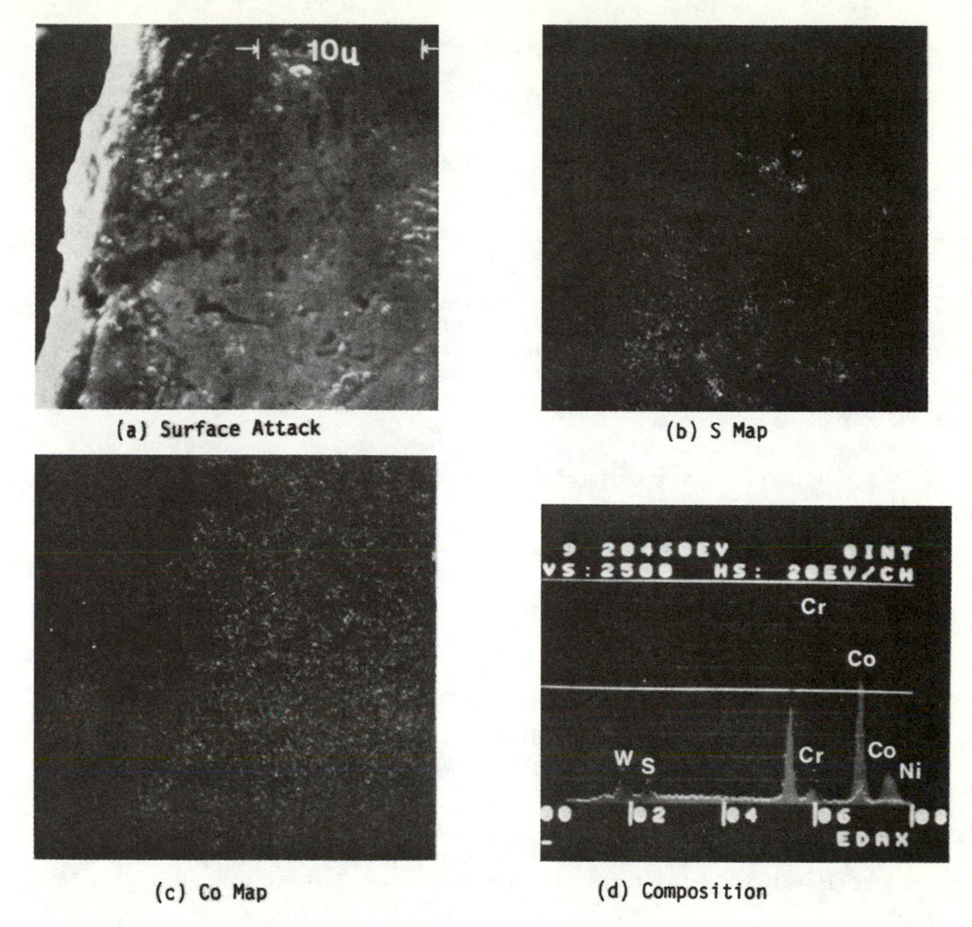

Figure 1 - Hot corrosion of M509 (a) surface attack, (b) sulfur map, (c) cobalt map.

all the alkali in the coal were effective in promoting sulfidation. Determination of the reasons for the orders-of-magnitude differences between coal and oil is essential to establishing the most effective approaches for inhibiting corrosion.

As a first principle, in order to eliminate sulfidation, the formation of active sulfates on the metal surfaces must be prevented. This can be accomplished by:

1. Maintaining the alkalis as condensed, inactive particulate matter;

2. Reacting alkali vapors to form inactive solid particles;

3. Reacting liquid condensates to form less active liquids or innocuous solids or;

4. Forming alkali compounds of such high volatility that no condensation occurs until temperatures are below those at which corrosion proceeds.

Since the preliminary testing at Combustion Power Company had indicated the corrosive potential from burning coals was substantially less than would be expected from oils containing only a minute fraction of the total alkali contained in the coal, some of these mechanisms of suppression must have been effective.

Analysis of ash particles from the flue gases by examination under the scanning electron microscope (SEM), and particles that were rich in potassium by energy-detection-analysis X-ray (EDAX), indicated the potassium-rich particles contained a complex feldspar $KAlSi_3O_8$, with calcium present as well. These results appear in Figure 2. Free-energy values from the literature[4] indicated that potassium feldspar and sodium feldspar ($NaAlSi_3O_8$) are more stable than simple silicates--and under some conditions can indeed be more stable than the respective alkali sulfates. While no data for calcium-rich feldspar were found in the literature, the fact that it was formed indicates that it is as stable as the simple potassium and sodium feldspars. Apparently some of the alkalis present were tied up as feldspars which inhibit the formation of the sulfates which are essential for hot corrosion to proceed.

Such findings were instrumental is determining the course of the test program.

THE TEST PROGRAM TO DATE

Testing was all related to the combustion of Illinois No. 6, a high-sulfur, moderately caking coal, with dolomite in the form of Kaiser "Dolowhite" as a sulfur suppressant. Analyses for the coal and dolomite are shown in Tables 1 and 2. Tests have been conducted in the model combustor at atmospheric pressure, with some testing done as well in the 4-atmosphere full-scale combustor.

The initial testing took the form of a parametric study, and consisted of a number of short-duration tests during which specimens of 22 alloys were exposed. The alloys and their compositions are shown in Table 3. Exposure temperatures were varied between 871 and 926°C nominal. On occasions temperature excursions to levels as high as 971°C were experienced. Additives intended to inhibt corrosion were used at levels of 0, 0.4%, and 0.8% of the coal weight. The additives were selected for their potential capability to tie up alkalis in

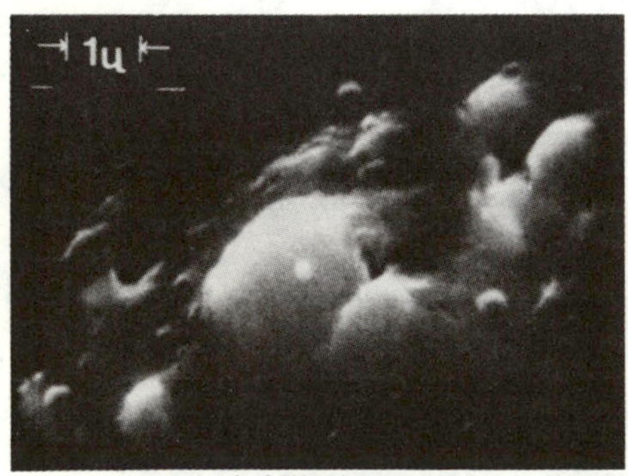

a) SEM Photograph

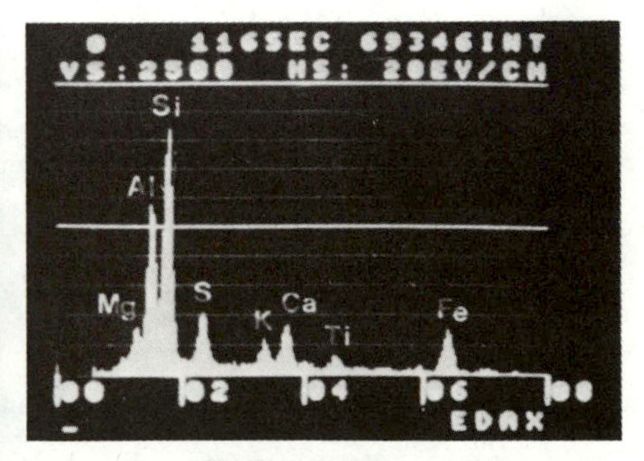

b) Edax Analysis

Figure 2 - Potassium-rich particles found in the hot-gas system (a) SEM photograph and (b) EDAX analysis.

TABLE 1

ANALYSIS OF ILLINOIS NO. 6 COAL
MIX OF 221 SAMPLES

PROXIMATE ANALYSIS	As received	Dry basis
% Moisture	9.28	xxxxx
% Ash	13.11	14.45
% Volatile	33.40	36.82
% Fixed Carbon	44.21	48.73
	100.00	100.00
Btu	10396	11459
% Sulfur	3.27	3.60
% Alk. as Na_2O	xxxxx	0.40

ULTIMATE ANALYSIS	% Weight As received	% Weight Dry basis
Moisture	9.28	xxxxx
Carbon	59.97	66.11
Hydrogen	4.18	4.61
Nitrogen	1.14	1.26
Chlorine	0.16	0.18
Sulfur	3.27	3.60
Ash	13.11	14.45
Oxygen (diff)	8.89	9.79
	100.00	100.00

MINERAL ANALYSIS	% Wt. Ignited Basis
Phos. pentoxide, P_2O_5	0.19
Silica; SiO_2	47.60
Ferric oxide, Fe_2O_3	16.64
Alumina, Al_2O_3	17.58
Titania, TiO_2	0.96
Lime, CaO	6.20
Magnesia, MgO	1.88
Sulfur trioxide, SO_3	5.35
Potassium oxide, K_2O	1.77
Sodium oxide, Na_2O	1.62
Undetermined	0.21
	100.00

SILICA VALUE = 65.82
T250 = 2370 F

TABLE 2

TYPICAL ANALYSIS OF KAISER DOLOMITE

	As Rec'd	Dry
% Moisture	0.88	xxxxx
% Ash	55.21	55.70
% L.O.I.	43.91	44.30

WATER SOLUBLE ALKALIES, Dry Basis

% Na_2O = 0.009

% K_2O = 0.000

	DRY BASIS Analysis	DRY BASIS Computed
Ignition Loss	44.30*	0.83**
Carbon dioxide	43.47	
CaO	32.80	5.50
MgO	20.20	
Phos. pentoxide	0.02	0.02
Silica	0.77	0.77
Ferric oxide	0.50	0.50
Alumina	0.57	0.57
Titania	0.00	0.00
$CaCO_3$		48.72
$MgCO_3$		42.25
Sulfur trioxide	0.14	0.14
Potassium oxide	0.28	0.28
Sodium oxide	0.12	0.12
Undetermined	0.30	0.30

*Including CO_2 **Not Including CO_2

stable compounds other than sulfates; those employed were clay ($Al_2O_3 \cdot 2SiO_2$) in the form of Burgess No. 10, a commercial pigment, and silica (SiO_2) in the form of Johns Manville Celite FC diatomaceous earth. Specimen exposure times in these initial tests were between 80 and 90 hours.

Specific test conditions are shown in Table 4, along with summary results for seven alloys of primary interest. From the data shown, it is evident that test conditions can produce widely varying degrees of surface attack and sulfidation. There is, for example, a sharp contrast in results for the first five alloys between the baseline test M-401 and test M-402A and B.

The contrast is made pictorially evident by comparison of (a) and (b) in Figure 3.* Sulfidation is evidenced in (a) by the dark spots located as much as 70 microns beneath the surface. These are readily identified in Inconel 600 by probing with EDAX. The EDAX analyses for the two specimens are shown as (c) and (d). In (c), which relates to the spot indicated by the arrow in (a), a distinctive sulfur peak is in evidence. In the analysis for (b), shown as (d), the sulfur peak is absent.

The clay pigment appears to be effective as a corrosion inhibitor even at the higher temperature of test M-402B. Indeed that was the conclusion initially drawn. Further inspection of the data, however, reveals an inability to repeat the results of M-402, either with double the amount of clay additive or with the silica additive used instead. Ultimately it became evident that the primary variable in all the tests was temperature--and in particular, temperature excursions. This observation was at first obscured by the seeming absence of any such excursions in test M-401. Later inspection of the data from the test, however, revealed that a temperature excursion had indeed occurred, but before coal had been introduced into the system; i.e., with only oil and dolomite present. It is now seen that traces of alkali--or the calcium itself--in the dolomite and sulfur in the oil are sufficient to cause sulfidation at temperatures above 1780 F.

The several illustrations in Figure 4 are for Inconel 601, following its exposure to temperature excursions as high as 954 F, during test M-403D. Normally this alloy is quite resistant to oxidation corrosion. The catastrophic rate of corrosion that resulted is typical for conditions highly conducive to sulfidation. Figure 4(a) is a secondary electron image, which is similar to an optical image but with somewhat different contrasts; the extent to which metal integrity is lost is easily seen.

The sulfur map in (d) covers the same field of view as the image directly above it; it discloses the front of sulfur moving into the metal ahead of the surface damage, a phenomenon characteristic of classical sulfidation. The chromium and nickel maps in (b) and (c) show the concentrations of these two elements to be complementary. Bright areas of the chromium map correspond to dark areas of the nickel map, and vice versa.

The series of short-duration tests (Table 4) were to have established the efficacy of the two additives used, and to have provided information as to the quantities required. The effects produced by occasional temperature excursions, however, were of such magnitude as to obscure the effects of parametric variation. In order to reduce the incidence and range of such excursions, the coal-feed system was at this point modified to give a quicker response to variations in bed temperature, and the size of the coal being fed was reduced from a maximum of 1/4" to a maximum of 1/8".

---

*A small array of marks, just below and to the right of center in (b), can be seen on close inspection to appear as well in (a). It is, in fact, a slight flaw in the photographic glass, and can be identified in most of the subsequent photomicrographs also.

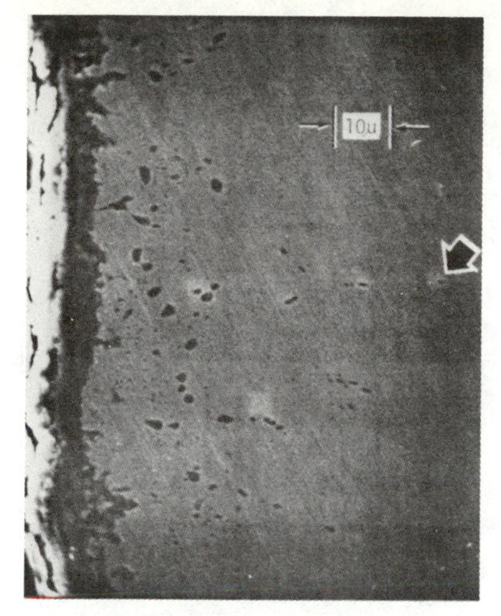

a) Test M-401, 1580 F,
No Additive

b) Test M-402B, 1675 F,
Clay Additive

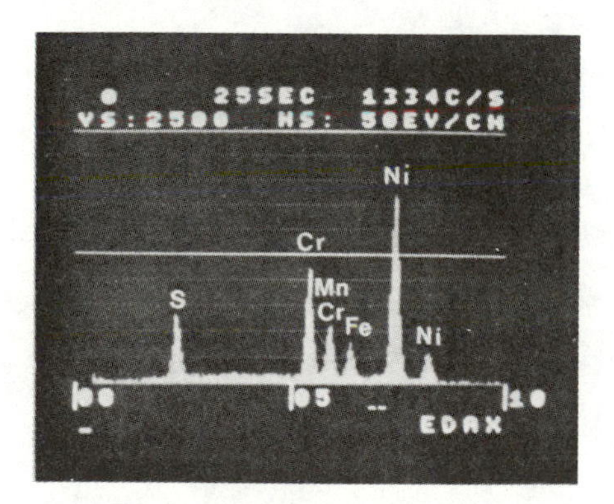

c) EDAX Analysis of
"Dark" Spot in (a)

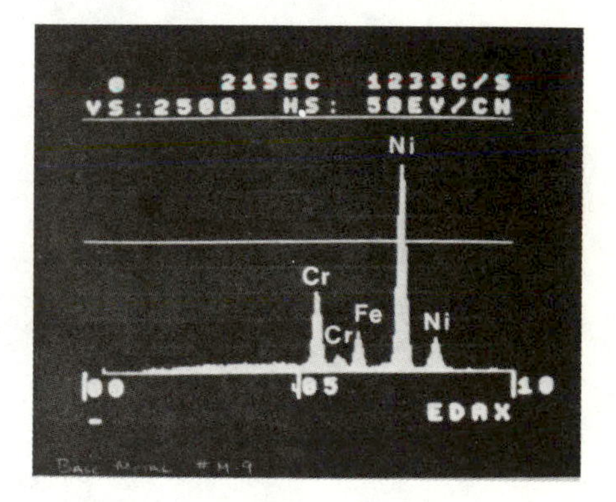

d) EDAX Analysis of
Base Metal

Figure 3 - Inconel 600 exposed 85 hours with and without additive (a) test M-401, 1580 F, no additive, (b) test M-402B, 1675 F, clay additive, (c) EDAX analysis of "dark" spot in (a), and (d) EDAX analysis of base metal.

Following the changes, an extended test of just over 1000 hours was carried out with excellent temperature control and no excursions. Test conditions are shown in Table 5, along with corrosion result for all 22 alloys. Since all exposures were for the full duration of the test, the penetrations listed may be interpreted directly as rates; i.e., microns per thousand hours.

No sulfidation was observed. Damage in general was minimal, but comparison of the numbers does show a ranking of alloys in terms of their corrosion resistance that further indicates sulfidation was not the chief mechanism operative in this test. IN 713, which contains aluminum and is quite susceptible to sulfidation in turbines, was shown here to be more resistant than FS 414, a cobalt-based superalloy developed specifically to resist sulfidation.

TABLE 3

TYPICAL CORROSION-TEST CANDIDATE MATERIALS

Alloy	Probable Use	Composition										
		Ni	Co	Cr	Fe	Al	Ti	Mo	W	Cb	Zr	C
SS 304	Nonturbine	10	--	20	Bal	--	--	--	--	--	--	0.05
SS 309	Nonturbine	14	--	23	Bal	--	--	--	--	--	--	0.05
SS 310	Nonturbine	20	--	25	Bal	--	--	--	--	--	--	0.05
SS 316	Nonturbine	10	--	18	Bal	--	--	2.5	--	--	--	0.05
SS 330	Nonturbine	35	--	19	Bal	--	--	--	--	--	--	0.05
SS 333	Nonturbine	48	3	25	18	--	--	3	3	--	--	0.05
SS 446	Nonturbine	0.7	--	25	74	--	--	--	--	--	--	0.10
Inconel 600	Nonturbine	76	--	15.5	8.0	--	--	--	--	--	--	0.08
Inconel 601	Nonturbine	61	--	23	14.1	1.35	--	--	--	--	--	0.05
Inconel 690	Nonturbine	60	--	30.0	9.5	--	--	--	--	--	--	0.03
Inconel 706	Nonturbine	42	--	16	40	0.20	1.75	--	--	--	--	0.03
N155	Stator	20	20	21	33	--	--	3	2.5	--	--	0.15
Incoloy 800	Nonturbine	32.5	--	21.0	45	0.38	0.38	--	--	--	--	0.05
Incoloy 825	Nonturbine	44	--	21.5	30.0	0.10	0.90	3.0	--	--	--	0.03
Nimonic 80A	Stator	76	--	19.5	--	1.3	2.5	--	--	--	--	0.06
Rene 77	Rotor	59	15.0	14.0	--	4.3	3.35	4.2	--	--	0.04	0.07
Inco 713C	Rotor	74	--	12.5	--	6.1	0.8	4.1	--	2.0	0.10	0.12
U-700	Stator	55	17.0	15.0	--	4.0	3.5	5.0	--	--	--	0.06
IN 738	Turbine	64	8.5	16	--	3.4	3.4	1.8	2.6	--	--	0.17
HS 31	Turbine	10	56	25.0	--	--	--	--	7.5	1.0	--	0.50
FS 414	Turbine	10	52	29.0	1.0	--	--	--	7.5	--	--	0.25
M 509	Turbine	10	58	23.5	--	--	0.2	--	7	--	0.5	--
PWA 68	Coating	--	Bal	19-25	--	12-15	--	--	--	--	--	--

TABLE 4
PARAMETRIC TESTS IN THE MODEL COMBUSTOR

Test	M401	M402A	M402B	M402C	M402D	M403A	M403B	M403C	M4[illegible]
Duration, hr.	87.4	84.9	84.2	84.3	84.2	83.9	84.6	84.4	84.[illegible]
Freeboard Temp. (°F) Nominal	1600	1600	1700	1600	1700	1600	1700	1600	17[illegible]
Average	1580	1570	1675	1570	1660	1580	1670	1570	1670
Maximum	1670	1590	1680	1730	1780	1730	1710	1580	1750
Corrosion Additive	None	0.4% Burg. #10 Pig.	0.4% Burg. #10 Pig.	0.8% Burg. #10 Pig.	0.8% Burg. #10 Pig.	0.4% J.M. Celite FC	0.4% J.M. Celite FC	0.8% J.M. Celite FC	0.8 J.[illegible] Celite FC
Alloy	Penetration+ In Microns								
FS414	6,13S	None	9	10,80S	10, 150S	20	25	18	50
Inconel 600	70S	4	10	27	1300	8	40	10	600
Nimonic 80A	30,45S	4	10	16	300S	14,20S	35,55S	17	29
IN 713C*	10	None	4	5	7	6	10	5	18
IN 738*	12	None	None	5	18	7	20	10	23
IN 601	24	16	33	20	360	15	16	13	300
IN 690	23	13	6	23	320	6	27	22	140

* The presence of molybdenum in these alloys precludes identification of low-level sulfidation by EDAX.

+ Numbers followed by "S" designate sulfur penetrations beyond visible surface damage.

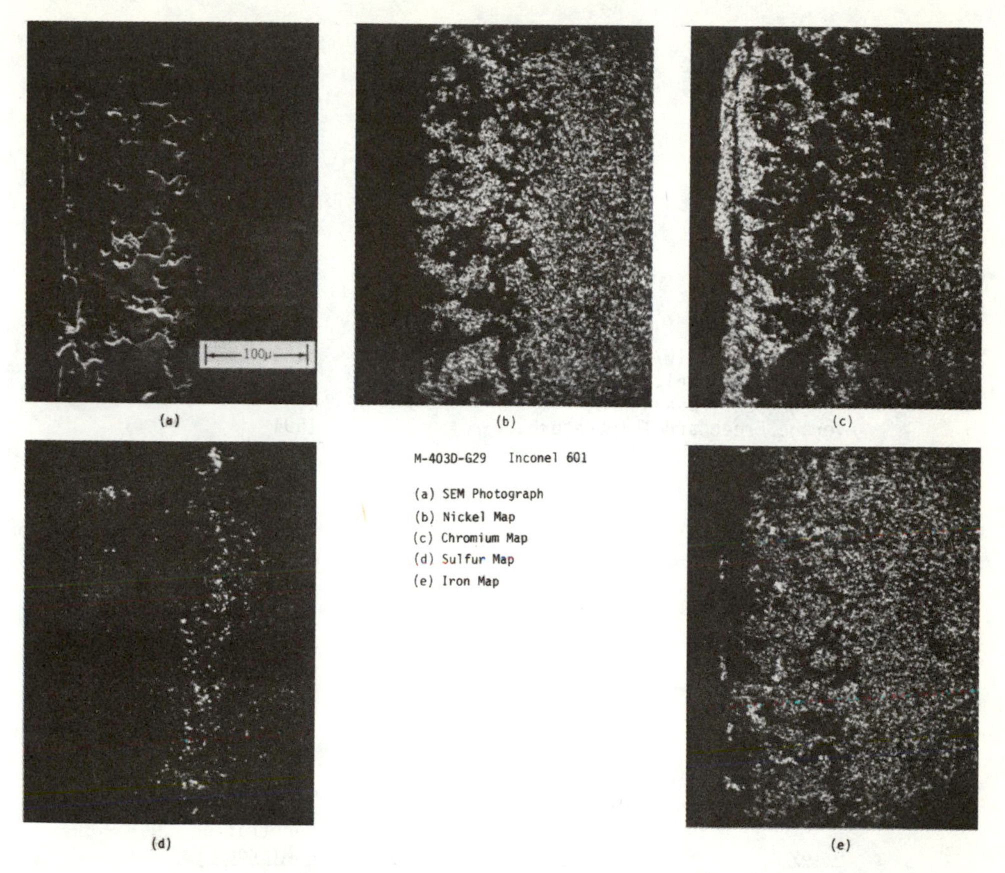

Figure 4 - Inconel 601 exposed to coal combustion 85 hours at 1700 F, (a) SEM photograph, M-403D-G29 Inconel 601, (b) nickel map, M-403D-G29 Inconel 601, (c) chromium map, M-403D-G29 Inconel 601, (d) sulfur map, M-403D-G29 Inconel 601, (e) iron map, M-403D-G29 Inconel 601.

Aside from the effects of temperature excursions, the tests to this point had shown low levels of sulfidation activity. A proper evaluation of the role of additives--including a better differentiation among alloys--would require a longer baseline test. The first 400 hours of the next planned long-duration test (M-406) were therefore set aside for further evaluation of the no-additive condition. This new baseline test, designated M-406A, was conducted at an average temperature of 882 F. Significantly, no temperature excursions were experienced. As a result, extensive corrosion was not found, confirming that the M-401 results reflected an effect of temperature that occurred before the start of coal combustion.

The final 568 hours of model testing, designated M-406B, were run with a freeboard temperature of 1670 F, in order to determine more accurately the temperature needed to produce sulfidation. Clay pigment was fed as additive at 0.4% of the coal weight to combine with any excess alkalis, and MgO was added to minimize fouling. Conditions of the test are summarized in Table 6, along with corrosion results for the turbine materials and selected Inconels. No sulfidation was found in the state-of-the-art turbine materials, and only trace indications were observed in the more susceptible materials.

TABLE 5

M-405 TEST SUMMARY

TEST CONDITIONS

Duration, hr	1003
Dolomite/Coal, lb/lb	.24
Clay Pigment/Coal, lb/lb	.004
Average Bed Temperature, F	1650
Average Freeboard Temperature, F	1594

CORROSION RESULTS

Alloy	Max. Penetration, Microns	Alloy	Max. Penetration, Microns
SS 304	17	Incoloy 800	47
SS 309	24	Incoloy 825	35
SS 310	27		

Alloy	Max. Penetration, Microns	Alloy	Max. Penetration, Microns
		Nimonic 80A	70
SS 316	18	Rene 77	28
SS 330	12	Inco 713C	18
SS 333	30		
SS 446	3	U-700	30
		IN 738	45
Inconel 600	70		
Inconel 601	65	HS 31	30
Inconel 690	50	FS 414	60
Inconel 706	18	M 509	22
N155	20	PWA 68	0

TABLE 6

M-406B TEST SUMMARY

TEST CONDITIONS

Duration, hr	568
Dolomite/Coal, lb/lb	.24
Clay Pigment/Coal, lb/lb	.004
Magnesia/Coal, lb/lb	.001
Average Bed Temperature, F	1685
Average Freeboard Temperature, F	1670

CORROSION RESULTS

Alloy	Penetration in Microns	
	Maximum	Average
IN 600	20	15
IN 601	30	20
IN 690	20	17
IN 713	25	13
IN 738	40	25
FS 414	60	50
Nimonic 80A	80	50
IN 800	80*	50
IN 825	55*	24

*Sulfidation identified

The attack sustained by four different types of turbine alloys is shown in Figures 5 and 6. Of these four types, each appears to be attacked in its own characteristic pattern, IN 713--(a) in both illustrations exhibits the least overall penetration and highly localized variations. The dark "crystallites" in 5(a) are rich in aluminum and show no sulfidation. IN 713 (b) in both illustrations is more uniformly attacked, with deeper penetrations resulting from the longer run at higher temperature. Nimonic 80A, (c), is the most severely attacked of the four alloys (note the different scales). FS 414, (d), a cobalt-based superalloy, is attacked almost as extensively as Nimonic 80A and much more than IN 738.

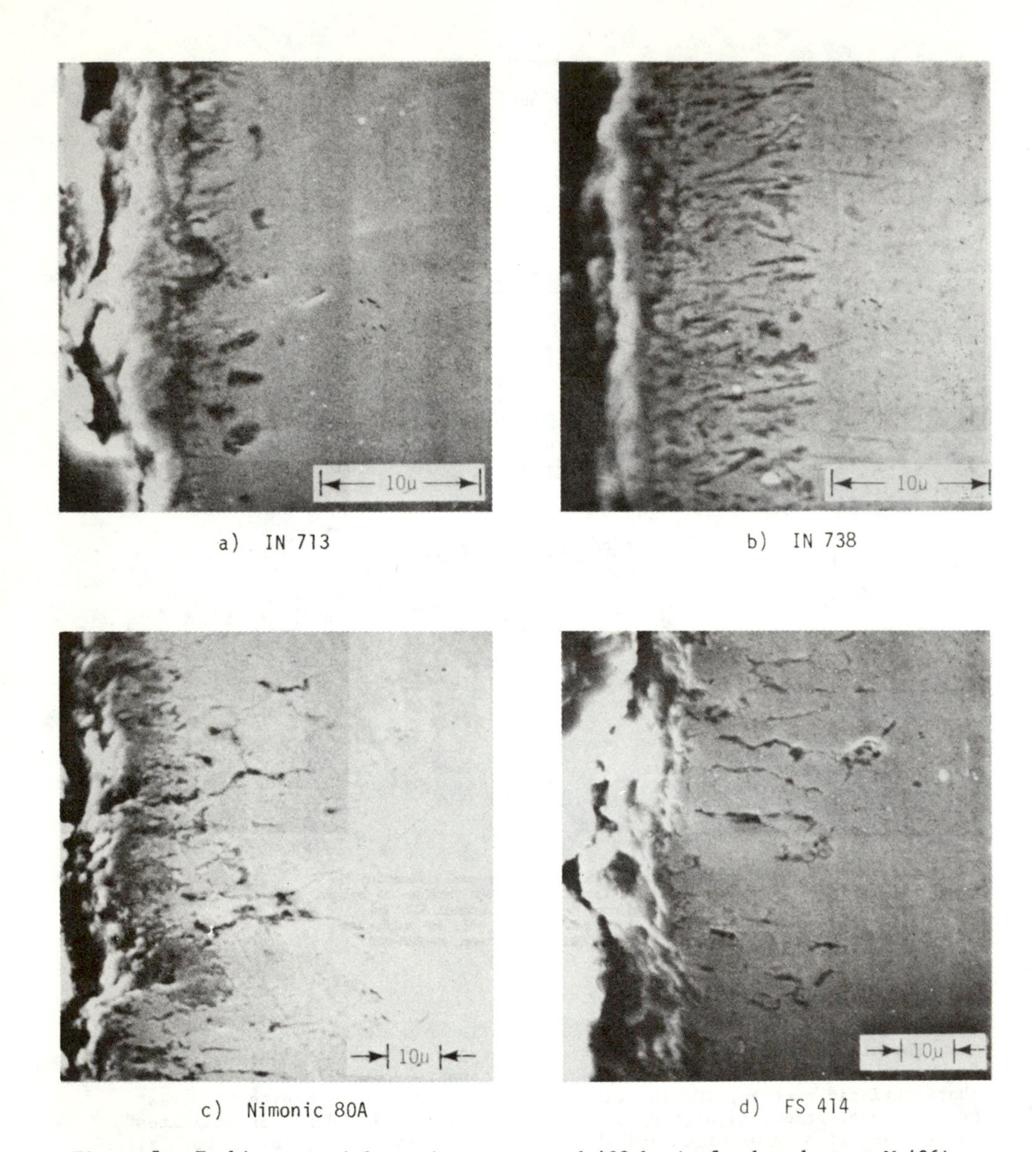

a) IN 713

b) IN 738

c) Nimonic 80A

d) FS 414

Figure 5 - Turbine-material specimens, exposed 400 hr in freeboard, test M-406A, (a) IN 713, (b) IN 738, (c) Nimonic 80A, and (d) FS 414.

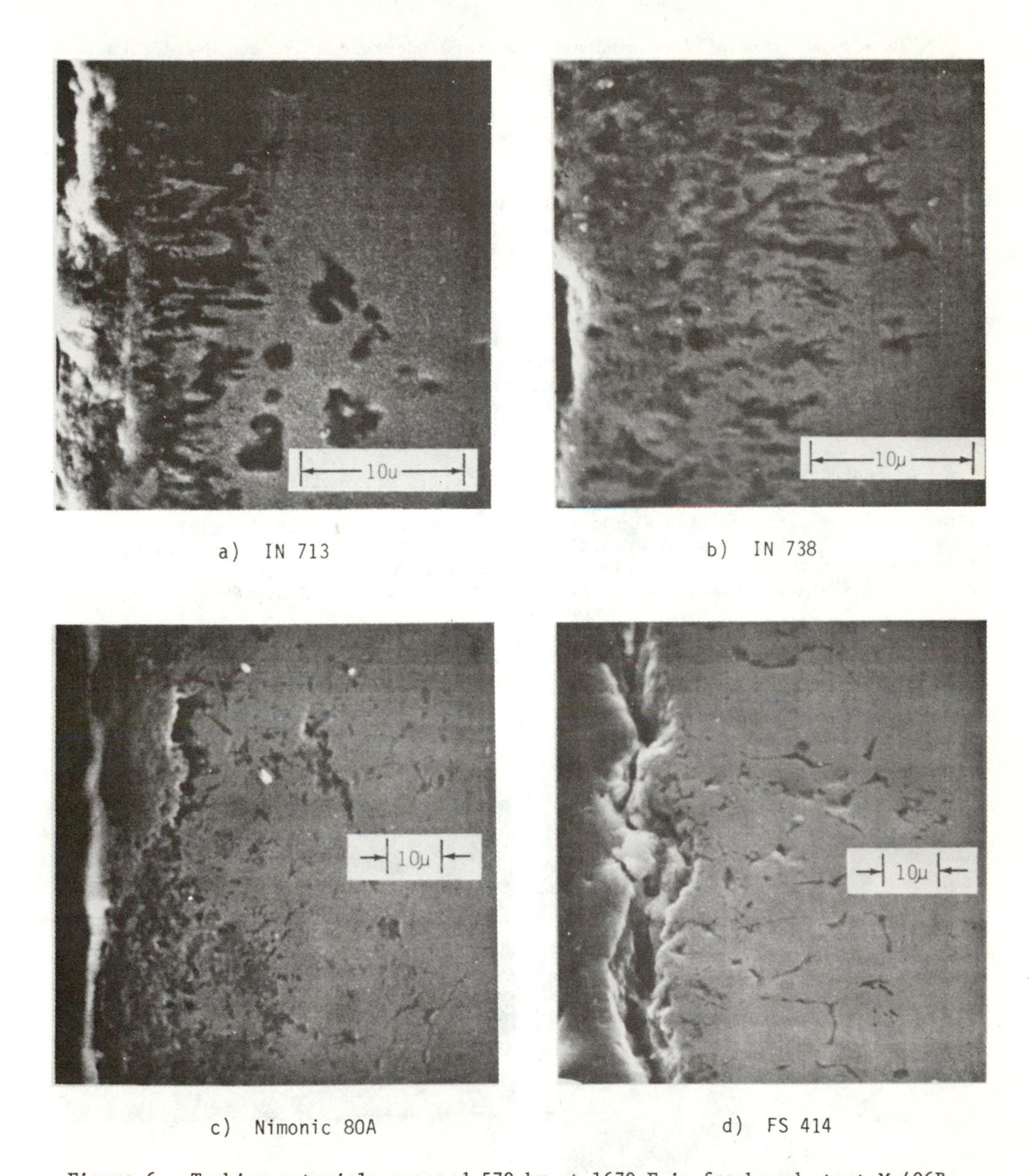

Figure 6 - Turbine materials exposed 570 hr at 1670 F in freeboard, test M-406B, (a) IN 713, (b) IN 738, (c) Nimonic 80A, and (d) FS 414.

DISCUSSION OF HOT-CORROSION RESULTS

The absence of sulfidation when temperature excursions are avoided indicates that no liquid sodium sulfate is condensing on the metal specimens. Analysis of the ash at various collection points in the system, shown in Table 7, shows essentially no water-soluble sodium or potassium in ash that is removed from the system hot. The insoluble alkali salts are assumed to be feldspars, while the solubles are sulfates, chlorides, or simple silicates.

The concentration of soluble alkalis in ash collected by the granular filter* indicates that this ash would be highly corrosive at high temperatures. A SEM image and EDAX analysis are shown in Figure 7. In tests at Solar,[5] specimens were painted with a slurry made from this ash and then reheated. Corrosion rates at 910 C were from four to six times those experienced by specimens exposed in the combustor freeboard. The operative factor is collection temperature. Because of system heat losses inherent in the test setup, the granular filter was working at a relatively low temperature (648 C). As a result, the material it collected is richer in alkali sulfates than would be the case of collection took place at higher temperature. The reason for this is best illustrated by considering the equilibrium lines shown in Figure 8.

To the right and below the diagonal lines, the stable solid phase is a feldspar ($KAlSi_3O_8$ or $NaAlSi_3O_8$) to the left and above the lines, the condensed phases are the corresponding alkali sulfates. At 871 F, and the partial pressures of SO_2 and O_2 prevailing (10^{-4} atm), alkali vapors will react to form feldspars. If these particles are not removed from the system and the gas/particle mixture is cooled, the equilibrium lines are crossed by moving to the

*A moving-bed granular filter is used as a final gas-cleanup stage, to remove particulate too fine for collection by conventional tangential-entry cyclones.

(a) SEM Image

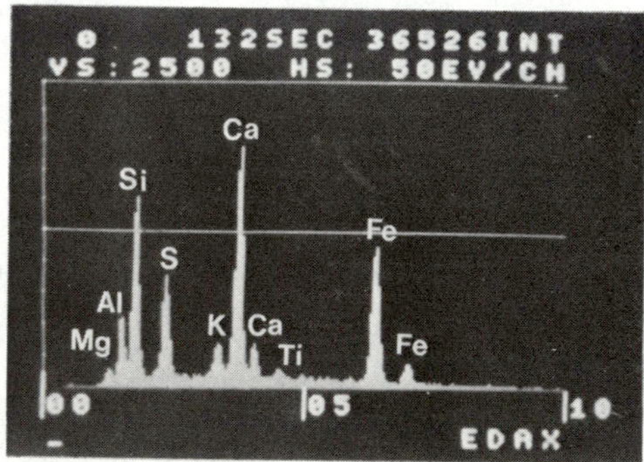

(b) EDAX Analysis

Figure 7 - Ash collected by granular filter and separated by media circulation and cleaning system (a) SEM image and (b) EDAX analysis.

TABLE 7

ASH COMPOSITION FROM SELECTED SOURCES

Test	M-406A					M-406B				
Source of Ash	Coal	Bed Drain	#2 Spec. Holder	2nd Stage	G/F	Coal	Bed Drain	#2 Spec. Holder	2nd Stage	G/F
Composition										
P	.08	.02	.03	.05	.04	.08	.02	.03	.04	.02
Si	22.43	7.31	9.77	10.54	10.68	22.72	6.38	10.54	9.80	15.30
Fe	12.50	1.29	6.58	6.47	6.84	12.88	1.77	4.14	6.17	8.09
Al	8.68	1.66	3.88	4.24	3.92	8.21	1.77	3.69	3.91	8.48
Ti	.49	.07	.20	.17	.13	.40	.07	.17	.13	.47
Ca	3.57	23.30	14.01	16.30	14.69	3.72	23.59	19.01	16.19	9.20
Mg	.90	12.30	7.06	7.69	6.33	.90	12.67	9.05	7.78	4.27
S	2.33	9.53	11.43	8.90	10.31	2.29	9.75	8.33	9.84	5.63
Total K	1.47	.52	.67	.67	.72	1.47	.52	.64	.61	1.21
K^+ (soluble)	.02	.03	.05	.05	.10	.04	.03	.01	.02	.26
Total Na	1.26	.72	.80	.80	.80	1.37	.60	.72	.72	.96
Na^+ (soluble)	.11	.05	.01	.08	.23	.13	.05	.03	.05	.29
K^+/K, %	1.36	6.03	.74	7.28	13.2	2.72	5.55	2.34	3.29	21.6
Na^+/Na, %	8.73	6.49	1.57	9.63	28.7	9.49	8.15	4.33	7.11	30.1

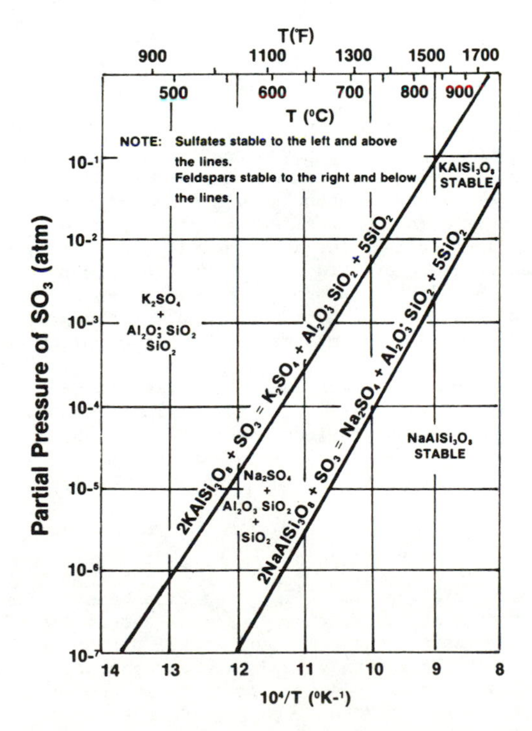

Figure 8 - Relative stability of alkali sulfates & feldspars.

left. The fine feldspar particles are converted to sulfates by reaction with SO_2 and O_2. When these sulfate salts are placed on samples in a system that prevents vaporization, as is being done at Solar, the material is highly corrosive. To account for the transfer of Na_2SO_4 as vapor would require a vapor pressure in excess of 13 ppm, compared to a saturating value of 50 ppb.

The relative stability of feldspar as against alkali sulfates was not predictable with the thermodynamic data available from Knacke.[6] However, experimental results at both Westinghouse and Combustion Power Company indicated that clay pigment was a most effective "absorber" for alkali salts. Thermodynamic data from a more recent source[4] confirm this theoretically. A Westinghouse plot based on these data, showing the fraction of potassium vapor remaining "unabsorbed", is presented here as Figure 9. It indicates that at one atmosphere total pressure and 877°C only 1.6% of the normal potassium activity would be present in the vapor if equilibrium with feldspar prevailed; at 10 atmospheres, 5% would be present. Calculations of this type will be used to arrive at the dew point of alkali sulfates which, if hot corrosion is to be completely avoided, must remain below the temperature regime of hot corrosion for the turbine expansion cycles.

Analysis of the combustion system by Westinghouse suggests that chlorides deriving from the coal and dolomite may play a role in preventing the condensation of Na_2SO_4 by forming the more volatile NaCl species. Such mechanisms are so dependent upon operating variables and exact thermochemical data that experimental work in their turbine-simulator rigs is planned. These tests will use precisely controlled amounts of sodium, chlorine, and sulfur to determine the range and extent of any operating windows in which hot corrosion does not occur.

TURBINE LIFE PROJECTION

Figure 10 provides an example of the effect of operational temperature on turbine life. This figure summarizes the corrosion data in terms of maximum penetration achieved on IN 738 alloy at Combustion Power Company. The line for 871°C indicates a rate of attack of about 43 microns per thousand hours. At this temperature, there is no apparent effect attributable to the additive. If one assumes that the allowable loss for a blade initially 0.125" thick is 20 percent per side, then the rate of attack at 871°C projects to a turbine life of about 14,000 hours. At 910°C, the projected life is about 9,000 hours, but however, if the temperature is reduced to 843°C, basing the projection on the single point available, a life of as much as 50,000 hours might be realized.

The results indicate that if operational temperatures are suitably limited, Illinois No. 6 coal contains sufficient clay materials to neutralize the effects of its alkali content. Corrosion problems for IN 738 will be tolerable at attainable SO_2 retention levels, provided temperature excursions are avoided and solids are removed prior to entry of the combustion products into the turbine.

EROSION AND DEPOSITION

The ash collected by the granular filter has also been studied by Solar to determine, as a function of temperature, its characteristics with respect to erosion and deposition. Their results are summarized by the graph of Figure 11. The erosion rate is seen to be strongly temperature-dependent. The data further indicate that the erosion problem becomes more acute in the cooler regions of the turbine (second, third, and fourth stages) where the relative particle velocity is still large.

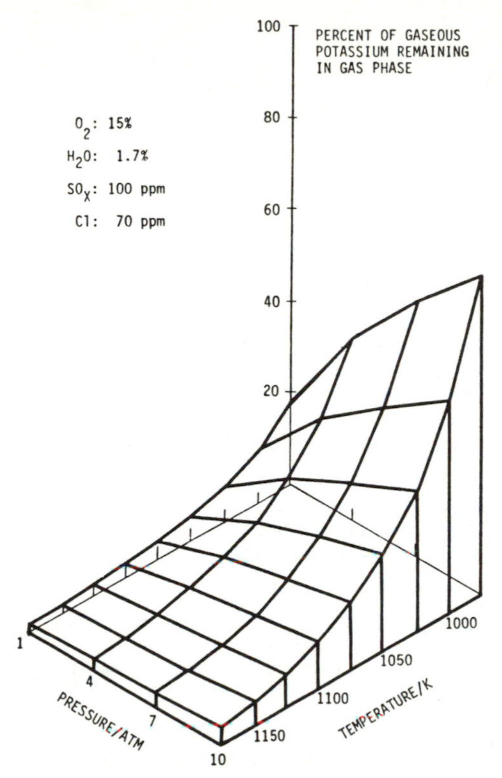

FIGURE 9 - Thermodynamics of gettering action of aluminosilicates in reducing gaseous potassium emissions.

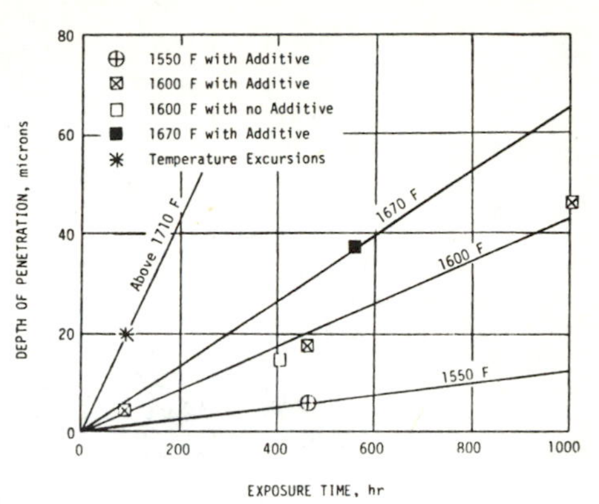

FIGURE 10 - Effect of temperature on corrosion rate for IN 738.

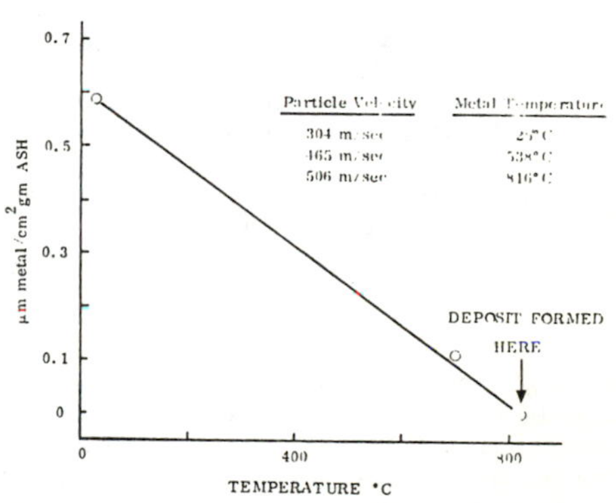

FIGURE 11 - Erosion by granular-filter ash as a function of temperature.

A potentially more serious problem is indicated by the fact that the ash forms an adherent deposit on hot (816 C) metal surfaces. In an investigation of the deposition process, Solar has determined that the amount of ash deposited is a linear function of time and particulate loading for times up to 2 hours and loadings up to 500 mg/m^3 (0.22 gr/ft^3). The data are presented in Figure 12, along with a projected curve for 70 mg/m^3, estimated as the maximum output loading for the full-size granular filter. It is seen that even this loading will deposit over 16 microns of ash per hour on hot metal surfaces.

The indication from past testing at Combustion Power Company is that these deposits grow to a particular thickness and then spall, which may result in large particles of high relative velocity hitting the rotating buckets and causing severe erosion. The ash-deposition problem is a serious one, and not yet well understood. It is probable, though, that reheated ash that contains a high concentration of alkali sulfates (used in these experiments) is more sticky than ash containing none of the lower-melting alkalis.

CONCLUSIONS

Temperature, is the single most important factor in determining whether or not sulfidation will occur in a coal-combustor system. At a temperature of 910 C with clay and magnesium oxide as additives, or at 870 C with or without additives, no evidence of sulfidation was found in turbine-blade materials.

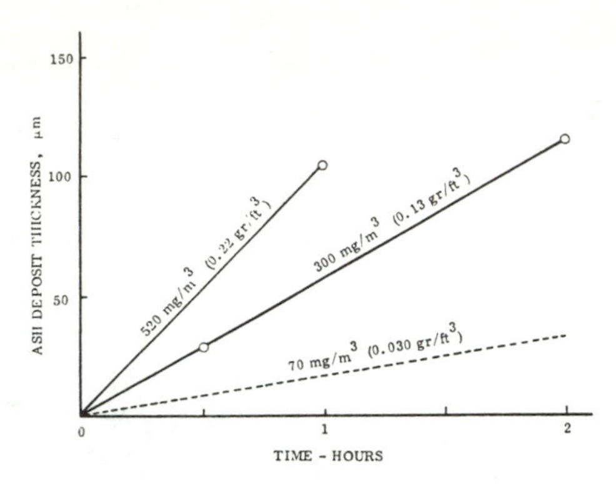

FIGURE 12 - Ash deposition at 816 C as a function of time and loading.

On the other hand, temperature excursions above 932 C for as little as one hour produced detectable sulfidation in some alloys; similar excursions to 954 C produced easily detectable sulfidation in turbine-blade materials and catastrophic sulfidation in more susceptible alloys.

Illinois No. 6 coal contains enough available aluminum silicate in the ash to suppress the alkali materials as feldspar at a temperature of 870 C. Corrosion results indicate for this temperature a potential turbine life of about 14,000 hours.

Taking into consideration the absence of metal sulfidation in the coal-burning system within temperature ranges associated with maximum sulfidation rates in oil-burning experiments; the generally low rates of attack; and the resistance ranking of the alloys, it is concluded that oxidation rather than sulfidation is the mechanism controlling the corrosion rates at temperatures below 910 C. At higher temperatures the ash constituents, including calcium sulfate and metal oxides, supply a liquid-like environment necessary to produce sulfidation. Calculations based on ash compositions and thermodynamic data indicate that higher temperature combustion gases would not cause sulfidation in a turbine, the solid particles could be removed prior to turbine entry. That is, liquid sodium sulfates or potassium sulfates would not condense from the vapor if the alkali feldspars were removed as solids before the temperature of the gas was reduced by heat losses or expansion in the turbine.

The turbine alloys most resistant to this atmosphere are those containing high aluminum concentrations, such as IN 713 and IN 738. The most resistant non-turbine materials are those high in chromium (e.g, SS 446). Both stainless-steel types SS 310 and SS 316 perform very well.

To establish whether corrosive alkali salts can be removed from the exhaust gas in the form of the feldspars that exist at high temperature, it will be necessary to conduct tests in which fine particulate is removed--as by a moving-bed granular filter--before the exhaust gas is allowed to cool.

In order to determine the range of applicability of the results of the tests here reported, direct test observations should be made on other combinations of coal and sulfur suppressant. These should be supported by specific bench-scale and turbine-rig testing and laboratory research provided by major turbine manufacturers.

Sulfidation conditions, when present, are easily detectable. Short runs (i.e., about 80 hours) can be effectively used for process-variable screening. Tests to determine relative material behavior, particularly when corrosion is inhibited, require longer runs.

ACKNOWLEDGEMENT

The experimental program from which the information presented in this paper was drawn was carried out under the sponsorship of the United States Energy Research and Development Administration, Contract E(49-18)-1536. The data and conclusions presented in this paper are essentially those of the Contractor and are not necessarily endorsed by ERDA, nor does mention of trade names or commercial products constitute endorsement or recommendation for use.

REFERENCE

1. Pressurized Fluidized Bed Combustion, British Coal Utilization Research Association, Ltd., Final Report to Office of Coal Research, Department of the Interior, November 1973.

2. Sims, Chester M., Keynote Paper for Third International Symposium on the Superalloys, September 1976.

3. Hart, A.B., and Cutler, A.J.B. (ed), Deposition and Corrosion in Gas Turbines, John Wiley and Sons, 1973; pp. 306-309.

4. Robie, R.A., and Waldbaum, D.R., Thermodynamic Properties of Minerals and Related Substances at 298.15°K (25.0°C) and One Atmosphere (1.013 Bars) Pressure and at Higher Temperatures, United States Geological Survey, Bulletin 1259.

5. Corrosion and Erosion Evaluation of Turbine Material in an Environment Simulating the CPU-400 Combustor Operating on Coal, Solar Division of International Harvester Company, Quarterly Technical Progress Report No. 2, ERDA Contract E(49-18)-1536, Subcontract No. 1536-03, 28 July 1976.

6. Barin, I., and Knacke, O., Thermochemical Properties of Inorganic Substances, Springer-Verlag, Berlin/Hiedelberg, and Verlag Stahleisen m.b.H., Dusseldorf, 1973.

HOT CORROSION/EROSION MATERIALS PROGRAM AND EXPERIENCE IN THE PRESSURIZED FLUIDIZED BED COAL COMBUSTION MINIPLANT

M. S. NUTKIS

Exxon Research and Engineering Company
Linden, New Jersey, USA 07036

The Exxon Research Fluidized Bed Coal Combustion Miniplant has been in operation since 1974, during which time it has accumulated information concerning the behavior and durability of some of its materials routinely exposed to the FBC environment.

A current Exxon Research/ERDA program is to operate the pressurized coal combustion Miniplant for 1100 hours to provide a test site and environment for exposure of specimens of potential PFBCC fireside heat exchanger alloys (supplied by Westinghouse) and gas turbine materials (supplied by General Electric). The principal objectives of this ERDA study is to compile a sound engineering data base for the characterization of the corrosion/erosion lifetime behavior of a number of commercially available alloys when exposed to a pressurized fluidized bed coal combustion environment. These PFBCC exposure tests will provide materials corrosion/erosion data and comparisons for application to advanced power systems using coal-derived fuels.

DESCRIPTION OF PFBCC MINIPLANT

Under contracts sponsored by the EPA, Exxon Research and Engineering Company designed, constructed and operates a pressurized fluidized bed coal combustion facility, the Miniplant, to demonstrate the feasibility of this coal combustion technique for environmentally acceptable energy production (Figure 1).

The Exxon Miniplant combustor is a 32 ft (10 m) tall vessel with a 24 inch (61 cm) shell refractory lined to a 13 inch (33 cm) inside diameter. Main fluidizing air for the combustor is supplied at pressures to 10 atmos. and controlled flow rates of 600 to 1200 SCFM. Coal and sorbent are proportioned, premixed and continuously pneumatically injected into the combustor at a side port 11 inch (28 cm) above the distributor grid. The air passes up through the distributing grid where it fluidizes the bed material and provides the air required for combustion.

Heat extraction in the combustor is achieved by cooling coils through which water is circulated. The coils are fabricated from type 316 stainless steel 1/2 inch pipe and are arranged in a serpentine, vertical orientation. Each coil consits of 5.9 ft^2 (0.55 m^2 of surface area; the number of coils used can be varied from 1 to 10.

Flue gas exits the combustor and discharges through 2 stages of cyclones for solids removal. The solids collected by the first stage cyclone are returned to the combustor 20 inches above the grid. Solids, mainly fly ash, collected in second cyclone are discharged by means of a lock hopper system. The expanded bed height is controlled by continuously rejecting solids through a port at the 90 inch (2.3 m) elevation. Discharge gases from the second cyclone enter a particulate clean-up vessel containing granular bed filters. These filters are for final particulate removal before the flue gases enter the turbine test section.

OPERATING EXPERIENCE

The PFBCC Miniplant has operated for 1500 actual test hours, during which over 400,000 lbs of coal were combusted. The coal used was predominantly Champion coal, a Pittsburgh seam coal with nominally 2 percent sulfur. Operating conditions were generally diverse, since the tests were conducted to investigate the effect of various operating parameters on PFBCC and desulfurization performance. Most runs were at a pressure of 9-9.5 atmospheres, a temperature range of 1500° to 1800°F, superficial bed velocities of 5 to 7 ft/sec and excess air levels of 10 to 20 percent.

Most internal metal components of the Miniplant combustor are type 316 stainless steel. This material has survived in a satisfactory manner in many locations. In certain applications, especially thermowells in higher temperature zones in the combustor, bare 316 stainless steel or Inconel 600 thermocouple sheaths have not been durable.

Metallurgical analysis was performed on two sections removed from the second stage cyclone when it was modified to a small diameter in January 1977. One sample was a section of 316 stainless steel, 5 inch (schedule 80) pipe cut from the outlet tube of the second stage cyclone. It was installed in September 1974 and had been exposed to 1400 hours of PFBCC service at temperatures of 1300° to 1600°F. The other sample was a section from the tangential stainless steel inlet liner of the second stage cyclone. It had about 1100 hours of exposure to flue gases at temperatures of 1300-1600°F and velocities of 60 to 80 ft/sec.

Metallurgical examination revealed that the two cyclone samples had suffered only minor surface damage. Both inside and outside surfaces of the outlet tube were covered with a corrosion scale approximately 0.001-0.002 inches thick. The scale contained primarily iron oxide, with traces of sulfur and calcium. The oxidation had penetrated intergranularly into the base metal to a depth of 0.004-0.005 inches. The exposed surface of the cyclone inlet liner also had corrosion deposits 0.001-0.002 inches thick and the scale was also mainly iron oxide with traces of sulfur and calcium. There was little penetration of this corrosion into the underlying metal, and there was no evidence of severe wear or erosion of the specimen.

CORROSION/EROSION TEST PROGRAM

The PFBCC hot corrosion/erosion exposure tests will start with a 100-hour shakedown test to assess the performance of the heat exchanger material probes and the turbine test passage in conjunction with the granular bed gas filter. This will be followed by the extended exposure tests of the heat exchanger materials supplied by Westinghouse Research and the gas turbine materials supplied by General Electric Company. Total exposure time will be up to 1000 hours, (in 100-hour segments), with the tests interrupted at the 200 and 500 accumulated hour intervals to allow removal, inspection and re-inspection of the specimens.

The intended miniplant combustor operating conditions for the exposure runs are 9 atmospheres pressure, 1700-1750°F combustor bed temperature, superficial bed velocity of 6 ft/sec (1.8 m/sec) and 15-20% excess air. The coal will be Illinois #6 and the sorbent Tymochtee dolomite. Expected GE turbine section inlet conditions are 9 atmospheres pressure, 1550°F gas temperature and 0.72 to 0.92 #/sec gas mass flow rate.

The heat exchanger specimen alloy-temperature matrix for in-bed and above bed Exxon Miniplant PFBCC exposure tests is shown in Table I.

Table I

Heat Exchanger Materials
Alloy-Temperature Matrix

Alloy	Specimen Temperature (°F)			
	In-Bed Only	In-Bed and Above-Bed		
	1050°	1200	1400	1600
2 1/4 Cr - 1 Mo	X			
9 Cr - 1 Mo	X			
304 Stainless Steel		X		
Incoloy 800		X	X	
Hastelloy X			X	X
Haynes 188				X

The specimens prepared by Westinghouse for the FBC combustor fireside corrosion/erosion exposures are 1.25" OD X 1.00" ID by 3" long cylinders with thermocouples inserted in the walls. Specimens of 2 different alloys selected for a given test matrix temperature are welded together, plugged at the end and welded to a tube section to form a bayonet probe. This specimen probe consists of a central cooling tube thermocouple conduits, cooling air inlet and outlet and a flange for connection to the miniplant combustor port.

Separate temperature control loops can set and maintain the temperature of each specimen probe independently by regulating the amount of cooling air delivered to each probe. This will provide a reliable, independent and readily adjustable temperature control for the 21 specimens probes (each with 2 alloy specimens) to be inserted in the combustor. It permits control of 12 in-bed specimen probes at mean temperatures of 1050°F, 1200°F, 1400°F and 1600°F, and 9 above-bed specimen probes at temperatures of 1200°F, 1400°F and 1600°F, with 3 specimen probes at each temperature.

The turbine test section will provide a region with representative flow velocities to furnish engineering information on the corrosion/erosion deterioration of gas turbine materials exposed to the exhaust gas from a pressurized fluidized bed coal combustor. The General Electric turbine hot gas passage for gas turbine materials corrosion/erosion evaluation incorporates both bucket and nozzle test specimens. Enclosing the flow duct containing the airfoil specimens is a pressure shell which comprises the GE turbine test section. This section will be inserted in the Exxon Miniplant discharge piping immediately downstream of the granular bed particulate clean-up filter.

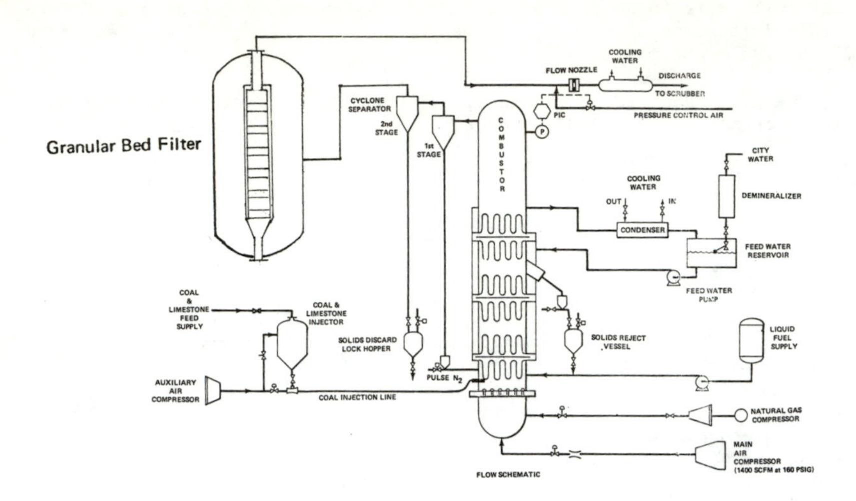

Figure 1. PFBCC Miniplant

PRELIMINARY REPORT ON CORROSION ANALYSIS OF HEAT EXCHANGER TUBES FROM A FLUIDIZED-BED COAL COMBUSTOR

R. H. COOPER and J. H. DeVAN

Metals and Ceramics Division
Oak Ridge National Laboratory
Oak Ridge, Tennessee, USA 37830

ABSTRACT

The Energy Division of the Oak Ridge National Laboratory is planning the construction of a fluidized-bed coal combustor pilot plant. This system (MIUS — Modular Integrated Utility System) was initially designed under joint funding provided by HUD and ERDA and called for a 7- to 10-MW multiple-train unit to supply power and heat to a large residential housing complex. In order to gain preliminary materials compatibility information, a small scale MIUS fluidized bed was constructed by the Fluidyne Corporation of Minneapolis, Minnesota and a testing program of candidate heat exchanger tube materials was initiated in this unit. These candidate materials included Incoloy 800, Inconel 600, types 304 and 316 stainless steel, and aluminized Incoloy 800 and 310 stainless steel. These candidate materials were exposed to the fluidized-environment at tube wall temperatures ranging from 820 to 875°C. This test unit and candidate materials have successfully completed a 500-hr run and preliminary analysis of these heat exchanger materials has been completed. Although analysis of these materials is not complete, some significant trends have been noted.

First, all tubes evaluated had a 25 to 50-μm-thick (1- to 2-mil) scale of $CaSO_4$ on the fire side. The presence of $CaSO_4$ scale indicates that bed material is depositing on the tube under operating conditions. Furthermore, a thin corrosion scale was observed below the $CaSO_4$ layer. Microprobe data indicated that this corrosion product was the result of oxidation, not sulfidation.

Secondly, no measureable corrosion rate was observed for any of the materials evaluated. However, in view of the short exposure time (500 hr) and measurement errors associated with wall thickness determination, linearly extrapolated corrosion rates less than 0.25 mm/year (10 mils/year) could not be resolved.

Third, results from the analysis of aluminized alloy 800 and type 310 stainless steel were not encouraging. These analyses revealed that the 310 and 800 materials had extensive cracking and internal attack in respective aluminized layers. In addition, the aluminized coating on the air side of the 310 materials was delaminated.

---

*Research sponsored by the Energy Research and Development Administration under contract with Union Carbide Corporation.

In general, the analysis has shown no "runaway" corrosion conditions for the combination of materials and bed conditions evaluated. However, because of short exposure time of this test, no extrapolation can be made regarding the successful applications of these test materials as heat exchanger tubing in a long-life commercial application. Clearly, long-term test data are needed to reliably answer this question.

INTRODUCTION

The Eenrgy Division at ORNL is planning the construction of a fluidized-bed coal combustor pilot plant at ORNL. This system (MIUS — Modular Integrated Utility System) was initially designed under joint funding provided by HUD and ERDA and called for a 7- to 10-MW multiple-train unit to supply power and heat to a large residential housing complex. The proposed single-train pilot plant will be used to study the performance of the fluidized-bed combustor design.

During the design of this system, the Metals and Ceramics Division of ORNL was aksed to identify candidate materials for use as the heat-exchanger tubing. The heat-exchanger tubes in a commercial system will be expected to last 20 years while operating at a maximum wall temperature of 856°C (1575°F) and carrying hoop stresses from 14 to 18 MPa (2 to 2.5 ksi). The tubing will contain air at temperatures ranging to 842°C (1550°F) and the outside surface will be exposed to the 899°C (1650°F) fluidized coal bed environment. These are obviously challenging and unknown conditions for any engineering material. To gain preliminary data, a small scale MIUS fluidized bed was constructed by the Fluidyne Corporation of Minneapolis, and a surveillance testing program of candidate heat exchanger tube material was initiated in this unit. This report summarizes the preliminary results of this surveillance testing program.

DESIGN AND OPERATION OF TEST FACILITY

The combustor used in this test was an atmospheric fluidized bed with an active fluidizing area of 0.209 m^2 (2.25 ft^2). (A schematic drawing is included as Fig. 1.) Temperature of the bed was controlled with an array of 40 heat exchanger tubes. These included two water-cooled tubes for coarse temperature control and 22 air-cooled trimmer tubes for fine control. The remaining 16 air-cooled sample tubes were assigned specific locations within the bed cross section (see Fig. 2). Each test tube was constructed of three spool pieces, to provide a total of 48 individual test conditions. Materials used as spool pieces were Incoloy 800H, Inconel 600, and stainless steel types 310H, 304, and 316. These materials were chosen as representing the range of high-temperature strength properties and corrosion resistance that could be expected from economically attractive heat exchanger alloys. The location of the 16 test tubes in the fluidized bed combustor is shown in Fig. 2, while Table 2 identifies the alloy used for each of the 48 spool pieces.

The fluidized bed operated with these test tubes for 500 hr at 890 ± 2°C (1634°F) with Illinois No. 6 coal and limestone (Table 2 lists the composition of the coal). Coal and limestone were bottom fed into the combustor at the rate of 22 and 13.5 kg/hr, respectively. With the Illinois No. 6 coal a Ca/S ratio of approximately 5.5 was maintained. The bed was fluidized with combustor air having a superficial velocity of 0.76 m/s. The oxygen content of the flue gas was constantly monitored and was used to adjust the coal feed rate to maintain a constant excess oxygen level in the flue gas of 2.5%. Additional analysis of the flue gas indicated that the average SO_2 content was 250—400 ppm, representing a sulfur capture ratio of 0.88.

Initial design of this system called for the wall temperature of the test tubes to be approximately 856°C (1575°F) during steady-state conditions. During operation the inside wall temperature of each test tube was monitored at the entrance, midsection, and exit (see Fig. 3). Monitoring of these thermocouples throughout the 500-hr test showed the average inlet and outlet temperatures for the 16 test tubes to be approximately 842 and 876°C (1550 and 1610°F), respectively.

ANALYSIS OF SURVEILLANCE COUPONS

In November 1976, the surveillance test tubes were removed from the Fluidyne bed after 500 hr of operation and shipped to ORNL. The exterior surfaces of the test tubes were covered with a straw-colored deposit. The scale adherence to the tube surfaces varies for each alloy; however, the scale appeared to be the most tenacious on the alloy 600 sections. The tubes as received from Fluidyne are shown in Figs. 4 and 5. Samples of this scale have been removed from the tube surfaces. A summary of a semiquantitative analysis of this scale is as follows:

Chemical Analysis of Scale

Element	wt %	Element	wt %
Al	4.6	K	0.03
Ca	>60	Na	0.03
Cr	1.0	Mo	0.02
Fe	11.0	Ni	0.3
Si	5.0	V	0.04

Additional analysis indicates the scale is made up primarily of $CaSO_4$, suggesting that bed material is depositing on the tubes during operation. More importantly, the presence of this scale is strong evidence that the tube material has not been eroded by bed material.

Initial test plans called for extensive eddy-current inspection of all test tubes and destructive examination of four test tubes after the first 500-hr exposure. Replacements for the destructively tested tubes and the original 12 tubes would later be reinserted in the fluidized bed for an additional 500-hr exposure.

The original eddy-current measurements of scale and tube wall thickness were to be taken on all 16 tubes removed from the bed. Similar measurements were to be made metallographically on the four destructively evaluated tubes. We anticipated that this information could be used to estimate scale and wall thicknesses for the remaining 12 tubes before their reinsertion into the fluidized bed. Technical problems in developing and demonstrating the eddy-current test equipment delayed the destructive evaluation of four tubes until March 1977, at which time test tubes 2-C, 9-D, 9-B, and 9-A were sectioned as shown in Fig. 6. These test sections plus samples of as-received alloy 800, alloy 600, and types 304 and 310 stainless steel were metallographically examined. Surface scale thickness, depth of intergranular attack, and wall thickness were measured for each material tested, and the results are summarized in Table 3 (also see Figs. 7–16). In addition, representative samples of each material were also tested with the microscope (see Figs. 17 and 18).

DISCUSSION OF RESULTS

Since analysis of these test materials is continuing, an in-depth assessment of the response of these materials to the fluidized-bed environment is not feasible at this time. Despite these limitations the following paragraphs will summarize the information available at this time on each of the material evaluated.

Incoloy 800. Intergranular attack 4 to 6 × 10 mm deep was noted on both the inside and outside contours of this alloy (Fig. 7). As a result, metallographic analysis of an as-received portion of the alloy 800 material was initiated (see Fig. 16). Analysis of this information revealed no conditions that would leave this material vulnerable to intergranular attack.

Microprobe analysis of areas of intergranular attack on the fire side of the alloy 800 material indicated that these areas were richer in Cr and Al than the matrix material. Note that Incoloy 800 exposed to impure helium environments have incompletely developed a protective surface corrosion scale, and Cr, Al, and Ti oxidized along subsurface grain boundaries.[1] Although there are insufficient data to prove that this mechanism applies to alloy 800 in the fluidized bed, the low oxygen potential associated with the fluidized environment and the microstructural evidence suggest that additional analysis is warranted.

Additional microprobe analysis, Fig. 18, revealed a uniform deposit of $CaSO_4$ on the surface of the Incoloy 800 material. A partially protective scale made of two layers—Fe-Cr oxide and Cr oxide—was identified between the $CaSO_4$ deposit and the base material. It should be noted that microprobe data showed no sign of sulfidation attack. Furthermore, wall thickness measurements showed no measurable metal loss.

Review of the inside contour of the Incoloy 800 samples showed significant intergranular attack. Exposure of this alloy to air at tube temperatures from 840 to 875°C would not have been expected to produce this type attack. A subsequent review of the cooling system indicated that cooling air was not filtered after being compressed and was apparently contaminated with compressor lubricant.

Type 304 Stainless Steel. Analysis of this material reveals considerable roughness of the outside surface, which could be the result of surface wastage (Fig. 8). However, within the limits of nominal wall thickness variation, comparison of wall thickness measurements of as-received and exposed material does not reveal measurable corrosion attack. Again, microprobe data revealed no indication of sulfidation attack; however, a duplex corrosion product was observed. The first layer was an Fe and Cr oxide over a second layer of Cr and Mn oxide.

Type 310 Stainless Steel. It is important to note that little or no intergranular attack was observed in this material (Fig. 10). Again surface roughness was noted on the outside contours; however, no measurable metal loss was observed. Microprobe analysis (Fig. 17) revealed a duplex corrosion product of Fe and Cr oxide over a Cr-rich scale.

Inconel 600. The outside surface of the alloy 600 had the thickest deposit of $CaSO_4$ of any of the materials evaluated (Fig. 9); however, only a thin layer of Cr_2O_3 was observed as a corrosion product. It should be emphasized that the surface deposit of the alloy 600 was continuous and showed no signs of exfoliation of $CaSO_4$ and/or corrosion product. The small quantity of corrosion product observed and the presence of a tightly adherent deposit suggest that alloy 600 had resisted this environment better than any of the other materials in this surveillance test. Obviously longer term data will be needed to substantiate this observation.

Alonized Type 310 Stainless Steel. In general, the response of the alonized material did not appear satisfactory (Fig. 11). Extensive cracking through the alonized coating was observed on the contour exposed to the fluidized bed. In addition, the alonized coating on the air side had delaminated.

Alonized Alloy 800. Significant internal attack had occurred in the coating (Fig. 12). One would anticipate that continued internal attack would lead to delamination of a significant portion of this alonized layer.

As indicated earlier, numerous wall thickness measurements were made metallographically on both surveillance samples and as-received tube materials. It was anticipated that a comparison of these wall thickness measurements could be used to determine the loss of metal resulting from exposure to the fluidized bed environment.

A review of the results of these wall thickness measurements (Table 3) indicates that in each case there is no statistically significant difference in wall thickness between as-received and exposed material. Although this observation is encouraging, one should note that the combination of short fluidized-bed exposure time and measurement errors associated with wall thickness determinations limited the minimum detectable linear corrosion rate to approximately 0.25 mm/year (10 mils/year).

CONCLUSION

As indicated earlier, analysis of these surveillance coupons is continuing; therefore, any observations made at this time should be considered preliminary. Nevertheless, some significant trends should be noted.

First, all tubes evaluated in this program had a 25 to 50 mm-thick (1–2 mil) deposit of $CaSO_4$ on the outside. The presence of $CaSO_4$ scale indicates that bed material is depositing on the tube under operating conditions. The presence of this scale suggests that erosive wear of these tubes by bed material is not a significant problem. It should also be noted that despite the presence of the tenacious $CaSO_4$ coating on the tubes, microprobe data indicate that corrosion is the result of oxidation, not sulfidation, as might be expected.[2]

Second, it must be emphasized that these surveillance samples were exposed for only 500 hr. In view of the metallographic measurement errors associated with the wall thickness determination, deterioration resulting from corrosion rates of 0.25 mm/year (10 mil/year) and less could not be identified. Recognizing that a tube life of 20 years is desired in commercial applications, capability of determining corrosion rates ranging from 25 to 130 mm/year (1–5 mils/year) is needed. Therefore, long-term surveillance tests will be needed to resolve this problem.

Third, results from the analysis of the alonized alloy 800 and type 310 stainless steel samples were not encouraging. The materials were extensively cracked and attacked internally in the alonized coating exposed to the fluidized bed. In addition, the alonized coating on the air side of the type 310 stainless steel was delaminated.

In general, the analysis has shown that there are no "runaway" corrosion conditions for the combination of materials and bed conditions evaluated. However, the short exposure time of this test precluded extrapolation regarding the successful application of these test materials as heat exchanger tubing in a long-life commercial application. Clearly long-term test data are needed to reliably answer this question.

REFERENCES

1. H.G.A. Bates, W. Betteridge, R. H. Cook, L. W. Graham, and D. F. Lupton, "The Influence of Impure Helium on Properties of Some Austenitic Steels," DP-Report-934 (May 1975).

2. R. D. LaNauze, J. C. Holder, and E. A. Rogers, "Investigation into Corrosion in Fluidized Bed Combustion Systems," paper given at the Conference on Corrosion and Deposition in Power Plants — June 1–2, 1977, ESSEN.

ACKNOLWEDGMENTS

The authors wish to acknowledge the contribution of M. D. Allen for his assistance in the metallographic analysis and Regina Collins for her assistance in preparing the manuscript.

Table 1. Test Matrix for Surveillance Testing in the Fluidized Bed Combustion Unit Built by Fluidyne Engineering Corporation

Row	Column	Position on Tube		
		a	b	c
2	C	Alloy 800	Alloy 800 Alonized	Alloy 800
5	D	Alloy 800	Alloy 800 Alonized	Alloy 800
8	D	310 SS	310 SS Alonized	310 SS
9	D	Alloy 800	310 SS	Alloy 600
3	C	310 SS	310 SS	310 SS
3	B	316 SS	316 SS	316 SS
4	C	316 SS	316 SS	316 SS
6	C	316 SS	316 SS	316 SS
7	B	304 SS	Alloy 600	Alloy 800
8	C	304 SS	Alloy 600	Alloy 800
9	C	304 SS	Alloy 600	Alloy 800
9	B	310 SS	310 SS Alonized	310 SS
2	B	Alloy 800	Alloy 800 Alonized	Alloy 800
6	B	310 SS	310 SS Alonized	310 SS
4	B	Alloy 800	Alloy 800 Alonized	Alloy 800
9	A	304 SS	Alloy 600	Alloy 800

Table 2. Composition and Properties of Illinois No. 6 Coal Used in Fluidyne Test

Ultimate Analysis	As Received	Dry
Moisture	10.41	
Carbon	60.20	67.20
Hydrogen	4.10	4.58
Nitrogen	0.61	0.68
Chlorine	0.01	0.01
Sulfur	3.42	3.82
Ash	13.55	15.12
Oxygen (by difference)	7.70	8.59
	100.00	100.00

Table 3. Summary of Metallographic Analysis of Tube Materials Removed From Fluidyne Surveillance Test

Material	Air Side		Coal Side		Wall Thickness, mm	
	Scale (μm)	Intergranular Attack (μm)	Deposit (μm)	Intergranular Attack (μm)	As Exposed	As Received
Alloy 800	6	4.5	3.5	3	1.88	1.87
304	15	0	5.0	0	1.70	1.69
Alloy 600	5	2.5	5.0	0	1.78	1.77
310	7.5	0.57	2.5	0	1.65	1.66
Alloy 800 Alonized	4.5	10.00[a]	6.2	13[a]	1.93	
310 Alonized	5.0	10.2[a]	7.5	20[a]	1.73	

[a]Depth of alonized layer.

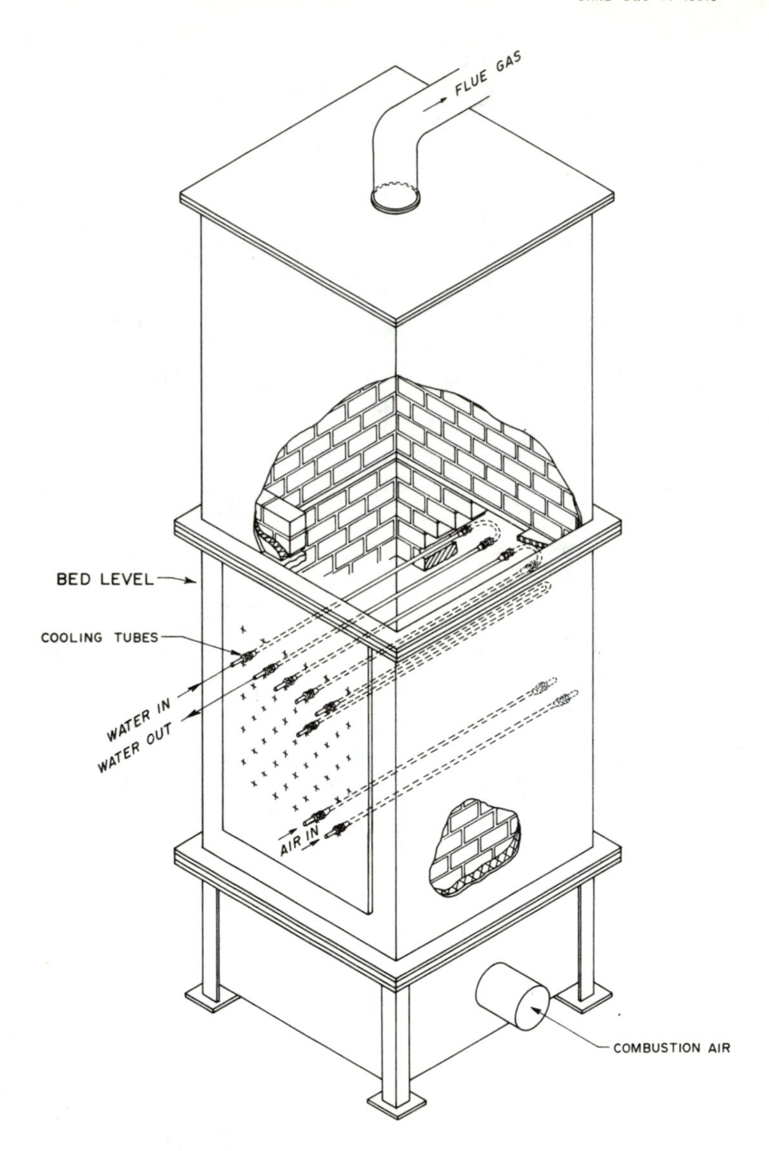

Fig. 1. The Fluidized Bed Built by Fluidyne, Showing the Location of the Cooling Coils in the Bed.

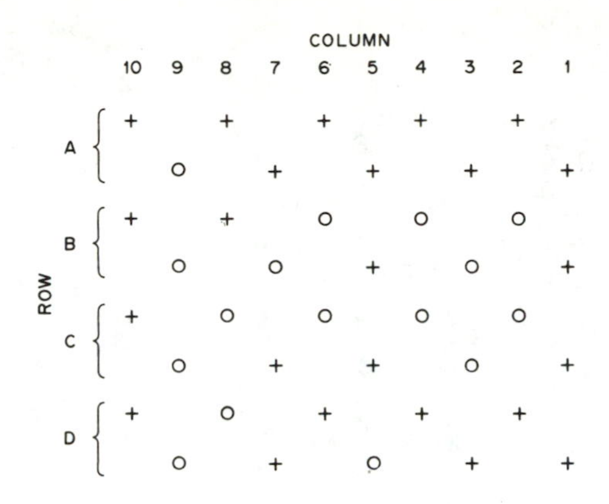

Fig. 2. Tube Array in the 0.46 by 0.46 m (18 × 18 in.) Fluidized-Bed Combustor Built by Fluidyne Engineering Corporation.

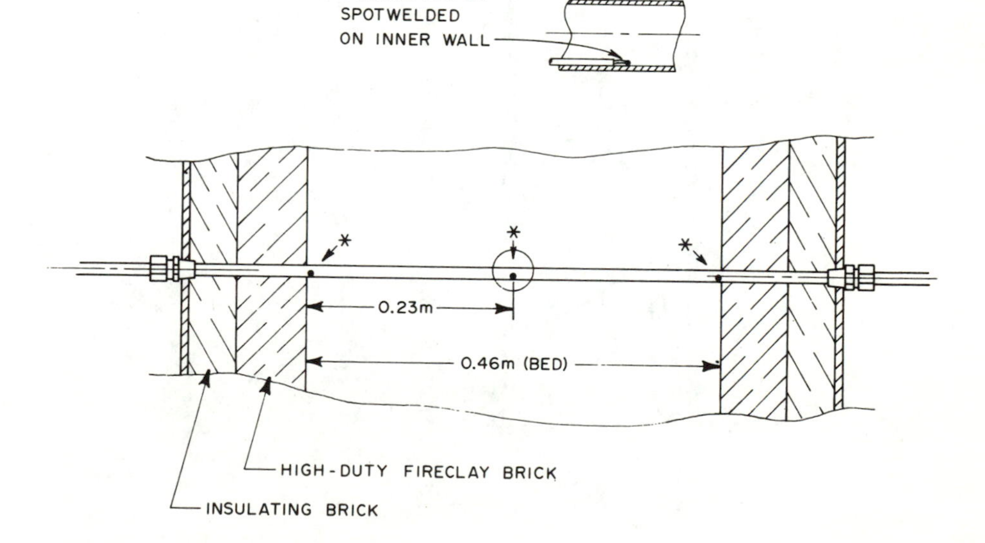

Fig. 3. ORNL Tube Installation in Combustor and Thermocouple Locations.

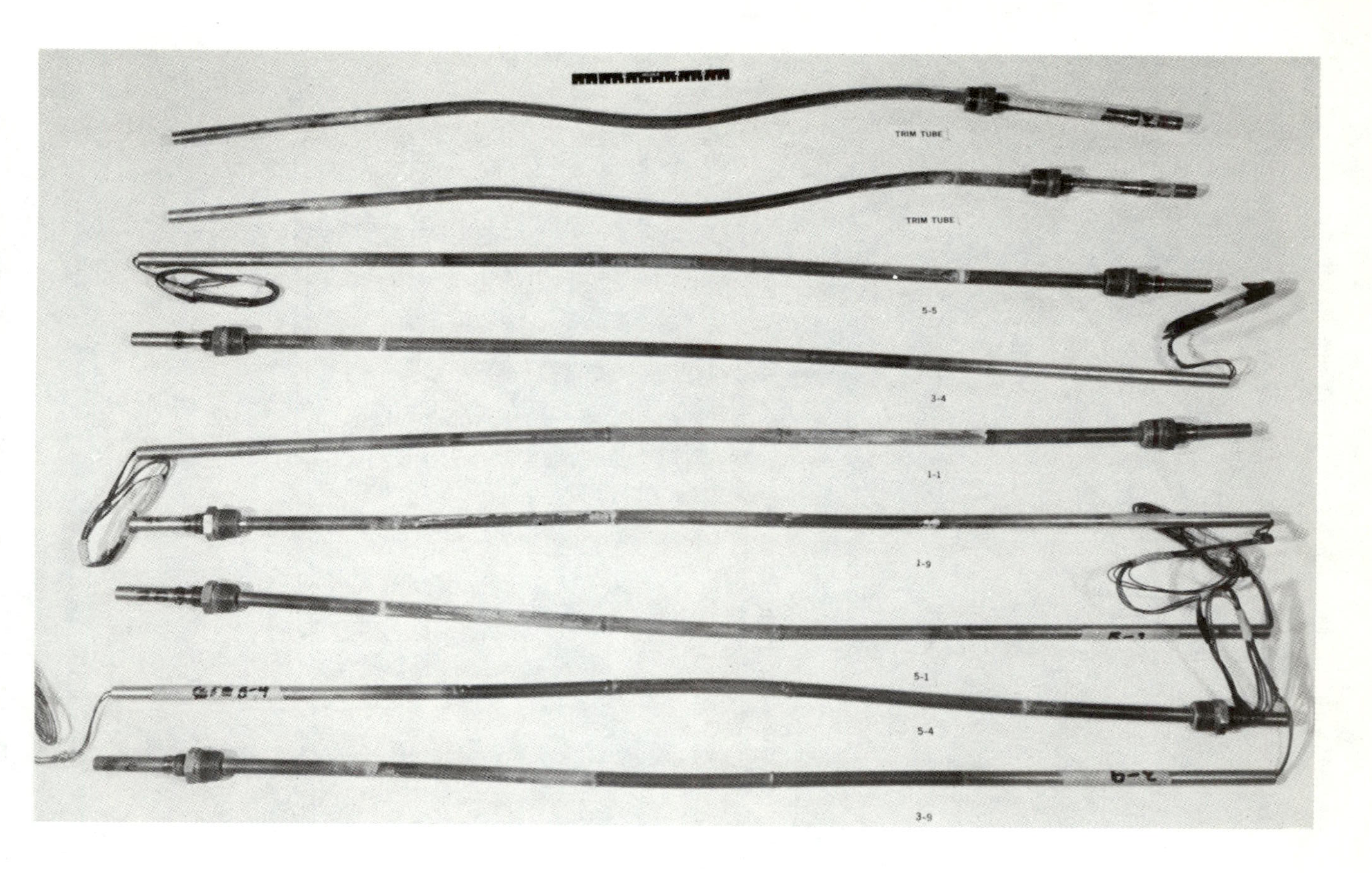

Fig. 4. Seven Surveillance Tubes and Two Trim Tubes.

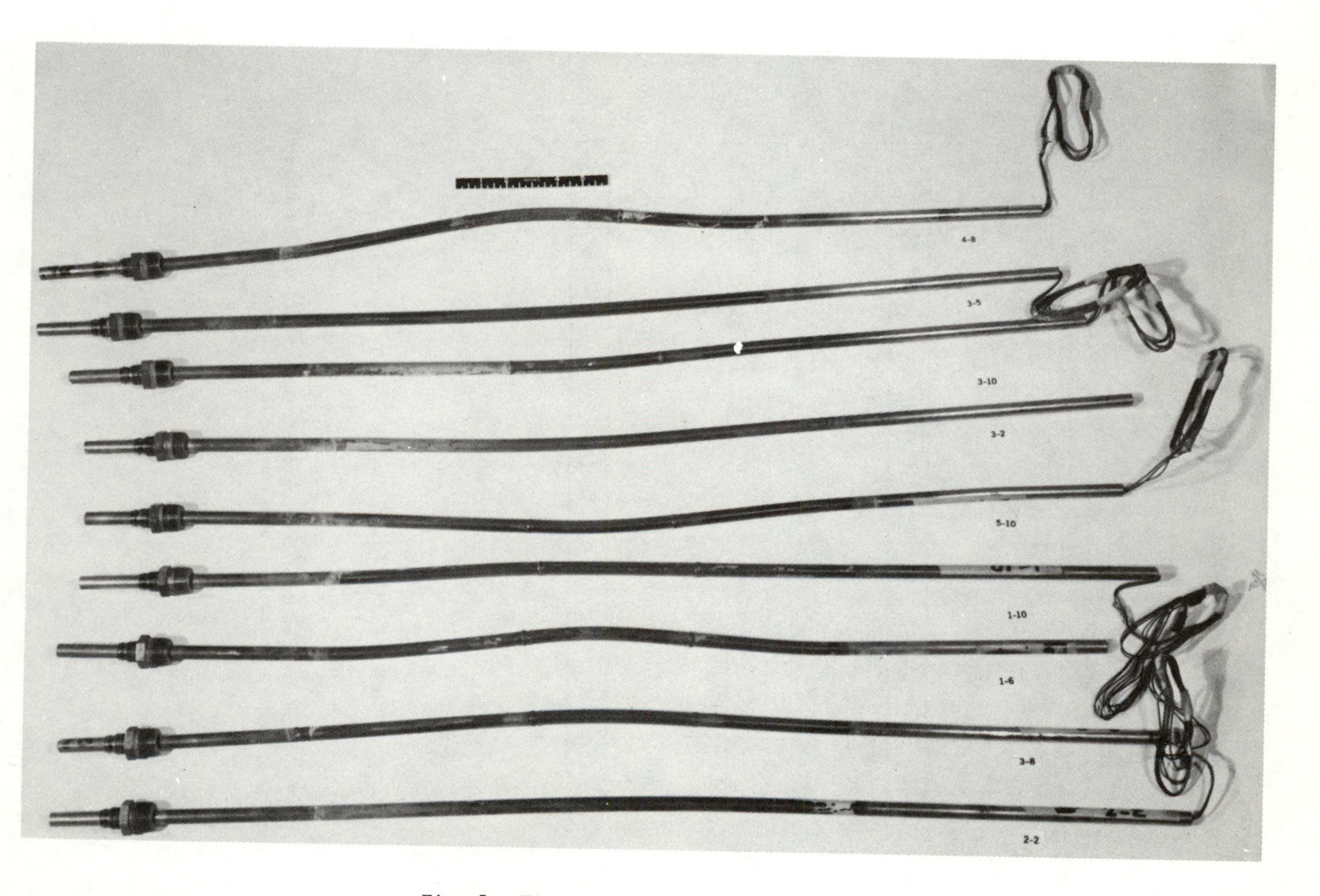

Fig. 5. Nine Surveillance Tubes.

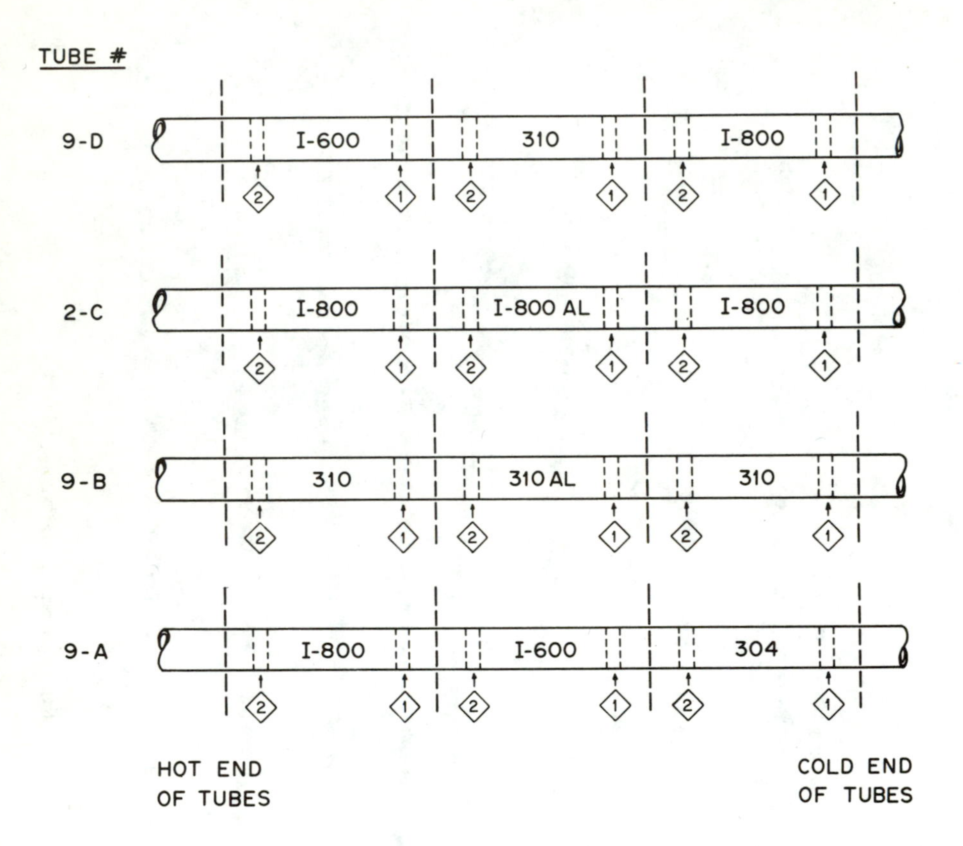

Fig. 6. Location of Metallographic Samples and Identification of Surveillance Tubes Exposed to Fluidized-Bed Coal Combustor.

Fig. 7. Cross Section of an Incoloy 800 Sample Taken from the Cold End of Tube 1-1. Note the intergranular attack associated with both the interior and exterior contours. 40×. As polished.

Fig. 8. Cross Section of Type 304 Stainless Steel Sample Taken from the Cold End of Tube 5-10. Note considerable surface roughness on the outside surface of the sample. This could be the result of surface wastage. 40×. As polished.

Fig. 9. Cross Section of an Inconel 600 Sample Taken from the Hot End of Tube 1-10. Note considerable surface roughness on the outside of this sample. 40×. As polished.

Fig. 10. Cross Section of Type 310 Stainless Steel Sample Taken from the Middle of Tube 1-10. Note minimal intergranular attack; however, considerable surface roughness is observable on the outside contour. 40×. As polished.

Fig. 11. Cross Section of an Alonized Type 310 Stainless Steel Sample Taken from the Middle of Tube 4-8. Note the delamination of a portion of the Alonized layer on this interior contour. Also note the extensive attack on a portion of the Alonized layer on the exterior contour. 40×. As polished.

Fig. 12. Cross Section of an Alonized Incoloy 800 Sample Taken from the Middle of Tube 1.1. Note the considerable internal attack in the Alonized layer. 40×. As polished.

Fig. 13. Cross Section of the Outside Contour of As-Received Type 310 Stainless Steel Tube Samples. 100×. Etched.

Fig. 14. Cross Section of the Outside Contour of As-Received Type 304 Stainless Steel Tube Samples. 100×. Etched.

Fig. 15. Cross Section of the Outside Contour of As-Received Inconel 600 Tube Samples. 100×. Etched.

Fig. 16. Cross Section of the Outside Contour of As-Received Incoloy 800 Tube Samples. 100×. Etched.

BACKSCATTERED ELECTRON IMAGE 500X

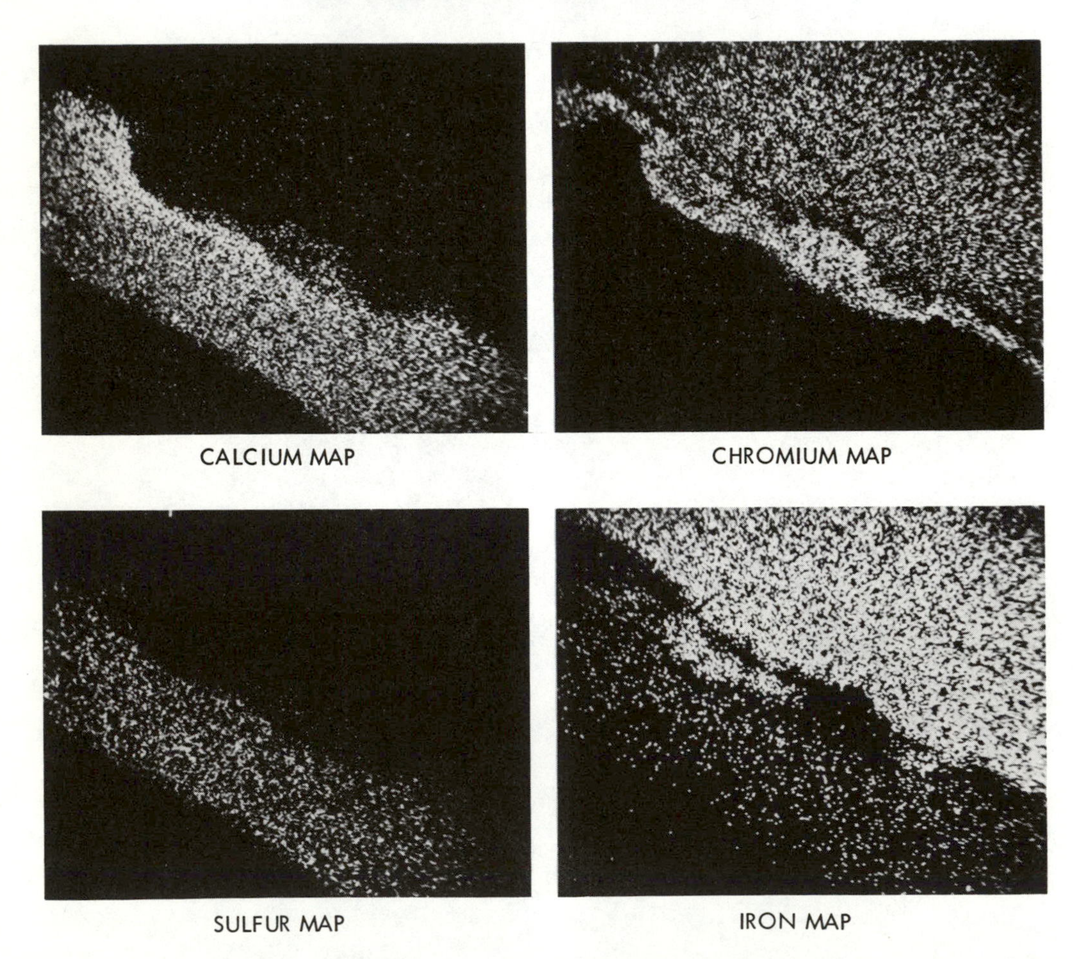

Fig. 17. Summary of Microprobe Analysis of 310 Stainless Steel.

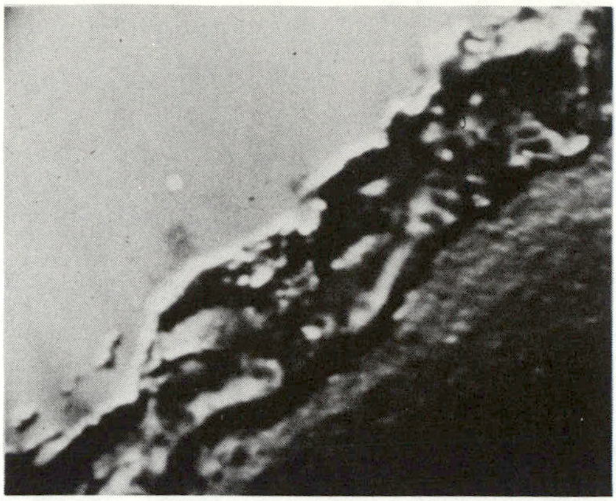

BACKSCATTERED ELECTRON IMAGE 500X

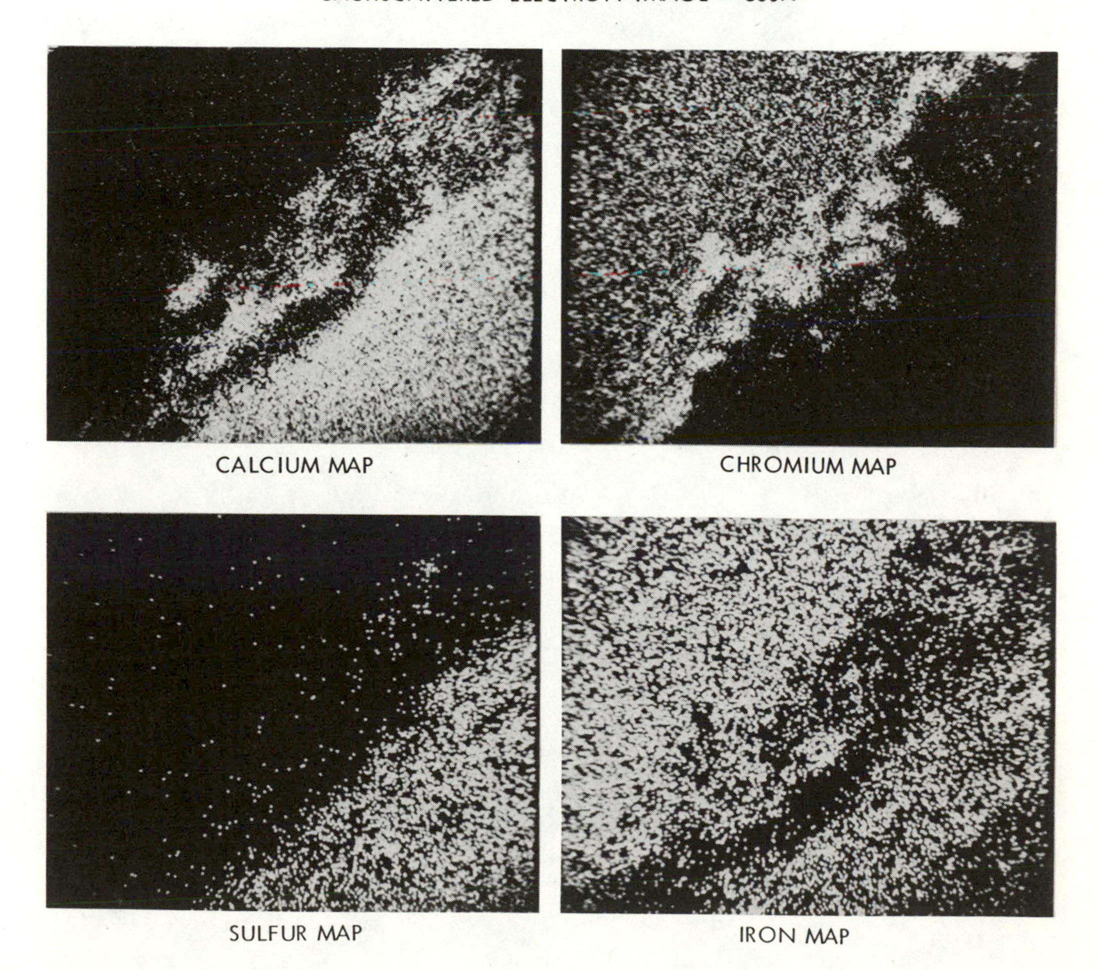

Fig. 18. Summary of Microprobe Analysis of Incoloy 800.

MATERIALS TESTS FOR AN INDIRECT AIR COOLED PFBC COMBINED CYCLE POWER PLANT

S. M. WOLOSIN

Curtiss-Wright Corporation
Wood-Ridge, New Jersey, USA 07075

ABSTRACT

This paper reports the results of a laboratory test program to evaluate alloys and coatings for use in a high sulfur coal fired PFB combined cycle system. Erosion testing at 871°C on nickel and cobalt base alloys and coatings indicated erosion of vertically oriented heat exchanger tubes in a PFB combustor will not be significant at low (0.82 m/sec) bed velocities. Combined hot corrosion erosion testing at 871°C on nickel and cobalt base turbine alloys and coatings revealed that

a. Cobalt base super alloys exhibit better resistance against hot corrosion-erosion than nickel base superalloys

b. Coatings provide excellent corrosion-erosion resistance in a particulate laden high velocity (91.4m/sec) gas stream.

1.0 INTRODUCTION

Pressurized-fluidized-bed combustion (PFBC) of coal in an integrated combined-cycle power plant is potentially an attractive means of producing clean, cost competitive electric power. Curtiss-Wright, under ERDA sponsored contract, is engaged in a program to design, construct, and operate a pilot electric power plant to evaluate the concept for use in the electric utility industry. The pilot plant will be constructed around an existing Total Energy Power Generating System at the Curtiss-Wright, Wood-Ridge Facility, and will have a PFBC with a bed diameter of about 3.8 meters and a bed depth of 4.9 meters. Plans are that it will be operational early in 1980.

High sulfur coal will be burned in the PFBC which will contain dolomite as a sulfur sorbent. The bed temperature will be maintained at 900°C by an air cooled heat exchanger within the bed. A gas turbine compressor will supply air to the bed for coal combustion as well as air to the heat exchanger located within the bed. One-third of the compressor discharge air will be used for coal combustion while two-thirds will flow through the heat exchanger to control bed temperature.

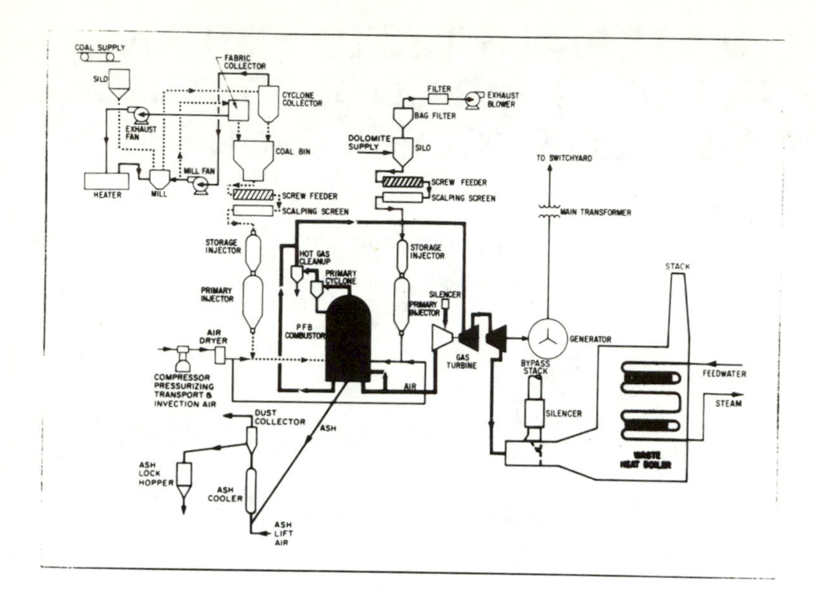

Figure 1. Combined Cycle Pilot Plant
Simplified Flow Diagram

The combustion gas stream will be cleaned of harmful particulate prior to being combined with the clean heat exchanger air stream for expansion through a gas turbine. Gases exiting the power turbine will be cycled through a conventional steam system for additional power generation.

A flow diagram of this system can be seen in Figure 1.

The five year, multi-phase program will address to a number of key technical issues, not the least of which is the durability of structural materials, both within the bed and within the gas turbine.

There is no doubt that the two key materials areas in the system involve the durability of the in-bed heat exchanger tubes and the gas turbine vanes and blades.

Because of the low fluidizing and particle velocities, and the vertical orientation of the tubes resulting in low impact angles, tube metal erosion is not believed to be a significant problem. Corrosion of the tubes, which will operate at or near bed temperature, is another matter. Localized conditions which may alternate between relatively oxidizing and relatively reducing or even carburizing conditions are a severe environment for most materials. Also, there is the concern posed by trace amounts of sodium and potassium salts in the coal and dolomite sulfur sorbent. Since metal temperatures are in the range where hot corrosion from these alkali metal sulfates may occur, test evaluations are required.

Even though the hot gases entering the turbine will be significantly diluted by bed coolant air, the anticipated particulate and alkali-metal content will probably be higher than similar turbines operating on petroleum fuels. Erosion and corrosion in the turbine is therefore another area requiring test evaluation.

This paper will report on two material evaluation test programs carried out in the laboratory to evaluate both turbine and heat exchanger alloy/coating systems for application in the PFB system.

2.0 TURBINE MATERIALS TEST PROGRAM

The turbine alloy evaluation was conducted in an aggressive simulated PFB gaseous environment utilizing a high temperature environmental test rig. Rod specimens 7.9 mm diameter x 83 mm long were utilized as test specimens. Alloys and coatings currently utilized in the PFB turbine as well as potential replacement alloys and coatings were included in the test program.

The turbine alloys tested were IN-738, U-500, U-700, U-710, IN-792, HS-31 and FSX-414. Coatings tested included aluminide, platinum aluminide, Co-Cr-Al-Y, Ni-Cr-Al-Y and Co-Cr-Ni-W systems. Coatings were applied by pack cementation, electrodeposition, electron beam vapor deposition and high velocity plasma spray processes.

Table 1 lists the compositions of the alloys and coatings tested.

TABLE 1

COMPOSITION OF ALLOYS & COATINGS TESTED IN TURBINE MATERIALS EVALUATION

Alloys	Cr	Ni	Co	Mo	W	Cb	Ti	Al	Fe	C	Mn	Si	B	Other
U-500	19.0	Bal.	18.0	4.0	-	-	2.9	2.9	4.0[1]	0.08	0.75[1]	0.75[1]	0.005	-
U-700	15.0	Bal.	18.5	5.2	-	-	3.5	4.3	1.0[1]	0.15[1]	-	-	0.05[1]	-
U-710	18.0	Bal.	15.0	3.0	1.5	-	5.0	2.5	-	0.07	-	-	0.20	-
IN-738	16.0	Bal.	8.5	1.7	2.6	0.9	3.4	3.4	0.5[1]	0.17	0.20[1]	0.30[1]	0.01	1.7 Ta
IN-792	12.7	Bal.	9.0	2.0	3.9	-	4.2	3.2	-	0.21	-	-	0.02	3.9 Ta
S-31	25.5	10.5	Bal.	-	7.5	-	-	-	2.0	0.50	1.0	0.50	0.01	-
FSX-414	29.5	10.5	Bal.	-	7.0	-	-	-	2.0	0.35	1.0	-	0.01	-
Coatings														
CW-3	-	-	-	-	-	-	-	100	-	-	-	-	-	-
RT-22	-	-	-	-	-	-	-	95	-	-	-	-	-	5 Pt.
ATD-1	35.3	Bal.	-	-	-	-	-	12.7	-	-	-	-	-	0.33Y
ATD-2A	21.9	-	Bal.	-	-	-	-	11.7	-	-	-	-	-	0.26Y
P-15[2]	X	-	X	-	-	-	-	X	-	-	-	-	-	X
P-15A[2]	X	-	X	-	-	-	-	X	-	-	-	-	-	X
45VF	25.5	10.5	Bal.	-	7.5	-	-	-	-	0.5	-	-	-	-

[1] Maximum

[2] Curtiss-Wright Proprietary Coating System

Figure 2. Hot Corrosion/Erosion Rig

Testing was carried out in a hot corrosion (sulfidation) erosion environment at 871°C utilizing a burner rig. The sulfidation environment was created through the introduction of high levels of sodium and potassium into combustion gases from a sulfur doped jet fuel. An erosion environment was created by introducing abrasive particulate into the rig heating chamber during elevated temperature exposure. Figure 2 is a view of the hot corrosion test rig showing the specimens, heating chamber and abrasive unit used to introduce abrasive particulate.

The particulate impinged upon the samples at a location 25 mm below the tip of the specimens causing metal erosion over an area 3 mm to 6.4 mm wide.

An SS-White Airbrasive/Unit, Laboratory Model K, was utilized to introduce the abrasive into the rig heating chamber. This unit is capable of delivering a gas propelled stream of finely graded abrasive particles against the test specimens. The abrasive was fed through a 3.2 mm diameter tube positioned five cm from the test specimens and impinged upon the specimens at a 90° angle. Prior to installing the abrasive unit into the hot corrosion rig, the particle velocity of the abrasive was determined at 621 kPa air pressure. Velocities were measured by the rotating disk method. Two parallel disks fixed to a shaft at a known separation are rotated at a constant speed. The disks are held in the particle stream with the direction of particle travel normal to the disks. Particles pass through a radial slot in the first disk to impinge on the second disk. An eroded image of the slot in the first disk is produced on the second disk which is displaced by an amount that depends on the velocity of the particles, the spacing between the disks, and the speed of rotation of the disks. The maximum velocity of the particles in the stream can then be determined by measuring the maximum angular displacement of the erosion image on the second disk.

Velocity calculations carried out using this procedure indicated a particle velocity of 76.2 m/sec.

A temperature profile of the test specimen was determined prior to testing by inserting a thermocouple through an axial hole in the center of a specimen and measuring the temperature at 6.4 mm intervals along the length of the specimen while the specimen was rotating in the combustion flame. A slip ring assembly was utilized to transmit the temperature to a recorder.

Although a temperature gradient existed along the length of the test specimen, the temperature 6.4 mm above the erosion zone was found to be the same as the specimen temperature at the erosion zone. Metallographic measurements taken at these two locations permitted corrosion effects and corrosion-erosion effects to be determined on each specimen after exposure.

Experiments were carried out utilizing ten micron alumina, ten micron silica and three micron silica as the abrasive particulate. Particulate flux levels were varied between .13 and .34 $mg/mm^2/sec$. An aqueous solution of sodium and potassium chloride was metered into the combustion gas stream to achieve concentration levels noted in Table 2. A petro-sulfur compound was added to the JP-5 fuel to achieve the sulfur concentration in the gas stream.

Table 2 contains a listing of the conditions of all the tests conducted.

TABLE 2

TURBINE MATERIALS EVALUATION PROGRAM

TEST ENVIRONMENT

	Rig	Expected PFB Turbine
Metal Temperature	871°C	871°C
Particle Size	10 Microns 3 Microns	$<$ 2 Microns
Particulate Composition	Al_2O_3 SiO_2	Mixture-SiO_2, Al_2O_3, MgO, CaO, $CaSO_4$, Fe_2O_3, C
L.E. Impact Velocity	91.4 M/Sec	$\sim$183-274 M/Sec
Particulate Flux	.13-34 $MG/MM^2/Sec$	$\sim$.01 $MG/MM^2/Sec$
Alkali Metal Content		
Sodium	5 PPM	$<$1.5 PPM
Potassium	2 PPM	$<$0.5 PPM
Sulfur	.015	.025%

After test exposure specimens were subjected to both metallographic examination and micrometer measurements to determine the extent of corrosion-erosion damage. These measurements allowed the alloys to be ranked in order of decreasing corrosion-erosion resistance.

The alloys were found to exhibit resistance to corrosion damage in agreement with results obtained in previous Curtiss-Wright and other investigations. The cobalt base alloys, Stellite-31 and FSX-414 exhibited the best resistance to hot corrosion while Udimet-700 experienced the greatest amount of attack. The order of the alloys in terms of decreasing resistance to hot corrosion between the extremes were Udimet 710, IN-738, U-500 and Inco 792.

The effect of the erosive particulate resulted in metal loss for many of the specimens. The nickel base alloys were found to erode at the highest rates while the aluminide coatings and the overlay vapor deposited coatings eroded at the lowest rates.

The cobalt base alloys along with the plasma spray coatings were found to erode at intermediate rates.

Table 3 is a ranking of the relative resistance of the alloys and coatings evaluated in this program. Table 4 is a ranking of the relative resistance of the alloys tested. Table 5 summarizes the results of this test program.

TABLE 3

TURBINE MATERIALS EVALUATION PROGRAM

CORROSION/EROSION RESISTANCE OF ALLOYS AND COATINGS

Coating (C) or Alloy (A)	Relative Resistance
RT-22(C)	1.00
CW-3(C)	.74
ATD-1+CW-3(C)	.57
ATD-2(C)	.42
ATD-1(C)	.38
S-31(A)	.22
FSX-414(A)	.20
45VF(C)	.20
P-15A(C)	.15
P-15(C)	.11
IN-738(A)	.09
IN-792(A)	.07
U-710(A)	.07
U-500(A)	.06
U-700(A)	.06

TABLE 4

TURBINE MATERIALS EVALUATION PROGRAM

CORROSION/EROSION RESISTANCE OF ALLOYS

Alloy	Relative Resistance
S-31	1.0
FSX-414	1.0
U-710	.5
IN-738	.4
U-500	.4
IN-792	.4
U-700	.3

TABLE 5

TURBINE MATERIALS EVALUATION PROGRAM

SUMMARY OF RESULTS

Particulate Size

10 μ - Severe Erosion
3 μ - Mild Erosion

Blade/Vane Materials

Cobalt-base more resistant than nickel-base

"Sulfidation-resistant" nickel-base (U-710, IN-738, U-500, IN-792) generally superior to "regular" nickel base (U-700)

Coatings

All coatings improve resistance

Very best RT-22, CW-3, ATD-1 + CW-3, ATD-1, ATD-2

Moderately effective CWP-15A, CWP-15, 45VF

3.0 HEAT EXCHANGER MATERIALS TEST PROGRAM

The heat exchanger tube and fin alloy evaluation was carried out to determine the erosion characteristics of candidate alloys exposed in an erosive and temperature environment simulating the actual PFB Combustor. Testing was carried out in a indirectly heated laboratory sized fluidized bed with alumina as the bed medium. Fluidizing velocity in the bed was set at .82 m/sec simulating the PFB design.

Three 1000 hour tests were conducted. Specimens utilized were 12.7 mm x 50.8 mm x 3.2 mm. In test number 1 candidate nickel and cobalt base tube alloys were exposed to the bed environment. In test number 2, braze areas and weld beads were exposed to the bed environment while in test number 3, samples coated with several types of coatings were subjected to the 1000 hour exposure. Table 6 lists the samples exposed in this test program.

The test specimens were weighed and checked for surface roughness before and after test. Samples were subjected to metallographic examination at 500 and 1000 hour intervals.

The test results revealed that all samples showed a slight degree of oxide buildup after test exposure. In a majority of cases the oxide was 0.003 mm or less in thickness. The surface roughness of the specimens increased from about 20 rms in the unexposed condition to between 80 and 120 rms after exposure. The increase in surface roughness was attributed to oxide formation and spalling when the samples were removed for examination. The test results indicated that metal loss due to erosion would not be significant on vertically oriented heat exchanger tubes in a PFB operating at low fluidizing velocities. The braze joints, weld joints and coated specimens similarly showed no effects of erosion.

TABLE 6

COMPOSITION OF ALLOYS & COATINGS TESTED IN HEAT EXCHANGER MATERIALS EVALUATION

Alloys	Cr	Ni	Co	Mo	W	Cb	Mn	Ta	Al	Fe	B	Y	Other
IN-800	22	33	-	-	-	-	0.8	-	0.4	Bal.	-	-	0.4 Ti
IN-601	23	Bal.	-	-	-	-	0.5	-	1.35	14	-	-	0.2 Si
IN-690	30	Bal.	-	-	-	-	-	-	-	9.5	-	-	
IN-617	22	Bal.	12.5	9	-	-	-	-	1	-	-	-	
H556	22	20	20	3	2.5	0.1	1.5	0.9	0.3	Bal.	-	-	0.02 La, 0.2N
Stellite 6B	30	3m	Bal.	1.5m	4.5	-	2m	-	-	3m	-	-	2Si, 1.2C
H188	22	22	Bal.	-	14	-	1.2m	-	-	1.5	-	-	0.08 La
Weld & Braze Alloys													
Inconel 625	21	Bal.	1.0m	9	-	3.5	.5m	-	.4m	5m	-	-	.4m Ti
Stellite 25	20	10	Bal.	-	15	-	1.5	-	-	-	-	-	
Colmonoy 150	15	Bal.	-	-	-	-	-	-	-	-	3.5	-	
Coatings													
Metco 45VF	25	10	Bal.	-	8	-	-	-	-	-	-	-	
P15*	X	-	X	-	-	-	-	-	X	-	-	-	X
Graded Zirconia	-	-	-	-	-	-	-	-	-	-	-	-	ZrO_2 + $CaCO_3$
ATD-2A	21.9	-	Bal.	-	-	-	-	-	11.7	-	-	0.26	

*Curtiss-Wright Proprietary Coating System

m = maximum

Corrosion conditions in a PFB Combustor burning high sulfur coal and using dolomite as a sulfur sorbent will be different than the oxidizing environment that existed in the laboratory test reported. The purpose of the test was to determine if erosion would significantly contribute to metal loss. Further testing will be carried out on candidate heat exchanger alloys in an actual PFBC rig.

HIGH TEMPERATURE CORROSION IN FLUIDIZED-BED COMBUSTION

JOHN STRINGER

Electric Power Research Institute
Palo Alto, California, USA

ABSTRACT

Because of the low combustion temperature, corrosion of the type observed on superheaters in conventional boilers is unlikely. However, it is shown that within the bed itself, the oxygen and sulfur activities are linked by the CaO/$CaSO_4$ equilibrium, and local low oxygen activities will result in local high sulfur activities. Low oxygen activities (of the order of 10^{-12} atm) have been observed in the bed in one combustor, implying sulfur activities of the order of 10^{-7} atm, sufficient to induce sulfidation of alloy components. Sulfidation/oxidation corrosion of in-bed components has indeed been observed in some cases. It is shown that quite small variations in bed chemistry may move the conditions away from the risk of this form of corrosion. Hot corrosion of turbine components due to the high alkali content of the combustion gases is also considered.

INTRODUCTION

The fireside corrosion of superheaters in conventional coal-fired boilers is associated with the deposition of alkali-rich salts on the metal surface: there is some disagreement about the details of the mechanism, but it seems clear that the salt may become molten beneath a sintered ash deposit. The process requires a high combustion temperature to release a relatively large proportion of the alkali metals in the coal, and a high-temperature gas stream to induce sintering of the ash deposit. It was correctly believed that this form of attack was unlikely in a fluidized-bed combustor.

However, there are two further forms of high temperature corrosion which must be considered. It is expected that the combustion gases from pressurized fluidized-bed combustors will be expanded through gas turbines, and here the possibility of alkali sulfate-induced accelerated oxidation, "hot corrosion," must be considered. Within the beds of both pressurized and atmospheric pressure combustors there is a possibility that regions of low oxygen partial pressure may exist, and these could induce a form of accelerated corrosion called sulfidation/oxidation attack.

The possibility of erosion, or the interaction of erosion and corrosion, must also be considered. As yet, high temperature erosion is not well understood; but for the sort of materials that make up the bed-coal, coal minerals, ash, limestone, calcium sulfate--and the sort of alloys used in the bed, typically, austenitic stainless steels--it seems likely that there is a threshold particle velocity below which erosion does not occur. There is some reason to believe that this velocity is of the order of 8 m/s (24 fps), but experiments now in progress[1] should shortly enable a more precise figure to be determined. Within the bed local particle velocities may be of the order of three times the fluidizing velocity, although under certain circumstances (bubbles flowing up in contact with vertical tubes) higher velocities may be obtained. Consequently, erosion should not be a problem within the bed for fluidizing velocities below 2.7 m/s (8 fps) or so. Indeed, most in-bed tubes appear to be covered with a stable deposit, which may be a thin black laquer in the case of lower temperature beds, or a dark reddish-brown layer up to 1 mm thick at 900 C. The presence of any deposit certainly implies that there is no erosion. In-bed erosion has been observed in the Exxon bed[2], but this is associated with the presence of long bertical tubes: the problem was eliminated by the use of baffles, although as will become clear later the presence of baffles may enhance the risk of sulfidation/oxidation attack.

Particle velocities in the freeboard should not significantly exceed those in the bed, and the particle loading will be lower, so that the risk of erosion should be less: any effect should be beneficial, preventing the accumulation of long-lived deposits on metal surfaces in the freeboard. There will be some acceleration entering the cyclones, but the problems should be no more severe than in other cyclones.

Potentially the most severe erosion conditions should be in the gas turbine following a pressurized fbc since the particle velocities will be of the order of 300 m/s (1000 fps). Erosion was a major problem in earlier studies of coal-burning gas turbines[3]. However, the ash particles from an fbc are very soft and friable, and harder particles such as the limestone and the coal minerals elutriated from the bed should be largely removed by the cyclones. Experiments at the Leatherhead Laboratories of the National Coal Board have shown no signs of erosion at particle loadings and velocities of the order anticipated[4]. However, this may depend initially on such factors as the depth of the bed, the fluidizing velocity and the efficiency of the cyclones. Experiments currently in progress[5] should provide more quantitative information on this aspect.

HOT CORROSION

The topic of hot corrosion in gas turbines has been reviewed at length recently[6]. Sodium salts present in the fuel or ingested in the air are deposited on the vanes and blades: the major constituent of the deposit is sodium sulfate. The most common source of the sodium is sea salt, and the major constituent of this is sodium chloride: in principle, sodium chloride will sulfate under the conditions encountered in the gas turbine, even with low sulfur fuel (0.1% S or so). However, the residence time in turbine is short--some 5 ms between entering the combustion chamber and striking the nozzle guide vanes--and this is generally regarded as too short to allow the sulfation reaction to occur. It is usually considered that the sodium sulfate condenses on the vanes and blades; but again the salt enters the engine as an aerosol, and the residence time is too short for even quite small particles to evaporate. It is possible therefore that the salt particles impact the metal surfaces. Hot corrosion may be encountered at very low salt levels--0.01 ppm sea salt in the intake air is usually specified as a safe limit--but it is possible that the salt is concentrated by being deposited within the compressor, these deposits breaking off from time to time (for example, during start-up).

Although sodium sufate is regarded as the aggressive constituent of the deposit, it typically contains significant amounts of magnesium and calcium sulfates, and sometimes silica as well. The mixed salt has a melting point significantly below that of pure sodium sulfate (880 C), and it is usually considered that it is necessary for the salt to be molten for severe hot corrosion to be encountered: this has not, however, been demonstrated, and for at least one aspect of the attack it would not be essential. In the laboratory it has been shown that quite small amounts of sodium chloride can have a major effect on the corrosion, producing a marked acceleration in attack. Generally, little or no chloride is detected in the deposits, although some is found in engines run in marine environments, but its effect cannot be discounted since sulfation is believed to be incomplete on kinetic grounds.

Simulator rig experiments suggest that there is a threshold temperature below which hot corrosion does not occur and a maximum temperature above which the attack diminishes sharply; and this general concept is supported by studies of blades corroded in operating engines. Originally, it was suggested that the lower threshold temperature was the melting point of the salt mixture, and the upper limit was the dew point of the salt in the atmosphere, but this now seems unlikely: both temperatures depend on the alloy composition. The value of the lower threshold temperature depends somewhat on the conditions: long term testing at low salt concentrations gives a lower threshold temperature than short term tests at higher salt concentrations.

This is probably connected with the kinetics of the corrosion. It seems likely that there is an incubation period during which there is little or no attack: at the end of this period the attack is initiated and subsequently propagates through the alloy. The less severe the conditions, the longer the incubation period; but eventually attack will occur. Some years ago it was generally thought that there would be little attack below 750 C (1382 F), but now it seems probable that the lower limit should be 700 C (1292 F) or so. Again, these values will depend to some extent on the alloy used, and on the detailed environment: the presence of sodium chloride probably lowers the apparent threshold temperature.

The presence of carbon in the deposit, due to incomplete combustion, appears to accelerate the attack. Thermal cycling is also harmful: this is probably because the initiation of the corrosion appears to be associated with the breakdown of the initially-formed protective oxide layer. For the same reason, it would be anticipated that erosive conditions would be deleterious, although this has not yet been demonstrated.

There have been some disagreements about the mechanism of the corrosion, but there now seems to be a greater degree of agreement among the various investigators. An important aspect is the ability of a molten salt layer to flux the protective scale, either as anionic species (basic fluxing) or a cationic species (acidic fluxing). This may be either an initiating step (removing the protective oxide) or a propagating mechanism. A second mechanism involves the sulfidation of the alloy by the salt; the sulfidized alloy then oxidizes rapidly. This is principally a propagation mechanism. It seems probable that both processes may be important under different circumstances.

Generally, high chromium contents increase the resistance of nickel-base alloys to hot corrosion, and this is true whether the alloy forms a Cr_2O_3 or an Al_2O_3 protective oxide. High molybdenum contents (above about 4 wt%) are usually harmful, although Hastelloy X, which contains 9% Mo, has excellent hot corrosion resistance. There is little agreement about the role of other elements. It is probable that higher chromium contents are also beneficial for cobalt-base alloys, but this has not been deomnstrated so clearly because most cobalt-base superalloys have high chromium contents anyway. There is relatively little information on the hot-corrosion of iron-base alloys.

Hot corrosion has been observed in the turbine test section of the pressurized fbc at Leatherhead[7]. Indeed, the aluminized X-40 nozzle guide vane segment used to redicrect the gas flow over the test pins showed very severe attack down the carbide networks, and this is normally regarded as a very resistant alloy: however, the conditions encountered by the vane may have been untypically severe because of the variety of experiments that were being performed on the rig and because of one or two considerable temperature excursions.

Alkali salt contents of 20 ppm in the exhaust gases are quoted, and this is very high. At a bed temperature of 850 C (1562 F) the equilibrium partial pressure of sodium sulfate would be well below 1 ppm[8]. The residence time in the bed is relatively long so any sodium chloride present in the coal should have been sulfated; there are of course other sodium-containing minerals in the coal, and at the low combustion temperature these may not be broken down. However, it seems probable that most of the sodium is present as an aerosol. This is of significance, since at these gas temperatures there is unlikely to be a large temperature difference between the gas and the metal of the first stage vanes and blades, so that condensation of salt is unlikely. However, at these relatively high alkali metal levels the deposition of salts by impaction must be considered.

In coal-fired gas turbines the accretion of material on the vanes has been a problem, and clearly, if there is a deposit, there is a real risk of hot corrosion from the alkali salts.

On the basis of earlier experience, it is probable that high chromium nickel-base alloys, or cobalt-base alloys such as X-40 or FS 414 would be the best materials to choose. Simple aluminization would probably not be very effective, but overlay coatings of the CoCrAlY type might be effective because of their high chromium content. Chromium-rich coatings might also be worth further study.

SULFIDATION/OXIDATION ATTACK

This is, as indicated above, one aspect of the propagation stage of hot corrosion; but it can occur in atmospheres containing sulfur and oxygen in the absence of any condensed salt; investigators have produced similar attack to that found in practice in the laboratory in SO_2/O_2 atmospheres[9] and by oxidizing specimens which had been presulfidized.

The sulfur is present as sulfides in the metal ahead of the oxidation front: the presence of those sulfides appears to prevent the alloy forming a continuous protective Cr_2O_3 or Al_2O_3 scale, and the alloy oxidizes rapidly. The sulfides themselves oxidize, and the sulfur released can either diffuse further into the metal forming sulfides, in which case the attack is self-propagating, or escape through the porous oxide to the atmosphere, in which case the attack is self-limiting, and will slow down and stop unless there is a fresh supply of sulfur.

In many cases the sulfides formed are chromium-rich. Originally, it was believed that the removal of the chromium to form the sulfide would allow the rapid oxidation of the depleted matrix, but in fact it appears that the chromium-rich sulfides themselves oxidize rapidly, forming a layer in which particles of metal depleted in chromium are embedded in a nonprotective chromium oxide matrix. Eventually, towards the outer part of the scale the particles of relatively noble metal oxidize. In other cases, the sulfur activity can rise sufficiently high that sulfides of the more noble metals form, and these may form liquid phases. Thus, the FeS/Fe eutectic is at 988 C (1810 F); the Co_4S_3/Co eutectic is at 877 C (1610 F); and the Ni_3S_2/Ni eutectic is at 645 C (1193 F). The appearance of a liquid phase within the metal appears to result in accelerated breakdown, the liquid phase penetrating rapidly along grain boundaries. However, it must be emphasized that unacceptably rapid corrosion can take place even if the sulfur activity never rises high enough to form these liquid phases.

The important part of this process is that the formation of the sulfide prevents the development of the normal continuous protective oxide layer. It is by no means clear that the collection of (for example) the chromium into a chromium-rich second phase necessarily precludes the formation of a continuous external Cr_2O_3 layer: El Dahshan, et al, have shown that a protective scale can form during the oxidation of Co-Cr-C alloys[10]. The oxidation of two-phase materials has recently been discussed in general terms[11] from which it appears that a coarse interconnected second phase is more likely to result in nonprotective oxidation than a fine discrete distribution.

Whether or not the presence of sulfide is deleterious is thus a matter connected with the details of the reaction mechanism and cannot be predicted a priori; however, the thermodynamic conditions determining whether or not a sulfide will be stable can be calculated in absolute terms.

THERMODYNAMICS OF THE CORROSION PROCESSES: STABILITY DIAGRAMS AND REACTION PATHS

It has been useful to present the thermodynamics of both hot corrosion and sulfidation/oxidation corrosion by using simple diagrams showing fields of stability of the different phases as a function of oxygen and sulfur partial pressures. This technique was first introduced for hot corrosion by Quets and Dresher[12], although it had been used in extractive metallurgy much earlier.

Figure 1 shows such a diagram for the system Cr-S-O at 871 C (1600 F) calculated by Hemmings and Perkins[13]. At sulfur partial pressures below 10^{-13} atm, CrS will not form on chromium metal; at oxygen partial pressures below 10^{-26} atm, Cr_2O_3 will not form. However, consider a situation where the initial partial pressure is 10^{-15} atm and the sulfur partial pressure is 10^{-5} atm. Clearly,

Cr O is the stable phase in equilibrium with the atmosphere, and an outer layer of oxide should form on the metal. (For simplicity this discussion neglects composition changes in the atmosphere, although in practice these may be very important[14].) Now, as one approaches the metal surface through the oxide layer, the activities of both oxygen and sulfur will change: the oxygen activity will presumably fall towards the equilibrium value of 10^{-26} atm. One may distinguish two extreme situations for the sulfur: if the oxide is wholly impervious to sulfur, the sulfur activity will fall abruptly to very low values immediately inside the scale where the oxygen activity is still high: this could be represented on the diagram as a vertical line. In this case, the sulfide cannot be stable anywhere in the system, and the oxide layer will be in contact with the metal. On the other hand, if the oxide is wholly transparent to the sulfur, the sulfur activity will remain essentially constant through the oxide as the oxygen activity falls: this could be represented on the diagram as a horizontal line. When the oxygen pressure has fallen to 10^{-20} atm, the line will cross the boundary between Cr_2O_3 and CrS, and consequently, an inner layer of sulfide can be expected to form. If diffusion of oxygen in the sulfide was rapid, the effect would be that the mean composition would move along the Cr_2O_3/CrS boundary and a two-phase layer would develop in contact with the metal, with an outer layer of Cr_2O_3.

This method of describing the sequence of phases in a reacting system by considering the changes in activity within the system and plotting them as a line on the stability diagram is qualitatively interesting: unfortunately, no detailed quantitative analyses have yet been performed. The line on the diagram representing the activity changes is called the "reaction path."

These diagrams are calculated assuming unit activity of all condensed phases, and neglect mutual solubility between different phases. Reducing the activity of a phase by alloying or by compound formation has the effect of expanding the field of stability of that phase, but the effect is not great. Several authors have calculated diagrams appropriate for the activities of the metallic components in particular alloys[15]; while these are interesting, they must be used with caution, since the selective oxidation or sulfidation of components of the alloys will alter the composition of the underlying metal.

It is common practice to analyze the behavior of complex alloys by superimposing diagrams of the type shown in Figure 1. This, too, is a practice which must be treated with caution, since it neglects the activity changes in the solid phases and the possibility of components involving more than one of the metallic components, such as spinels or thiospinels. Nevertheless, it is acceptable for determining the risk that unacceptable phases, such as liquid sulfides, may appear. Figure 2, also taken from Hemmings and Perkins[13], shows the superimposed diagrams for Mn, Cr, Fe and Ni at 1093 C (2000 F). Although manganese sulfide is more stable than nickel sulfide, the important factor in avoiding the formation of nickel sulfide at the interior of the scale is to displace the oxide/sulfide boundary as far to the left as possible, and this criterion suggests that chromium will be a more effective alloying addition.

In the case of corrosion by molten salts, it is sometimes convenient to use a representation in which the two independent variables are the activities of ionic species rather than the activities of the elements: thus, in the case of sodium sulfate corrosion, the oxygen activity, P_{O_2}, and the SO_3 activity have been used by Pettit and coworkers[16]. The SO_3 activity is then equivalent to the pH in the case of an aqueous solution, and the oxygen activity is equivalent to a potential, so the diagram becomes equivalent to the Pourbaix diagram used in aqueous corrosion. The two representations are exactly equivalent: it is possible to draw lines of constant sulfur pressure on the second type, and to transform it to the first type. Figure 3 shows a diagram representing the stability fields for Al_2O_3 in molten sodium sulfate, taken from Pettit, et al[16].

CONDITIONS WITHIN THE FLUIDIZED BED

If a fluidized bed contains an acceptor, then the bed material typically consists of a mixture of CaO, $CaSO_4$, ash, and uncombusted coal: the first two constituents form the large majority. It is probable, therefore, that local atmosphere conditions will be consistent with the CaO/$CaSO_4$ equilibrium, although there are two other possibilities: if all the CaO particles were covered with a

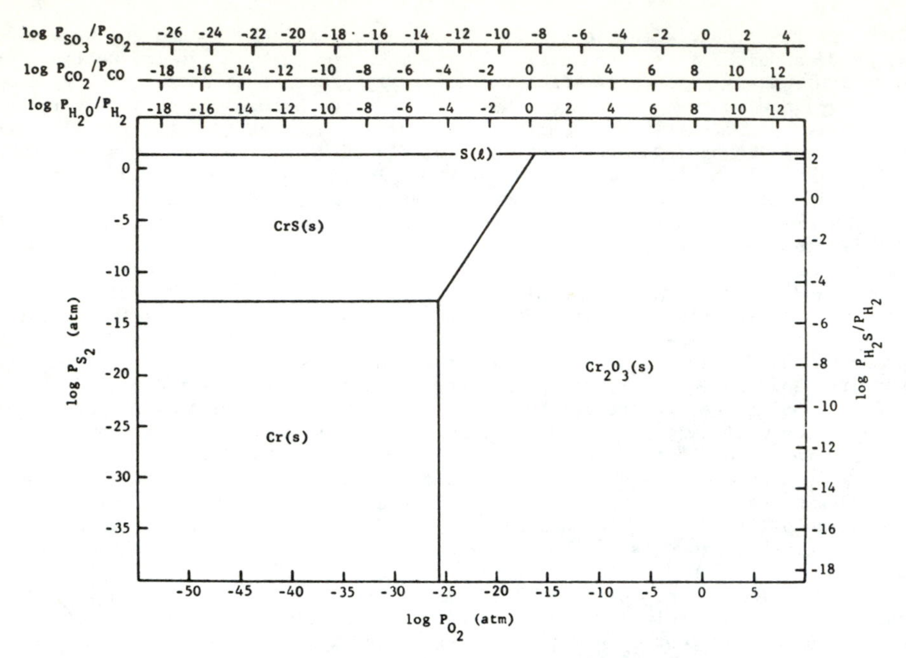

Figure 1. Phase stability diagram for the system Cr-S-O at 871C (1600F). From Hemmings and Perkins[13].

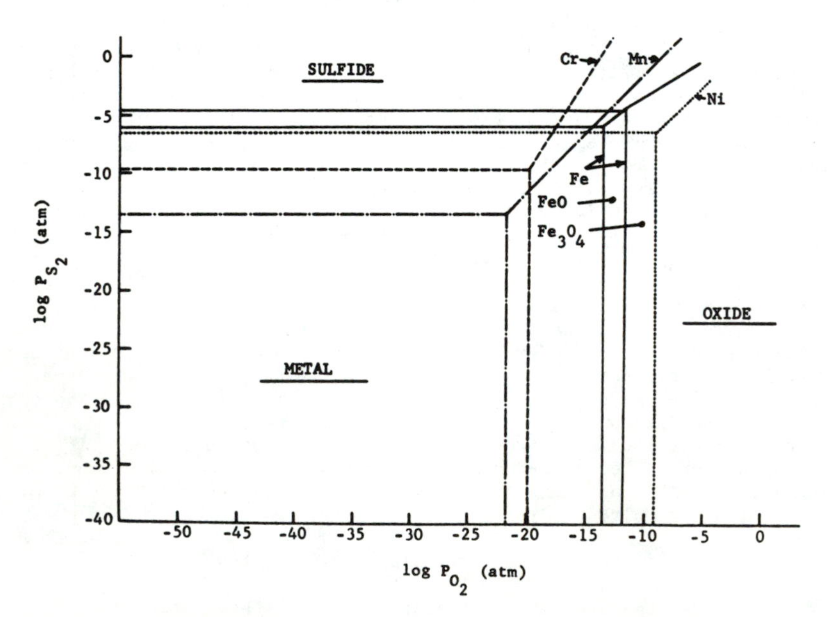

Figure 2. Superimposed phase stability diagrams for the metals Mn, Cr, Fe and Ni with S and O at 1093C (2000F). From Hemmings and Perkins[13].

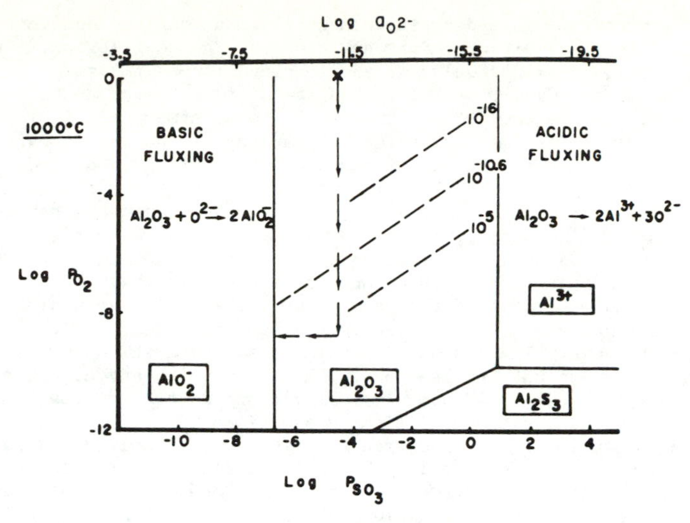

Figure 3. A phase stability diagram representing the phase in the Al-O-S system that are stable in Na_2SO_4 at 1000C (1832F). From Goebel et al.[16]

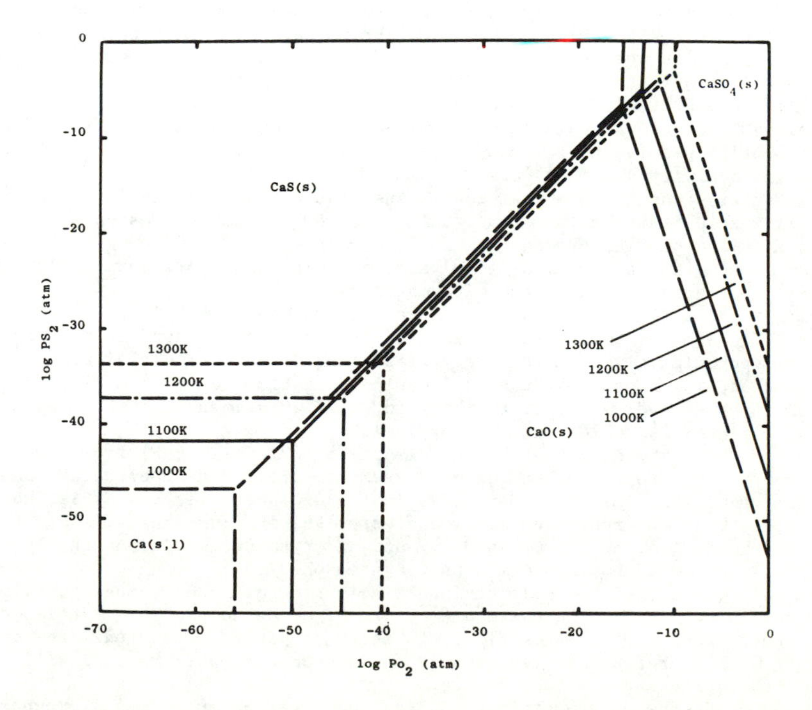

Figure 4. Phase stability diagram for the system Ca-S-O for four temperatures.

layer of $CaSO_4$, or if a stable deposit formed allowing complete sulfidation to take place, then local conditions might deviate from the equilibrium. It will be adequate, however, to consider only the simple situation at first.

Figure 4 shows the Ca-O-S stability diagram at four temperatures. In order to determine where on the $CaO/CaSO_4$ boundary the local atmosphere is located, it is necessary to have an estimate of either the sulfur or the oxygen partial pressure within the bed. The air is blown into the bottom of the bed, and bubbles pass upward, exchanging gas with the surroundings as they rise; the oxygen content should fall as the bubble rises, but it is not clear how fast this happens. Each coal particle is in principle surrounded by an envelope in which the oxygen activity falls from its value in the surrounding gas to a value equal to the C/CO equilibrium at the surface temperature, but it is probable that this envelope is quite small in extent, because of the rapid heat exchange in the bed. At anytime the bed will contain perhaps 1% coal, so on average each coal particle is 4-5 diameters from the next. The gas outside these small combustion envelopes will exchange with the air in the bubbles, and will thus have an oxygen activity below that of the overall system (0.03 atm, typically); but in the absence of more information, the actual activity is difficult to estimate.

One direct measurement of the oxygen activity in a fluidized bed has been reported: Cook, et al[17], used a solid electrolyte probe to determine the oxygen activity in the 0.3 m x 0.3 m combustor at the Coal Research Establishment. The bed temperature was 850 C, and the fluidizing velocity was 1 m/s. One probe was located 0.3 m above the top of the bed, and another 0.15 m below. The above-bed probe gave oxygen activities of some 0.02 atm, close to nominal, with momentary excursions to values as low as 10^{-6} atm, associated with high coal feed rate, or perhaps with the approach of burning coal particles to the probe. However, under substoichiometric combustion conditions, a value of 10^{-14} atm was obtained. A gas chromatograph showed there to be 16.1% CO_2, 2.85% CO and 0.4% oxygen in the gas; this CO/CO_2 ratio is roughly equivalent to an oxygen partial pressure of 10^{-15} atm, and these data suggest that equilibrium had not been established between these different species; it is interesting that the probe reported an oxygen activity similar to that expected for the CO_2/CO equilibrium.

Within the bed, the probe indicated an oxygen activity of 10^{-12} atm in normal operation, with occasional excursions to higher values which the authors regarded as due to bubbles passing up through the bed past the probe. In substoichiometric operation the activity was 10^{-14} atm.

It is interesting to note that the exhaust gas typically contains a partial pressure of 2 x 10^{-2} atm oxygen, and 2 x 10^{-2} atm SO_2. This corresponds approximately to a sulfur partial pressure of 10^{-30} atm or so; the $CaO/CaSO_4$ line would correspond to a sulfur activity of 10^{-34} atm, at this oxygen pressure, which is remarkably good agreement under the circumstances.

Examination of Figure 4 shows that at a temperature of 850 C, an oxygen activity of 10^{-12} atm for the $CaO/CaSO_4$ equilibrium would correspond to a sulfur activity of approximately 10^{-7} atm, very close indeed to the limit of stability of CaO: indeed, it is possible that CaS would be stable under these circumstances. These activities are very similar to those in gasifier atmospheres where values of $P_{O_2} = 10^{-15}$ atm, $P_{S_2} = 10^{-6}$ atm are typical.

Figure 5 shows the Ni-S-O stability diagram at 621 C (1150 F) from Hemmings and Perkins[13]. This represents a low metal temperature for a superheater, and under these conditions the atmosphere lies in the NiO phase field. It is, however, possible that the reaction path would enter the sulfide phase field. Figure 6 shows the diagram at 871 C (1600 F), and under these conditions the liquid sulfide is the phase in equilibrium with the atmosphere.

This analysis shows that sulfidation of alloys at elevated temperatures is entirely possible. At lower temperatures there might be no problems: this depends on the direction of the reaction path which cannot be predicted. Low excess oxygen will increase the risk, and it seems probable that the risk will increase with increasing bed temperature.

Regions where the oxygen activity is reduced, even by quite small amounts, may greatly increase the risk since the conditions lie so close to the phase boundaries. The region near the coal feed ports must be at risk, and excessively high coal feed rates are clearly to be avoided. In the CRE tests, as reported by LaNauze, et al[18], a relatively thick deposit formed on the in-bed tubes. Even if

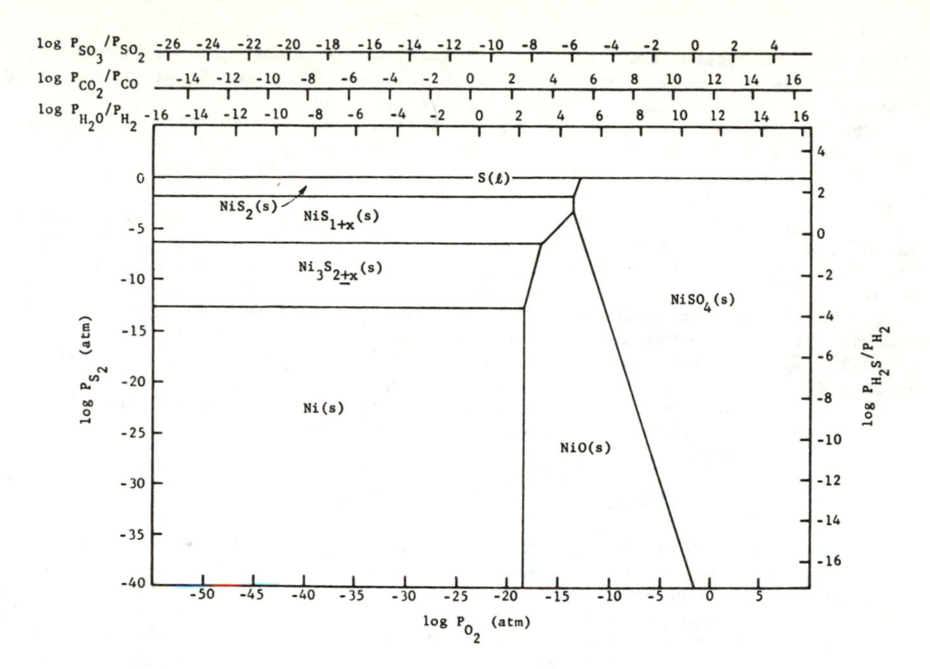

Figure 5. Phase stability diagram for the system Ni-S-O at 621C (1150F). From Hemmings and Perkins[13].

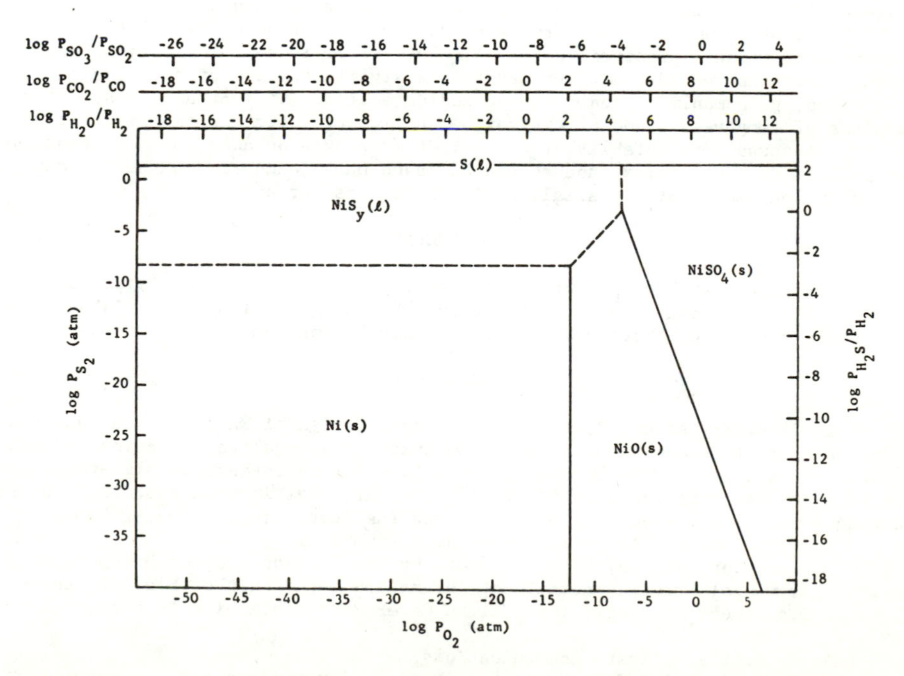

Figure 6. Phase stability diagram for the system Ni-S-O at 871C (1600F). From Hemmings and Perkins.[13]

this deposit was porous and only loosely sintered, it would act as a barrier to the flow of gas and could result in changes in the activities of the constituents of the atmosphere. In fact, it appears to be quite dense.

Other possible sources of activity variations are the variations in the voidage around the tube: the voidage is a minimum at the top of the tube[19] which means that in that vicinity there is rather less movement in the bed material, which could result in a lower oxygen activity.

Recent laboratory experiments[20] have shown that it is possible to reproduce many of the aspects of the sulfidation/oxidation corrosion observed in the CRE tests in a simple crucible experiment using reagent grade calcium sulfate, with the oxygen activity reduced to an unspecified value by the addition of graphite: the graphite is oxidized slowly and is replaced periodically. The sulfidation/oxidation attack is particularly severe on IN 671, a 50% Ni-50% Cr-type alloy, and on IN 597, a nickel-base alloy containing 25% Cr. Type 316 stainless steel was relatively resistant at 750 C, but showed some regions of severe attack at 900 C. As yet, the factors which determine the resistance of alloys to this form of corrosion are not well understood; but it now seems probable that techniques for laboratory simulation will be fairly easy to develop.

COMBATTING FLUIDIZED BED CORROSION

It now appears that the conditions in the bed are borderline for sulfidation/oxidation corrosion. It is obviously desirable to study the effect of changing the operating variables--bed temperature, fluidizing velocity, excess air--on the local sulfur and oxygen activities within the bed, with the aim of optimizing the operation conditions to minimize the risk of corrosion. The best alloys for resistance to this type of attack must be selected, and it seems probable at the present time that for the last stages of the superheaters these will be of the austenitic stainless steel type containing relatively small amounts of nickel. Some conservatism in the choice of materials for components with lower metal temperatures would also be advisable, and special attention should be given to the selection of materials for uncooooled support members within the bed.

It is still unclear how serious a problem hot corrosion of gas turbine components following a pressurized fbc will be. There is some indication that relatively inexpensive bed additives may have a significant effect on reducing corrosion, presumably by reducing the sodium salt activity[21], and conservative choices of temperature and alloys may be sufficient. Further experimentation is clearly necessary to establish the risk: this should be done for long test times at realistic temperatures, and should be based on an accurate characterization of the form and amount of the alkali salts in the gas stream.

ACKNOWLEDGEMENTS

This paper has been prepared with the support of the University of Liverpool and the Electric Power Research Institute. Discussions with Dr. R. LaNauze and his colleagues at CRE have been helpful in developing some of the ideas presented.

REFERENCES

1 I. G. Wright and R. B. Herchenroeder, Design of Materials for use under erosion/corrosion conditions at high temperatures in coal gasification and coal combustion systems. Task 1: design of alloys for resistance to high-temperature erosion-corrosion by low velocity (≤ 75 fps) particles in both reducing and oxidizing atmospheres, Electric Power Research Institute Project No. RP 589-1.

2 M. S. Nutkis, paper in the present symposium.

3 See, for example, "The Coal-Burning Gas Turbine Project," report of the interdepartmental steering committee, Australian Department of Minerals and Energy, Department of Supply (Australian Government Publishing Service, Canberra, 1973).

4 H. R. Hoy, private communications.

5 J. A. Goebel and F. S. Pettit, Design of Materials for use under erosion/corrosion conditions at high temperatures in coal gasification and coal combustion systems. Task 2: design of alloys for resistance to high temperature erosion-

corrosion by high velocity (1000 fps) particles. Electric Power Research Institute Project No. RP 543-1.

6 J. Stringer, in Proc. symp. on props of high temperature alloys, Z. A. Foroulis and F. S. Pettit, eidtors, Electrochemical Society, Princeton, New Jersey, 1976, 513.

7 British Coal Utilisation Research Association R and D, Report to the Office of Coal Research, September 1973, and General Electric Supplementary report.

8 I. I. Bessen and R. E. Fryxell, Proceedings of gas turbine materials conference, Naval Ship Engineering Center, Hyattsville, October 1972. See also reference 6.

9 See, for example, P. Hancock, Proceedings 1st International Congress on metallic corrosion, London, 1961, 193.

10 M. E. El-Dahshan, J. Stringer and D. P. Whittle, Cobalt, 1974.4, 86.

11 J. Stringer, P. S. Corkish and D. P. Whittle, "Stress effects and the oxidation of metals," edited by J. V. Cathcart, AIME, New York, 1975, 75.

12 J. M. Quets and W. H. Dresher, J. Materials 4, 1969, 583.

13 P. C. Hemmings and R. A. Perkins, Lockheed Palo Alto Research Laboratories, report to the Electric Power Research Institute on Project No. RP 716-1, March 1977.

14 A. Rahmel, Corr. Sci., 13, 1973, 125.

15 See, for example, K. Natesan and O. K. Chopra, in Proceedings Symposium on properties of high-temperature alloys, Z. A. Foroulis and F. S. Pettit, editors, Electrochemical Society, Princeton, New Jersey, 1976, 493.

16 J. A. Goebel, F. S. Pettit and G. W. Goward, Met. Trans., 4, 1975, 261.

17 M. J. Cooke, A. J. B. Cutler and E. Raask, J. Inst. Fuel, 1972, 153.

18 J. C. Holder, R. D. LaNauze, E. A. Rogers and G. G. Thurlow, paper in the present symposium.

19 N. M. Rooney, PhD thesis, University of Cambridge, 1974.

20 J. Stringer and D. P. Whittle, unpublished results: see preliminary data in Proc. Int. VGB conference on corrosion and deposits in power plants, Essen, June 1977.

21 K. E. Phillips, Combustion Power Company. Various reports to ERDA on Contract No. E(49-18)-1536, "Energy Conversion from Coal Utilizing CPU-400 Technology."

SLAGGING, FOULING, AND CORROSION PROBLEMS ANTICIPATED IN ADVANCED CYCLES

Session Chairman: **W. E. Young**
Westinghouse Electric Corporation
Co-Chairman: **George C. Wiedersum**
Philadelphia Electric Company

THE DESIGN OF ELECTRODE SYSTEMS AND EVALUATION OF ELECTRODE MATERIALS AND SEMI-HOT WALL MHD CHANNELS

L. H. CADOFF, J. A. DILMORE,* J. LEMPERT, B. R. ROSSING,
H. D. SMITH, A. B. TURNER, W. E. YOUNG,* and S. WAY**

Westinghouse Research and Development Center,
Electric Corporation
Pittsburgh, Pennsylvania, USA 15235

ABSTRACT

MHD generators operating with hot walls, over 1500°C, are of importance in reducing thermal losses and improving overall efficiency. A reduction in wall heat loss from 20% to 10% of the generated MHD power gives about 1.2 percentage points increase in efficiency for the power plant.

The advantages of hot wall generators are not limited to benefits for thermal efficiency. With high wall heat losses, which tend to accompany cold walls, the electrical conductivity of the gases is reduced because of the lower temperature and, consequently, the generator tends to become longer. Also, condensed layers of seed and slag on the insulating surfaces are reduced or eliminated with hot walls. Hot electrode walls also substantially reduce the potential drops in the interface between the electrodes and the plasma, and reduce arc spots.

Hot wall generators, while offering many attractive features, also present difficult problems. Among these are material durability, retention and attachment of wall elements, transfer of heat from the exposed surface to the cooled wall structure, thermal stress cracking, and provision for effective electrical insulation, or electrical conductance where required. The electrode and wall structures must function and exhibit durability in the presence of combustion product gases bearing seed and slag. In the design considerations discussed herein, it will be assumed that we are dealing with combustion products of coal having about 10% ash carryover from the combustor and a maximum gas temperature around 2600°K.

The laboratory screening tests described herein were run in the most corrosive environment to which electrode systems can be subjected - when the seed and slag are in the molten state. Therefore, to aid in the selection of the best material, static immersion and electrochemical corrosion tests were conducted on three candidate electrode materials in slag-seed mixtures at 1400°C. These tests indicate that the selection, testing and use of electrode materials should always be carried out with serious attention to the electrochemical reactions occurring at the electrode/slag interfaces.

---

*Member, ASME.
**Fellow, ASME.

INTRODUCTION

A magnetohydrodynamic (MHD) generator transforms heat energy directly into electricity using the motion of an electrically conducting fluid in the presence of electric and magnetic fields. It is a very simple and compact device with no mechanically moving parts, and might be likened to firing a rocket jet between the poles of a magnet. For example a large rocket engine could easily produce 10,000 megawatts of MHD power with an additional amount available from a conventional steam bottoming plant. It is necessary to combine the MHD topping unit with either a steam or gas turbine as the exhaust gases issuing from the MHD duct are still at a temperature of 2000°C. The resulting combined plant efficiency is in the neighborhood of 50%, and thermodynamically the generator operates with a non-condensing working fluid using a Brayton cycle.

An MHD generator has electrical characteristics similar to conventional generators or batteries i.e., maximum voltage is developed at no load, dropping to zero under short circuit conditions. Electrodes contacting the hot gas stream serve as current collectors, replacing the brushes of conventional D. C. rotating machines. An expression for power delivered to the load per unit volume of the generator is:

$$P = K(1-K)\sigma\mu^2B^2$$

where B is the magnetic field strength, μ is the gas stream velocity, σ is the gas electrical conductivity, and K a generator loading factor, usually taken as about 0.75. In order to obtain a reasonable length of generating duct, say 15 meters for central station proportions, with a 5 Tesla magnet and a gas stream velocity of 800 m/sec, the gas electrical conductivity should be 4 or 5 mhos/m. Generally gas electrical conductivities are very much less than metal and even at combustion flame temperatures it is necessary to seed the gas with an easily ionized alkali salt. With a 1% potassium addition the generator inlet temperature should be about 2500°C, a temperature not easily attained without oxygen addition to the combustor or the use of a ceramic air preheater.

Containing gases at 2500°C presents an immediate problem which is aggravated by the presence of the alkali seed and any fuel (coal) ash combustion products. These materials combine and react to produce low melting point eutectics which in turn penetrate the channel electrode and insulating materials and destroy their integrity. The reacted product and fluxed combustor and/or channel lining materials may then be carried downstream into the air preheater and steam generator where they may deposit, erode, or corrode heat transfer surfaces. The problem was first noted in 1960 when Westinghouse ran the first successful combustion driven MHD generator(1). It was determined at that time that MgO and ZrO_2 were possible candidates for insulators and current collectors respectively but that further development would be necessary. MgO tended to volatilize and the ZrO_2, not a completely adequate conductor even at MHD temperatures, tended to lose its stability and crack during thermal cycling.

In order to deal with the problem of deleterious reactions and degradation of the electrode system, three different approaches can be taken; each is based upon the operation of channel walls within a specific temperature regime: a) cold wall, i.e., well below the slag-seed solidification temperature, $T < 1000°C$, b) semi-hot wall ($1200 < T < 1500°C$) where the slag-seed forms a liquid layer on the wall; and c) hot wall ($T > 1500°C$) where little of the mineral ash or seed condenses on the surface.

By operating in the cold wall mode, the chemical degradation is minimized because the solid slag is slow to react with wall materials. Also metals and non-oxide ceramics can be used as electrodes. However, serious instabilities in the current transfer process occur due to arc mode current transfer and to the poor electrical conductivity of solid slag. In addition, the high heat losses associated with this mode of operation can adversely effect cycle efficiency.

On the other hand, operation in the hot wall mode with the rejection of most (>80%) of the slag in the combustor will minimize the condensation of slag. While current transfer is through a diffuse mode at these temperatures and arcing is avoided, the selection of materials is limited to a short list of refractory oxides having low vaporization rates and corrosion resistance to slag-seed condensates. Analysis of materials tested under these conditions indicate that the slag-seed condensates are in the form of isolated droplets[2] that recession rates are dramatically reduced over those predicted from tests in molten slag-seed mixtures[2], and that several oxides show resistance to this environment up to at least 1973°K[3].

In the intermediate temperature range, the semi-hot wall channel, which allows for liquid slag-seed contact with channel materials will presumably reduce the tendency for arc current transfer but will maximize the chemical dissolution of these materials. Refractory corrosion is generally a diffusion-limited process; thus the rate limiting diffusion process in the slag-seed and the chemical driving forces are important considerations. Thus, the primary mechanism for chemical compatibility in this mode of operation is to control the compositions of the gas phase, the slag and the electrode (insulator) to minimize concentration gradients to reduce the chemical driving forces. The Westinghouse experience has been predominately in the areas of hot and semi-hot wall operation.

Following the experimental program of the early 1960's at which time MHD concepts and principles were confirmed, several feasibility studies were made. The first of these was supported by the Office of Coal Research under contract 14-01-001-476 and culminated in a paper titled MHD Combination.[4][5] The most recent was the ECAS study[6] sponsored by the NASA Lewis Lab. In these studies it was concluded that MHD combined systems were competitive with conventional systems. However, several problem areas associated with gas electrical conductivity, high temperature combustion equipment, air preheater, seed recovery and processing, the superconductivity magnet, and durable duct construction required further investigation and development. At the present time the "durable duct construction" appears to be the last and most difficult problem standing in the way of successful large scale MHD generators and specifically this includes the electrodes and their associated insulation and lead outs. The electrode assemblies must be improved and developed to preserve their physical and electrical integrity in combustion product atmospheres and under MHD conditions.

Meanwhile experimental work was continuing. Under ERDA contracts[7][8] two pieces of experimental equipment were built, one to study the electrical conductivity of combustion product gas, and the other a component test facility located at the Waltz Mill site. This facility could shed light on other aspects of MHD systems such as combustion and fuel handling, heat exchangers, seed recovery and reprocessing, but was intended mainly to study and develop the channel construction. During the second contract period the program was expanded to include basic laboratory evaluation of materials and a project to cooperate with the Soviet Union in channel development was added. During 1976 this work resulted in the preparation of five papers.[3][9][10][11][12]

The National Program for MHD Coal-Burning Power Generation was prescribed in 1975 by the 94th Congress in Public Law S.744 which "authorized a vigorous Federal Program of research, development, and demonstration to assure the utilization of MHD (magnetohydrodynamics) to assist in meeting our national energy needs, and for other purposes." The program must culminate in the design, construction, and operation of a 500 MWe plant by the late 1980's. This plant will be preceded by two facilities, the ETF (Engineering Test Facility) and the CDIF (Component Development and Integration Facility), both to be located in Butte, Montana. Work has begun on the CDIF, a 50 MWt facility with a total flow rate of 20 lb/sec, burning six tons/hr. of coal. It will be a complete power plant with the exception of a steam turbine which in the case of a full-scale commercial plant extracts energy in a temperature range where the MHD process is not feasible. Initial operation is planned for late 1978.

In support of the National Program an overall R&D priority has been established and six general component problem areas (some such as the Effluent Control System being "state of the art") have been identified and assigned to various university and industry teams. Westinghouse was assigned the task of operating a high temperature materials test facility for the purpose of development of materials and construction for hot walls and electrodes in a channel which may run at temperatures of 1000-1800°C and contain a slag-laden gas stream passing through the channel at near sonic velocity, and a plasma temperature of 2600°C. To carry out this task Westinghouse operates several pieces of laboratory equipment for the selection of candidate materials. Tests are made to determine resistance to slag/seed corrosion under static and/or dynamic conditions, with and without the flow of electric current. Materials shown to be promising in these bench type tests are evaluated as electrode/insulator structures in the MTF (Materials Test Facility). This facility simulates MHD conditions but is not equipped with a magnet. It was intended that a final step in the evaluation be performed at Waltz Mill in the 50 KWe generating channel. All steps in the overall program are reiterative with information for modification sent back to the bench scale tests. Finally, supplementing the above study, a cooperative program is being carried out with scientists of the Soviet Union and with scientists of other American institutions in which American materials and structures are evaluated in the USSR U-02 Facility.

In this paper we will summarize work relating to the design of electrode systems for operation under hot wall conditions and then describe laboratory scale screening experiments for selection of materials for semi-hot wall operating conditions.

DESIGN CONSIDERATIONS

Material Selection

To realize good designs of electrode and wall elements, one should select materials that meet the basic requirements of high melting point, thermal shock resistance, durability in presence of seed and slag, and compatibility with surrounding materials. Then, it is important to further ascertain information on the various physical properties and their temperature dependence, particularly thermal and electrical conductivities. Also, one should know the chemical composition, porosity and microstructure. Possible candidate materials for electrodes are: Y_2O_3/ZrO_2, CeO_2/ZrO_2, $LaCrO_3$, $MgCr_2O_4$ and Pt/Rh (80-20). For insulating materials, there are favorable prospects for: MgO, $MgAl_2O_4$, and $SrZrO_3$.

Favorable preliminary results have been obtained for ZrO_2/Y_2O_3 electrodes and for MgO, $MgAl_2O_4$ and $SrZrO_3$ insulator materials. We have reason to expect also that serviceable hot electrodes may be constructed using Pt/20%Rh. The yttria stabilized zirconia has some limitations in the cathode because of the strong component of ionic conductivity. It is believed that if oxygen can be retained in the surrounding environment, if appropriate electric lead attachment is used, and if current densities are not excessive, the yttria zirconia will be a viable material. Further discussion of material properties and characteristics is given in Reference (13).

Heat Absorption and Removal

The convective heat transfer to the wall is given with a reasonable accuracy by the expression

$$q_c = N_s \rho u \,(h_{ad} - h_w)$$

where ρu is mass flow per unit area in the main stream, N_s is the Stanton number, and $h_{ad}-h_w$ is the difference in specific enthalpies at adiabatic and actual wall conditions. Also, we may, for the present purpose, assume a simple relation between N_s and the friction factor, such as $N_s = 0.6\ f$.

The radiant heat flux is usually much smaller than the convective portion, since we have usually a non-luminous type of radiation. For the present purpose, it is sufficiently accurate to simply allow for radiation by the assumption $q_r = 0.05\ q_c$. Thus, we may use the approximate formulation

$$q = 1.05\ N_s \rho u \,(h_{ad} - h_w) \tag{1}$$

For a thermally conducting wall with hot face temperature T_w and cool face T_c, constant thermal conductivity K and thickness c,

$$q = K\,(T_w - T_c)/c \tag{2}$$

If K is K(T), we have expressions for heat flux and wall thickness as follows

$$q = K\,dT/dy \tag{3}$$

and

$$c = \frac{1}{q}\int_{T_c}^{T_w} K(T)\,dT \tag{4}$$

Formulations (1) and (4) may be used to construct the heat flux chart shown in Figure 1. Here, we have used f = 0.004, and have taken the enthalpy/temperature relations appropriate for coal combustion products. For a different f value, we simply apply a correction factor to q on the chart. The K(T) relations used in Figure 1 were based on References (14), (15). The chart in Figure 1 can also be used where we have several wall materials in layers.

If a material of high thermal conductivity is used, the above relations call for a large thickness if we wish to obtain T_w above 1700°C. Rather than increasing the thickness, it is better practice to use a backup insulator behind this material of lower thermal conductivity, like MgO or low density ZrO_2, which acts as a thermal barrier.

Geometrical Considerations

For the electrode walls, decisions must be made, as in any segmented Faraday generator, concerning the ratio of electrode pitch s to channel width d, and electrode face width t to pitch s. The segmentation fineness ratio s/d is preferably kept less than 0.05 to avoid electrical losses, while the optimum ratio t/s ranges from 0.5 to 0.67 for Hall parameter between ∞ and 1.00. For cool walls s/d as small as 0.02 is sometimes recommended. The fineness of segmentation, given by ratio s/d, need not be so small in hot electrode wall designs in which the electrode bars are enclosed by insulating holders with $t/s \approx 0.5$. For hot electrode walls, it is suggested that t/s be 0.4 to 0.5; this will reduce the likelihood of arcing between electrodes and will increase interelectrode leakage resistance. Within the above mentioned constraints, every effort should be made to realize the most rugged possible construction. Sites of tensile stress should be avoided, particularly with stress concentrations in fillets and corners.

Electrical Insulation

Aside from breakage and loss of wall lining material, nothing is more damaging to the performance of the MHD generator than failure of the insulating capability of the walls.

The insulating ceramic blocks must be joined by a continuous layer of sealant, or by brazing, to their water-cooled supporting metal elements. The proportions of these water-cooled elements must be small enough to avoid excessive potential differences between them. However, the situation is more favorable than in peg wall constructions where the surfaces of the elements are exposed directly to the plasma. In the hot wall construction, the metal elements are protected by an outer facing of ceramic blocks. The metal elements should be completely covered with a silastic coating. With such construction, a difference of potential of 150 volts between adjacent elements is regarded as acceptable. If we have a field of 30 volts per cm, ceramic block element dimensions of 5 cm might be appropriate. However, the ability of insulator wall blocks having 5 cm dimensions to consistently sustain electric fields, in the plasma, of the order of 30 volts/cm requires further experimental verification. It may be necessary to reduce the 5 cm size.

There is another factor, however, which bears on the proportions of the insulating wall blocks. The thickness will be determined mainly by heat flux and maximum surface temperature considerations, and it is unwise to make transverse dimensions more than about 2.5 times the thickness for ceramic blocks with high thermal gradients. Otherwise, the tendency to assume curvature in the presence of restraints may cause lateral cracking.

The electrodes must be insulated from each other and from ground. Insulation from ground can be realized either by use of glass reinforced plastic (such as G-7) for the casing, or complete internal insulation with continuous sheets of fiberglass or glass mica. The use of an insulating casing greatly simplifies the problem of providing adequate insulation, particularly for smaller generator channels. The insulation of the electrodes from their neighbors calls for use of interposed ceramic having good resistivity at high temperature in the presence of seed. Analysis shows it is sufficient if inter-electrode insulation layers which are at high temperature have a resistivity of 200 ohm-cm. In addition, it is important that the material be able to withstand voltage gradients of 50 volts per cm or higher in the plasma environment without dielectric breakdown. In this way, the Hall current j_x will be made very small compared to the useful current density j_y. A number of insulating ceramics are available that would have resistivity as high as 200 ohm-cm at 1900°C. The real problem exists at the lower temperature regions (1000-1300°C) where seed compounds may be condensed. It is necessary either to prevent the seed from entering these regions by use of an impervious material, or to construct the insulating members so that the leakage current will be effectively interrupted. The use of air gaps between adjacent parts, and air infiltration in such gaps can be helpful.

The insulating wall blocks also should have a resistivity of at least 100 ohm-cm. For a small generator of 100 cm^2 flow cross-section, the heated portion of the walls should have cross-section about 25 cm^2. Consequently, with 100 ohm-cm resistivity, the walls would have a resistance of 4 ohms per cm of generator length. The plasma column on the other hand, with resistivity above 20 ohm cm would have resistance 0.2 ohm per cm of generator length. The ratio of wall resistance to plasma resistance per cm of length is therefore 20, and this is sufficient to reduce axial current leakage and Hall voltage degradation to acceptable levels. Although 100 ohm cm is sufficient, we would favor 200 ohm cm for the wall material resistivity. The same consideration regarding seed effects apply here as are mentioned in the preceding paragraph. Gaps of 1 to 2 mm are recommended between adjacent insulating wall blocks. Air transpiration can be effectively used on the insulating walls to minimize penetration of seed and/or slag into the spaces between the blocks. In larger generator channels, the wall insulation problem is less serious, because with increasing size, the wall resistance decreases less rapidly than the plasma column resistance.

In the construction of insulating walls, one may arrange the ceramic block pattern in a checkerboard array. The blocks could be staggered or in line alternately, the "checkerboard" pattern could be set diagonally to the passage so that the joint lines match the equipotentials in the plasma. Our experiments will indicate which of these installations is preferred.

Retention and Attachment

With hot wall construction, we generally have the problem of securely anchoring blocks of heated ceramic to water-cooled bars while providing for continuous heat removal. The latter requirement dictates that we avoid air gaps that interfere with the flow of heat. The attaching means must be compatible also with the electrical insulation requirements. There are three main approaches that have been used. These are metal clips or hangers, sealant adhesives, and brazing. Overlap of adjacent parts can also be used to assist in retention.

Electrode insulator blocks may be brazed to small metal plates which are in turn screwed or soldered to the water-cooled bars. To prevent electrical breakdown and reduce axial leakage, it is essential to insulate the water-cooled metal bars from each other. This can be accomplished with coatings and layers of silicone sealant.

The insulator blocks for either electrodes or side walls can be held in place by gripping clips extending 3 to 4 mm into slots, or around the edges of the ceramic. The blocks should also be sealed with silicone adhesive sealer having a 300°C operating temperature, the layer of sealant being not over 0.4 mm so as not to obstruct heat flow. Use of silicone adhesive alone, without clips or retaining overlap by neighboring parts, is a possibility.

The introduction of silicone adhesives is a technical development of greatest importance to MHD technology. The capability of these materials to operate continuously at temperatures of 260°C plus their flexibility and ability to accommodate thermal distortions and to provide reliable seals make them extremely useful and valuable.

Electrical Contact to Hot Electrodes

Materials like stabilized zirconia become conductive at temperatures over 1200°C. The temperatures required are too high for use of ferrous metal contacts. Some success has been experienced with the use of platinized surfaces on the zirconia electrode bars. Platinum contact wires can then be joined to this platinized surface. The current collection zone on zirconia electrodes should be on the downstream face of the anode bars and the upstream face of the cathode bars, to minimize the current concentrations that tend to take place at the edges.

It is possible that composite electrodes may be developed that will have zirconia properties on the hot face with good conductivity on the cool face. Lanthanum chromite as an electrode material offers the great advantage of possessing low temperature conductivity. With metallic electrodes of Pt/Rh, there is no problem with current collection, but top temperatures are limited to about 1750°C.

Suggested Design Approaches

Hot electrodes of stabilized zirconia, or Pt/Rh, may be installed in ceramic holders which are in turn attached to water-cooled copper electrode mounting bars.

Several possible designs are shown in Figure 2(a)(b)(c). In (a), we have an yttria/zirconia electrode bar held by zirconia cement in a slot cut in a magnesia or $MgAl_2O_4$ holder. This holder is, in turn, held to the water-cooled bar by a combination of metal clips and silicone adhesive. The electrical contact is not shown. By use of design information of the type shown in Figure 1, this electrode assembly can be designed to operate at face temperatures up to 1850°C. A modification of (a), not shown, is to make the holders of double width and to mount two electrodes in each. This gives a stronger installation. In (b), we have a similar mounting system but with the electrode holders brazed to metal strips, which are in turn soldered to the copper mounting bars. In (c), we show a possible hot electrode installation with Pt/Rh. The illustration is self explanatory.

Hot side walls may be constructed as shown in Figure 3. Here, the block is retained by recessed metal clips plus silicone adhesive. Transpiration air may be introduced between adjacent blocks, but we are not convinced this is necessary. Such walls can be designed to operate at 1900°C. The material should have a resistivity in alkali-seeded atmospheres of 100 to 200 ohm cm, as well as good dielectric strength in the presence of seed.

Experiments have been performed with hot zirconia electrodes at 1800°C to 1850°C, mounted in sets of five in MgO holders. Retention after 40 hours of operation appeared satisfactory. See Figure 4(a). Some bridging of the electrodes by the seed/slag mixture will be noted. The MgO side wall blocks - in excellent condition - are shown in Figure 4(b).

SEMI-HOT WALL CORROSION EXPERIMENTS

In this section, several laboratory screening tests are described in which materials are subjected to corrosion by liquid eastern type coal slags, with and without the flow of direct current and with and without movement of slag. The results of these experiments will be discussed not only in terms of promise of the electrode materials under evaluation but will also address the usefulness of these tests in evaluating electrode materials under semi-hot wall generator conditions.

The descriptive information on the three electrode materials tested in this study are shown in Table 1. Corrosion testing was performed in two synthetic slags which simulate eastern coal ash compositions. These slag formulations, given in Table 2, were prepared by ball mixing powders of reaction grade component oxides or their carbonates (i.e., $CaCO_3$, $MgCO_3$) for eight hours, pressing the mixed powders into pellets and melting these in a platinum crucible at 1400°C. The viscosities of these slags at 1400°C, listed in Table 2, were calculated according to the method of Shaw.[16] E-03 and E-01 can be considered as representing an eastern and a low silica eastern slag, respectively.

Static Immersion Test

The static immersion test simply involves submerging accurately dimensioned circular or rectangular plate sample coupons under a liquid slag; the variables being slag composition, immersion time and temperature. There is no intentional stirring of the melt, but certainly, slight convective currents due to compositional and thermal gradients are present. After testing, the samples are carefully aligned, sectioned and metallographically mounted for further analyses. These include measurements of sample thickness, reaction layer thickness, penetration depths, chemistry and phase changes, concentration gradients, etc., using optical microscopy, SEM, EDAX and x-ray diffraction characterization techniques. It was found early in this study that several corrosion reactions--surface dissolution, intergranular penetration, erosion, reaction layer formation, etc., were acting at the same time. The extent of corrosion was, therefore, difficult to define for comparison between materials and tests. To provide a standard measure adaptable to all materials tested, the extent of corrosion, ΔW, was defined as:

$$-\Delta W = \frac{W_o - W_f}{2} \qquad (5)$$

where W_o = original sample thickness, W_f = thickness of sample after test, measured between regions where slag constitutes less than 20 volume percent of the sample. The 20 v/o figure is arbitrary, but certainly, a material containing that much slag would have little structural integrity in an MHD channel operating under liquid slagging conditions.

For those materials which formed a protective or semi-protective reaction scale with the slag, a scale thickness, d, was also measured. The corrosion extent, ΔW, listed for materials with protective scales does not include this scale thickness but only the underlying material.

Static Electrochemical Test

Figure 5 is a schematic drawing of a test setup designed to investigate electrochemical reactions between liquid slag and potential electrode materials under static conditions. The electrochemical cell consists of high density MgO or $MgAl_2O_4$ square cross-section crucibles, into which, on opposing walls, are inserted tight fitting (≈0.6 cm diameter) rods of the experimental electrode materials. In operation, the crucibles are filled with solid slag granules raised to the desired test temperature in a furnace and a dc voltage is impressed across the electrodes. Voltage, current and temperature are continuously monitored during the test.

All experiments were run under constant current conditions (>1 A/cm^2). Interelectrode distances were about 2.5 cm. In general, the heating rate of the furnace used was approximately 400°C/hr so that in effect, the electrodes are subjected to a short time immersion test in addition to the electrochemical test. At the conclusion of a test, the cell is removed from the hot furnace and allowed to air cool. The cell is then sectioned along a vertical or horizontal plane containing the electrodes, metallographically mounted and polished and the corrosion reactions investigated.

Rotating Electrochemical Test

Figure 6 shows schematically the test arrangement to test electrode materials immersed and rotated in a slag bath with the passage of direct current. The slag is contained in a large slip cast silica crucible. The rotating electrode support system (see Figure 6) is essentially a high temperature ceramic chuck. This ceramic chuck fastens into a brush assembly which passes current while the chuck rotates. The electric circuit is completed with platinum leads which are wrapped around one end of the electrodes which have previously been metallized with a platinum paste. The electrodes are approximately 0.6 cm in diameter and greater than 2.5 cm long. A variable speed motor allows for variation in rotation speed. Electrode separation was from 2.5 cm to 4.5 cm. The entire electrode support assembly can be smoothly lowered and raised into and from the melt while the slag bath remains at test temperature. Before the surface of the slag is reached, a voltage (∿25 volts) is impressed between electrodes. When contact with the slag is made, current immediately begins to flow. The electrodes are lowered 0.6 cm below the surface. At this time, rotation begins (i.e., 30 rpm or 240 cm/min for a 2.5 cm separation). Tests were either at constant voltage or current for durations of 25 minutes and 2 hours.

RESULTS

Static Immersion Tests

3 $MgAl_2O_4$: 1 Fe_3O_4. This material demonstrated relatively high resistance to slag E-03. Although there was no chemical phase change in this material, there was interdiffusion of elements (ions) between it and the slag. Also, slag did penetrate along certain isolated preferential paths similar to that shown in Figure 8(e). The slag appears as a veinlet or river-like structure and these preferential paths are either grain boundaries and/or cracks in the material. For the short duration tests (<9 hours) the corrosion behavior of this material in slag E-01 appears to be very similar. However, at longer times this more fluid slag penetrates the grain boundaries throughout the entire sample. This is followed by recrystallization, grain growth reactions, and floating off (erosion) of particles into the slag.

The size of the samples remains relatively constant, due to two compensating processes--erosion at the surface and swelling due to slag penetration at the grain boundaries (see Table 3). Energy dispersive x-ray analysis (EDAX) indicates Fe diffusing into this spinel. To a lesser extent, Mg and Al diffuse into the slag from the sample. However, the spinel structure can tolerate large variations in composition without significant lattice parameter changes that could lead to a loss in structural integrity.

$La_{.95}Mg_{.05}CrO_3$. $LaCrO_3$ shows much poorer corrosion resistance to these eastern slags than the spinel described above. Several modes of attack are evident. First, there is direct chemical reaction between slag and sample. In the case of slag E-01 this attack is slowed by the formation of a semi-adherent Fe,Cr(Al) spinel reaction product layer which grows parabolically with time (Table 3). This reaction product is not a true protective layer as evidenced by swelling and penetration of slag into the grain boundaries and pores behind this layer and the interior of the unreacted $LaCrO_3$. Once inside the sample chemical reactions of the slag with grains continues with disruption of the structure and even more extensive slag attack. In the case of the eastern slag, E-03, corrosion is much more intense. Instead of an integral reaction layer, a large liquid/solid zone surrounds a relatively unreacted portion of the sample. This zone appears to be more than 50 v/o liquid (slag) and remains in place only because of the high viscosity of this liquid at temperature. The principal reaction products are a Fe,Cr,(Al) spinel and a lanthanum silicate (with some Ca). Finally, large quantities of La are found dissolved in the slag.

SnO_2. SnO_2 corrodes simply through the diffusion of tin into the surrounding slag. There is no significant diffusion of elements from the slag into the material, nor is there any indication of crystals being fluxed from the surface. For the slags investigated, there is rapid and extensive penetration of the slag at the grain boundaries and pores; the amount of intergranular slag increases with time. This causes recrystallization and a tendency for spheroidization of the SnO_2 grains. Although the material does hold together microscopic examination after long-term tests suggests that shear forces in a real MHD channel would rapidly cause disintegration.

Electrochemical Corrosion Tests

The corrosion tests in which direct current was passed on the three electrode materials are summarized in Table 4. In general, the results obtained in both the rotating and the static electrochemical tests were in agreement. For the sake of brevity, this discussion of electrochemical tests will draw from the results of both types of tests. Primarily, the corrosion of cathodes is much more severe than that of anodes. As shown, electrochemical tests indicate a factor of 5 to 18 more corrosion at cathodes over anodes. Except for some small differences, the reactions and corrosion rates at anodes was very similar to that observed in the simple immersion tests. Therefore, the remainder of this discussion will center on reactions observed at the cathodes of these candidate electrode materials.

3 $MgAl_2O_4$: 1 Fe_3O_4 (Spinel) Cathode. The severe reactions that occur with this material are shown schematically in a sample tested in a rotating test in Figure 7 along with micrographs of the corresponding areas in Figure 8. Below the slag line there is a general loss of material in the slag. In areas b and a (Figure 8(b) and (a)) metallic iron reaction products are found in the slag adhering to the sample. The iron deposits as globules in the slag and on the surface. While a metallic phase is formed in the slag, another phase, wustite (FeO), identified by x-ray is formed in the spinel itself. This phase, appearing as a light colored phase in Figure 8(b) and (c) (areas b, c and the upper part of d), is formed by the migration and reduction of iron oxide from the spinel. It is always found at grain-boundaries or pores, regions that have been intruded into by the liquid slag. Analysis by SEM-EDAX also shows that FeO is only formed below the slag-air interface. Figure 8(d) shows the sharp change in microstructure and microchemistry at this boundary. This indicates the formation of FeO is related to electrochemical reactions. Above the slag-air interface slag has covered the surface by 'wicking', however, corrosion here is similar to that found at the anode and in simple immersion tests.

$La_{.95}Mg_{.05}CrO_3$ Cathode. The lanthanum chromite cathode was penetrated by and chemically reacted with the eastern slag. Two distinct zones are shown in Figure 9. The outer zone is relatively coarse grained, loosely packed iron chromite spinel fluxing off the surface. The inner zone is a region of slag penetrated lanthanum chromite and intergranular iron chromite spinel. Note absence of open pores in this zone. The inner zone grades directly into material that resembles "as received" material, but with small amounts of grain boundary slag penetration. An iron deposit (not shown in the figure) was found in the slag near the sample surface. These results are similar to but more severe than those found in the static immersion test.

In Figure 10, the effect of rotation on the electrical resistance is illustrated. Here the resistance decreases with an increased rate of rotation. The exact functional relationship between rotation rate and resistance has not been worked out for the system. It is believed that the effect of rotation is to restrict the concentration polarization to a boundary layer of definite thickness. This is consistent with the resistance dependence.

SnO_2 Cathode. SnO_2 is unique in that no metallic Fe is formed at the cathode. Tin oxide has dissolved into the slag at an extremely accelerated rate but there is no evidence of SnO_2 particles being fluxed off the surface. In the slag below the cathode some SnO_2 particles have crystallized from the melt, as if the slag has been saturated with tin ions. Figure 11 is a photograph of the sectioned cell, illustrating these effects. Note especially the severe irregular corrosion at the cathode compared to the relatively unaffected anode.

Slag Electrolysis

For the voltage and current density conditions under which these experiments were run, both the E-01 and E-03 slags experienced electrolysis. This was manifested by a continuous decrease in the voltage drop (resistance) across the slag layer with time (Figure 10, Table 4) and by the following two reactions: a) voltage fluctuations and bubble formation, and b) iron formation at the cathode. Under constant current conditions, the fluctuations were about ±2 volts with a frequency of about 5-10 sec at the start of the experiment and droped to ±1 volt and ≈20-30 sec at the conclusion. Bogdanska, et.al.,[17] observed similar fluctuations in measuring conductivity of slags. Usually in electrolysis, fluctuations are associated with bubble formation at the electrodes and the formation of free O_2 has indeed been observed at the anodes.

Free α-iron has been determined to be depositing in the slag near and on the cathode. This is true both for spinel and $LaCrO_3$ cathodes for both types of electrochemical tests, but not for the experiment where SnO_2 electrodes were used. Recent rotating electrochemical experiments confirm the absence of iron. Since the oxygen partial pressure must be about 10^{-10} atm to form free Fe from FeO, the oxygen partial pressure at the cathode must be even lower.

Discussion of Results

In the static tests of the materials described above and in additional tests on other materials with different slags[18] two predominant types of reactions have been observed; one involving simple diffusional exchange between sample and slag and a second that consists of chemical reaction between sample and slag with the formation of new phase(s). In general, from the standpoint of corrosion resistance, the simple diffusional mode is preferred, especially if the sample is in quasi-equilibrium with the slag, i.e., where concentration gradients are small, or if the inter-diffusion coefficients are small. Reactions producing new phases are less desirable from the standpoint of introducing new sources of mechanical strain due to crystallographic incompatibility between product and matrix phases. This could result in cracking of the matrix with further increase in slag penetration. In the static immersion tests and in the anodes of the electrochemical tests, $3MgAl_2O_4{:}Fe_3O_4$ spinel and SnO_2 exhibited attack via the diffusional mode. The recession rates of the spinel were lower than the SnO_2 most probably due to the lower concentration gradients between slag and sample. The $LaCrO_3$ reacted with slag producing reaction products and displayed the highest recession rates.

From a materials point of view, the MHD generator must be considered an electrochemical device. Therefore, corrosion in the absence of an electric field gives only a partial picture of the nature and extent of the interactions between slag and electrodes. As has been demonstrated, highly reducing conditions at the cathodes and presumably highly oxidizing conditions at the anodes can be generated as a result of ionic conduction through the liquid slag layer. Electrode materials should therefore be selected on the basis of their oxidation and reduction resistance, generally ruling out most metals and carbides for use as anodes and many oxides as cathodes. The decomposition of spinel to FeO and $MgAl_2O_4$ is an example of an electrochemically induced cathodic phase change.

The SnO_2 electrode cell experiment provides a very striking example of the importance of electrochemical reactions in selecting materials for MHD electrodes. Again, the anode exhibits low corrosion, comparable to the immersion tests, but the cathode has been very severely reacted (Figure 11). Significantly, there is no evidence for free Fe formation although the voltage drop across the cell should be sufficient for this reaction.* The explanation for both of the above observations must simply be based on thermodynamic/electrochemical considerations. Presumably two principal reactions are in competition at the cathode; these are $Sn^{+4} + 2e \rightarrow Sn^{+2}$ (slag) and $Fe^{+2} + 2e \rightarrow Fe$. Thermodynamics favors the first reaction over the second by ≈8 Kcal/mole and free iron will not form, but the SnO_2 electrode will be forced to sacrificially dissolve as suggested by the first reaction. (This argument should also be stated in terms of electrochemical redox potentials, but unfortunately this information isn't available for the complex slag-ceramic system considered here.) Conceivably, in the other electrochemical experiments, similar sacrificial reactions may be occurring and this could explain the highly accelerated corrosion rates of the cathodes vs anodes. Understanding of these reactions however is hindered by the fact that $LaCrO_3$ and spinel do not (at least in the electrochemical experiments) classically dissolve off into slags like SnO_2 does (i.e., simple ion diffusion) but corrode or erode largely by the fluxing off of solid reaction products. Thus with time, a liquid/solid boundary zone of multiphase reaction products and of unknown electrical conductivities are built up, making for extremely complex half-cell reactions.

The electrochemical reactions noted here are in reality not unlike those discussed by Heywood, et. al.,(19) regarding electrolysis which occurs in ZrO_2, an ionically conducting electrode. In addition, such reactions would not be unexpected in so-called "capped electrodes" where high temperature ionic materials are used to cover and "protect" less refractory electronic conductors (i.e., ZrO_2-Y_2O_3 on $LaCrO_3$). The important thing to be aware of is that at any ionically conducting/electronically conducting interface, reduction reactions will occur; the extent being dependent on current density, voltage and temperature. The driving force (voltage drop) will be greater if the ionic material is solid but the kinetics will of course be accelerated if the ionic material is liquid as in liquid slag.

Analogously, electrochemical reactions must also be expected to occur at anodes. These reactions involve loss of electrons (oxidation) and formation of free O_2. However, most oxides are resistant to such reactions and will, in general, not undergo deleterious reactions. The relatively good corrosion resistance of the three ceramic anodes tested is consistent with this interpretation.

Electrochemical reactions may be reduced by any process which lowers the voltage drop (i.e., decreases resistance) and reduces ionic transference across the slag layer.** Thus increasing the conductivity, especially the electronic component, of the slag through appropriate doping and/or decreasing of thickness of the slag layer would be constructive ways of minimizing the electrochemical component of corrosion. Another way of reducing electrochemical corrosion is to reduce the current density. This is not a very viable option but does serve to underline the fact that an insulating material should not, in theory, undergo electrochemical corrosion.

A basic question that dominates this study is what is the relevance of screening tests in selecting electrode/insulator materials that will survive in an MHD channel under hot slagging wall conditions? There is as yet an insufficient data base to provide answers to this question. However, it should be recognized that laboratory type screening tests are by their very nature limited to studying a few variables and can't hope to duplicate the complex conditions that actually exist in an MHD channel. Their primary value is not in generating the same absolute rates of corrosion as occurs in a channel but in defining and clarifying what factors are important in the corrosion process, so that appropriate materials development and design work can be instituted to improve material survivability.

Of the screening tests investigated herein, static immersion tests are experimentally the easiest and least expensive to run and provide insights as

---

*Thermodynamically, the decomposition reaction, $FeO \rightarrow Fe + 1/2O_2$ requires ≈0.8 volts at 1400°C.

**This can be interpreted in terms of the Faraday equation, $W = \frac{ItA}{nF}$ where W = weight of material dissolved or deposited at electrodes, I = ionic component of the electrical current, t = time, A = atomic weight, n = valency state and F = Faraday constant.

to intrinsic and extrinsic corrosion reactions. In addition, since quasi-chemical equilibrium conditions can be approximated at the surface or in the pores of the sample, phase relationships between complex slags and electrode/insulator materials may be obtained. This information is important in designing thermodynamically compatible electrode (insulator)/slag systems. Rotating immersion tests, although not included in this study, are more difficult to run experimentally. Phase relationships can't be readily determined, mechanisms of the reactions can be obscured, but this test can provide "hard" engineering type numbers as to the extent of corrosion of a specific material. In rotation, there is no question as to the extent of corrosion as can occur in a static test where the sample might be eroding and swelling at the same time. In a rotating immersion test, any part of the sample that hasn't sheared away can be considered as having survived, or at least, having "passed" the test.

In light of the findings in this study, electrochemical screening tests are an essential tool in evaluating corrosion reactions of electrode materials with slags. Although these tests, both static and rotating, are more complex and expensive to run than simple immersion tests, they provide data that can't be obtained any other way. The above comments on the static vs rotational immersion tests should apply equally well as to the utility of static vs rotational electrochemical tests.

Finally, from the results of both electrochemical and immersion testing, it is possible to make a few tentative remarks on the survivability of the electrode materials tested. Spinel ($3MgAl_2O_4{:}Fe_3O_4$) exhibited the least amount of intrinsic corrosion in eastern slags. However, its sensitivity to decomposition in a reducing electrochemical environment would disqualify it for use as a cathode, at least under conditions similar to those of our tests. Spinel, however, could function adequately as an anode. SnO_2 exhibits even poorer cathodic stability although it could also be used as an anode. $LaCrO_3$ exhibited about the best electrochemical stability but is more extensively attacked in eastern slags than the others.

Electrochemical reactions between slags and electrodes are a basic phenomenon intrinsic to the operation of a coal fired MHD channel. These reactions can be expected to occur at any ionically conducting/electronically conducting interface. The severity of these electrochemical reactions should increase with increasing current density, voltage and ionicity of the slag. Under conditions of this study, cathodic corrosion of oxide ceramics was about an order of magnitude higher than anodic corrosion. The nature and extent of anodic corrosion for these materials was about the same as that which occurred in simple immersion tests.

Static tests, both immersion and electrochemical are especially suited for investigating corrosion mechanisms. Rotating tests can provide superior engineering corrosion loss data. Data from immersion tests provide information on the <u>best</u> corrosion resistance of a material in a given slag. Corrosion resistance of materials electrochemically tested should, in general, be worse.

In the absence of electrochemical effects, $3MgAl_2O_4$ spinel exhibited better corrosion resistance in eastern slags than SnO_2 which in turn was superior to $LaCrO_3$. Under electrochemical testing, $LaCrO_3$ was the most stable. Spinel decomposed to $MgAl_2O_4$ and FeO while SnO_2 experienced catastrophic dissolution.

REFERENCES

1 S. Way, S.M. DeCorso, R. L. Hundstad, G. A. Kemeny, W. A. Stewart, W. E. Young: "Experiments with MHD Power Generation." Journal of Engineering for Power, Vol. 83, Series A, No. 4. October, 1961. pp. 397-408.

2 B. R. Rossing et al., Corrosion Resistance of MHD Generator Materials to Seed/Slag Mixtures, 6th Intl. Conf. on MHD Electrical Power Generation, Washington, D.C. June (1975).

3 SLAG - REFRACTORY BEHAVIOR UNDER HOT WALL GENERATOR CONDITIONS, B. R. Rossing, J. A. Dilmore, H. D. Smith, W. E. Young, Westinghouse Research Laboratories, Pittsburgh, Pennsylvania.

4 S. Way, W. E. Young, I. S. Tuba, R. L. Chambers: "Fuels for Advanced Power Generation Systems." Amer. Society of Mechanical Engineers Annual Meeting, Philadelphia, November, 1963. (Westinghouse Sci. Paper 63-118-265-P4a)

5 W. E. Young, S. Way, T. C. Tsu, D. Q. Hoover, N. P. Cochran:"The MHD Combination", Mechanical Engineering, Nov. 1967. pp. 50-57.

6 "Energy Conversion Alternatives Study (ECAS)" Contract Number: NAS 3-19407 - NASA Lewis Research Center - Feb. 12, 1976.

7 "Magnetohydrodynamic Investigation for Coal Fired Open Cycle Systems" - Contract Number 14-32-0001-1540 - United States Energy Research and Development Administration (ERDA) 1974-1975

8 "Development, Testing and Evaulation of MHD Materials and Component Designs" - Contract Number E (49-18) 2248 - United States Energy Research and Development Administration (ERDA) 1976-1978

9 JOINT TEST OF A U.S. ELECTRODE SYSTEM IN THE U.S.S.R. U-02 MHD FACILITY, W. D. Jackson, Energy Research Development Administration, Washington, D.C.; S. J. Schneider, National Bureau of Standards, Washington, D.C.; W. E. Young, Westinghouse Electric Corporation, Pittsburgh, Pennsylvania; L. Bates, Battelle Northwest, Seattle, Washington, A. E. Sheindlin, G. P. Telegin, and D. K. Burenkov, Institute of High Temperatures, Moscow, U.S.S.R.

10 A U.S. ELECTRODE WALL FOR TESTING IN THE U.S.S.R. U-02 MHD FACILITY, R. J. Wright, B. R. Rossing, W. E. Young, Westinghouse Research Laboratories, Pittsburgh, Pennsylvania.

11 RECENT MHD GENERATOR STUDIES AT WESTINGHOUSE, A. B. Turner, S. Way, J. Lempert, W. E. Young, J. A. Dilmore, and B. R. Rossing, Westinghouse Research Laboratory, Pittsburgh, Pennsylvania.

12 THE DURABILITY OF MHD CHANNEL MATERIALS UNDER COAL-FIRED CONDITIONS, B. R. Rossing, J. A. Dilmore, H. D. Smith, S. Schneider, L. H. Cadoff and W. E. Young.
Proceedings of the 3rd US/USSR Colloquium on MHD Power Generation - Institute of High Temperatures, Academy of Sciences - Moscow, USSR - October 17-27, 1976.

13 Rossing, B. R. and Bowen, H. F., "Materials for Open Cycle MHD Channels," Critical Material Problems in Energy Production, C. Stein, Ed., Academic Press, May 1976.

14 Telegen, G. and Schneider, S., et al, Results of Joint US/USSR MHD Channel Tests, 5th European Conference on Thermophysical Properties of Materials, Moscow, May 1976.

15 Engineering Properties of Selected Ceramic Materials, American Ceramic Society, Columbus, Ohio, 1966.

16 H. R. Shaw, "Viscosities of Magmatic Silicate Liquids: An Empirical Method of Prediction," Amer. Journal of Science, Vol. 272, No. 9, p. 870-893, 1972.

17 M. Bogdanska, et al, "Effects of Slag on MHD Generator Performance," Proc. 6th Intern. Conf. on MHD Power Generation, Wash. D.C., June 1975.

18 Development, Testing and Evaluation of MHD Materials and Component Designs, Quarterly Report on ERDA Contract EX-76-C-01-2248, Westinghouse Electric Corp., April, 1977.

19 Heywood and Womack, Open Cycle MHD Power Generation, Pergamon Press, 1969.

Table 1. Characteristics of Test Materials

Sample I.D.	Material	Manu-facturer	Density, % Theoretical	Por-osity,* % Open	Purity	Comments
LC-1	$La_{.95}Mg_{.05}CrO_3$	Westing-house	92.8	1.5	99.4+	Major impuri-ties Al,Si,Fe. Some La sili-cates and $MgCr_2O_4$ at grain boundaries
FS-1	$3MgAl_2O_4$ $1Fe_3O_4$	BNW	92.6	0.15	99.5+	Major impuri-ties Si,Cr,Ni
T-1	SnO_2	Corning	90.6	3.4	98+	Large amount of copper oxide.

*Determined by water immersion.

Table 2. Slag Compositions

Slag I.D.	Composition (wt %)								Calculated, Viscosity, 1400°C poise
	SiO_2	Al_2O_3	Fe_2O_3	CaO	MgO	TiO_2	Na_2O	K_2O	
E-01*	33	26	21	8	1.1	0.8	1.8	4.8	4.8
E-03**	48	20	21	8	1	0.8	1.1	1.5	88.6

*Chemically analyzed.
**As formulated.

Table 3. Summary of Static Immersion Tests Run at 1400°C

Material	Slag I.D.	Immersion Time, hrs	Corrosion, mm ΔW (d)	Corrosion Mode
$3MgAl_2O_4$: $1Fe_3O_4$	E-01	1	+.025	Veinlet intrusion
		4	-.025	Veinlet intrusion
		8	0	Complete slag intrusion and recrystallization, grain fluxing
		24.5	0	
	E-03	25	Sample thermally shocked	Veinlet intrusion
$La_{.95}Mg_{.05}CrO_3$	E-01	1	+.012(.04)	Predominantly chemical reaction. FeCr spinel forms semi-adherent layer. La dissolves in slag.
		4	+.025(.076-.10)	
		8	-.038(.137)	
		24.5	-(.254)	
$La_{.95}Mg_{.05}CrO_3$	E-03	25	+.782	No spinel layer. Severe corrosion and swelling.
SnO_2	E-03	25	-.065	Surface dissolution. Intergranular slag.

Table 4. Summary of Electrochemical Corrosion Tests, Static (S) and Rotating (R)

Material	Spinel	Spinel	$LaCrO_3$	$LaCrO_3$	SnO_2
Type Test	S	R	S	R	S
Slag	E-01	E-03	E-01	E-03	E-03
Temperature, °C	1400	1370	1460	1375	1460
Duration, min	38	120	60	25	55
Electrode Separation, cm	2.5	2.5	2.5	2.5	2.5
Current Density A/cm^2	1.07	0.33	1.1	0.32	1.07
Voltage drop across slag, calculated. Start	14		30		22
End	8		16		10
Corrosion, ΔW, mm					
Cathode, ΔW_c	-1.14		-2.11		-2.79*
Anode, ΔW_a	-0.076		-0.45		-0.15
$\Delta W_c/\Delta W_a$	15		4.7		18.3

*Volume averaged to account for irregular corrosion (Fig. 11)

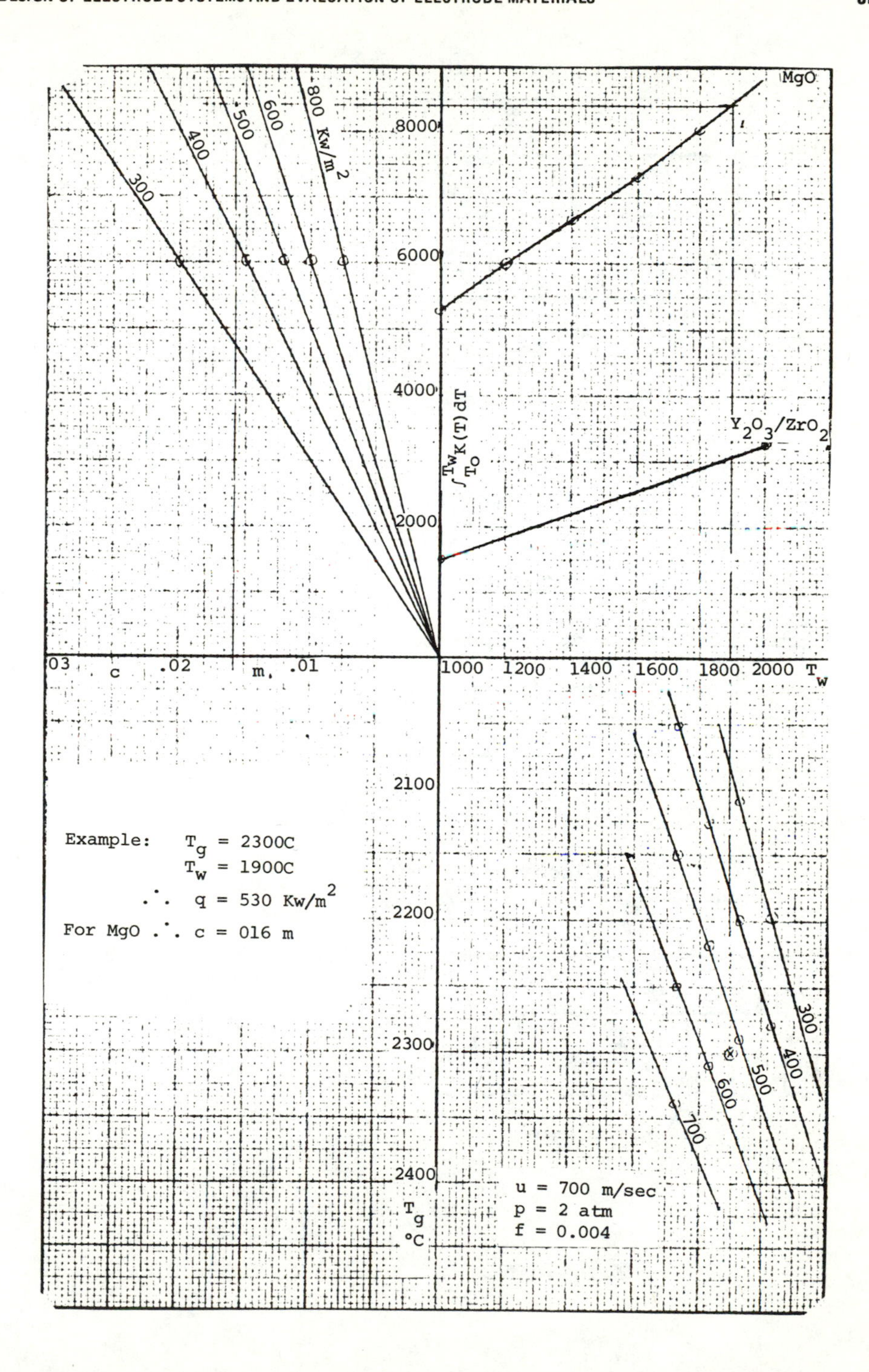

Fig. 1 - Wall Design for Heat Flux for T_c = 200C

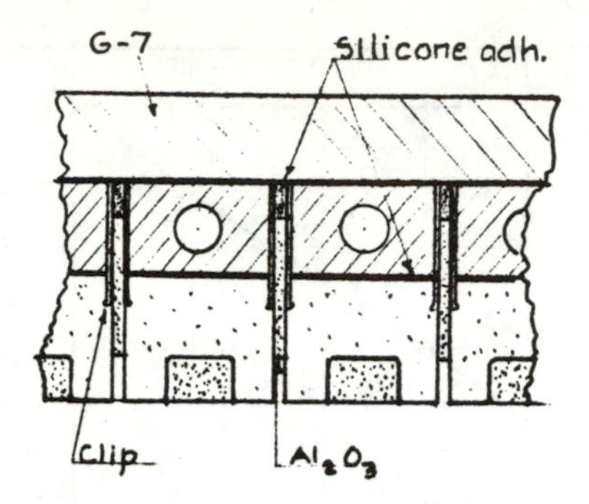

(a)

ZrO_2/Y_2O_3 electrodes
set in MgO holders

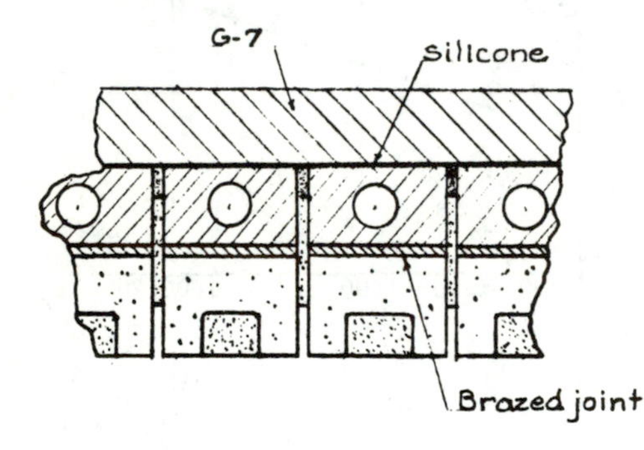

(b)

ZrO_2/Y_2O_3 electrodes
MgO holders
Braze Mounting

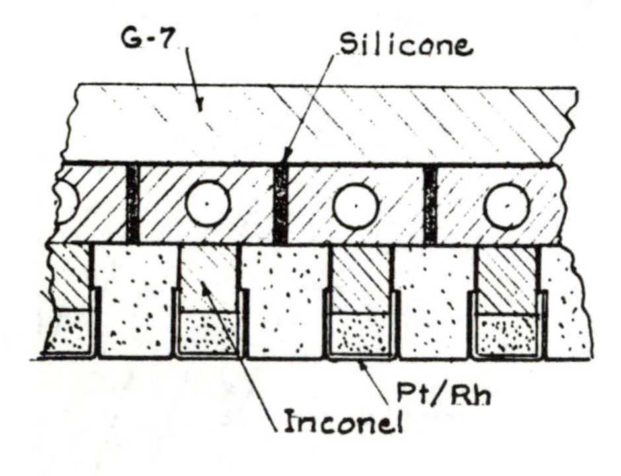

(c)

Metal electrodes
Pt/Rh

Fig. 2 - Examples of Hot Electrodes

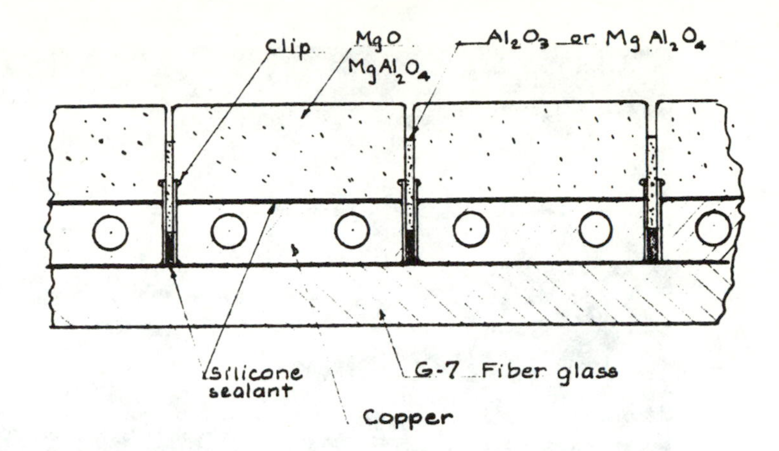

Fig. 3 - Insulating Wall Example of Hot Wall Construction

Fig. 4(a) - View of the Electrode Wall After 40 Hours of Test at Approximately 1800°C.

Fig. 4(b) - View of the Insulating Wall After 40 Hours of Test at Approximately 1800°C

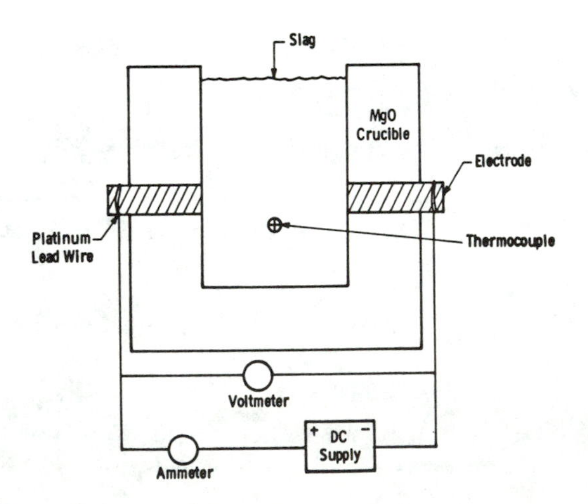

Fig. 5 - Schematic Diagram of Static Electrochemical Test Apparatus

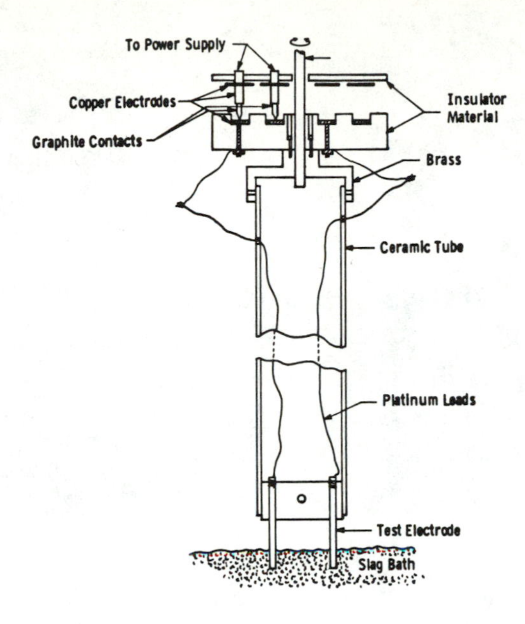

Fig. 6 - Schematic Section of Rotating Electrochemical Test Apparatus

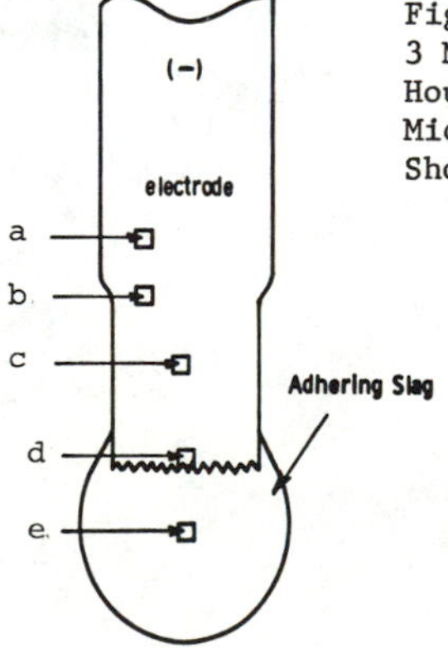

Fig. 7 - Schematic Cross-Section of 3 $MgAl_2O_4$: 1 Fe_3O_4 Cathode After Two Hours Rotating Electrochemical Tests. Micrographs of Lettered Areas Are Shown in Fig. 8

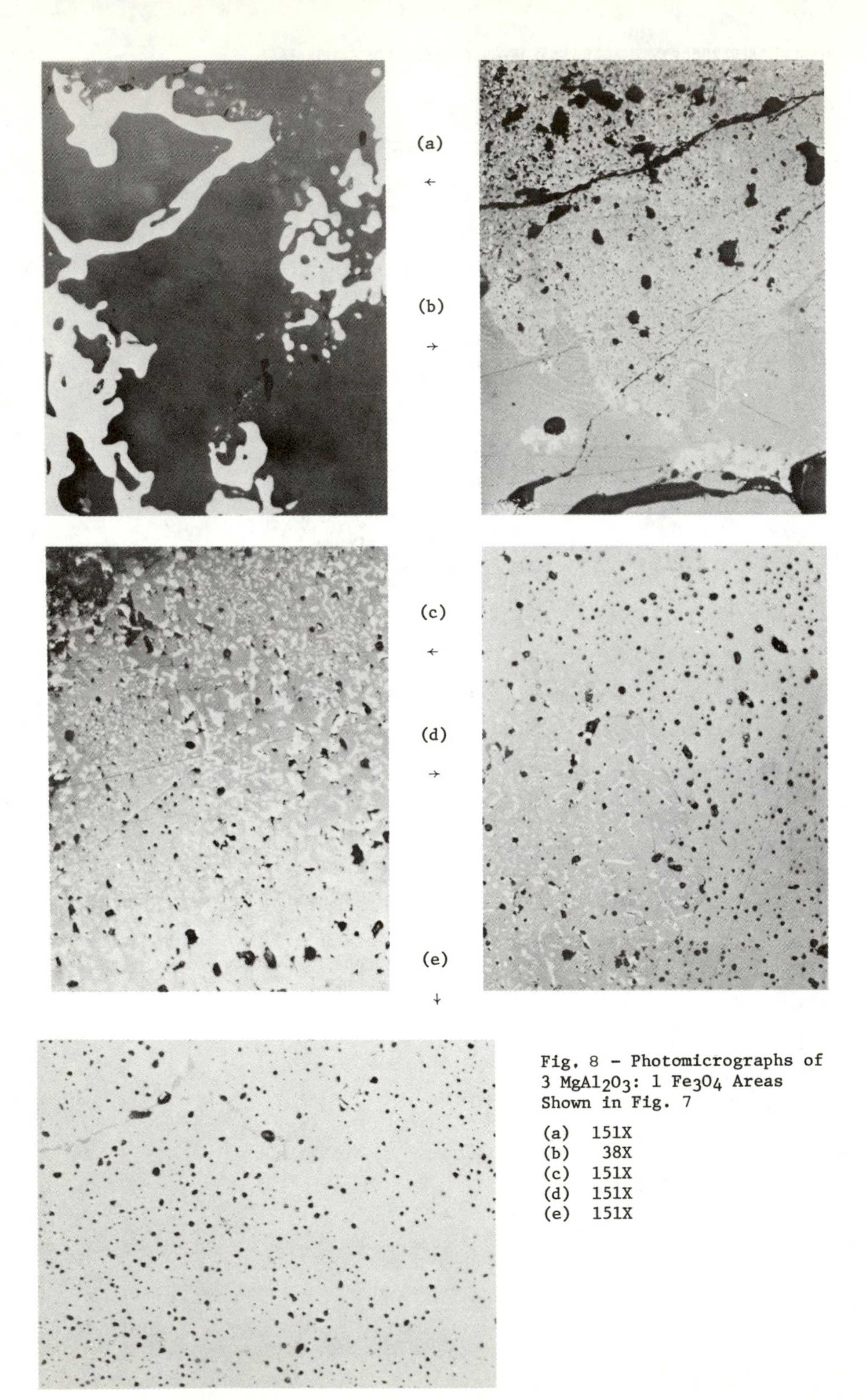

Fig. 8 - Photomicrographs of 3 $MgAl_2O_3$: 1 Fe_3O_4 Areas Shown in Fig. 7

(a) 151X
(b) 38X
(c) 151X
(d) 151X
(e) 151X

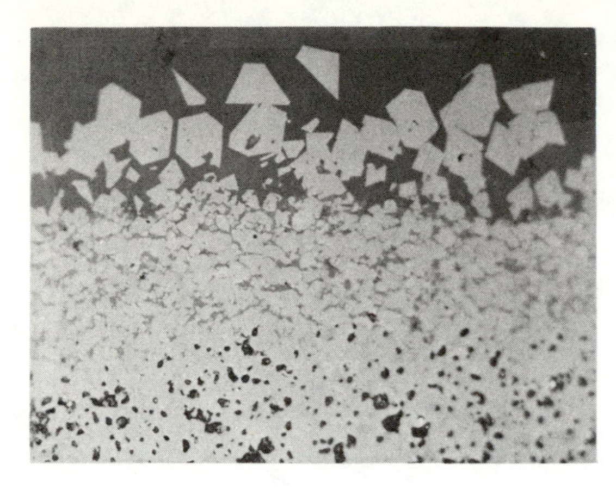

Fig. 9 - Photograph of Surface of $La_{.95}$ $Mg_{.05}$ CrO_3 After Rotating Electrochemical Test (755X)

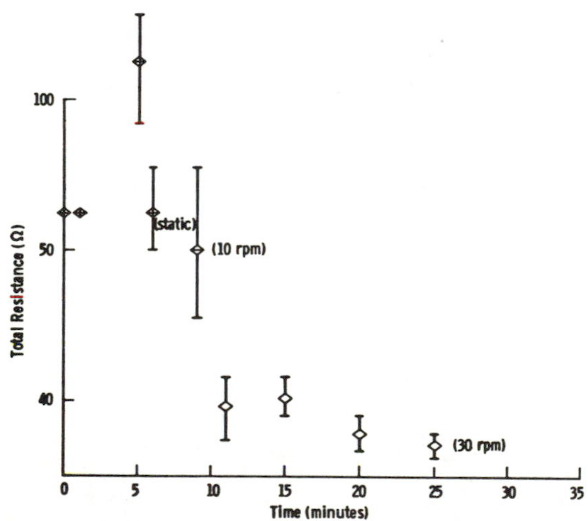

Fig. 10 - Effect of Rotation on Electrical Resistance of Electrochemical Test System-$LaCrO_3$ Electrodes

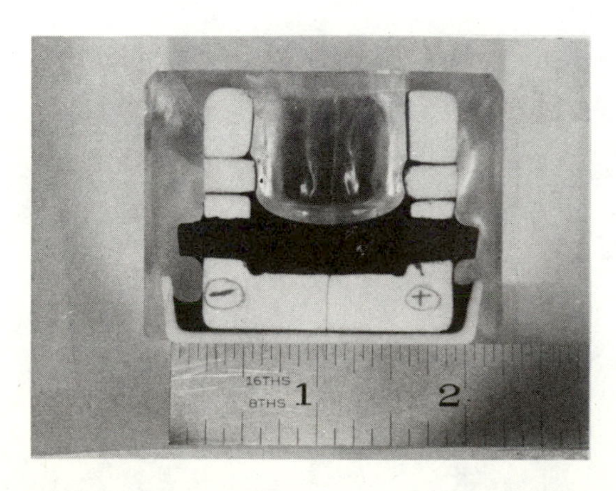

Fig. 11 - Polished Cross-Section of Static Electrochemical Test Cell-SnO_2 Electrodes, 1.51X

COAL SLAG EFFECTS IN MHD GENERATORS

J. K. KOESTER, M. E. RODGERS, R. M. NELSON, and R. H. EUSTIS

Stanford University
Stanford, California, USA 94305

ABSTRACT

Experiments have been conducted on the formation and properties of slag coatings resulting from the direct firing of coal under simulated MHD generator conditions.

The hydrodynamic behavior of the thin slag coatings were studied for various coals and types of channel wall. Stable, continuous coatings were achieved over suitable substrates. Slag layer characteristics, such as thickness, surface temperature, and development time were measured and compared with theory.

The response of slag layers to applied transverse and axial discharges under MHD generator duct conditions was investigated for various metal electrodes and two different coals. Axial slag breakdown and both diffuse and arc mode transverse discharges were observed with cinephotography.

The critical current density for diffuse mode operation was an order of magnitude larger for mild steel electrodes than for stainless steel or nickel electrodes. Measured values of slag layer resistance increased rapidly with decreasing electrode temperature. A large contact resistance was measured for the slag layer over the nickel electrode. The excellent slag contact with the steel electrode was traced to chemical interaction of the slag with an iron oxide film on the electrode surface. The large variations in diffuse mode performance for different electrode materials is shown to correlate well with a function of slag layer resistance.

The erosion rates of low carbon steel electrodes operating diffusely under slag were measured. Steel cathodes were not attacked; however, steel anodes were severely eroded which suggests electrochemical corrosion due to ionically conducting species in the slag layer. Large differences in slag microstructure over the cathode versus the anode were observed.

NOMENCLATURE

A = wetted area of channel
a = ash mass fraction
B = magnetic field
d = diameter of slag droplet
f_i = fraction of current carried by ions
G = dimensionless current density
H = dimensionless current density
g = dimensionless layer resistance
h = channel height
J = current density
k = slag thermal conductivity
$\dot{m}$ = mass flowrate
M_ℓ = mass of slag in coating
n = exponent in conductivity power law
$\dot{Q}_s$ = heat flux to slag surface
r_ℓ = layer resistance
s = slag layer thickness
t = time
T = temperature
u = gas velocity
v_{dep} = deposition velocity
v_+ = dimensionless deposition velocity
V = voltage
x = axial coordinate
y = transverse coordinate

α = slag thermal diffusivity
β = Hall parameter
γ = ratio of slag surface to wall temperatures
Γ = mass flux to wall
δ = electrode recession
$\dot{\delta}$ = erosion rate
ε_i = mass fraction of ith component
θ = dimensionless time
μ = gas viscosity
ρ = density
σ = electrical conductivity
τ = characteristic time
τ_w = gas shear stress at wall
τ_+ = dimensionless relaxation time
ϕ = fuel/oxidizer equivalence ratio
ψ = current stream function

Subscripts

c = critical
comb = combustor
d = development
el = electrode
g = gas
K = potassium
ℓ = layer
s = slag surface
sl = slag
t = total
th = thermionic
w = channel wall

INTRODUCTION

For the past several years, there has been an increasing interest in the direct firing of coal in magnetohydrodynamic (MHD) generator systems. It has been felt that there may be economic advantages of direct firing with the elimination of extensive coal processing and that coal slag deposits on electrodes and insulators may offer protection.

Programs have been initiated ranging from those in which 100% of the coal ash is carried through the MHD channel (1) to a combustion chamber development (2) in which combustion occurs in multiple stages to eliminate a large fraction (up to 95%) of the slag from the MHD duct. Depending on the amount of ash rejection, the ash mass fraction in the combustion products entering the MHD channel will range from 0.05% to 1% since most U.S. coals contain about 10% mineral matter. By comparison, the combustion products will contain around 1% potassium seed added to achieve a sufficient electrical conductivity.

Most of the ash components condense out below a temperature of ≈ 2100 K (3) in an MHD combustion gas environment; the volatile exception (aside from the alkali metals) is silica which doesn't begin heavy condensation until below 1900 K. Since the duct walls are likely to operate at a temperature below 2100 K, some fraction of the ash vapor and liquid droplets will condense out on the duct walls. This condensate builds up a slag layer until the slag surface temperature is high enough that the slag becomes fluid and "runs-off" due to the gas boundary layer shear force. When the slag layer "run-off" balances the slag transport from the gas, the layer reaches an equilibrium thickness, typically on the order of a millimeter.

This thin layer of slag may protect the generator walls by providing a barrier to oxidation as well as a thermal barrier from the high temperature combustion products. The coating of slag over the electrode and insulator elements has two primary effects on electrical performance: first, the transverse resistivity of the slag layer impedes the collection of current at the electrodes, and secondly, the axial conductance of the layer tends to short out the interelectrode insulators. However, the electrical conductivity range of most slags coupled with the thinness of the layer result in acceptable performance for large slag coated generators.

The diffusion of such ionic species as oxygen, iron, and potassium through the slag layer can both modify the slag properties and lead to electrochemical attack of the channel walls. The plasma composition (especially the oxygen and potassium concentrations) at the slag surface provides a source for the ion transport to the channel-slag interface.

The strong dependance of slag electrical conductivity on temperature results in limitations on diffuse mode current transfer and current transfer through the slag often occurs via small arcs.

This paper describes experimental and theoretical results on slag layer hydrodynamics, the electrical effects due to slag coatings, the chemical characteristization of slag layers, and the effect of slag on electrode erosion.

SLAG EXPERIMENTAL TESTS

Flow-Train. The present slag experimental facility of the High Temperature Gasdynamics Laboratory consists of a fuel and oxidant flow system, a combustor, a slag deposition test section, a slagging electrode module, a diagnostic section, and an exhaust system with cooling and scrubbing hardware. Important elements of the flow-train are shown schematically in Fig. 1.

Potassium-seeded ethanol and pulverized coal in an ethanol slurry are burned with oxygen and nitrogen and the resulting slag-laden plasma passes from a plenum chanber through an inlet nozzle into an experimental duct having a 3.8 × 7.6 cm cross-section and a length of 50 cm. All duct surfaces, including the inlet nozzle, are constructed of copper or mild steel with transverse ceramic-filled grooves in the deposition surface. The diagnostic section immediately downstream of the test duct acts as a radiation shield and provides ports for observation of the slag layer through the exit plane of the test duct. A purged optical port provides a 1.3 cm diameter view of the lower 7.6 cm wall approximately 20 cm from the test section inlet.

This test rig simulates the conditions expected in a central station MHD generator: high subsonic flow, plasma temperature 2200-2800 K, potassium seed fraction 0.5-2 percent, ash carryover fraction 0.05-1 percent, and average current densities of 1 amp/cm^2. The facility is steady state and has been operated up to six hours in duration.

Electrode Module. Slag discharge experiments were performed with an electrode module (shown schematically in Fig. 2) placed at the end of the slag deposition test section. Slag transverse and axial discharges were studied by applying voltage to the primary electrodes and to the auxiliary electrodes respectively. Platinum sidewall voltage probes were used to measure the slag layer voltage drop and plasma conductivity. The electrodes were cooled by water or by a variable flow of nitrogen so that electrode surface temperatures could be controlled over the range 500 to 1500 K. The electrode temperature was measured by two Pt/Pt13Rd thermocouples installed at different distances below the surface. The surface temperature was calculated from these measurements using an extrapolation scheme and the temperature dependent thermal conductivity of the electrode material. Electrodes constructed of steel 1018, nickel 200, and stainless steel 304 were tested.

The slag layer over the bottom electrodes was photographed through the exit jet with a cine camera. The slag surface temperature was measured by an optical pyrometer.

Data Recording. Reactant flowrates, channel temperatures, currents, and voltages were measured by standard techniques. Discrete sampling and data reduction were made with a scanning unit and integrating digital voltmeter interfaced with a minicomputer. Key signals were continuously recorded on an eight channel strip-chart and synchronized with the cine film by marker blips.

Test Conditions. Experiments on slag hydrodynamics and the response of slag layers to applied electric fields were carried out in an ongoing test program using various sources of ash. The test conditions used are summarized in Table 1. All of these tests were run at a N_2/O_2 ratio of 0.5 which results in a flame temperature of about 2750 K. The test section exit velocity was about 500 m/sec. The synthetic mix was composed of equal parts of silica, alumina, ferric oxide, and lime hydrate powders. The Pittsburgh coal was a bituminous of unknown origin; the Montana coal was a sub-bituminous from the Rosebud seam; the Illinois #6 coal was a bituminous. All of the coals were pulverized to 100% through 200 mesh.

Table 1. Experimental Test Conditions

Test #	Ash Source	$\dot{m}_t$ (kg/sec)	a (wt %)	ε_K (wt %)	ϕ	τ_{comb} (msec)
5	Synthetic Mix	0.136	1.5	0	1.05	18
6	Synthetic Mix	0.136	1.5	1.0	1.05	18
8	Pittsburgh Coal	0.136	0.22	0	1.05	18
9	Pittsburgh Coal	0.136	0.38	0.39-0.69	0.0-1.1	18
10	Montana Coal	0.145	0.84	0.57	1.3	18
11	Montana Coal	0.134	0.46	0.61	1.0	39
12	Illinois Coal	0.136	0.5	0.6	1.05	39
16,17,18	Montana Coal	0.136	0.5	0.6	1.05	39
19,20	Illinois Coal	0.136	1.0	0.91	1.10	40
21	Illinois Coal	0.136	1.0	0.6	1.05	39
22	Illinois Coal	0.136	1.0	0.6	1.05	39

where: $\dot{m}_t$ = total mass flowrate
a = ash mass fraction
ε_K = potassium seed mass fraction
ϕ = equivalence ratio (>1 for fuel rich)
τ_{comb} = combustor residence time

SLAG HYDRODYNAMICS

The slag flow on duct walls can be characterized as the very low Reynolds number shear flow of a fluid whose viscosity is a strong inverse function of temperature. Mass, momentum and heat are transported to the slag from the overlaying plasma boundary layer. In steady state, the mass flux in the layer is equal to the net deposition of slag to the portion of the layer upstream of the location of interest. The surface temperature adjusts to provide a viscosity level and temperature gradient in the slag which produce a velocity profile reflecting the correct mass flow and imposed shear stress (see Fig. 3).

A model for this phenomenon was first described by Rosa (4) and has been extended by workers at Stanford and elsewhere (5-8) to include effects of freezing, non-Newtonian flow near the freezing temperature, Joulean heating caused by normal to streamwise current transport, and transient layer build-up. For typical values of slag properties and flow parameters, the model can be used to investigate the effects of a variety of parameters on the slag layer performance. Two measurable layer characteristics which are important in MHD applications are the surface temperature and the layer thickness. For a typical slag, the effects of layer mass flux, wall shear stress, heat flux, and substrate temperature are shown in Figs. 4-7. Note that large changes in all of these parameters lead to relatively modest changes in thickness and particularly surface temperature, the latter quantity being most strongly influenced by the slag viscosity level. More viscous slags yield thicker layers and higher surface temperatures, while slags of higher thermal conductivity yield thinner layers and higher surface temperatures.

Quantitative comparisons of the model predictions with experimental results require reasonable estimates of slag viscosity, thermal conductivity, freezing point, and density, as well as measurements or estimates of the slag mass flux, boundary layer shear stress, substrate temperature, and plasma core temperature.

The slag deposition experiments which have been carried out in this laboratory have been aimed initially at obtaining qualitative information about slag deposition and flow on channel walls.

Slag Layer Development

The temporal development of the slag coating on the grooved stainless steel slab near the exit of the test duct has been recorded during several tests by cinephotography. Typical stages of layer development are shown in Fig. 8 by selected cinefilm frames from test #8. Slag droplets begin to form on the hot ceramic stripes and these droplets begin to coalesce into globules (frames a,b). As the globules grow and attain a higher surface temperature, they elongate in the downstream direction (frame c) and bridge over the metal stripes to adjacent ceramic stripes (frame d). This behavior is similar to the preferential deposition on ceramic stripes and metal bridging observed by Stickler (6). This bridging may advance to a state where much of the wall surface is coated without any significant slag upstream run-off. In some cases, upstream run-off appears early in the form of isolated rivulets. In frame e, a single "river" of slag appeared after 121 minutes of operation with coal. This "river" is close to the conditions of an equilibrium flowing slag layer, with material flowing along the surface and depositing slag which freezes and builds up to the equilibrium thickness before advancing further. In this case, the "river" propagated at a speed of 1.2 cm/min. The "river", from its first appearance, exhibited transverse ripples which propagated downstream at a rate about 50 times greater than the "head" of the "river". After propagating to the channel exit, the river appeared almost stable for a considerable time period (frame f). The "river" slowly broadens to complete the coverage or in some cases, several rivulets appear and slowly merge together.

The development and flow of slag layers on upward-and downward-facing horizontal surfaces, as well as on vertical surfaces, has been observed to be qualitatively similar under the conditions studied, but diagonal rivulets on vertical channel walls have suggested that gravity has some effect and may be important in large-scale devices or in regions of low wall shear.

To quantify and compare the temporal development of the layers in the various tests, the times required to attain particular stages in the development sequence are normalized with respect to ash flowrate and layer equilibrium thickness. The amount of slag in an equilibrium layer is:

$$M_{\ell} = \rho_{sl} A \langle s \rangle \qquad (1)$$

where ρ_{sl} is the slag density, A is the area of interior walls of the duct, and $<s>$ is the average layer thickness. The total feed rate of slag into the channel is given by

$$\dot{m}_{sl} = \dot{m}_t \, a \tag{2}$$

where a is the ash mass fraction. Since the deposition of slag on the walls is, in the non-flowing stages of development, proportional to the feedrate of ash, then a characteristic development time can be defined as:

$$\tau_d = M_\ell / \dot{m}_{sl} = \frac{\rho_{sl} \, A \, <s>}{\dot{m}_t \, a} \tag{3}$$

This time will be smaller than the actual development time of a uniform slag layer. There are two events in the slag layer development whose time of occurrence can be roughly determined: the bridging of slag from ceramic stripe to ceramic stripe, and the appearance of the propagating river. These times can then be normalized with τ_d to obtain a measure of the relative rate of the development:

$$\theta_{event} \equiv t_{event} / \tau_d \tag{4}$$

The characteristic time, τ_d, and the normalized event times for the various tests are presented in Table 2. The river formation time for test #5 and both river formation time and bridging time for test #8 are longer than in the comparable tests (#6 and #9) with seed. This suggests that the potassium seed "fluxes" the slag droplets and enhances the rate of layer development.

Table 2. Summary of Slag Layer Development at Channel Exit

Test #	τ (min)	θ_{bridge}	θ_{river}	$\theta_{duration}$	% coated	% flowing
5	12.7	-	2.8	3.7	75	40
6	8.9	<1.2	1.2	8.5	100	100
8	35.2	2.7	3.5	4.6	90	25
9	20.1	2.0	2.3	6.8	100	60
10	13.7	0.9	1.3	6.7	100	60
11	16.8	1.0	1.8	10.2	100	70
19	7.0	-	5.1	38.4	100	100

The state of development at the end of each test is also summarized in Table 2 in terms of the percentage of the duct wall coated and the percentage of the slag surface that was flowing. There is rough correlation between normalized run duration and percent surface flow. For test #19, a 100% flowing slag layer was obtained for a normalized time of ~ 15.

The rate of deposition of slag droplets onto duct walls in turbulent plasmas is strongly dependent on the slag droplet size distribution. Experiments have been performed by Holve (9) to measure the mean droplet size in the plasma at the end of the experimental test duct described above. The technique involves the measurement of the percent extinction of a He-Ne laser beam directed across the plasma free jet. This measurement, with an estimate of the mass loading of ash in the plasma and assumed size distribution, yields a line-of-sight averaged value of the mean particle size. Under the conditions outlined in Table 2, the measured mass mean droplet size was:

$$d = 1.6 \ \mu m \pm 48\% \tag{5}$$

This mean droplet size can be used to obtain estimates of the time required for the development of an equilibrium layer. Experimental data acquired by Liu and Agawall (10) provide a correlation between droplet size and deposition rate in terms of dimensionless quantities defined as follows.

$$\tau_+ = \frac{\rho_{sl}\, d^2\, \tau_w}{18\, \mu^2} \qquad (6)$$

$$v_+ = v_{dep} \sqrt{\rho/\tau_w} \qquad (7)$$

where ρ_{sl} and d are the particle density and diameter, respectively, τ_w is the wall shear stress, μ and ρ are the mean gas viscosity and density, respectively, and v_{dep} is the deposition velocity. The mass flux to the wall is given by

$$\Gamma_{sl} = \rho_{sl}\, a\, V_{dep} \qquad (8)$$

where a is the mass fraction of slag in the flow.

A curve fit for the experimental data is shown in Fig. 9. For conditions appropriate to the experiments described above, 1.6 μm particles lead to a deposition rate of 2×10^{-4} kg/m^2•s at an ash fraction of .005. For a typical equilibrium layer thickness of $s = 2$ mm and a slag density $\rho_{sl} \simeq 2.5$ g/cm^3, a development time can be crudely calculated:

$$t_d \sim \rho_{sl}\, s/\Gamma_{sl} \sim 430 \text{ min} \qquad (9)$$

This time is considerably longer than typical development times observed for the present experimental configuration. The distribution range of droplet sizes about the mass mean value is not known. However, the saturation of the particle deposition velocity to a value of $v_+ \approx 0.1$ for large values of τ_+ (corresponding to $d > 4$ μm here) allow the estimation of a more realistic development time. Assuming one half of the particles are large (> 4 μm), a development of 86 minutes results. This is less than the 105 minutes observed experimentally, but in quite reasonable agreement considering the unknown distribution range and the sensitivity of deposition rage to particle size.

Characteristics of Flowing Slag Layers

Narrow transverse ripples, relatively widely spaced, were observed on the slag surface for all cases in which the layer was flowing. The ripples were observed on both horizontal as well as vertical surfaces. Since the ripples appear brighter than the surrounding slag, the ripple crests are evidently higher in temperature. Measurements made with an optical pyrometer indicate that the temperature difference between ripple crests and troughs is 25-40 K. Due to the strong temperature dependence of slag viscosity, the ripples may play a role in slag transport. Average values for ripple propagation speed and ripple spacing are presented in Table 3. The ripple spacing did not change much, but there is a large variation in ripple speeds. Both ash flowrate and slag viscosity were varied, but the effect of these parameters on ripple speed is not well understood.

Also shown in Table 3 is the average slag surface temperature near the exit plane of the test duct. Comparison of tests #8 and 9 and of tests #19 and 21 show that the presence of potassium in the slag causes the surface temperature to increase by over 100 K. This is consistent with the measurements at the National Bureau of Standards which show that slag viscosity increases with increasing potassium content up to a limiting value. Comparison of tests #18 and 19 shows that the higher viscosity of the Illinois #6 coal compared to the Montana Rosebud also results in higher surface temperature (11).

Table 3. Flowing Slag Layer Characteristics

Test #	Ripple Speed (cm/sec)	Ripple Separation (cm)	Slag Surface Temperature (°K) Bottom Wall
6	6.0	-	1920
8	0.8	0.4	1700
9	1.8	0.3 - 0.6	1860
10	14.1	0.5 - 1.0	-
11	5.1	0.5 - 0.8	1750
16	2.6	.5 - 1	1770
18	2.1	.5 - 1	1800
19	-	-	1873
21	-	-	1740

Recent experiments (12) have been aimed at quantitative comparison of experimental data with the slag layer flow model. In these experiments, the flowing slag thickness was measured directly using a laser ranging instrument capable of resolving thickness changes of 0.3 mm and tracking ripples on the slag surface, Fig. 10. The temperature of the wall surface beneath the slag layer was varied from 400 to 1400 K by using a special gas-cooled wall module. Slag properties were estimated from published correlations, (11) and plasma-side boundary conditions were calculated from standard boundary layer correlations and a system heat balance. To obtain closure for the model, the slag run-off was directly measured using a deflector/hopper collection system. Figs. 11-13 summarize the results of these experiments. The solid lines represent the predictions based on the smooth layer model described above. For both types of coal, the surface temperature remains essentially constant for wall temperatures from 400 to 1400 K, a result in good agreement with the model predictions. For the Illinois coal the layer thickness measurements also agree well with predictions for wall temperatures greater than 800 K. The discrepancy at low wall temperature may be caused by a high thermal contact resistance between the slag and the wall.

ELECTRICAL EFFECTS OF SLAG COATINGS

A coal fired central station MHD power plant (13,19) will probably operate at high subsonic velocities (600-900 m/sec) and a magnetic field of 5 to 6 tesla. The combustion products will enter the MHD duct at a gas temperature of around 2600 K and with a potassium seed fraction of about 1 percent, the gas electrical conductivity in the MHD channel will range from about 10 down to 1 mho/m at the exit. Due to the large Hall parameter (1.5 to 4) of this working fluid, the generator electrodes must be segmented into narrow (around 1 cm) elements and the MHD channel must withstand axial electric fields of 4-8 kv/m. The average current density flowing transversely through the working fluid is about 0.5 to 1 amp/cm^2.

The thin (on the order of 1 mm) coating of conducting slag on the generator walls affects the electrical performance in two basic ways. First, current must pass through the slag layer from the electrodes to the working fluid. Secondly, the slag layer tends to short out the axial electric field which balances the Hall effect.

Rosa (4) has presented some simple expressions for estimating the deleterious effects of slag on electrical performance. Rosa assumes a power law for slag conductivity ($\sigma_{s1} \sim T^n$), diffuse current conduction, and linear temperature profiles in the slag. By requiring the voltage drop across the slag layers to be less than 5 percent of the generator open circuit voltage (uBh), the slag conductivity at the electrode-slag interface must be greater than a minimum value:

$$\sigma_w \geq \frac{2(1-n)}{(n-1).05} \sigma_g \left(\frac{s}{h}\right)\left(\frac{T_w}{T_s-T_w}\right)\left[1-\left(\frac{T_w}{T_s}\right)^{n-1}\right] \quad (10)$$

This relation reveals that the minimum value for σ_w varies inversely with generator size. For typical values of the above parameters with T_w = 1200 K and h = 1 m, σ_w should exceed 0.003 mho/m.

By prescribing that the axial conductance of the slag coatings be less than 5 percent of the axial interval conductance of the gas, Rosa finds a maximum value for the slag conductivity at the slag-plasma interface:

$$\sigma_s \leq \frac{0.05(1+n)}{4(1+\beta^2)} \sigma_g \left(\frac{h}{s}\right)\left(\frac{T_s-T_w}{T_s}\right)\Bigg/\left[1-\left(\frac{T_w}{T_s}\right)^{n+1}\right] \quad (11)$$

Hence, a large generator can tolerate a larger slag surface conductivity. The maximum allowable σ_{s1} is strongly decreased by large Hall parameter values. At a β of 3 and T_w = 1200, σ_s must typically be less than about 20 mho/m.

Most coal slags (except those with very large fractions of iron oxide) operating between a temperature of 1200 K to 1900 K have electrical conductivities within the above limits (11,14).

Slag coatings can have a beneficial effect on electrical performance. The relatively high surface temperature of the molten slag layer results in a reduction of the gas boundary layer voltage drop and in an increase in thermionic electron emission at the cathode wall. The resistive nature of the coating tends to reduce concentrations of current density at the electrode edges.

Although most of the above comments have assumed a diffuse current discharge through the slag, slags are electrothermally unstable at sufficiently large current densities due to the strong temperature dependence (n = 6-17) of slag conductivity. The electrothermal instability leads to current conduction through small arc columns in the slag.

Arc mode discharge transverse through the slag layer can be beneficial electrically in that a low impedance path between the electrode and gas can be maintained while the slag layer has a large impedance in the axial direction. However, the high temperature arcs tend to accelerate locally the erosion of electrodes. The development of axial arcs shorts out the axial electrical field (with a corresponding decrease in generator performance) and usually results in a failure of the interelectrode insulators.

EFFECTS OF SLAG ON CURRENT DISTRIBUTION

Current distributions in the plasma and slag regions have been calculated at magnetic fields up to 5 T. Diffuse mode current transport is assumed. From Maxwell's equations and the generalized Ohm's law, the current stream function, ψ, for the two-dimensional generator model is governed by the equation:

$$\frac{\partial^2\psi}{\partial x^2} + \frac{\partial^2\psi}{\partial y^2} + \frac{\partial(\ell n\sigma)}{\partial y}\left[\beta\frac{\partial\psi}{\partial x} - \frac{\partial\psi}{\partial y}\right] - \frac{\partial\beta}{\partial y}\frac{\partial\psi}{\partial x} = 0 \quad (12)$$

This equation along with periodic end conditions and ideal electrode and insulator conditions was solved numerically using given property distributions. The profiles of conductivity and Hall parameter used are shown in Fig. 14. The current distribution for a segmented Faraday generator at a magnetic field of 5 T is shown in Fig. 15.

In reality, the energy equation is strongly coupled into the present equations through the Joulean heating terms and will eventually be included in the analysis, but the results of the present calculations indicate problem areas. As can be seen in Fig. 15, the current concentrations on the electrodes is not as severe as in cases without slag. The present results show a current density concentration factor of about 1.5 for the upstream end of the anode and downstream end of the cathode, whereas Celinski and Fisher (16) found a concentration factor of about 7 for the case without slag. Current does tend to concentrate at the slag surface where the conductivities are high giving a concentration factor of about 11 there. Even though the maximum current concentrations are on the slag surface, the present results indicate about 6 times more Joulean heating in the cool highly resistive layer near the electrode surface than in the hot conductive region near the slag-plasma interface.

Transverse Discharge Results

The slag electrothermal stability and the slag-electrode chemical interactions have been studied by applied voltage discharges transverse to the slag layer (5,17). A magnetic field was not present in these experiments. Two types of coal and three electrode materials were tested. The primary variables were current density, electrode temperature, and anode versus cathode. In test #11, the bottom primary electrode was 1018 steel and tested as both anode and cathode. The bottom primary electrode in test #12 was stainless steel 304 operated as both anode and cathode. The upstream auxiliary electrode (steel 1018) was tested in the anode configuration. In test #19, the bottom primary electrode was split into two separate electrodes so that nickel 200 and steel 1018 could be compared side by side. These electrodes were operated as anodes only.

Visual Observations. The mode of electrical conduction was determined by cinephotography of the slag surface. Both the diffuse mode and the arc mode were observed for all the electrodes tested. A side by side comparison of the different modes of conduction is shown by the cine film reproduced in Fig. 16. The mild steel anode on the left is operating diffusely at a current density of 1.31 amps/cm^2 and an electrode (surface) temperature of 1450 K. The slag surface over the steel anode appears similar to the slag over the channel wall without current flow. The nickel anode on the right is operating in the arc mode although the electrode temperature is nearly the same as the steel electrode and the current density is 30 percent lower. Three quasi-stationary arc spots are visible on the edges of the nickel electrode (edges are radiused to 1.6 mm). At the start of the unstable discharge, many smaller and highly mobile arc spots appeared randomly distributed over the electrode surface. These smaller arc spots develop into a few larger arc spots which become entrenched on the electrode edges. Some slag layer damage from the arcs is visible, but the slag run-off repaired these voids before the coverage was significantly decreased.

During the initial period of applied current, the slag layer appeared to be bubbling, especially over the steel electrode. No hot spots were observed in conjunction with this bubbling.

Slag Discharge Mode

The critical current density for transition from the diffuse mode to the arc mode was sought for by applying a four level current schedule; the electrode current was held for five minutes at each level, then roughly doubled in value for the next level. The results of applying these current schedules at three different electrode temperatures are shown in Fig. 17 for the nickel anode. The experimental curve for this critical current density separates the current density versus electrode temperature plots into two regions, a diffuse mode region at low values of current density and an unstable arc mode region at large values of current density. The critical current density is a sensitive function of electrode temperature.

The results of the anode electrothermal stability tests for various coals and materials are summarized in Fig. 18. Not only is the critical current density strongly dependent on electrode temperature, but also on the type of electrode material. The mild steel electrode was almost an order of magnitude more stable than the nickel electrode for both the Montana Rosebud and the Illinois #6 coals. The critical current density for the steel anode increases more rapidly with electrode temperature than with the nickel case. The stainless steel anode was much less stable than the mild steel and only slightly better than nickel.

Slag Layer Polarization. The voltage drop across the slag layer (and part of the plasma boundary layer) was measured for each electrode in test #19 by a platinum sidewall probe about 0.8 cm from the electrode surface. Typical voltage drop and electrode current signals for the nickel anode are shown in Fig. 19 versus time. At the lower current, the voltage drop increased from about 8 to 80 volts over a time period of about 3 minutes. At a temperature of 1430 K, the polarization time for the nickel anode decreased to about 1 minute. Similar polarization behavior was also observed at the mild steel anode. At the beginning of the discharge experiments, the nickel anode (1240 K temperature) did not polarize completely until three current levels (at 5 minutes per level) had been applied. During the low temperature (1040 K) runs, the mild steel anode did not completely polarize after 5 minutes in three runs.

In Fig. 19, the current was increased to about 0.27 amps (J_{el} = 0.095 amp/cm^2 after 5 minutes of low current operation. The discharge immediately (on a time scale of seconds) began arcing. The arc voltage remained fairly constant at about 50 volts with a fluctuation level of about ± 5 volts. Arc voltage drops at the nickel anode ranged from 50 to 70 volts over the temperature range: 1020-1420 K and the current density range: 0.085-0.92 amps/cm^2. The arc voltage drop did not exhibit a significant dependence on electrode temperature in this range of temperatures.

Voltage-Current Characteristic. The polarized value of the electrode voltage drop is plotted versus electrode current (and average current density) in Fig. 20. The voltage-current characteristics indicate a resistive behavior for the diffuse mode discharges. The slope of the V-I characteristics increases steeply as the electrode temperature is decreased, indicating the expected large increase in slag layer resistance at low electrode temperatures.

The arc mode voltage drop is more constant voltage than resistive in nature; the arc voltage drop is only 62 volts at a current density of 5.6 amps/cm^2.

Slag Layer Resistance. The measured slag layer resistance is plotted versus electrode temperature in Fig. 21 for the mild steel and the nickel anodes. These values of slag layer resistance are for diffuse discharges, polarized layers, and small current densities (layer not modified by Joule heating). The layer resistance is expressed in ohm-cm^2 so that the voltage drop can be determined by multiplying layer resistance times current density. The layer resistance for the nickel anode is more than an order of magnitude greater than at the mild steel anode. The layer resistance in both cases decreases rapidly with increasing electrode temperature.

Diffuse Slag Layer Resistance. The slag voltage drop, V_{sl}, for the diffuse mode is equal to $r_\ell J_{el}$ where the layer resistance is: $r_\ell = \int_0^s dy/\sigma_{sl}$ and s is the slag layer thickness. The slag conductivity, σ_{sl}, is a sensitive function of temperature which requires that the energy equation for the slag layer be solved in order to evaluate r_ℓ. For wall temperatures, T_w, below the slag temperature of critical viscosity, the slag surface temperature T_s, is not affected by changes in T_w and only slightly affected by J_{el} (5,12). The slag resistance for the conditions of test #19 was calculated by using the measured value of T_s = 1870 K and a slag thermal conductivity, k, value of 1.3 watts/m-K determined from a frozen slag thickness measurement and the equation: $s = k(T_s - T_w)/\dot{Q}_s$ where $\dot{Q}_s$ is the heat flux. The slag electrical conductivity measurements by NBS (13) on an Illinois #6 coal slag sample from an MHD generator were used. The calculated slag voltage drop-current density characteristics are shown in Fig. 22. The voltage drop increases rapidly with a decrease in electrode temperature. The V-J characteristics decrease in slope at large values of J_{el} due to Joule heating; however, the diffuse discharge becomes unstable before this effect is significant. The boundary curve in Fig. 22 for diffuse-arc mode was calculated from a semi-empirical correlation described below. The calculated slag layer resistance is compared (see Fig. 21) with the resistance measurements of test #19 by adding on the plasma resistance between the slag surface and the voltage probe (estimated as 12 ohm-cm^2). Above T_w = 1200 K, the slag resistance at the mild steel electrode is lower than calculated, probably due to the high conductivity of iron rich phases found at the steel-slag interface. Below T_w = 1100 K, the larger value of measured resistance at the steel may be caused by contact resistance. The resistance at the nickel electrode is considerably larger than calculated from slag properties and indicates a large temperature dependent contact resistance at the nickel-slag interface.

Slag Discharge Stability. The strong temperature dependence of slag electrical conductivity results in electrothermal instability of the slag layer at some critical value of current density. The temperature dependence may be characterized by a power law exponent, $n = d\ \ell n\ \sigma / d\ \ell n\ T$, where n ranges from 6 to 17 for slags. The characteristic time for electrothermal breakdown is s^2/α where α is a slag thermal diffusivity. Using the α measurements of Bates (18), the characteristic breakdown time is a few seconds. A correlation, based on dimensional analysis, for the critical current density (5) may be expressed as:

$$J_c = \sqrt{k\,\sigma_w\,T_w/s^2}\;\;G(n,\gamma) \tag{13}$$

where $\gamma = T_s/T_w$. Since σ_w varies as T_w^n, J_c is a strong function of temperature, varying as $T_w^{(n+1)/2}$. Results from a linear perturbation analysis of the electrothermal instability show that $G(n,1) \sim 1/\sqrt{n}$ and suggest that $G \sim 1/\sqrt{n}$ for $\gamma \neq 1$. Since σ_w can vary by orders of magnitude, the primary factor governing the critical current density is the slag conductivity near the electrode-slag interface.

The slag layer resistance can be expressed as $r_\ell = s/\sigma_w\ g(n,\gamma)$ where g varies from 0.16 to 0.24 over the range γ = 1.2-1.8. Since σ_w is closely related to r_ℓ, the critical current density correlation can be expressed in terms of r_ℓ:

$$J_c = \sqrt{k\,T_w/r_\ell\,s}\;\;H(n,\gamma) \tag{14}$$

This correlation should apply to all electrode materials and slags. If the slag resistance is a fairly strong function of temperature, $1/r_\ell \sim T^6 - T^{17}$, then the effect of n and γ on J_c should be second order effect. This correlation is evaluated over a wide range of resistances by using the measured values shown in Fig. 21. J_c is plotted versus $\sqrt{k\,T_w/r_\ell\,s}$ in Fig. 23. On a log-log plot, the correlation predicts a straight line of unity slope should separate the arc mode points from the diffuse mode points for both nickel and steel electrodes. As shown by Fig. 23, this correlation works well over a range of about three orders of magnitude with H = 0.44.

Cathode Results. A stability map for a mild steel cathode is shown in Fig. 24. The anode stability boundary for mild steel is included for reference. Above a current density value of about 1.8 amps/cm^2, the cathode operated in a constricted discharge mode for all electrode temperatures investigated. This suggested that the cathode may be limited by thermionic emission at the slag layer surface. Based on a slag surface temperature of 1750 K, the effective work function is 2.9 eV. This value is similar to that of metal surfaces with an adsorbed layer of potassium. Below the thermionic limiting current density, J_{th}, diffuse cathode discharges were observed at electrode temperatures above 1300 K and constricted hot-spot discharges at 550 K temperature in test #10. However, the present number of cathode data points below J_{th} are insufficient to define the electrothermal stability boundary.

Axial Slag Breakdown Experiments

The breakdown strength of the slag layer was measured in test #12 with Montana coal by applying an axial voltage across two 0.32 cm long alumina insulators with a central floating metal segment 1.27 cm in length (see Fig. 25). The discharge before and after breakdown are shown by selected frames from the cine film in Fig. 27 (the location of electrodes and insulators are indicated by an overlay described in Fig. 26). At V = 52 volts, the discharge had no visible effect on the slag surface. After breakdown, the voltage dropped from 150 volts to 54 volts. In this discharge mode, nearly all the current is transferred through an intense localized discharge across each ceramic gap.

The voltage-current characteristic for the axial discharge is shown in Fig. 28. Stable discharges were observed at voltages up to 115 volts with breakdown observed at about 145 volts. These results indicate a slag axial breakdown strength (based on the ceramic gap) between 18 and 23 kv/m.

CHEMICAL CHARACTERIZATION OF SLAG LAYERS

The bulk composition of the slag deposits from various coals and over various surfaces are presented in Table 4. The composition was determined by elemental atomic absorption measurements. The mineral analyses of the coal used in the corresponding test is given for comparison.

Table 4. Composition of Slag Layer over Various Surfaces with Different Coals

Oxide (wt. %)	SiO_2	Al_2O_3	Fe_2O_3	CaO	MgO	K_2O	SO_3	C	NiO
Pittsburgh Coal Ash	54.1	32.2	5.9	1.8	0.8	1.2	0.4	-	-
TEST #8									
Deposition Channel Wall	51.4	27.5	5.2	1.5	5.8	3.2			
TEST #9									
Deposition Channel Wall	36.7	33.0	3.2	0.8	4.2	20.2	0.7	0.2	-
TEST #12									
Montana Coal Ash	42.9	17.2	4.4	13.3	3.3	0.5	17.3	-	-
Copper Channel Wall	38.0	27.0	3.4	12.9	3.6	21.3	-	-	-
SS 304 Electrode	34.6	25.2	10.0	10.6	5.3	21.4	-	-	-
Mild Steel Electrode	23.1	21.0	43.4	7.9	3.5	8.8	-	-	-
TEST #19									
Illinois #6 Coal Ash	54.1	18.7	13.0	5.2	-	2.4	3.1	-	-
Copper Channel Wall	52.0	24.6	7.7	0.4	-	22.6	.03	-	0.04
Nickel Anode	44.5	18.9	9.4	0.2	-	24.8	.03	-	2.2
Mild Steel Anode	41.8	18.9	10.5	0.4	-	25.2	-	-	-
Steel Cathode, Near Plasma	45.0	18.1	9.5	0.3	-	28.6	-	-	-
Steel Cathode, Near Electrode	42.9	-	15.8	-	-	32.2	-	-	-
TEST #20									
Steel Anode	-	17.2	13.4	-	-	19.2	-	-	-
Steel Cathode	-	19.7	19.8	-	-	28.3	-	-	-

Deposits from several locations (upstream, downstream, copper sidewall, stainless steel lower wall) were analyzed for tests #8 and #9, but no large variations observed. Large amounts of potassium oxide (~ 20%) was absorbed into the slag layer for all tests except #8, which was run without seed. The amount absorbed is in rough agreement with the equilibrium results of reference (19) based on the slag surface temperature.

The slag over the cold copper channel wall in test #12 contained a small amount of iron oxide similar to the value in the Montana coal ash. The slag over the stainless steel 304 electrode contained about 3 times as much iron oxide as the copper wall sample while the slag at the 1018 steel electrode contained almost 31 times that in the copper sample.

The mild steel electrodes were uniformly coated with an adherent slag layer after test #12. A nonuniform, weakly adherent slag coating was observed at the stainless steel electrode.

In test #19 (with Illinois #6 coal), the effect of polarity on composition was assessed by running the bottom electrodes as anodes only. The amount of iron oxide found on the nickel anode, the mild steel anode, and the steel cathode slag near the plasma was comparable to the amount in the coal ash when normalized by the large amount of potassium oxide absorbed into the slag. A small increase in iron oxide was measured at the steel anode. However, a factor of 1.7 extra iron oxide was observed in the slag near the steel interface at the cathode; a factor of about 1.3 increase in K_2O was also observed in this sample. This transport of positive ions to the cathode surface was observed after 0.7 amp-hour/cm^2 of charge transferred through the cathode.

The steel side half of the slag coating was bonded extremely well to the cathode surface. The slag coating at the steel anode was adherent. The slag layer at the nickel anode was not adherent. Although the combustor was run 10 percent fuel rich in test #19 to minimize the oxidation of the metal electrodes, a thin greenish film of nickel oxide was found at the nickel-slag interface.

Test #20 was operated under the conditions of constant electrode temperature (1400 K), constant current density, as well as the same polarity. As in test #19, extra iron and potassium was observed at the cathode.

Polished slag cross-sections were prepared and studied with the scanning electron microscope at the Stanford University Center for Materials Research. The composition distributions through the slag layers were measured with an electron beam microprobe. A typical distribution of iron through the slag is shown in Fig. 30 for a sample from test #12. The distribution revealed an iron rich (≃ 30 percent) region, near the electrode interface, which extends about 25 percent through the slag layer. The iron content in the plasma side of the slag layer was uniform and low (≃ 1.8 percent) roughly corresponding to the amount in the ash.

The processes which form the iron rich region over steel electrodes have been investigated by a detailed study of the slag-electrode microstructure performed at the Metallurgy and Composites Laboratory of the Lockheed Research Laboratory, Palo Alto, Calif. The results of an analysis of a slag coated mild steel electrode from test #12 are summarized in Fig. 30. The outer zone consists of a glassy high density slag near the high temperature slag-plasma interface. The next zone consists of an iron rich devitrified second phase in a glassy matrix of potassium-aluminum-silicate. The third region consists of three phases: Wustite and Fayalite crystals in a potassium-aluminum-iron-silicate slag. The steel electrode was covered with a thin layer of Wustite with intergranular and interfacial penetration of slag.

These observations indicate that oxygen from the plasma boundary layer had diffused through the slag to form a thick scale of Wustite (FeO) on the mild steel electrode. The iron oxide scale was then attacked by the slag to form large crystals of Fayalite (2 $FeO-SiO_2$) in a fused silica mass.

Photomicrographs of the slag layers formed over steel electrodes during the constant conditions of test #20 are shown in Fig. 31.

The slag microstructure at the steel anode is typical in that the slag near the plasma was glassy and homogeneous while near the cooler electrode, the slag contained devitrified second phases. The slag microstructure at the cathode was quite different: the glassy region was absent over almost the entire slag cross-section; this layer was considerably more porous; small regions of pure iron were found more or less uniformly distributed across the entire slag layer. The largest regions of pure iron phase were about 20 microns in size. The pure iron regions were found at a number density (based on slag cross-section area) of about 280 mm^2. These iron regions could have formed from ejected electrode material or from reduction of slag due to oxygen ion transport out of the cathode slag layer.

The presence of crystallized phases in slag should increase the viscosity of slag and hence the slag layer thickness. The slag layer at the cathode was 30 percent thicker than at the anode which is consistent with the microstructure observations.

ELECTRODE EROSION

As discussed above, current transfer to MHD electrodes of several amps/cm^2 may be achieved with voltage drops of less than one hundred volts regardless of electrode temperature or mode of current conduction. For large scale (~ 1,000 MWe) MHD-steam power plants, electrode voltage drops of ≈ 100 volts result in a minimal decrease in plant performance. The erosion of MHD electrodes for long duration power plant operation is the limiting factor rather than that of electrical performance.

Since an MHD power plant generator has no moving parts and lateral dimensions on the order of a few meters, substantial erosion of the electrode wall might be tolerated. In the worse case, the erosion must be less than the distance to the electrode cooling passage. This distance could range from ≈ 200 mm for copper to ≈ 10 mm for stainless steel to values of only a few millimeters for the low thermal conductivity ceramic electrode materials. For cheap metal electrodes, other considerations such as large localized erosion, disturbance to the gas boundary layer flow, or contamination of downstream components may be the limiting factor. For the purpose of this discussion, a maximum erosion of 10 mm is assumed. For precious metals, economics is the limiting factor and the erosion of these materials must be less than a fraction of a millimeter.

A practical lifetime for a power plant MHD generator is in the range of 5,000 - 10,000 hours (20). Based on these considerations, erosion rates of $\approx 2 \times 10^{-3}$ mm/hr for cheap materials and $\approx 1 \times 10^{-4}$ mm/hr for precious materials are required. Despite the importance of the erosion problem, quantitative values for erosion rates are not known for most operating conditions. The most extensive and careful study of erosion rates were made in the British MHD program (19) of the sixties. However, this study did not include the effect of coal slag. The erosion of several materials operating in the arc mode under coal slag have been reported by Petty, et al (21). Erosion of some ceramic electrodes operating diffusely with coal slag in a static test were reported by Cadoff et al (22). The erosion of low carbon steel electrodes operating diffusely with coal slag was measured in our slag test #20. Several tests have been made of high temperature ceramic electrodes without coal slag (19,23,24), but these results will not be discussed here.

The British Results (19)

The British studied the erosion of various metal electrodes in a simulated MHD duct environment using potassium carbonate or potassium sulphate as the seed material. The erosion increased linearly with time and roughly linear with current density over the range 1 to 5 amps/cm^2. For the purpose of comparison and extrapolation, an electrode erosion rate, $\dot{\delta}_{el}$, is defined as:

$$\dot{\delta}_{el} = \frac{\delta}{J \tau} \tag{15}$$

The electrode erosion rate was found to depend strongly on electrode temperature, material, polarity, and seed deposits (carbonate vs sulphate). Variations over four orders of magnitude were observed for materials including: copper, mild steel, stainless steel, nickel, gold, silver, platinum, rhodium, and palladium. The electrode erosion rate was only weakly dependent on the equivalence ratio and almost independent of heat flux, velocity, seed concentration above 0.5 wt % K, and hall parameter up to 1.

Selected erosion results are given in Table 5. The experimental conditions were: $u = 200$ m/sec, $\dot{Q}_s = 130$ w/cm^2, $T_g = 3000$ K, $T_{el} = 420$ K, $\varepsilon_K = 0.5\%$, $B = 3$ T.

Table 5. Erosion Rates of Cold Metal Electrodes as Measured in the British Program.

K_2CO_3 Seed	I = 0 $\dot{\delta}$ (mm/hr)	Anode $\dot{\delta}_{el}$ (mm/amp-hr/cm^2)	Cathode
Cu	$< 10^{-5}$	$\sim 9 \times 10^{-3}$	2.2×10^{-4}
Au	-	-	2.0×10^{-2}
Pt	$< 10^{-5}$	5.3×10^{-5}	2.7×10^{-3}
K_2SO_4 Seed			
Cu	3.2×10^{-4}	4.6×10^{-1}	6.3×10^{-4}
Au	-	3.1×10^{-2}	9.6×10^{-3}
Pt	2×10^{-4}	4.0×10^{-4}	1.1×10^{-3}

For copper, the erosion at the anode is greater than at the cathode despite the much more concentrated discharge observed at the cathode (electron emitter). The British conclude that cathode erosion is primarily caused by metal evaporation at the arc root. This damage can be minimized by a large thermal diffusivity material which explains the good erosion resistance of copper cathodes. Anode erosion is caused by chemical attack at the high temperature of the concentrated discharge and electrolytic attack by surface deposits. The sulphate deposits result in a large increase in the erosion rate at the anode. The use of noble metals such as gold and platinum were found effective in decreasing the anode erosion rate.

Since a current density of about 1 amp/cm^2 is required, values of $\dot{\delta}_{el} = 2 \times 10^{-3}$ and 1×10^{-4} mm/amp-hr/cm^2 are desirable for cheap electrodes and precious electrodes respectively. For clean fuel with potassium carbonate seed, copper cathodes and platinum clad copper anodes meet the erosion criteria. The erosion performance in the presence of sulfur is not sufficient.

AERL Results (21)

Several metal alloy electrodes were tested with and without coal slag at AERL under electrical conditions approaching those expected in a power plant MHD generator. The nominal test conditions were $M = 1.5$, $B = 2.3$ T, $E_x = 1.8$ kV/m,

J_y = 0.7 amps/cm^2, $\beta \approx 2$. Electrodes were operated for about 20 hours and checked for damage by visual observation. Assuming that erosion on the order of 0.1 mm is detectable by visual observation, erosion rates greater than 7×10^{-3} mm/amp-hr/cm^2 were measured qualitatively.

OFHC copper and tungsten-copper (57 W - 43 Cu) cathodes operating at 700 K and 420 K respectively were rated as "excellent" (no material loss) for both with and without slag. K-Monel (66 Ni, 30 Cu, 3 Al, 1 Fe) and Hastelloy B (66 Ni, 28 Mo, 5 Fe, 1 Cr) operating at about 900 K were rated as "poor". Tantung (45 Co, 27 Cr, 14 W), TD nickel (98 Ni, 2 ThO_2) and Inconel (76 Ni, 16 Cr, 8 Fe) were rated from "very good" to "fair" respectively.

Anode erosion was more severe than at the cathode and was increased by slag. Inconel anodes operated at about 900 K and 0.75 amp/cm^2 were uniformly eroded on the upstream edges in correspondence with the predicted current density concentrations on the upstream edge due to the Hall effect. From data presented, the electrode erosion rates can be estimated as $\approx 2 \times 10^{-2}$ mm/amp-hr/cm^2 without slag and $\approx 4 \times 10^{-2}$ mm/amp-hr/cm^2 with slag. The enhanced erosion with slag was blamed on oxygen transport in the slag. About four percent ionic transport in the slag accounts for this erosion. Copper anodes performed better than Inconel with some upstream erosion and a few local areas with larger chips missing. More erosion resistant anodes are required and copper clad with a noble metal are logical candidates for improvement.

Westinghouse Results (22)

Static erosion measurements of ceramic electrodes in contact with coal slag were made at high temperatures (~ 1700 K) and diffuse current transport. The results are shown in Table 6. The relatively oxidation resistant ceramics performed much better as anodes than cathodes, but have an unacceptably large erosion rate in either case. Even at 1700 K, several volts were required to drive the current across the liquid slag; hence, very stable materials are required to withstand electrolytic attack by molten slag.

Table 6. Erosion Rates of Moderately High Temperature Ceramic Electrodes in Contact with Slag.

	$\dot{\delta}$ (mm/hr)	Anode $\dot{\delta}_{el}$ (mm/amp-hr/cm^2)	Cathode
Spinel 3 $MgAl_2O_4$-1 Fe_3O_4 1670 K	-	1.1×10^{-1}	1.68
La Cr O_3 1730 K	-	4.1×10^{-1}	1.92
Sn O_2 1730 K	2.6×10^{-3}	1.5×10^{-1}	2.85

Stanford Results

The low carbon steel electrodes were shown above to perform well electrically with slag. Large diffuse currents were achieved at the relatively low electrode temperatures of 1400 K. The steel 1018 electrodes were operated for two hours in test #20 under the steady conditions of T_w = 1400 K, and $J_{el} \approx$ 1 amp/cm^2. The amp-hours accumulated by the steel electrodes during the two tests are given in Table 7.

The steel anode was altered from a rectangular cross section of 12.7 mm width to a curved top with a radius of about 12.4 mm. The average recession of the steel anode surface was 2.0 mm which results in an erosion rate of 6.8×10^{-1} mm/amp-hr/cm^2. This heavy attack suggests that electrochemical corrosion

Table 7. Electrode Amp-hours

	J τ (amp-hr/cm^2)	
	Test #19	Test #20
Steel Anode	1.33	1.62
Steel Cathode	0.70	0.78

is occurring and that some of the current in the slag is conducted by ionic species. Applying Faraday's law to divalent ions (either $O^{=}$ or Fe^{++}), the recession rate of the steel electrode is:

$$\dot{\delta}_{el} = 1.35 f_i \tag{16}$$

The corresponding value of f_i for the steel anode experiment is 0.5.

The steel cathode did not change dimensions by a significant amount which implies a cathode erosion rate of less than 7×10^{-2} mm/amp-hr/cm^2.

CONCLUSIONS

For experiments with the channel wall temperatures below the slag liquidus temperature, stable, continuous coatings of slag were formed over suitable substrates. The slag surface temperature is independent of the channel wall temperature and the slag layer acts as a thermal barrier. The slag layer absorbs a large (~ 25% K_2O) amount of potassium from the seeded combustion plasma.

The slag layer conducts large current densities (several amps/cm^2) at a reasonable voltage drop (less than 100 volts) independent of the electrode temperature. The mode of current conduction (arc versus diffuse) is a very sensitive function of electrode temperature and material. Electrode temperatures of at least 1400 K are required for diffuse conduction of a few amps/cm^2. For diffuse conduction, the electrode must be in good electrical "contact" with the slag layer. Mild steel makes good contact with slag due to the chemical interaction of the iron oxide scale with the slag. Nickel and stainless steel make poor contact with slag.

The primary MHD electrode problem is that of erosion. Electrode erosion is roughly a linear function of current density and depends strongly on electrode temperature, material, polarity, and surface deposits. Metal anodes are eroded more rapidly than cathodes due to electrolytic attack by either seed or slag deposits. For ceramic electrodes, the cathodes are attacked the greatest.

Chemical attack may be minimized by operating the electrode wall at low temperatures in the arc mode. The electrode materials must have a very large value of thermal diffusivity to dissipate the thermal energy of the arcs. Cold copper cathodes and platinum clad copper anodes appear capable of operating with slag for a few thousand hours.

The erosion resistance of high temperature diffuse mode electrodes is degraded by the strong electrolytic attack caused by the slag layer. Metal cathodes may possess sufficient erosion resistance in the moderate temperature diffuse mode; however, anode materials considerably more stable than any material tested previously are required for practical diffusively operating anodes. Slag compatible ceramic materials and noble metal clad electrodes are candidates for a long duration diffuse mode anode.

REFERENCES

1. Dicks, J. B., et al., "The Direct-Coal-Fired MHD Generator System," 14th Symposium on Engineering Aspects of MHD, Tullahoma, Tenn., April 1974.

2. Lacey, J. J., Demeter, J. J., and Bienstock D., "Production of a Clean Working Fluid for Coal-Burning, Open Cycle MHD Power Generation," 12th Symposium on Engineering Aspects of MHD, Argonne, Ill., March, 1972.

3. Feldmann, H. F., Simons, W. H., and Bienstock, D., "Design Data for Coal-Burning Open-Cycle MHD Systems: 1. Effect of Slag Formation on Duct Operation and Seed Recovery," Advances in Energy Conversion Engineering, Proceedings of the Intersociety Energy Conversion Engineering Conference, Miami Beach, Fla., August 1967, pp. 423-429.

4. Rosa, R. J., "Design Considerations for Coal-Fired MHD Generator Ducts," Fifth International Conference on MHD Electrical Power Generation, Munich 1971.

5. Koester, J. K., Rodgers, M. E. and Eustis, R. H., "In-Channel Observations on Coal Slag," 15th Symposium on Engineering Aspects of MHD, Philadelphia, May 1976.

6. Stickler, D. B. and DeSaro, R., "Replenishment Analysis and Technology Development," Sixth International Conference on MHD Electric Power Generation, Washington, D.C. 1975.

7. Stickler, D. B. and DeSaro, R., "Replenishment Processes and Flow Train Interaction," 15th Symposium on Engineering Aspects of MHD, Philadelphia, May 1976.

8. Crawford, L., et al., "Investigation of Slag Deposits and Seed Absorption in a Direct-Fired MHD Power Generator," Sixth International Conference on MHD Electric Power Generation, Washington, D.C., 1975.

9. Holve, D., "Optical Measurements of Mean Particle Size in the Exhaust of a Coal-Fired MHD Generator," Memorandum Report #4, HTGL, Mech. Eng. Dept., Stanford, Ca.

10. Liu, B. Y. H., and Agarwal, J. K., <u>Aerosol Science</u>, Vol. 5, pp. 145-155, 1974.

11. Capps, W., "Some Properties of Coal Slags of Importance to MHD," 16th Symposium on Engineering Aspects of MHD, Pittsburgh, May 1977.

12. Rodgers, M. E., Ariessohn, P. C. and Kruger, C. H., "Comparison of Measurements and Predictions of the Fluid Mechanics and Thermal Behavior of MHD Channel Slag Layers," 16th Symposium on Engineering Aspects of MHD, Pittsburgh, May 1977.

13. Jackson, W. D., et al, "Status of the Reference Dual-Cycle MHD-Steam Power Plant," 16th Symposium on Engineering Aspects of MHD, Pittsburgh, May 1977.

14. Bates, J. L., "Electrical Conductivity of Molten Coal Slags Containing Potassium Seed," 16th Symposium on Engineering Aspects of MHD, Pittsburgh May 1977.

15. Armstrong, A. J., Hosler, W. R., and Frederiske, H. P. R., "Development, Testing, and Evaluation of MHD-Materials," ERDA Contract E(49-1)-3800 Report, NBS, September 30, 1976.

16. Celinski, Z. N., and Fischer, F. W., "Effect of Electrode Size in MHD Generators with Segmented Electrodes," AIAA Journal, March, 1966.

17. Koester, J. K. and Nelson, R. M., "Electrical Behavior of Slag Coatings in Coal-Fired MHD Generators," 16th Symposium on Engineering Aspects of MHD, Pittsburgh, May 1977.

18. Bates, J. L., "Properties of Molten Coal Slag Relating to Open Cycle MHD," NSF-RANN Grant GI-44100 Report, Battelle Pacific Northwest Laboratories, February, 1975.

19. Heywood, J. B. and Womack, G. J. (eds.), Open Cycle M.H.D. Power Generation Pergamon Press, 1969.

20. Zygielbaum, P. S., Electric Power Research Institute, Palo Alto, Ca., Personal communication, June 1977.

21. Petty, S, Demirjian, A., and Solbes, A., "Electrode Phenomena in Slagging MHD Channels," 16th Symposium on Engineering Aspects of MHD, Pittsburgh, May 1977.

22. Cadoff, L. H., Smith, H. D., and Rossing, B. R., "The Evaluation of Electrode Materials for Slag Coated MHD Channels," 16th Symposium on Engineering Aspects of MHD, Pittsburgh, May 1977.

23. Gokhshteyn, Ya. P., Safanov, A. A. and Lyubimov, V. P., On the Physical and Chemical Behavior of the Electrodes of an MHD Generator Made of ZrO_2-Y_2O_3 and ZrO_2-CeO_2, in "MHD Methods of Producing Electrical Energy," (Ed. by V. A. Kirillin and A. Y. Sheindlin).

24. Rudins, G. et al, "The Second Joint Test of a U.S. Electrode System in the U.S.S.R. U-02 Facility," 16th Symposium on Engineering Aspects of MHD, Pittsburgh, May 1977.

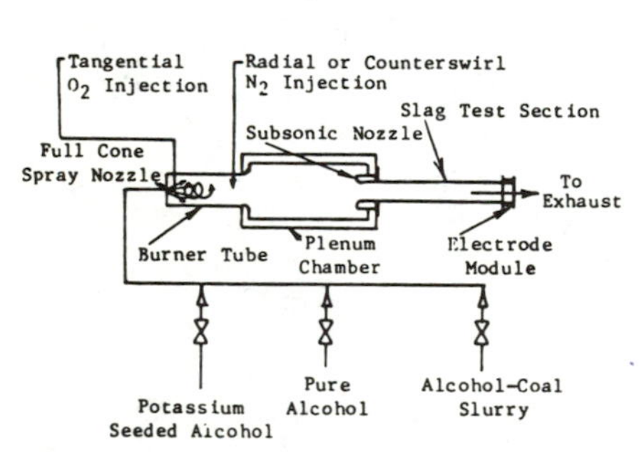

Fig. 1. Slag Experimental Facility - Flow Train Schematic.

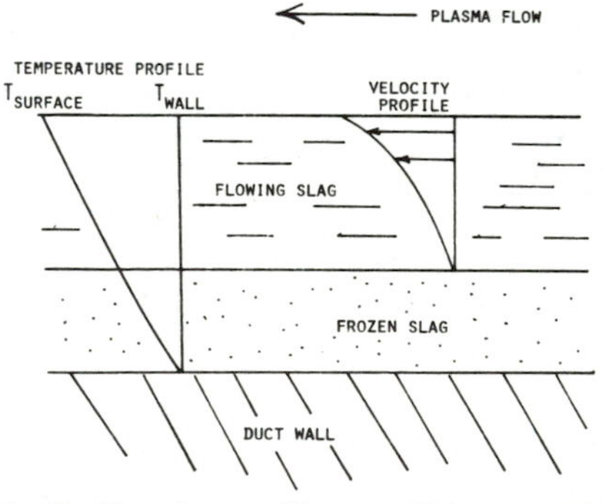

Fig. 3. Slag Layer Flow on MHD Duct Walls.

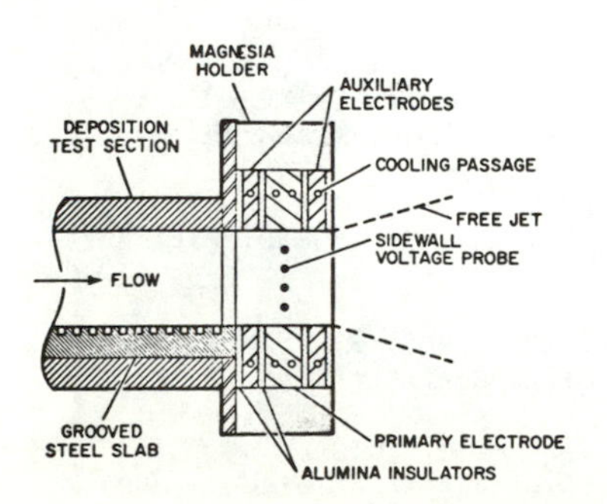

Fig. 2. Schematic of Electrode Module.

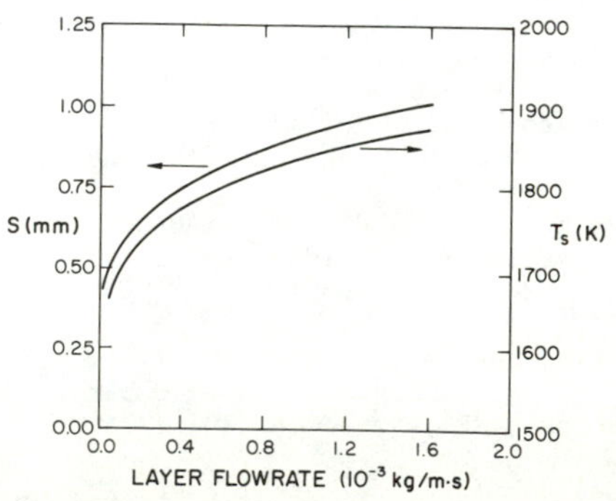

Fig. 4. Effect of Slag Flowrate on Surface Temperature and Layer Thickness.

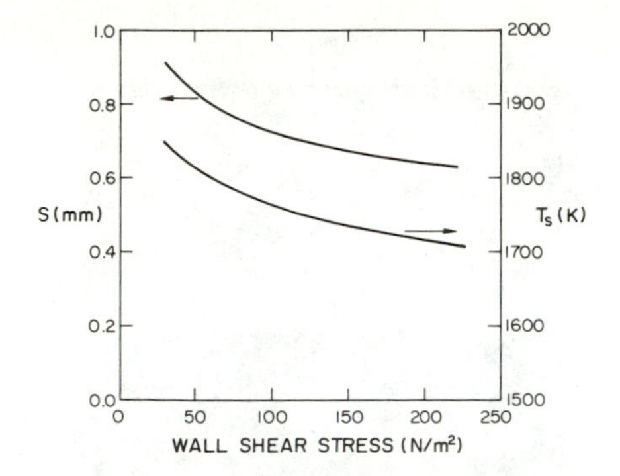

Fig.5. Effect of Wall Shear Stress on Surface Temperature and Layer Thickness.

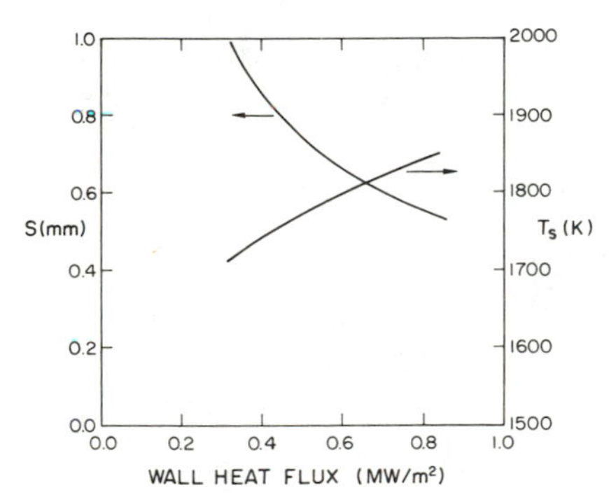

Fig.6. Effect of Heat Flux on Surface Temperatures and Layer Thickness.

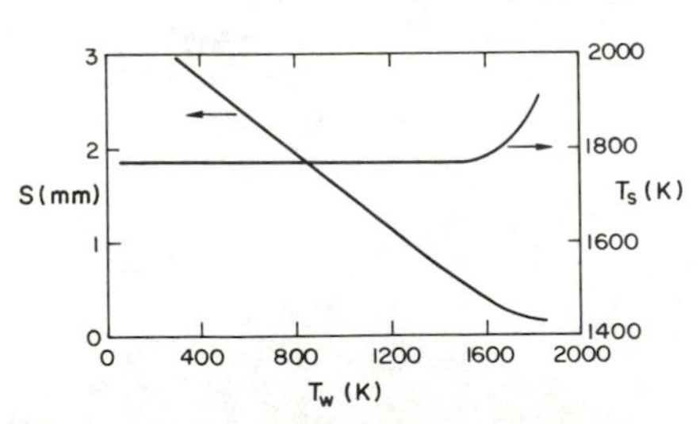

Fig.7. Effect of Wall Temperature on Surface Temperature and Layer Thickness.

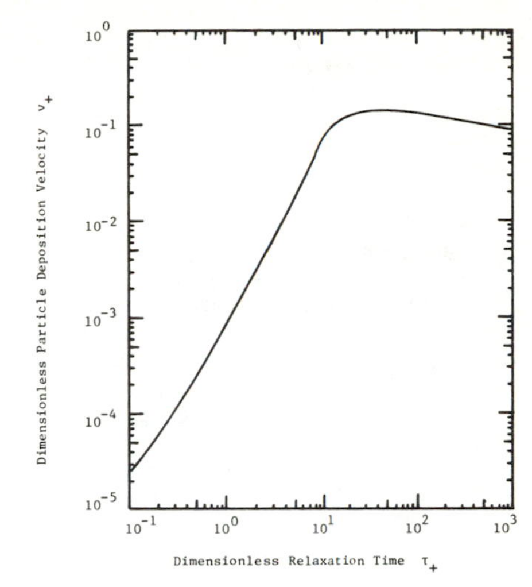

Fig.9. Dimensionless Particle Deposition Velocity vs Dimensionless Relaxation Time.

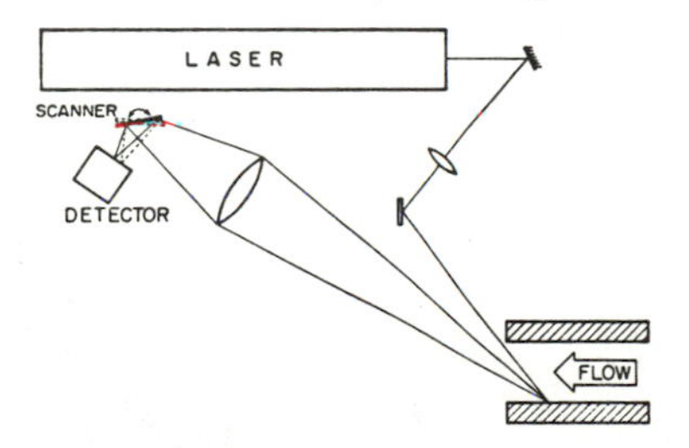

Fig.10. Laser Ranging Instrument for Slag Thickness Measurement.

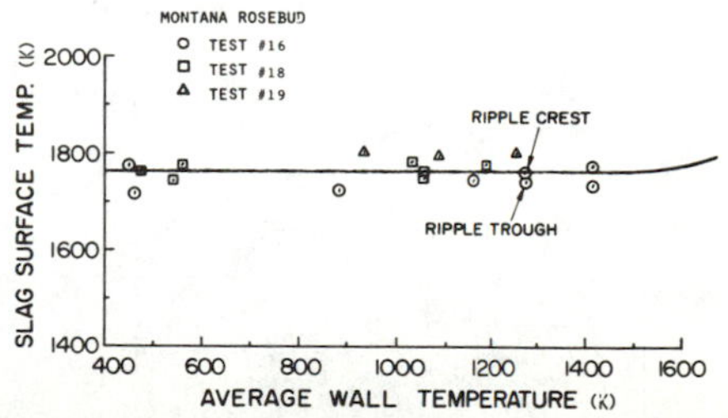

Fig.11. Slag Surface Temperature vs Wall Temperature.

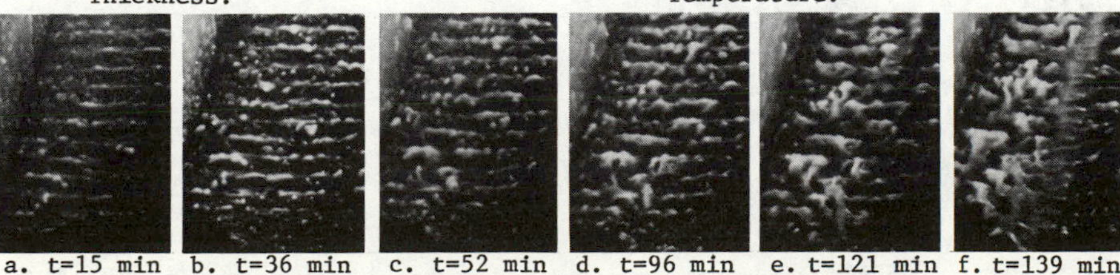

a. t=15 min b. t=36 min c. t=52 min d. t=96 min e. t=121 min f. t=139 min

Figs. 8a-f. Development of Slag Layer on the Grooved Stainless Steel Surface During Test #8.

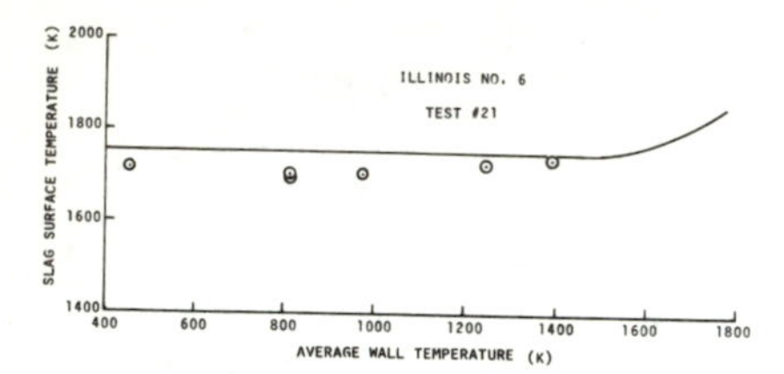

Fig. 12. Slag Surface Temperature versus Wall Temperature.

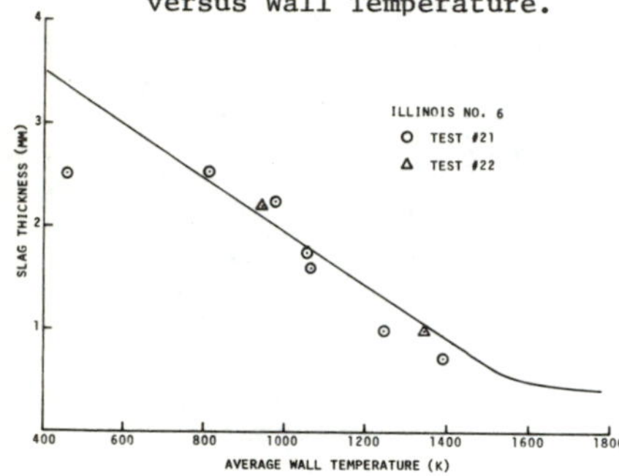

Fig. 13. Slag Thickness versus Wall Temperature.

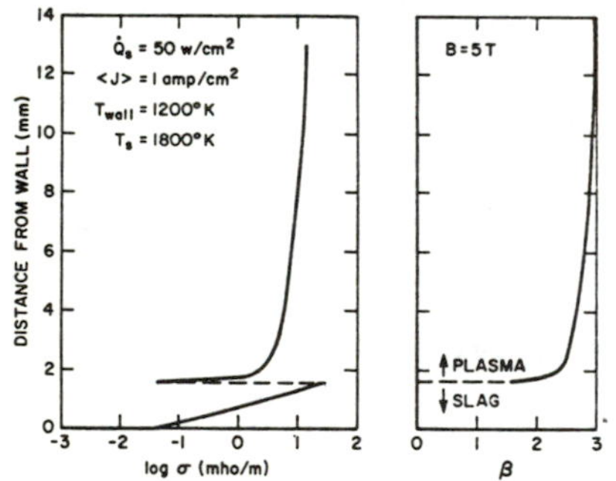

Fig. 14. Conductivity and Hall Parameter Profiles.

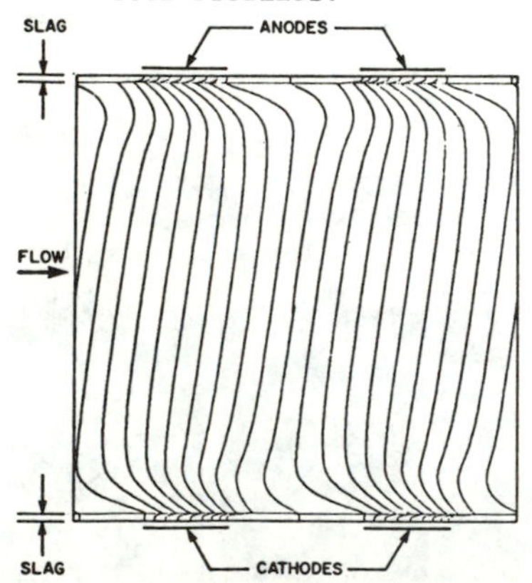

Fig. 15. Current Streamlines for a Slag Coated Segmented Faraday Generator at B = 5T.

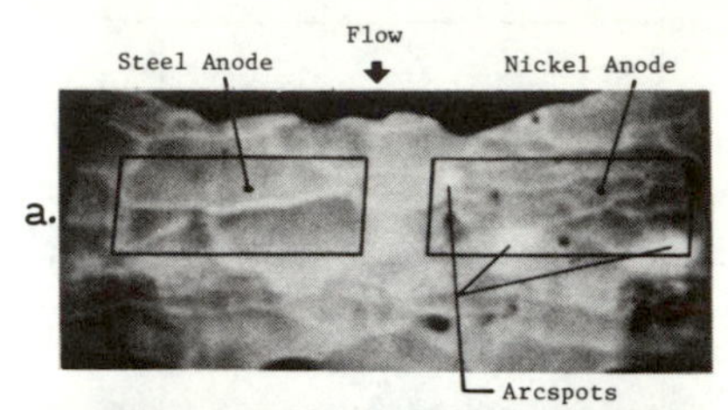

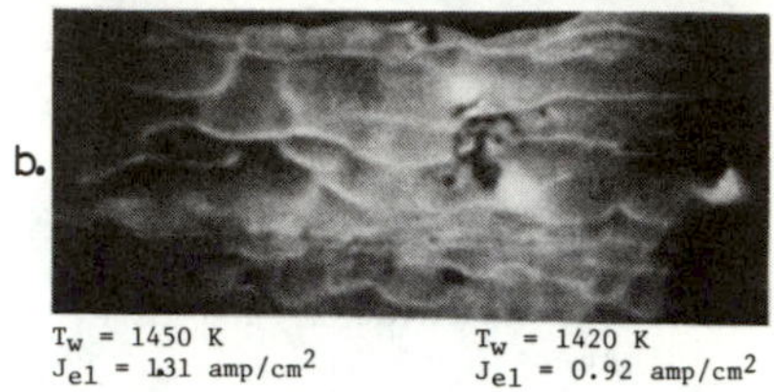

Fig. 16. Side by Side Comparison of Electrode Operating in the Diffuse Mode and the Arc Mode.

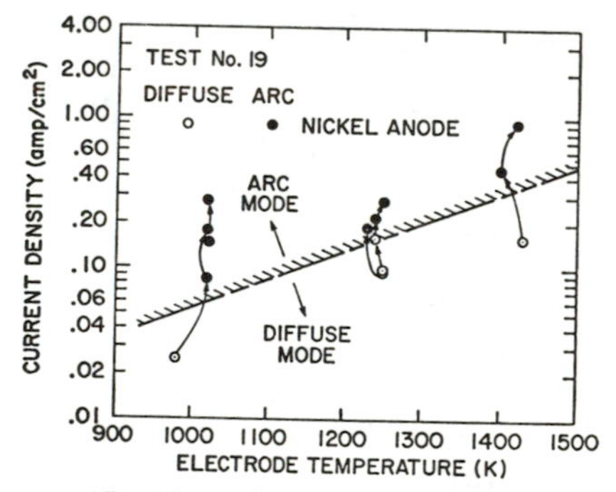

Fig. 17. Slag Electric Discharge Stability Map.

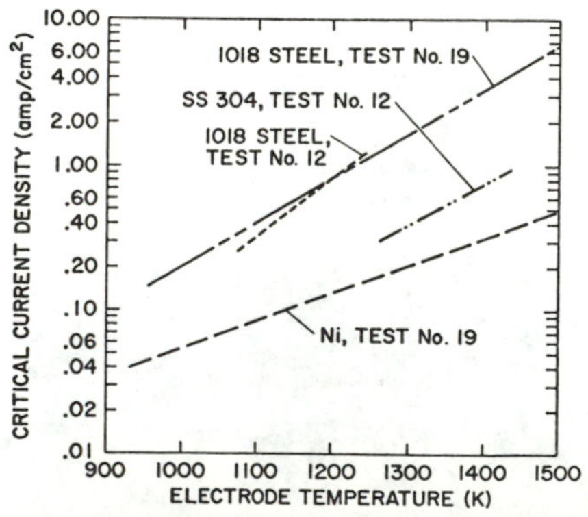

Fig. 18. Critical Current Density for Diffuse to Arc Mode Transition as Function of Electrode Temperature.

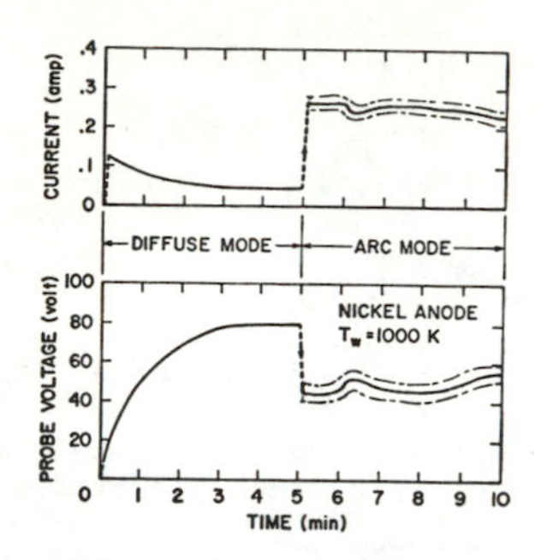

Fig. 19. Electrode Current and Voltage Drop Vs Time for Two Modes of Electric Discharge.

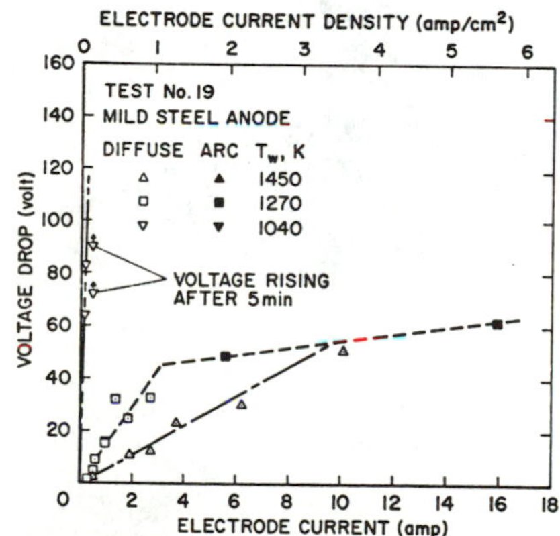

Fig. 20. Electrode Voltage Drop Vs. Current for Various Electrode Temperatures.

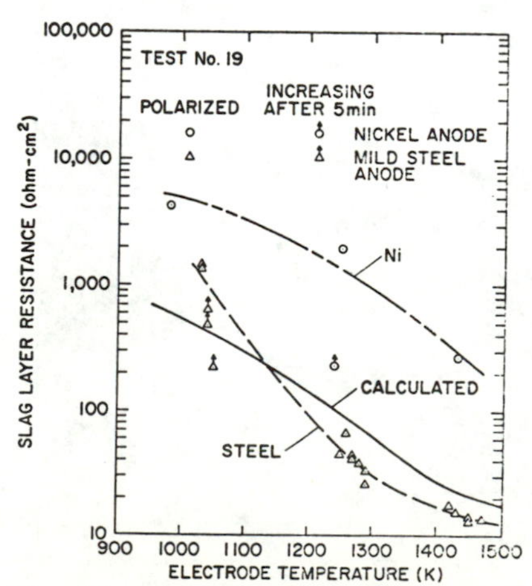

Fig. 21. Slag Layer Resistance versus Electrode Temperature for Two Electrode Materials.

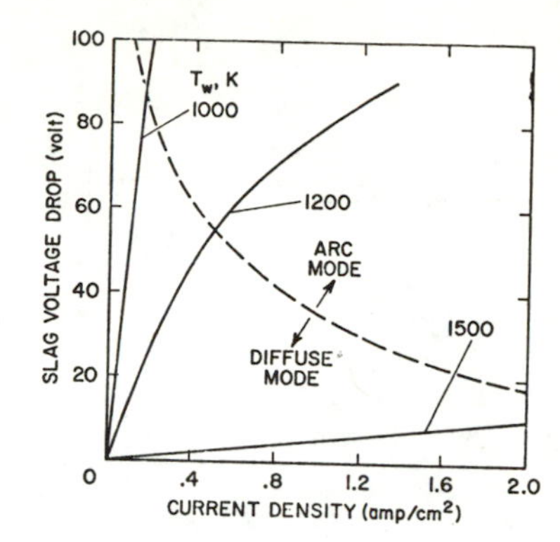

Fig. 22. Slag Layer Voltage-Current Characteristics for Various Electrode Temperatures.

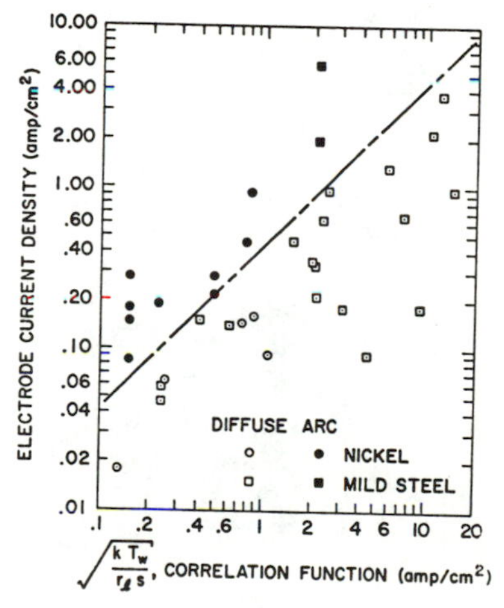

Fig. 23. Correlation of Critical Current Density with Slag Layer Resistance.

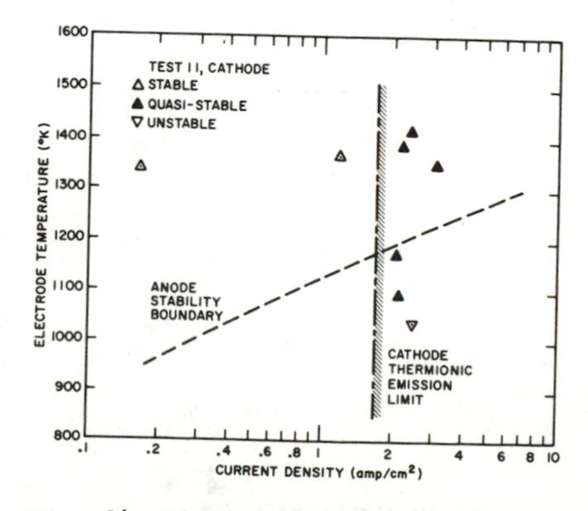

Fig. 24. Slag Electric Discharge Stability Map for the Cathode.

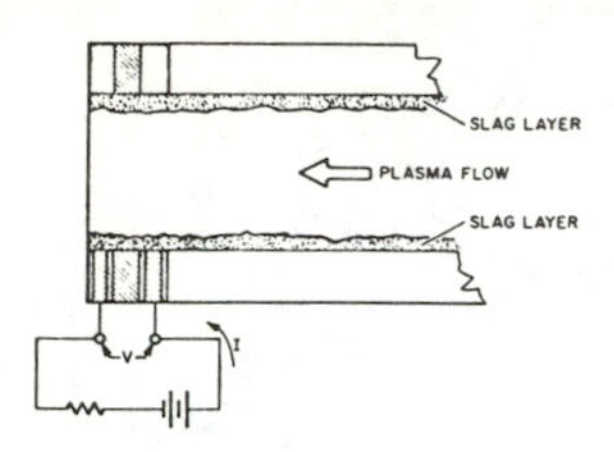

Fig. 25. Axial Slag Discharge Configuration.

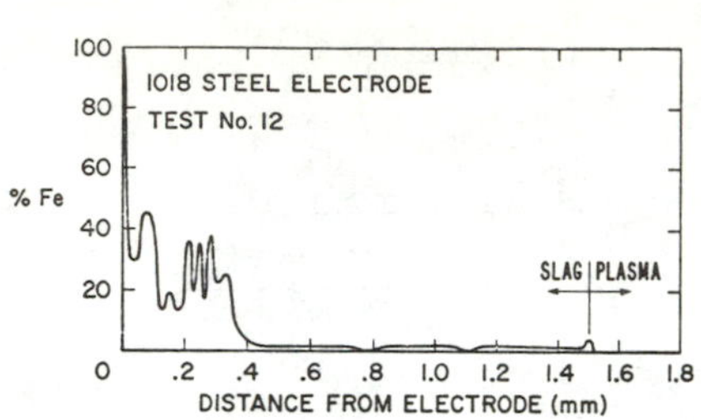

Fig.29. Distribution of Iron in Slag Over Mild Steel Electrodes.

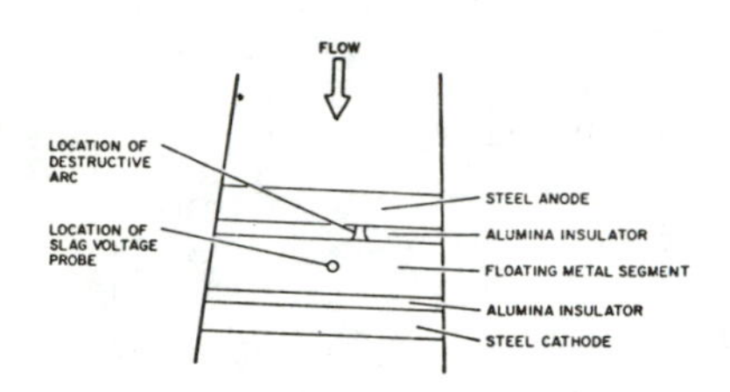

Fig. 26. Overlay on Fig.16 Showing Location of Electrodes

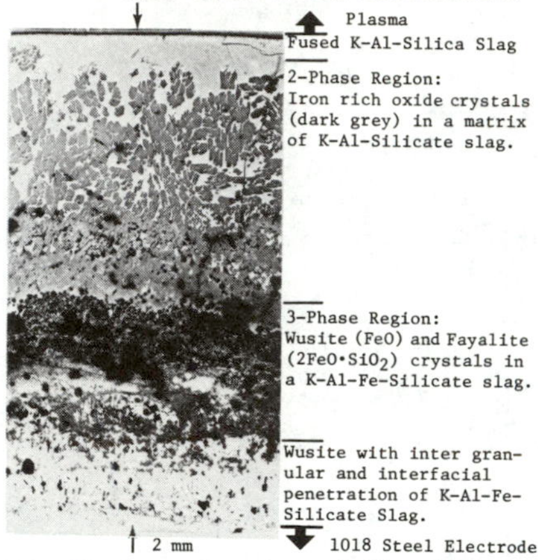

Fig.30. Cross Section of Slag Layer on Mild Steel Electrode, Test #12.

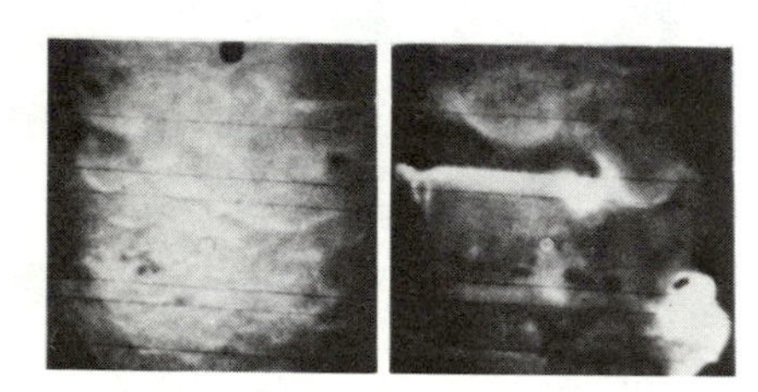

I = 1.8 amp V = 52 v. I = 10.5 amp V = 54 v.

Fig. 27. Axial Discharge Before and After Breakdown.

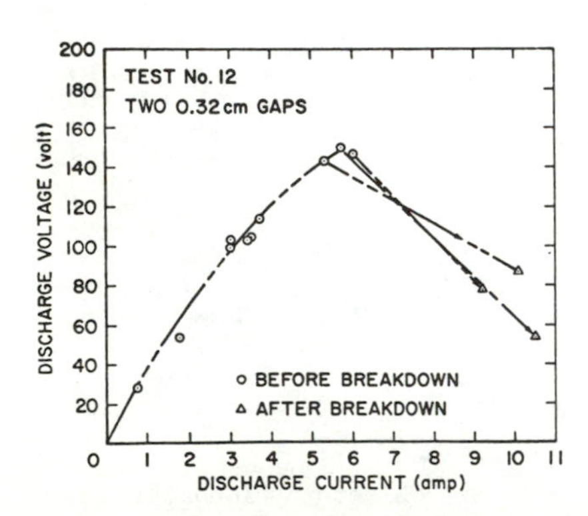

Fig. 28. Voltage-Current Characteristic for Axial Slag Discharge.

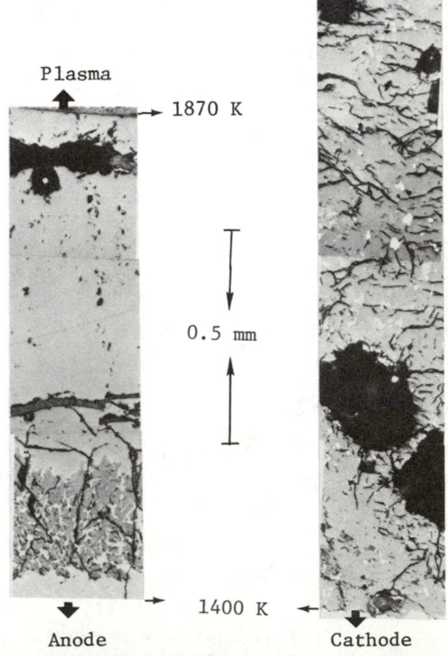

Fig.31. Cross Section of Slag Layer on 1018 Steel Electrodes, Test #20.

CORROSION AND DEPOSITS IN MHD GENERATOR SYSTEMS

J. B. DICKS, L. W. CRAWFORD, C. K. PETERSEN, and M. S. BEATON

The University of Tennessee Space Institute
Energy Conversion Division
Tullahoma, Tennessee, USA 37388

ABSTRACT

Experiences in deposition of and corrosion-erosion by ash constituents and ash-additive mixtures in The University of Tennessee Space Institute MHD Laboratory are described. Particular components which are discussed include the combustor, the MHD generator and gas ducts, high temperature heat transfer surface, and low temperature heat transfer surface. Effects of the deposits on metallic and ceramic materials are considered.

MHD POWER GENERATION

The MHD Process And Central Power Generation

Magnetohydrodynamic (MHD) power generation occurs when a conducting fluid passes through a magnetic field. The interaction gives rise to a potential difference which may be connected in an external circuit to draw electrical (dc) power from the moving fluid. The usual thermodynamic cycle employed is the Brayton cycle, in which cool fluids are pressurized, heated, and caused to achieve high velocities (several hundred m/s) in a channel or duct (the MHD generator), the fluid flowing normal to a magnetic field. Various configurations are used to collect the power generated. A number of heating means and working fluids have been conceived of and tested in small scale. The most mature variant is open-cycle MHD power generation, in which the heating of the pressurized fluid is caused by combustion of a fuel, and in which the fluid is made conducting by addition of a relatively small amount of a compound of an easily ionizable element (seed). (For large scale and long durations, potassium is generally recognized to be the most practical seed material.) At suitable combustion temperatures, the equilibrium ionization of the seed makes the combustion products sufficiently conducting for significant levels of power to be generated when the products are passed through the generator.

Over the past several years, numerous research and development activities have been carried out in the USA, supported mainly by funds of US Government agencies, toward the objective of using open-cycle MHD power generation for central power stations, especially based on coal or coal-derived products as a fuel. Such power stations are projected to have thermal efficiencies in the range of 50 percent for first-generation commercial plants, with the possibility of higher efficiencies in subsequent generations. For commercial plants, it will be necessary to use some of the generator-exhaust thermal energy (or separately fire) to preheat air to obtain appropriate combustor and channel temperatures. Thermal energy from the generator exhaust and other components will be used for a conventional steam plant heat source. The thermal efficiency referred to above is for the combined cycle (MHD-steam turbine).

The coexistence of coal ash (slag) materials and seed leads to problems of at least two types, namely, the corrosiveness of the slag, the seed, or the mixture, and the economical separation of seed from slag. A major attribute of the seed in the MHD process is the possibility of injecting seed in the combustor as hydroxide or carbonate, whereby the solid condensation product is the sulfate, thus leading to sulfur removal from the exhaust gases. It is necessary, however, to recover the largest portion of the seed, and to regenerate it into suitable form for re-injection in the combustor. Therefore, in the apparatus and work described below, major components include those for slag separation from seed and recovery of the seed particulate, as well as combustor, nozzle, generator, diffuser, and other components.

Research and development in open-cycle MHD have been and are being carried out in other countries, with the largest current efforts in the USSR and Japan. However, in the USSR, the major fuel used is natural gas, while a fuel oil is used in Japan.

MHD Research And Development At UTSI

At The University of Tennessee Space Institute (UTSI) in recent years, the major effort in power generation has involved direct-coal firing with no overt attempt to remove slag in the combustor. Thus, all the slag formed from the coal ash is carried into the generator and downstream components. In such a situation, it is necessary to separate seed (mainly vapor) and slag (mainly condensed) at high temperature near the generator exit; otherwise, the potassium may become bound as condensed alumino-silicates from which it is difficult to recover the seed. In the existing laboratory, one flow leg (UTSI-III) contains a cyclone separator to remove the particles larger than about 5 μm. Such a cyclone or bank of cyclones for slag-seed separation is an essential part of projected large-scale plants with full slag carryover. (Other concepts, not currently studied experimentally at UTSI, involve separation of the major portion of the slag in a cyclone combustor, or by prior gasification of the coal.)

Studies carried out at UTSI include combustion, MHD power generation, diffuser performance, high temperature heat transfer, cyclone separation, low temperature heat transfer, particulate removal, and gaseous pollutant production and control. In conjunction with these, relatively short-time corrosion and erosion observations on a number of materials in various parts of the equipment have been made.

THE UTSI MHD LABORATORY

Components Of Flow Trains

The UTSI MHD Laboratory includes three flow legs supported (not simultaneously) by common oxidizer and pulverized fuel supplies, utilities, and data acquisition system. The generator leg (UTSI-I) includes combustor, nozzle, generator in a 2T magnet, and diffuser, exhausting to the atmosphere. The materials test leg (UTSI-II) has a short channel to hold materials (electrode, insulator, and seal), no diffuser, and no magnet. The third leg (UTSI-III) includes a combustor, nozzle, channel (aerodynamically similar to the generator of UTSI-I, but with no magnet), diffuser, high temperature heat transfer surface, slag-separation cyclone, low temperature heat transfer surface, and a baghouse. The coal-seed supply is contained in pressurized hoppers, and is sufficient (1800 kg) to operate 1 - 2 hours at usual operating conditions. The oxidizer reservoir contains pressurized liquid oxygen, which is vaporized and warmed to near-ambient temperatures.

Total mass flow of combustion products is in the range 0.7-1.4 kg/s, with flow being determined by objectives of a given test. Because of the use of oxygen as oxidizer, ash constituents make up about five percent of the total mass flow. Significant quantitites of slag, with some seed admixture, coat most surfaces of the flow train as far as the slag-separator cyclone. Seed particulate tends to deposit downstream.

Combustion Product Conditions

Table I, below, indicates the approximate combustion product conditions at various points in the UTSI apparatus complex. The component referred to as "radiant boiler" is in the position where such a component might be in a full-scale plant, although in UTSI's apparatus, much of the heat is transferred convectively. In previous parts of this paper, the component was referred to as high temperature heat transfer surface, to avoid this explanation at a more awkward point.

The considerable ranges presented for the components reflect various mass flows and component sizes.

TABLE I

Combustion Product Conditions

Component	Temperature K	Pressure atm. absolute	Velocity m/s
Combustor	2700 - 3000	2 - 5	20 - 50
Generator or Channel Entrance	2500 - 2700	1 - 2	900 - 1200
Generator or Channel Exit	2300 - 2500	0.5 - 1	900 - 1400
Radiant Boiler Exit	1600 - 1800	1+	30 - 60
Cyclone Exit	800 - 1200	1+	30 - 40
Baghouse Entrance	350 - 450	1+	15 - 30

DEPOSITION OF CONDENSED SUBSTANCES IN THE UTSI APPARATUS

It is evident that gravity, centrifugal force, inertia, or various other phenomena may cause particles to deposit on exposed surfaces. In turbulent flow parallel to surfaces, random motions imparted to the particles probably lead to impingement. Concentration gradients of condensible vapors (such as seed substances) lead to deposition, also.

Some calculations of particle deposition rates in parallel flow have been made (1,2). In regions of high gas temperature, the phenomena of deposition, re-entrainment, and layer flow are probably extremely complex. A more-or-less steady state is often reached in which very viscous or essentially solid slag adheres to a cold wall, while fluid slag flows over the colder layer. The thickness of layer depends on many factors (2,3,4), such as gas temperature, cold-wall temperature, slag viscosity variation with temperature, and shear stress and rate of heat transfer (both of which depend on gas mass flow). In UTSI's apparatus, most surfaces exposed to the combustion products are relatively cool (from near ambient to a few hundred deg C although in several cases described below, hot surfaces have been tested).

Up to the entrance to the cyclone, slag at the hot interface and in the gas stream remains fluid. In components of the flow train with relatively low velocities (e.g. combustor) considerable pools of molten slag build up with time. In components, with high velocities (near 1000 m/s) slag layer thickness commonly range from less than one to a few mm, with little apparent effect of gravity. In the cyclone and downstream components, particles are solid. In the cyclone, quantities of dust (from 5 μm up) are collected, but some larger chunks that have broken loose from upstream components also appear. Downstream of the cyclone, particles exist mainly as fine potassium sulfate crystals (with some slag components which escaped the cyclone) with dimensions down to sub-micron size. Considerable quantities accumulate on low-temperature convective heat transfer surfaces prior to the bag-house. Sufficient material accumulates on these surfaces to affect heat transfer rates, and to indicate the need for intermittent removal with longer operation times.

It should be mentioned that as originally conceived at UTSI (5) the slag-separation cyclone should operate at liquid slagging conditions, to improve separation of seed and slag and decrease heat loss. It would be necessary in the present apparatus to pre-coat the cyclone walls with ceramic or run longer times to achieve a fluid slag. This has not been done as yet.

CORROSION AND EROSION OBSERVATIONS

Combustion Chamber Liner

In usual UTSI practice, combustion chambers consist of cylindrical smooth cooled metallic walls or helical cooled tubes with refractory cement between tubes. In one series of experiments, however a liner of curved silicon carbide bricks was tested to investigate the effect on reducing heat loss. The bricks were cemented into the metallic shell, to a length of 30 cm and subjected to combustion conditions in UTSI-II in several tests for a cumulative time of 86 min. Figure 1 shows an erosion record as increase in diameter of the chamber due to erosion of the SiC brick. The largest erosion occurred at the downstream end of the chamber, where the last course of bricks was not protected. Considerable erosion occurred 10 cm upstream of the end of the brick. Much smaller corrosion (not shown) occurred further upstream.

Slag samples after two of the runs were examined using SEM-EDX. The major slag constituent was a glassy substance of varying density (due to entrapped gas bubbles). The x-ray energy spectrum for this material Figure 2 shows silicon to be the most abundant element. Potassium is present in intermediate amounts with calcium and iron in minor amounts. Calcium was found by x-ray analysis in the silicon carbide bricks used to line the combustor. Iron is present in the coal, both as a combined element and as magnetic iron-iron oxide chunks discovered after the runs in the coal barrels.

Spheres were found on some surfaces of the glass slag samples. Figure 3 exhibits a large number of these spheres ranging in size from about 5 μm to about 180 μm. Figure 4 is a higher magnification of the largest sphere in Figure 3. The bottom x-ray energy spectrum in Figure 5 shows that the spheres are essentially pure iron. The top spectrum is due to the surrounding material, and is very similar to Figure 2.

In one slag sample taken from the bottom of the combustor a layer of coke one to four mm thick was deposited on the combustor wall at the beginning of the run, due evidently to initially incomplete combustion of the coal. This layer was covered with a dense glassy slag at least two mm thick. Figure 6 shows part of this sample extending from the combustor wall. Four x-ray energy spectra were collected at the points marked in Figure 6 to indicate the distribution of potassium and sulfur. The numbered spectra in Figure 7 correspond to the numbers in Figure 6; S and K denote the energy peaks for sulfur and potassium. Point 4 in the coke layer and point 3 at the edge of the coke layer have substantial concentrations of both potassium and sulfur. Points 2 and 1 in the glassy region have much less potassium and sulfur which is barely detectable with the available apparatus. This shows that the seed material is deposited on the combustor wall along with partially burned coal at the beginning of the run.

Several chunks of yet another material were found within the slag. The substance appeared in spheroidal forms suspended in the glassy slag. The chunks exhibited a silver-blue metallic lustre, dendritic solidification structure, hardness approaching that of the silicon carbide liner brick, and low electrical resistivity. The dendritic structure of one such chunk is shown in Figure 8, and at higher magnification in Figure 9. X-ray energy spectra (Figure 10) were collected at the points indicated on Figure 9. The spectra collected from the primary dendrite (closed symbol) shows the presence of silicon only, while the interdendritic area marked by the open symbol shows a large amount of iron present along with the silicon. The high silicon content of this material strongly suggests that it was derived from the silicon carbide liner bricks. The substantial iron content and the presence of iron-iron oxide chunks in the coal supply further suggest that an alloying reaction occurred in the silicon carbide. Such a reaction would lower the melting point of the material substantially causing melting to occur as much as 200° C below the melting point of silicon carbide.

Metallic And Ceramic Specimens Exposed To Channel Flow

While copper electrodes have been found to be satisfactory for the relatively short times of running at UTSI (6), and apparently have proved to be the most slag-resistant material in larger duration tests (7), it is intuitively appealing to search for a material which can operate at higher surface temperatures, in order, possibly, to promote diffuse electric currents rather than arc mode electrical transport, and possibly cut down on electrical losses in the slag layer.

A number of metallic and ceramic specimens have been exposed to combustion product flow in UTSI-II, without electrical current, to determine erosion and corrosion behavior. Two configurations of samples have been used. In one configuration, a plate 2.5 cm by 5.0 cm is held by screws in a special copper frame, or is held down at the ends with copper bars. In the second configuration, a number of bevelled ceramic pieces which hold a small trapezoidal (approximately 1 cm square) metallic or ceramic specimen, are held down by copper bars. One or more temperatures in the specimens of both types are measured. Figure 11 shows a sketch of the second configuration. In each configuration, the sample is flush with surfaces of upstream and downstream copper frames.

In the first configuration, several metals have been tested at temperatures up to several hundred deg C. The metals included copper, stainless steels, inconels, molybdenum, titanium, and tantalum. Most metals, including copper, showed considerable erosion at the higher temperatures, although it is apparent that some of them, such as molybdenum and titanum, may have been oxidized while copper may have melted. Tantalum was found to be resistant to erosion, corrosion, or oxidation at the temperatures used.

In the second configuration (small trapezoidal specimens) several metals and some ceramics have been tested. A summary of the results is contained in Table II, which has been displayed in part previously (8). In the majority of tests discussed here, the pulverized coal was obtained from TVA's Widows Creek Plant and had the ash composition (nominal) given in Table III. In certain of the tests mentioned in Table II, the slag was rendered more basic by addition of sufficient limestone to the fuel to bring the CaO content of the slag up to 20 more percent.

In many of the specimens presented in Table II, a rather narrow reaction zone between slag and sample was apparent between slag and sample, even if a considerable thickness of sample had disappeared. It appears obvious that a low-melting reaction product formed, and was swept away until a stable temperature at the interface was reached. Microscopic and SEM-EDX observation generally revealed more penetration along grain boundaries then into the grains themselves. In certain alloys, especially those containing nickel, considerable diffusion of the metal into the slag layer occurred.

Downstream Components

For the most part, negligible corrosion has been observed in components downstream of the diffuser. The only exception is the radiant boiler, in which tubes have occasionally ruptured. These tubes operate at only several atmospheres of water pressure. In spite of the name "boiler", the unit is ordinarily operated at high enough water flow to prevent boiling. In some of the early runs, some tubes were apparently weakened by over-heating due to poor distribution of

TABLE II. Observations On In Channel Corroded Samples

Sample	Max. Slag-Sample Interface Temp(°C)	Corrosion Time(min)	Corrosion (mm)
TZM*	1100	10	less 0.1
No diffusion of slag into sample, small wetting of slag on surface.			
Nb+1%*Zr	1600	10	0.4 to 1.2
Slag wet sample, some small diffusion of slag into sample			
In-601*	1200 to 1400	10	0.8 to 2.9
Slag has wet surface, no apparent diffusion of slag into sample. Large diffusion of sample into slag.			
W-5% Re*	1200 to 1400	30	less 0.1
Slag has wet sample, no diffusion of slag into sample.			
Al_2O_3 (99% dense)	1200 to 1600	60	less 0.1 to 3.0
Only small diffusion of slag into sample, less than 200 microns for up to 3 mm corrosion. Slag wet sample. Basic slag is more corrosive than acidic slag. Below 1300°C less than 0.1 mm corrosion.			
Cu	200 to 900	60	less 0.1
Diffusion of sample into slag. No grain growth at interface. Small diffusion of slag into copper approximately 50 microns.			
Si_3N_4 (70-80% dense)		60	.5 to 5
Slag penetration is small, less than 100 microns. Samples appear eroded more than corroded.			
SiC (Hot Pressed)	1700		3.0
Sample completely corroded away with basic slag.			
$CeO_2 \cdot ZrO_2$* Hot-Pressed	1700	10	0.1 to 0.4
Slag has wet sample, small diffusion of slag into sample. Thermal cracked during cool-down.			
W*	1300	10	less 0.1
Slag has not wet surface, no diffusion of slag into sample. Some thin layer-wise separation of sample at sample-slag interface.			
Ta*	1300	10	less 0.1
Slag has wet surface, no diffusion of slag into sample. Some reaction with Pt thermocouple.			
HS-188*	1200	10	less 0.1
Slag has wet surface, no diffusion of slag into sample			
310SS*	1300	10	0.1 to 0.6
Possible alloying with Pt thermocouple, slag has wet surface, no diffusion of slag into sample. Large diffusion of sample into slag layer.			

*Sample supplied by Oak Ridge National Lab

TABLE III

Nominal Analysis of Ash From TVA Coal

Equivalent Oxide	Percent by Mass
SiO_2	50
Al_2O_3	18
Fe_2O_3	20
CaO	6
MgO	2
Na_2O	1
K_2O	3
TiO_2	2

Note: Individaul analyses may show variations of several relative percent.

the water. We have attributed failure to chromium carbide formation at excessive temperatures.

FUTURE WORK

An additional MHD Laboratory complex is to be started-up at UTSI in a year or so. The new laboratory will be capable of somewhat greater mass flows and much longer continuous operating times. Components, including combustor, generator, cyclone, heat transfer surfaces, and final particulate removal, will be operated over tons of hours rather than fractions of hours as at present. It is anticipated that more subtle effects of corrosion and erosion will be observable under these conditions. The work at UTSI is aimed at obtaining engineering data on performance of all components, including corrosion of materials, for incorporation in design of larger MHD facilities in the future.

ACKNOWLEDGMENTS

The work reported here has been supported under contracts: 14-32-0001-1213, Office of Coal Research, U. S. Department of the Interior, and the Tennessee Valley Authority, EX-76-C-01-1760, U. S. Energy Research and Development Administration.

REFERENCES

1. Dicks, J. B. et al "The Direct-Coal-Fired MHD Generator System" Proceedings of the 14th Symposium on Engineering Aspects of Magnetohydrodynamics, pp II.1.1-10, Tullahoma, TN, April 8-10, 1974.

2. Stickler, D. B., and DeSaro, R. "Replenishment Analysis and Technology Development" Sixth International Conference on Magnetohydrodynamic Electrical Power Generator, Vol II, pp 31-49, Washington, DC, June 9-13, 1975.

3. Crawford, L. et al "Investigation of Slag Deposits and Seed Absorption in a Direct-Fired MHD Generator" ibid. pp 51-65.

4. Koester, J. K. et al "In-Channel Observations on Coal Slag" Proceedings of the 15th Symposium on Engineering Aspects of Magnetohydrodynamics, pp I.6.1-10, Philadelphia, PA, May 24-26, 1976.

5. Dicks, J. B. et al "MHD Direct Energy Conversion" UTSI 8th Quarterly Report, Office of Coal Research, August 2, 1973.

6. Crawford, L. W. et al "Slag Layers in Direct Coal-Fired MHD Power Generation" Proceedings of the 15th Symposium on Engineering Aspects of Magnetohydrodynamics, pp I.8.1-6, Philadelphia, PA, May 24-26, 1976.

7. Petty, S. et al "Electrode Phenomena on Slagging MHD Channels" Proceed-

ings of the 16th Symposium, Engineering Aspects of Magnetohydronamics, pp. VIII .1.1-12, Pittsburgh, PA May 16-18, 1977.

8. Crawford, L. W. et al "Generator Wall Slag Coating and Materials Corrosion Experiments" ibid pp IV.5.29-34.

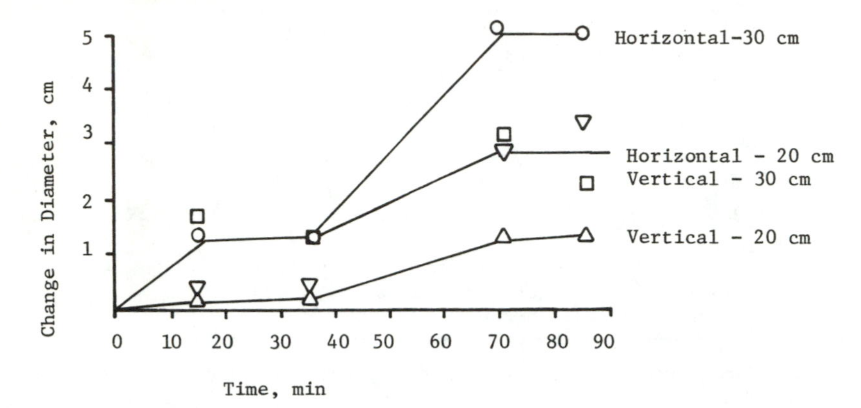

Figure 1. Erosion of SiC brick at and near the downstream end of the combustor.

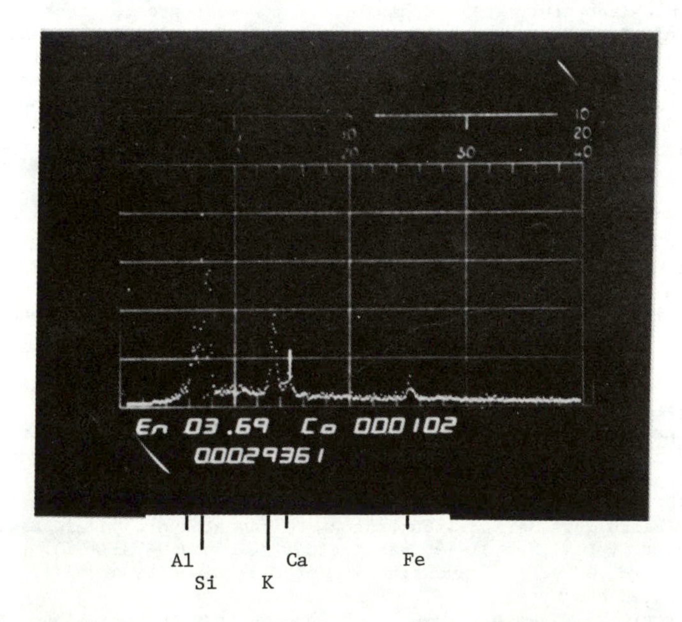

Figure 2. X-ray Energy Spectrum from Glass Slag Component of Combustor

100 Microns

Figure 3. Iron Spheres on Surface of Combustor Slag - 100X

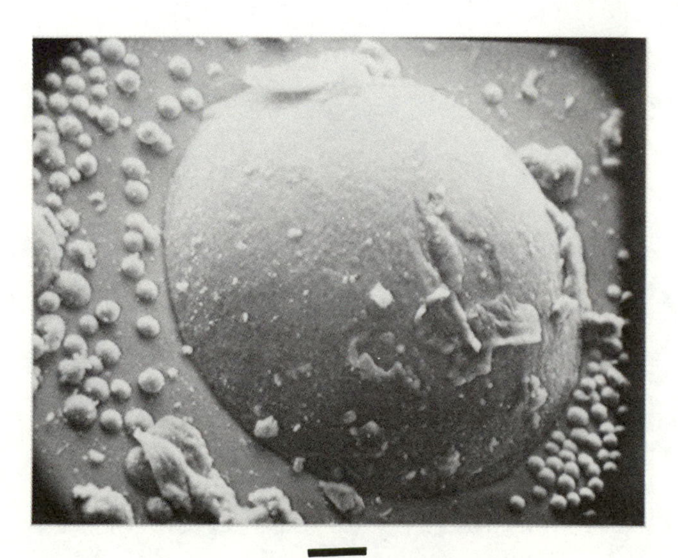

20 Microns

Figure 4. Iron Spheres on Surface of Combustor Slag - 500X.

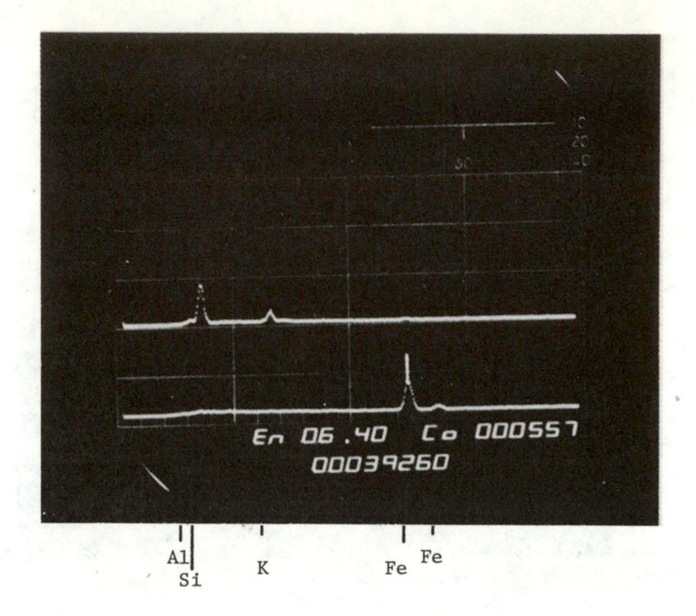

Figure 5. X-ray Energy Spectra from Figure 4: Top Spectrum from Glassy Slag Material; Bottom Spectrum from Sphere (Iron Only).

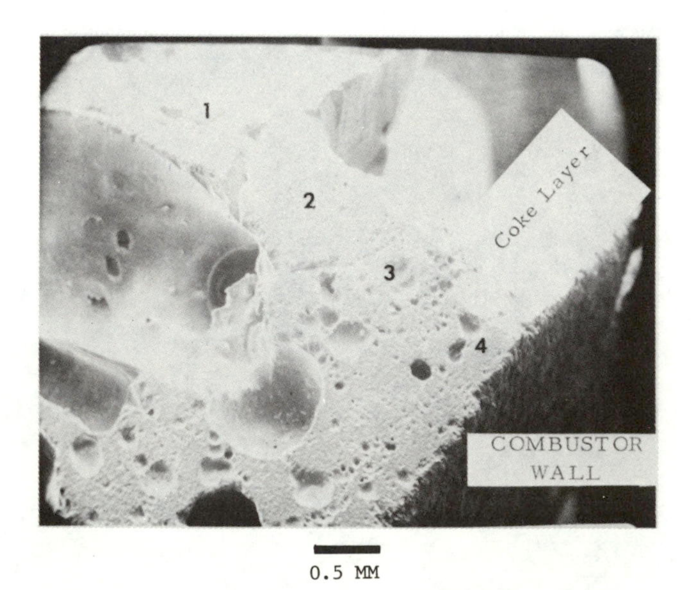

Figure 6. Cross Section of Slag Sample Taken from Bottom of Combustor, Run II 930 (20X).

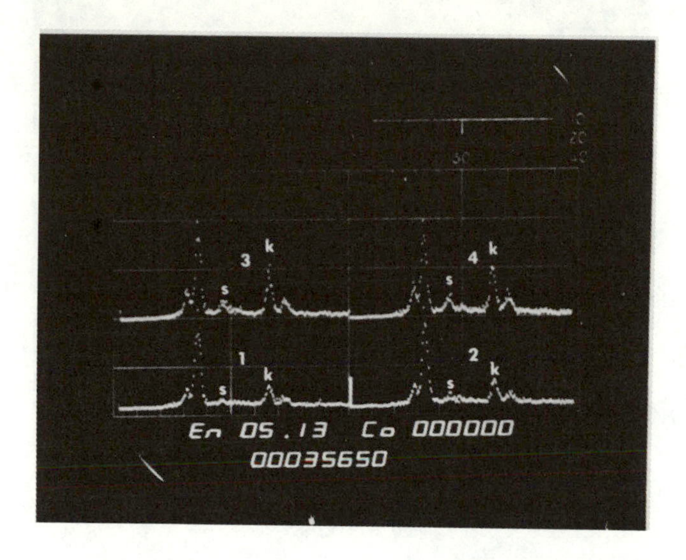

Figure 7. X-ray Energy Spectra from the Numbered Locations in Figure 6. Sulfur and Potassium Peaks are Marked.

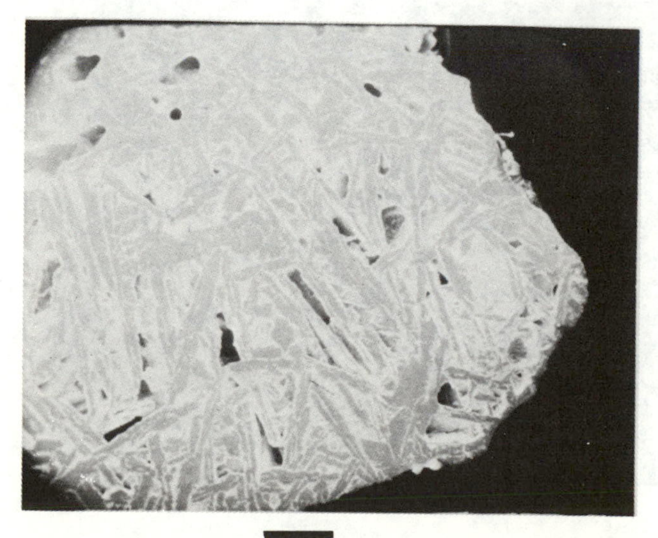

Figure 8. Polished Cross Section of Silicon-Based, Metal-Like Material from Combustor, Run II 930 (12X).

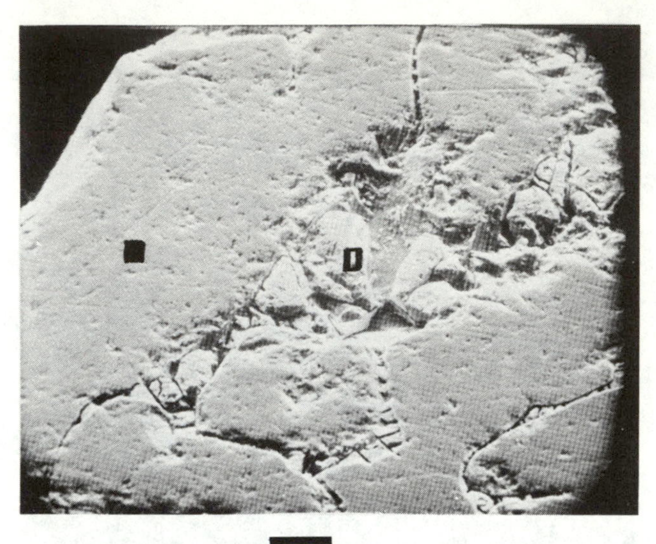

50 Microns

Figure 9. Higher Magnification of Polished Cross Section in Figure 8 Showing Locations of X-ray Energy Spectra (200X)

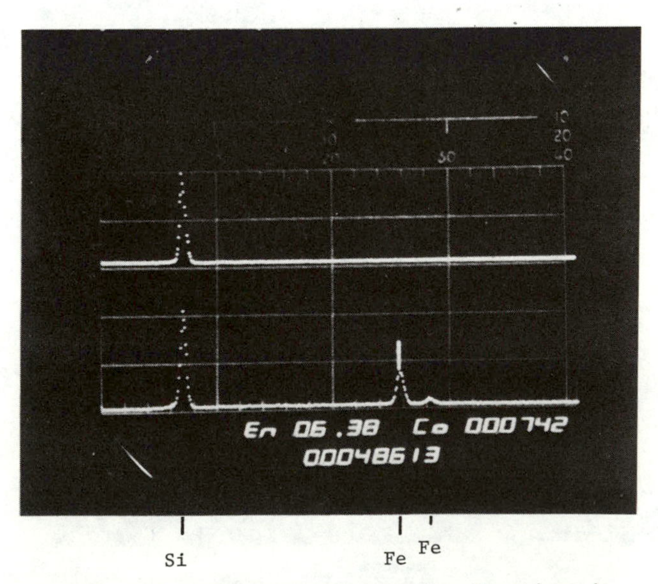

Figure 10. X-ray Energy Spectra for Two Locations in Figure 9. Top Spectrum: Primary Dendrite, Silicon Only. Bottom Spectrum: Interdendritic Region, Silicon Plus Iron.

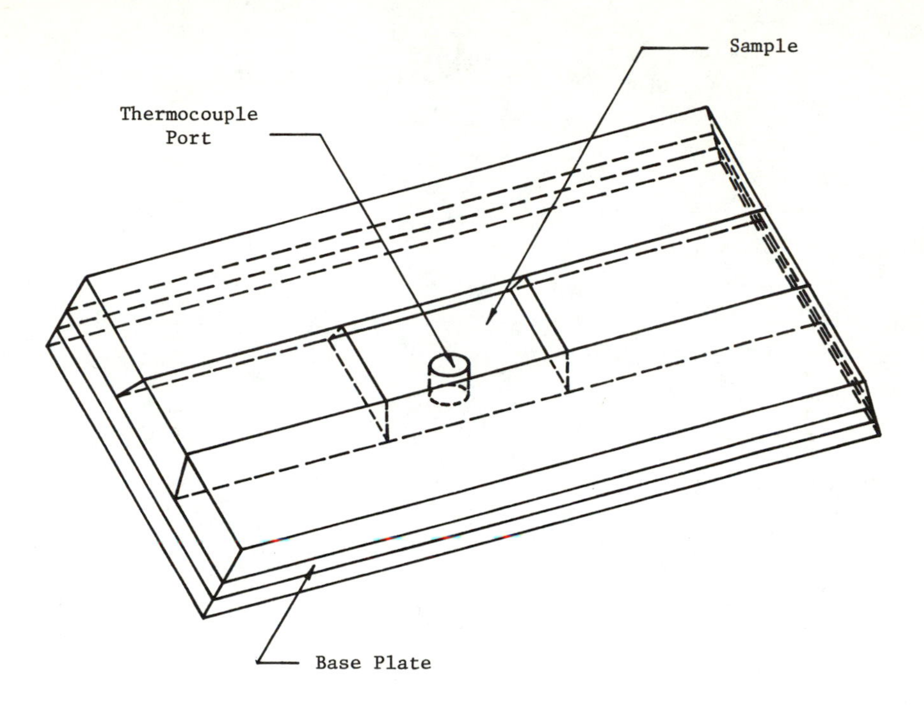

Figure 11. Sketch of Ceramic Sample Holder

CONTROLLED UTILIZATION OF COAL SLAG IN THE MHD TOPPING CYCLE

D. B. STICKLER and R. DeSARO

Avco Everett Research Laboratory, Inc.
Everett, Massachusetts, USA 02149

ABSTRACT

Slag coating development for structural protection from gas dynamic erosion, and from current concentrations in the MHD channel, is described. Experimental and analytical study of the behavior of coal ash on the wall structures of MHD systems has resulted in an understanding of controlling transport processes. Experimental measurements under both subsonic and supersonic conditions indicate that on a wettable wall structure, operating at temperatures well below T_{250}, slag develops by deposition from the gas flow at a predictable rate. A layer equilibrates with a viscosity-temperature distribution defined by balancing of local mass flow in the layer with skin friction and pressure gradient forces driving it. Typically, at 0.3 wgt percent mineral matter in a mach 1.4 flow, equilibration time is about forty minutes. This is somewhat greater than predicted by simple theory, due to the existence of a finite rate of re-entrainment prior to equilibration. The steady state layer is on the order of one mm thick, with surface temperature about 1800 K. Effects of ash composition, combustion stoichiometry, flow field, and wall structure are discussed, together with corresponding analyses.

INTRODUCTION

Development of open cycle magnetohydrodynamic (MHD) power generation technology for use in the United States is based upon coal as fuel. A schematic of a combined cycle MHD-steam power plant is given in Fig. 1. The MHD generator of the topping cycle gives DC electrical power, output directly from the interaction of the high velocity plasma and magnetic field. Subsequently, the output gas is used for steam production in radiant and convective boilers. It may also be used in a heat exchanger system to provide air preheat for the coal combustor. As indicated in Fig. 2, combined cycle plant efficiency is projected to range from 50 to 60 percent (1).

Effective mineral matter removal from the coal prior to combustion, or as slag in a gasification-combustion process, is unattractive in terms of basic cycle efficiency, or generated power cost. The working fluid of the high temperature MHD cycle consequently contains coal mineral matter as vapor and droplets. Consequently, the MHD flow train components are required to function in a slagging environment. Long duration service of open cycle MHD component surfaces and minimum heat transfer loss from the generator working fluid may best be obtained in the presence of slag or ceramic surface layer. Since this surface is itself subject to vaporization, flow, and spalling, replenishment of it is necessary. This is obtained by deposition of ceramic materials as liquid droplets or vapor from the hot gas flow. These materials may be added deliberately, or, in a coal-fired system, be present naturally as part of the fuel. The equilibrium thermal and physical condition of such a coating interacts strongly with flow transport processes, notably electrical current flow in the generator gas boundary layer and wall structure.

The end state desired is a stable, continuous layer of slag or ceramic covering the flow train surfaces. Thermal and electrical properties and conditions should preferably be controlled to give suitable temperature and composition distribution for diffuse current transport in both gas and condensed material. Also, axial current leakage must be minimized.

The work reported here has had as its goal the development of a scientific and engineering base for such component operation. The approach chosen is based on the concept of promoting the development of a stable flowing slag layer, coating all exposed surfaces of the MHD topping cycle components.

Use of coal slag as a wall coating material derives from work by Louis (2). He proposed that electrodes used for current extraction in the generator be maintained in situ by continuous deposition of vaporized zirconia from the plasma flow. This approach was subsequently demonstrated in a clean fuel fired system, operating with zirconia addition for one hundred hours (3). The descriptive term employed for this process was replenishment, which has since become a generic term in the MHD field for in situ wall coating processes. A predictive model describing such a process was developed by Rosa (4), who applied it to both zirconia and coal slag. The first demonstration of MHD generator operation with coal slag in the flow was done at University of Tennessee Space Institute (UTSI) (5, 6). Mineral matter concentration in the plasma was quite high (2-4%) relative to potential commercial systems, and test duration limited by feed systems to about one hour. Longer duration generator development work (ten to one hundred hours) at electrical power extraction levels of 200 to 600 kilowatts, has been underway at Avco Everett Research Laboratory (AERL) under ERDA sponsorship (7, 8). This effort has been concentrated on generator hardware technology, operating in a slagging environment. The mineral matter results from addition of coal, fly ash, or fly ash plus additives, to the combustion products of fuel oil and oxygen enriched air. The resulting gas flow, seeded with partially ionized potassium for electrical conductivity, contains about 0.2% (wgt) mineral matter, as vapor and droplets. This corresponds to the output of a coal combustor operating with 80% slag rejection, burning bituminous or sub-bituminous coal with air preheated to about 1800 K. Research aimed specifically at understanding processes of slag layer development and flow, in support of the generator development effort, has been carried out both at Stanford University (9) and AERL.

The work reported here has been concentrated on description of the fluid dynamic and thermal aspects of slag behavior. Initially, transport processes and wall surface conditions necessary for development of a stable slag layer were studied. This led to establishment of requirements on wall structure, gas flow field history, and slag physical state. Effects of wall structure, gas flow behavior, mineral matter composition, and combustion stoichiometry on both transient and steady state slag layer properties have been investigated.

Interaction of the slag layer inside the generator with electrical power extraction gives some further constraints on wall structure and slag quantity and composition.

The results of this work, as well as extensive generator development, indicate that it is both feasible and desirable to utilize slag in the MHD cycle. A stable flowing layer of slag can be developed, coating all components from the combustor, through the nozzle, the generator, and the supersonic and subsonic diffusers. It functions as a self renewing protective coating on the entire internal structure.

EXPERIMENTAL OBSERVATIONS

Test Conditions

Measurements of slag layer development and properties have been made in both subsonic and supersonic flows, with a variety of wall structures and slag feed rates. Test conditions are summarized in Table I, and are designed to simulate baseload generator conditions. Combustor, nozzle, and generator fluid dynamic conditions have been simulated, as well as generator slag studies with an externally imposed electric field. The test duct used for generator simulation is circa 0.8 meter long, with a fixed cross dimension of 6 cm, and adjustable height, entrance and exit geometry, to control mach number distribution. Interchangeable blocks are used for the experimental wall structure. The test structure exposed to the two phase product flow normally includes two or three different ceramic filled test wall structures, bare copper, and nickel plated copper, the latter for erosion resistance. Stainless steel 304 is used as end blocks in the supersonic configuration. In subsonic flow experiments, the contraction section between combustor and channel was of nickel plated cold copper. Supersonic generator flow simulation used both a short, heavily cooled nozzle, and a longer nozzle designed for slag layer optimization.

Slag layer development and steady state condition is strongly affected by the design of the wall structure, in terms of materials and operating temperature, as well as the chemical composition of the mineral matter itself. The potassium added to provide an electron source in the gas also interacts to give a slag with typically 20% K_2O content. Combustor stoichiometry affects slag flow behavior through its viscosity dependence on iron oxidation state.

In general, it is found that an equilibrium slag layer develops with surface temperature on the order of 1800 K, for test conditions of interest. The corresponding thickness is about a millimeter, depending on local gas dynamic transport conditions and metal wall temperature. Bonding is to exposed ceramic, with bridging over exposed metal. The surface is generally rippled on a depth scale of circa 0.1 mm, due to capillary wave propagation.

Aside from general flow train conditions, diagnostics include local calorimetry, optical bolometers with narrow passbands centered at 0.9 and 2.2 μm, scanning over the wall, and still photography. The optical units look through a common one inch aperture, air purged quartz viewport in most cases. Observations have also been made using two viewports at different axial locations, with videotape recording of slag layer surface appearance. Also, specialized optics have been used for observation of slag layer behavior with current transport driven by imposed electric fields. Test channel pressure distribution is taken on gauges via static wall taps, with one serving as source for on-line gas analysis (CO, CO_2, O_2). Primary recording is on a CEC galvanometer recorder, with data logging from gauge readouts. Input to a HP minicomputer system is also available.

Results are given below first in terms of general slag layer development and equilibrium behavior. The effect of mineral matter composition is then addressed. Finally, some observations on current transport interaction are given.

Table I

Test Conditions-Combustor Feed

Primary feed	O_2-N_2-$CH_{1.8}$	
Mach No.	0.6-0.7	1.1-1.6
Combustor	MK VI	EPRI
τ_c ms	15	8
P_o psia	30-35	55-63
T_o K	2800-2900	2500-2600
$\dot{m}$ Kgm/sec	1.0	0.64-0.70
Replen. feed	Seacoal*	Fly ash†
ϕ_∞ wgt frac.	0.001-0.0025	0.0005
ϕ_k wgt frac.	.007	0

*See Table II for properties

†See Table III for properties

Table II

Replenishment Feed Material Properties

	(1) Seacoal	(2)
Volatile	34.4%	35%
Fixed C.	58.0	51
Moisture	2.0	2.5
Ash	5.7	11.5
SiO_2	54.1	52
Al_2O_3	32.2	24
Fe_2O_3	5.9	16
MgO	0.75	1
CaO	1.75	3
TiO_2	2.6	1
K_2O	1.17	2
Na_2O	0.33	<1
P_2O_5, SO_3, etc.	bal.	bal.
Size Distribution (% < μm)		87%-74 68%-44 20%-10

(1) Inconel wall, M < 1
(2) Cu wall, M < 1

Combustor-Nozzle Processes

Physical understanding of the replenishment process breaks down naturally into various fairly distinct physical processes, when the time history of injected coal or ceramic is followed, Figure 3 outlines some of the pertinent phenomena. In the case of a ceramic particle feed, the material undergoes heating and state change, primarily in the combustor. Modeling of particle heat transfer from combustion gas has been carried out for a low particle number density flow. Effects such as velocity lag in combustor and nozzle, radiation, and state change were considered (10). Figure 4 gives some typical results for the AERL Mk VI combustor conditions with Al_2O_3 additive.

Table III

Mineral Feed Compositions (Wgt %)

	Western Ash	Standard Eastern Ash	Additive Silica	High Silica Ash*	CaO Ash I	CaO Ash II	Sulphur-Ash
SiO_2	55.0	41.6	100	59.1	41.6	36.1	37.6
Al_2O_3	19.5	23.7		16.6	17.8	17.0	21.4
Fe_2O_3	5.4	20.2		14.1	15.1	14.4	18.2
MgO	4.0	1.3		0.9	1.0	0.9	1.2
CaO	10.0	5.6		3.9	19.0	26.1	5.1
TiO_2	0.8	0.8		0.6	0.6	0.6	0.7
K_2O	.66	1.9		1.3	1.4	1.4	1.7
Na_2O	.66	1.5		1.0	1.1	1.1	1.4
ZrO_2	--	--		--	--	--	--
S	--	--		--	--	--	--
P_2O_5, SO_3, etc.	Bal.	Bal.		Bal.	Bal.	Bal.	Bal.
Distribution: Size (% < μm)		100%-20 90%-15 80%-10 16%-5	100%-20 100%-15 100%-10 10%-5	100%-20 93%-15 86%-10 14%-5			

*70% Std., 30% SiO_2

The salient point to be observed is that a strong upper bound on feed particle size exists, if molten droplets are to be produced in the combustor flow. Analysis of velocity lag in the nozzle flow suggests that while it gives strongly increased heat flux to the particles, the time scale is too short for significant heat up or fusion over that which occurred in the combustor. Due to the low number density of particles, nozzle agglomeration appears not to be a significant factor.

Coal combustion systems, or clean fueled systems with coal used as an additive, have a more complex and at present poorly defined ceramic/slag history. Rapid coal particle heating and high volatile loss in excess of sixty weight percent on a DAF basis characterizes initial coal particle combustion (11,12,13). Residual char (carbon) burnout is controlled by heterogeneous reaction, primarily dependent on oxygen at ordinary combustor flame temperatures, but with carbon dioxide and steam of potential importance in MHD combustors. This step proceeds at high enough particle temperature that the varied minerals dispersed in the initial coal particle are probably fused, but not necessarily homogenized. Final char burnout may result in multiple mineral matter droplets of varying composition from a given coal particle, or in agglomeration to a single droplet (14).

Due to the short combustor residence time in clean fueled systems, typically of the order of that for fuel droplet vaporization, ceramic vaporization appears unrealistic as a major process. This is associated with the low vapor pressure of most metal oxides at stagnation conditions, relative to a fuel droplet boiling at the combustion pressure. Coal fired systems may be found to give somewhat different results, in that a residence time in the range of twenty to fifty milliseconds has been quoted as desirable for char burnout for a typical power plant grind bituminous coal in turbulent plug flow (13,15).

The ceramic or slag output from an MHD combustor appears most probably to be in the form of a dispersion of small droplets. For clean fueled systems, they originate as a finely pulverized metal oxide input, and for coal fired as the mineral content of the coal. In both cases, some degree of chemical species homogenization may occur by vapor diffusion or agglomeration in the turbulent combustor flow, but a heterogeneous species droplet distribution and vapor species probably dominate transport to the wall.

Wall Impact and Bonding

Conditions necessary for adhesion of replenishment material to exposed surfaces can be considered in terms of ceramic or slag properties in the flow and the wall surface condition. In terms of elementary wall interaction, it is necessary that either the replenishment ceramic be liquid or gaseous, or that the wall surface be molten. For practical applications, deposition of solid particles on a molten wall surface is unattractive. This derives from two considerations. First, the continuous existence of a molten wall cannot be guaranteed, due to possible spalling of large areas of slag or long term operation without replenishment feed. Secondly, as indicated above, solid phase replenishment material is expected to be present in the gas flow as large particulates. Approximate analysis of boundary layer particle transport is outlined in Ref. (10) and results plotted in Fig. 5. It indicates that particles of circa 30 μm diameter and larger may impact the wall at sufficient tangential velocity to deform exposed metal or splash molten ceramic.

The effect of initial wall surface condition on bonding of molten droplets of slag has been observed both in the Mk VI generator flow train and in the slag replenishment research facility. Qualitatively, strongest surface bonding is obtained with a combination of low thermal diffusivity, high temperature wall material, readily wettable by the slag. Figure 6 illustrates the face of a grooved copper wall block used in the replenishment test channel, with the grooves filled with a phosphate bonded cast zirconia. This general form of wall structure, combining well cooled metal with ceramic, is found experimentally to function effectively as a slagging wall. A typical sequence of local slag layer development, starting with an uncoated wall, is shown in Fig. 7. The field of view is approximately 2.5 cm, with the groove structure repeating on a 0.6 cm interval. This particular experiment used coal feed, and was performed at subsonic flow conditions. Gas flow is from left to right.

A cross section of a slag sample removed after the test run is shown in Fig. 11. It is typical of observations made. Adhesion is found to be to the exposed ceramic only. Definite counter diffusion of slag and zirconia is observed, to a depth in the zirconia representative of an isotherm for slag crystallization. In general, the gas side region of the slag is found to be glassy, and the region close to the metal crystalline, following rapid cooldown during the test system shutdown transient. It is also observed that the streamer structures of Figs. 7b and 9b are frozen into the slag layer cold side structure.

Combustor and Nozzle Slagging

In order to establish more realistic source flow conditions for study of wall slag flow, both combustor liners and nozzle have been modified to accept a slag coating. The combustor liners were changed from aluminum to grooved copper filled with a ceramic. A new nozzle was fabricated, grooved and lengthened to improve continuity of slag wall flow. Pertinent properties of both are shown in Table IV. Slag has developed on the combustor liners and nozzle as shown in Figs. 18 and 19. In Fig. 18 slag layer development started at about 4-6 inches downstream of the injector plate and coated the remainder of the liners. The first 4-6" did not develop a slag coating. Analysis of heat transfer to ash particles in the combustor flow indicates that this corresponds to the spatial locus of significant melting. Streamers can be noticed in Fig. 18 flowing axially and spaced relatively uniformly around the circumference. Some downflow is evident, probably due to gravitational body force, although it is certainly not a major competing mechanism under these combustion conditions.

Table IV

Combustor, Nozzle Flow Parameters

	Modified Slagging	Original Non-Slagging
Combustor:		
τ_c (ms)	8	8
P_o (atmos.)	3.7-4.3	3.7-4.3
T_o^o (K)	2500-2600	2500-2600
$\dot{m}$ (Kgm/ sec)	.64-.70	.64-.70
Liners	Grooved Cu	Al
Ceramic in Grooves	ZrO_2	None
Nozzle:		
M_e	1.2	1.2
$\Delta P/\Delta X$ (psi/ in)	-2.0	-4.5
Base Material	Cu, Ni plated	Cu, Ni plated
Ceramic in Grooves	ZrO_2	None
Slagging Walls	One*	None

*Compatible with channel test section wall

The purpose of the slagging combustor and nozzle is to increase the slag droplet transport rate in the upstream boundary layer, and to give a source wall shear flow at the channel entrance. This results in decreased equilibration time for downstream components, and enhanced coverage on the first upstream test block. This is in agreement with theoretical results (17). Figure 20 shows the effect of upstream source shear flow. The slag layer thickness increases at entrance (x = 0) for increased $\dot{M}_o$. This predicted result is also found experimentally, both in the research system and the generator development hardware. Heat loss from the slagging nozzle is larger than for the non-slagging one due to the longer length needed to keep the pressure gradient low. This is partially offset by the thermal insulation afforded by the slag itself. This effect is artificial, in that a full scale power plant would operate under pressure gradient conditions compatible with stable nozzle slagging in any case. In the combustor, the total heat flux loss decreases due to the slag insulating effect. During the first run using the grooved liners, the heat flux loss decreased by about 25% from an unslagged to an equilibrium coated condition.

Slag Layer Evolution

In general, growth behavior and steady state conditions are independent of both wall surface structure and free stream mach number in the range investigated. The sole notable exception is that a flame spray coated wall exhibits relatively uniform local slag development, with strong bonding obtained. A strong dependence on mineral matter physical state, i.e., vapor or droplet flow, and on gas acceleration history is found to exist.

A copper wall structure, grooved and cast with zirconia, has been used as a control surface. Subsonic gas flow layer development is shown in Fig. 7 and a supersonic gas flow case in Fig. 9.* It is clear that the slag layer

*t = run time, t_f = replenishment on time,
t_k = potassium on time,
ϕ_∞, ϕ_k = wgt % ash, seed in flow.

growth behavior is qualitatively similar. Initial bonding is to the exposed ceramic. Development of local bumps with high shear loading leads to small scale streamer runoff across the exposed copper, followed by broader curtain like structures. Eventually, the copper is bridged locally and a more nearly continuous coating develops. In most cases observed, final development of a steady state condition occurs as a result of a larger scale shear flow in the streamwise direction, initiated near the upstream end of the coated wall. The end state is found to be a shear flow of slag, with transversely rippled surface, with local mass flow rate presumably equal to the integrated deposition over the upstream surface. Differences in final surface appearance in the flowing regions probably reflect surface wave structure dependence on skin friction and the magnitude of the slag shear flow. Sequence photographs at six second intervals show a surface wave motion and flow to exist in the supersonic case, as well as in the subsonic. In the early stages of local layer development, e.g. Figs. 7b and 9b, the crossover structure is apparently almost isothermal, based on bolometric signal and photograph density. It remains relatively isothermal with time until essentially complete bridging is obtained.

The time required for development of an equilibrium slag layer is plotted in Fig. 12 as a function of mineral concentration in the combustor flow. In the limit of zero re-entrainment from wall to gas, the time scale would be expected to be inversely proportional to the concentration (16), so that $\phi_{ash} t_e$ would be constant. In fact, as indicated in Fig. 15 this is not the case. Low ϕ_{ash} requires relatively long equilibration time, and is interpreted as indicating significant competition between deposition and re-entrainment. Re-entrainment has been observed using a real time video monitor. An example of it can be seen in Fig. 9b. The local streamers appearing with bright downstream tips and elongated lengths function as sources of slag droplets, which originate at the tips. Fluid slag is fed to the tip via shear flow on the upstream structure. The spherical tip region is observed to cyclicly grow and abruptly disappear, on a time scale of circa one to ten seconds. Presumably, large spacing between bands of exposed wettable ceramic, high shear, and low ϕ_{ash} would result in equilibrium between deposition and re-entrainment, without full wall coverage.

Slag Feed Composition-Flow Behavior

The effect on slagging properties of ash composition is considered. Generally, altering the composition of the ash will affect the viscosity, thermal and electrical conductivity of the resulting slag. To first order, slag layer composition parallels that of the ash feed. This section deals primarily with the viscosity response to composition.

In some MHD coal combustor design approaches, chemical fractionation of the coal mineral content may occur with low vapor pressure species being retained and higher vapor pressure material convecting with the plasma flow. This would lead to a channel slag layer enhanced, for example, in silica and iron oxide. A high silica content ash has been investigated in the replenishment facility. Silica was added to the standard ash (Eastern coal type ash) normally used, to produce the mix shown in Table III. CaO was also added to the standard ash (see Table III) and its effect on the slag layer was observed.

Coal ash viscosity-temperature characteristics have been most conveniently correlated in the past using, among others, two parameters -- the silica factor and the base-acid ratio defined as

$$SF = \frac{SiO_2}{SiO_2 + Fe_2O_3 + CaO + MgO}$$

and

$$B/A = \frac{Fe_2O_3 + CaO + MgO + NA_2O + K_2O}{SiO_2 + Al_2O_3 + TiO_2}$$

respectively, where

SiO_2, Fe_2O_3, etc. = % by wgt of constituent in ash

As SF increases and B/A decreases for different slags the temperature for a given viscosity rises. The values of SF and B/A for typical eastern and western coal ashes and for the various ash mixtures discussed in this paper are shown in Table III.

For a high SiO_2 content ash one would expect the slag viscosity at a given temperature to increase and correspondingly force the equilibrium layer temperature higher since the gas shear loading remains constant. Also, since the slag layer becomes thicker to accommodate the higher surface temperature, the heat flux through the slag layer should decrease. Figure 13 shows these trends. Figure 13 is slag surface temperature measured using an optical pyrometer filtered for 2.2 μm and a disappearing filament pyrometer. The disappearing filament pyrometer was not corrected for transmissivity while the 2.2 μm bolometer was. Correcting the disappearing filament for the correct transmissivity would push the temperatures higher. Emissivity of the slag layer was assumed to be unity. The ash feed varied so that

$$.30\% < \phi_{ash} < .52\%$$

Surface temperature increases and heat flux decreases as % SiO_2 in ash increases, agreeing with theoretical results. Measured heat flux through the slag layer during a run also confirms theory with the heat flux decreasing as the high SiO_2 ash is fed.

A western coal is calcia rich and iron oxide poor relative to an eastern coal. This trend was investigated by adding calcia* and silica to the standard ash resulting in the high CaO mixes shown in Table III. It was desired to investigate the results the CaO would have independent of the SiO_2 so the resulting mix #I was made holding the % SiO_2 constant. Mix #II was mixed with a higher CaO content than mix #I, further decreasing its SF number. The slag surface temperature decreases as shown in Fig. 13 where SF = .54 corresponds to the high CaO ash mix #I and SF = .47 for mix #II. No measurable change is observed in the heat flux corresponding to the high CaO ash.

The theoretical curves of the foregoing are shown in Fig. 14 based on an analysis from Ref. (17). The slag viscosity was modeled as

$$\mu_s = \exp\left[C + DT_s + ET_s^2\right]$$

with the constants chosen from Ref. (18) as a function of the SF of the ash feedstocks. Values of the constants are shown in Table IV. The result of theoretical prediction for surface temperature and heat flux is given in Fig. 14 as a function of SF. They are in general quantitative agreement with experimental observations, as given in Fig. 13. Consistent with the observed slag surface ripple structure, a rough wall was assumed in the fluid dynamic transport modeling. An effective wall roughness factor is defined as RF = C_f/C_{fo}, where C_f is the skin friction coefficient for boundary layer transport on a rough wall, and C_{fo} that predicted for a smooth wall. Use of

*Fisher Scientific Company, Chemical Manufacturing Division, Fair Lawn, New Jersey, Order #C-114

an RF value taken from rough wall studies gives realistic results for predicted heat flux, as well as surface temperature.

Table V

Viscosity-Temperature Parameters.
Silica Factor and Base/Acid Ratio

$$\mu_s = \exp[C + DT_s + ET_s^2]$$

	SF	B/A	C	-D	$E \times 10^6$
Eastern Coal Ash	.6	.51	39.1	.032	6.62
Western Coal Ash	.7	.28			
Test Runs:					
Standard	.61	.46	39.1	.032	6.62
High SiO_2	.76	.28	71.6	.064	14.83
High CaO I	.54	.63	32.2	.026	5.08
High CaO II	.47	.82			
High S	.61	.46	39.1	.032	6.62

To test the effect of sulphur compounds on slag flow behavior, a mix of standard ash with powdered sulphur[†] added was run (Table III and IV). During cycling from standard ash to the high sulphur/ash mix, no changes in slag surface temperature, heat flux or appearance were noticed. It is concluded that the gross fluid dynamic behavior of slag flow, under reducing combustion conditions, is insensitive to equilibrium sulphur compounds in the combustion product.

Potassium Seeding and Transient Effects

Seed input to the flow system gives vapor phase potassium species, which transport to the slag layer at a relatively high rate. As in the case of vaporized mineral species, this occurs at a higher rate at the inlet end of the nozzle and generator. End state concentration in the wall slag layer is typically 18% (wgt) as K_2O. This decreases slag thermal conductivity (19), driving a surface temperature increase, and a corresponding strong viscosity decrease. Under the gas dynamic shear conditions, this low viscosity slag is driven off, presumably in a steep front as predicted for vapor deposition of slag. Replenishment channel tests under both subsonic and supersonic flow conditions have shown this process. Sequence photographs are given in Figs. 7d, e, f and 8d, e and f. Observations of the evolution of local current transport stability in the Mk VI system may also be compatible with this mechanism. It is anticipated that feed of seed during the slag layer growth and equilibration period would minimize this phenomenon, except as associated with vapor phase mineral species transport. Seed may also affect the viscosity temperature characteristics of the slag, effectively increasing the B/A number since seed is an alkali. After the initial runoff, the surface temperature has been observed to decrease with seed feed on. Figure 15 shows results of slag surface temperature measurements. Slag surface temperature decreases each time seed feed is turned on and increases with seed feed off. With the shear loading remaining constant, a lower slag surface temperature is required for a higher B/A (corresponding to a higher alkali concentration) slag to keep the viscosity unchanged, thereby balancing the various transport terms. The points beyond 130 min are somewhat higher than expected due to the fact that the bolometers are scanned axially and are not always stopped at exactly the

---

[†] Fisher Scientific Company, Chemical Manufacturing Division, Fair Lawn, New Jersey, Order #S-593.

same spot. The bolometers were stopped possibly at a hotter spot at 130 min than at previous times. Figure 16 shows measured heat flux data indicating the total thermal impedance changes taking place. In general, the heat flux increases with seed feed on and decreases with seed feed off, indicating a decrease and increase respectively in total thermal impedance. Changes are on the order of 15-20%.

Potassium radiation is not important since the measurements are taken directly at the surface of the slag layer and through the slag layer. Looking at the heat flux equation

$$\dot{q}_s = \frac{(T_s - T_c)}{\frac{S}{K_s} + \frac{t}{K_m}} \qquad (1)$$

where $\dot{q}_s$ = heat flux through slag
T_s = surface temperature of slag
S = thickness of slag
K_s = thermal conductivity of slag
t = thickness of metal test wall = const.
K_m = thermal conductivity of metal test wall = const.

Measurements show that T_s decreases and $\dot{q}_s$ increases so that K_s/S must increase.

Another transient mechanism is that of ash cycling leading to a wave runoff. Figure 17 is a tracing of a strip chart record during a typical runoff. Shown are the two bolometer traces. When the ash feed is turned off the temperature of the slag layer decays. About 35 seconds after the ash feed is again turned on a temperature jump appears at the viewport, most probably associated with a slag wave runoff. This is probably a result of the integrated mass flux falling onto the slag layer upstream of the viewport, preferentially on the upstream portion of the channel, running axially due to its lower viscosity, as a steep fronted wave.

Current Transport Interaction

The interaction of generator current transport with the slag layer has been addressed at AERL primarily in the generator development program (20, 21). In general, it has been found that generator operation is compatible with the presence of a flowing slag layer. Both beneficial and deleterious effects are observed, the latter being primarily enhanced anode metal electrochemical oxidation, and increased current distribution non-uniformity.

The observed slag surface temperature range, controlled primarily by viscosity, suggests that current transport from generator core flow to the slag surface occurs predominantly as local current concentrations. Observations by Vasil'yev et al. (22) and by Koester (9) place some bounds on diffuse versus arc transport in terms of core current density, surface temperature and current vector (anode/cathode). A prime purpose of a flowing replenishment coating is of course to sacrificially vaporize in such local hot regions, while protecting the backing electrode structure. An upper bound on local arc current is definable in terms of either arc burn-through of the slag layer, or such high local heat input that the backing wall is locally damaged while retaining a continuous coating. Current transport within the slag to the metal is in this case dependent on local temperature rise throughout the slag, to the metal interface, so that a local, heat flux driven high electrical conductivity may result in the slag, without dependence on significant electrothermal instability within the slag. At the other extreme, an extremely cold, high thermal conductance wall can force secondary electrothermal instability in the slag or ceramic close to the cold metal. This was observed on the MK VI A channel, using zirconia as electrode ceramic, to cause metallic electrode failure.

An upper bound on local T_m, imposed by slag layer thickness, has been treated in a very crude sense (23). Arc column properties are based upon the results of Hsu (24). Two limits may exist, based upon layer thickness necessary to prevent arc column burn-through, and shear flow required to provide a mass source for stationary arc column slag vaporization.

As a first approximation to slag layer working limits, the results are shown in Figs. 21 and 22. Both figures are plotted in terms of local layer thickness S versus T_m, for various axial locations, and corresponding transport behavior, in a generator. Figure 21 refers to zero slag layer shear carryover at the generator entrance, while Fig. 22 is based on an assumed ceramic faced low pressure gradient nozzle. In both cases, S is seen to decrease with increasing T_m. Initially, this is solely a result of decreasing insulation thickness, with $T_s \sim$ constant. As T_m is further increased, however, T_s is more strongly affected. At $T_m >> T_{250}$, a very thin, fluid slag layer is predicted.

The requirements for nondestructive arc mode current transport are indicated in terms of S required for various I. Further, the $\dot{M}$ limit is indicated on Fig. 21. In the case represented by Fig. 22, the local $\dot{M}$ at S is in all cases shown greater than the minimum required for steady state arc transport.

The bounds indicated for arc transport wall protection are more nearly speculative than defined. However, in terms of the gas phase problem, they do suggest a need for considerable slag coverage. Also, the existence of requirements on both layer thickness and slag flow is indicated, suggesting a potential bound for arc intensity with stationary arcs. Secondary arcing within a slag layer is a further problem, as mentioned above. A strong coupling with the mode of current transport, i.e. ionic or electronic, has also been observed (9,21).

Power loss due to axial current leakage in the wall slag layers was initially addressed by Rosa (4). He concluded that a slag based on the overall content of an eastern bituminous coal did not represent a significant leakage path for realistic scale channels. However, as indicated above, wall slag composition may depend both on coal feed and combustor ash fractionation. Experimental verification of slag layer surface temperature dependence on silica fraction is discussed above. This implies also higher axial leakage conductance in the slag, as indicated qualitatively in Fig. 23. The iron oxide also contributes to increased slag conductivity. Consequently, in the limit of low power dissipation, enhanced axial current leakage is projected, relative to the earlier estimates by Rosa, or measurements with unmodified coal slag. The result of experimental test of this projection is shown in Fig. 24 as curves of hall voltage variation on the Mk VIC channel. Output at steady state slagging conditions was initially established with fly ash feed (Table III, column 2, 2×10^{-3} wgt fraction nominal). Following this, the feed was switched to a high silica content feed (Table III, column 4, 2×10^{-3} wgt fraction nominal). There resulted a net hall field and power loss, which recovered when the mineral feed was terminated. Subsequent feed of fly ash gave a slight decrease, indicating finite but small leakage with it. Analysis of these measurements, and comparison with measurements of slag electrical conductivity, suggests that surface electrothermal breakdown was triggered by the high silica concentration slag, rather than simple current leakage directly controlled by system fluid dynamics.

Preliminary treatment of scaling laws suggests that in the linear limit of low leakage power dissipation, fractional power loss by axial leakage may not be dependent on system scale for constant slag composition. The observations reported on the effect of calcia on slag layer properties suggest that western coals are attractive in this context, and that dolomite addition to eastern coal derived slag carryover could be effective.

MODELING

Steady State

Slag layer behavior has been described mathematically in order to quantitatively correlate experimental observations with physical phenomena. Following this correlation, the model is employed for predictive extrapolation to other flow conditions and power plant scale hardware. Both steady state and transient processes have been treated with reasonable success. The details of the mathematical approach are given elsewhere (4, 12, 25, 26). Physically, as indicated in Fig. 25, a local steady state slag layer mass flow is defined as the integral of all mass deposited on the wall upstream. The force available to move this flow is defined as the local boundary layer skin friction and pressure gradient. This force works against the drag associated with viscous flow of the slag. Slag viscosity is represented as a function of temperature, and the temperature distribution through the layer derived from heat transfer modeling. Iteration leads to definition of local surface temperature and layer thickness for which the viscous force distribution, at the required local mass flow rate, just balances the driving forces. Typically, a velocity profile in the slag layer as given in Fig. 26 results. Predicted slag layer dependence on mineral concentration in the gas flow is shown in Fig. 27 as a representative result.

The effect of the backing wall structure temperature on slag layer properties was described above, with reference to Figs. 21 and 22. Increased T_m decreases the slag layer thickness, without significantly affecting the surface flow temperature, as long as $T_m < T_{250}$. The reason for this is apparent from Fig. 26. Changing T_m basically affects only the thickness of the lower, stationary part of the slag layer. As T_m reaches a range for which slag viscosity is relatively low, the surface temperature is projected to increase. In general, however, to drive the slag surface temperature significantly above that for flow over a cold wall, the backing wall structure must reach or exceed that flow temperature, e.g. 1800 K. This has been done experimentally, for example in ceramic lined high temperature coal combustors (13, 15). However, the slag flow rapidly attacks and removes the ceramic, until it becomes thin enough that the ceramic-slag interface is stationary, and the slag surface temperature again becomes independent of the interface temperature.

It has been shown that a typical generator wall structure, comprising alternate ceramic and metal exposed surfaces, develops a coating through initial slag bonding to exposed ceramic, followed by streamer and subsequently curtain flow over the metal. One limit of slag layer development is that for which such streamers are unstable to cross the metal, but instead re-entrain into the gas flow. As a first approximation, the equilibrium slag layer thickness, predicted for a given T_m, can be compared with a re-entrained droplet diameter scale, defined in terms of surface tension stability. Such a scale is associated with Weber number, defined as

$$W_e = \frac{\rho u^2 d}{\sigma} \tag{2}$$

Setting $W_e \sim 1$ as a stability bound, and using such slag surface tension data as exist (27), one finds the bounds shown on Figs. 21 and 22. The implication drawn from this is that upper bounds exist on metal wall temperature, above which a continuous slag coating will not develop. This is clearly not applicable in the case of exposed ceramic walls, which are wettable by slag. In this case, however, other phenomena, such as relations between slag boiloff and droplet arrival rate, could control development of a continuous layer.

Transient Processes

Evolution of the slag layer following flow initiation proceeds as a competition among several processes. It is treated here on an overall mass balance basis, without consideration of the detailed local structure. Also, a quasi-one-dimensional system is assumed. Boundary layer slag species transport is the primary local source term. Re-entrainment associated with droplet removal and surface vaporization compete. Shear driven slag flow is also a local sink or source term. During the period of slag layer growth on moderate temperature substrates, the surface temperature is probably low enough that vapor loss is not a major process. Some initial surface conditions may result in a surface structure for which droplet re-entrainment is important, but these are of marginal interest to slagging electrode system operation. Consequently, a first approximation to the local layer growth size is based on the phenomena indicated in Fig. 25. The appropriate equation is

$$\frac{\partial S}{\partial t} = \frac{1}{\rho}\left(\dot{m} - \frac{\partial \dot{M}}{\partial x}\right) \tag{3}$$

where S is the local instantaneous slag layer thickness, ρ the slag material density, $\dot{m}$ the local deposition rate as droplets and vapor condensate, and $\dot{M}$ the local instantaneous shear driven slag flow. Input shear flow, e.g. from a contoured nozzle, is represented by $\dot{M}_o$. The relation of these terms of gas-dynamic and slag properties, together with assumptions made, were given previously (4, 9, 24, 25). Numerical integration of the equation system gives a prediction of layer thickness, surface temperature, heat flux, and shear flow distribution in space and time. The ratio of local layer shear flow to integrated upstream deposition is used to define the approach to steady state as

$$F = \dot{M}(x) \Big/ \left(\int_o^x \dot{m}\, dx + M_o\right) \tag{4}$$

where F = 0 corresponds to an initial condition of a bare wall, and F = 1 implies equilibrium at x. The overall equilibrium condition is defined by F = 1 at all x.

Calculation of slag layer development in generator flow fields yields two distinct overall growth sequences. These depend upon the physical state of the coal mineral content in the flow field. In the limit of droplet carryover, the turbulent transport model gives increasing diffusivity with increasing boundary layer thickness, and correspondingly with axial position in the generator. Figure 28 shows the result of this in terms of slag layer surface temperature evolution. It is predicted that growth and equilibration are most rapid at the exit end, and slowest at the inlet. Also, no discontinuities are found.

In the limit of ash carryover solely as vapor species, turbulent boundary layer transport is again assumed the limiting step for wall deposition. However, in this case, local transport rate scales with skin friction, and is correspondingly a maximum at the duct inlet. Figure 29 gives predicted surface temperature development for this case, and Fig. 30 the slag layer shear flow development. Growth of layer thickness and surface temperature is most rapid at the inlet end, and results ($t \sim 1000$ sec) in local equilibrium between deposition and shear driven runoff. Subsequently, continued deposition over the front region provides a mass source for continued runoff, which results in a steep front in thickness, temperature, and flow propagating down the length of the duct. Upstream of this, the layer is equilibrated, while downstream, growth by vapor deposition continues. Under steady state operating conditions, it is clear that this front passage is a one time event, rather than a repetitive, cyclic process.

While the model presents the discontinuity as a well defined spatially propagating event, in the real system, a stream dominated structure, with a large length scale, may well exist. However, the transverse interfaces of such streamers, e.g. Figs. 7c and 10e, do themselves represent quite significant temperature discontinuities.

A case of combined droplet and vapor deposition is given in Fig. 31. Here the temperature step associated with the runoff is weaker than predicted for the pure vapor case, and, due to transport and concentration constants chosen, the time scale shorter. This case is more closely related to the operation of the MK VI generator facility than are the others, in terms of absolute carryover, vapor fraction, and transport parameters.

The slag runoff surface discontinuity predicted for vapor deposition, and observed in replenishment experiments, may also be driven by two other mechanisms, associated with generator operation. These are seed feed following slag layer development, and joule dissipation in the wall region associated with initiation of power extraction.

Seed input to the flow system gives vapor phase potassium species, which transport to the slag layer at a relatively high rate. As in the case of vaporized mineral species, this occurs at higher rate at the inlet end of the nozzle and generator. End state concentration in the wall slag layer is typically 18% (wgt) as K_2O. This decreases slag thermal conductivity, driving a surface temperature increase, and a corresponding strong viscosity decrease. Under the gas dynamic shear conditions, this low viscosity slag is driven off, presumably in a steep front as predicted for vapor deposition of slag. Replenishment channel tests under subsonic flow conditions have shown this process. Sequence photographs are given in Figs. 2d, e, f and 3d, e, f. Observations of the evolution of local current transport stability in the MK VI system may also be compatible with this mechanism. It is anticipated that feed of seed during the slag layer growth and equilibration period would minimize this phenomena, except associated uniquely with vapor phase mineral species transport.

Power extraction from a generator is observed to increase net heat flux to the electrode system. Consequently, presupposing an initial equilibrium seed-slag wall coating, turn on of the magnetic field would be expected to drive the slag surface temperature up and viscosity down. In most cases, this is again probably more pronounced toward the front of the duct, and should result in a final slag runoff event. Shutdown of either seed or magnetic field would presumably allow a gradual re-equilibration with runoff reoccurring following turn on.

As mentioned above, some generator electrical phenomena have been observed which may correlate with a seed-induced slag layer re-equilibration. Given the magnitude of the temperature step predicted for vapor phase slag transport, this is probably to be expected. In a practical sense, the propagating surface temperature discontinuity may result in locally high and non-uniform current densities and associated voltage gradients. This may cause excessive electrical and thermal loading of the electrode system.

DISCUSSION AND CONCLUSIONS

Observations of slag layer development, steady state, and transient behavior show dependence on test conditions generally consistent with physical expectations.

Slag surface temperature appears to be weakly dependent on ash feed rate. Other variables such as combustor stoichiometry, ash composition and measurement uncertainties seem to dominate. The nominal combustion product test condition employed is chosen as weakly reducing. Typically, fuel flow is adjusted to give circa 4% CO, 1.5% O_2 in the products. More strongly reducing conditions have negligible effect on slag flow observations. However, for the standard ash feed, transition to stoichiometric or slightly oxidizing conditions changes the slag layer structure radically, from a uniformly flowing, transversely rippled structure, to a relatively static surface appearance dominated by a diamond shaped luminosity pattern. This has a characteristic dimension of about one cm, and appears stable. Similar experiments with high calcia ash gave no significant departure from the reducing atmosphere flowing structure, even at circa 90% of stoichiometric fuel. It appears possible to interpret these observations in terms of the concentration of iron oxides in the slag, and the dependence of slag viscosity on its oxidation state. Alternatively, the effect of calcia as a fluxant may dominate.

Use of a slagging combustor liner results in decreased heat loss, and possibly enhanced slag layer growth in the channel section. A low pressure gradient transonic nozzle further enhances slagging development and measurably improves layer axial uniformity. Steady state slag conditions are found to depend in a consistent manner on viscosity, as related to silica fraction, with lower SF slags equilibrating at lower surface temperature. Flow dependence on combustor stoichiometry is related to iron oxide fraction or base-acid ratio. Sulphur feed does not measurably affect slag behavior. In terms of slag behavior in a generator system, it appears qualitatively desirable to operate arc mode channels with a moderate silica, low iron oxide carryover. Required carryover mass flow can be decreased by use of a low acceleration nozzle. This is generally compatible with, for example, Montana Rosebud coal as a mineral matter source, but may be strongly perturbed by combustor slag rejection behavior.

Development testing of generators, under simulated base load conditions, in the range of ten to one hundred hours of continuous power extraction suggests the validity of a slagging wall system design approach for long duration commercial units. Further development presently underway include refinement of the electrode structure and coupling of power extraction with slag layer behavior.

ACKNOWLEDGMENTS

The authors wish to express their appreciation to their colleagues in the MHD program group at AERL. In particular, thanks are due Mr. Frank Aronno for expert technical support, and Ms. Karen Cardello for final typing. Also, the activities of Mr. Paul Zygielbaum of EPRI, in defining technical questions and maintaining program coordination, were of signal value.

REFERENCES

1 "Energy Conversion Alternatives Study", (ECAS) General Electric Phase II Final Report, NASA-CR134949 Volume I, December, 1976.

2 Fourth Annual Report of the MHD Generator Development Program, May, 1964, Avco Everett Research Laboratory, Inc., Everett, Massachusetts.

3 Fifth Annual Report of the MHD Generator Development Program, 1965, Avco Everett Research Laboratory, Inc., Everett, Massachusetts.

4 Rosa, R. J., "Design Considerations for Coal Fired MHD Generator Ducts", Fifth International Conference on MHD Power Generation, Munich, 1971.

5 Dicks, J. B., et al., "The Direct-Coal-Fired MHD Generator System", 14th Symposium on Engineering Aspects of MHD, UTSI, Tullahoma, Tenn., 1974.

6 Wu, J. C. L., et al., "Experimental and Theoretical Investigation on a Direct Coal Fired MHD Generator", Sixth International Conference on Magnetohydrodynamic Electrical Power Generation, Washington, D.C., 1975.

7 Petty, S., et al., "Developments With The Mk VI Long Duration MHD Generator", Sixth International Conference on Magnetohydrodynamic Electrical Power Generation, Washington, D.C., 1975.

8 Petty, S., et al., "Progress on the Mk VI Long-Duration MHD Generator", 15th Symposium on Engineering Aspects of MHD, University of Pennsylvania, Philadelphia, 1976.

9 Koester, J. K., and Nelson, R. M., "Electrical Behavior of Slag Coatings in Coal-Fired MHD Generator", 16th Symposium on Engineering Aspects of MHD, University of Pittsburgh, May, 1977.

10 "Collaborative Program for MHD Power Generation", Annual Report; February 15, 1973-February 14, 1974, Avco Everett Research Laboratory, Inc., Everett, Massachusetts.

11 Stickler, D. B., Gannon, R. E., and Kobayashi, H., "Rapid Devolatilization Modeling of Coal", Eastern States Section Meeting, Combustion Institute, Johns Hopkins University, A.P.L., November, 1974.

12 Ubhayakar, S., Stickler, D. B., Gannon, R. E., and von Rosenberg, Jr., C., "Rapid Devolatilization of Pulverzied Coal in Hot Combustion Gases", 16th Symposium (International) on Combustion, M.I.T., Cambridge, Massachusetts, August, 1976.

13 Stickler, D. B., Ubhayakar, S., and Becker, F. E., "Ignition and Combustion Behavior of Pulverized Coal Jets in Hot Oxidizing Atmospheres" 16th Symposium on Engineering Aspects of MHD, University of Pittsburgh, May, 1977.

14 Kobayashi, H., Padia, A., and Chomiak, J., "Effects of Devolatilization Kinetics and Ash Behavior on Coal Fired MHD Combustor Design", 15th Symposium on Engineering Aspects of MHD, University of Pennsylvania, Philadelphia, May, 1976.

15 Teno, J., et al., "A Coal Combustion System for MHD Generators", presented at Intersociety Energy Conversion Engineering Conference, Las Vegas, 1970.

16 Stickler, D.B., DeSaro, R., Quarterly Progress Report, March 1-May 30, 1976, p. 11, EPRI Contract RP 322-1.

17 Stickler, D. B., and DeSaro, R., "Slag Interaction Phenomena on MHD Generator Electrodes", AIAA Journal of Energy, Vol. I No. 3, June, 1977.

18 Sage, W. K., Mc Ilroy, J. B., "Relationship of Coal-Ash Viscosity to Chemical Composition", transactions of ASME, Journal of Engineering for Power, Series A, 82, 11. 145-155, 1960.

19 Bates, J. L., "Thermal Diffusivity and Electrical Conductivity of Molten and Solid Coal Slag", Sixth International Conference on MHD Electrical Power Generation, Washington, D.C., 1975.

20 Solbes, A., et al., "Mark VI MHD Generator Studies", 16th Symposium on Engineering Aspects of MHD, University of Pittsburgh, May 1977.

21 Petty, S., Demirjian, A., and Solbes, A., "Electrode Phenomena in Slagging MHD Channels", 16th Symposium on Engineering Aspects of MHD, University of Pittsburgh, May, 1977.

22 Vasil'yev, N. N. et al., "Parameters of Near Electrode Arcing Processes in MHD Generators", Sixth International Conference on MHD Electrical Power Generation, Washington, D. C., June, 1975.

23. Progress Report, June 1-September 1, 1975, EPRI Contract RP 322-1.

24 Hsu, M. S. S., "Thermal Instabilities and Arcs in the MHD Boundary Layers", 13th Symposium on MHD, Stanford University, 1973.

25 Stickler, D. B., and DeSaro, R., "Replenishment Analysis and Technology Development", Sixth International Conference on MHD Power Generation, Washington, D. C., June, 1975.

26 Stickler, D. B., and DeSaro, R., "Replenishment Processes and Flow Train Interaction", Proceedings of the 15th Symposium Engineering Aspects of MHD, University of Pennsylvania, 1976.

27 Clover, G. M., and Raask, Open Cycle MHD Power Generation, Pergamon Press, Oxford, p. 52 (1969). Edited by J. B. Heywood and G. J. Womack.

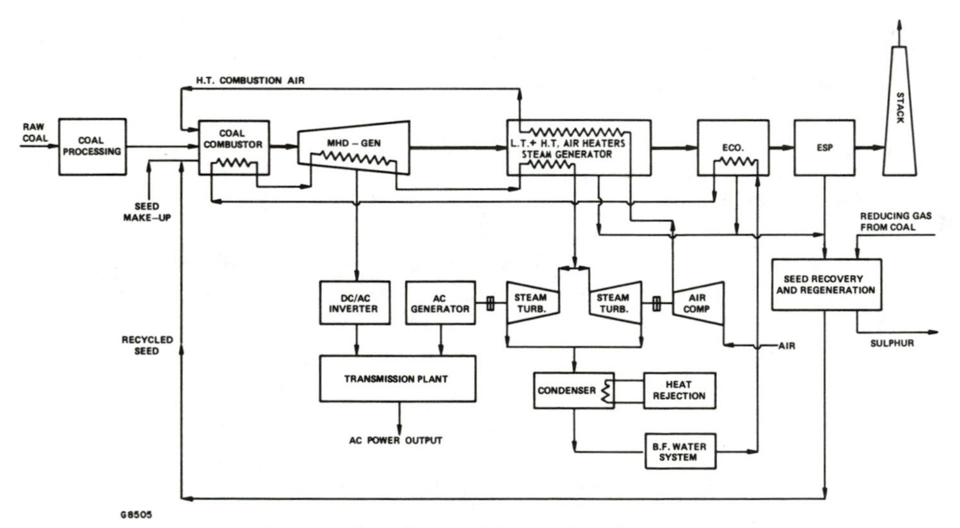

Fig. 1 Simplified Schematic of Developed MHD Steam Power Cycle

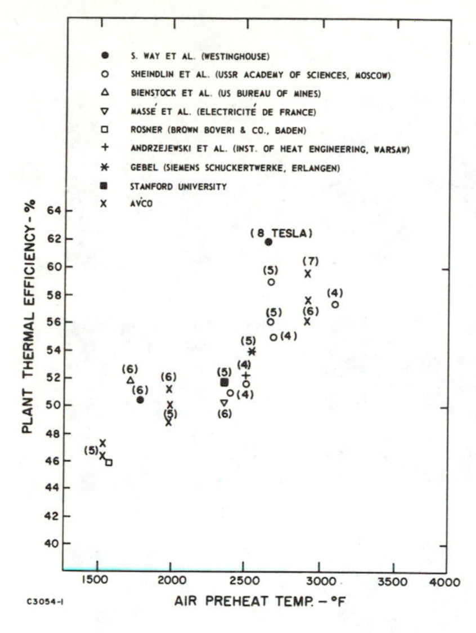

Fig. 2 Baseload Combined MHD Power Cycle Efficiency

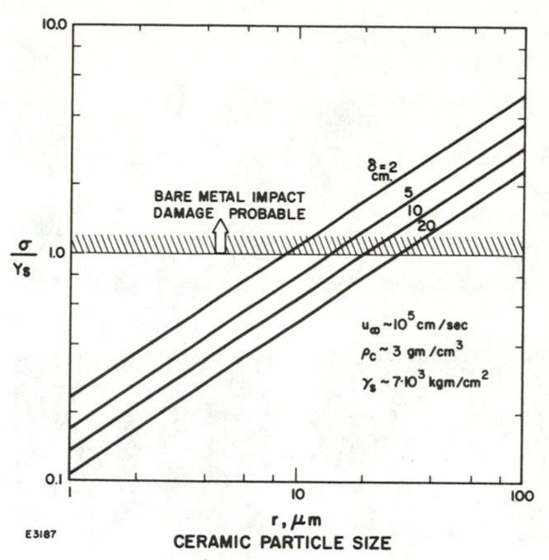

Fig. 5 Particle Impact Stress On Protruding Wall Structures

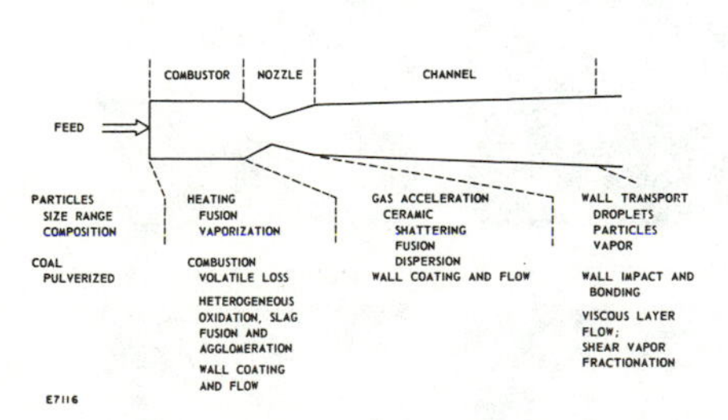

Fig. 3 Coal Slag and Ceramic Material History in Flow Train

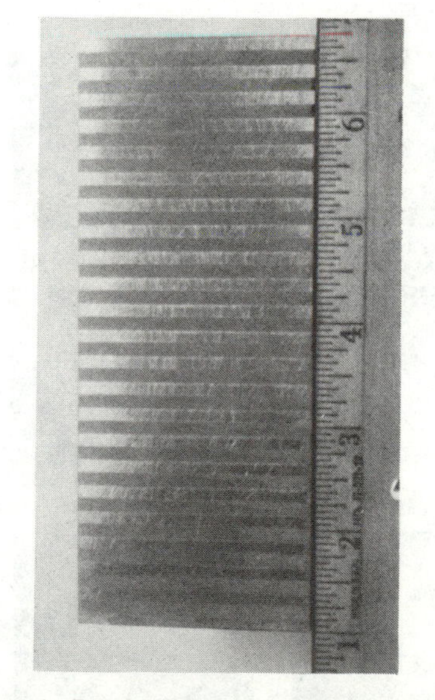

Fig. 6 Replenishment Channel Wall Appearance Prior to Deposition, Grooved Copper Wall With Zirconia Castable

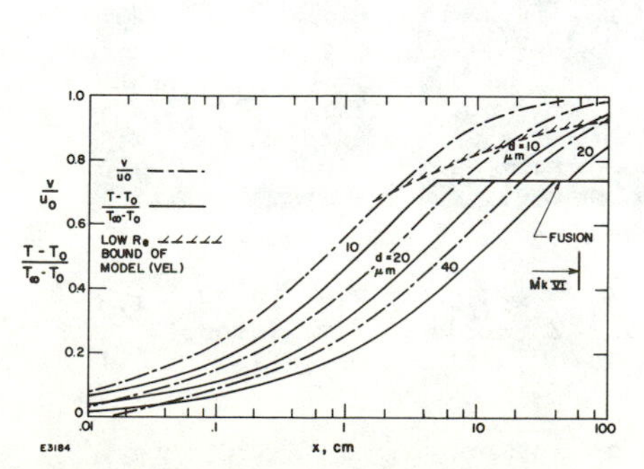

Fig. 4 Particle Temperature and State History, Al_2O_3 in Mk VI Combustor

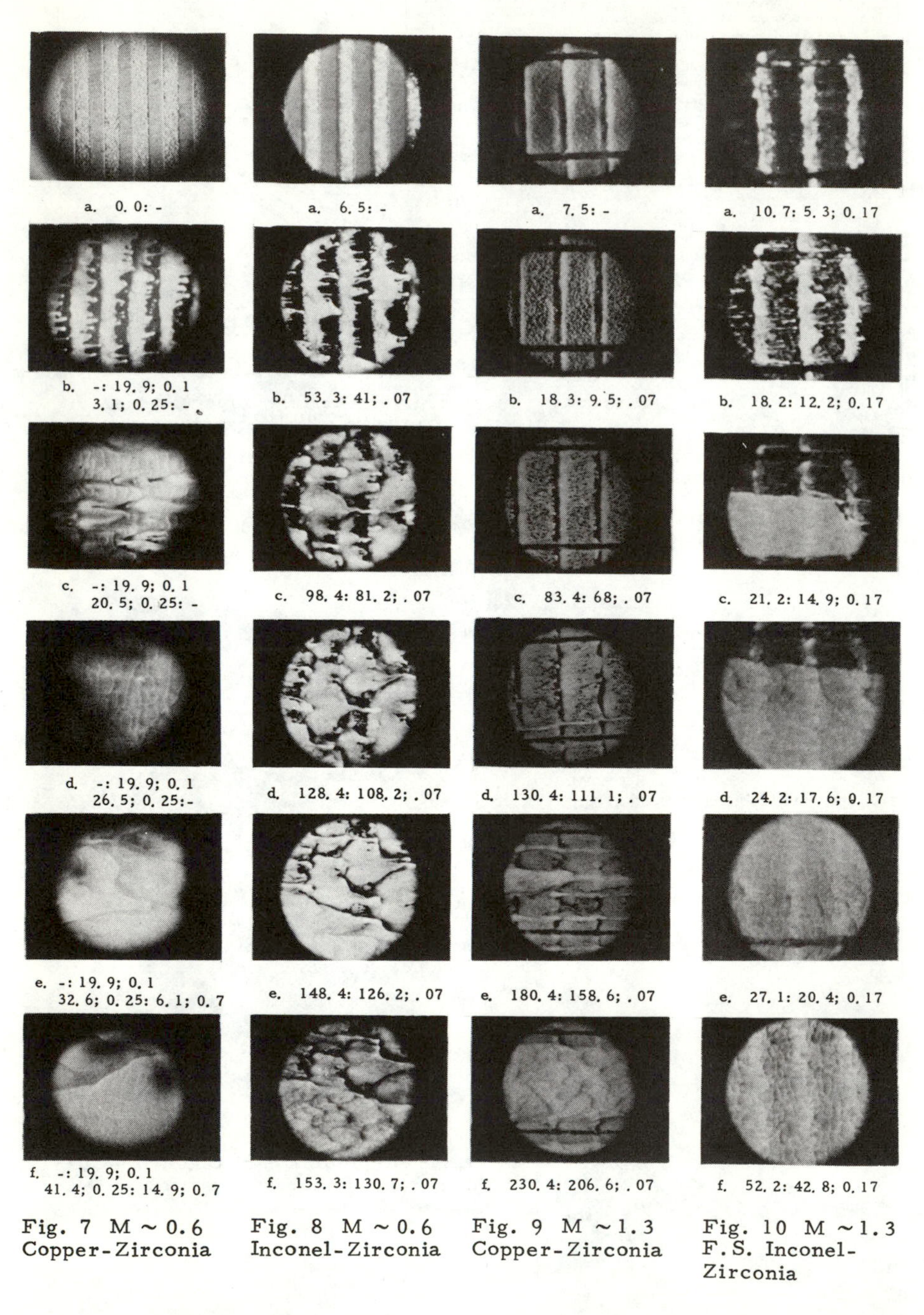

Fig. 7 M ~ 0.6 Copper-Zirconia

Fig. 8 M ~ 0.6 Inconel-Zirconia

Fig. 9 M ~ 1.3 Copper-Zirconia

Fig. 10 M ~ 1.3 F.S. Inconel-Zirconia

SLAG LAYER GROWTH SEQUENCE PHOTOGRAPHS

Key to Captions: t: t_f; ϕ_∞ (ϕ = wgt%)
t_f; ϕ_∞: t_k; ϕ_k

Fig. 11 Composite Cross Section Photomicrograph of Slag Layer From Copper-Zirconia Wall

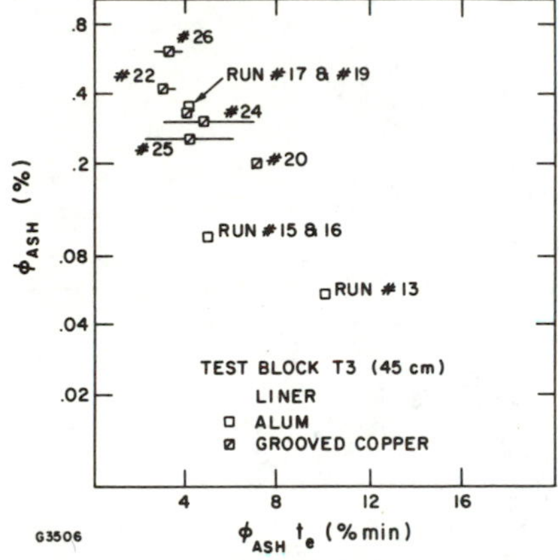

Fig. 12 Slag Layer Equilibration Time Dependence on Slag Fraction in Flow

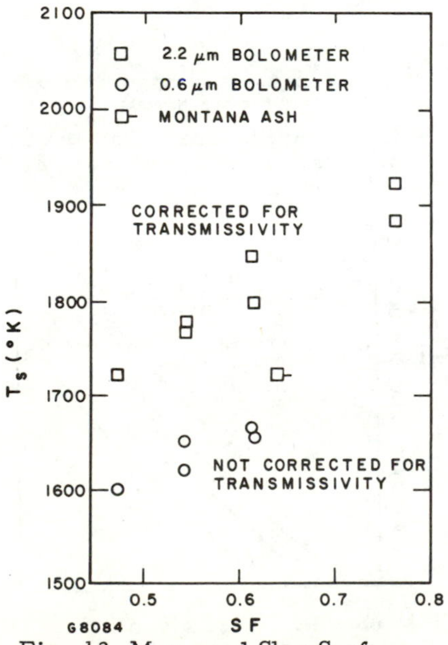

Fig. 13 Measured Slag Surface Temperature of a Grooved Inconel Block

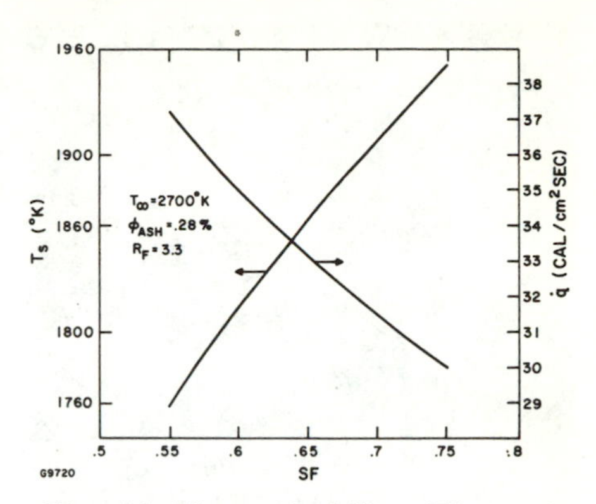

Fig. 14 Theoretical Heat Flux and Slag Surface Temperature as a Function of Slag Composition

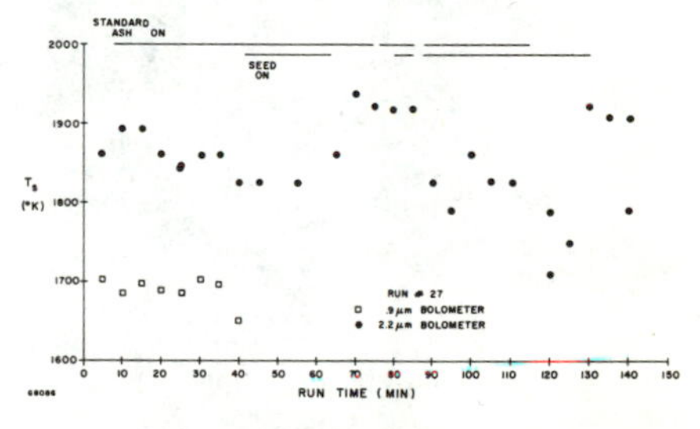

Fig. 15 Measured Surface Temperature Development with Time

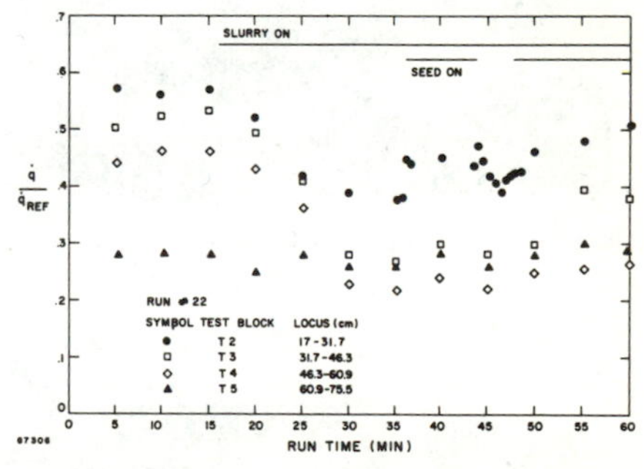

Fig. 16 Dimensionless Heat Flux Development at Various Axial Locations and Wall Structures

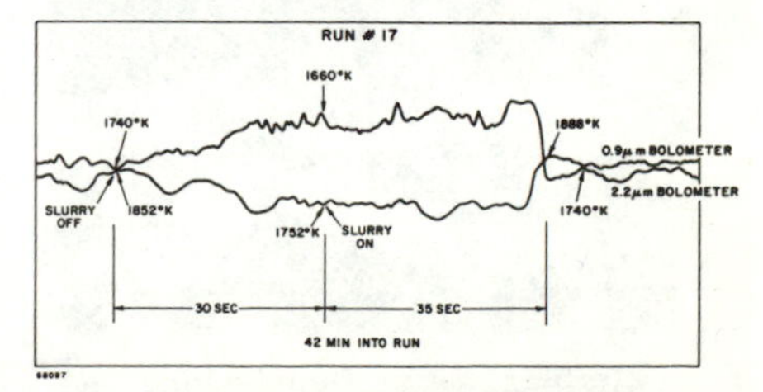

Fig. 17 Tracing of Strip Chart Record During Typical Runoff for Run 17

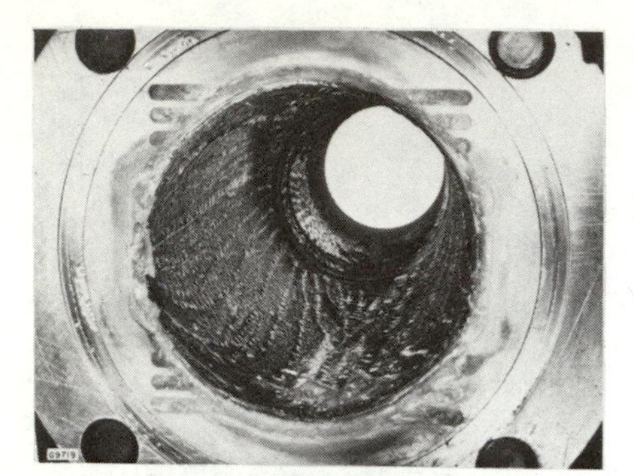

Fig. 18 End State Photo of Combustor Can After Run 37. Inside diameter 14 cm.

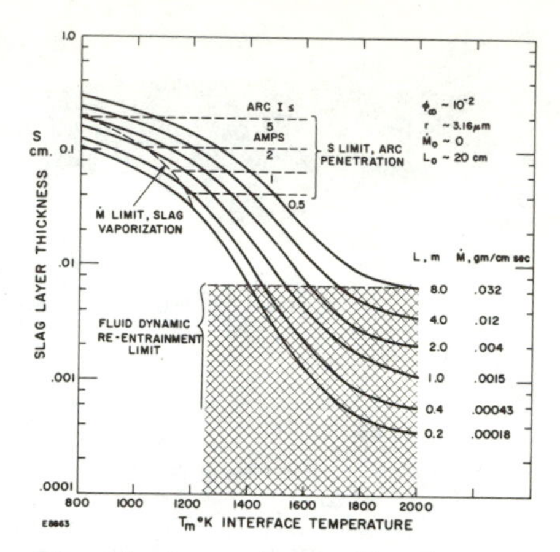

Fig. 21 Approximate Slag Layer Thermal Response Without Nozzle Wall Flow

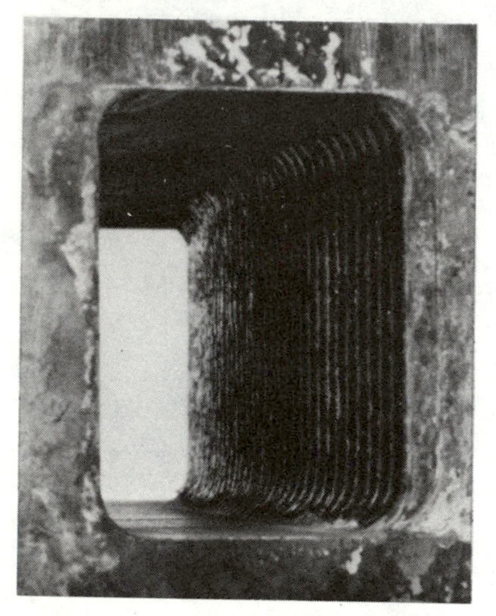

Fig. 19 End State Photo of Downstream Portion of Nozzle After Run 37

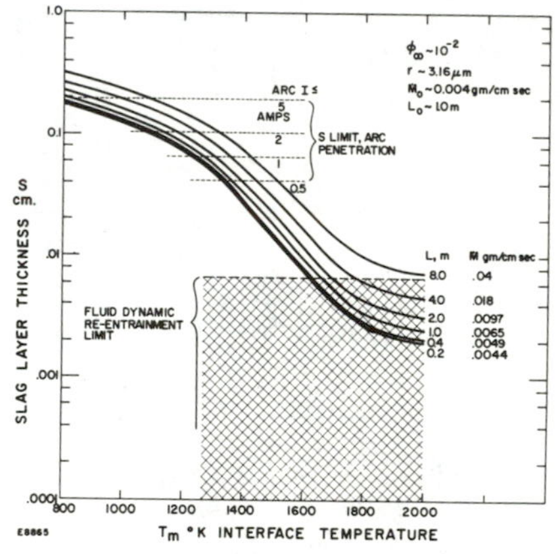

Fig. 22 Approximate Slag Layer Thermal Response With Nozzle Wall Flow

Fig. 20 Slag Layer Dependence on Slag Layer Shear Flow Rate at Generator Entrance

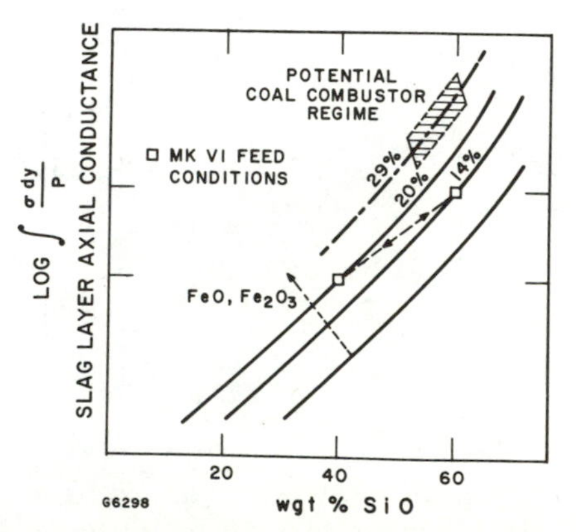

Fig. 23 Qualitative Dependence of Axial Leakage on Slag Composition

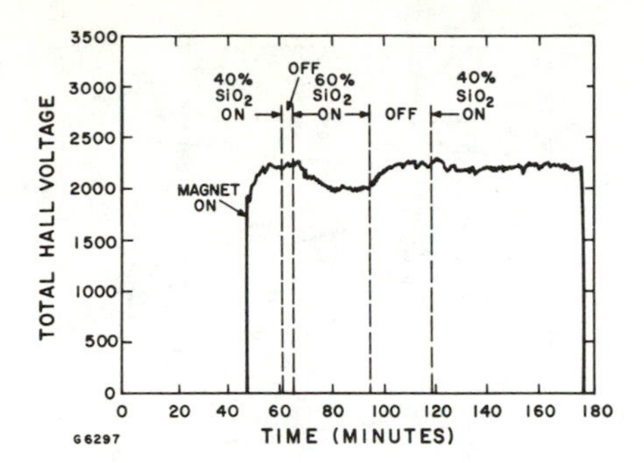

Fig. 24 Hall Field Time History with Slag Feed Composition Changes

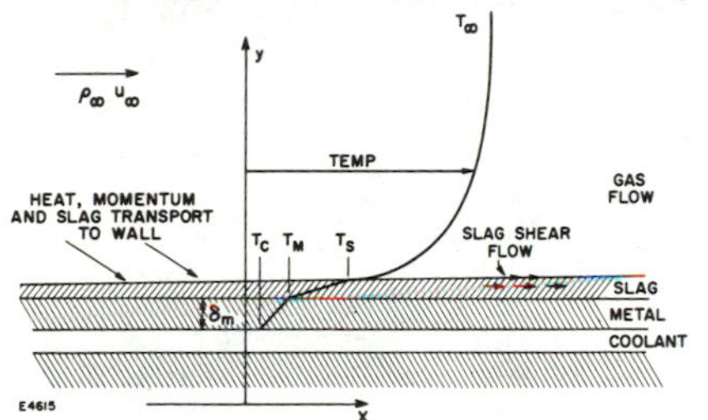

Fig. 25 Model Flow Field and Wall Geometry

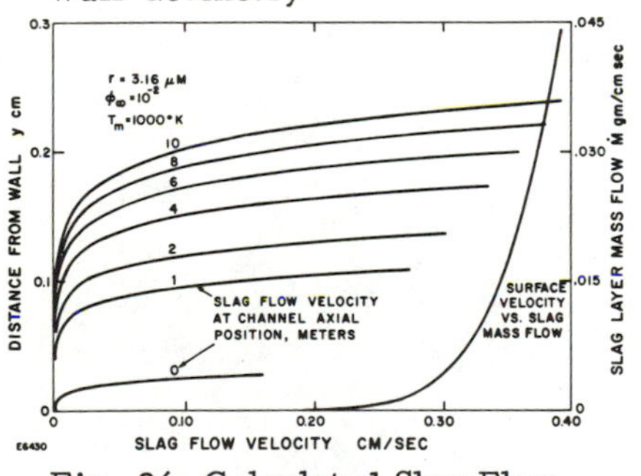

Fig. 26 Calculated Slag Flow Velocity in Wall Layer

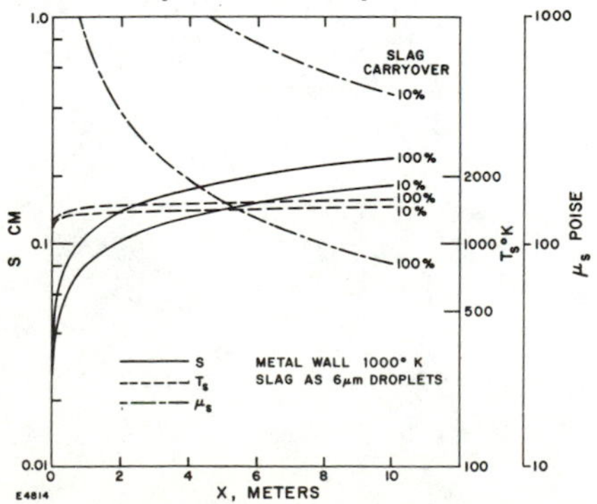

Fig. 27 Slag Layer Dependence on Free Stream Slag Concentration

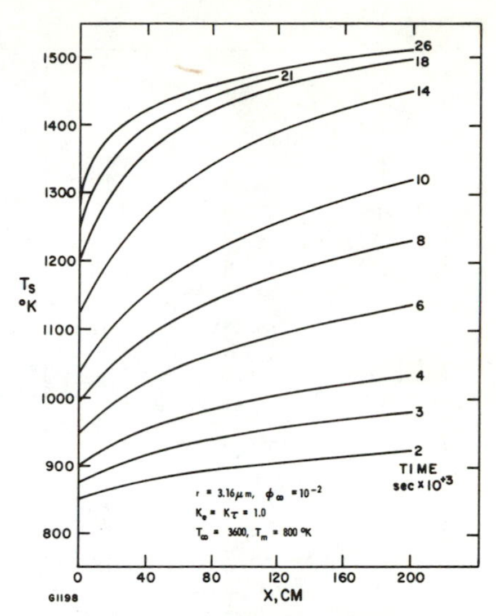

Fig. 28 Predicted Slag Layer Growth (Surface Temperature) For Droplet Transport

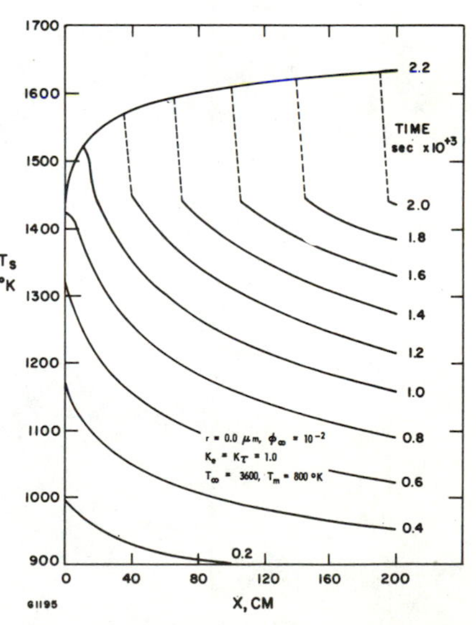

Fig. 29 Predicted Slag Layer Growth (Surface Temperature) For Vapor Transport

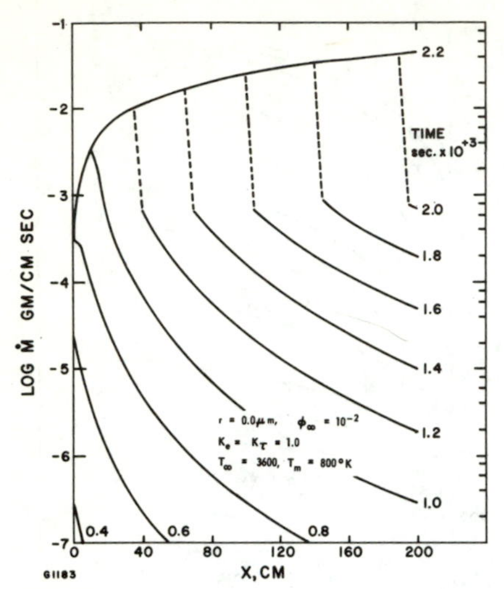

Fig. 30 Predicted Slag Layer Shear Flow Development For Vapor Transport

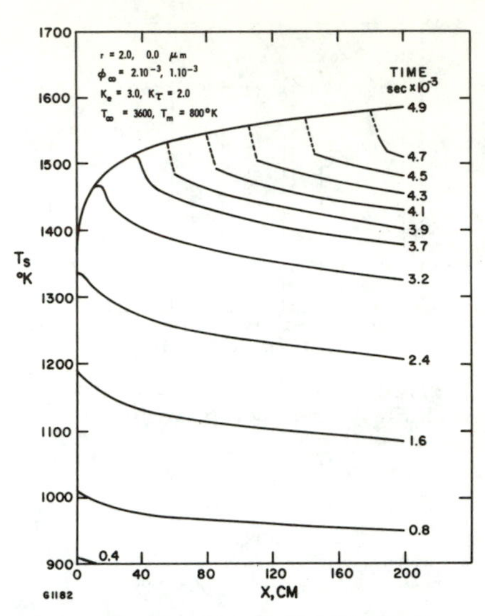

Fig. 31 Predicted Slag Layer Development in Generator Simulation, Short, Cold Wall Transonic Nozzle

INDEX